FRANK H. NETTER, MD

EDITION
8

NETTER ATLAS
of HUMAN
ANATOMY

Classic Regional Approach

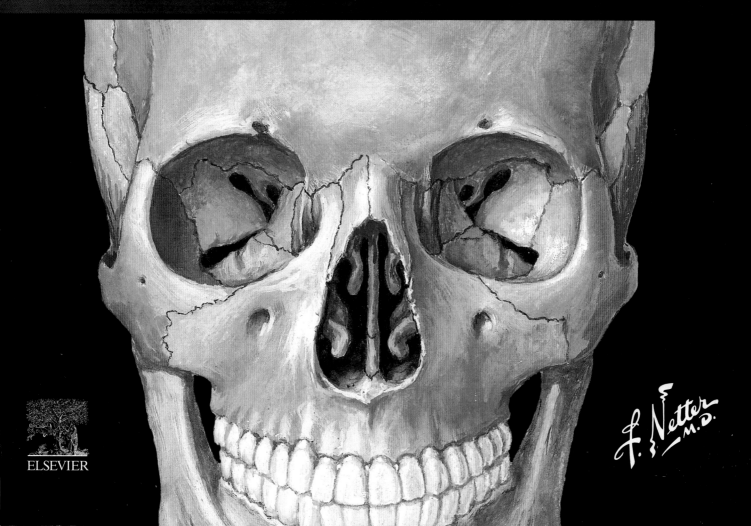

ELSEVIER

F. Netter M.D.

ELSEVIER
1600 John F. Kennedy Blvd.
Ste. 1600
Philadelphia, PA 19103-2899

ATLAS OF HUMAN ANATOMY: CLASSIC REGIONAL APPROACH, EIGHTH EDITION

Standard Edition: 978-0-323-68042-4
Professional Edition: 978-0-323-79373-5
International Edition: 978-0-323-79374-2

Notices

International Standard Book Number: 978-0-323-68042-4

Publisher: Elyse O'Grady
Senior Content Strategist: Marybeth Thiel
Publishing Services Manager: Catherine Jackson
Senior Project Manager/Specialist: Carrie Stetz
Book Design: Renee Duenow

Printed in China

9 8 7 6 5 4 3 2

Working together to grow libraries in developing countries

www.elsevier.com • www.bookaid.org

CONSULTING EDITORS

EDITORS OF PREVIOUS EDITIONS

First Edition
Sharon Colacino, PhD

Second Edition
Arthur F. Dalley II, PhD

Third Edition
Carlos A.G. Machado, MD
John T. Hansen, PhD

Fourth Edition
Carlos A.G. Machado, MD
John T. Hansen, PhD
Jennifer K. Brueckner, PhD
Stephen W. Carmichael, PhD, DSc
Thomas R. Gest, PhD
Noelle A. Granger, PhD
Anil H. Waljii, MD, PhD

Fifth Edition
Carlos A.G. Machado, MD
John T. Hansen, PhD
Brion Benninger, MD, MS
Jennifer K. Brueckner, PhD
Stephen W. Carmichael, PhD, DSc
Noelle A. Granger, PhD
R. Shane Tubbs, MS, PA-C, PhD

Sixth Edition
Carlos A.G. Machado, MD
John T. Hansen, PhD
Brion Benninger, MD, MS
Jennifer Brueckner-Collins, PhD
Todd M. Hoagland, PhD
R. Shane Tubbs, MS, PA-C, PhD

Seventh Edition
Carlos A.G. Machado, MD
John T. Hansen, PhD
Brion Benninger, MD, MS
Jennifer Brueckner-Collins, PhD
Todd M. Hoagland, PhD
R. Shane Tubbs, MS, PA-C, PhD

OTHER CONTRIBUTING ILLUSTRATORS

Rob Duckwall, MA (DragonFly Media Group)
Kristen Wienandt Marzejon, MS, MFA
Tiffany S. DaVanzo, MA, CMI
James A. Perkins, MS, MFA

INTERNATIONAL ADVISORY BOARD

PREFACE

The illustrations comprising the *Netter Atlas of Human Anatomy* were painted by physician-artists, Frank H. Netter, MD, and Carlos Machado, MD. Dr. Netter was a surgeon and Dr. Machado is a cardiologist. Their clinical insights and perspectives have informed their approaches to these works of art. The collective expertise of the anatomists, educators, and clinicians guiding the selection, arrangement, labeling, and creation of the illustrations ensures the accuracy, relevancy, and educational power of this outstanding collection.

You have a copy of the **regionally organized** 8th edition with English language terminology. This is the traditional organization and presentation that has been used since the first edition. Also available is a Latin terminology option that is also regionally organized, as well as an option with English terminology organized by body system. In all cases, the same beautiful and instructive Art Plates and Table information are included.

New to this Edition

New Art

More than 20 new illustrations have been added and over 30 art modifications have been made throughout this edition. Highlights include new views of the temporal and infratemporal fossa, pelvic fascia, nasal cavity and paranasal sinuses, plus multiple new perspectives of the heart, a cross-section of the foot, enhanced surface anatomy plates, and overviews of many body systems. In these pages you will find the most robust illustrated coverage to date for modern clinical anatomy courses.

Terminology and Label Updates

This 8th edition incorporates terms of the *Terminologia Anatomica* (2nd edition), as published by the Federative International Programme on Anatomical Terminology in 2019 (https://fipat.library.dal.ca/ta2) and adopted by the International Federation of Associations of Anatomy in 2020. A fully searchable database of the updated *Terminologia Anatomica* can be accessed at https://ta2viewer.openanatomy.org. Common clinical eponyms and former terminologies are selectively included, parenthetically, for clarity. In addition, a strong effort has been made to reduce label text on the page while maximizing label information through the use of abbreviations (muscle/s = m/mm.; artery/ies = a./aa.; vein/s = v./vv.; and nerve/s = n./nn.) and focusing on the labels most relevant to the subject of each Plate.

Nerve Tables

The muscle tables and clinical tables of previous editions have been so positively received that new tables have been added to cover four major nerve groups: cranial nerves and the nerves of the cervical, brachial, and lumbosacral plexuses.

Future of the Netter Anatomy Atlas

As the Netter Atlas continues to evolve to meet the needs of students, educators, and clinicians, we welcome suggestions! Please use the following form to provide your feedback:

https://tinyurl.com/NetterAtlas8

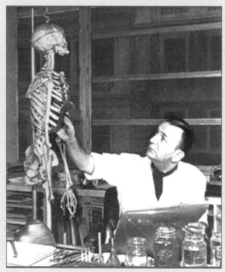

To my dear wife, Vera

PREFACE TO THE FIRST EDITION

I have often said that my career as a medical artist for almost 50 years has been a sort of "command performance" in the sense that it has grown in response to the desires and requests of the medical profession. Over these many years, I have produced almost 4,000 illustrations, mostly for *The CIBA* (now *Netter*) *Collection of Medical Illustrations* but also for *Clinical Symposia.* These pictures have been concerned with the varied subdivisions of medical knowledge such as gross anatomy, histology, embryology, physiology, pathology, diagnostic modalities, surgical and therapeutic techniques, and clinical manifestations of a multitude of diseases. As the years went by, however, there were more and more requests from physicians and students for me to produce an atlas purely of gross anatomy. Thus, this atlas has come about, not through any inspiration on my part but rather, like most of my previous works, as a fulfillment of the desires of the medical profession.

It involved going back over all the illustrations I had made over so many years, selecting those pertinent to gross anatomy, classifying them and organizing them by system and region, adapting them to page size and space, and arranging them in logical sequence. Anatomy of course does not change, but our understanding of anatomy and its clinical significance does change, as do anatomical terminology and nomenclature. This therefore required much updating of many of the older pictures and even revision of a number of them in order to make them more pertinent to today's ever-expanding scope of medical and surgical practice. In addition, I found that there were gaps in the portrayal of medical knowledge as pictorialized in the illustrations I had previously done, and this necessitated my making a number of new pictures that are included in this volume.

In creating an atlas such as this, it is important to achieve a happy medium between complexity and simplification. If the pictures are too complex, they may be difficult and confusing to read; if oversimplified, they may not be adequately definitive or may even be misleading. I have therefore striven for a middle course of realism without the clutter of confusing minutiae. I hope that the students and members of the medical and allied professions will find the illustrations readily understandable, yet instructive and useful.

At one point, the publisher and I thought it might be nice to include a foreword by a truly outstanding and renowned anatomist, but there are so many in that category that we could not make a choice. We did think of men like Vesalius, Leonardo da Vinci, William Hunter, and Henry Gray, who of course are unfortunately unavailable, but I do wonder what their comments might have been about this atlas.

Frank H. Netter, MD
(1906–1991)

FRANK H. NETTER, MD

Frank H. Netter was born in New York City in 1906. He studied art at the Art Students League and the National Academy of Design before entering medical school at New York University, where he received his Doctor of Medicine degree in 1931. During his student years, Dr. Netter's notebook sketches attracted the attention of the medical faculty and other physicians, allowing him to augment his income by illustrating articles and textbooks. He continued illustrating as a sideline after establishing a surgical practice in 1933, but he ultimately opted to give up his practice in favor of a full-time commitment to art. After service in the United States Army during World War II, Dr. Netter began his long collaboration with the CIBA Pharmaceutical Company (now Novartis Pharmaceuticals). This 45-year partnership resulted in the production of the extraordinary collection of medical art so familiar to physicians and other medical professionals worldwide.

Icon Learning Systems acquired the Netter Collection in July 2000 and continued to update Dr. Netter's original paintings and to add newly commissioned paintings by artists trained in the style of Dr. Netter. In 2005, Elsevier Inc. purchased the Netter Collection and all publications from Icon Learning Systems. There are now over 50 publications featuring the art of Dr. Netter available through Elsevier Inc.

Dr. Netter's works are among the finest examples of the use of illustration in the teaching of medical concepts. The 13-book *Netter Collection of Medical Illustrations,* which includes the greater part of the more than 20,000 paintings created by Dr. Netter, became and remains one of the most famous medical works ever published. *The Netter Atlas of Human Anatomy,* first published in 1989, presents the anatomic paintings from the Netter Collection. Now translated into 16 languages, it is the anatomy atlas of choice among medical and health professions students the world over.

The Netter illustrations are appreciated not only for their aesthetic qualities, but, more importantly, for their intellectual content. As Dr. Netter wrote in 1949 "clarification of a subject is the aim and goal of illustration. No matter how beautifully painted, how delicately and subtly rendered a subject may be, it is of little value as a *medical illustration* if it does not serve to make clear some medical point." Dr. Netter's planning, conception, point of view, and approach are what inform his paintings and what make them so intellectually valuable.

Frank H. Netter, MD, physician and artist, died in 1991.

ABOUT THE EDITORS

Carlos A.G. Machado, MD was chosen by Novartis to be Dr. Netter's successor. He continues to be the main artist who contributes to the Netter collection of medical illustrations.

Self-taught in medical illustration, cardiologist Carlos Machado has contributed meticulous updates to some of Dr. Netter's original plates and has created many paintings of his own in the style of Netter as an extension of the Netter collection. Dr. Machado's photorealistic expertise and his keen insight into the physician/patient relationship inform his vivid and unforgettable visual style. His dedication to researching each topic and subject he paints places him among the premier medical illustrators at work today.

Learn more about his background and see more of his art at: https://netterimages.com/artist-carlos-a-g-machado.html

Paul E. Neumann, MD was clinically trained in anatomical pathology and neuropathology. Most of his research publications have been in mouse neurogenetics and molecular human genetics. In the past several years, he has concentrated on the anatomical sciences, and has frequently written about anatomical terminology and anatomical ontology in the journal *Clinical Anatomy*. As an officer of the Federative International Programme for Anatomical Terminology (FIPAT), he participated in the production of Terminologia Anatomica (2nd edition), Terminologia Embryologica (2nd edition), and Terminologia Neuroanatomica. In addition to serving as the lead Latin editor of the 8th edition of Netter's Atlas, he was a contributor to the 33rd edition of *Dorland's Illustrated Medical Dictionary*.

R. Shane Tubbs, MS, PA-C, PhD is a native of Birmingham, Alabama and a clinical anatomist. His research interests are centered around clinical/surgical problems that are identified and solved with anatomical studies. This investigative paradigm in anatomy as resulted in over 1,700 peer reviewed publications. Dr. Tubbs' laboratory has made novel discoveries in human anatomy including a new nerve to the skin of the lower eyelid, a new space of the face, a new venous sinus over the spinal cord, new connections between the parts of the sciatic nerve, new ligaments of the neck, a previously undescribed cutaneous branch of the inferior gluteal nerve, and an etiology for postoperative C5 nerve palsies. Moreover, many anatomical feasibility studies from Dr. Tubbs' laboratory have gone on to be used by surgeons from around the world and have thus resulted in new surgical/clinical procedures such as treating hydrocephalus by shunting cerebrospinal fluid into various bones, restoration of upper limb function in paralyzed patients with neurotization procedures using the contralateral spinal accessory nerve, and harvesting of clavicle for anterior cervical discectomy and fusion procedures in patients with cervical instability or degenerative spine disease.

Dr. Tubbs sits on the editorial board of over 15 anatomical journals and has reviewed for over 150 scientific journals. He has been a visiting professor to major institutions in the United States and worldwide. Dr. Tubbs has authored over 40 books and over 75 book chapters. His published books by Elsevier include *Gray's Anatomy Review*, *Gray's Clinical Photographic Dissector of the Human Body*, *Netter's Introduction to Clinical Procedures*, and *Nerves and Nerve Injuries* volumes I and II. He is an editor for the 41st and 42nd editions of the over 150-year-old *Gray's Anatomy*, the 5th through 8th editions of *Netter's Atlas of Anatomy*, and is the editor-in-chief of the journal *Clinical Anatomy*. He is the Chair of the Federative International Programme on Anatomical Terminologies (FIPAT).

Jennifer K. Brueckner-Collins, PhD is a proud Kentucky native. She pursued her undergraduate and graduate training at the University of Kentucky. During her second year of graduate school there, she realized that her professional calling was not basic science research in skeletal muscle biology, but was instead helping medical students master the anatomical sciences. She discovered this during a required teaching assistantship in medical histology, where working with students at the 10-headed microscope changed her career path.

The next semester of graduate school, she assisted in teaching dissection-based gross anatomy, although she had taken anatomy when the lab component was prosection based. After teaching in the first lab, she knew that she needed to learn anatomy more thoroughly through dissection on her own, so she dissected one to two labs ahead of the students that semester; that was when she really learned anatomy and was inspired to teach this discipline as a profession. All of this occurred in the early 1990s, when pursuing a teaching career was frowned upon by many; it was thought that you only pursued this track if you were unsuccessful in research. She taught anatomy part-time during the rest of her graduate training, on her own time, to gain requisite experience to ultimately secure a faculty position.

Dr. Brueckner-Collins spent 10 years at the University of Kentucky as a full-time faculty member teaching dissection-based gross anatomy to medical, dental, and allied health students. Then, after meeting the love of her life, she moved to the University of Louisville and has taught medical and dental students there for more than a decade. Over 20 years of teaching full time at two medical schools in the state, her teaching efforts have been recognized through receipt of the highest teaching honor at each medical school in the state, the Provost's Teaching Award at University of Kentucky, and the Distinguished Teaching Professorship at University of Louisville.

Martha Johnson Gdowski, PhD earned her BS in Biology cum laude from Gannon University in 1990, followed by a PhD in Anatomy from the Pennsylvania State University College of Medicine in 1995. She completed postdoctoral fellowships at the Cleveland Clinic and Northwestern University School of Medicine prior to accepting a faculty position in the Department of Neuroscience at the University of Rochester School of Medicine and Dentistry in 2001. Previous research interests include the development of an adult model of hydrocephalus, sensorimotor integration in the basal ganglia, and sensorimotor integration in normal and pathological aging.

Her passion throughout her career has been in her service as an educator. Her teaching has encompassed a variety of learning formats, including didactic lecture, laboratory, journal club, and problem-based learning. She has taught for four academic institutions in different capacities (The Pennsylvania State University School of Medicine, Northwestern University School of Medicine, Ithaca College, and The University of Rochester School of Medicine and Dentistry). She has taught in the following curricula: Undergraduate and Graduate Neuroscience, Graduate Neuroanatomy, Graduate Human Anatomy and Physiology for Physical Therapists, Undergraduate Medical Human Anatomy and Histology, and Undergraduate and Graduate Human Anatomy. These experiences have provided an opportunity to instruct students that vary in age, life experience, race, ethnicity, and economic background, revealing how diversity in student populations enriches learning environments in ways that benefit everyone. She has been honored to be the recipient of numerous awards for her teaching and mentoring of students during their undergraduate medical education. Martha enjoys gardening, hiking, and swimming with her husband, Greg Gdowski, PhD, and their dogs, Sophie and Ivy.

Virginia T. Lyons, PhD is an Associate Professor of Medical Education and the Associate Dean for Preclinical Education at the Geisel School of Medicine at Dartmouth. She received her BS in Biology from Rochester Institute of Technology and her PhD in Cell Biology and Anatomy from the University of North Carolina at Chapel Hill. Dr. Lyons has devoted her career to education in the anatomical sciences, teaching gross anatomy, histology, embryology, and neuroanatomy to medical students and other health professions students. She has led courses and curricula in human gross anatomy and embryology for more than 20 years and is a strong advocate for incorporating engaged pedagogies into preclinical medical education. Dr. Lyons has been recognized with numerous awards for teaching and mentoring students and was elected to the Dartmouth chapter of the Alpha Omega Alpha Honor Medical Society. She is the author of *Netter's Essential Systems-Based Anatomy* and co-author of the Human Anatomy Learning Modules website accessed by students worldwide. Dr. Lyons also serves as the Discipline Editor for Anatomy on the Aquifer Sciences Curriculum Editorial Board, working to integrate anatomical concepts into virtual patient cases that are used in multiple settings including clerkships and residency training.

Peter J. Ward, PhD grew up in Casper, Wyoming, graduating from Kelly Walsh High School and then attending Carnegie Mellon University in Pittsburgh, Pennsylvania.

He began graduate school at Purdue University, where he first encountered gross anatomy, histology, embryology, and neuroanatomy. Having found a course of study that engrossed him, he helped teach those courses in the veterinary and medical programs at Purdue. Dr. Ward completed a PhD program in anatomy education and, in 2005, he joined the faculty at the West Virginia School of Osteopathic Medicine (WVSOM) in Lewisburg, West Virginia. There he has taught gross anatomy, embryology, neuroscience, histology, and the history of medicine. Dr. Ward has received numerous teaching awards, including the WVSOM Golden Key Award, the Basmajian Award from the American Association of Anatomists, and has been a two-time finalist in the West Virginia Merit Foundation's Professor of the Year selection. Dr. Ward has also been director of the WVSOM plastination facility, coordinator of the anatomy graduate teaching assistants, chair of the curriculum committee, chair of the faculty council, creator and director of a clinical anatomy elective course, and host of many anatomy-centered events between WVSOM and two Japanese Colleges of Osteopathy. Dr. Ward has also served as council member and association secretary for the American Association of Clinical Anatomists. In conjunction with Bone Clones, Inc., Dr. Ward has produced tactile models that mimic the feel of anatomical structures when intact and when ruptured during the physical examination. He created the YouTube channel Clinical Anatomy Explained! and continues to pursue interesting ways to present the anatomical sciences to the public. Dr. Ward was the Senior Associate Editor for the three volumes of *The Netter Collection: The Digestive System*, 2nd Edition, a contributor to *Gray's Anatomy*, 42nd Edition, and is author of *Netter's Integrated Musculoskeletal System: Clinical Anatomy Explained.*

Brion Benninger, MD, MBChB, MSc currently teaches surgical, imaging, and dynamic anatomy to medical students and residents in several countries (United States, New Zealand, China, Japan, Korea, The Caribbean, Mexico). He develops, invents, and assesses ultrasound probes, medical equipment, simulations, and software while identifying dynamic anatomy. He enjoys mixing educational techniques integrating macro imaging and surgical anatomy. Dr. Benninger developed the teaching theory of anatomy deconstruction/reconstruction and was the first to combine ultrasound with Google Glass during physical examination, coining the term "triple feedback examination." An early user of ultrasound, he continues to develop eFAST teaching and training techniques, has developed and shares a patent on a novel ultrasound finger probe, and is currently developing a new revolutionary ultrasound probe for breast screening. He is a reviewer for several ultrasound, clinical anatomy, surgical, and radiology journals and edits and writes medical textbooks. His research interests integrate clinical anatomy with conventional and emerging technologies to improve training techniques in situ and simulation. Dr. Benninger pioneered and coined the term "dynamic anatomy," developed a technique to deliver novel contrast medium to humans, and was the first to reveal vessels and nerves not previously seen using CT and MRI imaging. He has mentored more than 200 students on over 350 research projects presented at national and international conferences and has received numerous awards for projects related to emergency procedures, ultrasound, sports medicine, clinical anatomy, medical simulation, reverse translational research, medical education, and technology. He is proud to have received medical teaching awards from several countries and institutions, including being the first recipient in more than 25 years to receive the Commendation Medal Award from the Commission of Osteopathic Accreditation for innovative clinical anatomy teaching that he designed and facilitated in Lebanon, Oregon. Dr. Benninger has received sports medicine accolades from Sir Roger Bannister regarding his medical invention on shoulder proprioception. He is also Executive Director of the Medical Anatomy Center and collaborates with colleagues globally from surgical and nonsurgical specialties. He is also an invited course speaker for surgical anatomy in New Zealand. Dr. Benninger collects medical history books, loves mountains and sports, and is an anonymous restaurant critic. British mentors directly responsible for his training include Prof. Peter Bell (surgery), Prof. Sir Alec Jeffreys (genetic fingerprinting), Profs. David deBono and Tony Gershlick (cardiology), Prof. Roger Greenhalgh (vascular surgery), Profs. Chris Colton, John Webb, and Angus Wallace (orthopaedics), Prof. Harold Ellis CBE (surgery and clinical anatomy), and Prof. Susan Standring (Guys Hospital/Kings College).

Todd M. Hoagland, PhD is Clinical Professor of Biomedical Sciences and Occupational Therapy at Marquette University in the College of Health Sciences. Previously he was Professor of Anatomy at the Medical College of Wisconsin (MCW). Prior to MCW, Dr. Hoagland was at Boston University School of Medicine (BUSM) and he still holds an adjunct faculty position at Boston University Goldman School of Dental Medicine. Dr. Hoagland is a passionate teacher and is dedicated to helping students achieve their goals. He believes in being a strong steward of the anatomical sciences, which involves teaching it to students while contemporaneously developing resources to improve the transfer of knowledge and preparing the next generation to be even better teachers. While at BUSM, Dr. Hoagland was a leader for the Carnegie Initiative on the Doctorate in Neuroscience and helped develop the Vesalius Program (teacher training) for graduate students. The program ensures that graduate students learn about effective teaching, receive authentic experiences in the classroom, and understand how to share what they learn via scholarship.

Dr. Hoagland's dedication to health professions education has been richly rewarded by numerous teaching awards from the University of Notre Dame, BUSM, and MCW. Dr. Hoagland received the Award for Outstanding Ethical Leadership in 2009, was inducted into the Alpha Omega Alpha Honor Medical Society in 2010, received the American Association of Anatomists Basmajian Award in 2012, and was inducted into the Society of Teaching Scholars in 2012 and was their director from 2016–2020.

Dr. Hoagland's scholarly activity centers on (1) evaluating content and instructional/learning methodology in Clinical Human Anatomy and Neuroanatomy courses, especially as relevant to clinical practice, (2) translating basic anatomical science research findings into clinically meaningful information, and (3) evaluating professionalism in students to enhance their self-awareness and improve patient care outcomes. Dr. Hoagland is also consulting editor for *Netter's Atlas of Human Anatomy,* co-author for the digital anatomy textbook *AnatomyOne,* and lead author for *Clinical Human Anatomy Dissection Guide.*

ACKNOWLEDGMENTS

Carlos A. G. Machado, MD

With the completion of this 8th edition, I celebrate 27 years contributing to the Netter brand of educational products, 25 years of which have been dedicated to the update—seven editions—of this highly prestigious, from birth, *Atlas of Human Anatomy*. For these 25 years I have had the privilege and honor of working with some of the most knowledgeable anatomists, educators, and consulting editors—my treasured friends—from whom I have learned considerably.

For the last 16 years it has also been a great privilege to be part of the Elsevier team and be under the skillful coordination and orientation of Marybeth Thiel, Elsevier's Senior Content Development Specialist, and Elyse O'Grady, Executive Content Strategist. I thank both for their friendship, support, sensibility, and very dedicated work.

Once more I thank my wife Adriana and my daughter Beatriz for all their love and encouragement, and for patiently steering me back on track when I get lost in philosophical divagations about turning scientific research into artistic inspiration—and vice-versa!

It is impossible to put in words how thankful I am to my much-loved parents, Carlos and Neide, for their importance in my education and in the formation of my moral and ethical values.

I am eternally grateful to the body donors for their inestimable contribution to the correct understanding of human anatomy; to the students, teachers, health professionals, colleagues, educational institutions, and friends who have, anonymously or not, directly or indirectly, been an enormous source of motivation and invaluable scientific references, constructive comments, and relevant suggestions.

My last thanks, but far from being the least, go to my teachers Eugênio Cavalcante, Mário Fortes, and Paulo Carneiro, for their inspiring teachings on the practical application of the knowledge of anatomy.

Paul E. Neumann, MD

It has been a privilege to work on the English and Latin editions of *Netter's Atlas of Human Anatomy*. I thank the staff at Elsevier (especially Elyse O'Grady, Marybeth Thiel and Carrie Stetz), Dr. Carlos Machado, and the other editors for their efforts to produce a new, improved edition. I am also grateful to my wife, Sandra Powell, and my daughter, Eve, for their support of my academic work.

R. Shane Tubbs, MS, PA-C, PhD

I thank Elyse O'Grady and Marybeth Thiel for their dedication and hard work on this edition. As always, I thank my wife, Susan, and son, Isaiah, for their patience with me on such projects. Additionally, I thank Drs. George and Frank Salter who inspired and encouraged me along my path to anatomy.

Jennifer K. Brueckner-Collins, PhD

Reba McEntire once said "To succeed in life, you need three things: a wishbone, a backbone and a funny bone."

My work with the *Netter Atlas* and the people associated with it over the past 15 years has played an instrumental role in helping me develop and sustain these three metaphorical bones in my professional and personal life.

I am forever grateful to John Hansen, who believed in my ability to serve as an editor starting with the 4th edition.

I extend my sincere thanks to Marybeth Thiel and Elyse O'Grady for not only being the finest of colleagues, but part of my professional family as well. Thanks to you both for your professionalism, support, patience, and collegiality.

To Carlos Machado, you continue to amaze me and inspire me with your special gift of bringing anatomy to life through your art.

For this edition, I also count in my blessings, the ability to work closely with the talented team of educational leaders, including Martha Gdowski, Virginia Lyons, and Peter Ward. It is humbling to work with such brilliant and dedicated teachers as we collectively assembled the systems-based *Netter Atlas* concept.

Finally, I dedicate my work on this edition with unconditional and infinite love to Kurt, Lincoln, my Dad in Heaven, as well as my dog boys, Bingo and Biscuit.

Martha Johnson Gdowski, PhD

I am grateful for the honor to work with the team of editors that Elsevier has selected for the preparation of this 8th edition; they are exceptional in their knowledge, passion as educators, and collegiality. I especially would like to thank Elyse O'Grady and Marybeth Thiel, who have been outstanding in their expertise, patience, and guidance. I am grateful to John T. Hansen, PhD, for his guidance, mentorship, and friendship as a colleague at the University of Rochester and for giving me the opportunity to participate in this work. He continues to be an outstanding role model who has shaped my career as an anatomical sciences educator. Special thanks to Carlos Machado for his gift for making challenging anatomical dissections and difficult concepts accessible to students of anatomy through his artistry, research of the details, and thoughtful discussions. I am indebted to the selfless individuals who have gifted their bodies for anatomical study, the students of anatomy, and my colleagues at the University of Rochester, all of whom motivate me to work to be the best educator I can be. I am most grateful for my loving husband and best friend, Greg, who is my greatest source of support and inspiration.

Virginia T. Lyons, PhD

It has been a joy to work with members of the editorial team on the iconic *Atlas of Human Anatomy* by Frank Netter. I would like to thank Elyse O'Grady and Marybeth Thiel for their expert guidance and ability to nourish the creative process while also keeping us focused (otherwise we would have reveled in debating anatomy minutiae for hours!). I am amazed by the talent of Carlos Machado, who is able to transform our ideas into beautiful, detailed illustrations that simplify concepts for students. I appreciate the patience and support of my husband, Patrick, and my children, Sean and Nora, who keep me sane when things get busy. Finally, I am grateful for the opportunity to teach and learn from the outstanding medical students at the Geisel School of Medicine at Dartmouth. I am fulfilled by their energy, curiosity, and love of learning.

Peter J. Ward, PhD

It is a thrill and honor to contribute to the 8th edition of *Netter's Atlas of Human Anatomy*. It still amazes me that I am helping to showcase the incomparable illustrations of Frank Netter and Carlos Machado. I hope that this atlas continues to bring these works of medical art to a new generation of students as they begin investigating the awesome enigma of the human body. Thanks to all the amazing contributors and to the hardworking team at Elsevier, especially Marybeth Thiel and Elyse O'Grady, for keeping all of us moving forward. Thank you especially to Todd Hoagland for recommending me to the team. I have immense gratitude to James Walker and Kevin Hannon, who introduced me to the world of anatomy. They both seamlessly combined high expectations for their students along with enthusiastic teaching that made the topic fascinating and rewarding. Great thanks to my parents, Robert and Lucinda Ward, for their lifelong support of my education and for the many formative museum trips to stare at dinosaur bones. Sarah, Archer, and Dashiell, you are all the reason I work hard and try to make the world a slightly better place. Your love and enthusiasm mean everything to me.

Brion Benninger, MD, MBChB, MSc

I thank all the healthcare institutions worldwide and the allopathic and osteopathic associations who have provided me the privilege to wake up each day and focus on how to improve our knowledge of teaching and healing the anatomy of the mind, body, and soul while nurturing humanism. I am grateful and fortunate to have my lovely wife, Alison, and thoughtful son, Jack, support my efforts during late nights and long weekends. Their laughs and experiences complete my life. I thank Elsevier, especially Marybeth Thiel, Elyse O'Grady, and Madelene Hyde for expecting the highest standards and providing guidance, enabling my fellow coeditors to work in a fluid diverse environment. Many thanks to Carlos Machado and Frank Netter: the world is proud. I thank clinicians who trained me, especially my early gifted surgeon/anatomist/teacher mentors, Drs. Gerald Tressidor and Harold Ellis CBE (Cambridge & Guy's Hospital); Dr. S. Standring and Dr. M. England, who embody professionalism; Drs. P. Crone, E. Szeto, and J. Heatherington, for supporting innovative medical education; my past, current, and future students and patients; and clinical colleagues from all corners of the world who keep medicine and anatomy dynamic, fresh, and wanting. Special thanks to Drs. J.L. Horn, S. Echols, J. Anderson, and J. Underwood, friends, mentors and fellow visionaries who also see "outside the box," challenging the status quo. Heartfelt tribute to my late mentors, friends, and sister, Jim McDaniel, Bill Bryan, and Gail Hendricks, who represent what is good in teaching, caring, and healing. They made this world a wee bit better. Lastly, I thank my mother for her love of education and equality and my father for his inquisitive and creative mind.

Todd M. Hoagland, PhD

It is a privilege to teach clinical human anatomy, and I am eternally grateful to all the body donors and their families for enabling healthcare professionals to train in the dissection laboratory. It is my honor to work with occupational therapy and health professions students and colleagues at Marquette University. I am grateful to John Hansen and the professionals of the Elsevier team for the opportunity to be a steward of the incomparable *Netter Atlas*. Marybeth Thiel and Elyse O'Grady were especially helpful and a pleasure to work with. It was an honor to collaborate with the brilliant Carlos Machado and all the consulting editors. I thank Dave Bolender, Brian Bear, and Rebecca Lufler for being outstanding colleagues, and I thank all the graduate students I've worked with for helping me grow as a person; it is such a pleasure to see them flourish. I am deeply appreciative of Stan Hillman and Jack O'Malley for inspiring me with masterful teaching and rigorous expectations. I am indebted to Gary Kolesari and Richard Hoyt Jr for helping me become a competent clinical anatomist, and to Rob Bouchie for the intangibles and his camaraderie. I am most grateful to my brother, Bill, for his unwavering optimism and for always being there. I thank my mother, Liz, for her dedication and love, and for instilling a strong work ethic. I am humbled by my three awesome children, Ella, Caleb, and Gregory, for helping me redefine love, wonder, and joy. Olya, ty moye solntse!

CONTENTS

7th Edition to 8th Edition Plate Number Conversion Chart Available Online at https://tinyurl.com/Netter7to8conversion

SECTION 2   HEAD AND NECK • Plates 22–177

Surface Anatomy • Plates 22–24

Bones and Joints • Plates 25–47

Neck • Plates 48–58

Larynx and Endocrine Glands • Plates 103–109

Eye • Plates 110–120

Ear • Plates 121–126

Brain and Meninges • Plates 127–142

SECTION 3   BACK • Plates 178–201

SECTION 5 ABDOMEN · Plates 267–351

SECTION 6   PELVIS · Plates 352–421

Surface Anatomy · Plate 352

Bony Pelvis · Plates 353–357

Pelvic Diaphragm and Viscera · Plates 358–368

SECTION 7 UPPER LIMB • Plates 422–490

SECTION 8   LOWER LIMB • Plates 491–556

INTRODUCTION 1

ELECTRONIC BONUS PLATES

BP 1 Pilosebaceous Apparatus

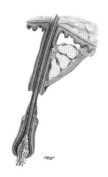

BP 2 Major Body Cavities

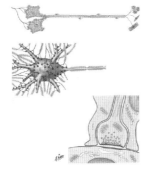

BP 3 Neurons and Synapses

BP 4 Features of a Typical Peripheral Nerve

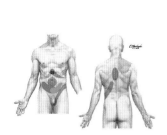

BP 5 Sites of Visceral Referred Pain

BP 6 Sympathetic Nervous System: General Topography

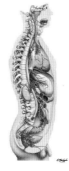

BP 7 Parasympathetic Nervous System: General Topography

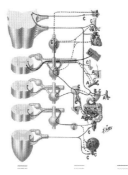

BP 8 Cholinergic and Adrenergic Synapses: Schema

ELECTRONIC BONUS PLATES—*cont'd*

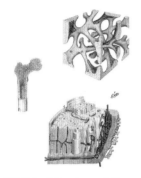

BP 9 Architecture of Bone

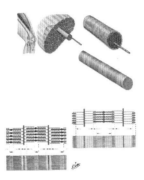

BP 10 Muscle Structure

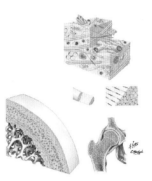

BP 11 Joints: Connective Tissues and Articular Cartilage

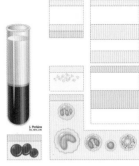

BP 12 Cardiovascular System: Composition of Blood

BP 13 Arterial Wall

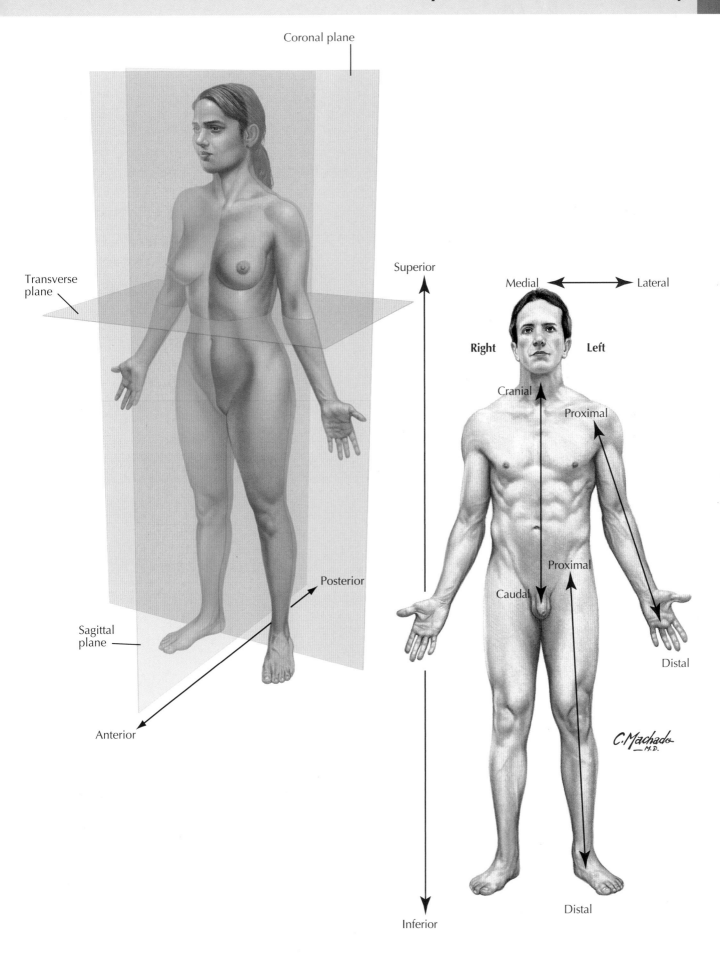

Coronal plane

Transverse plane

Superior

Medial Lateral

Right Left

Cranial

Proximal

Proximal

Caudal

Posterior

Distal

Sagittal plane

Anterior

Inferior

Distal

C. Machado
—M.D.

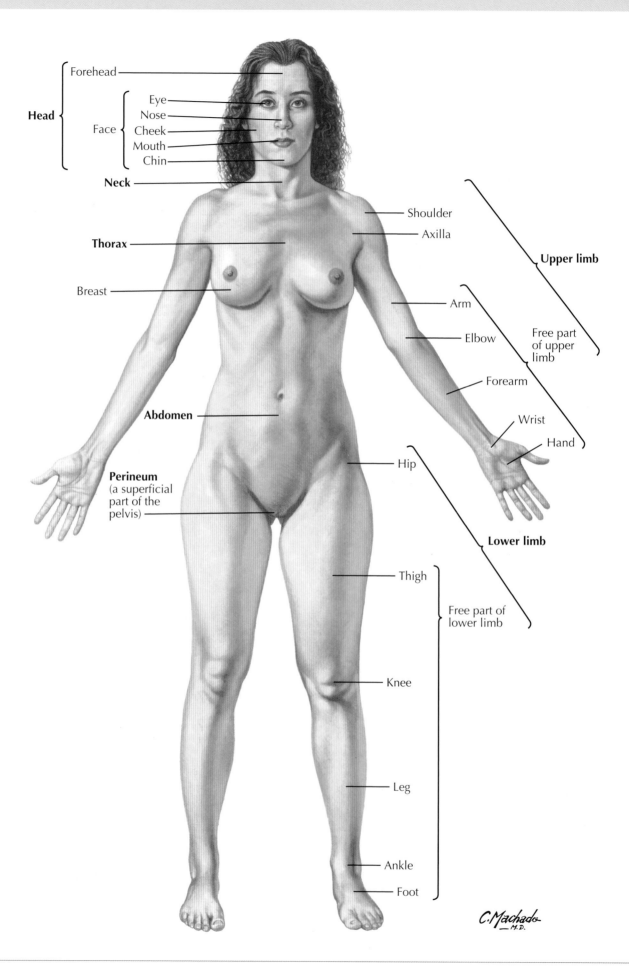

Head

Forehead

Face
Eye
Nose
Cheek
Mouth
Chin

Neck

Thorax

Breast

Abdomen

Perineum
(a superficial
part of the
pelvis)

Shoulder

Axilla

Arm

Elbow

Forearm

Wrist

Hand

Hip

Thigh

Knee

Leg

Ankle

Foot

Upper limb

Free part
of upper
limb

Lower limb

Free part of
lower limb

Plate 2

General Anatomy

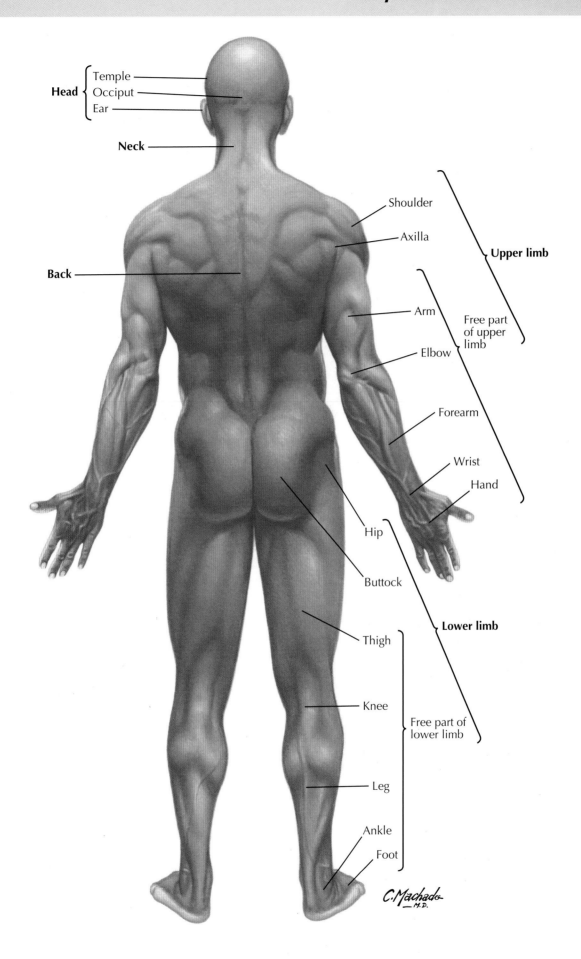

Head { Temple
       Occiput
       Ear

Neck

Shoulder

Axilla

Upper limb

Back

Arm

Free part of upper limb

Elbow

Forearm

Wrist

Hand

Hip

Buttock

Lower limb

Thigh

Knee

Free part of lower limb

Leg

Ankle

Foot

C. Machado
— M.D.

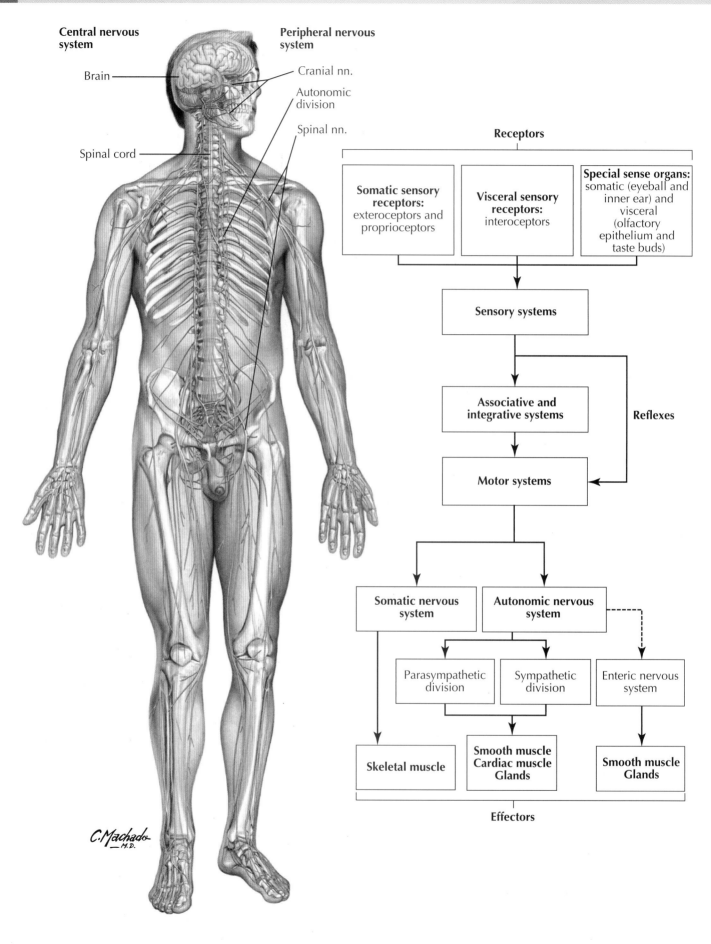

Central nervous system

Peripheral nervous system

Brain

Cranial nn.

Autonomic division

Spinal nn.

Spinal cord

Receptors

| Somatic sensory receptors: exteroceptors and proprioceptors | Visceral sensory receptors: interoceptors | **Special sense organs:** somatic (eyeball and inner ear) and visceral (olfactory epithelium and taste buds) |

Sensory systems

Associative and integrative systems

Reflexes

Motor systems

| Somatic nervous system | Autonomic nervous system |

| Parasympathetic division | Sympathetic division | Enteric nervous system |

| **Skeletal muscle** | **Smooth muscle Cardiac muscle Glands** | **Smooth muscle Glands** |

Effectors

Plate 4

Systematic Anatomy

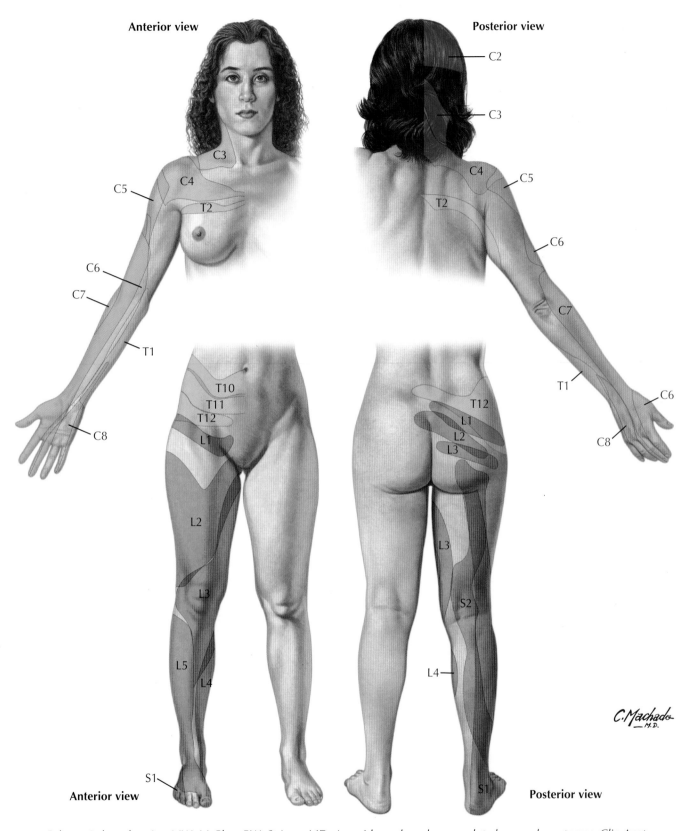

Anterior view

Posterior view

C2

C3

C3

C4

C4

C5

C5

C5

C6

T2

T2

C6

T2

C6

C7

C7

T1

T1

T10

T12

T11

L1

T12

L2

L1

L3

L2

C8

C6

L2

L3

C8

L3

S2

L3

L5

L4

L4

S1

S1

Anterior view

Posterior view

C.Machado
_M.D._

Schematic based on Lee MW, McPhee RW, Stringer MD. An evidence-based approach to human dermatomes. Clin Anat. 2008;21(5):363–373. doi: 10.1002/ca.20636. PMID: 18470936. Please note that these areas are not absolute and vary from person to person. S3, S4, S5, and Co supply the perineum but are not shown for reasons of clarity. Of note, the dermatomes are larger than illustrated as the figure is based on best evidence; gaps represent areas in which the data are inconclusive.

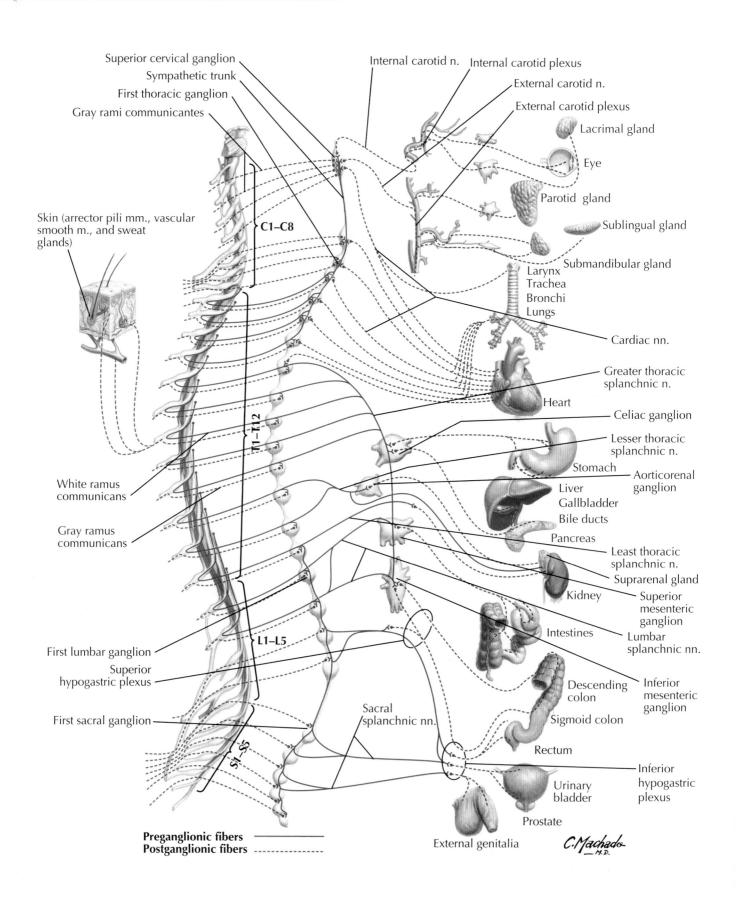

Superior cervical ganglion

Sympathetic trunk

First thoracic ganglion

Gray rami communicantes

Skin (arrector pili mm., vascular smooth m., and sweat glands)

C1–C8

Internal carotid n.

Internal carotid plexus

External carotid n.

External carotid plexus

Lacrimal gland

Eye

Parotid gland

Sublingual gland

Submandibular gland

Larynx
Trachea
Bronchi
Lungs

Cardiac nn.

Greater thoracic splanchnic n.

Heart

Celiac ganglion

Lesser thoracic splanchnic n.

Stomach

Aorticorenal ganglion

Liver
Gallbladder
Bile ducts

Pancreas

White ramus communicans

Gray ramus communicans

T1–T12

Least thoracic splanchnic n.

Suprarenal gland

Kidney

Superior mesenteric ganglion

Intestines

Lumbar splanchnic nn.

First lumbar ganglion

Superior hypogastric plexus

L1–L5

Descending colon

Inferior mesenteric ganglion

First sacral ganglion

Sacral splanchnic nn.

Sigmoid colon

Rectum

Inferior hypogastric plexus

S1–S5

Urinary bladder

Prostate

External genitalia

Preganglionic fibers ————
Postganglionic fibers - - - - - - - -

C. Machado
—M.D.

Plate 6

Systematic Anatomy

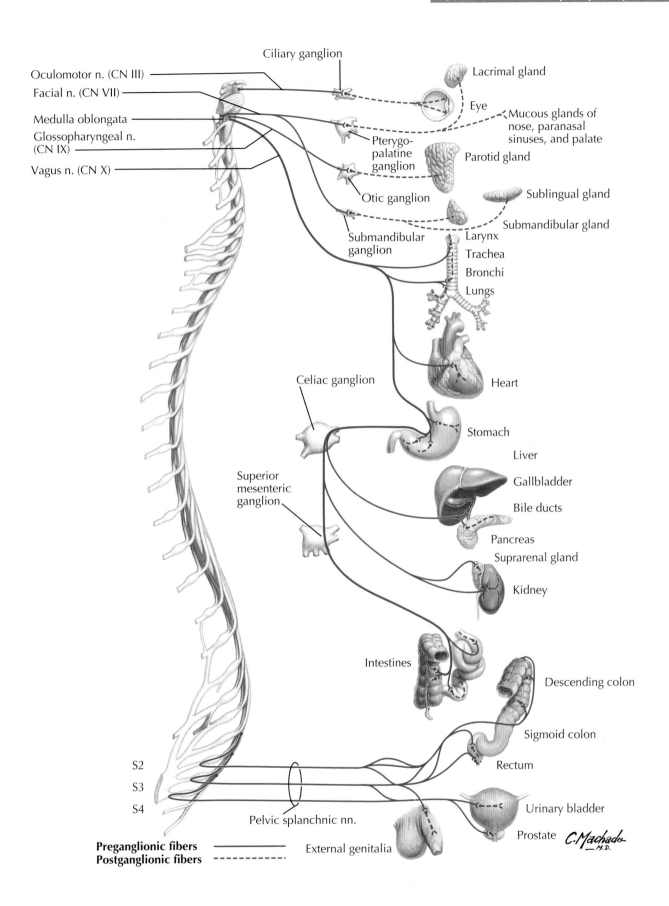

Ciliary ganglion

Oculomotor n. (CN III)

Lacrimal gland

Facial n. (CN VII)

Eye

Medulla oblongata

Mucous glands of
nose, paranasal
sinuses, and palate

Glossopharyngeal n.
(CN IX)

Pterygo-
palatine
ganglion

Parotid gland

Vagus n. (CN X)

Otic ganglion

Sublingual gland

Submandibular gland

Submandibular
ganglion

Larynx

Trachea

Bronchi

Lungs

Heart

Celiac ganglion

Stomach

Liver

Gallbladder

Bile ducts

Superior
mesenteric
ganglion

Pancreas

Suprarenal gland

Kidney

Intestines

Descending colon

Sigmoid colon

Rectum

S2

S3

S4

Urinary bladder

Pelvic splanchnic nn.

Prostate

Preganglionic fibers
Postganglionic fibers

External genitalia

C. Machado
_M.D._

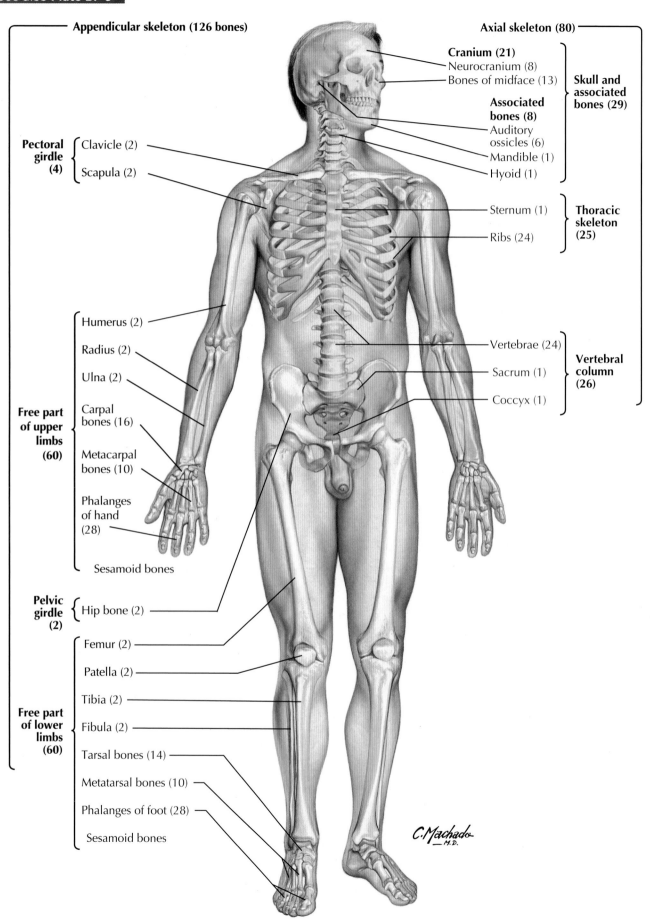

Appendicular skeleton (126 bones)

Axial skeleton (80)

Cranium (21)
Neurocranium (8)
Bones of midface (13)

Skull and associated bones (29)

Associated bones (8)
Auditory ossicles (6)
Mandible (1)
Hyoid (1)

Pectoral girdle (4)
Clavicle (2)
Scapula (2)

Sternum (1)

Ribs (24)

Thoracic skeleton (25)

Free part of upper limbs (60)
Humerus (2)
Radius (2)
Ulna (2)
Carpal bones (16)
Metacarpal bones (10)
Phalanges of hand (28)
Sesamoid bones

Vertebrae (24)

Sacrum (1)

Coccyx (1)

Vertebral column (26)

Pelvic girdle (2)
Hip bone (2)

Free part of lower limbs (60)
Femur (2)
Patella (2)
Tibia (2)
Fibula (2)
Tarsal bones (14)
Metatarsal bones (10)
Phalanges of foot (28)
Sesamoid bones

C. Machado
_M.D.

Plate 8

Systematic Anatomy

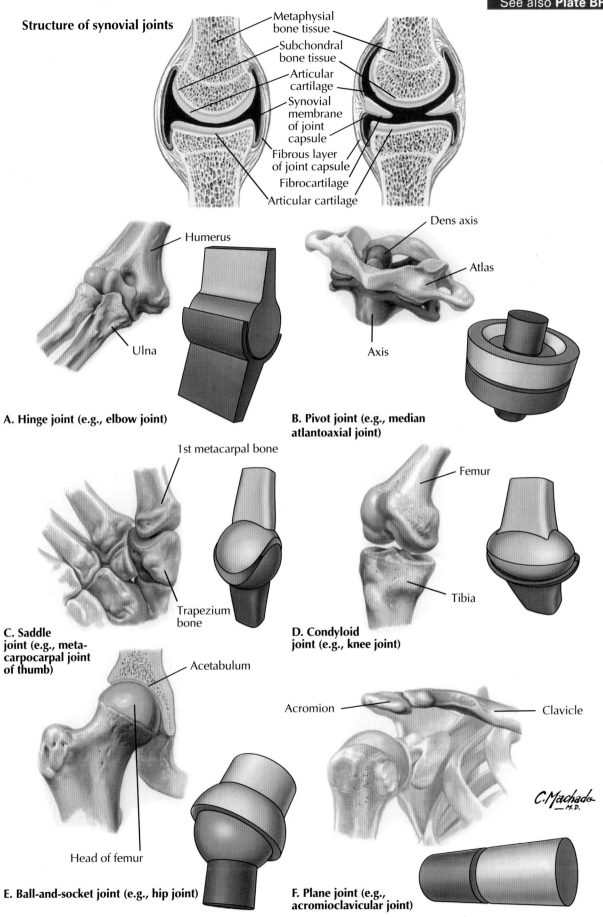

Structure of synovial joints

Metaphysial bone tissue
Subchondral bone tissue
Articular cartilage
Synovial membrane of joint capsule
Fibrous layer of joint capsule
Fibrocartilage
Articular cartilage

Humerus
Ulna

A. Hinge joint (e.g., elbow joint)

Dens axis
Atlas
Axis

B. Pivot joint (e.g., median atlantoaxial joint)

1st metacarpal bone
Trapezium bone

C. Saddle joint (e.g., metacarpocarpal joint of thumb)

Femur
Tibia

D. Condyloid joint (e.g., knee joint)

Acetabulum
Head of femur

E. Ball-and-socket joint (e.g., hip joint)

Acromion
Clavicle

F. Plane joint (e.g., acromioclavicular joint)

C.Machado
_M.D.

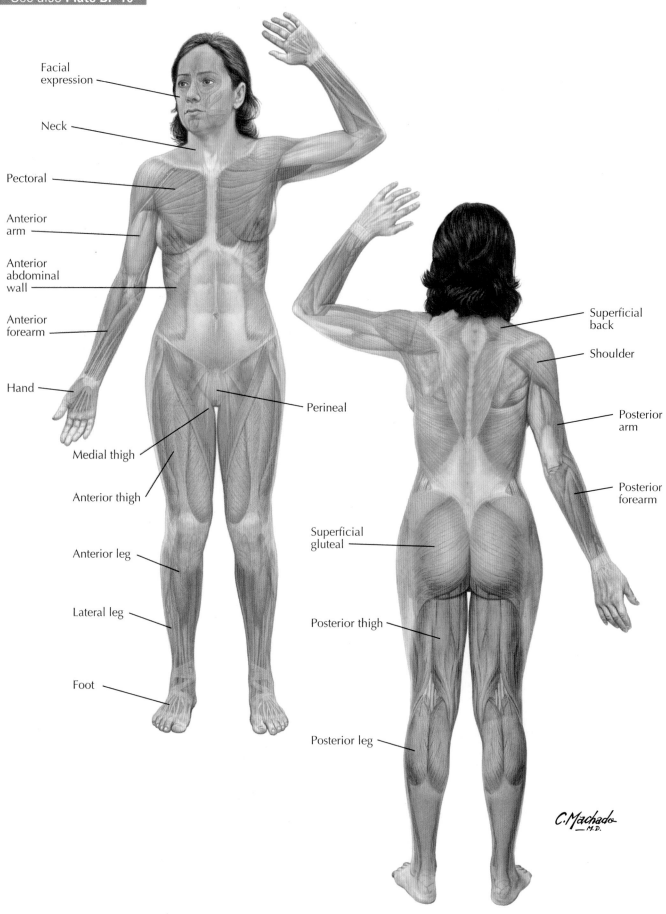

Facial expression

Neck

Pectoral

Anterior arm

Anterior abdominal wall

Anterior forearm

Hand

Medial thigh

Anterior thigh

Anterior leg

Lateral leg

Foot

Perineal

Superficial back

Shoulder

Posterior arm

Posterior forearm

Superficial gluteal

Posterior thigh

Posterior leg

C. Machado
M.D.

Plate 10

Systematic Anatomy

Segmental innervation of upper limb movements

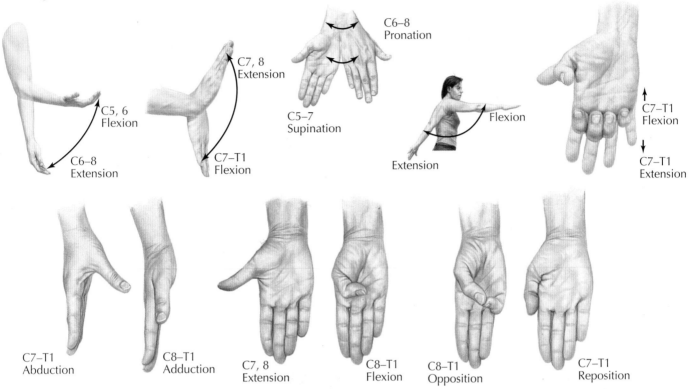

C6–8
Pronation

C7, 8
Extension

C5, 6
Flexion

C6–8
Extension

C5–7
Supination

C7–T1
Flexion

Flexion

Extension

C7–T1
Flexion

C7–T1
Extension

C7–T1
Abduction

C8–T1
Adduction

C7, 8
Extension

C8–T1
Flexion

C8–T1
Opposition

C7–T1
Reposition

Segmental innervation of lower limb movements

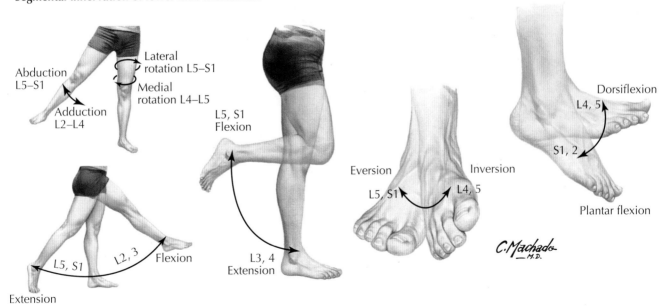

Abduction
L5–S1

Adduction
L2–L4

Lateral
rotation L5–S1

Medial
rotation L4–L5

L5, S1
Flexion

L5, S1
Extension

L2, 3
Flexion

L3, 4
Extension

Eversion
L5, S1

Inversion
L4, 5

Dorsiflexion
L4, 5

S1, 2

Plantar flexion

C.Machado
M.D.

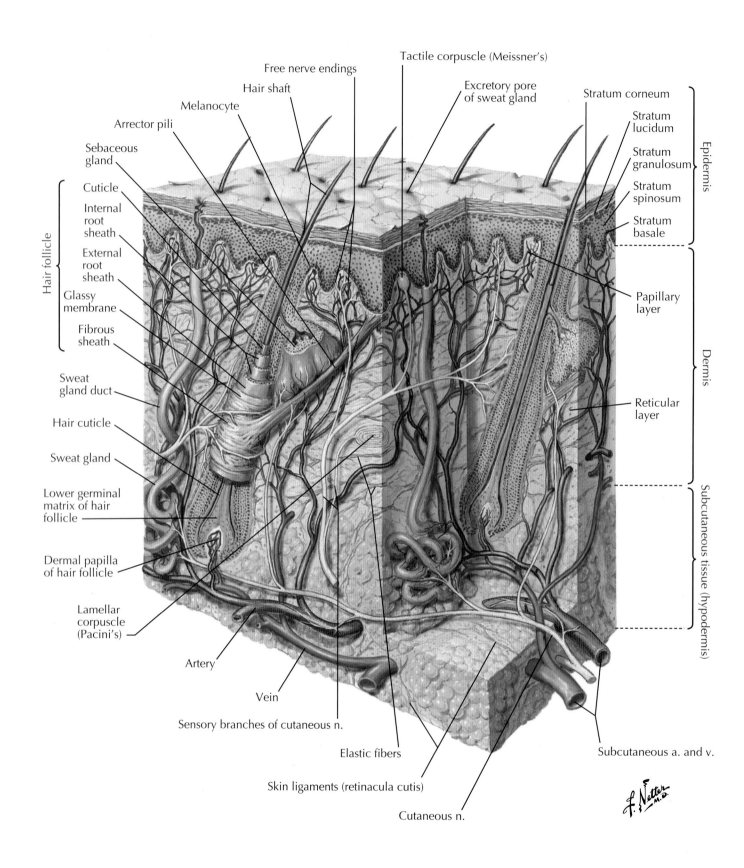

Free nerve endings

Tactile corpuscle (Meissner's)

Hair shaft

Excretory pore of sweat gland

Stratum corneum

Melanocyte

Stratum lucidum

Arrector pili

Stratum granulosum

Sebaceous gland

Stratum spinosum

Cuticle

Stratum basale

Internal root sheath

Epidermis

External root sheath

Papillary layer

Glassy membrane

Hair follicle

Fibrous sheath

Dermis

Sweat gland duct

Reticular layer

Hair cuticle

Sweat gland

Lower germinal matrix of hair follicle

Dermal papilla of hair follicle

Subcutaneous tissue (hypodermis)

Lamellar corpuscle (Pacini's)

Artery

Vein

Subcutaneous a. and v.

Sensory branches of cutaneous n.

Elastic fibers

Skin ligaments (retinacula cutis)

Cutaneous n.

Plate 12

Systematic Anatomy

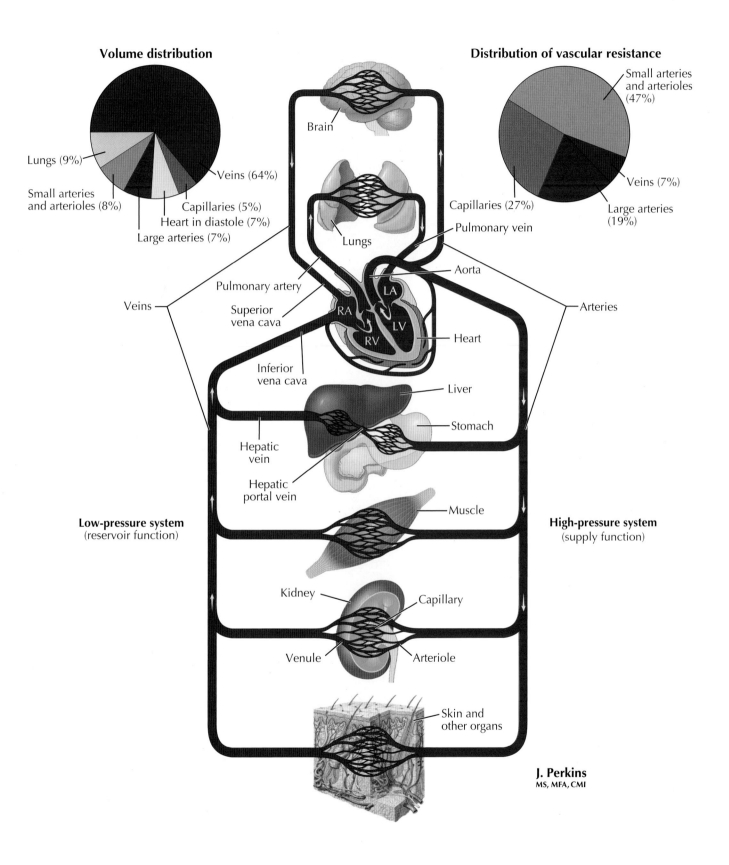

Volume distribution

Lungs (9%)

Veins (64%)

Small arteries and arterioles (8%)

Capillaries (5%)

Heart in diastole (7%)

Large arteries (7%)

Distribution of vascular resistance

Small arteries and arterioles (47%)

Veins (7%)

Capillaries (27%)

Large arteries (19%)

Brain

Lungs

Pulmonary vein

Aorta

LA

RA

LV

RV

Heart

Veins

Pulmonary artery

Superior vena cava

Arteries

Inferior vena cava

Liver

Stomach

Hepatic vein

Hepatic portal vein

Muscle

Low-pressure system
(reservoir function)

High-pressure system
(supply function)

Kidney

Capillary

Venule

Arteriole

Skin and other organs

J. Perkins
MS, MFA, CMI

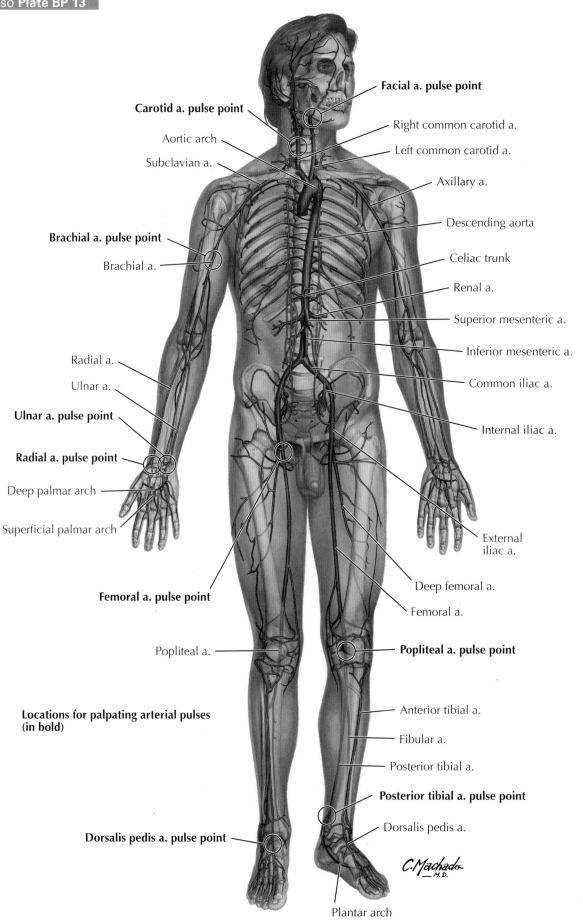

Facial a. pulse point

Carotid a. pulse point

Right common carotid a.

Aortic arch

Left common carotid a.

Subclavian a.

Axillary a.

Descending aorta

Brachial a. pulse point

Celiac trunk

Brachial a.

Renal a.

Superior mesenteric a.

Inferior mesenteric a.

Radial a.

Common iliac a.

Ulnar a.

Ulnar a. pulse point

Internal iliac a.

Radial a. pulse point

Deep palmar arch

Superficial palmar arch

External
iliac a.

Deep femoral a.

Femoral a. pulse point

Femoral a.

Popliteal a. pulse point

Popliteal a.

Anterior tibial a.

**Locations for palpating arterial pulses
(in bold)**

Fibular a.

Posterior tibial a.

Posterior tibial a. pulse point

Dorsalis pedis a. pulse point

Dorsalis pedis a.

Plantar arch

Plate 14

Systematic Anatomy

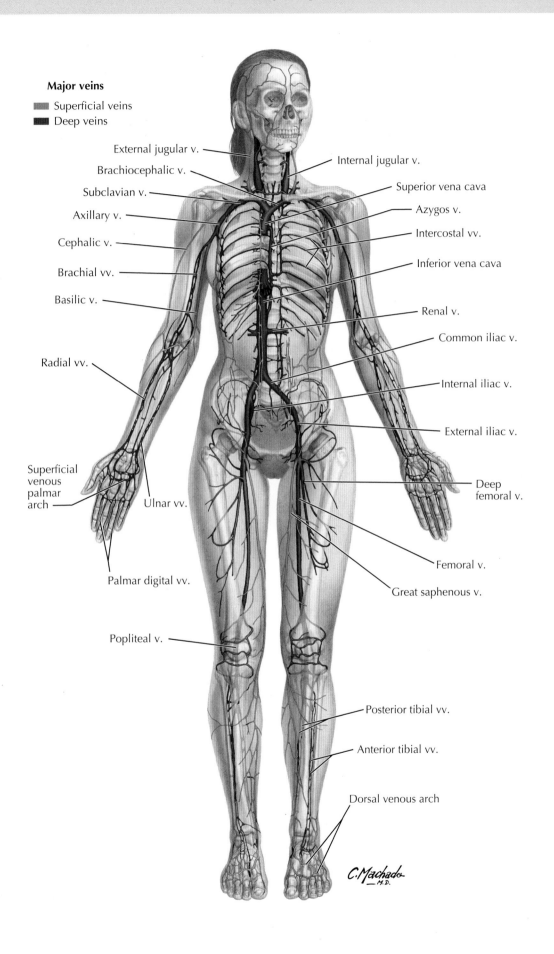

Major veins

Superficial veins

Deep veins

External jugular v.

Brachiocephalic v.

Subclavian v.

Axillary v.

Cephalic v.

Brachial vv.

Basilic v.

Radial vv.

Superficial venous palmar arch

Ulnar vv.

Palmar digital vv.

Popliteal v.

Internal jugular v.

Superior vena cava

Azygos v.

Intercostal vv.

Inferior vena cava

Renal v.

Common iliac v.

Internal iliac v.

External iliac v.

Deep femoral v.

Femoral v.

Great saphenous v.

Posterior tibial vv.

Anterior tibial vv.

Dorsal venous arch

C. Machado
M.D.

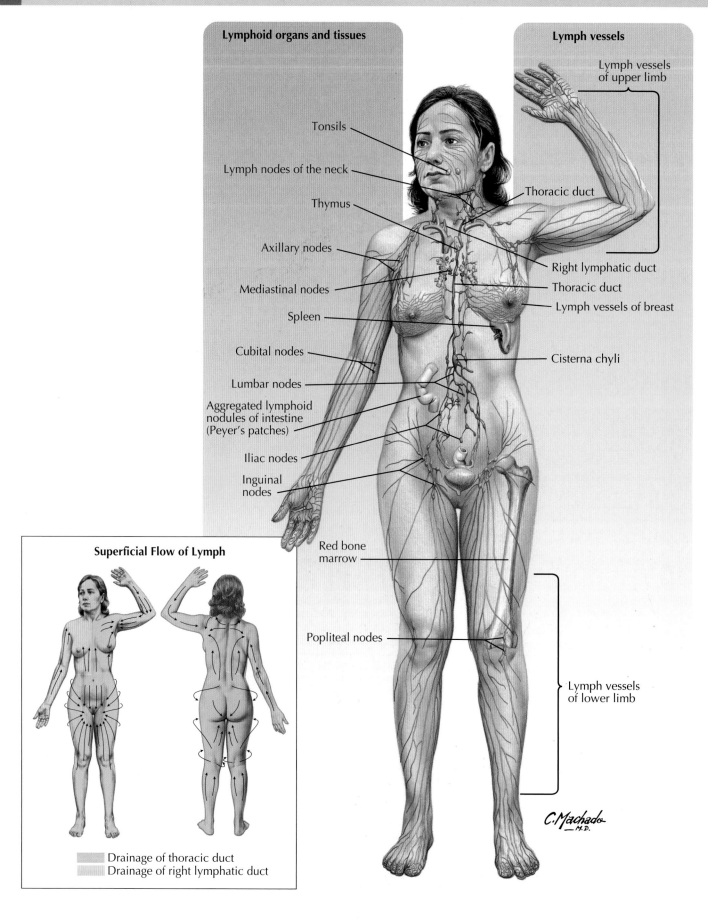

Lymphoid organs and tissues

Lymph vessels

Lymph vessels of upper limb

Tonsils

Lymph nodes of the neck

Thymus

Axillary nodes

Mediastinal nodes

Spleen

Cubital nodes

Lumbar nodes

Aggregated lymphoid nodules of intestine (Peyer's patches)

Iliac nodes

Inguinal nodes

Thoracic duct

Right lymphatic duct

Thoracic duct

Lymph vessels of breast

Cisterna chyli

Red bone marrow

Popliteal nodes

Lymph vessels of lower limb

Superficial Flow of Lymph

Drainage of thoracic duct
Drainage of right lymphatic duct

C. Machado M.D.

Plate 16

Systematic Anatomy

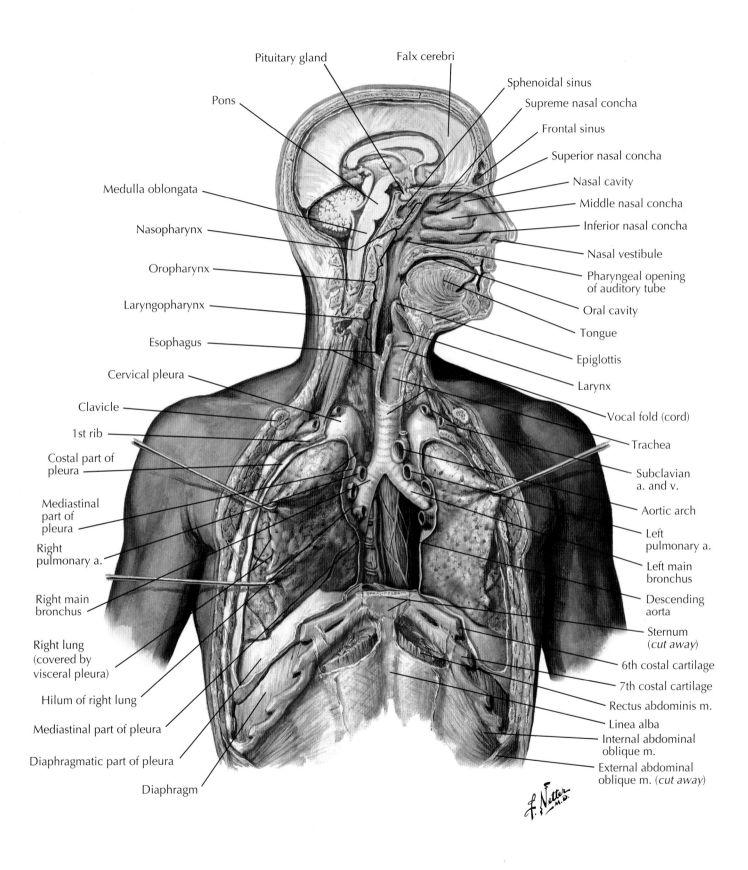

Pituitary gland

Falx cerebri

Pons

Sphenoidal sinus

Supreme nasal concha

Frontal sinus

Superior nasal concha

Medulla oblongata

Nasal cavity

Middle nasal concha

Nasopharynx

Inferior nasal concha

Oropharynx

Nasal vestibule

Pharyngeal opening
of auditory tube

Laryngopharynx

Oral cavity

Esophagus

Tongue

Cervical pleura

Epiglottis

Larynx

Clavicle

Vocal fold (cord)

1st rib

Trachea

Costal part of
pleura

Subclavian
a. and v.

Mediastinal
part of
pleura

Aortic arch

Left
pulmonary a.

Right
pulmonary a.

Left main
bronchus

Right main
bronchus

Descending
aorta

Right lung
(covered by
visceral pleura)

Sternum
(cut away)

6th costal cartilage

7th costal cartilage

Hilum of right lung

Rectus abdominis m.

Mediastinal part of pleura

Linea alba

Internal abdominal
oblique m.

Diaphragmatic part of pleura

External abdominal
oblique m. (cut away)

Diaphragm

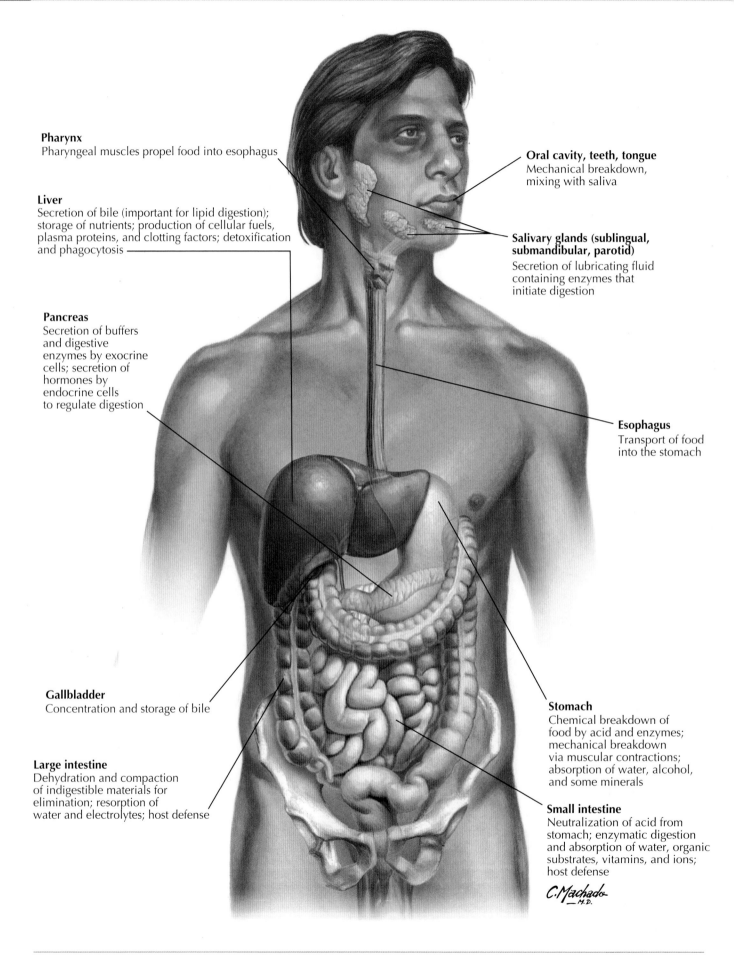

Pharynx
Pharyngeal muscles propel food into esophagus

Liver
Secretion of bile (important for lipid digestion); storage of nutrients; production of cellular fuels, plasma proteins, and clotting factors; detoxification and phagocytosis

Pancreas
Secretion of buffers and digestive enzymes by exocrine cells; secretion of hormones by endocrine cells to regulate digestion

Oral cavity, teeth, tongue
Mechanical breakdown, mixing with saliva

Salivary glands (sublingual, submandibular, parotid)
Secretion of lubricating fluid containing enzymes that initiate digestion

Esophagus
Transport of food into the stomach

Gallbladder
Concentration and storage of bile

Large intestine
Dehydration and compaction of indigestible materials for elimination; resorption of water and electrolytes; host defense

Stomach
Chemical breakdown of food by acid and enzymes; mechanical breakdown via muscular contractions; absorption of water, alcohol, and some minerals

Small intestine
Neutralization of acid from stomach; enzymatic digestion and absorption of water, organic substrates, vitamins, and ions; host defense

Plate 18

Systematic Anatomy

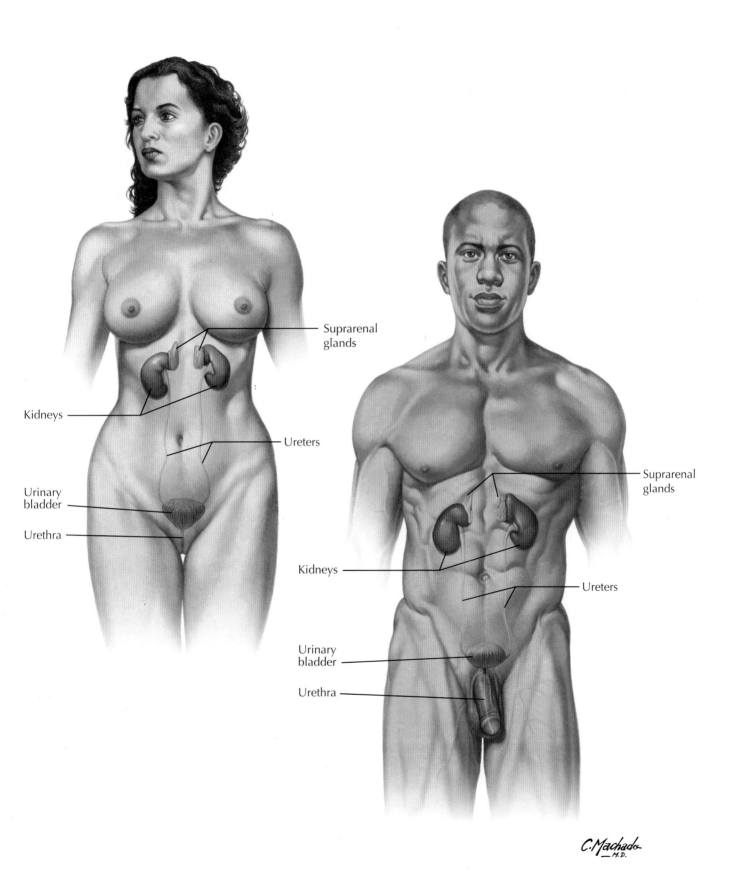

Suprarenal glands

Kidneys

Ureters

Urinary bladder

Urethra

Suprarenal glands

Kidneys

Ureters

Urinary bladder

Urethra

C.Machado
_M.D.

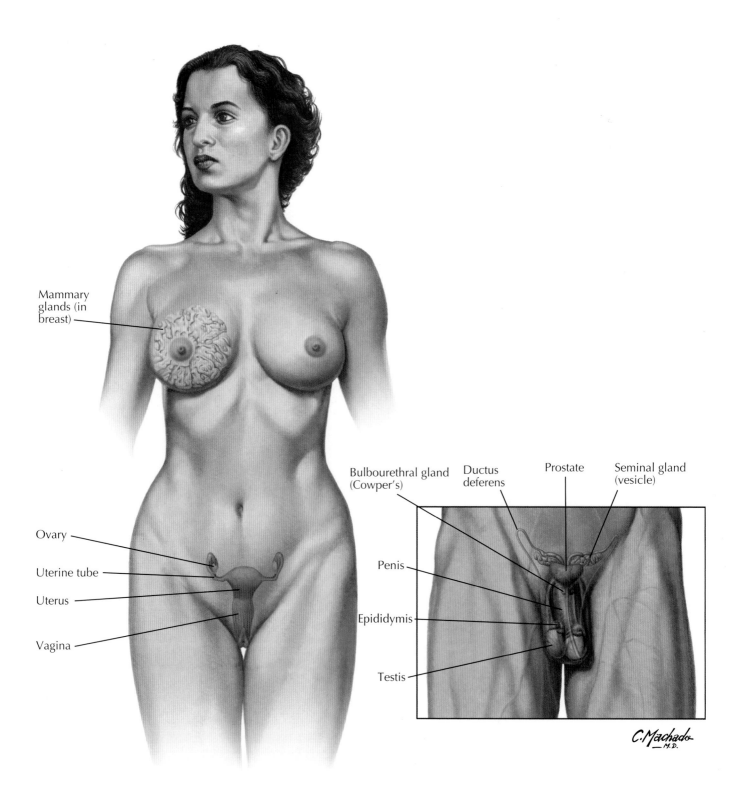

Mammary glands (in breast)

Bulbourethral gland (Cowper's)

Ductus deferens

Prostate

Seminal gland (vesicle)

Penis

Ovary

Uterine tube

Uterus

Epididymis

Vagina

Testis

C. Machado
M.D.

Plate 20

Systematic Anatomy

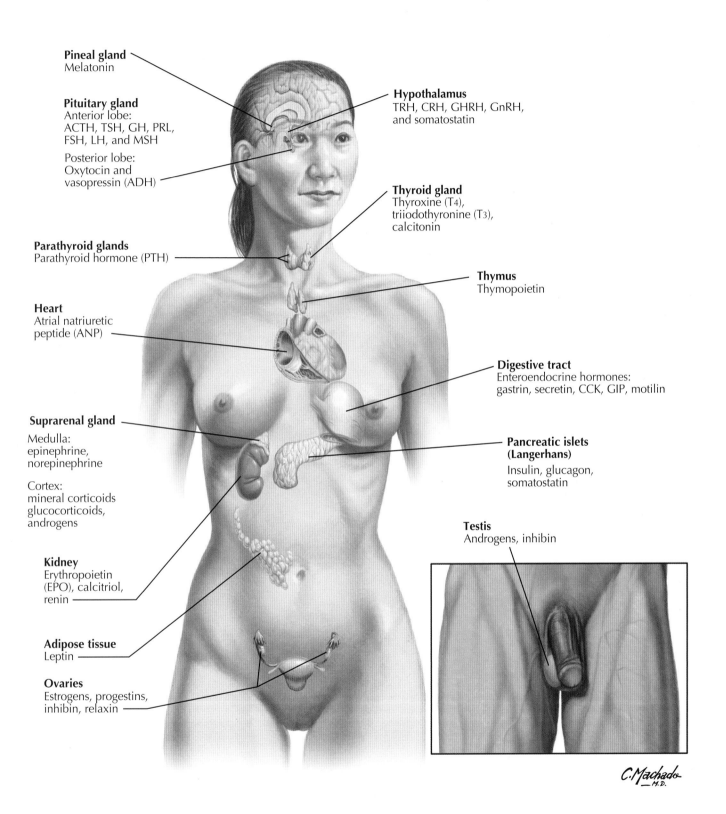

Pineal gland
Melatonin

Pituitary gland
Anterior lobe:
ACTH, TSH, GH, PRL,
FSH, LH, and MSH

Posterior lobe:
Oxytocin and
vasopressin (ADH)

Hypothalamus
TRH, CRH, GHRH, GnRH,
and somatostatin

Thyroid gland
Thyroxine (T4),
triiodothyronine (T3),
calcitonin

Parathyroid glands
Parathyroid hormone (PTH)

Thymus
Thymopoietin

Heart
Atrial natriuretic
peptide (ANP)

Digestive tract
Enteroendocrine hormones:
gastrin, secretin, CCK, GIP, motilin

Suprarenal gland

Medulla:
epinephrine,
norepinephrine

Cortex:
mineral corticoids
glucocorticoids,
androgens

**Pancreatic islets
(Langerhans)**

Insulin, glucagon,
somatostatin

Testis
Androgens, inhibin

Kidney
Erythropoietin
(EPO), calcitriol,
renin

Adipose tissue
Leptin

Ovaries
Estrogens, progestins,
inhibin, relaxin

C. Machado
_M.D.

HEAD AND NECK 2

ELECTRONIC BONUS PLATES

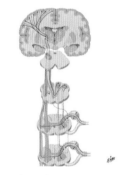

BP 14 Somatosensory System: Trunk and Limbs

BP 15 Pyramidal System

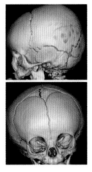

BP 16 3D Skull Reconstruction CTs

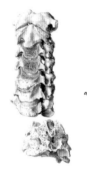

BP 17 Degenerative Changes in Cervical Vertebrae

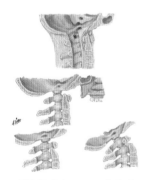

BP 18 Atlantooccipital Junction

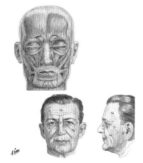

BP 19 Muscles of Facial Expression: Anterior View

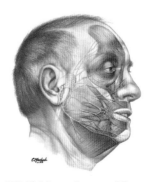

BP 20 Musculature of Face

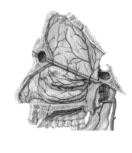

BP 21 Arteries of Nasal Cavity: Bony Nasal Septum Turned Up

ELECTRONIC BONUS PLATES—*cont'd*

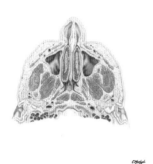

BP 22 Nose and Maxillary Sinus: Transverse Section

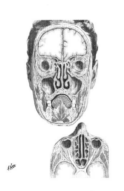

BP 23 Paranasal Sinuses

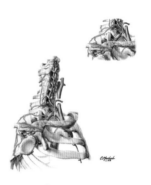

BP 24 Subclavian Artery

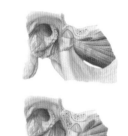

BP 25 Opening of the Mandible

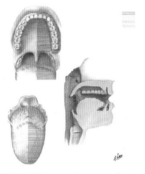

BP 26 Afferent Innervation of Oral Cavity and Pharynx

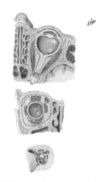

BP 27 Fasciae of Orbit and Eyeball

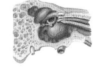

BP 28 Tympanic Cavity: Medial and Lateral Views

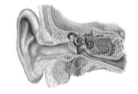

BP 29 Anatomy of the Pediatric Ear

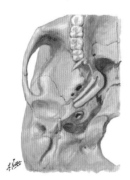

BP 30 Auditory Tube (Eustachian)

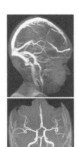
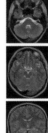

BP 31 Cranial Imaging (MRV and MRA)

BP 32 Axial and Coronal MRIs of Brain

See also **Plates 18, 188**

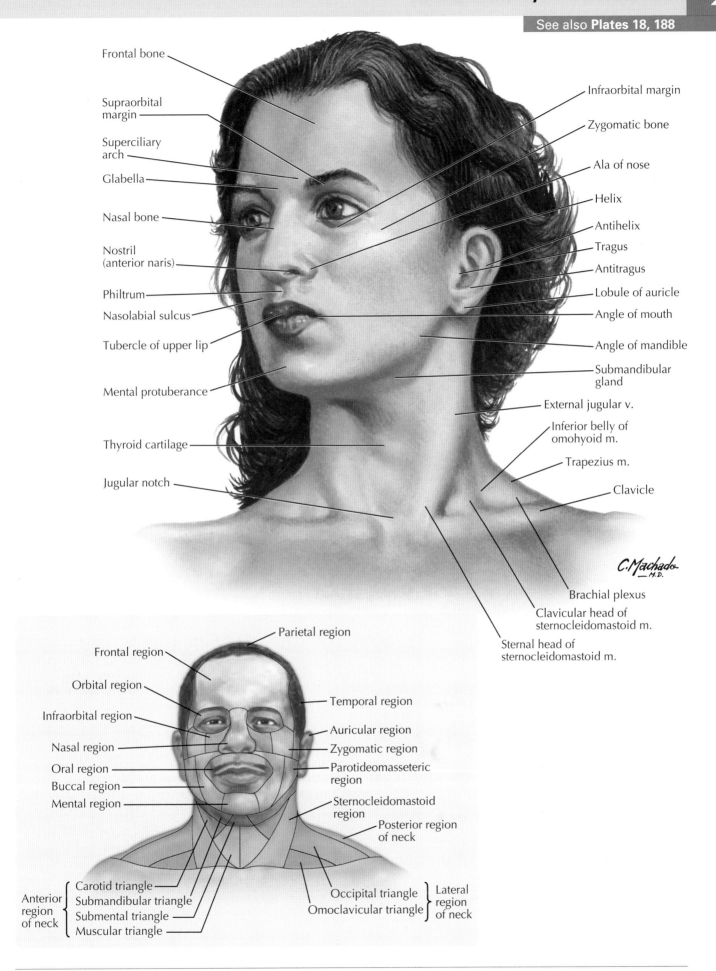

Frontal bone

Supraorbital margin

Superciliary arch

Glabella

Nasal bone

Nostril (anterior naris)

Philtrum

Nasolabial sulcus

Tubercle of upper lip

Mental protuberance

Thyroid cartilage

Jugular notch

Infraorbital margin

Zygomatic bone

Ala of nose

Helix

Antihelix

Tragus

Antitragus

Lobule of auricle

Angle of mouth

Angle of mandible

Submandibular gland

External jugular v.

Inferior belly of omohyoid m.

Trapezius m.

Clavicle

Brachial plexus

Clavicular head of sternocleidomastoid m.

Sternal head of sternocleidomastoid m.

C. Machado
M.D.

Parietal region

Frontal region

Orbital region

Infraorbital region

Nasal region

Oral region

Buccal region

Mental region

Temporal region

Auricular region

Zygomatic region

Parotideomasseteric region

Sternocleidomastoid region

Posterior region of neck

Anterior region of neck {
Carotid triangle
Submandibular triangle
Submental triangle
Muscular triangle

Occipital triangle
Omoclavicular triangle

Lateral region of neck

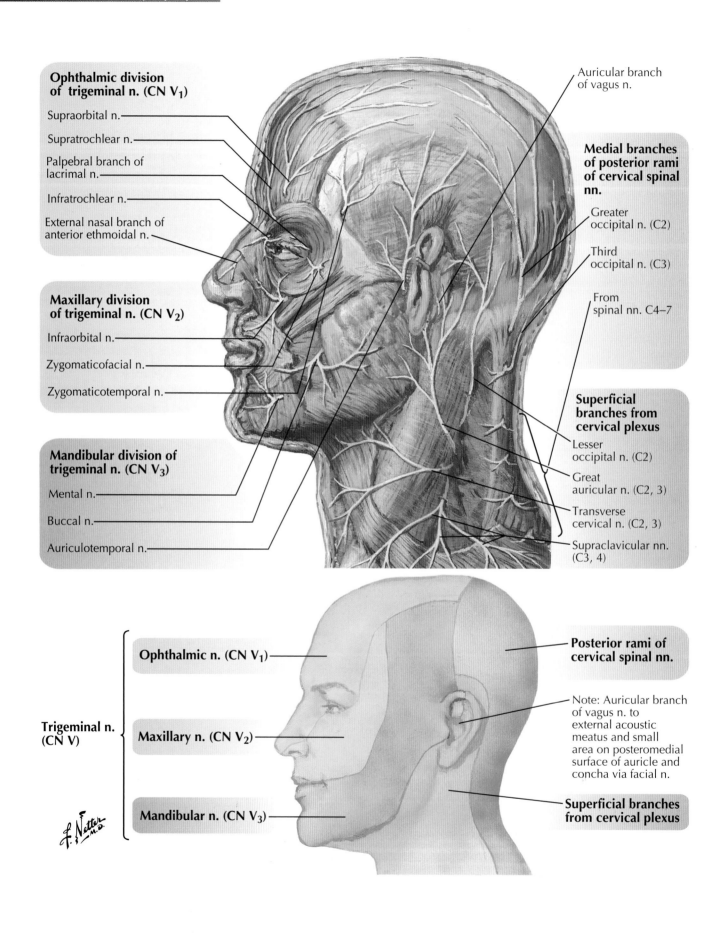

Ophthalmic division of trigeminal n. (CN V₁)

Supraorbital n.

Supratrochlear n.

Palpebral branch of lacrimal n.

Infratrochlear n.

External nasal branch of anterior ethmoidal n.

Maxillary division of trigeminal n. (CN V₂)

Infraorbital n.

Zygomaticofacial n.

Zygomaticotemporal n.

Mandibular division of trigeminal n. (CN V₃)

Mental n.

Buccal n.

Auriculotemporal n.

Auricular branch of vagus n.

Medial branches of posterior rami of cervical spinal nn.

Greater occipital n. (C2)

Third occipital n. (C3)

From spinal nn. C4–7

Superficial branches from cervical plexus

Lesser occipital n. (C2)

Great auricular n. (C2, 3)

Transverse cervical n. (C2, 3)

Supraclavicular nn. (C3, 4)

Ophthalmic n. (CN V₁)

Trigeminal n. (CN V)

Maxillary n. (CN V₂)

Mandibular n. (CN V₃)

Posterior rami of cervical spinal nn.

Note: Auricular branch of vagus n. to external acoustic meatus and small area on posteromedial surface of auricle and concha via facial n.

Superficial branches from cervical plexus

Plate 23

Surface Anatomy

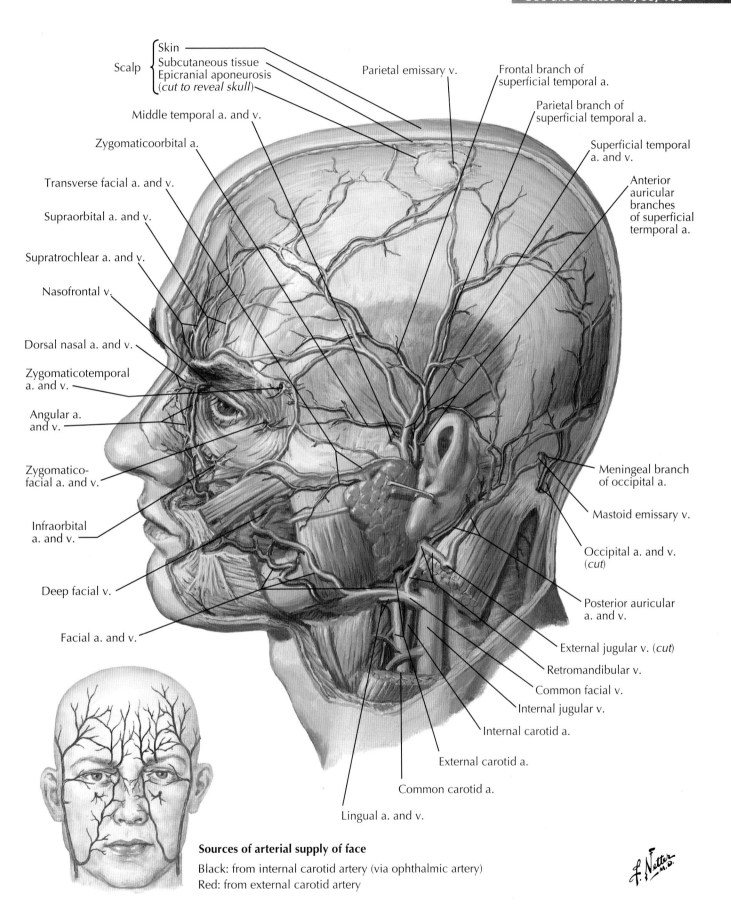

Scalp
- Skin
- Subcutaneous tissue
- Epicranial aponeurosis
- (*cut to reveal skull*)

Middle temporal a. and v.

Zygomaticoorbital a.

Transverse facial a. and v.

Supraorbital a. and v.

Supratrochlear a. and v.

Nasofrontal v.

Dorsal nasal a. and v.

Zygomaticotemporal a. and v.

Angular a. and v.

Zygomatico-facial a. and v.

Infraorbital a. and v.

Deep facial v.

Facial a. and v.

Parietal emissary v.

Frontal branch of superficial temporal a.

Parietal branch of superficial temporal a.

Superficial temporal a. and v.

Anterior auricular branches of superficial termporal a.

Meningeal branch of occipital a.

Mastoid emissary v.

Occipital a. and v. (*cut*)

Posterior auricular a. and v.

External jugular v. (*cut*)

Retromandibular v.

Common facial v.

Internal jugular v.

Internal carotid a.

External carotid a.

Common carotid a.

Lingual a. and v.

Sources of arterial supply of face

Black: from internal carotid artery (via ophthalmic artery)
Red: from external carotid artery

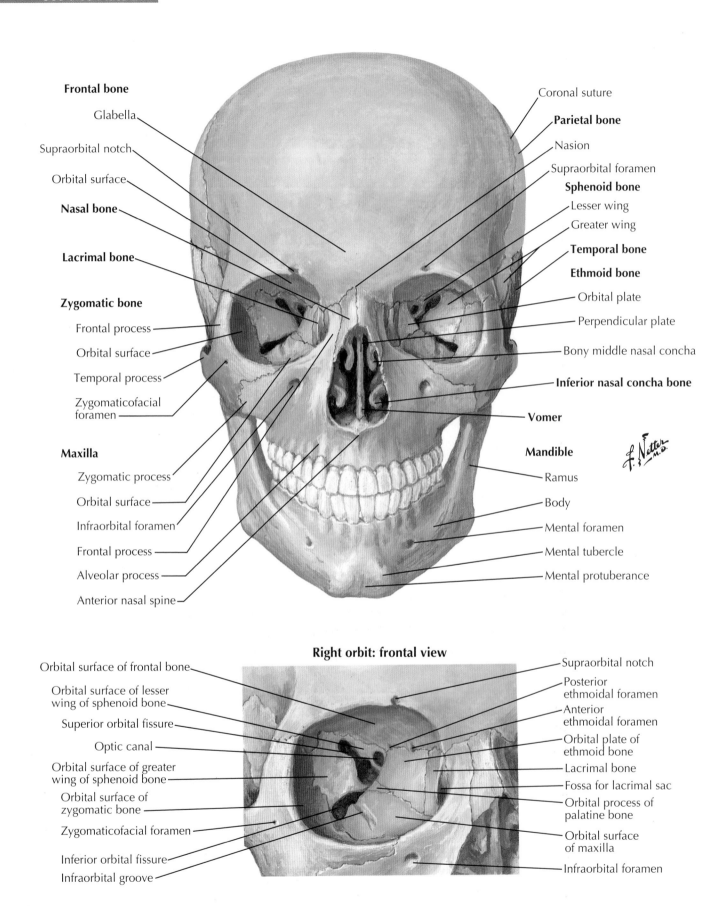

Frontal bone

Glabella

Supraorbital notch

Orbital surface

Nasal bone

Lacrimal bone

Zygomatic bone

Frontal process

Orbital surface

Temporal process

Zygomaticofacial foramen

Maxilla

Zygomatic process

Orbital surface

Infraorbital foramen

Frontal process

Alveolar process

Anterior nasal spine

Coronal suture

Parietal bone

Nasion

Supraorbital foramen

Sphenoid bone

Lesser wing

Greater wing

Temporal bone

Ethmoid bone

Orbital plate

Perpendicular plate

Bony middle nasal concha

Inferior nasal concha bone

Vomer

Mandible

Ramus

Body

Mental foramen

Mental tubercle

Mental protuberance

Right orbit: frontal view

Orbital surface of frontal bone

Orbital surface of lesser wing of sphenoid bone

Superior orbital fissure

Optic canal

Orbital surface of greater wing of sphenoid bone

Orbital surface of zygomatic bone

Zygomaticofacial foramen

Inferior orbital fissure

Infraorbital groove

Supraorbital notch

Posterior ethmoidal foramen

Anterior ethmoidal foramen

Orbital plate of ethmoid bone

Lacrimal bone

Fossa for lacrimal sac

Orbital process of palatine bone

Orbital surface of maxilla

Infraorbital foramen

Plate 25

Bones and Joints

Posterior anterior view

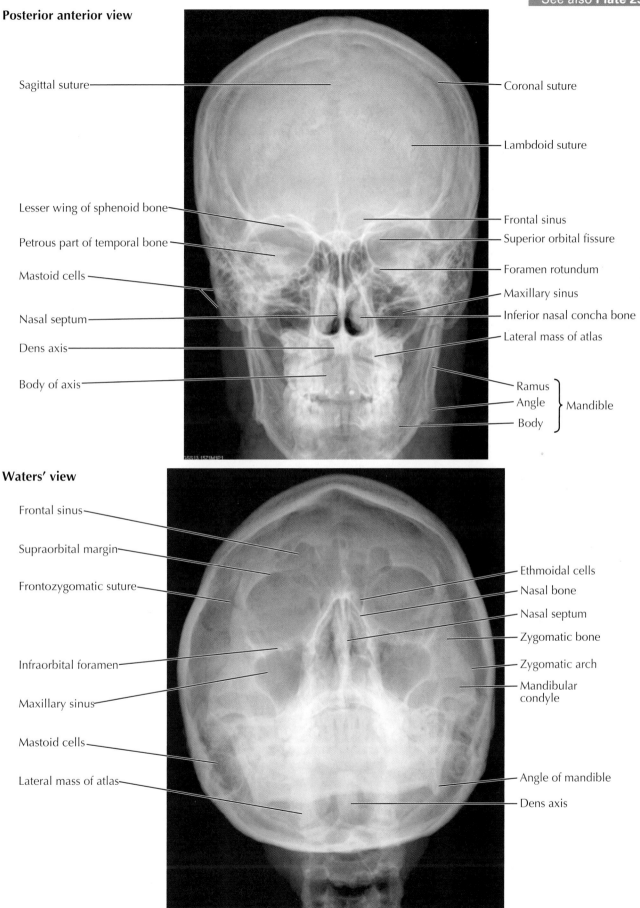

Sagittal suture

Coronal suture

Lambdoid suture

Lesser wing of sphenoid bone

Petrous part of temporal bone

Mastoid cells

Nasal septum

Dens axis

Body of axis

Frontal sinus

Superior orbital fissure

Foramen rotundum

Maxillary sinus

Inferior nasal concha bone

Lateral mass of atlas

Ramus ⎫
Angle ⎬ Mandible
Body ⎭

Waters' view

Frontal sinus

Supraorbital margin

Frontozygomatic suture

Infraorbital foramen

Maxillary sinus

Mastoid cells

Lateral mass of atlas

Ethmoidal cells

Nasal bone

Nasal septum

Zygomatic bone

Zygomatic arch

Mandibular condyle

Angle of mandible

Dens axis

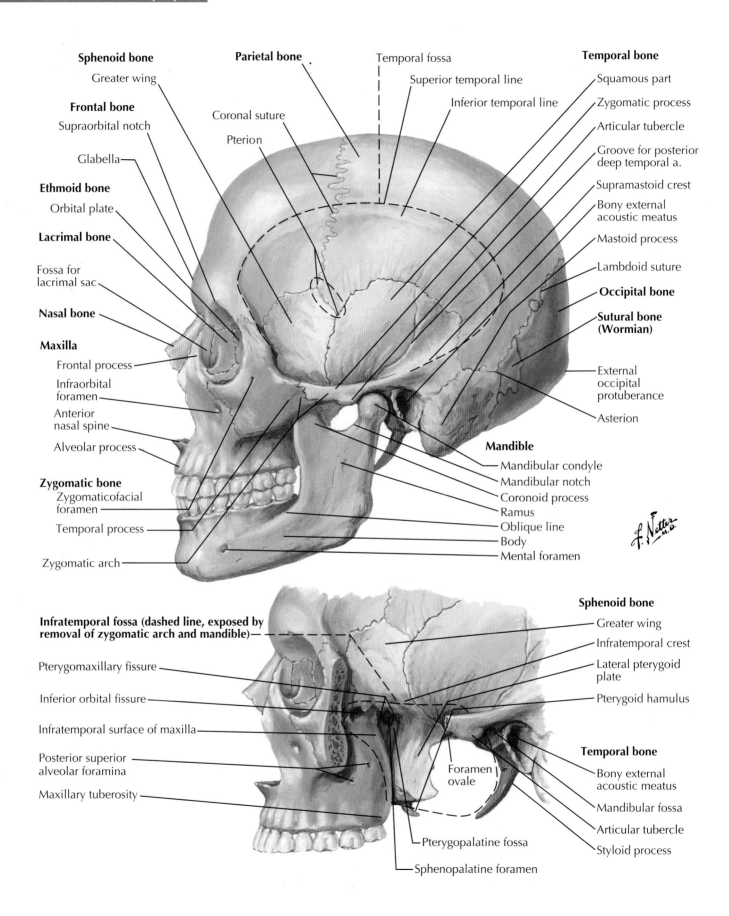

Sphenoid bone
Greater wing

Frontal bone
Supraorbital notch

Glabella

Ethmoid bone
Orbital plate

Lacrimal bone

Fossa for
lacrimal sac

Nasal bone

Maxilla
Frontal process

Infraorbital
foramen

Anterior
nasal spine

Alveolar process

Zygomatic bone
Zygomaticofacial
foramen

Temporal process

Zygomatic arch

Parietal bone

Coronal suture

Pterion

Temporal fossa
Superior temporal line
Inferior temporal line

Temporal bone
Squamous part
Zygomatic process
Articular tubercle
Groove for posterior
deep temporal a.
Supramastoid crest
Bony external
acoustic meatus
Mastoid process
Lambdoid suture
Occipital bone
**Sutural bone
(Wormian)**

External
occipital
protuberance

Asterion

Mandible
Mandibular condyle
Mandibular notch
Coronoid process
Ramus
Oblique line
Body
Mental foramen

Infratemporal fossa (dashed line, exposed by
removal of zygomatic arch and mandible)

Pterygomaxillary fissure

Inferior orbital fissure

Infratemporal surface of maxilla

Posterior superior
alveolar foramina

Maxillary tuberosity

Sphenoid bone
Greater wing
Infratemporal crest
Lateral pterygoid
plate
Pterygoid hamulus

Temporal bone
Bony external
acoustic meatus
Mandibular fossa
Articular tubercle
Styloid process

Foramen
ovale

Pterygopalatine fossa

Sphenopalatine foramen

Plate 27   **Bones and Joints**

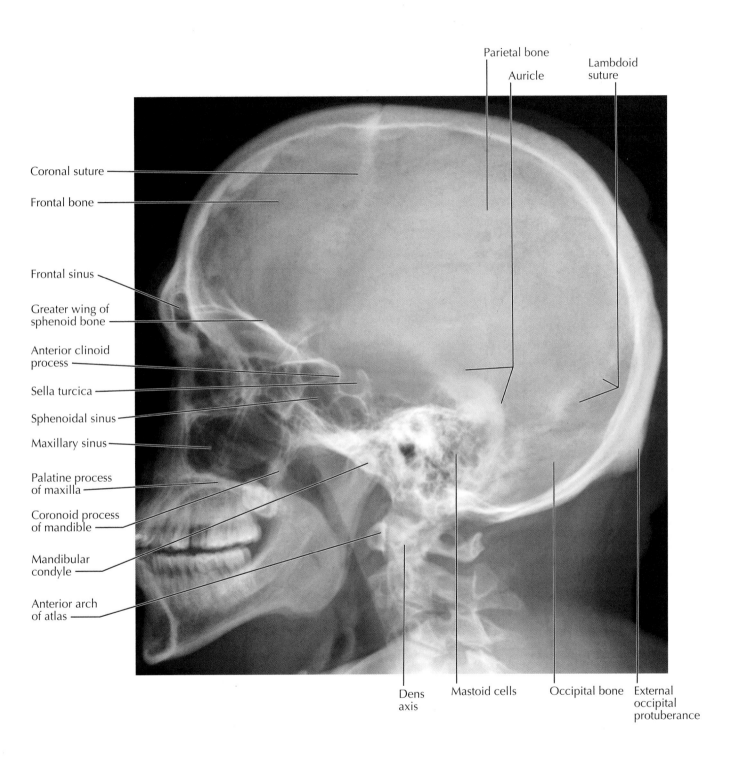

Parietal bone

Auricle

Lambdoid suture

Coronal suture

Frontal bone

Frontal sinus

Greater wing of sphenoid bone

Anterior clinoid process

Sella turcica

Sphenoidal sinus

Maxillary sinus

Palatine process of maxilla

Coronoid process of mandible

Mandibular condyle

Anterior arch of atlas

Dens axis

Mastoid cells

Occipital bone

External occipital protuberance

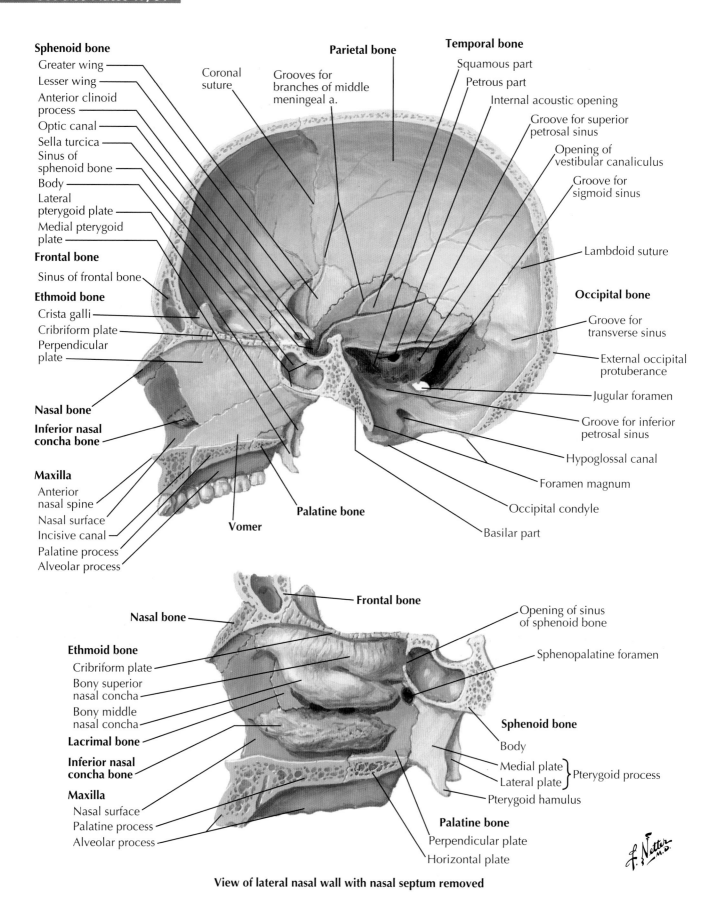

Sphenoid bone
Greater wing
Lesser wing
Anterior clinoid process
Optic canal
Sella turcica
Sinus of sphenoid bone
Body
Lateral pterygoid plate
Medial pterygoid plate

Frontal bone
Sinus of frontal bone

Ethmoid bone
Crista galli
Cribriform plate
Perpendicular plate

Nasal bone

Inferior nasal concha bone

Maxilla
Anterior nasal spine
Nasal surface
Incisive canal
Palatine process
Alveolar process

Coronal suture

Grooves for branches of middle meningeal a.

Parietal bone

Temporal bone
Squamous part
Petrous part
Internal acoustic opening
Groove for superior petrosal sinus
Opening of vestibular canaliculus
Groove for sigmoid sinus

Lambdoid suture

Occipital bone
Groove for transverse sinus
External occipital protuberance
Jugular foramen
Groove for inferior petrosal sinus
Hypoglossal canal
Foramen magnum
Occipital condyle
Basilar part

Palatine bone

Vomer

Nasal bone

Ethmoid bone
Cribriform plate
Bony superior nasal concha
Bony middle nasal concha

Lacrimal bone

Inferior nasal concha bone

Maxilla
Nasal surface
Palatine process
Alveolar process

Frontal bone

Opening of sinus of sphenoid bone

Sphenopalatine foramen

Sphenoid bone
Body
Medial plate
Lateral plate
Pterygoid hamulus

Palatine bone
Perpendicular plate
Horizontal plate

Pterygoid process

View of lateral nasal wall with nasal septum removed

Plate 29

Bones and Joints

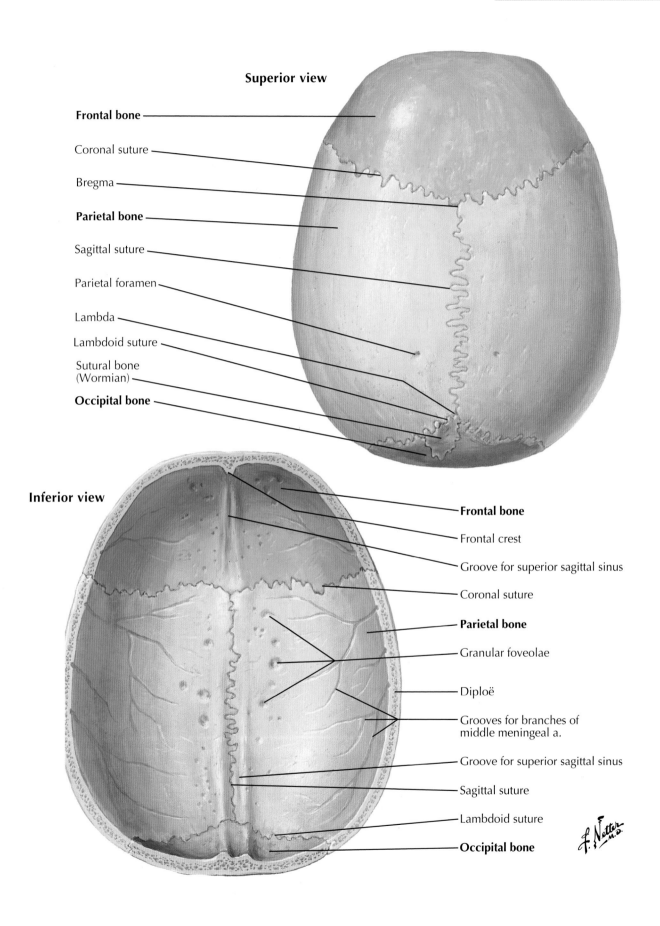

Superior view

Frontal bone

Coronal suture

Bregma

Parietal bone

Sagittal suture

Parietal foramen

Lambda

Lambdoid suture

Sutural bone
(Wormian)

Occipital bone

Inferior view

Frontal bone

Frontal crest

Groove for superior sagittal sinus

Coronal suture

Parietal bone

Granular foveolae

Diploë

Grooves for branches of
middle meningeal a.

Groove for superior sagittal sinus

Sagittal suture

Lambdoid suture

Occipital bone

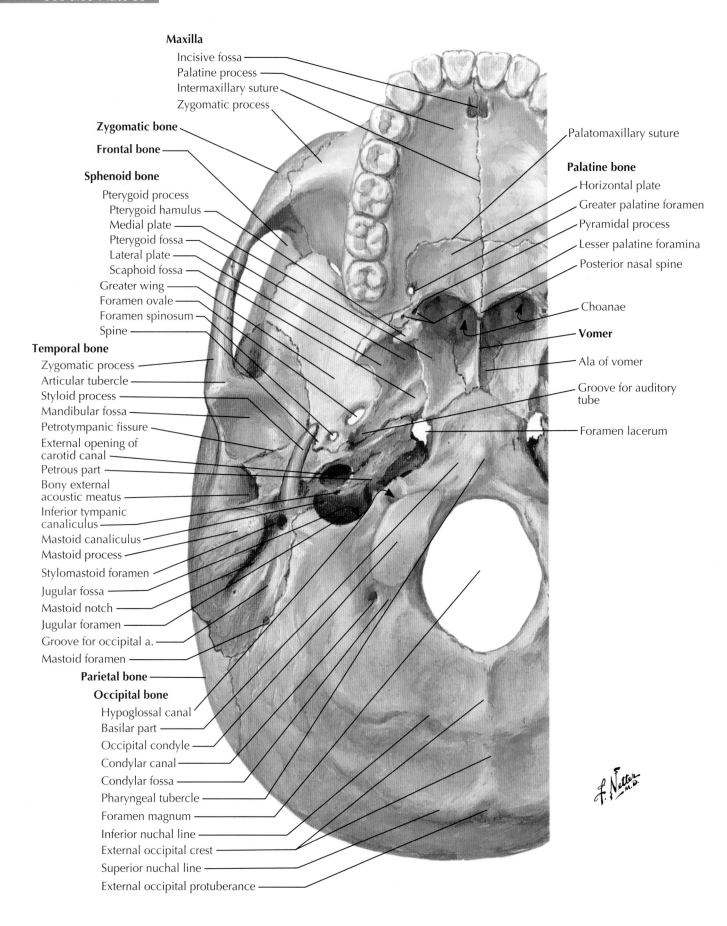

Maxilla
Incisive fossa
Palatine process
Intermaxillary suture
Zygomatic process

Zygomatic bone

Frontal bone

Sphenoid bone
Pterygoid process
Pterygoid hamulus
Medial plate
Pterygoid fossa
Lateral plate
Scaphoid fossa
Greater wing
Foramen ovale
Foramen spinosum
Spine

Temporal bone
Zygomatic process
Articular tubercle
Styloid process
Mandibular fossa
Petrotympanic fissure
External opening of carotid canal
Petrous part
Bony external acoustic meatus
Inferior tympanic canaliculus
Mastoid canaliculus
Mastoid process
Stylomastoid foramen
Jugular fossa
Mastoid notch
Jugular foramen
Groove for occipital a.
Mastoid foramen

Parietal bone

Occipital bone
Hypoglossal canal
Basilar part
Occipital condyle
Condylar canal
Condylar fossa
Pharyngeal tubercle
Foramen magnum
Inferior nuchal line
External occipital crest
Superior nuchal line
External occipital protuberance

Palatomaxillary suture

Palatine bone
Horizontal plate
Greater palatine foramen
Pyramidal process
Lesser palatine foramina
Posterior nasal spine

Choanae

Vomer

Ala of vomer

Groove for auditory tube

Foramen lacerum

Plate 31

Bones and Joints

Frontal bone
Groove for superior sagittal sinus
Frontal crest
Groove for anterior meningeal a.
Foramen cecum
Orbital part

Ethmoid bone
Crista galli
Cribriform plate

Sphenoid bone
Lesser wing
Anterior clinoid process
Greater wing
Groove for frontal branches
of middle meningeal a.
Body
Sphenoidal yoke
Chiasmatic sulcus
Tuberculum sellae
Sella   Hypophysial fossa
turcica   Dorsum sellae
Posterior clinoid process
Carotid sulcus
Clivus

Temporal bone
Squamous part
Petrous part
Groove for lesser petrosal n.
Groove for greater petrosal n.
Trigeminal impression
Arcuate eminence
Groove for superior petrosal sinus
Groove for sigmoid sinus

Parietal bone
Groove for parietal branches
of middle meningeal a.
Mastoid angle

Occipital bone
Clivus
Groove for inferior petrosal sinus
Basilar part
Groove for posterior meningeal aa.
Occipital condyle
Groove for transverse sinus
Groove for occipital sinus
Internal occipital crest
Internal occipital protuberance
Groove for superior sagittal sinus

Anterior
cranial
fossa

Middle
cranial
fossa

Posterior
cranial
fossa

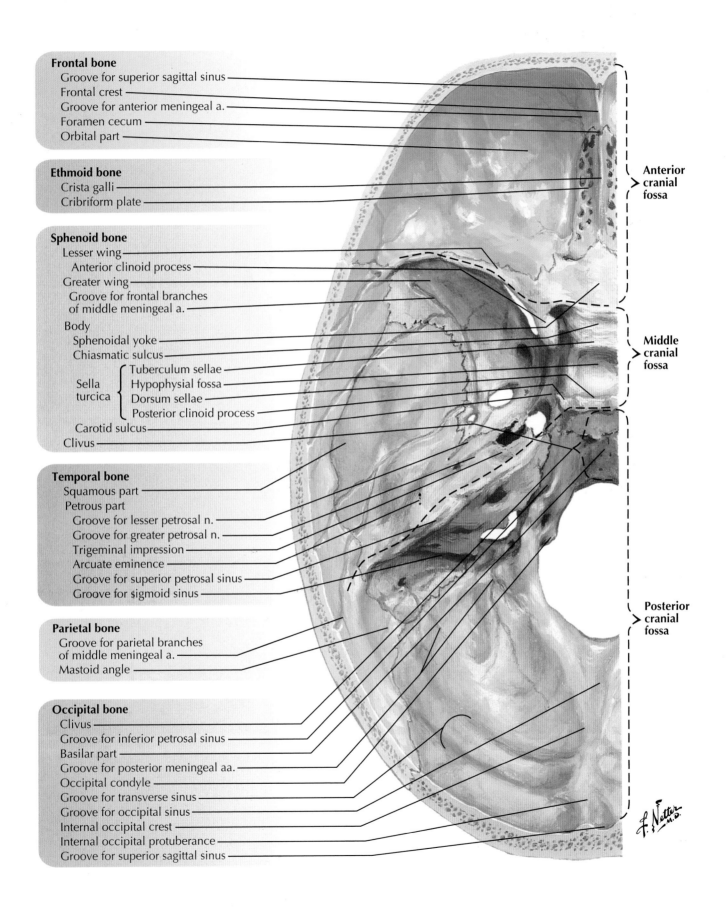

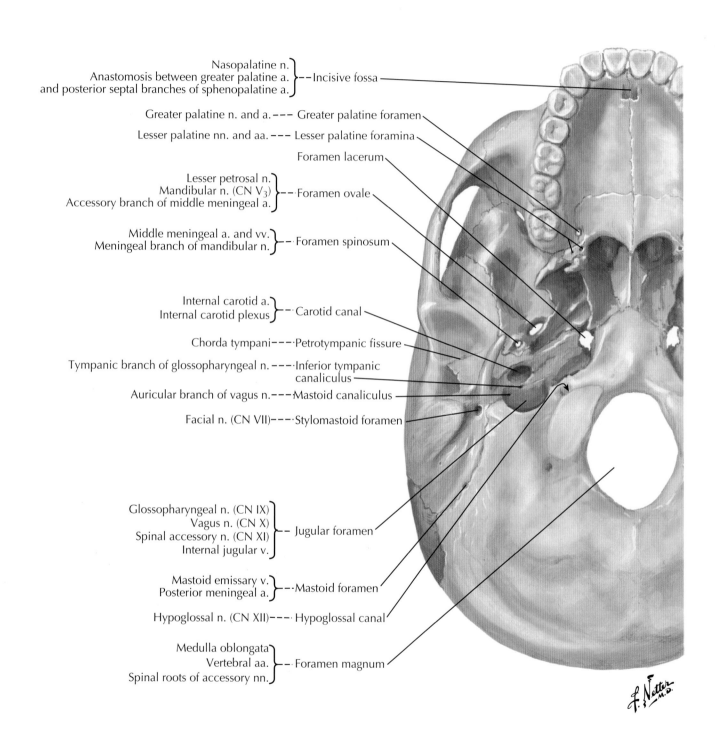

Nasopalatine n.
Anastomosis between greater palatine a.
and posterior septal branches of sphenopalatine a. } — Incisive fossa

Greater palatine n. and a. — — — Greater palatine foramen

Lesser palatine nn. and aa. — — — Lesser palatine foramina

Foramen lacerum

Lesser petrosal n.
Mandibular n. (CN V₃) } — — Foramen ovale
Accessory branch of middle meningeal a.

Middle meningeal a. and vv.
Meningeal branch of mandibular n. } — — Foramen spinosum

Internal carotid a.
Internal carotid plexus } — — Carotid canal

Chorda tympani — — — Petrotympanic fissure

Tympanic branch of glossopharyngeal n. — — — Inferior tympanic canaliculus

Auricular branch of vagus n. — — — Mastoid canaliculus

Facial n. (CN VII) — — — Stylomastoid foramen

Glossopharyngeal n. (CN IX)
Vagus n. (CN X)
Spinal accessory n. (CN XI) } — Jugular foramen
Internal jugular v.

Mastoid emissary v.
Posterior meningeal a. } — — Mastoid foramen

Hypoglossal n. (CN XII) — — — Hypoglossal canal

Medulla oblongata
Vertebral aa. } — — Foramen magnum
Spinal roots of accessory nn.

Plate 33

Bones and Joints

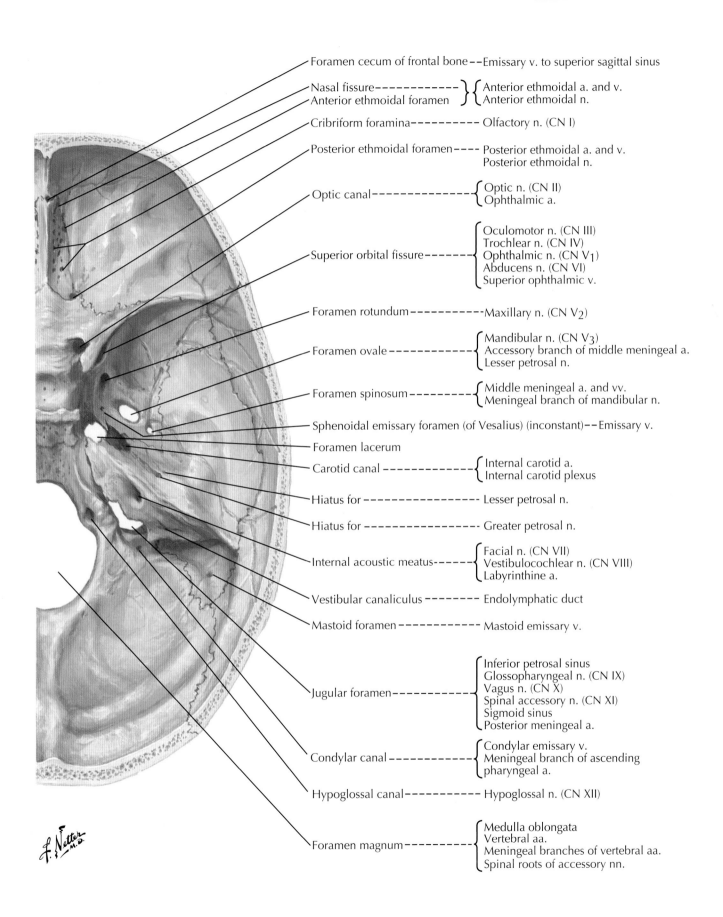

Foramen cecum of frontal bone--Emissary v. to superior sagittal sinus

Nasal fissure------------ } { Anterior ethmoidal a. and v.
Anterior ethmoidal foramen   { Anterior ethmoidal n.

Cribriform foramina--------- Olfactory n. (CN I)

Posterior ethmoidal foramen---- Posterior ethmoidal a. and v.
                                Posterior ethmoidal n.

Optic canal------------- { Optic n. (CN II)
                         { Ophthalmic a.

Superior orbital fissure------- { Oculomotor n. (CN III)
                                { Trochlear n. (CN IV)
                                { Ophthalmic n. (CN V$_1$)
                                { Abducens n. (CN VI)
                                { Superior ophthalmic v.

Foramen rotundum----------Maxillary n. (CN V$_2$)

Foramen ovale----------- { Mandibular n. (CN V$_3$)
                         { Accessory branch of middle meningeal a.
                         { Lesser petrosal n.

Foramen spinosum-------- { Middle meningeal a. and vv.
                         { Meningeal branch of mandibular n.

Sphenoidal emissary foramen (of Vesalius) (inconstant)--Emissary v.

Foramen lacerum

Carotid canal ----------- { Internal carotid a.
                          { Internal carotid plexus

Hiatus for --------------- Lesser petrosal n.

Hiatus for --------------- Greater petrosal n.

Internal acoustic meatus------ { Facial n. (CN VII)
                               { Vestibulocochlear n. (CN VIII)
                               { Labyrinthine a.

Vestibular canaliculus -------- Endolymphatic duct

Mastoid foramen----------- Mastoid emissary v.

Jugular foramen---------- { Inferior petrosal sinus
                          { Glossopharyngeal n. (CN IX)
                          { Vagus n. (CN X)
                          { Spinal accessory n. (CN XI)
                          { Sigmoid sinus
                          { Posterior meningeal a.

Condylar canal----------- { Condylar emissary v.
                          { Meningeal branch of ascending
                          { pharyngeal a.

Hypoglossal canal---------- Hypoglossal n. (CN XII)

Foramen magnum--------- { Medulla oblongata
                        { Vertebral aa.
                        { Meningeal branches of vertebral aa.
                        { Spinal roots of accessory nn.

f. Netter
M.D.

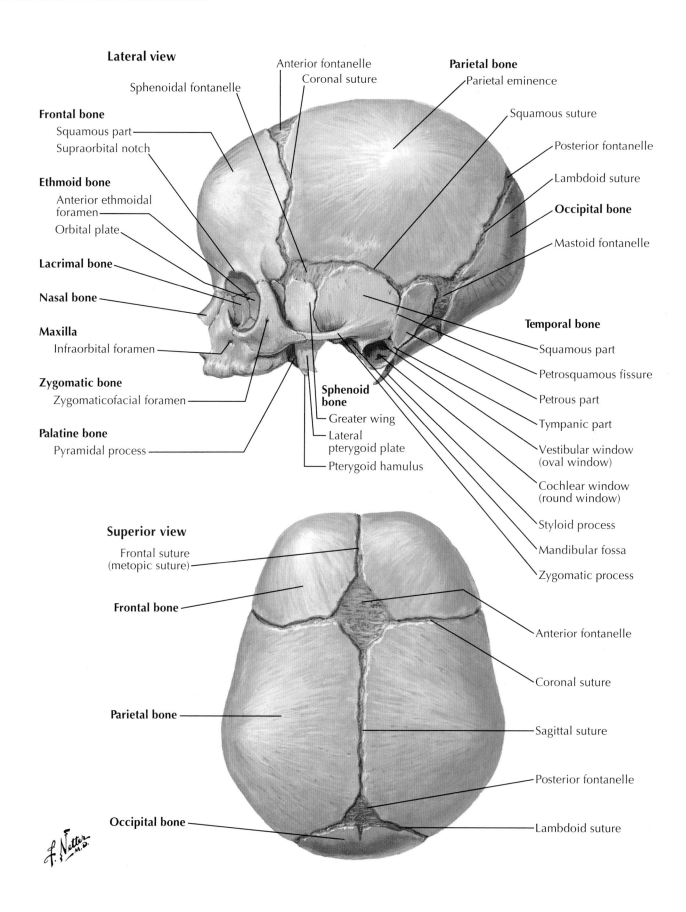

Lateral view

Sphenoidal fontanelle

Anterior fontanelle
Coronal suture

Parietal bone
Parietal eminence

Frontal bone
Squamous part
Supraorbital notch

Squamous suture

Posterior fontanelle

Lambdoid suture

Ethmoid bone
Anterior ethmoidal foramen
Orbital plate

Occipital bone

Mastoid fontanelle

Lacrimal bone

Nasal bone

Maxilla
Infraorbital foramen

Temporal bone
Squamous part
Petrosquamous fissure

Zygomatic bone
Zygomaticofacial foramen

Sphenoid bone
Greater wing
Lateral pterygoid plate
Pterygoid hamulus

Petrous part

Tympanic part

Vestibular window (oval window)

Cochlear window (round window)

Palatine bone
Pyramidal process

Styloid process

Mandibular fossa

Zygomatic process

Superior view
Frontal suture (metopic suture)

Frontal bone

Anterior fontanelle

Coronal suture

Parietal bone

Sagittal suture

Posterior fontanelle

Occipital bone

Lambdoid suture

Plate 35

Bones and Joints

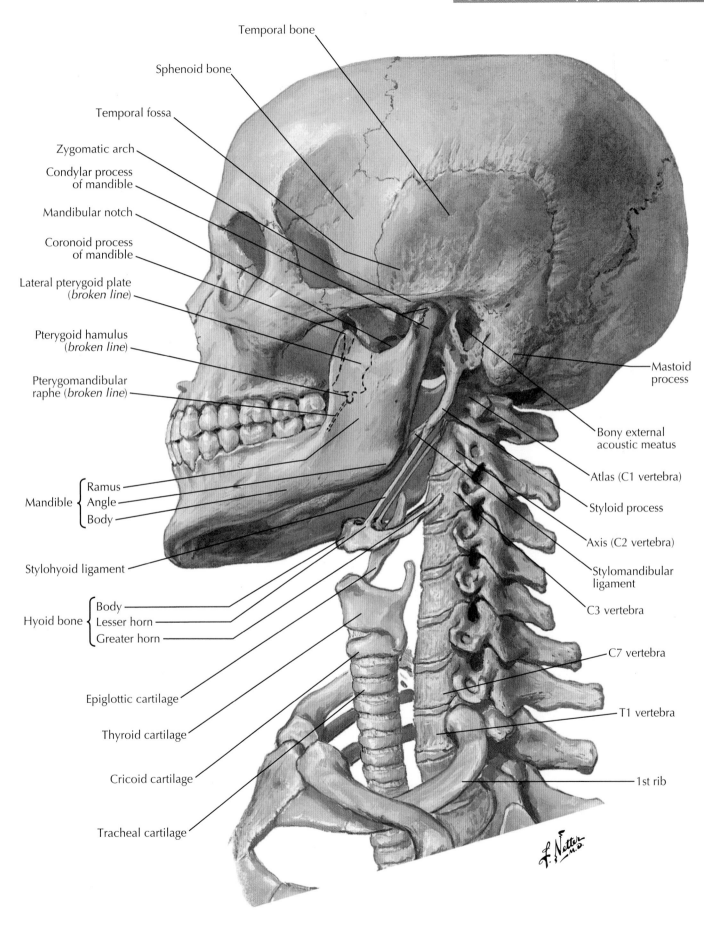

Temporal bone

Sphenoid bone

Temporal fossa

Zygomatic arch

Condylar process of mandible

Mandibular notch

Coronoid process of mandible

Lateral pterygoid plate (*broken line*)

Pterygoid hamulus (*broken line*)

Pterygomandibular raphe (*broken line*)

Mandible {
Ramus
Angle
Body
}

Stylohyoid ligament

Hyoid bone {
Body
Lesser horn
Greater horn
}

Epiglottic cartilage

Thyroid cartilage

Cricoid cartilage

Tracheal cartilage

Mastoid process

Bony external acoustic meatus

Atlas (C1 vertebra)

Styloid process

Axis (C2 vertebra)

Stylomandibular ligament

C3 vertebra

C7 vertebra

T1 vertebra

1st rib

Plane of section   1 Ethmoid bone
                       2 Inferior nasal
                                  concha bone

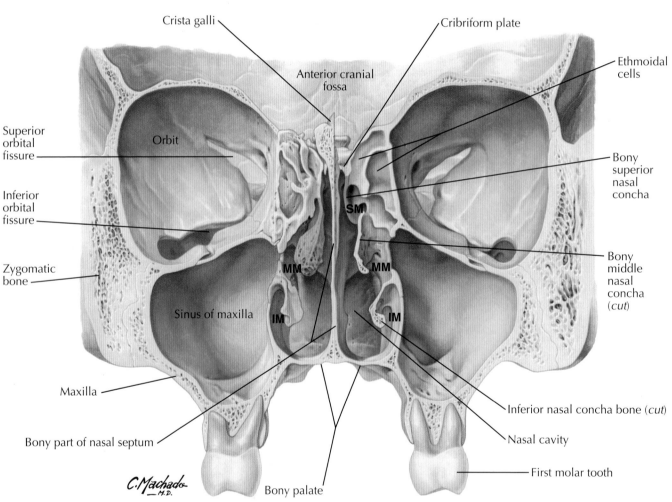

Crista galli

Cribriform plate

Anterior cranial fossa

Ethmoidal cells

Superior orbital fissure

Orbit

Bony superior nasal concha

Inferior orbital fissure

SM

Zygomatic bone

MM   MM

Bony middle nasal concha (*cut*)

Sinus of maxilla   IM   IM

Maxilla

Inferior nasal concha bone (*cut*)

Nasal cavity

Bony part of nasal septum

First molar tooth

Bony palate

Bony superior nasal concha   **SM**   Superior nasal meatus

Bony middle nasal concha   **MM**   Middle nasal meatus

Inferior nasal concha bone   **IM**   Inferior nasal meatus

Posterior view

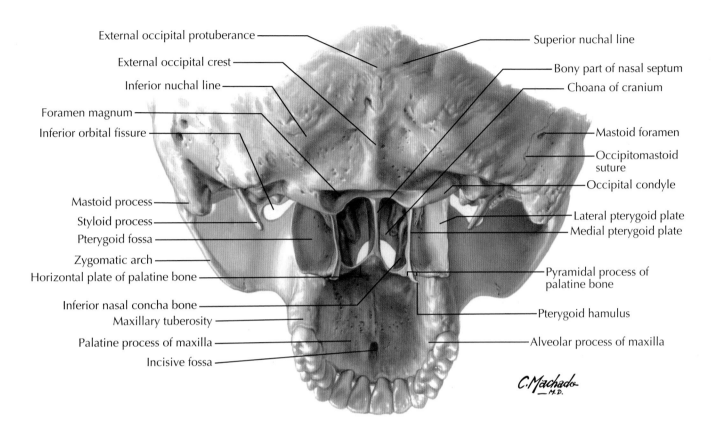

External occipital protuberance

External occipital crest

Inferior nuchal line

Foramen magnum

Inferior orbital fissure

Mastoid process

Styloid process

Pterygoid fossa

Zygomatic arch

Horizontal plate of palatine bone

Inferior nasal concha bone

Maxillary tuberosity

Palatine process of maxilla

Incisive fossa

Superior nuchal line

Bony part of nasal septum

Choana of cranium

Mastoid foramen

Occipitomastoid suture

Occipital condyle

Lateral pterygoid plate

Medial pterygoid plate

Pyramidal process of palatine bone

Pterygoid hamulus

Alveolar process of maxilla

C. Machado —M.D.

Lateral view

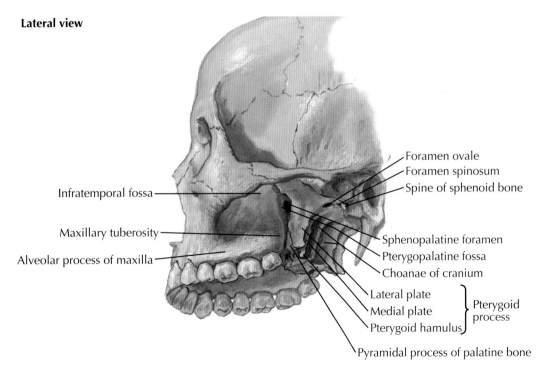

Infratemporal fossa

Maxillary tuberosity

Alveolar process of maxilla

Foramen ovale

Foramen spinosum

Spine of sphenoid bone

Sphenopalatine foramen

Pterygopalatine fossa

Choanae of cranium

Lateral plate

Medial plate — Pterygoid process

Pterygoid hamulus

Pyramidal process of palatine bone

F. Netter M.D.

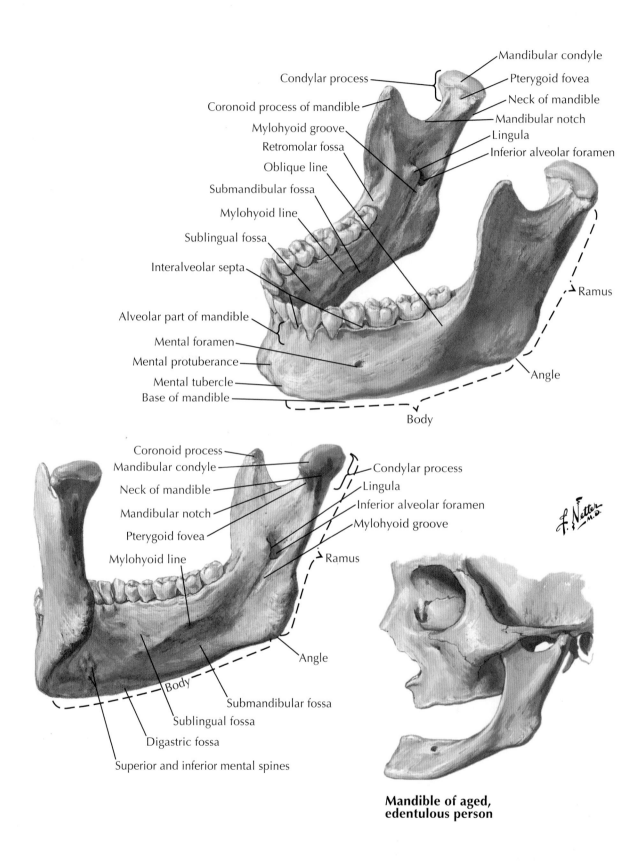

Condylar process
Mandibular condyle
Pterygoid fovea
Coronoid process of mandible
Neck of mandible
Mandibular notch
Mylohyoid groove
Lingula
Retromolar fossa
Inferior alveolar foramen
Oblique line
Submandibular fossa
Mylohyoid line
Sublingual fossa
Interalveolar septa
Ramus
Alveolar part of mandible
Mental foramen
Mental protuberance
Mental tubercle
Angle
Base of mandible
Body

Coronoid process
Mandibular condyle
Condylar process
Neck of mandible
Lingula
Mandibular notch
Inferior alveolar foramen
Pterygoid fovea
Mylohyoid groove
Mylohyoid line
Ramus
Angle
Body
Submandibular fossa
Sublingual fossa
Digastric fossa
Superior and inferior mental spines

**Mandible of aged,
edentulous person**

Plate 39

Bones and Joints

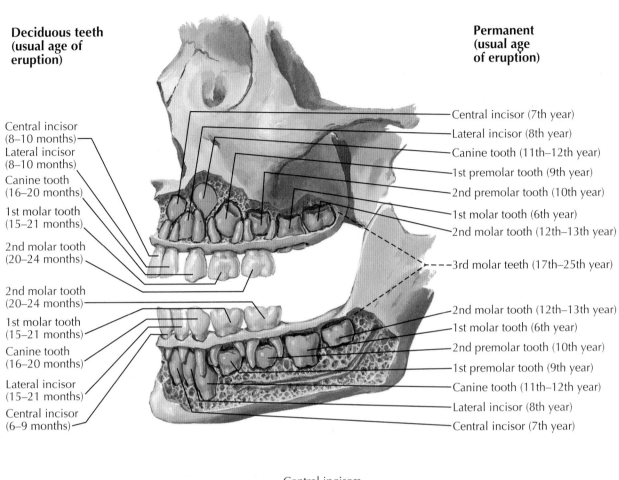

Deciduous teeth (usual age of eruption)

Central incisor (8–10 months)
Lateral incisor (8–10 months)
Canine tooth (16–20 months)
1st molar tooth (15–21 months)
2nd molar tooth (20–24 months)

2nd molar tooth (20–24 months)
1st molar tooth (15–21 months)
Canine tooth (16–20 months)
Lateral incisor (15–21 months)
Central incisor (6–9 months)

Permanent (usual age of eruption)

Central incisor (7th year)
Lateral incisor (8th year)
Canine tooth (11th–12th year)
1st premolar tooth (9th year)
2nd premolar tooth (10th year)
1st molar tooth (6th year)
2nd molar tooth (12th–13th year)

3rd molar teeth (17th–25th year)

2nd molar tooth (12th–13th year)
1st molar tooth (6th year)
2nd premolar tooth (10th year)
1st premolar tooth (9th year)
Canine tooth (11th–12th year)
Lateral incisor (8th year)
Central incisor (7th year)

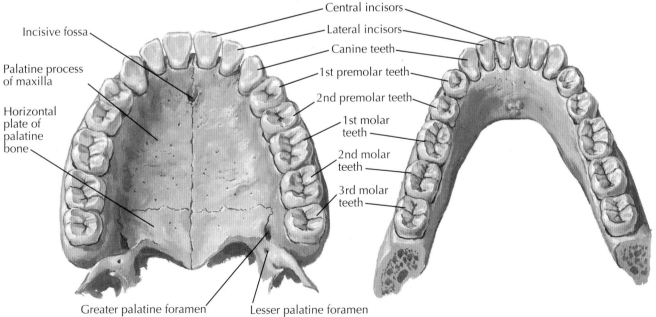

Incisive fossa

Palatine process of maxilla

Horizontal plate of palatine bone

Central incisors
Lateral incisors
Canine teeth
1st premolar teeth
2nd premolar teeth
1st molar teeth
2nd molar teeth
3rd molar teeth

Greater palatine foramen

Lesser palatine foramen

Upper permanent teeth

Lower permanent teeth

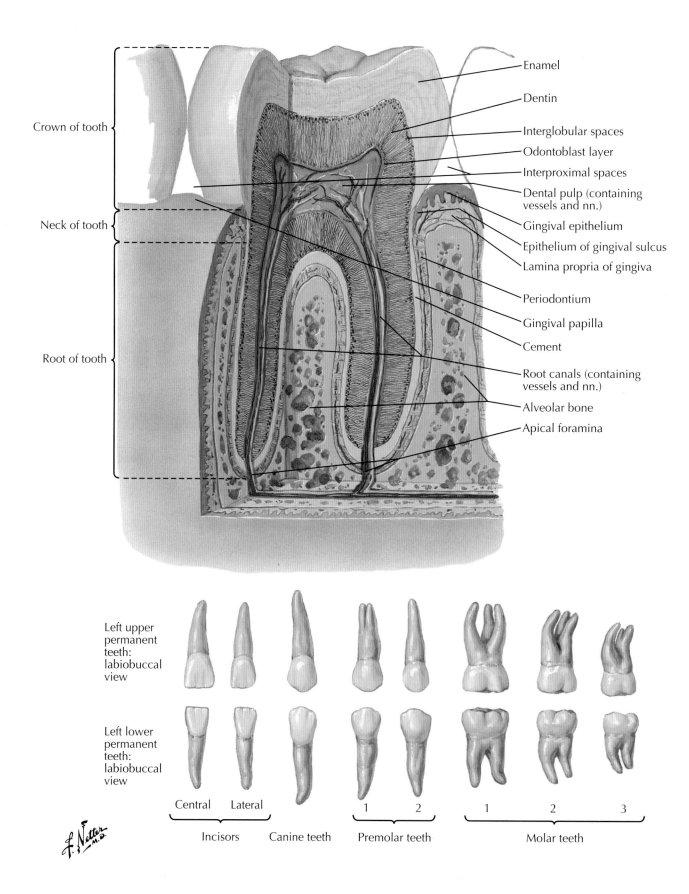

Enamel

Dentin

Interglobular spaces

Odontoblast layer

Interproximal spaces

Dental pulp (containing vessels and nn.)

Gingival epithelium

Epithelium of gingival sulcus

Lamina propria of gingiva

Periodontium

Gingival papilla

Cement

Root canals (containing vessels and nn.)

Alveolar bone

Apical foramina

Crown of tooth

Neck of tooth

Root of tooth

Left upper permanent teeth: labiobuccal view

Left lower permanent teeth: labiobuccal view

Central   Lateral

Incisors   Canine teeth

1   2

Premolar teeth

1   2   3

Molar teeth

Plate 41

Bones and Joints

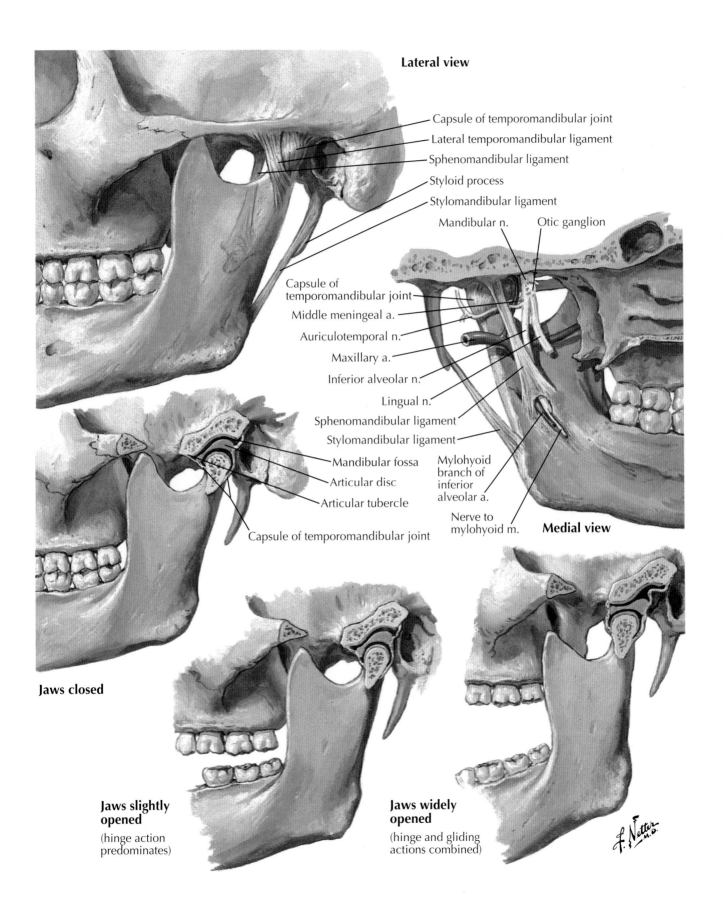

Lateral view

Capsule of temporomandibular joint
Lateral temporomandibular ligament
Sphenomandibular ligament
Styloid process
Stylomandibular ligament
Mandibular n.   Otic ganglion

Capsule of temporomandibular joint
Middle meningeal a.
Auriculotemporal n.
Maxillary a.
Inferior alveolar n.
Lingual n.
Sphenomandibular ligament
Stylomandibular ligament
Mandibular fossa
Articular disc
Articular tubercle
Capsule of temporomandibular joint

Mylohyoid branch of inferior alveolar a.
Nerve to mylohyoid m.   **Medial view**

Jaws closed

Jaws slightly opened

(hinge action predominates)

Jaws widely opened

(hinge and gliding actions combined)

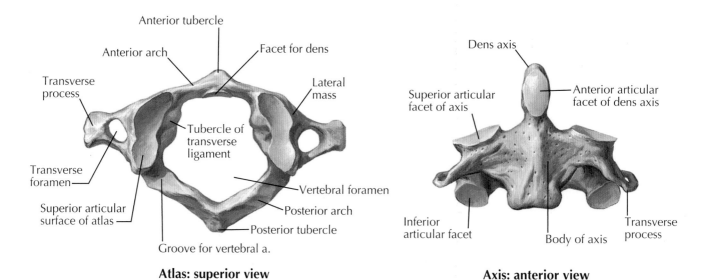

Anterior tubercle

Anterior arch

Facet for dens

Transverse process

Lateral mass

Tubercle of transverse ligament

Transverse foramen

Superior articular surface of atlas

Vertebral foramen

Posterior arch

Posterior tubercle

Groove for vertebral a.

Atlas: superior view

Dens axis

Superior articular facet of axis

Anterior articular facet of dens axis

Inferior articular facet

Body of axis

Transverse process

Axis: anterior view

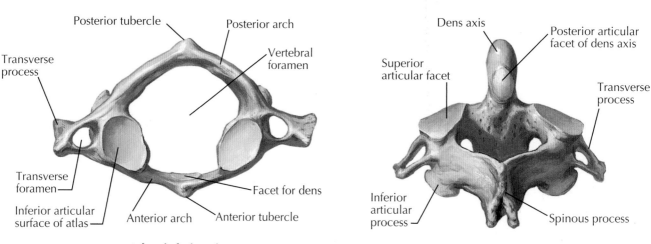

Posterior tubercle

Posterior arch

Transverse process

Vertebral foramen

Transverse foramen

Inferior articular surface of atlas

Facet for dens

Anterior arch

Anterior tubercle

Atlas: inferior view

Dens axis

Superior articular facet

Posterior articular facet of dens axis

Transverse process

Inferior articular process

Spinous process

Axis: posterosuperior view

Dens axis

Atlas (C1 vertebra)

Superior articular surface of atlas

Axis (C2 vertebra)

Posterior articular facet of dens axis

C3 vertebra

C4 vertebra

Upper cervical vertebrae: posterosuperior view

Plate 43

Bones and Joints

Inferior aspect of C3 vertebra and superior aspect of C4 vertebra showing the sites of the articular surfaces of the uncovertebral joints

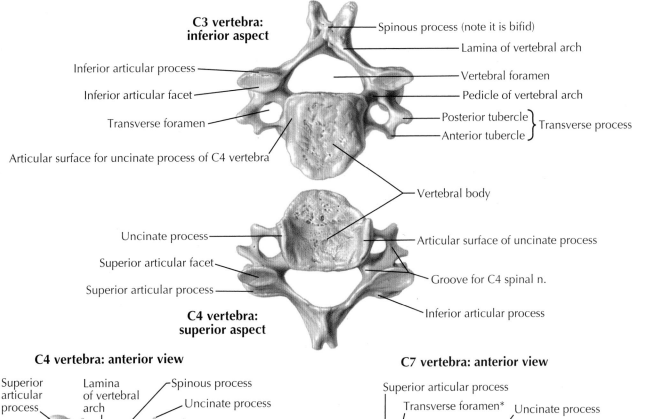

C3 vertebra: inferior aspect

Inferior articular process

Inferior articular facet

Transverse foramen

Articular surface for uncinate process of C4 vertebra

Spinous process (note it is bifid)

Lamina of vertebral arch

Vertebral foramen

Pedicle of vertebral arch

Posterior tubercle } Transverse process
Anterior tubercle }

Vertebral body

Uncinate process

Superior articular facet

Superior articular process

Articular surface of uncinate process

Groove for C4 spinal n.

Inferior articular process

C4 vertebra: superior aspect

C4 vertebra: anterior view

Superior articular process

Lamina of vertebral arch

Spinous process

Uncinate process

Articular surface

Posterior tubercle }
Anterior tubercle } Transverse process

Inferior articular facet

Transverse foramen

Vertebral body

C7 vertebra: anterior view

Superior articular process

Transverse foramen*

Uncinate process

Articular surface of uncinate process

Inferior articular facet

Vertebral body

Posterior tubercle }
Anterior tubercle } Transverse process

C7 vertebra (vertebra prominens): superior view

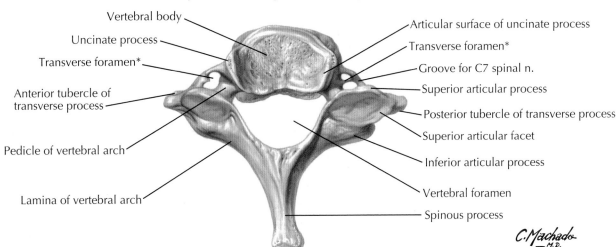

Vertebral body

Uncinate process

Transverse foramen*

Anterior tubercle of transverse process

Pedicle of vertebral arch

Lamina of vertebral arch

Articular surface of uncinate process

Transverse foramen*

Groove for C7 spinal n.

Superior articular process

Posterior tubercle of transverse process

Superior articular facet

Inferior articular process

Vertebral foramen

Spinous process

C.Machado
—M.D.

The transverse foramina of C7 transmit vertebral veins, but usually not the vertebral artery, and are asymmetrical in these drawings. Note the right transverse foramen is septated.

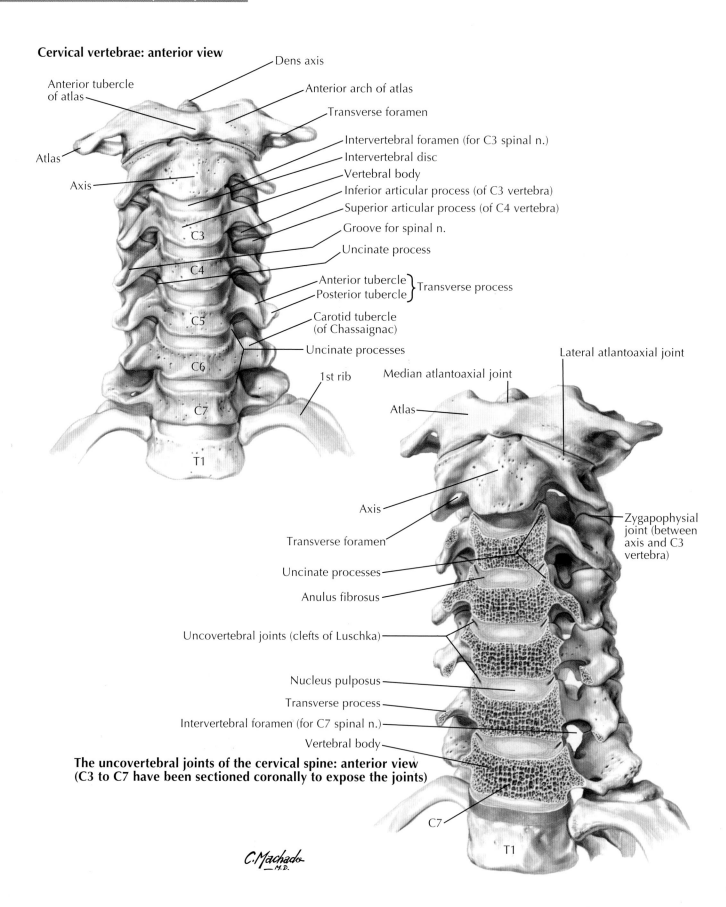

Cervical vertebrae: anterior view

Dens axis

Anterior tubercle of atlas

Anterior arch of atlas

Transverse foramen

Atlas

Intervertebral foramen (for C3 spinal n.)

Axis

Intervertebral disc

Vertebral body

Inferior articular process (of C3 vertebra)

Superior articular process (of C4 vertebra)

Groove for spinal n.

Uncinate process

C3

C4

Anterior tubercle
Posterior tubercle } Transverse process

Carotid tubercle (of Chassaignac)

C5

Uncinate processes

C6

1st rib

C7

T1

Lateral atlantoaxial joint

Median atlantoaxial joint

Atlas

Axis

Transverse foramen

Uncinate processes

Anulus fibrosus

Zygapophysial joint (between axis and C3 vertebra)

Uncovertebral joints (clefts of Luschka)

Nucleus pulposus

Transverse process

Intervertebral foramen (for C7 spinal n.)

Vertebral body

The uncovertebral joints of the cervical spine: anterior view (C3 to C7 have been sectioned coronally to expose the joints)

C7

T1

C. Machado
—M.D.

Plate 45

Bones and Joints

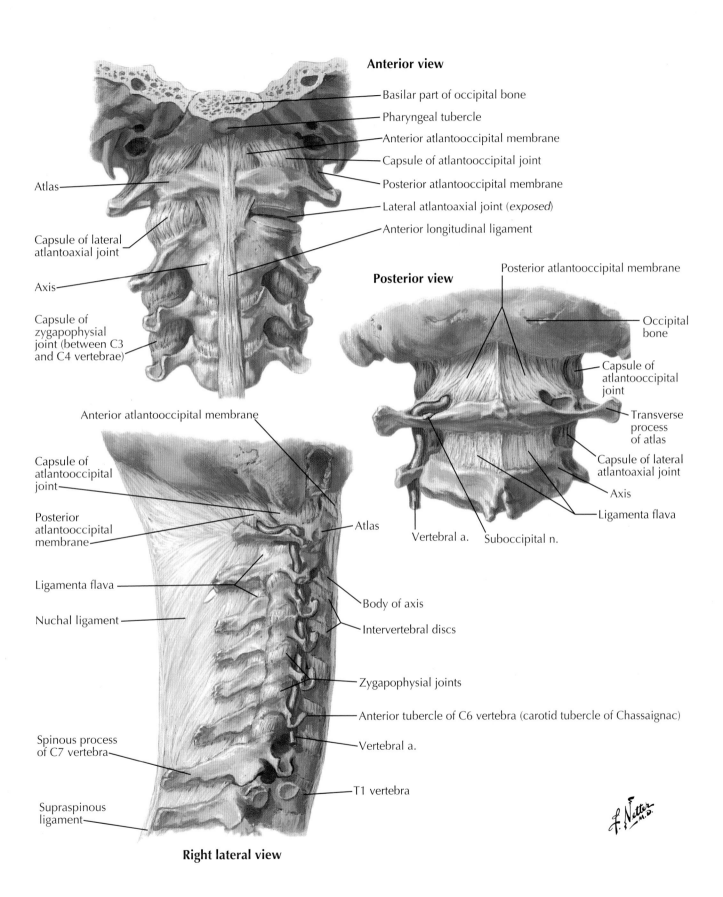

Anterior view

Basilar part of occipital bone

Pharyngeal tubercle

Anterior atlantooccipital membrane

Capsule of atlantooccipital joint

Posterior atlantooccipital membrane

Lateral atlantoaxial joint (*exposed*)

Anterior longitudinal ligament

Atlas

Capsule of lateral atlantoaxial joint

Axis

Capsule of zygapophysial joint (between C3 and C4 vertebrae)

Posterior view

Posterior atlantooccipital membrane

Occipital bone

Capsule of atlantooccipital joint

Transverse process of atlas

Capsule of lateral atlantoaxial joint

Axis

Ligamenta flava

Vertebral a.   Suboccipital n.

Anterior atlantooccipital membrane

Capsule of atlantooccipital joint

Posterior atlantooccipital membrane

Ligamenta flava

Nuchal ligament

Atlas

Body of axis

Intervertebral discs

Zygapophysial joints

Anterior tubercle of C6 vertebra (carotid tubercle of Chassaignac)

Vertebral a.

Spinous process of C7 vertebra

T1 vertebra

Supraspinous ligament

Right lateral view

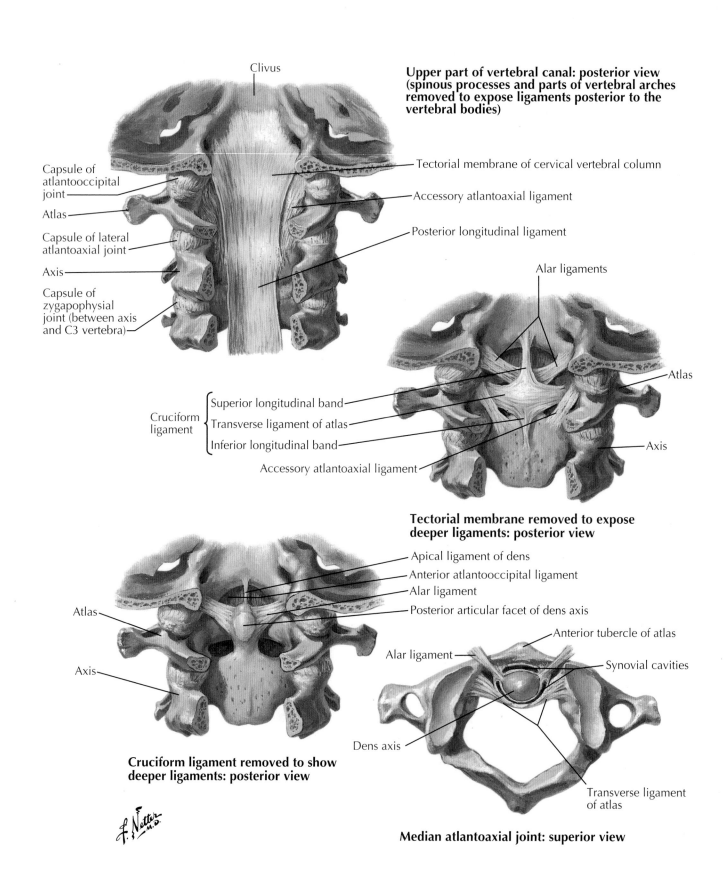

Clivus

Upper part of vertebral canal: posterior view (spinous processes and parts of vertebral arches removed to expose ligaments posterior to the vertebral bodies)

Capsule of atlantooccipital joint

Atlas

Capsule of lateral atlantoaxial joint

Axis

Capsule of zygapophysial joint (between axis and C3 vertebra)

Tectorial membrane of cervical vertebral column

Accessory atlantoaxial ligament

Posterior longitudinal ligament

Alar ligaments

Atlas

Axis

Cruciform ligament
- Superior longitudinal band
- Transverse ligament of atlas
- Inferior longitudinal band

Accessory atlantoaxial ligament

Tectorial membrane removed to expose deeper ligaments: posterior view

Atlas

Axis

Apical ligament of dens

Anterior atlantooccipital ligament

Alar ligament

Posterior articular facet of dens axis

Anterior tubercle of atlas

Alar ligament

Synovial cavities

Dens axis

Transverse ligament of atlas

Cruciform ligament removed to show deeper ligaments: posterior view

Median atlantoaxial joint: superior view

Plate 47

Bones and Joints

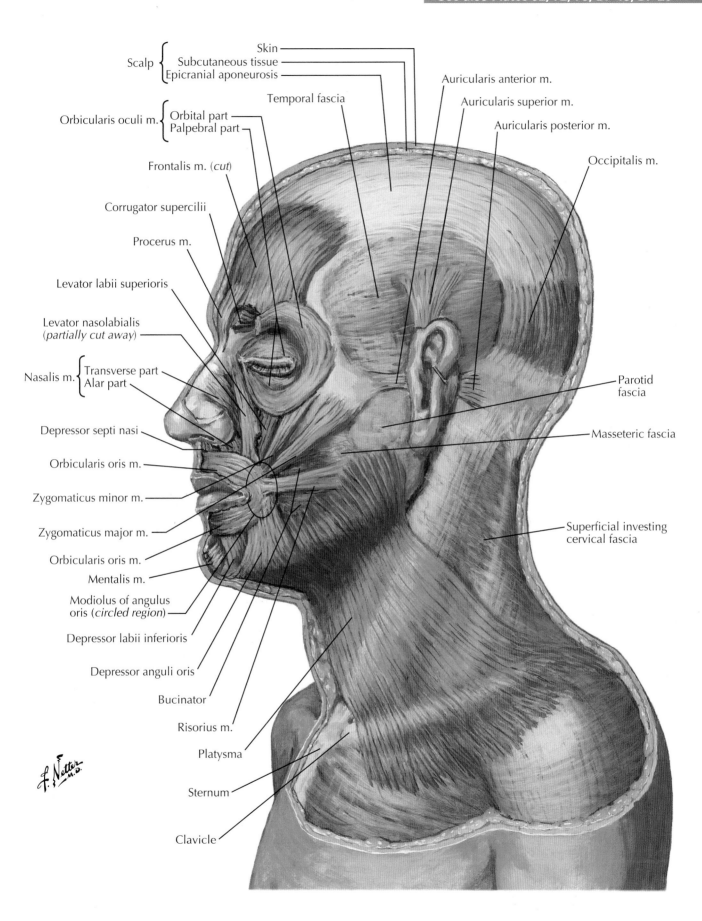

Scalp {
Skin
Subcutaneous tissue
Epicranial aponeurosis

Temporal fascia

Auricularis anterior m.

Auricularis superior m.

Auricularis posterior m.

Occipitalis m.

Orbicularis oculi m. {
Orbital part
Palpebral part

Frontalis m. (*cut*)

Corrugator supercilii

Procerus m.

Levator labii superioris

Levator nasolabialis
(*partially cut away*)

Nasalis m. {
Transverse part
Alar part

Depressor septi nasi

Orbicularis oris m.

Zygomaticus minor m.

Zygomaticus major m.

Orbicularis oris m.

Mentalis m.

Modiolus of angulus
oris (*circled region*)

Depressor labii inferioris

Depressor anguli oris

Bucinator

Risorius m.

Platysma

Sternum

Clavicle

Parotid
fascia

Masseteric fascia

Superficial investing
cervical fascia

f. Netter
M.D.

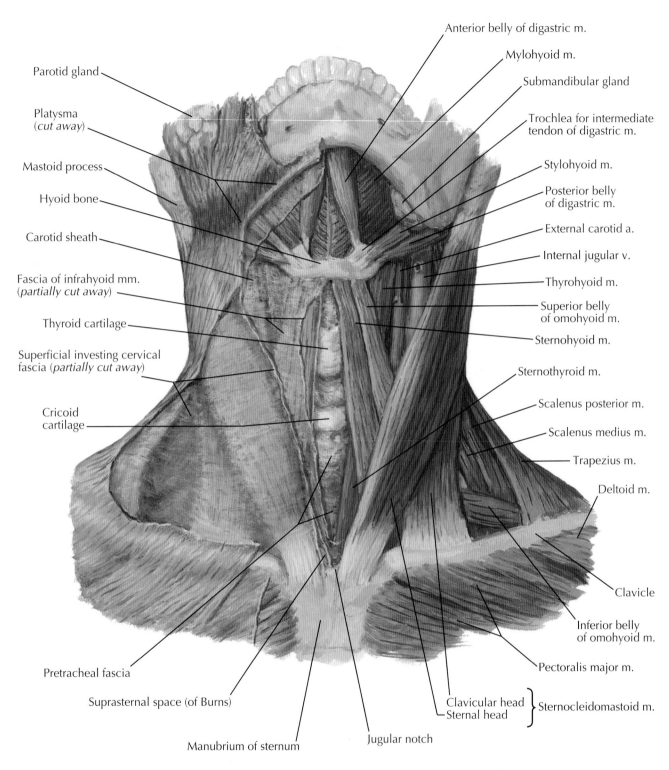

Anterior belly of digastric m.

Mylohyoid m.

Submandibular gland

Trochlea for intermediate tendon of digastric m.

Stylohyoid m.

Posterior belly of digastric m.

External carotid a.

Internal jugular v.

Thyrohyoid m.

Superior belly of omohyoid m.

Sternohyoid m.

Sternothyroid m.

Scalenus posterior m.

Scalenus medius m.

Trapezius m.

Deltoid m.

Clavicle

Inferior belly of omohyoid m.

Pectoralis major m.

Clavicular head
Sternal head } Sternocleidomastoid m.

Jugular notch

Manubrium of sternum

Suprasternal space (of Burns)

Pretracheal fascia

Cricoid cartilage

Superficial investing cervical fascia (*partially cut away*)

Thyroid cartilage

Fascia of infrahyoid mm. (*partially cut away*)

Carotid sheath

Hyoid bone

Mastoid process

Platysma (*cut away*)

Parotid gland

Plate 49   **Neck**

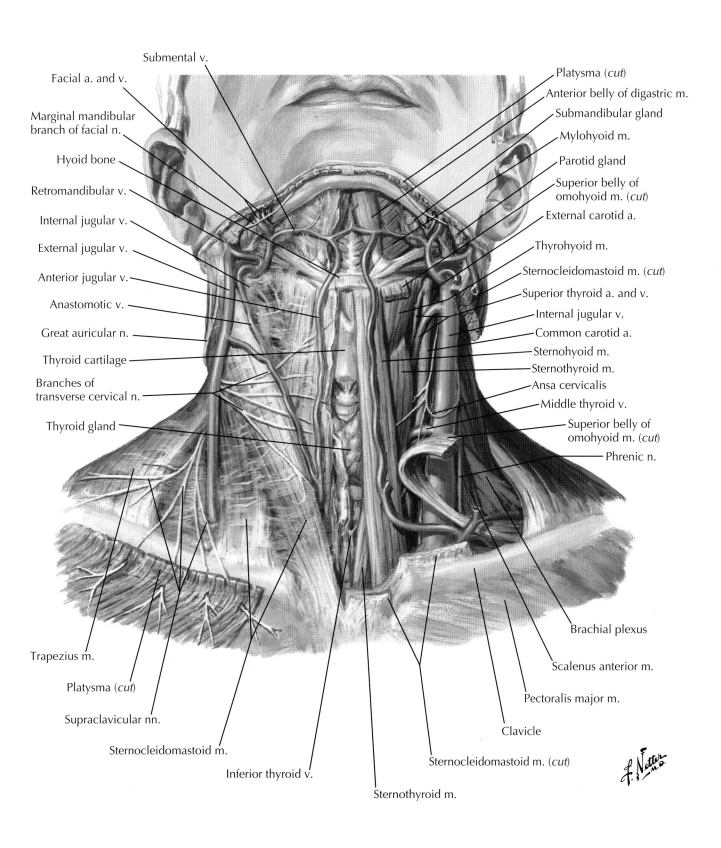

Submental v.

Facial a. and v.

Marginal mandibular
branch of facial n.

Hyoid bone

Retromandibular v.

Internal jugular v.

External jugular v.

Anterior jugular v.

Anastomotic v.

Great auricular n.

Thyroid cartilage

Branches of
transverse cervical n.

Thyroid gland

Trapezius m.

Platysma (*cut*)

Supraclavicular nn.

Sternocleidomastoid m.

Inferior thyroid v.

Sternothyroid m.

Platysma (*cut*)

Anterior belly of digastric m.

Submandibular gland

Mylohyoid m.

Parotid gland

Superior belly of
omohyoid m. (*cut*)

External carotid a.

Thyrohyoid m.

Sternocleidomastoid m. (*cut*)

Superior thyroid a. and v.

Internal jugular v.

Common carotid a.

Sternohyoid m.

Sternothyroid m.

Ansa cervicalis

Middle thyroid v.

Superior belly of
omohyoid m. (*cut*)

Phrenic n.

Brachial plexus

Scalenus anterior m.

Pectoralis major m.

Clavicle

Sternocleidomastoid m. (*cut*)

Cross section

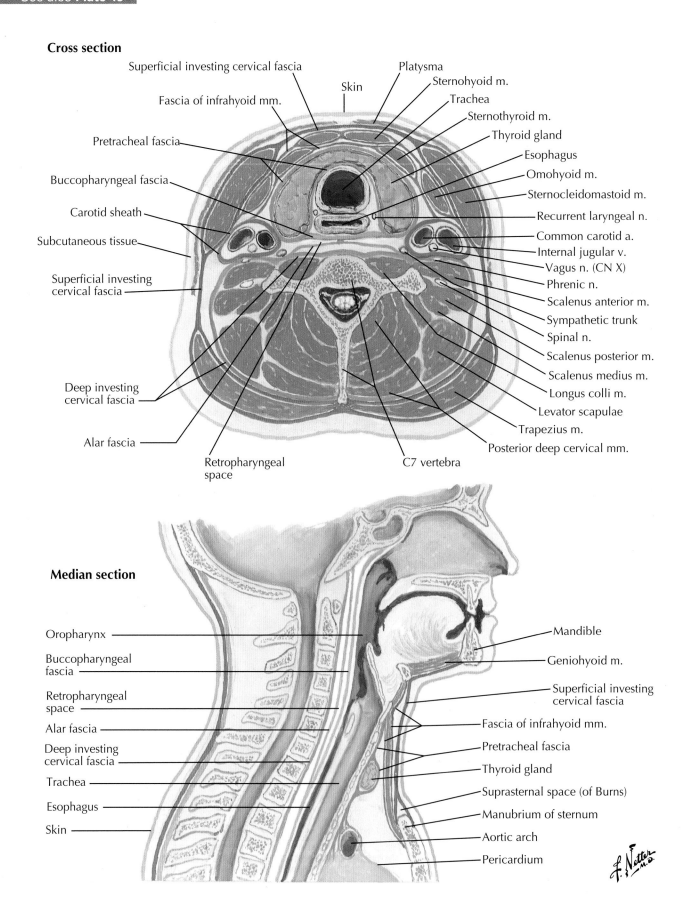

Superficial investing cervical fascia

Fascia of infrahyoid mm.

Pretracheal fascia

Buccopharyngeal fascia

Carotid sheath

Subcutaneous tissue

Superficial investing cervical fascia

Deep investing cervical fascia

Alar fascia

Retropharyngeal space

Skin

Platysma

Sternohyoid m.

Trachea

Sternothyroid m.

Thyroid gland

Esophagus

Omohyoid m.

Sternocleidomastoid m.

Recurrent laryngeal n.

Common carotid a.

Internal jugular v.

Vagus n. (CN X)

Phrenic n.

Scalenus anterior m.

Sympathetic trunk

Spinal n.

Scalenus posterior m.

Scalenus medius m.

Longus colli m.

Levator scapulae

Trapezius m.

Posterior deep cervical mm.

C7 vertebra

Median section

Oropharynx

Buccopharyngeal fascia

Retropharyngeal space

Alar fascia

Deep investing cervical fascia

Trachea

Esophagus

Skin

Mandible

Geniohyoid m.

Superficial investing cervical fascia

Fascia of infrahyoid mm.

Pretracheal fascia

Thyroid gland

Suprasternal space (of Burns)

Manubrium of sternum

Aortic arch

Pericardium

Plate 51

Neck

Superficial investing cervical fascia
Fascia of infrahyoid mm.
Pretracheal fascia
Buccopharyngeal fascia
Carotid sheath
Deep investing cervical fascia

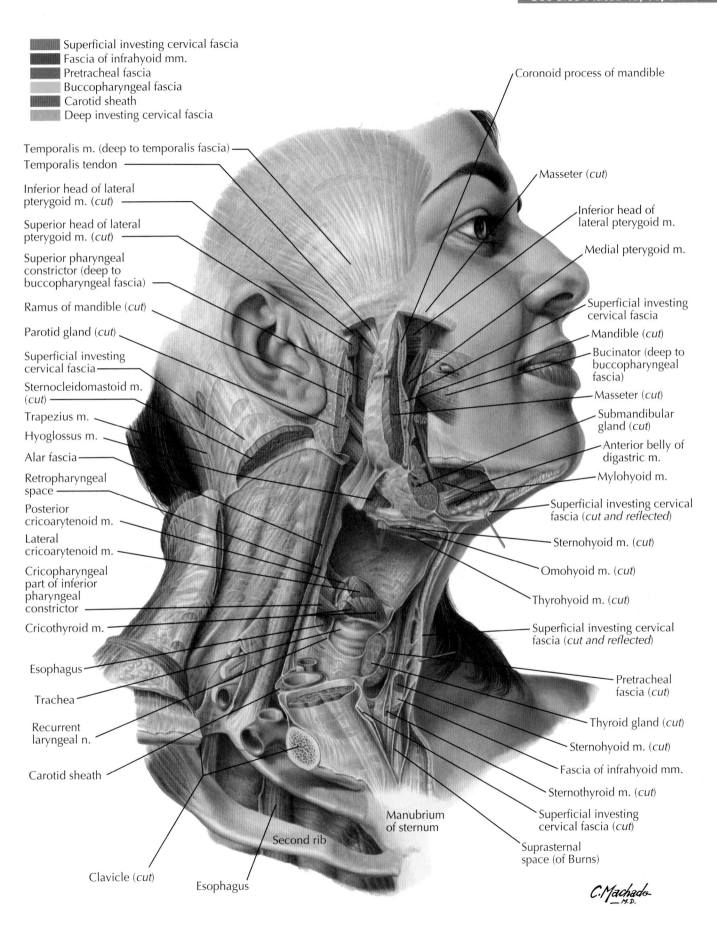

Coronoid process of mandible

Temporalis m. (deep to temporalis fascia)
Temporalis tendon
Inferior head of lateral pterygoid m. (*cut*)
Superior head of lateral pterygoid m. (*cut*)
Superior pharyngeal constrictor (deep to buccopharyngeal fascia)
Ramus of mandible (*cut*)
Parotid gland (*cut*)
Superficial investing cervical fascia
Sternocleidomastoid m. (*cut*)
Trapezius m.
Hyoglossus m.
Alar fascia
Retropharyngeal space
Posterior cricoarytenoid m.
Lateral cricoarytenoid m.
Cricopharyngeal part of inferior pharyngeal constrictor
Cricothyroid m.
Esophagus
Trachea
Recurrent laryngeal n.
Carotid sheath
Clavicle (*cut*)
Esophagus

Masseter (*cut*)
Inferior head of lateral pterygoid m.
Medial pterygoid m.
Superficial investing cervical fascia
Mandible (*cut*)
Bucinator (deep to buccopharyngeal fascia)
Masseter (*cut*)
Submandibular gland (*cut*)
Anterior belly of digastric m.
Mylohyoid m.
Superficial investing cervical fascia (*cut and reflected*)
Sternohyoid m. (*cut*)
Omohyoid m. (*cut*)
Thyrohyoid m. (*cut*)
Superficial investing cervical fascia (*cut and reflected*)
Pretracheal fascia (*cut*)
Thyroid gland (*cut*)
Sternohyoid m. (*cut*)
Fascia of infrahyoid mm.
Sternothyroid m. (*cut*)
Superficial investing cervical fascia (*cut*)
Suprasternal space (of Burns)

Manubrium of sternum
Second rib

C. Machado
— M.D.

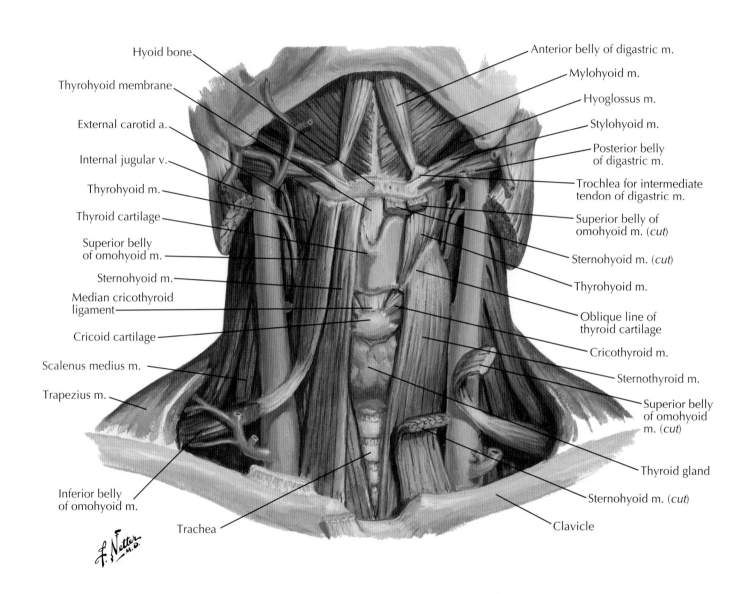

Hyoid bone

Thyrohyoid membrane

External carotid a.

Internal jugular v.

Thyrohyoid m.

Thyroid cartilage

Superior belly of omohyoid m.

Sternohyoid m.

Median cricothyroid ligament

Cricoid cartilage

Scalenus medius m.

Trapezius m.

Inferior belly of omohyoid m.

Trachea

Anterior belly of digastric m.

Mylohyoid m.

Hyoglossus m.

Stylohyoid m.

Posterior belly of digastric m.

Trochlea for intermediate tendon of digastric m.

Superior belly of omohyoid m. (*cut*)

Sternohyoid m. (*cut*)

Thyrohyoid m.

Oblique line of thyroid cartilage

Cricothyroid m.

Sternothyroid m.

Superior belly of omohyoid m. (*cut*)

Thyroid gland

Sternohyoid m. (*cut*)

Clavicle

Plate 53

Neck

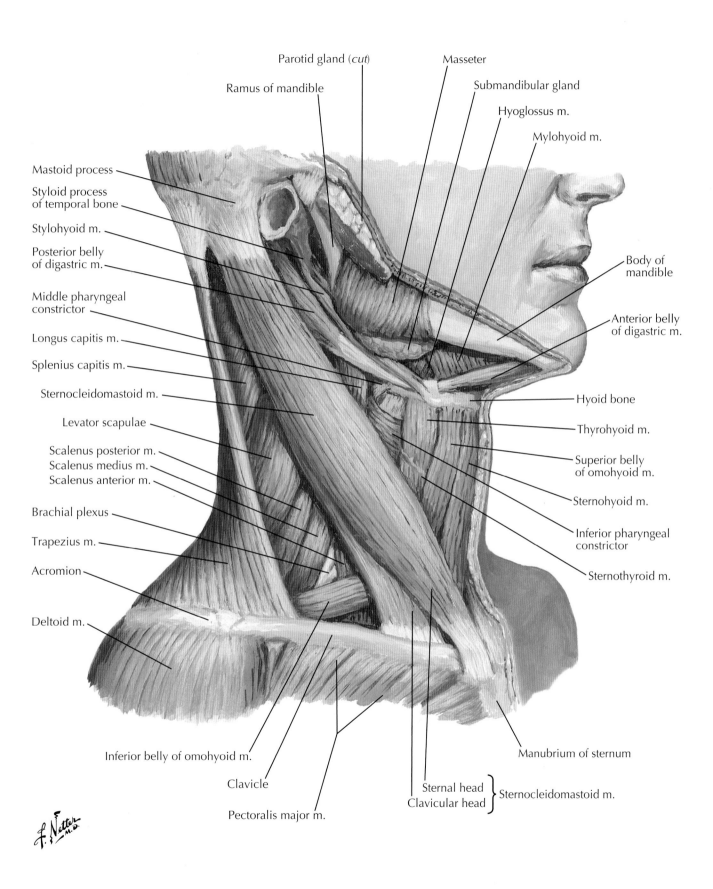

Parotid gland (*cut*)

Ramus of mandible

Masseter

Submandibular gland

Hyoglossus m.

Mylohyoid m.

Mastoid process

Styloid process
of temporal bone

Stylohyoid m.

Posterior belly
of digastric m.

Middle pharyngeal
constrictor

Longus capitis m.

Splenius capitis m.

Sternocleidomastoid m.

Levator scapulae

Scalenus posterior m.
Scalenus medius m.
Scalenus anterior m.

Brachial plexus

Trapezius m.

Acromion

Deltoid m.

Body of
mandible

Anterior belly
of digastric m.

Hyoid bone

Thyrohyoid m.

Superior belly
of omohyoid m.

Sternohyoid m.

Inferior pharyngeal
constrictor

Sternothyroid m.

Inferior belly of omohyoid m.

Clavicle

Pectoralis major m.

Sternal head
Clavicular head
} Sternocleidomastoid m.

Manubrium of sternum

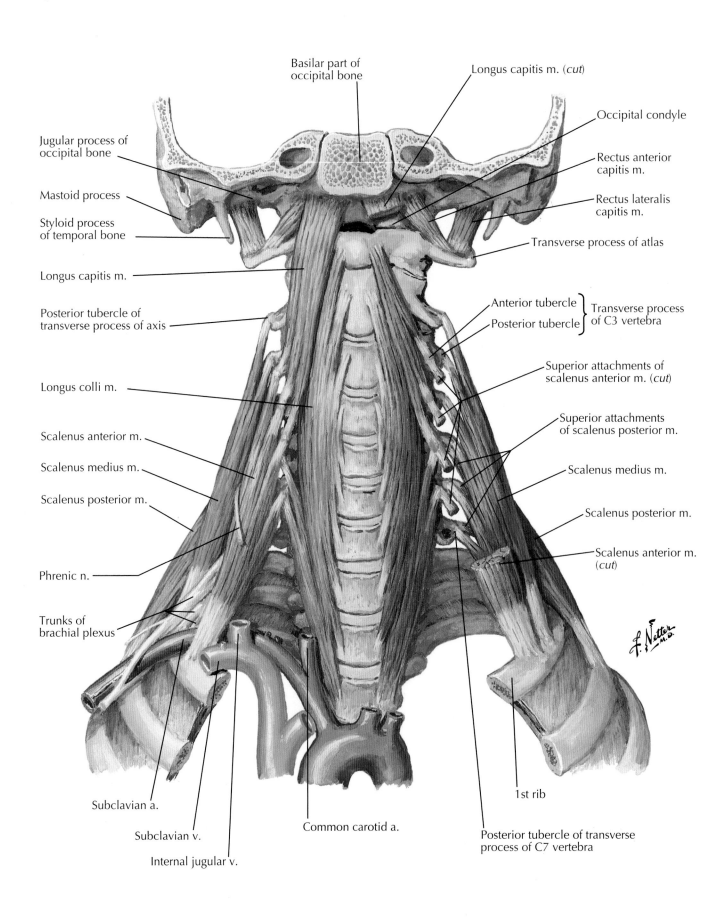

Basilar part of
occipital bone

Longus capitis m. (cut)

Occipital condyle

Jugular process of
occipital bone

Mastoid process

Styloid process
of temporal bone

Longus capitis m.

Posterior tubercle of
transverse process of axis

Longus colli m.

Scalenus anterior m.

Scalenus medius m.

Scalenus posterior m.

Phrenic n.

Trunks of
brachial plexus

Rectus anterior
capitis m.

Rectus lateralis
capitis m.

Transverse process of atlas

Anterior tubercle
Posterior tubercle
} Transverse process
of C3 vertebra

Superior attachments of
scalenus anterior m. (cut)

Superior attachments
of scalenus posterior m.

Scalenus medius m.

Scalenus posterior m.

Scalenus anterior m.
(cut)

Subclavian a.

Subclavian v.

Internal jugular v.

Common carotid a.

1st rib

Posterior tubercle of transverse
process of C7 vertebra

Plate 55

Neck

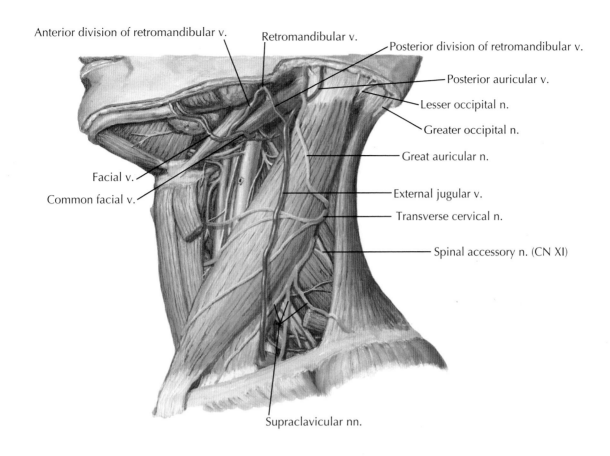

Anterior division of retromandibular v.

Retromandibular v.

Posterior division of retromandibular v.

Posterior auricular v.

Lesser occipital n.

Greater occipital n.

Great auricular n.

Facial v.

Common facial v.

External jugular v.

Transverse cervical n.

Spinal accessory n. (CN XI)

Supraclavicular nn.

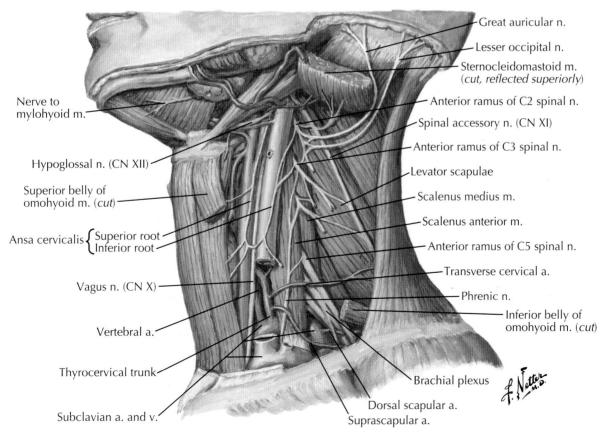

Great auricular n.

Lesser occipital n.

Sternocleidomastoid m.
(cut, reflected superiorly)

Nerve to
mylohyoid m.

Anterior ramus of C2 spinal n.

Spinal accessory n. (CN XI)

Anterior ramus of C3 spinal n.

Hypoglossal n. (CN XII)

Levator scapulae

Superior belly of
omohyoid m. (cut)

Scalenus medius m.

Scalenus anterior m.

Ansa cervicalis { Superior root
Inferior root

Anterior ramus of C5 spinal n.

Transverse cervical a.

Vagus n. (CN X)

Phrenic n.

Inferior belly of
omohyoid m. (cut)

Vertebral a.

Thyrocervical trunk

Brachial plexus

Dorsal scapular a.

Subclavian a. and v.

Suprascapular a.

Cervical plexus: schema

(S = gray ramus communicans from superior cervical ganglion)

Geniohyoid branch of hypoglossal n.

Thyrohyoid branch of hypoglossal n.

Transverse cervical n.

Nerve to superior belly of omohyoid m.

Ansa cervicalis { Superior root / Inferior root }

Nerve to sternothyroid m.

Nerve to sternohyoid m.

Nerve to inferior belly of omohyoid m.

Supraclavicular nn.

Hypoglossal n. (CN XII)

Spinal accessory n. (CN XI)

Great auricular n.

Lesser occipital n.

Nerves to rectus lateralis capitis, longus capitis, and rectus anterior capitis mm.

Nerves to longus capitis and longus colli mm.

Nerves to scalenus mm. and levator scapulae

Phrenic n.

S

C1

S   C2

S   C3

S   C4

C5

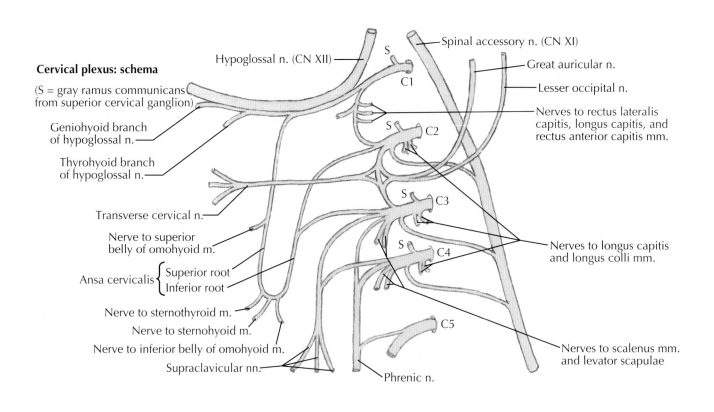

Right anterior view

Internal jugular v.

Common carotid a.

Ascending cervical a.

Phrenic n.

Scalenus anterior m.

Inferior thyroid a.

Transverse cervical a.

Brachial plexus

Suprascapular a.

Dorsal scapular a.

Costocervical trunk

Thyrocervical trunk

Subclavian a. and v.

Thyroid gland (*retracted*)

Middle cervical ganglion

Vagus n. (CN X)

Vertebral a.

Common carotid a.

Recurrent laryngeal n.

Brachiocephalic trunk

Internal jugular v. (*cut*)

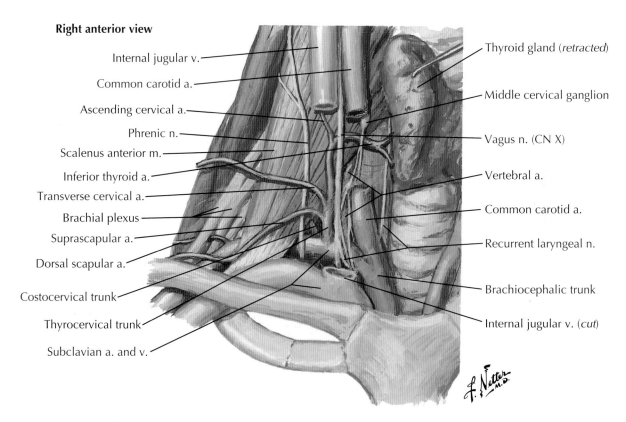

Plate 57   **Neck**

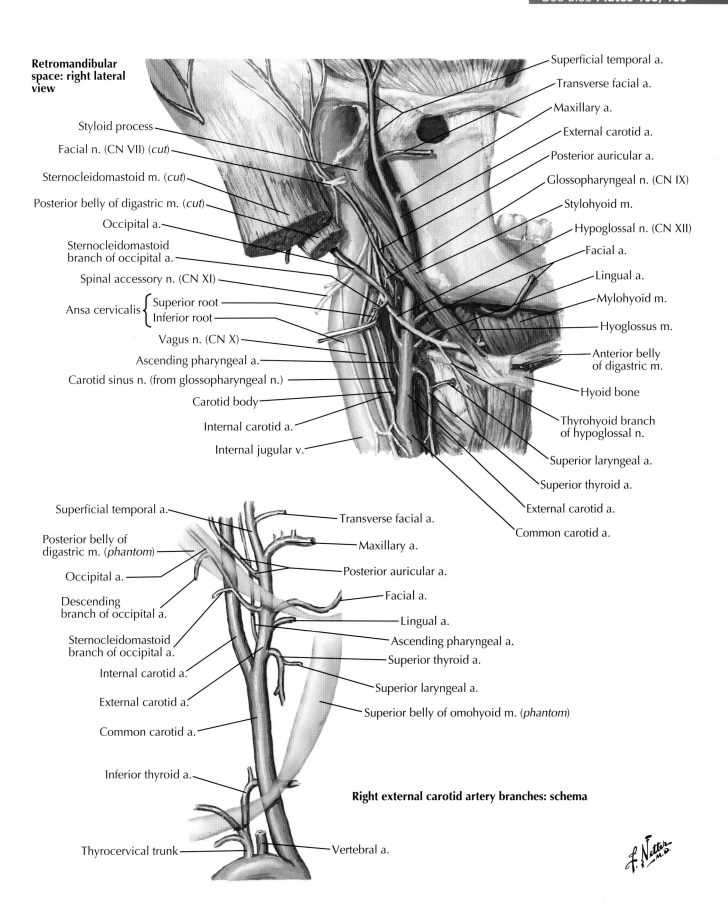

Retromandibular space: right lateral view

Styloid process

Facial n. (CN VII) (*cut*)

Sternocleidomastoid m. (*cut*)

Posterior belly of digastric m. (*cut*)

Occipital a.

Sternocleidomastoid branch of occipital a.

Spinal accessory n. (CN XI)

Ansa cervicalis { Superior root
Inferior root

Vagus n. (CN X)

Ascending pharyngeal a.

Carotid sinus n. (from glossopharyngeal n.)

Carotid body

Internal carotid a.

Internal jugular v.

Superficial temporal a.

Transverse facial a.

Maxillary a.

External carotid a.

Posterior auricular a.

Glossopharyngeal n. (CN IX)

Stylohyoid m.

Hypoglossal n. (CN XII)

Facial a.

Lingual a.

Mylohyoid m.

Hyoglossus m.

Anterior belly of digastric m.

Hyoid bone

Thyrohyoid branch of hypoglossal n.

Superior laryngeal a.

Superior thyroid a.

External carotid a.

Common carotid a.

Superficial temporal a.

Posterior belly of digastric m. (*phantom*)

Occipital a.

Descending branch of occipital a.

Sternocleidomastoid branch of occipital a.

Internal carotid a.

External carotid a.

Common carotid a.

Inferior thyroid a.

Thyrocervical trunk

Transverse facial a.

Maxillary a.

Posterior auricular a.

Facial a.

Lingual a.

Ascending pharyngeal a.

Superior thyroid a.

Superior laryngeal a.

Superior belly of omohyoid m. (*phantom*)

Right external carotid artery branches: schema

Vertebral a.

Anterolateral view

Frontal bone

Nasal bones

Frontal process of maxilla

Lateral processes of
septal nasal cartilage

Septal nasal cartilage

Minor alar cartilage

Accessory nasal cartilage

Major alar cartilage { Lateral crus

Medial crus

Septal nasal cartilage

Anterior nasal spine

Fibrofatty tissue of ala of nose

Infraorbital foramen

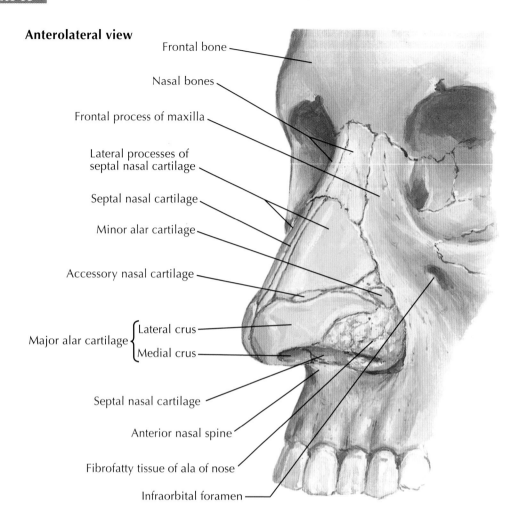

Inferior view

Major alar cartilage

Lateral
crus

Medial
crus

Fibrofatty tissue
of ala of nose

Anterior nasal
spine

Septal
nasal cartilage

Intermaxillary suture

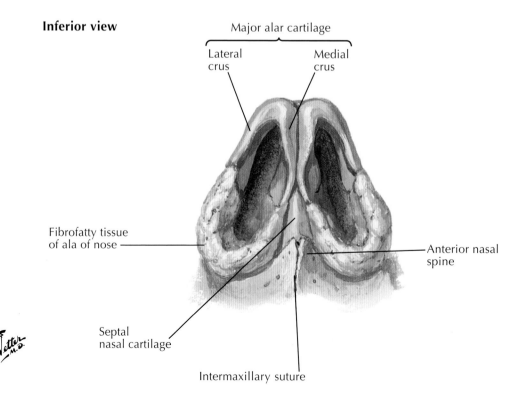

Plate 59

Nose

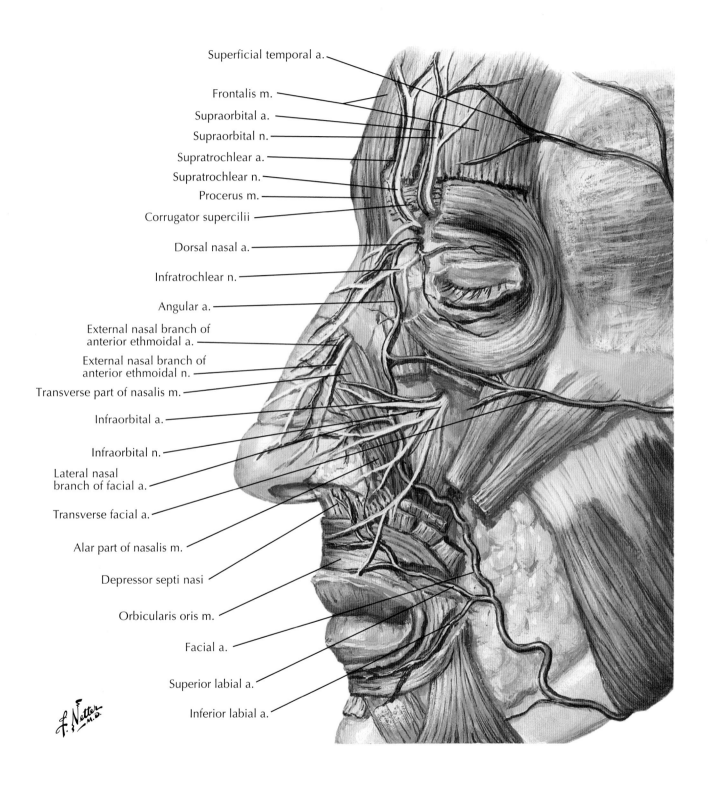

Superficial temporal a.

Frontalis m.

Supraorbital a.

Supraorbital n.

Supratrochlear a.

Supratrochlear n.

Procerus m.

Corrugator supercilii

Dorsal nasal a.

Infratrochlear n.

Angular a.

External nasal branch of anterior ethmoidal a.

External nasal branch of anterior ethmoidal n.

Transverse part of nasalis m.

Infraorbital a.

Infraorbital n.

Lateral nasal branch of facial a.

Transverse facial a.

Alar part of nasalis m.

Depressor septi nasi

Orbicularis oris m.

Facial a.

Superior labial a.

Inferior labial a.

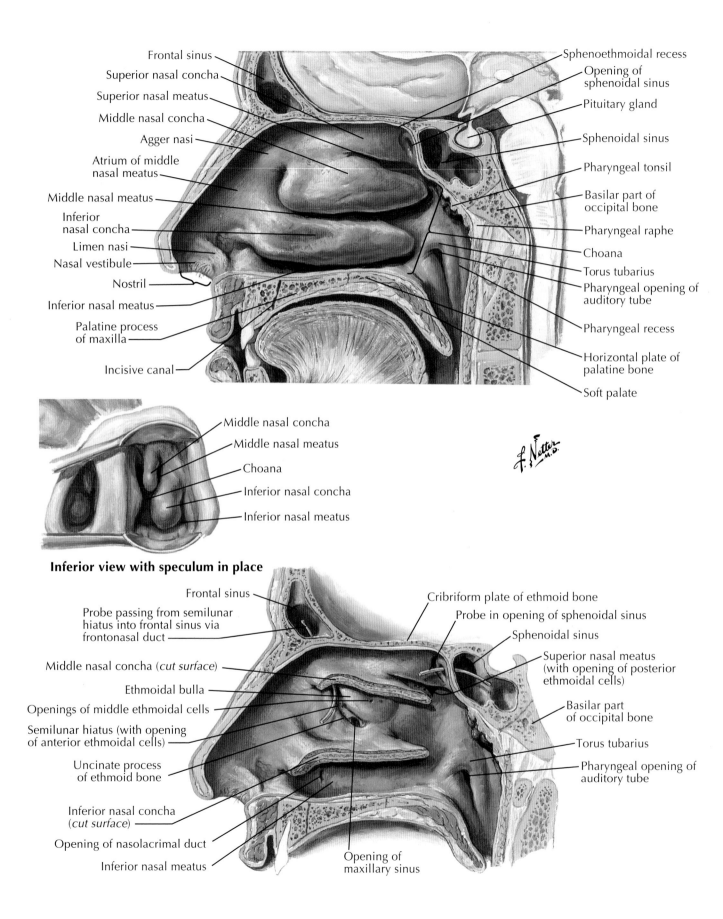

Frontal sinus

Superior nasal concha

Superior nasal meatus

Middle nasal concha

Agger nasi

Atrium of middle nasal meatus

Middle nasal meatus

Inferior nasal concha

Limen nasi

Nasal vestibule

Nostril

Inferior nasal meatus

Palatine process of maxilla

Incisive canal

Sphenoethmoidal recess

Opening of sphenoidal sinus

Pituitary gland

Sphenoidal sinus

Pharyngeal tonsil

Basilar part of occipital bone

Pharyngeal raphe

Choana

Torus tubarius

Pharyngeal opening of auditory tube

Pharyngeal recess

Horizontal plate of palatine bone

Soft palate

Middle nasal concha

Middle nasal meatus

Choana

Inferior nasal concha

Inferior nasal meatus

Inferior view with speculum in place

Frontal sinus

Probe passing from semilunar hiatus into frontal sinus via frontonasal duct

Middle nasal concha (*cut surface*)

Ethmoidal bulla

Openings of middle ethmoidal cells

Semilunar hiatus (with opening of anterior ethmoidal cells)

Uncinate process of ethmoid bone

Inferior nasal concha (*cut surface*)

Opening of nasolacrimal duct

Inferior nasal meatus

Cribriform plate of ethmoid bone

Probe in opening of sphenoidal sinus

Sphenoidal sinus

Superior nasal meatus (with opening of posterior ethmoidal cells)

Basilar part of occipital bone

Torus tubarius

Pharyngeal opening of auditory tube

Opening of maxillary sinus

Plate 61   **Nose**

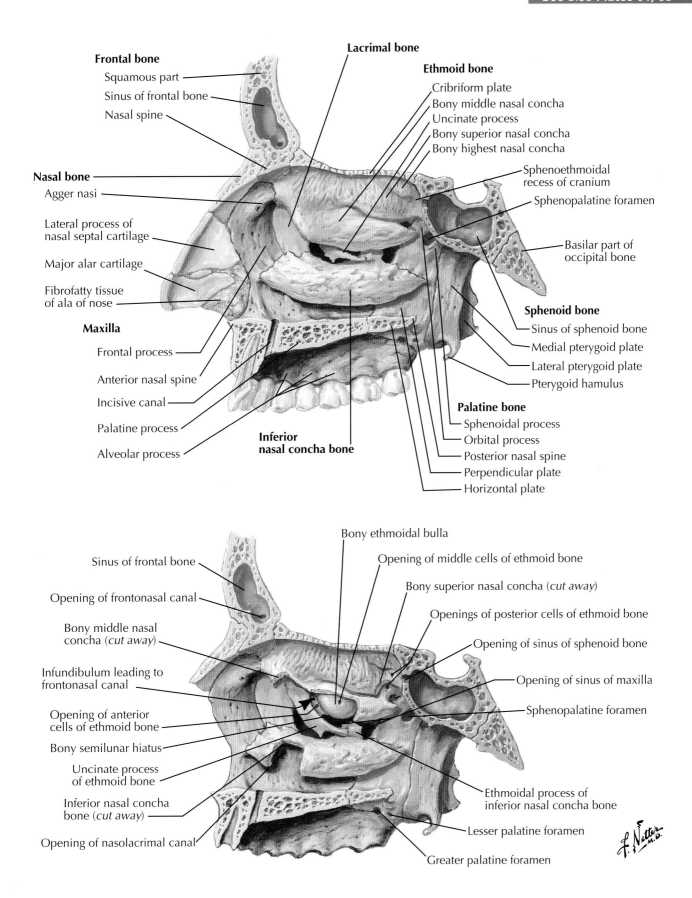

Frontal bone
Squamous part
Sinus of frontal bone
Nasal spine

Lacrimal bone

Ethmoid bone
Cribriform plate
Bony middle nasal concha
Uncinate process
Bony superior nasal concha
Bony highest nasal concha

Sphenoethmoidal recess of cranium

Sphenopalatine foramen

Nasal bone
Agger nasi

Lateral process of nasal septal cartilage

Major alar cartilage

Fibrofatty tissue of ala of nose

Basilar part of occipital bone

Sphenoid bone
Sinus of sphenoid bone
Medial pterygoid plate
Lateral pterygoid plate
Pterygoid hamulus

Maxilla
Frontal process
Anterior nasal spine
Incisive canal
Palatine process
Alveolar process

Inferior nasal concha bone

Palatine bone
Sphenoidal process
Orbital process
Posterior nasal spine
Perpendicular plate
Horizontal plate

Bony ethmoidal bulla

Opening of middle cells of ethmoid bone

Sinus of frontal bone

Opening of frontonasal canal

Bony middle nasal concha (*cut away*)

Infundibulum leading to frontonasal canal

Opening of anterior cells of ethmoid bone

Bony semilunar hiatus

Uncinate process of ethmoid bone

Inferior nasal concha bone (*cut away*)

Opening of nasolacrimal canal

Bony superior nasal concha (*cut away*)

Openings of posterior cells of ethmoid bone

Opening of sinus of sphenoid bone

Opening of sinus of maxilla

Sphenopalatine foramen

Ethmoidal process of inferior nasal concha bone

Lesser palatine foramen

Greater palatine foramen

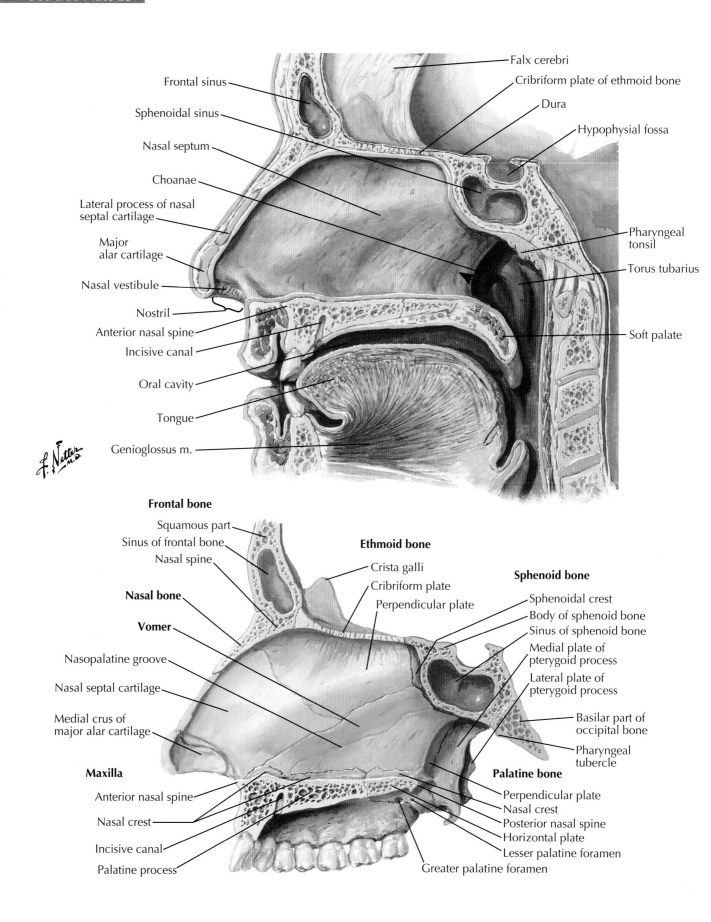

Falx cerebri

Frontal sinus

Cribriform plate of ethmoid bone

Sphenoidal sinus

Dura

Nasal septum

Hypophysial fossa

Choanae

Lateral process of nasal septal cartilage

Major alar cartilage

Pharyngeal tonsil

Nasal vestibule

Torus tubarius

Nostril

Anterior nasal spine

Incisive canal

Soft palate

Oral cavity

Tongue

Genioglossus m.

Frontal bone

Squamous part

Sinus of frontal bone

Nasal spine

Ethmoid bone

Crista galli

Cribriform plate

Perpendicular plate

Sphenoid bone

Nasal bone

Sphenoidal crest

Body of sphenoid bone

Vomer

Sinus of sphenoid bone

Nasopalatine groove

Medial plate of pterygoid process

Nasal septal cartilage

Lateral plate of pterygoid process

Medial crus of major alar cartilage

Basilar part of occipital bone

Pharyngeal tubercle

Maxilla

Palatine bone

Anterior nasal spine

Perpendicular plate

Nasal crest

Nasal crest

Posterior nasal spine

Incisive canal

Horizontal plate

Lesser palatine foramen

Palatine process

Greater palatine foramen

Plate 63

Nose

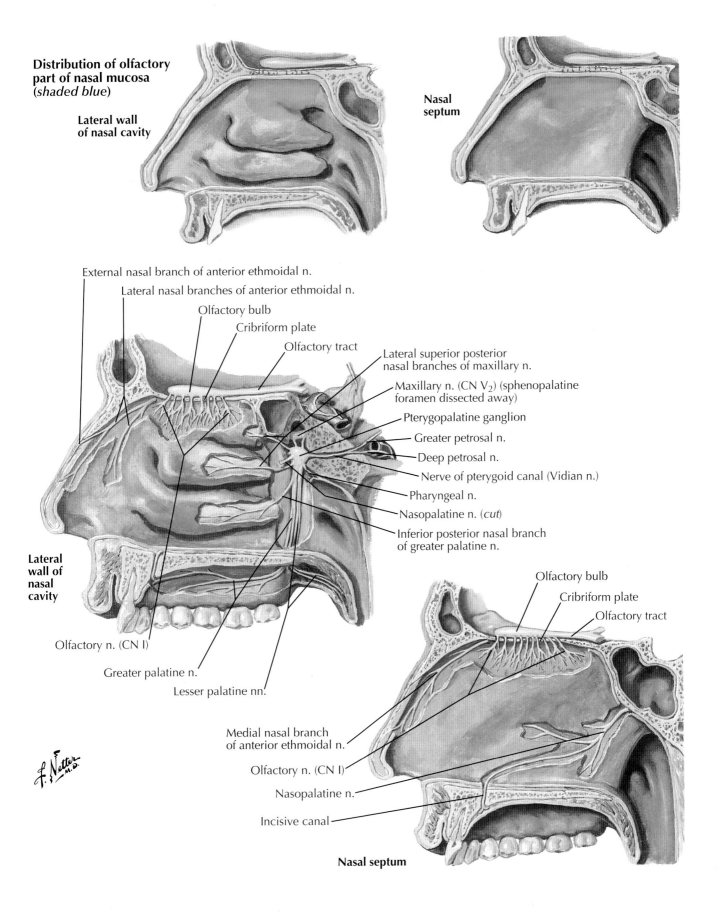

Distribution of olfactory part of nasal mucosa (*shaded blue*)

Lateral wall of nasal cavity

Nasal septum

External nasal branch of anterior ethmoidal n.

Lateral nasal branches of anterior ethmoidal n.

Olfactory bulb

Cribriform plate

Olfactory tract

Lateral superior posterior nasal branches of maxillary n.

Maxillary n. (CN V₂) (sphenopalatine foramen dissected away)

Pterygopalatine ganglion

Greater petrosal n.

Deep petrosal n.

Nerve of pterygoid canal (Vidian n.)

Pharyngeal n.

Nasopalatine n. (*cut*)

Inferior posterior nasal branch of greater palatine n.

Lateral wall of nasal cavity

Olfactory n. (CN I)

Greater palatine n.

Lesser palatine nn.

Medial nasal branch of anterior ethmoidal n.

Olfactory n. (CN I)

Nasopalatine n.

Incisive canal

Olfactory bulb

Cribriform plate

Olfactory tract

Nasal septum

Lateral wall of nasal cavity

Anterior ethmoidal a.

Anterior lateral nasal branch of anterior ethmoidal a.

Posterior ethmoidal a.

Ophthalmic a.

Internal carotid a.

Sphenopalatine a.

Posterior lateral nasal branches of sphenopalatine a.

Descending palatine a.

Maxillary a.

Alar branches of lateral nasal branch of facial a.

Greater palatine a.

Lesser palatine a.

External carotid a.

Internal carotid a.

Nasal septum

Anterior septal branches of anterior ethmoidal a.

Kiesselbach's plexus

Posterior septal branch of sphenopalatine a.

Greater palatine a.

Anastomosis between posterior septal branch of sphenopalatine a. and greater palatine a. in incisive canal

Nasal septal branches of superior labial a.

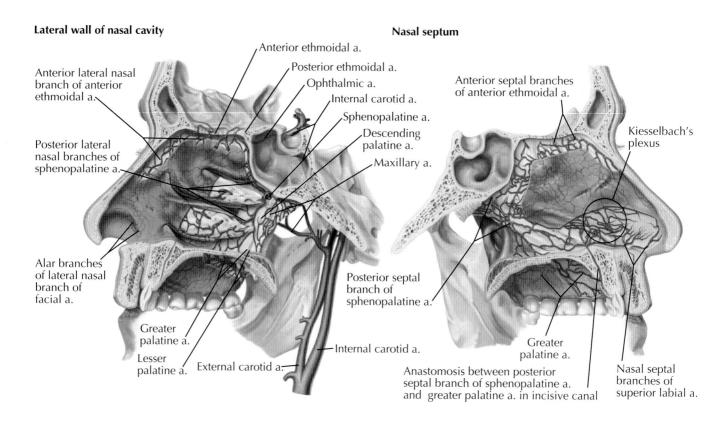

Lateral wall of nasal cavity

Anterior ethmoidal v.

Posterior ethmoidal v.

Superior ophthalmic v.

Inferior ophthalmic v.

Cavernous sinus

Nasofrontal v.

Anterior lateral nasal tributary of anterior ethmoidal v.

External nasal v.

Sphenopalatine v.

Pterygoid venous plexus

Posterior lateral nasal tributaries of sphenopalatine v.

Descending palatine v.

Posterior septal tributary of sphenopalatine v.

Facial v.

Retromandibular v.

Nasal septum

Communicating vv. between oral and nasal cavities in incisive canal

C. Machado
—M.D.

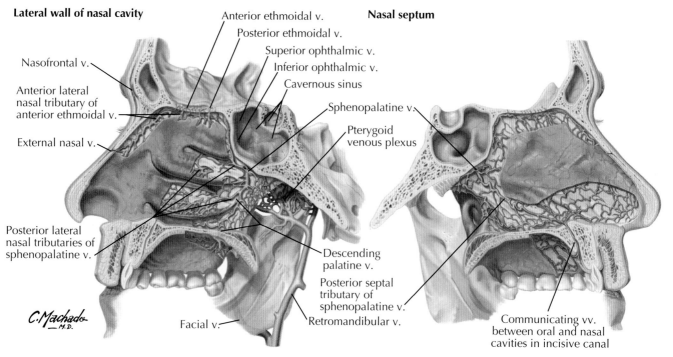

Plate 65

Nose

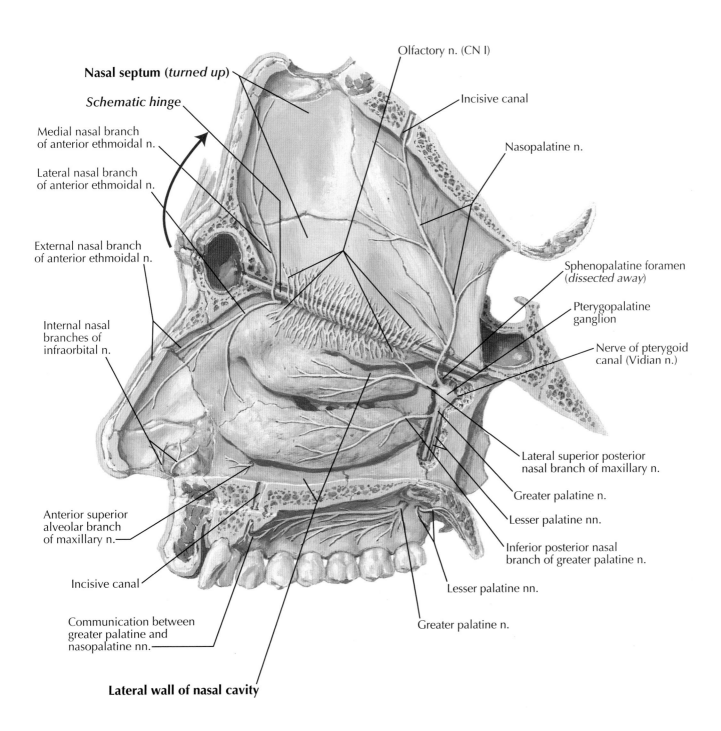

Olfactory n. (CN I)

Nasal septum (*turned up*)

Incisive canal

Schematic hinge

Nasopalatine n.

Medial nasal branch
of anterior ethmoidal n.

Lateral nasal branch
of anterior ethmoidal n.

External nasal branch
of anterior ethmoidal n.

Sphenopalatine foramen
(*dissected away*)

Pterygopalatine
ganglion

Internal nasal
branches of
infraorbital n.

Nerve of pterygoid
canal (Vidian n.)

Lateral superior posterior
nasal branch of maxillary n.

Greater palatine n.

Anterior superior
alveolar branch
of maxillary n.

Lesser palatine nn.

Inferior posterior nasal
branch of greater palatine n.

Incisive canal

Lesser palatine nn.

Communication between
greater palatine and
nasopalatine nn.

Greater palatine n.

Lateral wall of nasal cavity

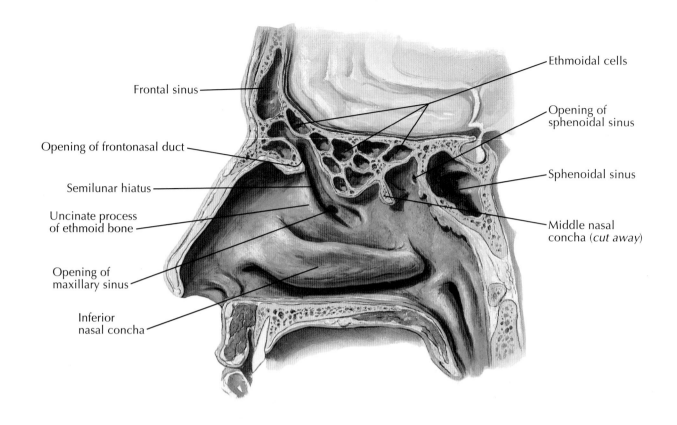

Frontal sinus

Opening of frontonasal duct

Semilunar hiatus

Uncinate process
of ethmoid bone

Opening of
maxillary sinus

Inferior
nasal concha

Ethmoidal cells

Opening of
sphenoidal sinus

Sphenoidal sinus

Middle nasal
concha (*cut away*)

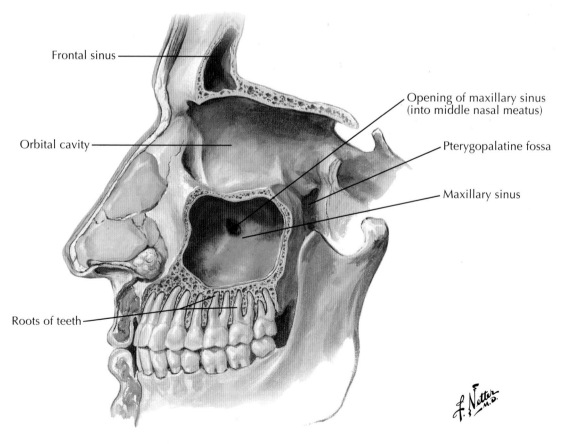

Frontal sinus

Orbital cavity

Roots of teeth

Opening of maxillary sinus
(into middle nasal meatus)

Pterygopalatine fossa

Maxillary sinus

Plate 67

Nose

Bones of nasal cavity and paranasal sinuses at birth

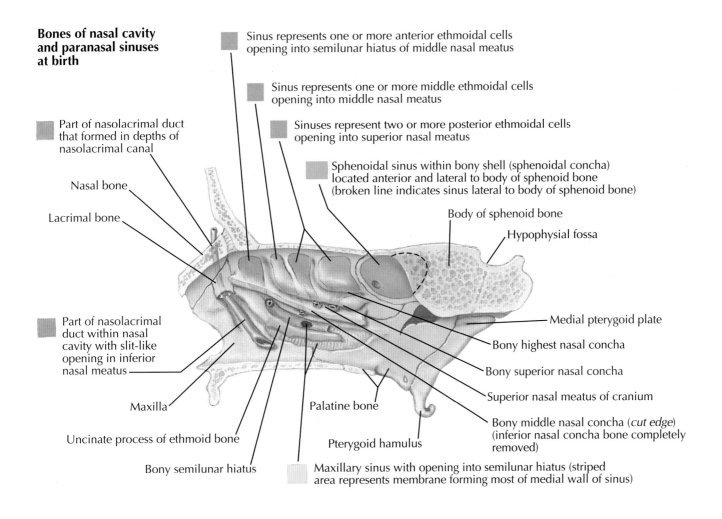

Sinus represents one or more anterior ethmoidal cells opening into semilunar hiatus of middle nasal meatus

Sinus represents one or more middle ethmoidal cells opening into middle nasal meatus

Sinuses represent two or more posterior ethmoidal cells opening into superior nasal meatus

Sphenoidal sinus within bony shell (sphenoidal concha) located anterior and lateral to body of sphenoid bone (broken line indicates sinus lateral to body of sphenoid bone)

Part of nasolacrimal duct that formed in depths of nasolacrimal canal

Nasal bone

Lacrimal bone

Body of sphenoid bone

Hypophysial fossa

Part of nasolacrimal duct within nasal cavity with slit-like opening in inferior nasal meatus

Medial pterygoid plate

Bony highest nasal concha

Bony superior nasal concha

Superior nasal meatus of cranium

Bony middle nasal concha (*cut edge*) (inferior nasal concha bone completely removed)

Maxilla

Uncinate process of ethmoid bone

Palatine bone

Pterygoid hamulus

Bony semilunar hiatus

Maxillary sinus with opening into semilunar hiatus (striped area represents membrane forming most of medial wall of sinus)

Growth of sinuses in frontal bone and maxilla throughout life

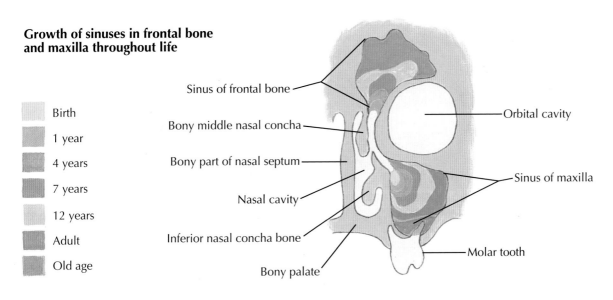

Birth

1 year

4 years

7 years

12 years

Adult

Old age

Sinus of frontal bone

Bony middle nasal concha

Bony part of nasal septum

Nasal cavity

Inferior nasal concha bone

Bony palate

Orbital cavity

Sinus of maxilla

Molar tooth

Nose

Plate 68

Coronal section

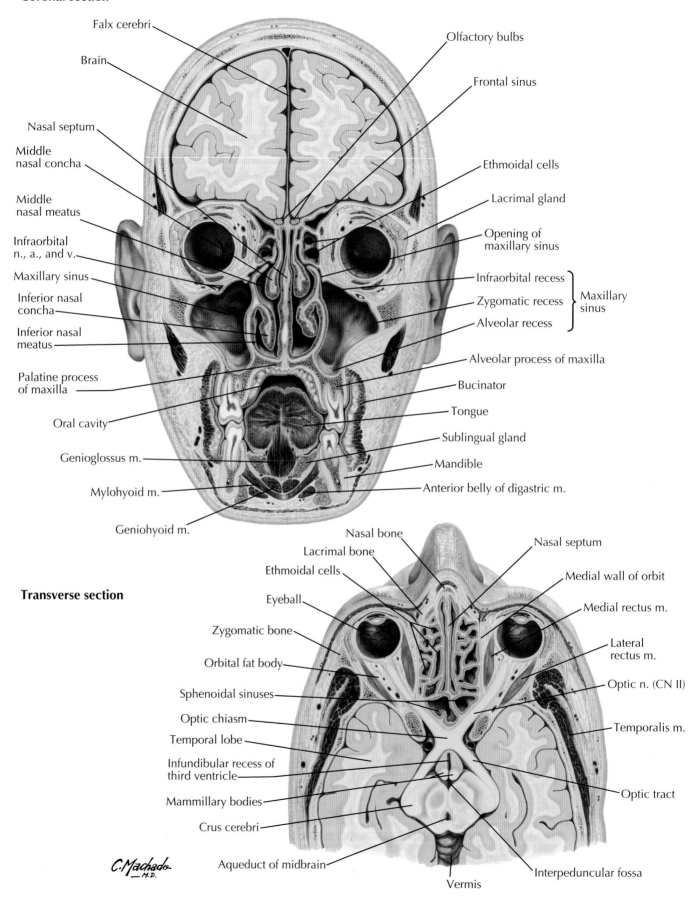

Falx cerebri

Brain

Nasal septum

Middle nasal concha

Middle nasal meatus

Infraorbital n., a., and v.

Maxillary sinus

Inferior nasal concha

Inferior nasal meatus

Palatine process of maxilla

Oral cavity

Genioglossus m.

Mylohyoid m.

Geniohyoid m.

Olfactory bulbs

Frontal sinus

Ethmoidal cells

Lacrimal gland

Opening of maxillary sinus

Infraorbital recess

Zygomatic recess

Alveolar recess

Maxillary sinus

Alveolar process of maxilla

Bucinator

Tongue

Sublingual gland

Mandible

Anterior belly of digastric m.

Transverse section

Nasal bone

Lacrimal bone

Ethmoidal cells

Eyeball

Zygomatic bone

Orbital fat body

Sphenoidal sinuses

Optic chiasm

Temporal lobe

Infundibular recess of third ventricle

Mammillary bodies

Crus cerebri

Aqueduct of midbrain

Nasal septum

Medial wall of orbit

Medial rectus m.

Lateral rectus m.

Optic n. (CN II)

Temporalis m.

Optic tract

Interpeduncular fossa

Vermis

C. Machado
—M.D.

Plate 69

Nose

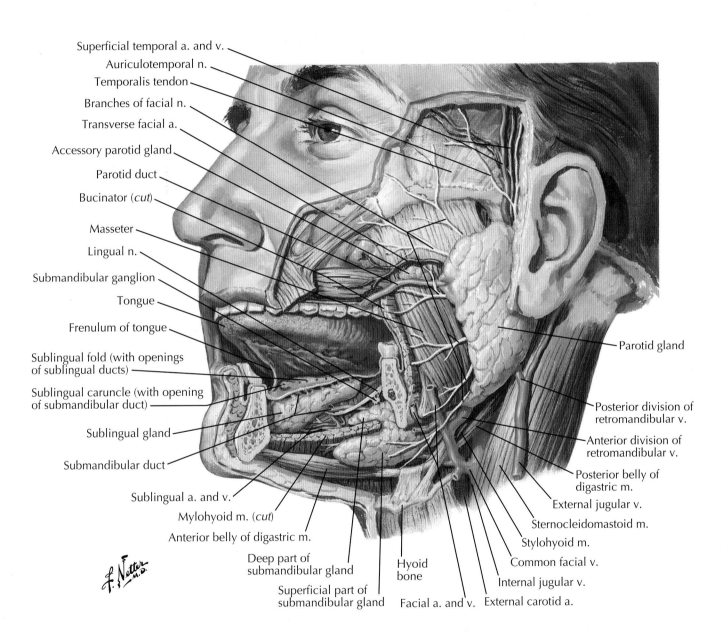

Superficial temporal a. and v.

Auriculotemporal n.

Temporalis tendon

Branches of facial n.

Transverse facial a.

Accessory parotid gland

Parotid duct

Bucinator (*cut*)

Masseter

Lingual n.

Submandibular ganglion

Tongue

Frenulum of tongue

Sublingual fold (with openings of sublingual ducts)

Sublingual caruncle (with opening of submandibular duct)

Sublingual gland

Submandibular duct

Sublingual a. and v.

Mylohyoid m. (*cut*)

Anterior belly of digastric m.

Deep part of submandibular gland

Superficial part of submandibular gland

Hyoid bone

Facial a. and v.

External carotid a.

Internal jugular v.

Common facial v.

Stylohyoid m.

Sternocleidomastoid m.

External jugular v.

Posterior belly of digastric m.

Anterior division of retromandibular v.

Posterior division of retromandibular v.

Parotid gland

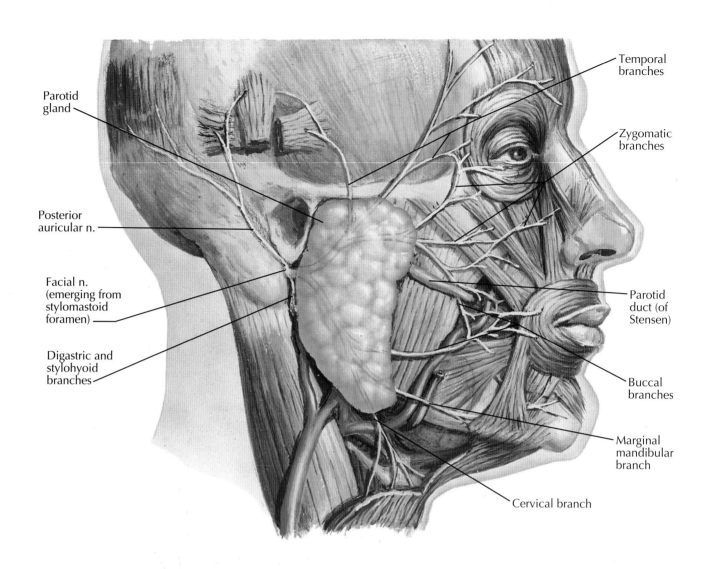

Parotid gland

Posterior auricular n.

Facial n. (emerging from stylomastoid foramen)

Digastric and stylohyoid branches

Temporal branches

Zygomatic branches

Parotid duct (of Stensen)

Buccal branches

Marginal mandibular branch

Cervical branch

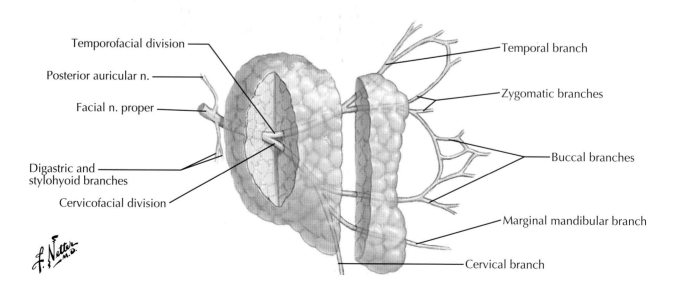

Temporofacial division

Posterior auricular n.

Facial n. proper

Digastric and stylohyoid branches

Cervicofacial division

Temporal branch

Zygomatic branches

Buccal branches

Marginal mandibular branch

Cervical branch

Plate 71

Nose

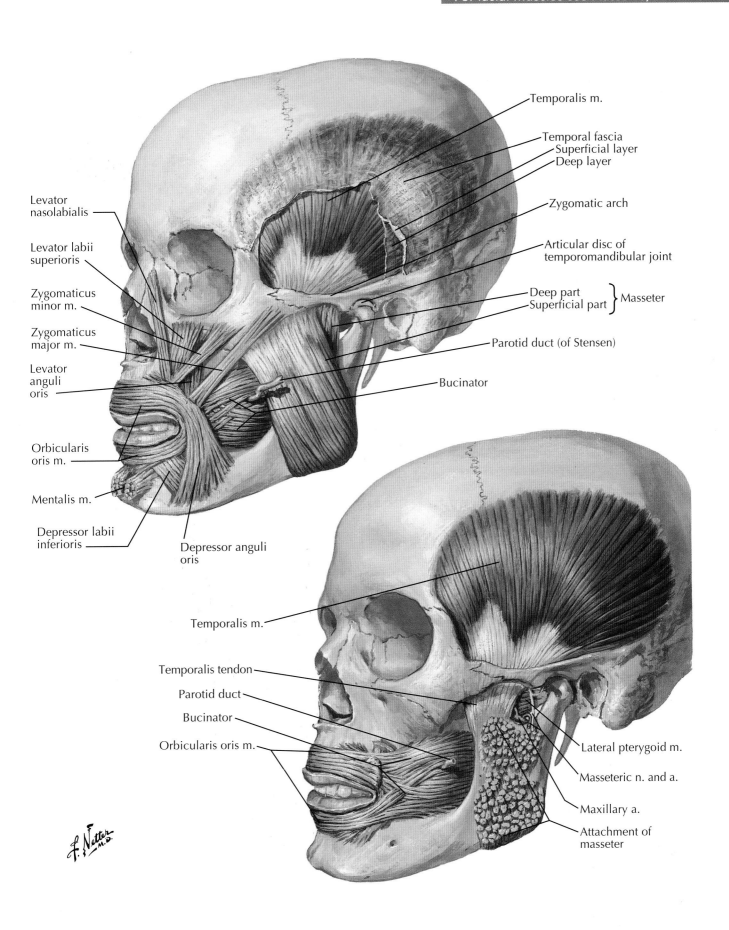

Temporalis m.

Temporal fascia
Superficial layer
Deep layer

Zygomatic arch

Articular disc of
temporomandibular joint

Deep part
Superficial part } Masseter

Parotid duct (of Stensen)

Bucinator

Levator
nasolabialis

Levator labii
superioris

Zygomaticus
minor m.

Zygomaticus
major m.

Levator
anguli
oris

Orbicularis
oris m.

Mentalis m.

Depressor labii
inferioris

Depressor anguli
oris

Temporalis m.

Temporalis tendon

Parotid duct

Bucinator

Orbicularis oris m.

Lateral pterygoid m.

Masseteric n. and a.

Maxillary a.

Attachment of
masseter

Nose

Plate 72

Lateral view

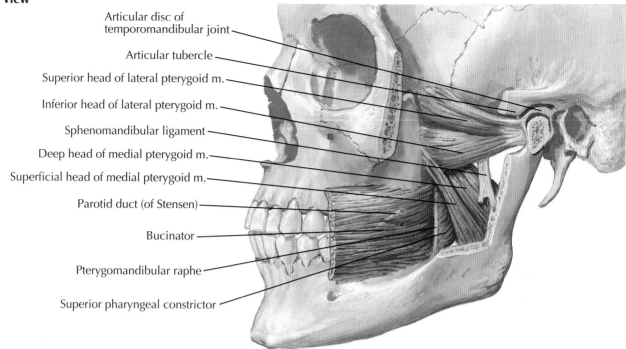

Articular disc of
temporomandibular joint

Articular tubercle

Superior head of lateral pterygoid m.

Inferior head of lateral pterygoid m.

Sphenomandibular ligament

Deep head of medial pterygoid m.

Superficial head of medial pterygoid m.

Parotid duct (of Stensen)

Bucinator

Pterygomandibular raphe

Superior pharyngeal constrictor

Posterior view

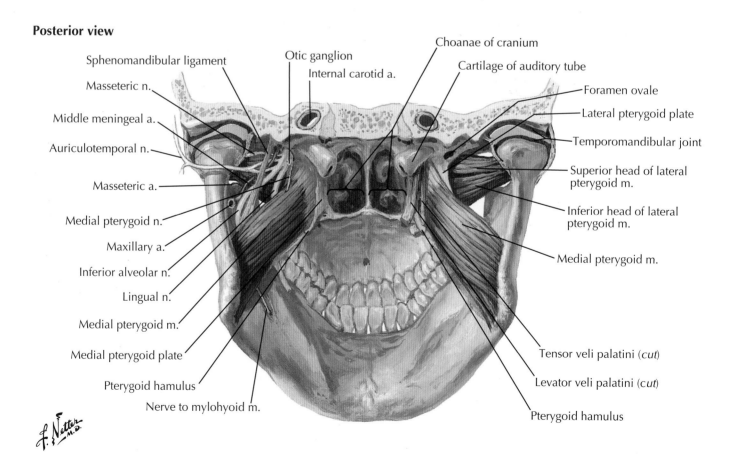

Sphenomandibular ligament

Masseteric n.

Middle meningeal a.

Auriculotemporal n.

Masseteric a.

Medial pterygoid n.

Maxillary a.

Inferior alveolar n.

Lingual n.

Medial pterygoid m.

Medial pterygoid plate

Pterygoid hamulus

Nerve to mylohyoid m.

Otic ganglion

Internal carotid a.

Choanae of cranium

Cartilage of auditory tube

Foramen ovale

Lateral pterygoid plate

Temporomandibular joint

Superior head of lateral
pterygoid m.

Inferior head of lateral
pterygoid m.

Medial pterygoid m.

Tensor veli palatini (*cut*)

Levator veli palatini (*cut*)

Pterygoid hamulus

Plate 73

Nose

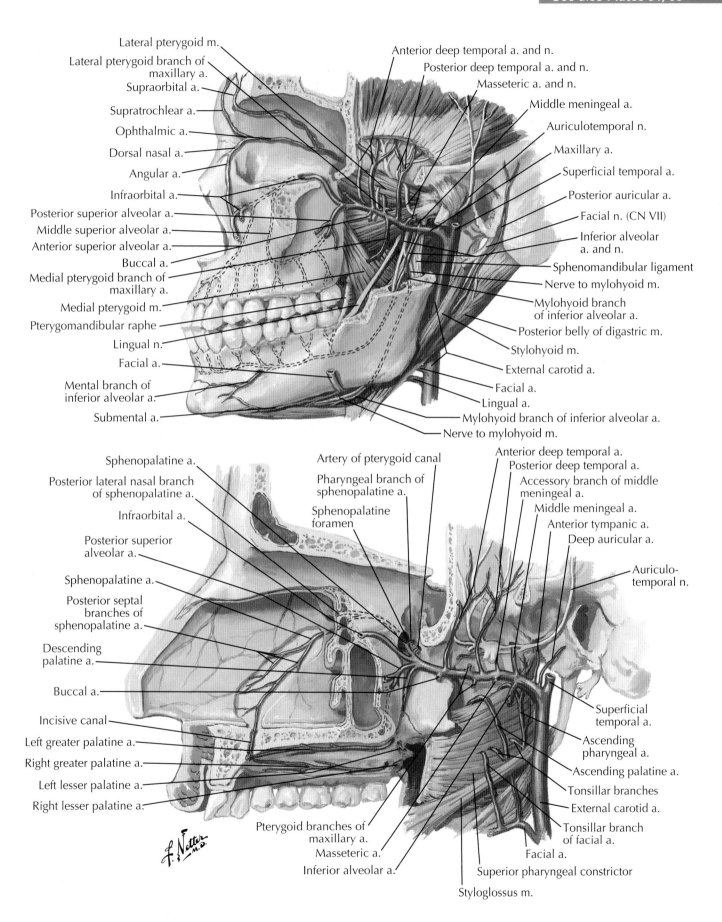

Lateral pterygoid m.
Lateral pterygoid branch of maxillary a.
Supraorbital a.
Supratrochlear a.
Ophthalmic a.
Dorsal nasal a.
Angular a.
Infraorbital a.
Posterior superior alveolar a.
Middle superior alveolar a.
Anterior superior alveolar a.
Buccal a.
Medial pterygoid branch of maxillary a.
Medial pterygoid m.
Pterygomandibular raphe
Lingual n.
Facial a.
Mental branch of inferior alveolar a.
Submental a.

Anterior deep temporal a. and n.
Posterior deep temporal a. and n.
Masseteric a. and n.
Middle meningeal a.
Auriculotemporal n.
Maxillary a.
Superficial temporal a.
Posterior auricular a.
Facial n. (CN VII)
Inferior alveolar a. and n.
Sphenomandibular ligament
Nerve to mylohyoid m.
Mylohyoid branch of inferior alveolar a.
Posterior belly of digastric m.
Stylohyoid m.
External carotid a.
Facial a.
Lingual a.
Mylohyoid branch of inferior alveolar a.
Nerve to mylohyoid m.

Sphenopalatine a.
Posterior lateral nasal branch of sphenopalatine a.
Infraorbital a.
Posterior superior alveolar a.
Sphenopalatine a.
Posterior septal branches of sphenopalatine a.
Descending palatine a.
Buccal a.
Incisive canal
Left greater palatine a.
Right greater palatine a.
Left lesser palatine a.
Right lesser palatine a.

Artery of pterygoid canal
Pharyngeal branch of sphenopalatine a.
Sphenopalatine foramen

Anterior deep temporal a.
Posterior deep temporal a.
Accessory branch of middle meningeal a.
Middle meningeal a.
Anterior tympanic a.
Deep auricular a.
Auriculo-temporal n.
Superficial temporal a.
Ascending pharyngeal a.
Ascending palatine a.
Tonsillar branches
External carotid a.
Tonsillar branch of facial a.
Facial a.

Pterygoid branches of maxillary a.
Masseteric a.
Inferior alveolar a.
Superior pharyngeal constrictor
Styloglossus m.

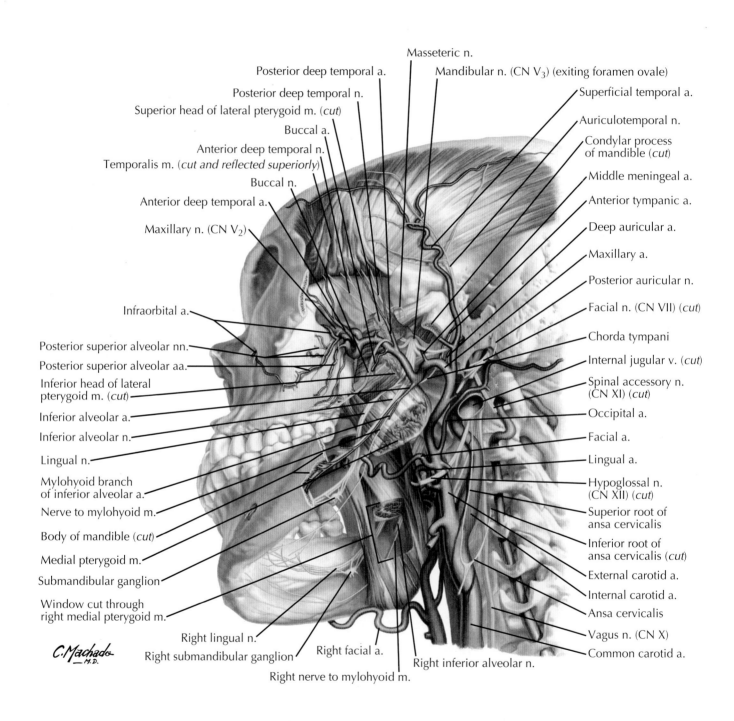

Masseteric n.

Posterior deep temporal a.

Mandibular n. (CN V₃) (exiting foramen ovale)

Posterior deep temporal n.

Superficial temporal a.

Superior head of lateral pterygoid m. (*cut*)

Auriculotemporal n.

Buccal a.

Condylar process of mandible (*cut*)

Anterior deep temporal n.

Temporalis m. (*cut and reflected superiorly*)

Middle meningeal a.

Buccal n.

Anterior tympanic a.

Anterior deep temporal a.

Deep auricular a.

Maxillary n. (CN V₂)

Maxillary a.

Posterior auricular n.

Infraorbital a.

Facial n. (CN VII) (*cut*)

Posterior superior alveolar nn.

Chorda tympani

Posterior superior alveolar aa.

Internal jugular v. (*cut*)

Inferior head of lateral pterygoid m. (*cut*)

Spinal accessory n. (CN XI) (*cut*)

Inferior alveolar a.

Occipital a.

Inferior alveolar n.

Facial a.

Lingual n.

Lingual a.

Mylohyoid branch of inferior alveolar a.

Hypoglossal n. (CN XII) (*cut*)

Nerve to mylohyoid m.

Superior root of ansa cervicalis

Body of mandible (*cut*)

Inferior root of ansa cervicalis (*cut*)

Medial pterygoid m.

External carotid a.

Submandibular ganglion

Internal carotid a.

Window cut through right medial pterygoid m.

Ansa cervicalis

Vagus n. (CN X)

Right lingual n.

Common carotid a.

Right submandibular ganglion

Right facial a.

Right inferior alveolar n.

Right nerve to mylohyoid m.

C. Machado —M.D.

Plate 75

Nose

Lateral view

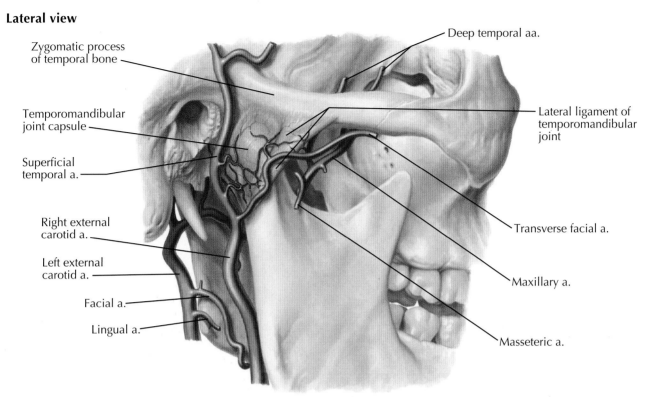

Deep temporal aa.

Zygomatic process
of temporal bone

Temporomandibular
joint capsule

Superficial
temporal a.

Right external
carotid a.

Left external
carotid a.

Facial a.

Lingual a.

Lateral ligament of
temporomandibular
joint

Transverse facial a.

Maxillary a.

Masseteric a.

Medial view

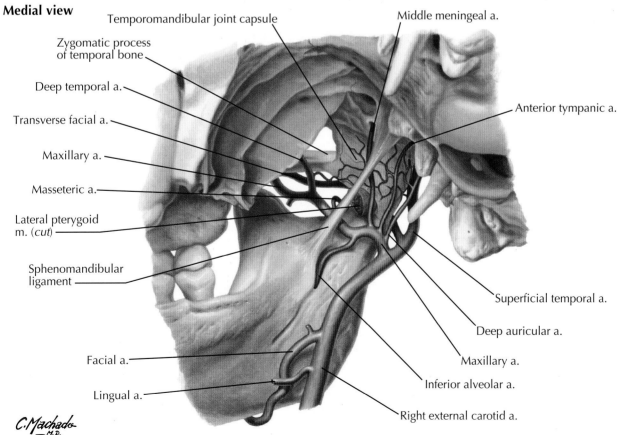

Temporomandibular joint capsule

Middle meningeal a.

Zygomatic process
of temporal bone

Deep temporal a.

Transverse facial a.

Maxillary a.

Masseteric a.

Lateral pterygoid
m. (*cut*)

Sphenomandibular
ligament

Facial a.

Lingual a.

Anterior tympanic a.

Superficial temporal a.

Deep auricular a.

Maxillary a.

Inferior alveolar a.

Right external carotid a.

C. Machado
_M.D.

Nose

Plate 76

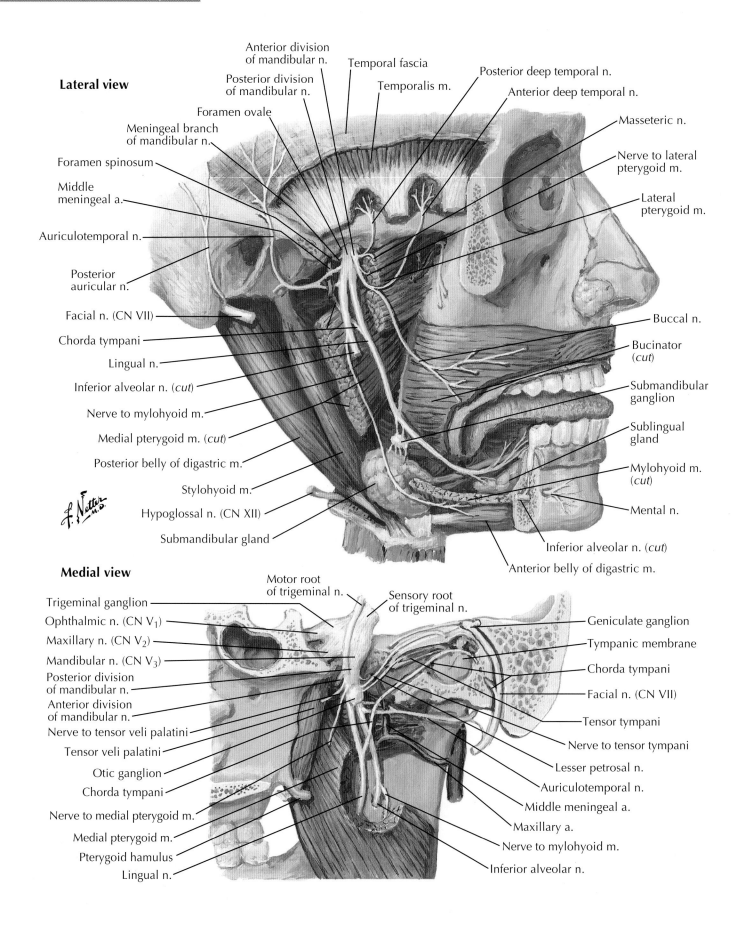

Lateral view

Anterior division of mandibular n.

Posterior division of mandibular n.

Temporal fascia

Temporalis m.

Posterior deep temporal n.

Anterior deep temporal n.

Foramen ovale

Meningeal branch of mandibular n.

Masseteric n.

Foramen spinosum

Nerve to lateral pterygoid m.

Middle meningeal a.

Lateral pterygoid m.

Auriculotemporal n.

Posterior auricular n.

Facial n. (CN VII)

Buccal n.

Chorda tympani

Bucinator (cut)

Lingual n.

Submandibular ganglion

Inferior alveolar n. (cut)

Nerve to mylohyoid m.

Sublingual gland

Medial pterygoid m. (cut)

Posterior belly of digastric m.

Mylohyoid m. (cut)

Stylohyoid m.

Hypoglossal n. (CN XII)

Mental n.

Submandibular gland

Inferior alveolar n. (cut)

Anterior belly of digastric m.

Medial view

Motor root of trigeminal n.

Sensory root of trigeminal n.

Trigeminal ganglion

Ophthalmic n. (CN V₁)

Geniculate ganglion

Maxillary n. (CN V₂)

Tympanic membrane

Mandibular n. (CN V₃)

Chorda tympani

Posterior division of mandibular n.

Facial n. (CN VII)

Anterior division of mandibular n.

Tensor tympani

Nerve to tensor veli palatini

Nerve to tensor tympani

Tensor veli palatini

Lesser petrosal n.

Otic ganglion

Auriculotemporal n.

Chorda tympani

Middle meningeal a.

Nerve to medial pterygoid m.

Maxillary a.

Medial pterygoid m.

Nerve to mylohyoid m.

Pterygoid hamulus

Inferior alveolar n.

Lingual n.

Plate 77

Nose

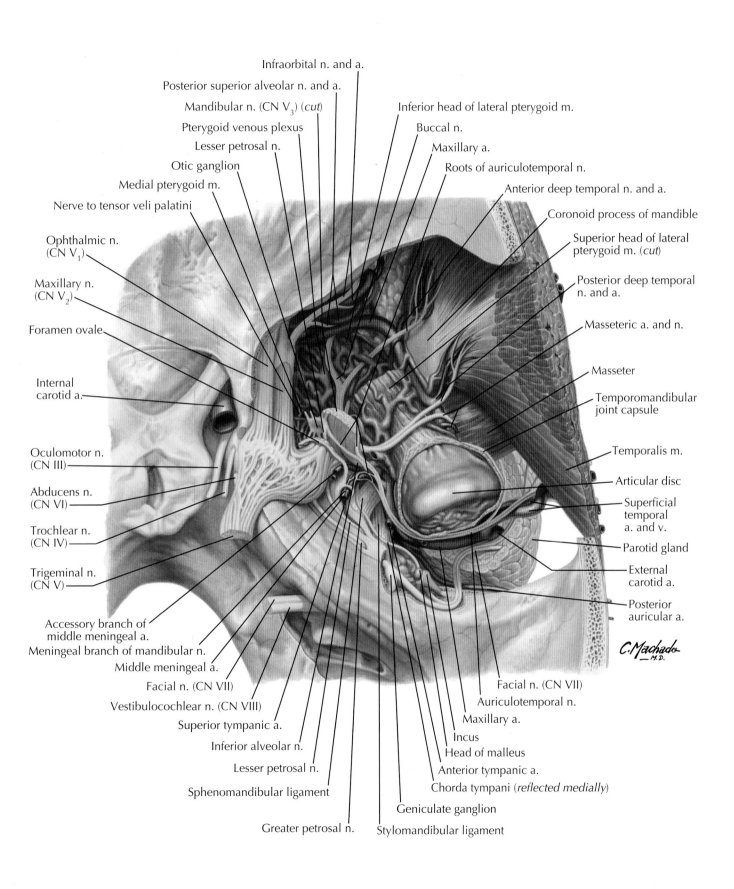

Infraorbital n. and a.

Posterior superior alveolar n. and a.

Mandibular n. (CN V₃) (cut)

Pterygoid venous plexus

Lesser petrosal n.

Otic ganglion

Medial pterygoid m.

Nerve to tensor veli palatini

Ophthalmic n. (CN V₁)

Maxillary n. (CN V₂)

Foramen ovale

Internal carotid a.

Oculomotor n. (CN III)

Abducens n. (CN VI)

Trochlear n. (CN IV)

Trigeminal n. (CN V)

Accessory branch of middle meningeal a.

Meningeal branch of mandibular n.

Middle meningeal a.

Facial n. (CN VII)

Vestibulocochlear n. (CN VIII)

Superior tympanic a.

Inferior alveolar n.

Lesser petrosal n.

Sphenomandibular ligament

Greater petrosal n.

Inferior head of lateral pterygoid m.

Buccal n.

Maxillary a.

Roots of auriculotemporal n.

Anterior deep temporal n. and a.

Coronoid process of mandible

Superior head of lateral pterygoid m. (cut)

Posterior deep temporal n. and a.

Masseteric a. and n.

Masseter

Temporomandibular joint capsule

Temporalis m.

Articular disc

Superficial temporal a. and v.

Parotid gland

External carotid a.

Posterior auricular a.

Facial n. (CN VII)

Auriculotemporal n.

Maxillary a.

Incus

Head of malleus

Anterior tympanic a.

Chorda tympani (reflected medially)

Geniculate ganglion

Stylomandibular ligament

C. Machado
—M.D.

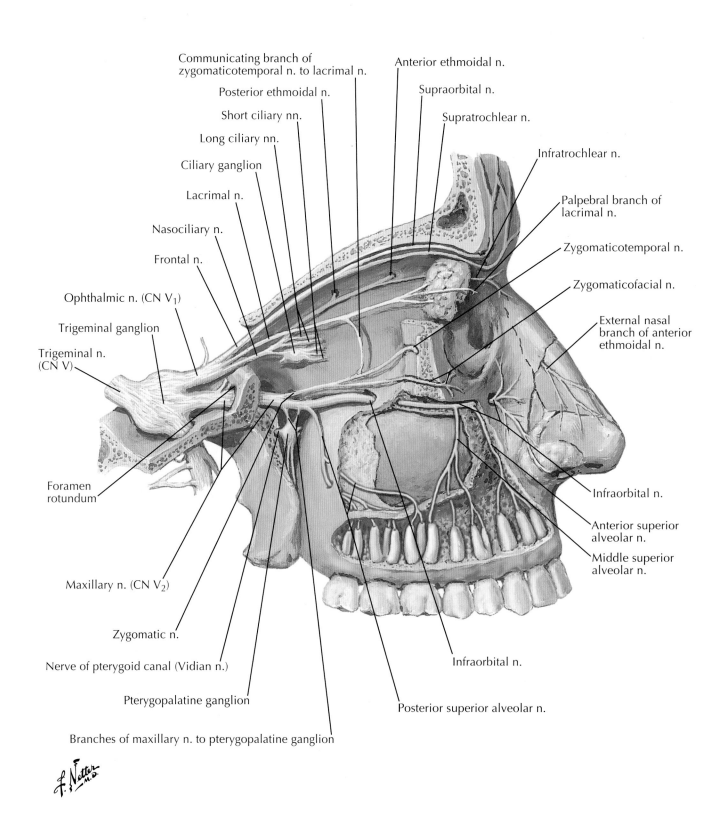

Communicating branch of
zygomaticotemporal n. to lacrimal n.

Posterior ethmoidal n.

Short ciliary nn.

Long ciliary nn.

Ciliary ganglion

Lacrimal n.

Nasociliary n.

Frontal n.

Ophthalmic n. (CN V₁)

Trigeminal ganglion

Trigeminal n.
(CN V)

Foramen
rotundum

Maxillary n. (CN V₂)

Zygomatic n.

Nerve of pterygoid canal (Vidian n.)

Pterygopalatine ganglion

Branches of maxillary n. to pterygopalatine ganglion

Anterior ethmoidal n.

Supraorbital n.

Supratrochlear n.

Infratrochlear n.

Palpebral branch of
lacrimal n.

Zygomaticotemporal n.

Zygomaticofacial n.

External nasal
branch of anterior
ethmoidal n.

Infraorbital n.

Anterior superior
alveolar n.

Middle superior
alveolar n.

Infraorbital n.

Posterior superior alveolar n.

Plate 79

Nose

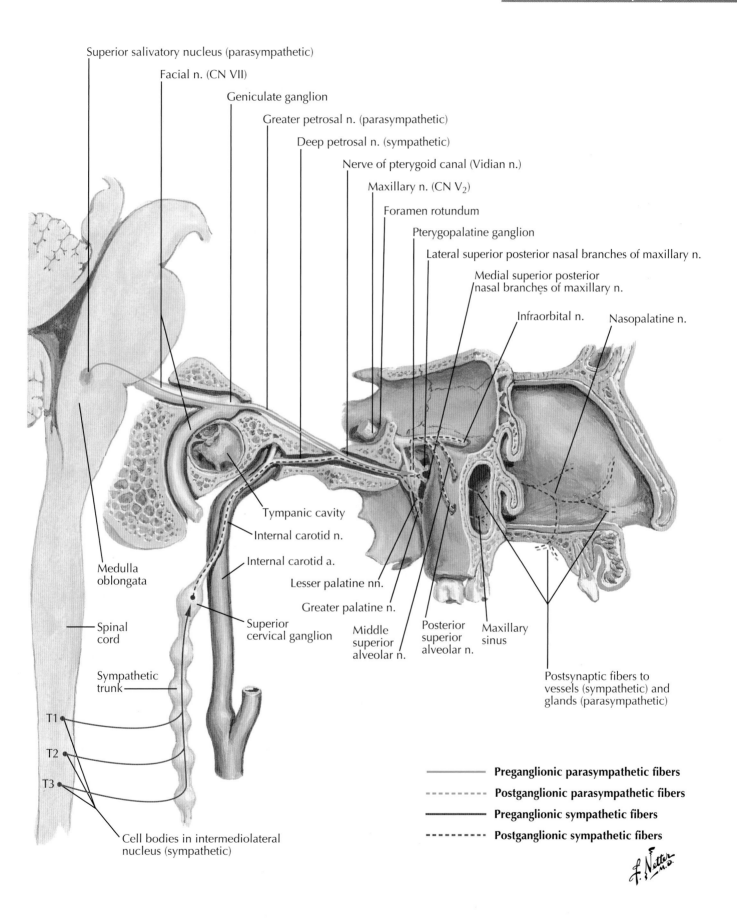

Superior salivatory nucleus (parasympathetic)

Facial n. (CN VII)

Geniculate ganglion

Greater petrosal n. (parasympathetic)

Deep petrosal n. (sympathetic)

Nerve of pterygoid canal (Vidian n.)

Maxillary n. (CN V₂)

Foramen rotundum

Pterygopalatine ganglion

Lateral superior posterior nasal branches of maxillary n.

Medial superior posterior nasal branches of maxillary n.

Infraorbital n.

Nasopalatine n.

Tympanic cavity

Internal carotid n.

Internal carotid a.

Lesser palatine nn.

Greater palatine n.

Medulla oblongata

Spinal cord

Sympathetic trunk

Superior cervical ganglion

Middle superior alveolar n.

Posterior superior alveolar n.

Maxillary sinus

Postsynaptic fibers to vessels (sympathetic) and glands (parasympathetic)

T1

T2

T3

Cell bodies in intermediolateral nucleus (sympathetic)

——— **Preganglionic parasympathetic fibers**

- - - - - **Postganglionic parasympathetic fibers**

━━━ **Preganglionic sympathetic fibers**

- - - - - **Postganglionic sympathetic fibers**

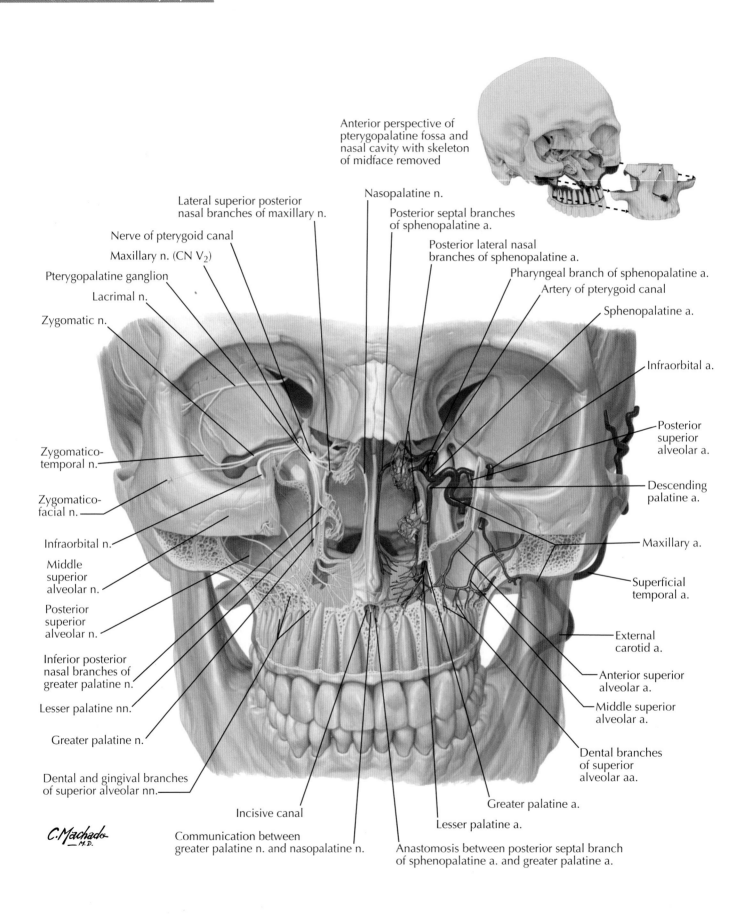

Anterior perspective of pterygopalatine fossa and nasal cavity with skeleton of midface removed

Lateral superior posterior nasal branches of maxillary n.

Nerve of pterygoid canal

Maxillary n. (CN V$_2$)

Pterygopalatine ganglion

Lacrimal n.

Zygomatic n.

Zygomatico-temporal n.

Zygomatico-facial n.

Infraorbital n.

Middle superior alveolar n.

Posterior superior alveolar n.

Inferior posterior nasal branches of greater palatine n.

Lesser palatine nn.

Greater palatine n.

Dental and gingival branches of superior alveolar nn.

Incisive canal

Communication between greater palatine n. and nasopalatine n.

Nasopalatine n.

Posterior septal branches of sphenopalatine a.

Posterior lateral nasal branches of sphenopalatine a.

Pharyngeal branch of sphenopalatine a.

Artery of pterygoid canal

Sphenopalatine a.

Infraorbital a.

Posterior superior alveolar a.

Descending palatine a.

Maxillary a.

Superficial temporal a.

External carotid a.

Anterior superior alveolar a.

Middle superior alveolar a.

Dental branches of superior alveolar aa.

Greater palatine a.

Lesser palatine a.

Anastomosis between posterior septal branch of sphenopalatine a. and greater palatine a.

C. Machado, M.D.

Plate 81   **Nose**

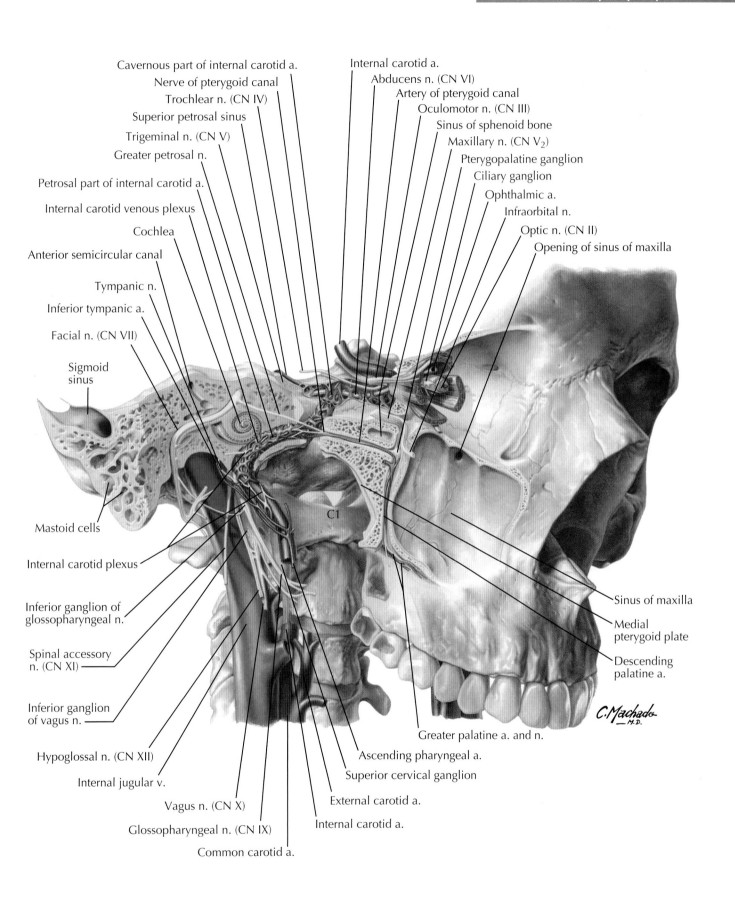

Cavernous part of internal carotid a.

Nerve of pterygoid canal

Trochlear n. (CN IV)

Superior petrosal sinus

Trigeminal n. (CN V)

Greater petrosal n.

Petrosal part of internal carotid a.

Internal carotid venous plexus

Cochlea

Anterior semicircular canal

Tympanic n.

Inferior tympanic a.

Facial n. (CN VII)

Sigmoid sinus

Mastoid cells

Internal carotid plexus

Inferior ganglion of glossopharyngeal n.

Spinal accessory n. (CN XI)

Inferior ganglion of vagus n.

Hypoglossal n. (CN XII)

Internal jugular v.

Vagus n. (CN X)

Glossopharyngeal n. (CN IX)

Common carotid a.

Internal carotid a.

Abducens n. (CN VI)

Artery of pterygoid canal

Oculomotor n. (CN III)

Sinus of sphenoid bone

Maxillary n. (CN V$_2$)

Pterygopalatine ganglion

Ciliary ganglion

Ophthalmic a.

Infraorbital n.

Optic n. (CN II)

Opening of sinus of maxilla

C1

Sinus of maxilla

Medial pterygoid plate

Descending palatine a.

C.Machado M.D.

Greater palatine a. and n.

Ascending pharyngeal a.

Superior cervical ganglion

External carotid a.

Internal carotid a.

Nose

Plate 82

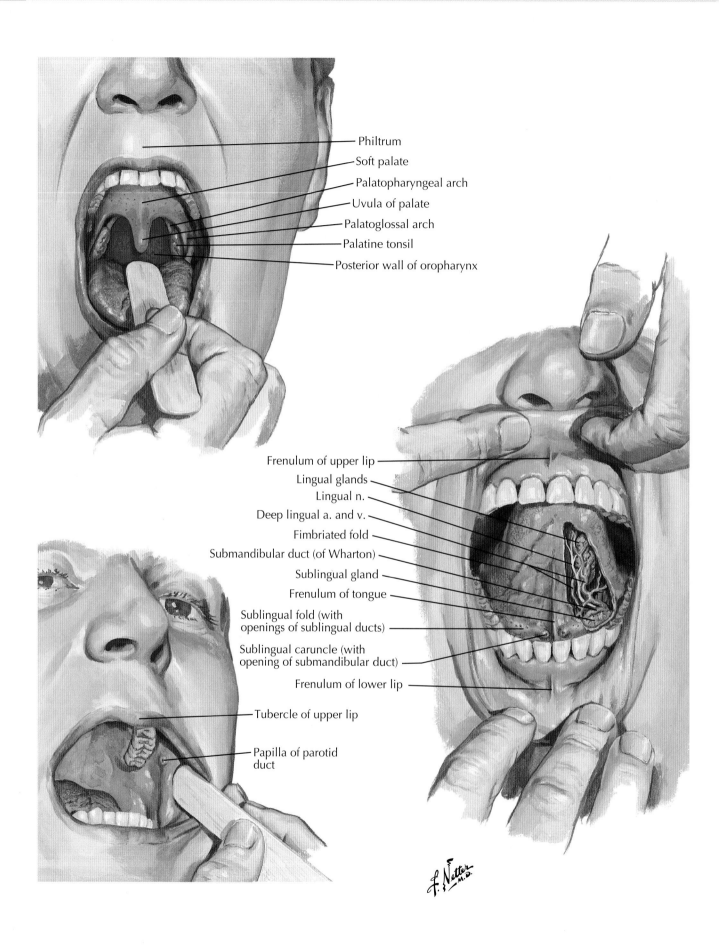

Philtrum

Soft palate

Palatopharyngeal arch

Uvula of palate

Palatoglossal arch

Palatine tonsil

Posterior wall of oropharynx

Frenulum of upper lip

Lingual glands

Lingual n.

Deep lingual a. and v.

Fimbriated fold

Submandibular duct (of Wharton)

Sublingual gland

Frenulum of tongue

Sublingual fold (with openings of sublingual ducts)

Sublingual caruncle (with opening of submandibular duct)

Frenulum of lower lip

Tubercle of upper lip

Papilla of parotid duct

Plate 83

Mouth

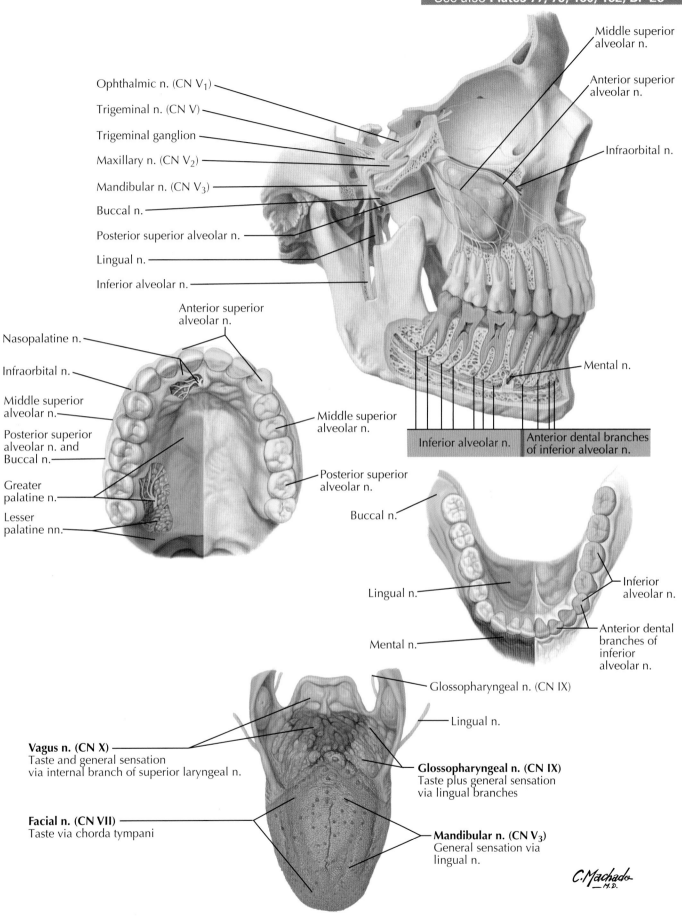

Middle superior alveolar n.

Anterior superior alveolar n.

Ophthalmic n. (CN V₁)

Trigeminal n. (CN V)

Trigeminal ganglion

Infraorbital n.

Maxillary n. (CN V₂)

Mandibular n. (CN V₃)

Buccal n.

Posterior superior alveolar n.

Lingual n.

Inferior alveolar n.

Anterior superior alveolar n.

Nasopalatine n.

Infraorbital n.

Middle superior alveolar n.

Posterior superior alveolar n. and Buccal n.

Greater palatine n.

Lesser palatine nn.

Mental n.

Middle superior alveolar n.

Posterior superior alveolar n.

Inferior alveolar n.

Anterior dental branches of inferior alveolar n.

Buccal n.

Inferior alveolar n.

Lingual n.

Anterior dental branches of inferior alveolar n.

Mental n.

Glossopharyngeal n. (CN IX)

Lingual n.

Vagus n. (CN X)
Taste and general sensation via internal branch of superior laryngeal n.

Glossopharyngeal n. (CN IX)
Taste plus general sensation via lingual branches

Facial n. (CN VII)
Taste via chorda tympani

Mandibular n. (CN V₃)
General sensation via lingual n.

C. Machado
M.D.

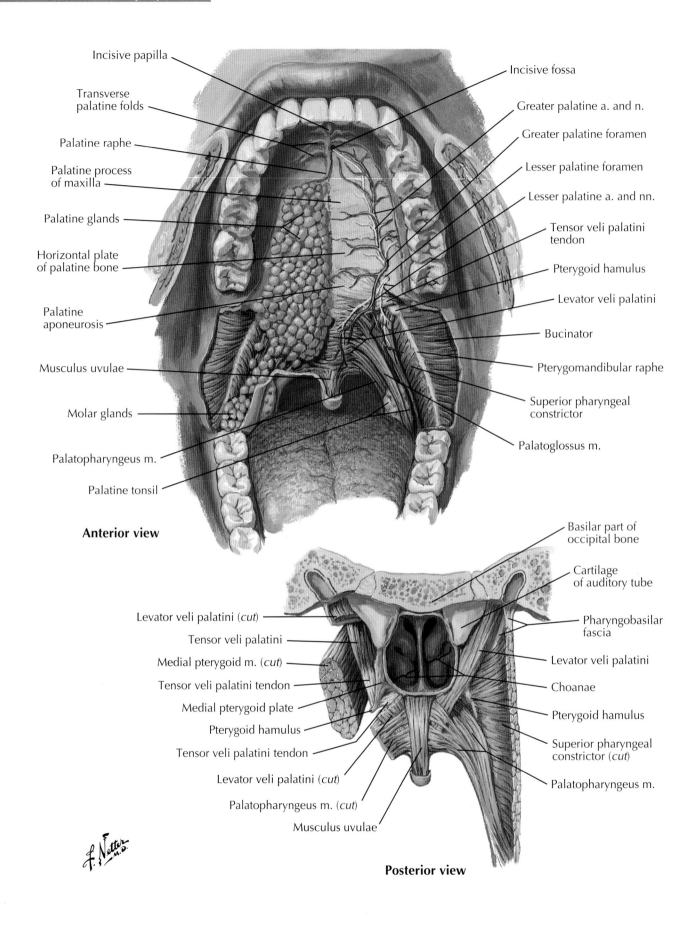

Incisive papilla

Transverse palatine folds

Palatine raphe

Palatine process of maxilla

Palatine glands

Horizontal plate of palatine bone

Palatine aponeurosis

Musculus uvulae

Molar glands

Palatopharyngeus m.

Palatine tonsil

Anterior view

Incisive fossa

Greater palatine a. and n.

Greater palatine foramen

Lesser palatine foramen

Lesser palatine a. and nn.

Tensor veli palatini tendon

Pterygoid hamulus

Levator veli palatini

Bucinator

Pterygomandibular raphe

Superior pharyngeal constrictor

Palatoglossus m.

Levator veli palatini (*cut*)

Tensor veli palatini

Medial pterygoid m. (*cut*)

Tensor veli palatini tendon

Medial pterygoid plate

Pterygoid hamulus

Tensor veli palatini tendon

Levator veli palatini (*cut*)

Palatopharyngeus m. (*cut*)

Musculus uvulae

Basilar part of occipital bone

Cartilage of auditory tube

Pharyngobasilar fascia

Levator veli palatini

Choanae

Pterygoid hamulus

Superior pharyngeal constrictor (*cut*)

Palatopharyngeus m.

Posterior view

Plate 85

Mouth

Horizontal section below lingula of mandible (superior view)

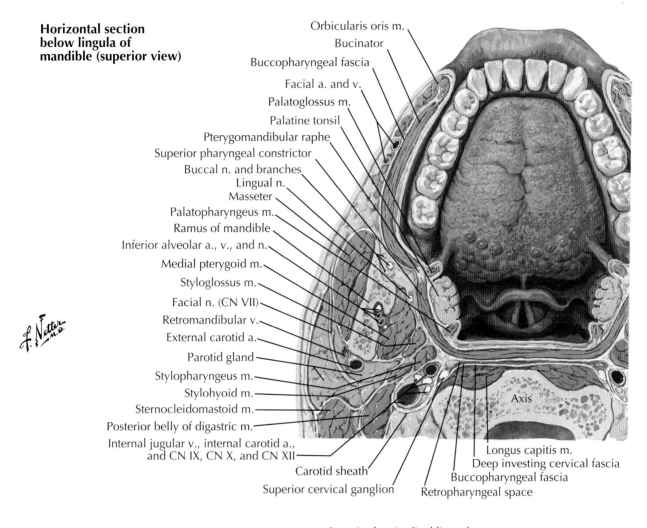

Orbicularis oris m.

Bucinator

Buccopharyngeal fascia

Facial a. and v.

Palatoglossus m.

Palatine tonsil

Pterygomandibular raphe

Superior pharyngeal constrictor

Buccal n. and branches

Lingual n.

Masseter

Palatopharyngeus m.

Ramus of mandible

Inferior alveolar a., v., and n.

Medial pterygoid m.

Styloglossus m.

Facial n. (CN VII)

Retromandibular v.

External carotid a.

Parotid gland

Stylopharyngeus m.

Stylohyoid m.

Sternocleidomastoid m.

Posterior belly of digastric m.

Internal jugular v., internal carotid a., and CN IX, CN X, and CN XII

Carotid sheath

Superior cervical ganglion

Axis

Longus capitis m.

Deep investing cervical fascia

Buccopharyngeal fascia

Retropharyngeal space

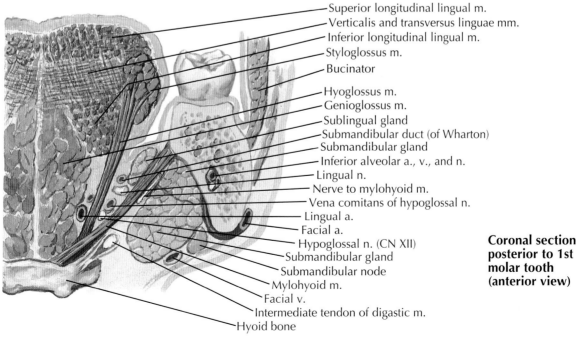

Superior longitudinal lingual m.

Verticalis and transversus linguae mm.

Inferior longitudinal lingual m.

Styloglossus m.

Bucinator

Hyoglossus m.

Genioglossus m.

Sublingual gland

Submandibular duct (of Wharton)

Submandibular gland

Inferior alveolar a., v., and n.

Lingual n.

Nerve to mylohyoid m.

Vena comitans of hypoglossal n.

Lingual a.

Facial a.

Hypoglossal n. (CN XII)

Submandibular gland

Submandibular node

Mylohyoid m.

Facial v.

Intermediate tendon of digastic m.

Hyoid bone

Coronal section posterior to 1st molar tooth (anterior view)

Mouth

Plate 86

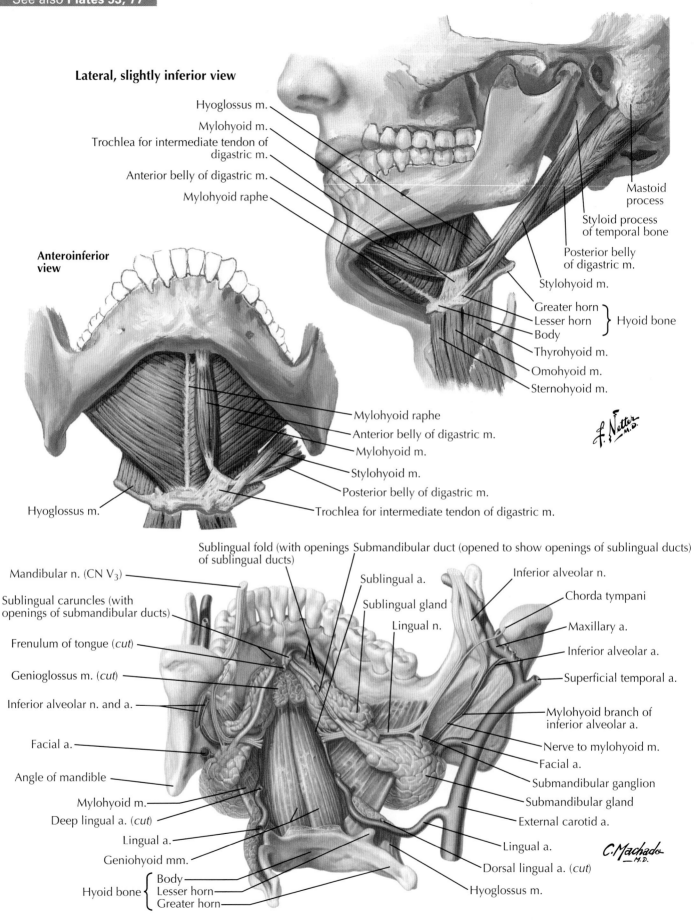

Lateral, slightly inferior view

Hyoglossus m.

Mylohyoid m.

Trochlea for intermediate tendon of digastric m.

Anterior belly of digastric m.

Mylohyoid raphe

Mastoid process

Styloid process of temporal bone

Posterior belly of digastric m.

Stylohyoid m.

Greater horn
Lesser horn } Hyoid bone
Body

Thyrohyoid m.

Omohyoid m.

Sternohyoid m.

Anteroinferior view

Mylohyoid raphe

Anterior belly of digastric m.

Mylohyoid m.

Stylohyoid m.

Posterior belly of digastric m.

Trochlea for intermediate tendon of digastric m.

Hyoglossus m.

Sublingual fold (with openings of sublingual ducts)

Submandibular duct (opened to show openings of sublingual ducts)

Sublingual a.

Inferior alveolar n.

Mandibular n. (CN V₃)

Sublingual caruncles (with openings of submandibular ducts)

Sublingual gland

Lingual n.

Chorda tympani

Maxillary a.

Frenulum of tongue (cut)

Inferior alveolar a.

Genioglossus m. (cut)

Superficial temporal a.

Inferior alveolar n. and a.

Mylohyoid branch of inferior alveolar a.

Facial a.

Nerve to mylohyoid m.

Facial a.

Angle of mandible

Submandibular ganglion

Mylohyoid m.

Submandibular gland

Deep lingual a. (cut)

External carotid a.

Lingual a.

Lingual a.

Geniohyoid mm.

Dorsal lingual a. (cut)

Body
Hyoid bone { Lesser horn
Greater horn

Hyoglossus m.

Plate 87

Mouth

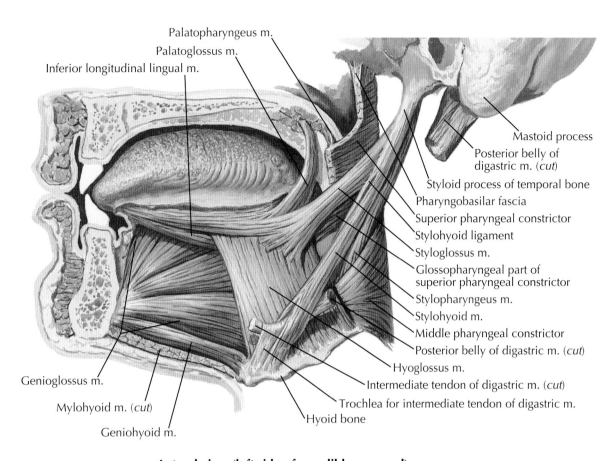

Palatopharyngeus m.

Palatoglossus m.

Inferior longitudinal lingual m.

Mastoid process

Posterior belly of digastric m. (*cut*)

Styloid process of temporal bone

Pharyngobasilar fascia

Superior pharyngeal constrictor

Stylohyoid ligament

Styloglossus m.

Glossopharyngeal part of superior pharyngeal constrictor

Stylopharyngeus m.

Stylohyoid m.

Middle pharyngeal constrictor

Posterior belly of digastric m. (*cut*)

Hyoglossus m.

Intermediate tendon of digastric m. (*cut*)

Trochlea for intermediate tendon of digastric m.

Genioglossus m.

Mylohyoid m. (*cut*)

Geniohyoid m.

Hyoid bone

Lateral view (left side of mandible removed)

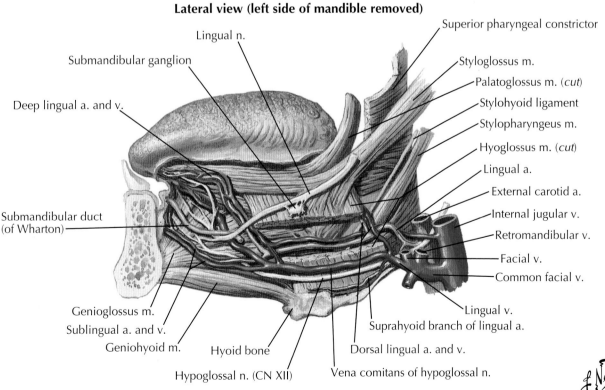

Lingual n.

Submandibular ganglion

Deep lingual a. and v.

Submandibular duct (of Wharton)

Superior pharyngeal constrictor

Styloglossus m.

Palatoglossus m. (*cut*)

Stylohyoid ligament

Stylopharyngeus m.

Hyoglossus m. (*cut*)

Lingual a.

External carotid a.

Internal jugular v.

Retromandibular v.

Facial v.

Common facial v.

Genioglossus m.

Sublingual a. and v.

Geniohyoid m.

Hyoid bone

Hypoglossal n. (CN XII)

Vena comitans of hypoglossal n.

Dorsal lingual a. and v.

Suprahyoid branch of lingual a.

Lingual v.

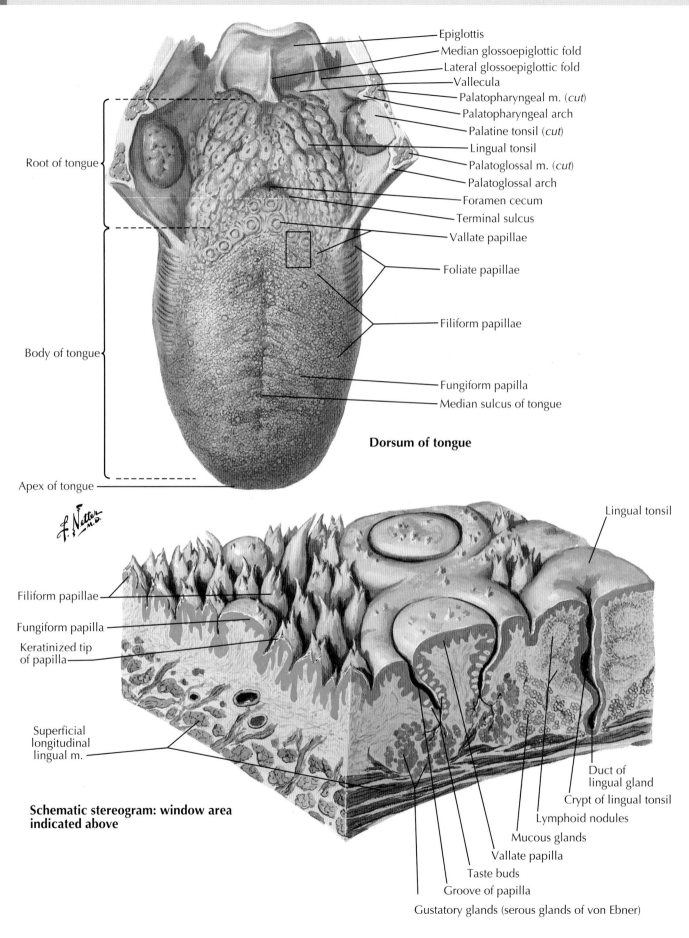

Epiglottis

Median glossoepiglottic fold

Lateral glossoepiglottic fold

Vallecula

Palatopharyngeal m. (*cut*)

Palatopharyngeal arch

Palatine tonsil (*cut*)

Lingual tonsil

Palatoglossal m. (*cut*)

Palatoglossal arch

Foramen cecum

Terminal sulcus

Vallate papillae

Foliate papillae

Filiform papillae

Fungiform papilla

Median sulcus of tongue

Root of tongue

Body of tongue

Apex of tongue

Dorsum of tongue

Filiform papillae

Fungiform papilla

Keratinized tip of papilla

Superficial longitudinal lingual m.

Lingual tonsil

Duct of lingual gland

Crypt of lingual tonsil

Lymphoid nodules

Mucous glands

Vallate papilla

Taste buds

Groove of papilla

Gustatory glands (serous glands of von Ebner)

Schematic stereogram: window area indicated above

Plate 89   **Mouth**

**Medial view
sagittal section**

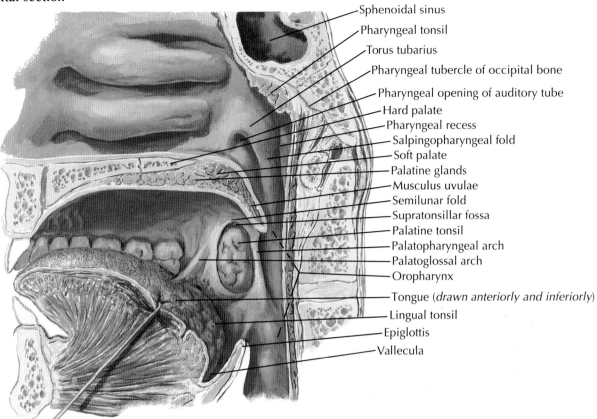

- Sphenoidal sinus
- Pharyngeal tonsil
- Torus tubarius
- Pharyngeal tubercle of occipital bone
- Pharyngeal opening of auditory tube
- Hard palate
- Pharyngeal recess
- Salpingopharyngeal fold
- Soft palate
- Palatine glands
- Musculus uvulae
- Semilunar fold
- Supratonsillar fossa
- Palatine tonsil
- Palatopharyngeal arch
- Palatoglossal arch
- Oropharynx
- Tongue (*drawn anteriorly and inferiorly*)
- Lingual tonsil
- Epiglottis
- Vallecula

Pharyngeal mucosa removed

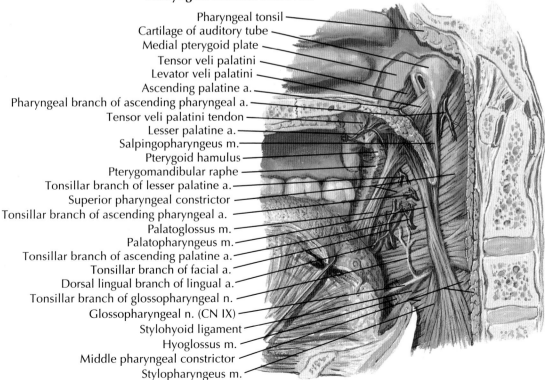

- Pharyngeal tonsil
- Cartilage of auditory tube
- Medial pterygoid plate
- Tensor veli palatini
- Levator veli palatini
- Ascending palatine a.
- Pharyngeal branch of ascending pharyngeal a.
- Tensor veli palatini tendon
- Lesser palatine a.
- Salpingopharyngeus m.
- Pterygoid hamulus
- Pterygomandibular raphe
- Tonsillar branch of lesser palatine a.
- Superior pharyngeal constrictor
- Tonsillar branch of ascending pharyngeal a.
- Palatoglossus m.
- Palatopharyngeus m.
- Tonsillar branch of ascending palatine a.
- Tonsillar branch of facial a.
- Dorsal lingual branch of lingual a.
- Tonsillar branch of glossopharyngeal n.
- Glossopharyngeal n. (CN IX)
- Stylohyoid ligament
- Hyoglossus m.
- Middle pharyngeal constrictor
- Stylopharyngeus m.

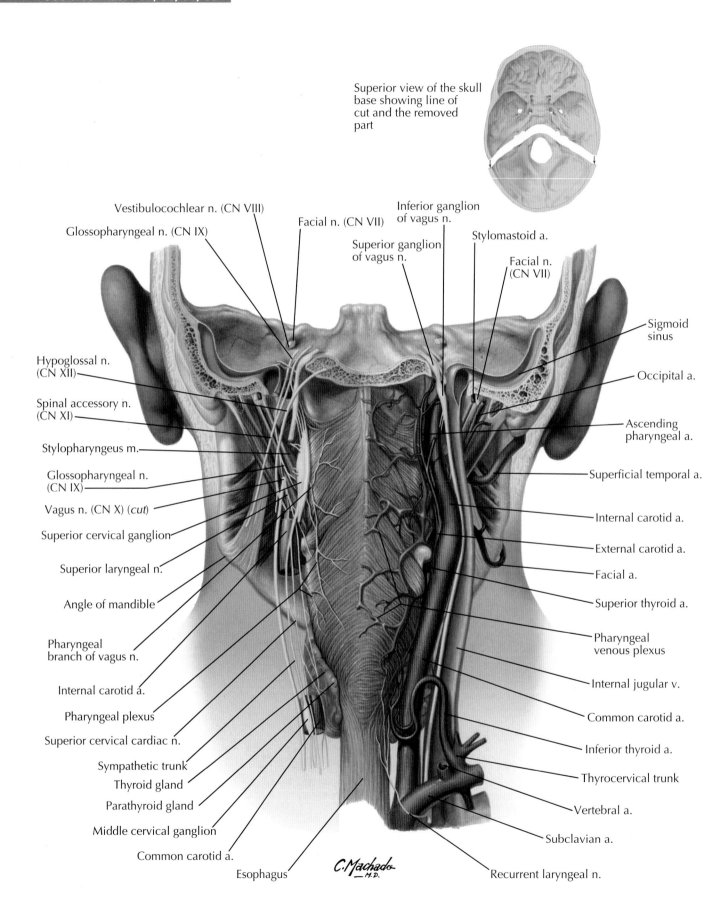

Superior view of the skull base showing line of cut and the removed part

Vestibulocochlear n. (CN VIII)

Glossopharyngeal n. (CN IX)

Facial n. (CN VII)

Inferior ganglion of vagus n.

Superior ganglion of vagus n.

Stylomastoid a.

Facial n. (CN VII)

Sigmoid sinus

Hypoglossal n. (CN XII)

Occipital a.

Spinal accessory n. (CN XI)

Ascending pharyngeal a.

Stylopharyngeus m.

Superficial temporal a.

Glossopharyngeal n. (CN IX)

Internal carotid a.

Vagus n. (CN X) (cut)

External carotid a.

Superior cervical ganglion

Facial a.

Superior laryngeal n.

Superior thyroid a.

Angle of mandible

Pharyngeal venous plexus

Pharyngeal branch of vagus n.

Internal jugular v.

Internal carotid a.

Common carotid a.

Pharyngeal plexus

Superior cervical cardiac n.

Inferior thyroid a.

Sympathetic trunk

Thyrocervical trunk

Thyroid gland

Parathyroid gland

Vertebral a.

Middle cervical ganglion

Subclavian a.

Common carotid a.

Recurrent laryngeal n.

Esophagus

C. Machado M.D.

Plate 91

Pharynx

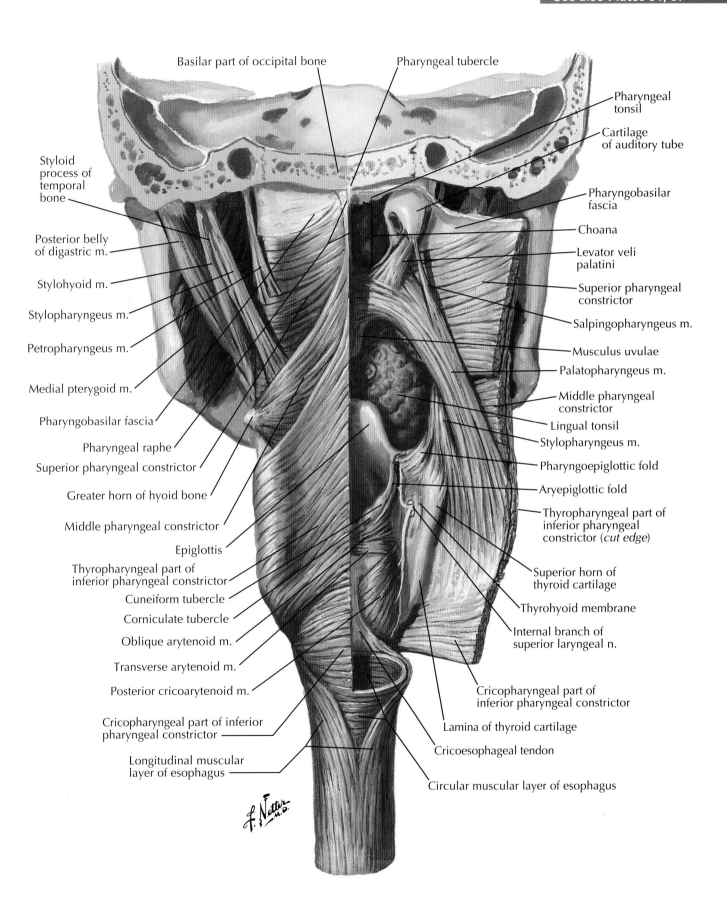

Basilar part of occipital bone

Pharyngeal tubercle

Pharyngeal tonsil

Cartilage of auditory tube

Styloid process of temporal bone

Pharyngobasilar fascia

Choana

Levator veli palatini

Posterior belly of digastric m.

Superior pharyngeal constrictor

Stylohyoid m.

Salpingopharyngeus m.

Stylopharyngeus m.

Musculus uvulae

Petropharyngeus m.

Palatopharyngeus m.

Medial pterygoid m.

Middle pharyngeal constrictor

Lingual tonsil

Pharyngobasilar fascia

Stylopharyngeus m.

Pharyngeal raphe

Pharyngoepiglottic fold

Superior pharyngeal constrictor

Aryepiglottic fold

Greater horn of hyoid bone

Thyropharyngeal part of inferior pharyngeal constrictor (*cut edge*)

Middle pharyngeal constrictor

Epiglottis

Superior horn of thyroid cartilage

Thyropharyngeal part of inferior pharyngeal constrictor

Thyrohyoid membrane

Cuneiform tubercle

Internal branch of superior laryngeal n.

Corniculate tubercle

Oblique arytenoid m.

Transverse arytenoid m.

Posterior cricoarytenoid m.

Cricopharyngeal part of inferior pharyngeal constrictor

Cricopharyngeal part of inferior pharyngeal constrictor

Lamina of thyroid cartilage

Longitudinal muscular layer of esophagus

Cricoesophageal tendon

Circular muscular layer of esophagus

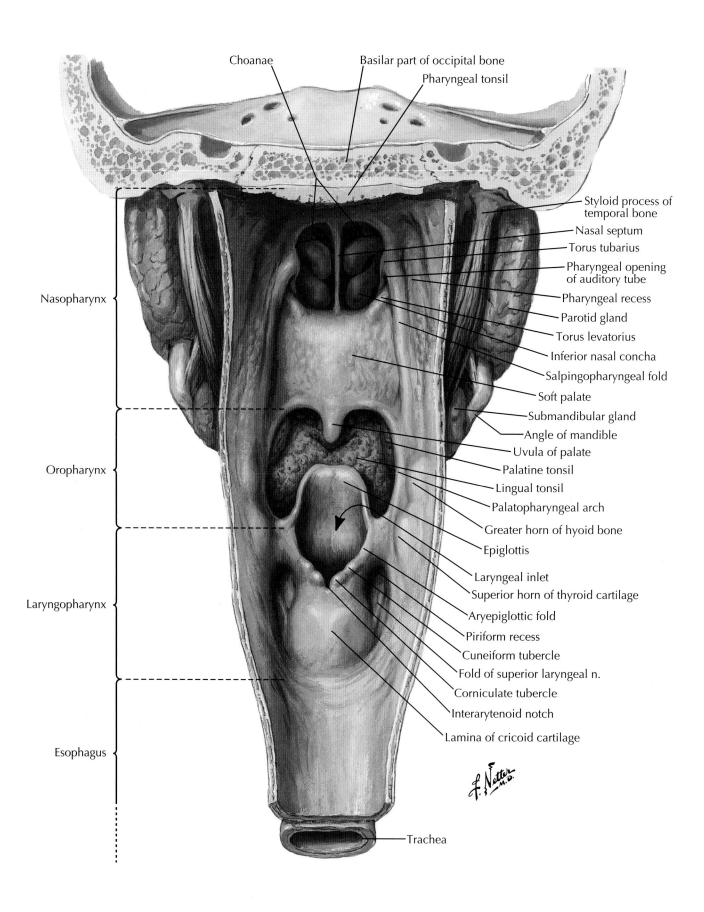

Choanae

Basilar part of occipital bone

Pharyngeal tonsil

Styloid process of temporal bone

Nasal septum

Torus tubarius

Pharyngeal opening of auditory tube

Pharyngeal recess

Parotid gland

Torus levatorius

Inferior nasal concha

Salpingopharyngeal fold

Soft palate

Submandibular gland

Angle of mandible

Uvula of palate

Palatine tonsil

Lingual tonsil

Palatopharyngeal arch

Greater horn of hyoid bone

Epiglottis

Laryngeal inlet

Superior horn of thyroid cartilage

Aryepiglottic fold

Piriform recess

Cuneiform tubercle

Fold of superior laryngeal n.

Corniculate tubercle

Interarytenoid notch

Lamina of cricoid cartilage

Nasopharynx

Oropharynx

Laryngopharynx

Esophagus

Trachea

Plate 93

Pharynx

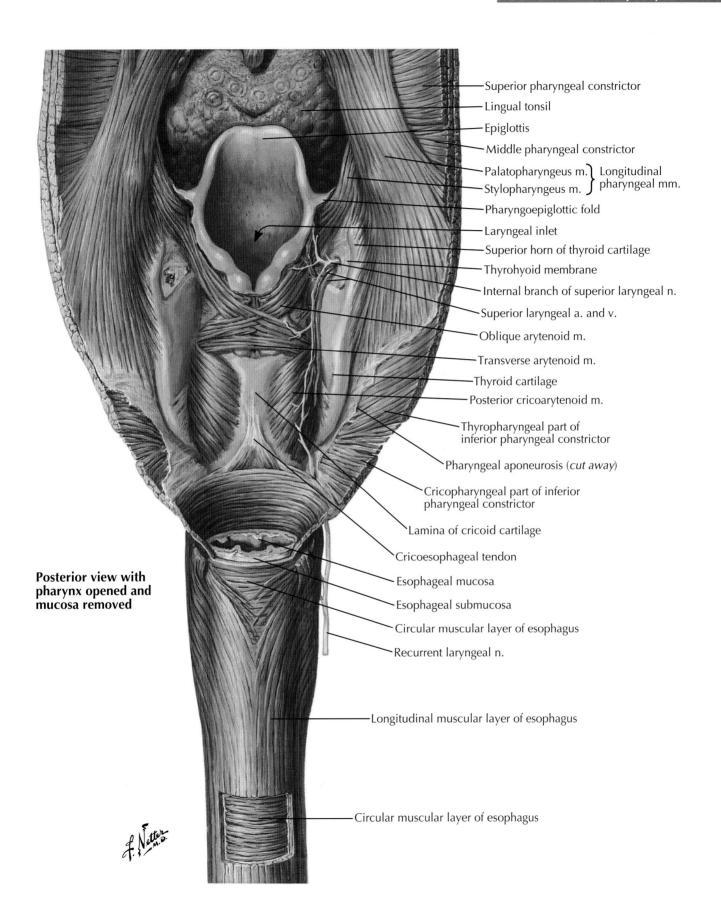

Superior pharyngeal constrictor

Lingual tonsil

Epiglottis

Middle pharyngeal constrictor

Palatopharyngeus m. } Longitudinal
Stylopharyngeus m. } pharyngeal mm.

Pharyngoepiglottic fold

Laryngeal inlet

Superior horn of thyroid cartilage

Thyrohyoid membrane

Internal branch of superior laryngeal n.

Superior laryngeal a. and v.

Oblique arytenoid m.

Transverse arytenoid m.

Thyroid cartilage

Posterior cricoarytenoid m.

Thyropharyngeal part of
inferior pharyngeal constrictor

Pharyngeal aponeurosis (*cut away*)

Cricopharyngeal part of inferior
pharyngeal constrictor

Lamina of cricoid cartilage

Cricoesophageal tendon

Esophageal mucosa

Esophageal submucosa

Circular muscular layer of esophagus

Recurrent laryngeal n.

Longitudinal muscular layer of esophagus

Circular muscular layer of esophagus

**Posterior view with
pharynx opened and
mucosa removed**

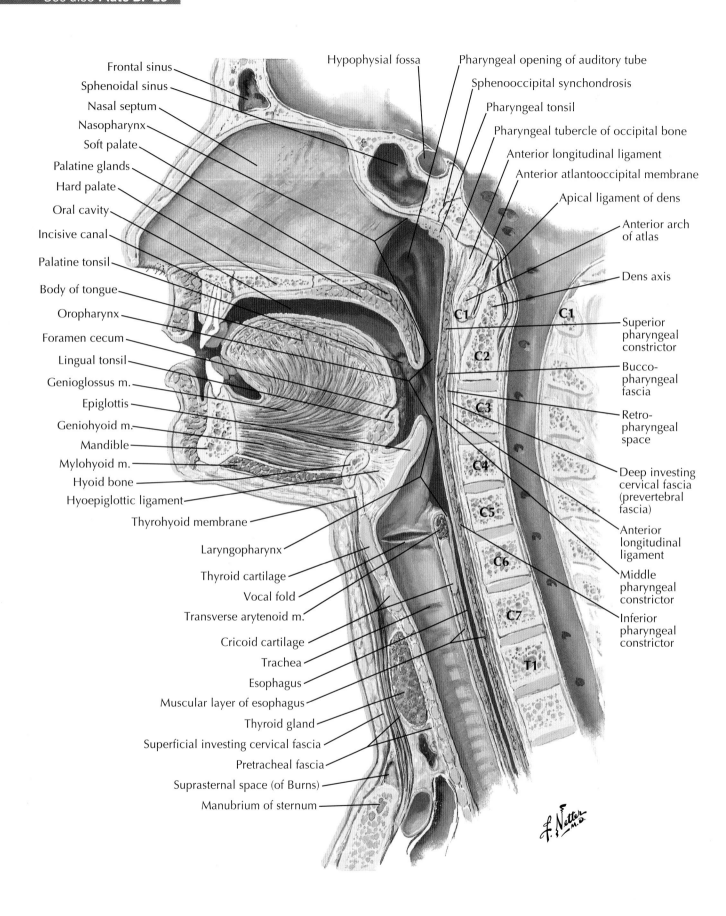

Frontal sinus

Sphenoidal sinus

Nasal septum

Nasopharynx

Soft palate

Palatine glands

Hard palate

Oral cavity

Incisive canal

Palatine tonsil

Body of tongue

Oropharynx

Foramen cecum

Lingual tonsil

Genioglossus m.

Epiglottis

Geniohyoid m.

Mandible

Mylohyoid m.

Hyoid bone

Hyoepiglottic ligament

Thyrohyoid membrane

Laryngopharynx

Thyroid cartilage

Vocal fold

Transverse arytenoid m.

Cricoid cartilage

Trachea

Esophagus

Muscular layer of esophagus

Thyroid gland

Superficial investing cervical fascia

Pretracheal fascia

Suprasternal space (of Burns)

Manubrium of sternum

Hypophysial fossa

Pharyngeal opening of auditory tube

Sphenooccipital synchondrosis

Pharyngeal tonsil

Pharyngeal tubercle of occipital bone

Anterior longitudinal ligament

Anterior atlantooccipital membrane

Apical ligament of dens

Anterior arch of atlas

Dens axis

Superior pharyngeal constrictor

Bucco-pharyngeal fascia

Retro-pharyngeal space

Deep investing cervical fascia (prevertebral fascia)

Anterior longitudinal ligament

Middle pharyngeal constrictor

Inferior pharyngeal constrictor

C1

C2

C3

C4

C5

C6

C7

T1

C1

C1

Plate 95

Pharynx

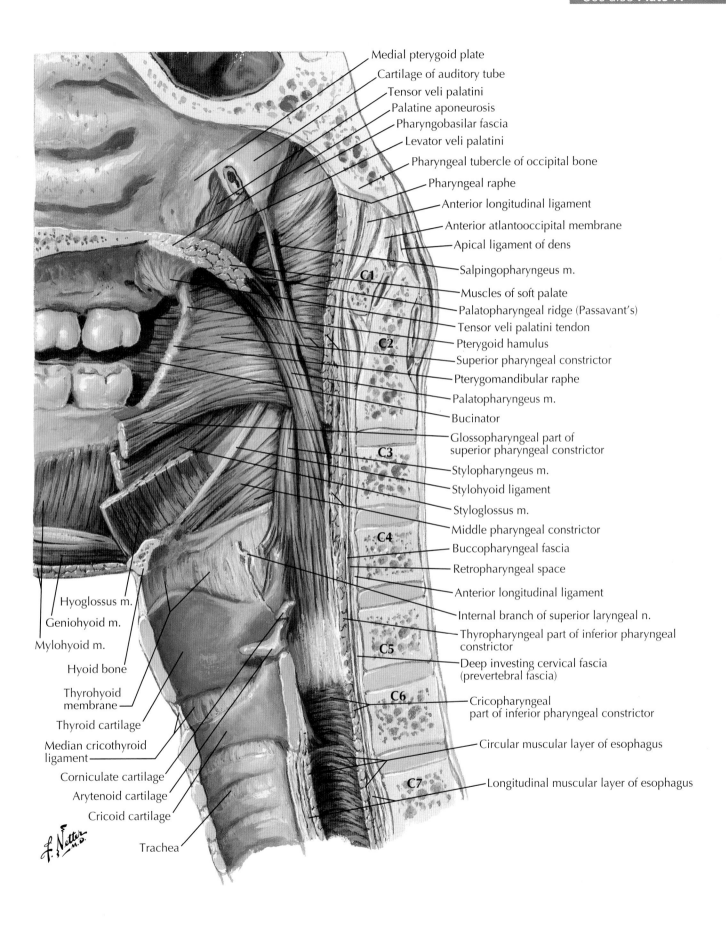

Medial pterygoid plate

Cartilage of auditory tube

Tensor veli palatini

Palatine aponeurosis

Pharyngobasilar fascia

Levator veli palatini

Pharyngeal tubercle of occipital bone

Pharyngeal raphe

Anterior longitudinal ligament

Anterior atlantooccipital membrane

Apical ligament of dens

C1

Salpingopharyngeus m.

Muscles of soft palate

Palatopharyngeal ridge (Passavant's)

Tensor veli palatini tendon

C2

Pterygoid hamulus

Superior pharyngeal constrictor

Pterygomandibular raphe

Palatopharyngeus m.

Bucinator

Glossopharyngeal part of superior pharyngeal constrictor

C3

Stylopharyngeus m.

Stylohyoid ligament

Styloglossus m.

Middle pharyngeal constrictor

C4

Buccopharyngeal fascia

Retropharyngeal space

Anterior longitudinal ligament

Internal branch of superior laryngeal n.

Thyropharyngeal part of inferior pharyngeal constrictor

C5

Deep investing cervical fascia (prevertebral fascia)

C6

Cricopharyngeal part of inferior pharyngeal constrictor

Circular muscular layer of esophagus

Longitudinal muscular layer of esophagus

C7

Hyoglossus m.

Geniohyoid m.

Mylohyoid m.

Hyoid bone

Thyrohyoid membrane

Thyroid cartilage

Median cricothyroid ligament

Corniculate cartilage

Arytenoid cartilage

Cricoid cartilage

Trachea

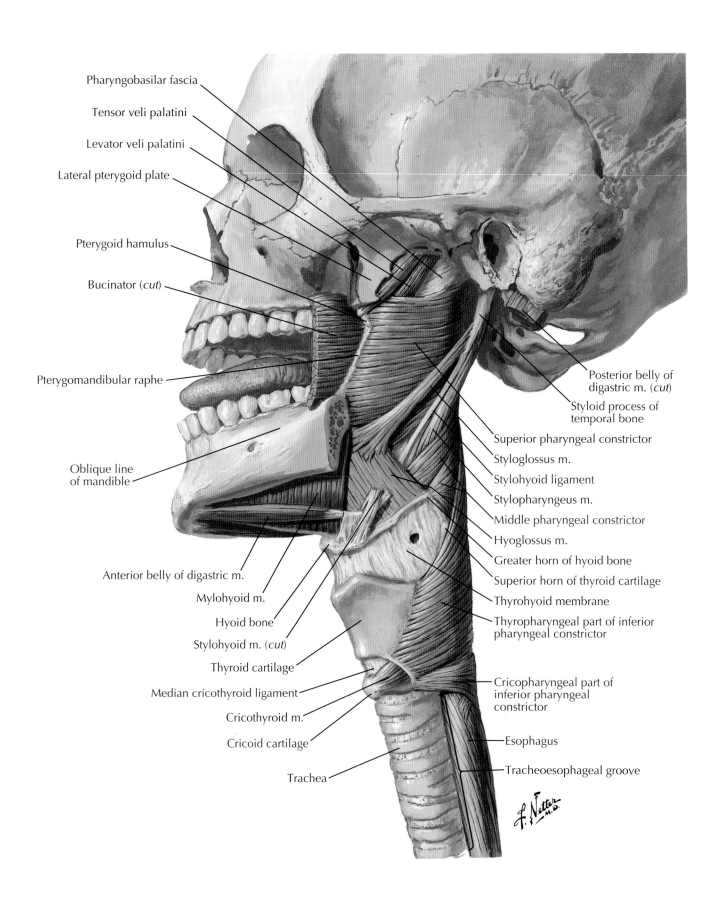

Pharyngobasilar fascia

Tensor veli palatini

Levator veli palatini

Lateral pterygoid plate

Pterygoid hamulus

Bucinator (*cut*)

Pterygomandibular raphe

Oblique line of mandible

Anterior belly of digastric m.

Mylohyoid m.

Hyoid bone

Stylohyoid m. (*cut*)

Thyroid cartilage

Median cricothyroid ligament

Cricothyroid m.

Cricoid cartilage

Trachea

Posterior belly of digastric m. (*cut*)

Styloid process of temporal bone

Superior pharyngeal constrictor

Styloglossus m.

Stylohyoid ligament

Stylopharyngeus m.

Middle pharyngeal constrictor

Hyoglossus m.

Greater horn of hyoid bone

Superior horn of thyroid cartilage

Thyrohyoid membrane

Thyropharyngeal part of inferior pharyngeal constrictor

Cricopharyngeal part of inferior pharyngeal constrictor

Esophagus

Tracheoesophageal groove

Plate 97

Pharynx

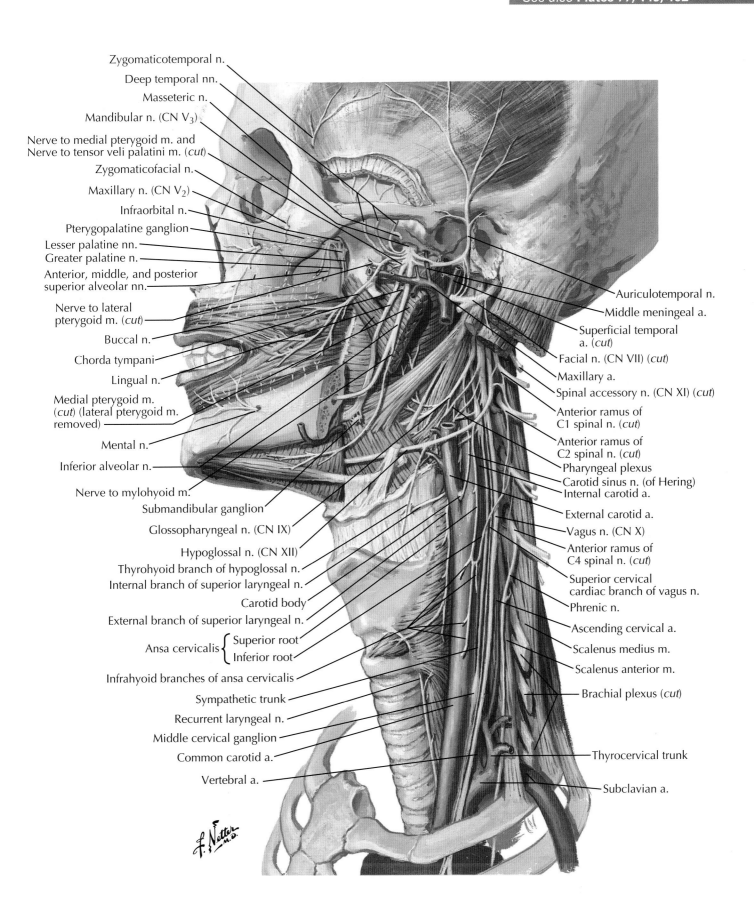

Zygomaticotemporal n.

Deep temporal nn.

Masseteric n.

Mandibular n. (CN V₃)

Nerve to medial pterygoid m. and
Nerve to tensor veli palatini m. (*cut*)

Zygomaticofacial n.

Maxillary n. (CN V₂)

Infraorbital n.

Pterygopalatine ganglion

Lesser palatine nn.

Greater palatine n.

Anterior, middle, and posterior
superior alveolar nn.

Nerve to lateral
pterygoid m. (*cut*)

Buccal n.

Chorda tympani

Lingual n.

Medial pterygoid m.
(*cut*) (lateral pterygoid m.
removed)

Mental n.

Inferior alveolar n.

Nerve to mylohyoid m.

Submandibular ganglion

Glossopharyngeal n. (CN IX)

Hypoglossal n. (CN XII)

Thyrohyoid branch of hypoglossal n.

Internal branch of superior laryngeal n.

Carotid body

External branch of superior laryngeal n.

Ansa cervicalis { Superior root / Inferior root

Infrahyoid branches of ansa cervicalis

Sympathetic trunk

Recurrent laryngeal n.

Middle cervical ganglion

Common carotid a.

Vertebral a.

Auriculotemporal n.

Middle meningeal a.

Superficial temporal
a. (*cut*)

Facial n. (CN VII) (*cut*)

Maxillary a.

Spinal accessory n. (CN XI) (*cut*)

Anterior ramus of
C1 spinal n. (*cut*)

Anterior ramus of
C2 spinal n. (*cut*)

Pharyngeal plexus

Carotid sinus n. (of Hering)

Internal carotid a.

External carotid a.

Vagus n. (CN X)

Anterior ramus of
C4 spinal n. (*cut*)

Superior cervical
cardiac branch of vagus n.

Phrenic n.

Ascending cervical a.

Scalenus medius m.

Scalenus anterior m.

Brachial plexus (*cut*)

Thyrocervical trunk

Subclavian a.

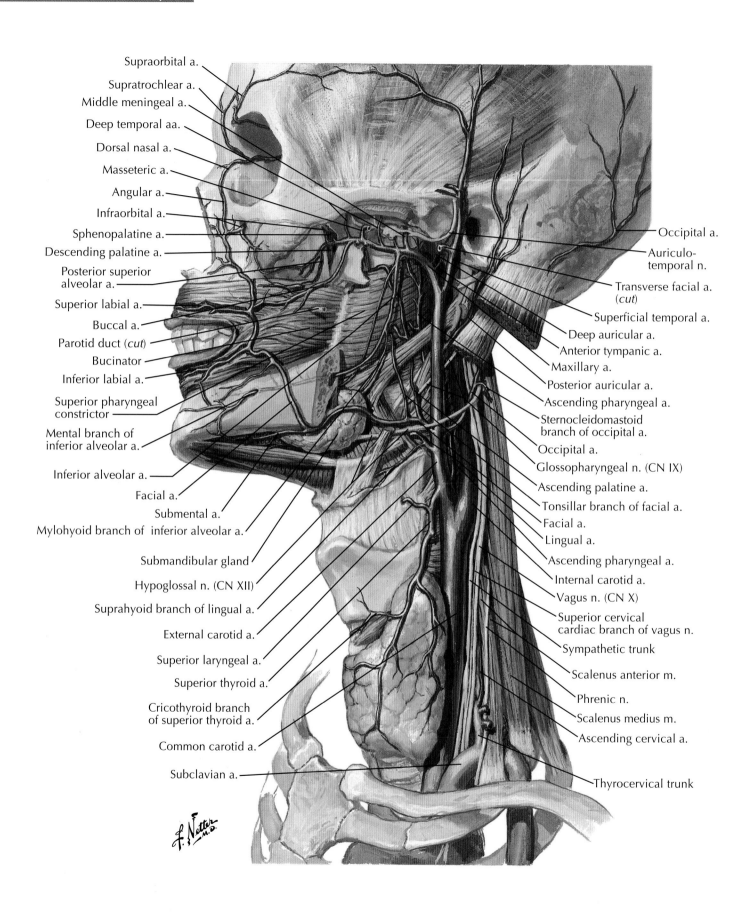

Supraorbital a.

Supratrochlear a.

Middle meningeal a.

Deep temporal aa.

Dorsal nasal a.

Masseteric a.

Angular a.

Infraorbital a.

Sphenopalatine a.

Descending palatine a.

Posterior superior alveolar a.

Superior labial a.

Buccal a.

Parotid duct (*cut*)

Bucinator

Inferior labial a.

Superior pharyngeal constrictor

Mental branch of inferior alveolar a.

Inferior alveolar a.

Facial a.

Submental a.

Mylohyoid branch of inferior alveolar a.

Submandibular gland

Hypoglossal n. (CN XII)

Suprahyoid branch of lingual a.

External carotid a.

Superior laryngeal a.

Superior thyroid a.

Cricothyroid branch of superior thyroid a.

Common carotid a.

Subclavian a.

Occipital a.

Auriculo-temporal n.

Transverse facial a. (*cut*)

Superficial temporal a.

Deep auricular a.

Anterior tympanic a.

Maxillary a.

Posterior auricular a.

Ascending pharyngeal a.

Sternocleidomastoid branch of occipital a.

Occipital a.

Glossopharyngeal n. (CN IX)

Ascending palatine a.

Tonsillar branch of facial a.

Facial a.

Lingual a.

Ascending pharyngeal a.

Internal carotid a.

Vagus n. (CN X)

Superior cervical cardiac branch of vagus n.

Sympathetic trunk

Scalenus anterior m.

Phrenic n.

Scalenus medius m.

Ascending cervical a.

Thyrocervical trunk

Plate 99

Pharynx

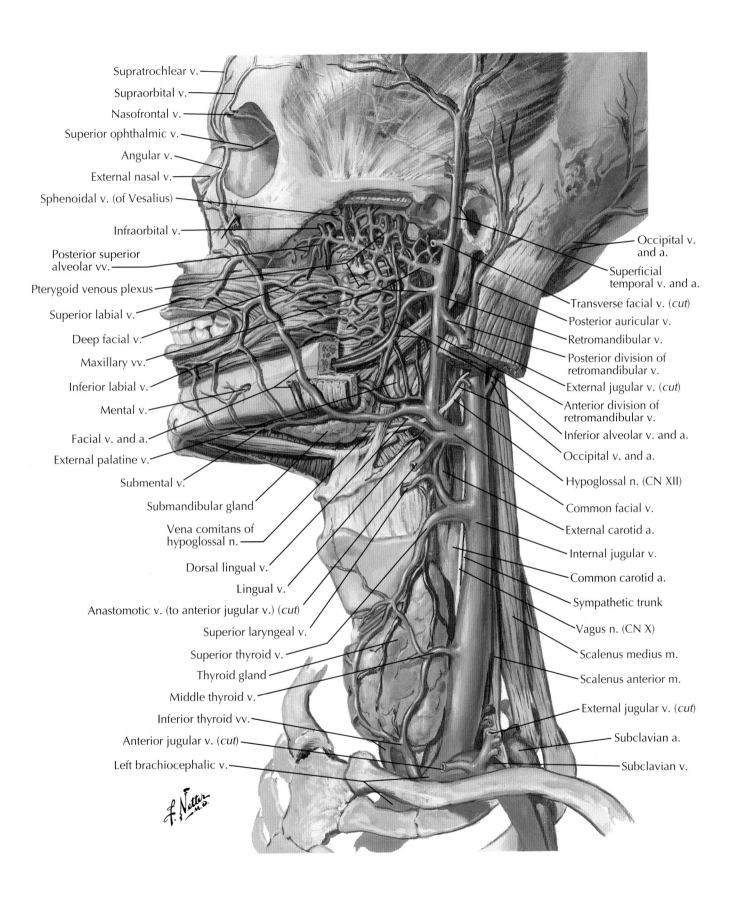

Supratrochlear v.

Supraorbital v.

Nasofrontal v.

Superior ophthalmic v.

Angular v.

External nasal v.

Sphenoidal v. (of Vesalius)

Infraorbital v.

Posterior superior alveolar vv.

Pterygoid venous plexus

Superior labial v.

Deep facial v.

Maxillary vv.

Inferior labial v.

Mental v.

Facial v. and a.

External palatine v.

Submental v.

Submandibular gland

Vena comitans of hypoglossal n.

Dorsal lingual v.

Lingual v.

Anastomotic v. (to anterior jugular v.) (cut)

Superior laryngeal v.

Superior thyroid v.

Thyroid gland

Middle thyroid v.

Inferior thyroid vv.

Anterior jugular v. (cut)

Left brachiocephalic v.

Occipital v. and a.

Superficial temporal v. and a.

Transverse facial v. (cut)

Posterior auricular v.

Retromandibular v.

Posterior division of retromandibular v.

External jugular v. (cut)

Anterior division of retromandibular v.

Inferior alveolar v. and a.

Occipital v. and a.

Hypoglossal n. (CN XII)

Common facial v.

External carotid a.

Internal jugular v.

Common carotid a.

Sympathetic trunk

Vagus n. (CN X)

Scalenus medius m.

Scalenus anterior m.

External jugular v. (cut)

Subclavian a.

Subclavian v.

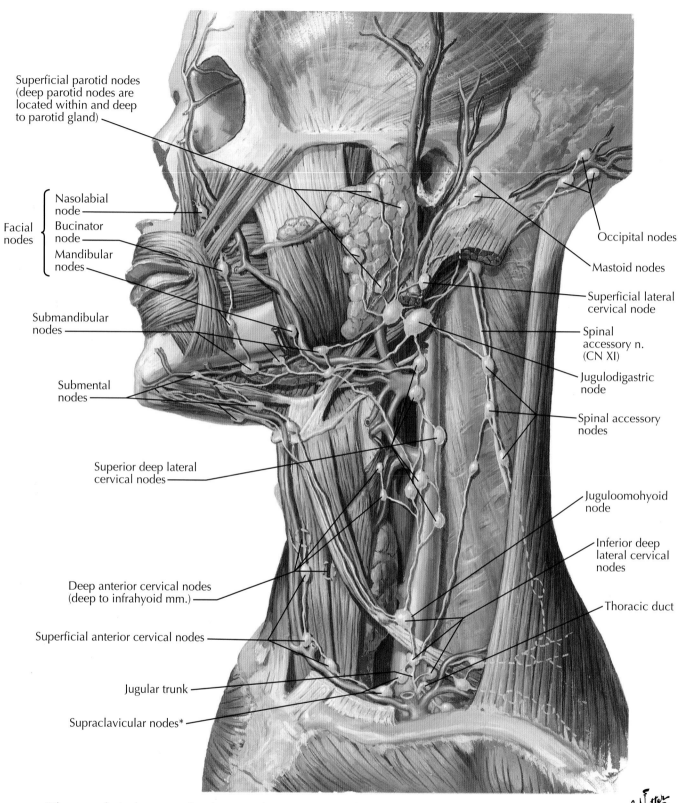

Superficial parotid nodes
(deep parotid nodes are
located within and deep
to parotid gland)

Facial nodes
Nasolabial node
Bucinator node
Mandibular nodes

Submandibular nodes

Submental nodes

Superior deep lateral cervical nodes

Deep anterior cervical nodes
(deep to infrahyoid mm.)

Superficial anterior cervical nodes

Jugular trunk

Supraclavicular nodes*

Occipital nodes

Mastoid nodes

Superficial lateral cervical node

Spinal accessory n. (CN XI)

Jugulodigastric node

Spinal accessory nodes

Juguloomohyoid node

Inferior deep lateral cervical nodes

Thoracic duct

*The supraclavicular group of nodes,
especially on the left, are also sometimes
referred to as the signal or sentinel lymph
nodes of Virchow or Troisier, especially when
sufficiently enlarged and palpable. These
nodes (or a single node) are so termed because
they may be the first recognized presumptive
evidence of malignant disease in the viscera.

Plate 101

Pharynx

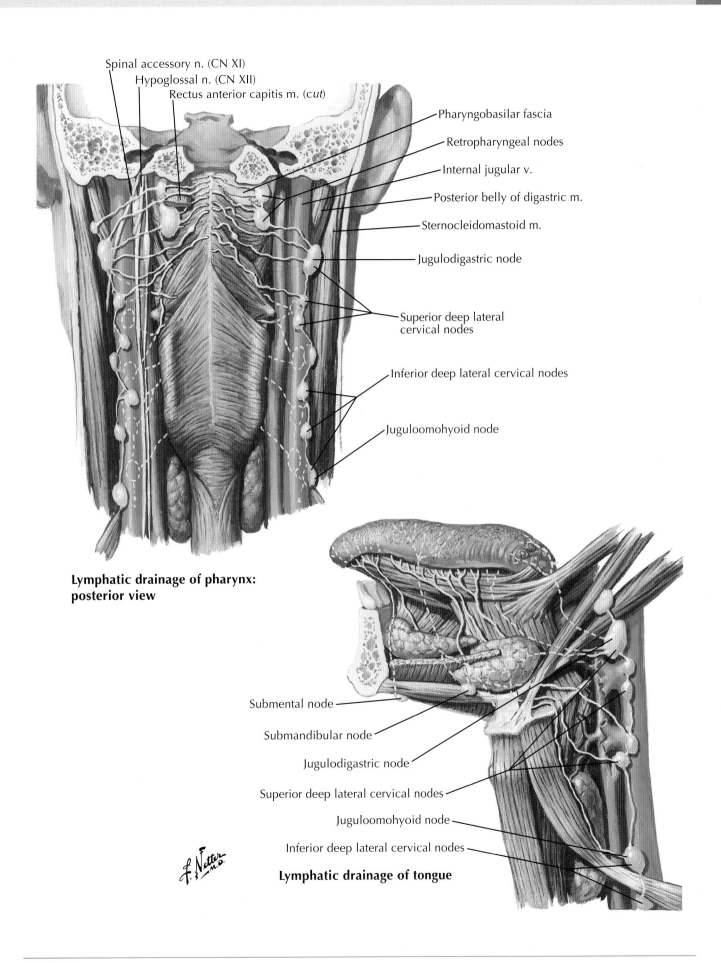

Spinal accessory n. (CN XI)
Hypoglossal n. (CN XII)
Rectus anterior capitis m. (cut)

Pharyngobasilar fascia

Retropharyngeal nodes

Internal jugular v.

Posterior belly of digastric m.

Sternocleidomastoid m.

Jugulodigastric node

Superior deep lateral cervical nodes

Inferior deep lateral cervical nodes

Juguloomohyoid node

Lymphatic drainage of pharynx: posterior view

Submental node

Submandibular node

Jugulodigastric node

Superior deep lateral cervical nodes

Juguloomohyoid node

Inferior deep lateral cervical nodes

Lymphatic drainage of tongue

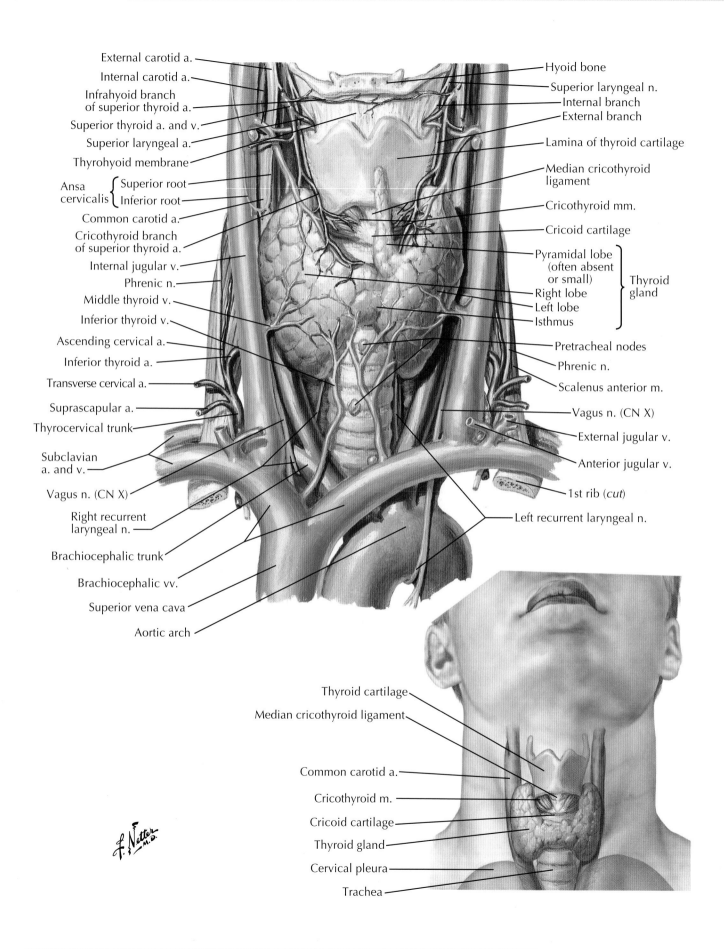

External carotid a.
Internal carotid a.
Infrahyoid branch of superior thyroid a.
Superior thyroid a. and v.
Superior laryngeal a.
Thyrohyoid membrane
Ansa cervicalis { Superior root
Inferior root }
Common carotid a.
Cricothyroid branch of superior thyroid a.
Internal jugular v.
Phrenic n.
Middle thyroid v.
Inferior thyroid v.
Ascending cervical a.
Inferior thyroid a.
Transverse cervical a.
Suprascapular a.
Thyrocervical trunk
Subclavian a. and v.
Vagus n. (CN X)
Right recurrent laryngeal n.
Brachiocephalic trunk
Brachiocephalic vv.
Superior vena cava
Aortic arch

Hyoid bone
Superior laryngeal n.
Internal branch
External branch
Lamina of thyroid cartilage
Median cricothyroid ligament
Cricothyroid mm.
Cricoid cartilage
Pyramidal lobe (often absent or small)
Right lobe
Left lobe
Isthmus
} Thyroid gland
Pretracheal nodes
Phrenic n.
Scalenus anterior m.
Vagus n. (CN X)
External jugular v.
Anterior jugular v.
1st rib (cut)
Left recurrent laryngeal n.

Thyroid cartilage
Median cricothyroid ligament
Common carotid a.
Cricothyroid m.
Cricoid cartilage
Thyroid gland
Cervical pleura
Trachea

Plate 103

Larynx and Endocrine Glands

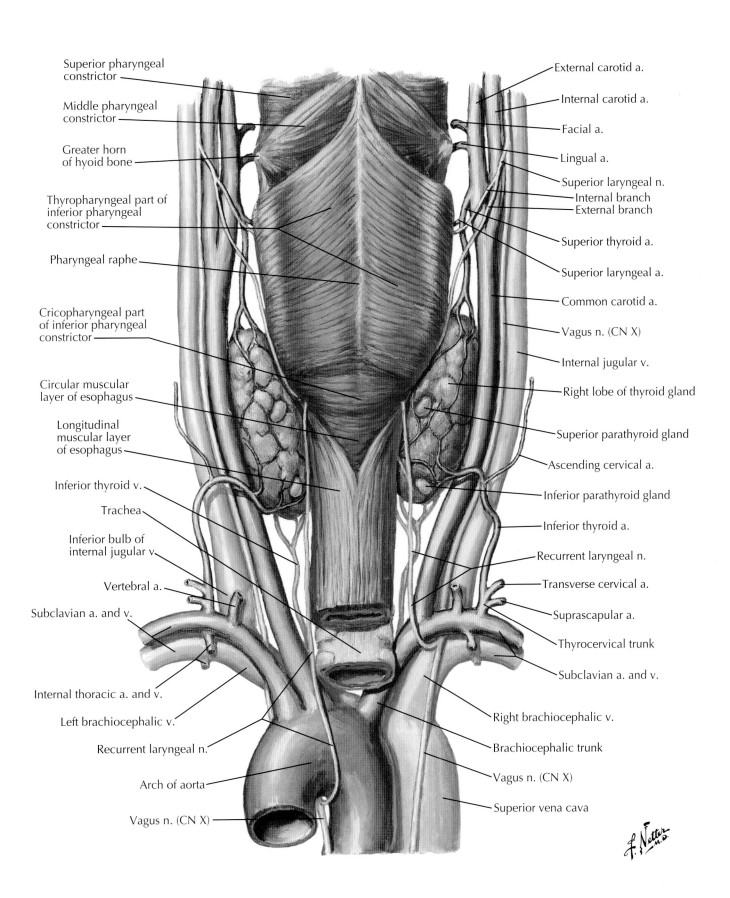

Superior pharyngeal constrictor

Middle pharyngeal constrictor

Greater horn of hyoid bone

Thyropharyngeal part of inferior pharyngeal constrictor

Pharyngeal raphe

Cricopharyngeal part of inferior pharyngeal constrictor

Circular muscular layer of esophagus

Longitudinal muscular layer of esophagus

Inferior thyroid v.

Trachea

Inferior bulb of internal jugular v.

Vertebral a.

Subclavian a. and v.

Internal thoracic a. and v.

Left brachiocephalic v.

Recurrent laryngeal n.

Arch of aorta

Vagus n. (CN X)

External carotid a.

Internal carotid a.

Facial a.

Lingual a.

Superior laryngeal n.
Internal branch
External branch

Superior thyroid a.

Superior laryngeal a.

Common carotid a.

Vagus n. (CN X)

Internal jugular v.

Right lobe of thyroid gland

Superior parathyroid gland

Ascending cervical a.

Inferior parathyroid gland

Inferior thyroid a.

Recurrent laryngeal n.

Transverse cervical a.

Suprascapular a.

Thyrocervical trunk

Subclavian a. and v.

Right brachiocephalic v.

Brachiocephalic trunk

Vagus n. (CN X)

Superior vena cava

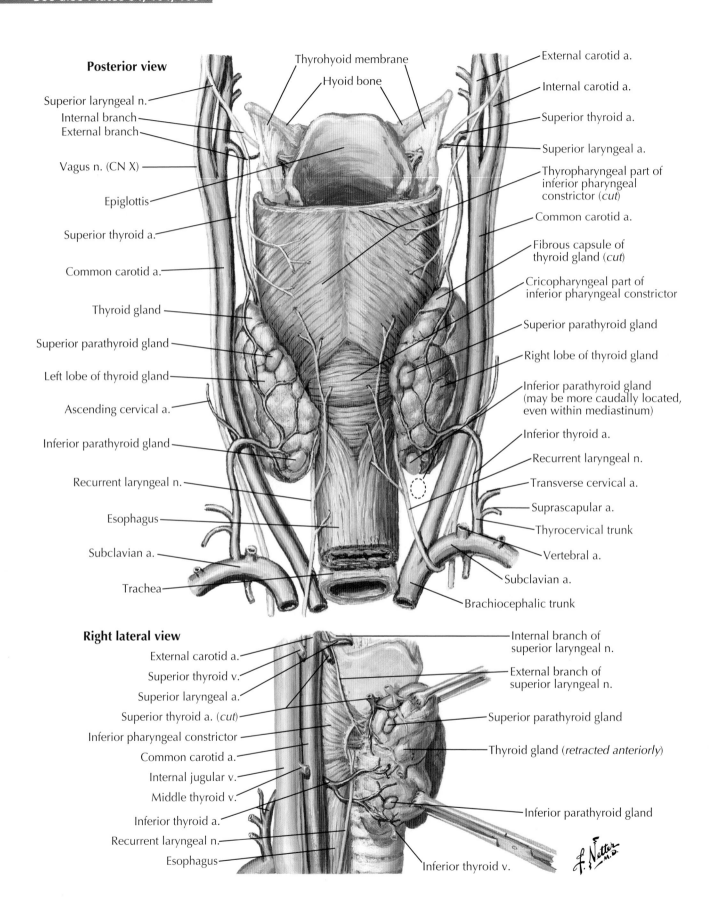

Posterior view

Superior laryngeal n.
Internal branch
External branch

Vagus n. (CN X)

Epiglottis

Superior thyroid a.

Common carotid a.

Thyroid gland

Superior parathyroid gland

Left lobe of thyroid gland

Ascending cervical a.

Inferior parathyroid gland

Recurrent laryngeal n.

Esophagus

Subclavian a.

Trachea

Thyrohyoid membrane
Hyoid bone

External carotid a.

Internal carotid a.

Superior thyroid a.

Superior laryngeal a.

Thyropharyngeal part of inferior pharyngeal constrictor (*cut*)

Common carotid a.

Fibrous capsule of thyroid gland (*cut*)

Cricopharyngeal part of inferior pharyngeal constrictor

Superior parathyroid gland

Right lobe of thyroid gland

Inferior parathyroid gland (may be more caudally located, even within mediastinum)

Inferior thyroid a.

Recurrent laryngeal n.

Transverse cervical a.

Suprascapular a.

Thyrocervical trunk

Vertebral a.

Subclavian a.

Brachiocephalic trunk

Right lateral view

External carotid a.
Superior thyroid v.
Superior laryngeal a.
Superior thyroid a. (*cut*)
Inferior pharyngeal constrictor
Common carotid a.
Internal jugular v.
Middle thyroid v.
Inferior thyroid a.
Recurrent laryngeal n.
Esophagus

Internal branch of superior laryngeal n.

External branch of superior laryngeal n.

Superior parathyroid gland

Thyroid gland (*retracted anteriorly*)

Inferior parathyroid gland

Inferior thyroid v.

Plate 105

Larynx and Endocrine Glands

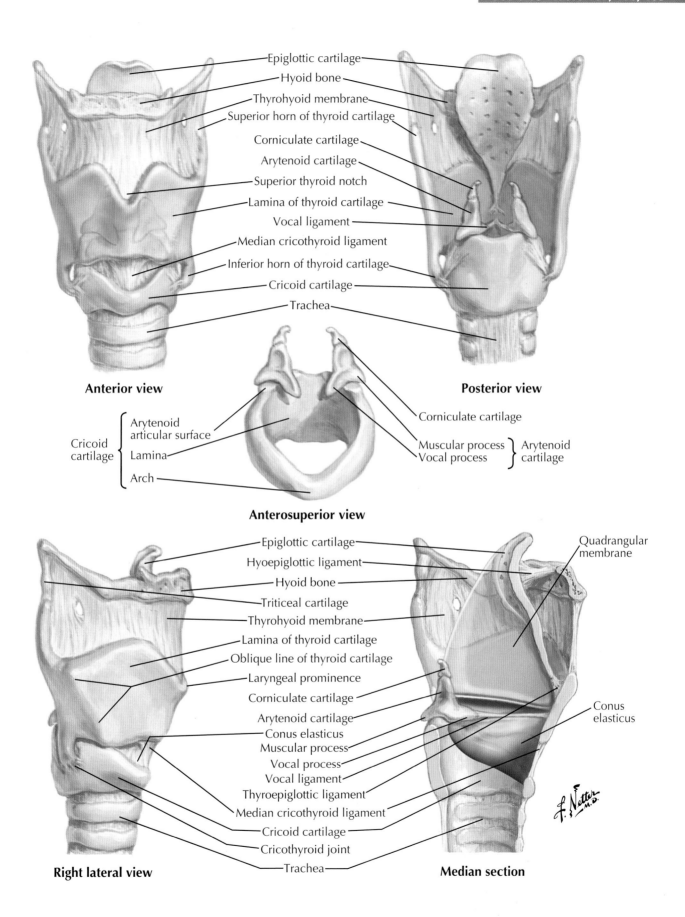

Epiglottic cartilage

Hyoid bone

Thyrohyoid membrane

Superior horn of thyroid cartilage

Corniculate cartilage

Arytenoid cartilage

Superior thyroid notch

Lamina of thyroid cartilage

Vocal ligament

Median cricothyroid ligament

Inferior horn of thyroid cartilage

Cricoid cartilage

Trachea

Anterior view

Posterior view

Cricoid cartilage { Arytenoid articular surface / Lamina / Arch

Corniculate cartilage

Muscular process / Vocal process } Arytenoid cartilage

Anterosuperior view

Epiglottic cartilage

Hyoepiglottic ligament

Hyoid bone

Triticeal cartilage

Thyrohyoid membrane

Lamina of thyroid cartilage

Oblique line of thyroid cartilage

Laryngeal prominence

Corniculate cartilage

Arytenoid cartilage

Conus elasticus

Muscular process

Vocal process

Vocal ligament

Thyroepiglottic ligament

Median cricothyroid ligament

Cricoid cartilage

Cricothyroid joint

Trachea

Quadrangular membrane

Conus elasticus

Right lateral view

Median section

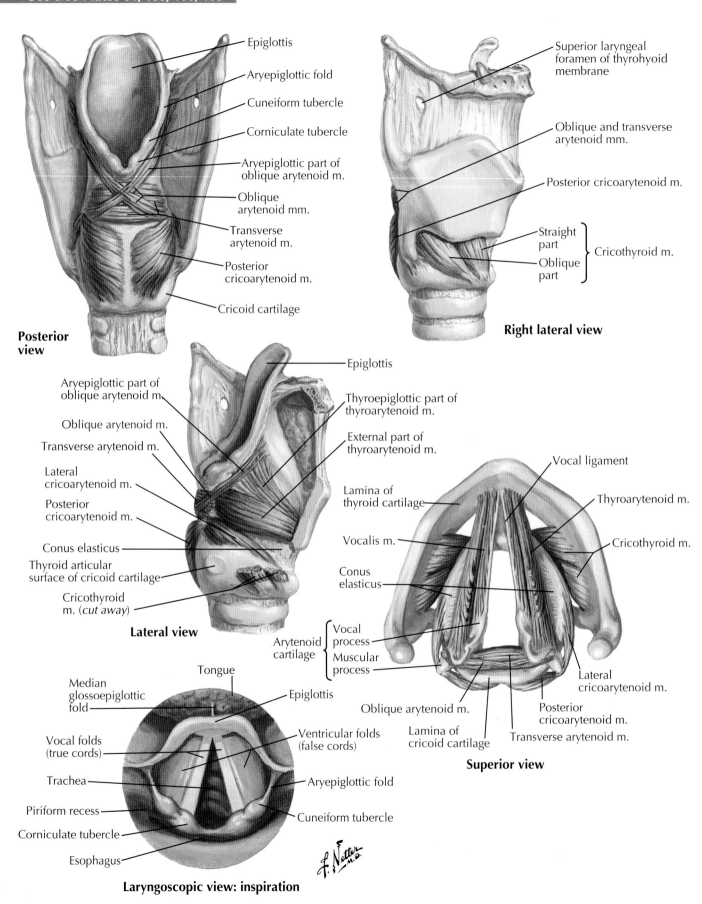

Epiglottis

Aryepiglottic fold

Cuneiform tubercle

Corniculate tubercle

Aryepiglottic part of oblique arytenoid m.

Oblique arytenoid mm.

Transverse arytenoid m.

Posterior cricoarytenoid m.

Cricoid cartilage

Posterior view

Superior laryngeal foramen of thyrohyoid membrane

Oblique and transverse arytenoid mm.

Posterior cricoarytenoid m.

Straight part
Oblique part
} Cricothyroid m.

Right lateral view

Aryepiglottic part of oblique arytenoid m.

Oblique arytenoid m.

Transverse arytenoid m.

Lateral cricoarytenoid m.

Posterior cricoarytenoid m.

Conus elasticus

Thyroid articular surface of cricoid cartilage

Cricothyroid m. (*cut away*)

Lateral view

Epiglottis

Thyroepiglottic part of thyroarytenoid m.

External part of thyroarytenoid m.

Lamina of thyroid cartilage

Vocalis m.

Conus elasticus

Arytenoid cartilage { Vocal process
Muscular process

Vocal ligament

Thyroarytenoid m.

Cricothyroid m.

Oblique arytenoid m.

Lamina of cricoid cartilage

Lateral cricoarytenoid m.

Posterior cricoarytenoid m.

Transverse arytenoid m.

Superior view

Tongue

Median glossoepiglottic fold

Epiglottis

Vocal folds (true cords)

Ventricular folds (false cords)

Trachea

Piriform recess

Aryepiglottic fold

Corniculate tubercle

Cuneiform tubercle

Esophagus

Laryngoscopic view: inspiration

Plate 107

Larynx and Endocrine Glands

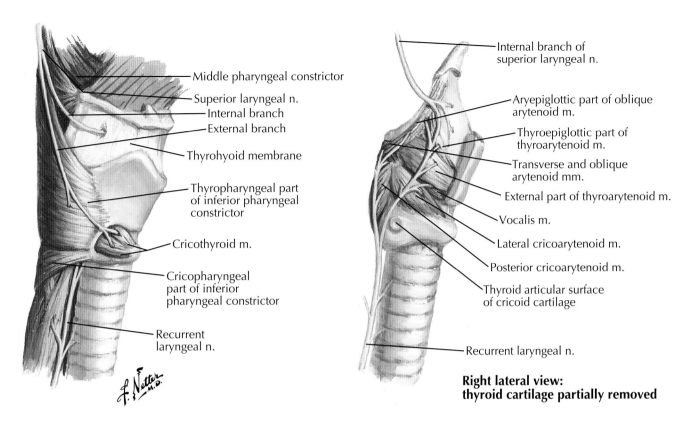

Middle pharyngeal constrictor

Superior laryngeal n.
Internal branch
External branch
Thyrohyoid membrane

Thyropharyngeal part of inferior pharyngeal constrictor

Cricothyroid m.

Cricopharyngeal part of inferior pharyngeal constrictor

Recurrent laryngeal n.

Internal branch of superior laryngeal n.

Aryepiglottic part of oblique arytenoid m.

Thyroepiglottic part of thyroarytenoid m.

Transverse and oblique arytenoid mm.

External part of thyroarytenoid m.

Vocalis m.

Lateral cricoarytenoid m.

Posterior cricoarytenoid m.

Thyroid articular surface of cricoid cartilage

Recurrent laryngeal n.

Right lateral view: thyroid cartilage partially removed

Coronal section through larynx

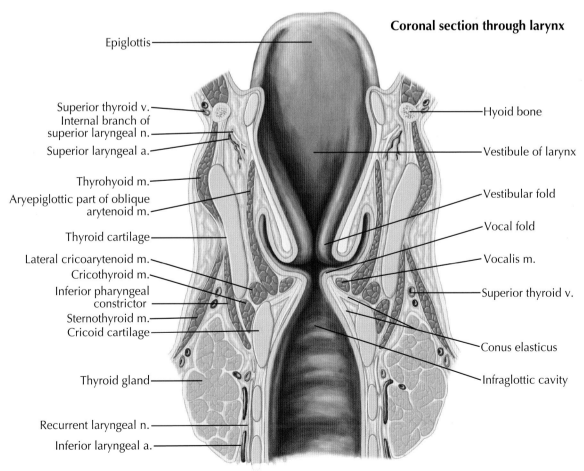

Epiglottis

Superior thyroid v.
Internal branch of superior laryngeal n.
Superior laryngeal a.

Thyrohyoid m.
Aryepiglottic part of oblique arytenoid m.

Thyroid cartilage

Lateral cricoarytenoid m.
Cricothyroid m.
Inferior pharyngeal constrictor
Sternothyroid m.
Cricoid cartilage

Thyroid gland

Recurrent laryngeal n.
Inferior laryngeal a.

Hyoid bone

Vestibule of larynx

Vestibular fold

Vocal fold

Vocalis m.

Superior thyroid v.

Conus elasticus

Infraglottic cavity

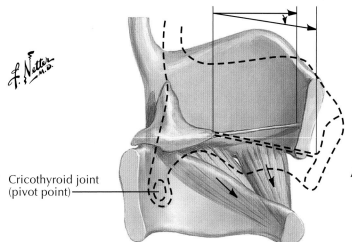

Cricothyroid joint (pivot point)

Action of cricothyroid muscles
Stretching of vocal ligaments

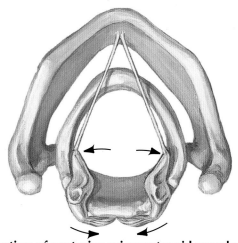

Action of posterior cricoarytenoid muscles
Abduction of vocal ligaments

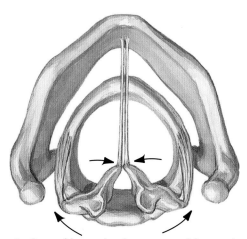

Action of lateral cricoarytenoid muscles
Adduction of vocal ligaments

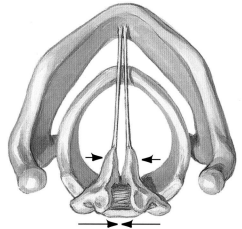

Action of transverse and oblique arytenoid muscles
Adduction of vocal ligaments

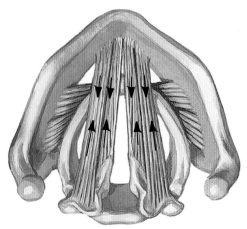

Action of vocalis and thyroarytenoid muscles
Shortening (relaxation) of vocal ligaments

Plate 109

Larynx and Endocrine Glands

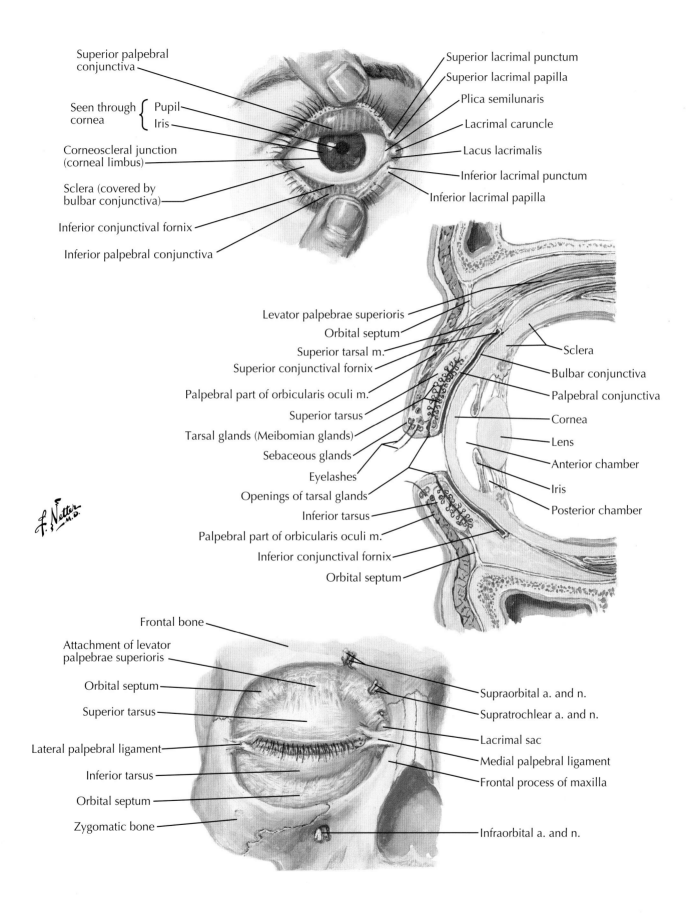

Superior palpebral conjunctiva

Seen through cornea { Pupil / Iris

Corneoscleral junction (corneal limbus)

Sclera (covered by bulbar conjunctiva)

Inferior conjunctival fornix

Inferior palpebral conjunctiva

Superior lacrimal punctum

Superior lacrimal papilla

Plica semilunaris

Lacrimal caruncle

Lacus lacrimalis

Inferior lacrimal punctum

Inferior lacrimal papilla

Levator palpebrae superioris

Orbital septum

Superior tarsal m.

Superior conjunctival fornix

Palpebral part of orbicularis oculi m.

Superior tarsus

Tarsal glands (Meibomian glands)

Sebaceous glands

Eyelashes

Openings of tarsal glands

Inferior tarsus

Palpebral part of orbicularis oculi m.

Inferior conjunctival fornix

Orbital septum

Sclera

Bulbar conjunctiva

Palpebral conjunctiva

Cornea

Lens

Anterior chamber

Iris

Posterior chamber

Frontal bone

Attachment of levator palpebrae superioris

Orbital septum

Superior tarsus

Lateral palpebral ligament

Inferior tarsus

Orbital septum

Zygomatic bone

Supraorbital a. and n.

Supratrochlear a. and n.

Lacrimal sac

Medial palpebral ligament

Frontal process of maxilla

Infraorbital a. and n.

Eye

Plate 110

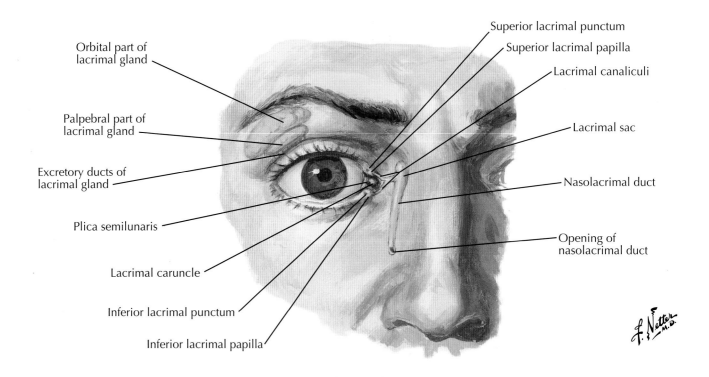

Orbital part of lacrimal gland

Palpebral part of lacrimal gland

Excretory ducts of lacrimal gland

Plica semilunaris

Lacrimal caruncle

Inferior lacrimal punctum

Inferior lacrimal papilla

Superior lacrimal punctum

Superior lacrimal papilla

Lacrimal canaliculi

Lacrimal sac

Nasolacrimal duct

Opening of nasolacrimal duct

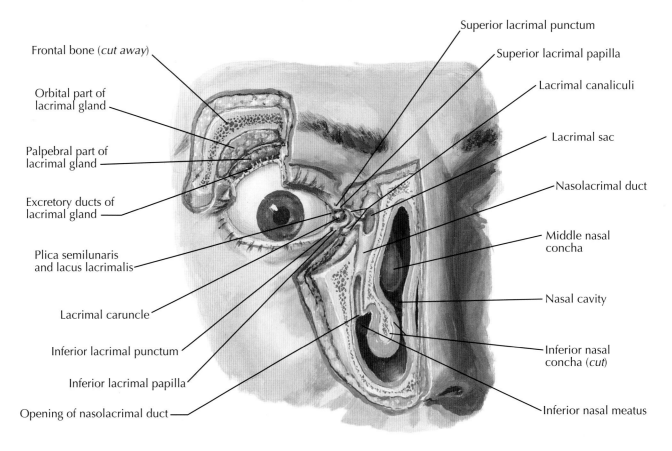

Frontal bone (*cut away*)

Orbital part of lacrimal gland

Palpebral part of lacrimal gland

Excretory ducts of lacrimal gland

Plica semilunaris and lacus lacrimalis

Lacrimal caruncle

Inferior lacrimal punctum

Inferior lacrimal papilla

Opening of nasolacrimal duct

Superior lacrimal punctum

Superior lacrimal papilla

Lacrimal canaliculi

Lacrimal sac

Nasolacrimal duct

Middle nasal concha

Nasal cavity

Inferior nasal concha (*cut*)

Inferior nasal meatus

Plate 111

Eye

Right lateral view

Superior oblique m.

Levator palpebrae superioris

Superior rectus m.

Medial rectus m.

Common tendinous ring (of Zinn)

Lateral rectus m. (*cut*)

Inferior rectus m.

Maxillary sinus

Trochlea

Optic n. (CN II)

Lateral rectus m. (*cut*)

Inferior oblique m.

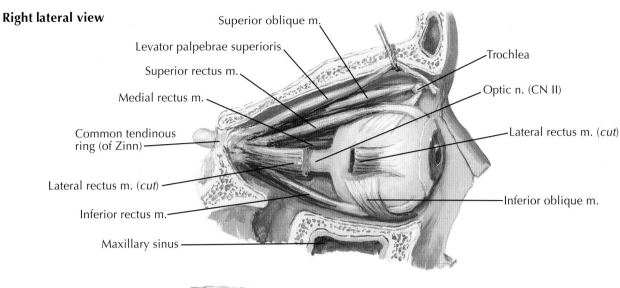

Superior view

Trochlea of superior oblique m.

Superior oblique m.

Medial rectus m.

Inferior rectus m.

Common tendinous ring (of Zinn)

Superior tarsus

Levator palpebrae superioris (*cut*)

Superior rectus m. (*cut*)

Lateral rectus m.

Optic n. (CN II)

Superior rectus m. (*cut*)

Levator palpebrae superioris (*cut*)

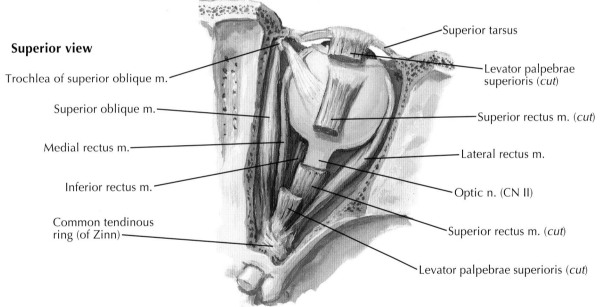

Innervation of extrinsic eye muscles: anterior view

Trochlea of superior oblique m.

Oculomotor n. (CN III)

Levator palpebrae superioris

Superior rectus m.

Medial rectus m.

Inferior rectus m.

Inferior oblique m.

Superior oblique m. { **Trochlear n. (CN IV)**

Lateral rectus m. { **Abducens n. (CN VI)**

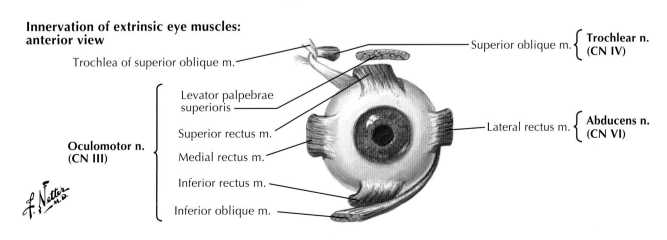

f. Netter.
M.D.

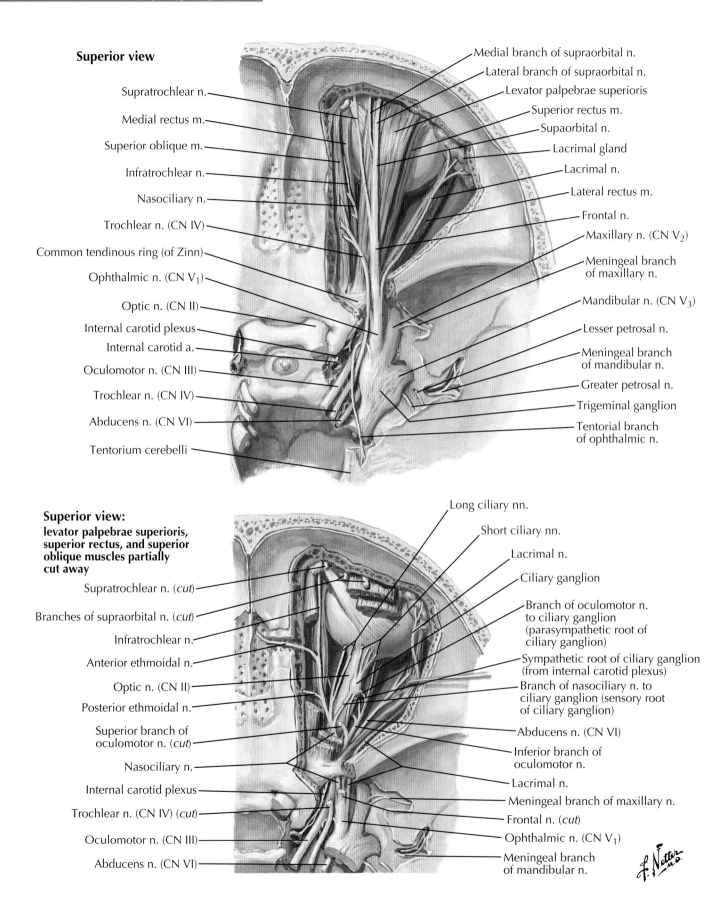

Superior view

Supratrochlear n.

Medial rectus m.

Superior oblique m.

Infratrochlear n.

Nasociliary n.

Trochlear n. (CN IV)

Common tendinous ring (of Zinn)

Ophthalmic n. (CN V₁)

Optic n. (CN II)

Internal carotid plexus

Internal carotid a.

Oculomotor n. (CN III)

Trochlear n. (CN IV)

Abducens n. (CN VI)

Tentorium cerebelli

Medial branch of supraorbital n.

Lateral branch of supraorbital n.

Levator palpebrae superioris

Superior rectus m.

Supaorbital n.

Lacrimal gland

Lacrimal n.

Lateral rectus m.

Frontal n.

Maxillary n. (CN V₂)

Meningeal branch
of maxillary n.

Mandibular n. (CN V₃)

Lesser petrosal n.

Meningeal branch
of mandibular n.

Greater petrosal n.

Trigeminal ganglion

Tentorial branch
of ophthalmic n.

**Superior view:
levator palpebrae superioris,
superior rectus, and superior
oblique muscles partially
cut away**

Supratrochlear n. (*cut*)

Branches of supraorbital n. (*cut*)

Infratrochlear n.

Anterior ethmoidal n.

Optic n. (CN II)

Posterior ethmoidal n.

Superior branch of
oculomotor n. (*cut*)

Nasociliary n.

Internal carotid plexus

Trochlear n. (CN IV) (*cut*)

Oculomotor n. (CN III)

Abducens n. (CN VI)

Long ciliary nn.

Short ciliary nn.

Lacrimal n.

Ciliary ganglion

Branch of oculomotor n.
to ciliary ganglion
(parasympathetic root of
ciliary ganglion)

Sympathetic root of ciliary ganglion
(from internal carotid plexus)

Branch of nasociliary n. to
ciliary ganglion (sensory root
of ciliary ganglion)

Abducens n. (CN VI)

Inferior branch of
oculomotor n.

Lacrimal n.

Meningeal branch of maxillary n.

Frontal n. (*cut*)

Ophthalmic n. (CN V₁)

Meningeal branch
of mandibular n.

Plate 113

Eye

Superior view of right orbit

Posterior

Abducens n. (CN VI)
Trigeminal ganglion
Ophthalmic n. (CN V₁)
Maxillary n. (CN V₂)
Mandibular n. (CN V₃)
Frontal n.
Lateral rectus m.
Abducens n.
Ciliary ganglion
Short ciliary nn.
Infraorbital n.
Medial rectus m.
Superior rectus m.
Lacrimal gland
Supratrochlear n.
Levator palpebrae superioris
Supraorbital n.:
Medial branch
Lateral branch
Superior oblique m.
Superior oblique tendon
Inferior oblique tendon
Anterior ethmoidal cells
Internal nasal branches
of anterior ethmoidal n.

Oculomotor n. (CN III)
Trochlear n. (CN IV)
Optic n. (CN II)
Sphenoidal sinus
Anterior clinoid process
Posterior ethmoidal n.
Long ciliary n.
Inferior rectus m.
Nasociliary n.
Skin
Anterior ethmoidal n.
Left anterior ethmoidal n.
Frontalis m.
Supraorbital n.
Lateral branch
Medial branch
Supratrochlear n.
Right and left infratrochlear nn.
Lateral nasal branch of anterior ethmoidal n.
External nasal branch
of anterior ethmoidal n.
External nasal branch
of left anterior ethmoidal n.
Medial nasal branch
of anterior ethmoidal n.

C. Machado
—M.D.

Anterior view of right orbit
(eyeball transparent)

Anterior

Posterior
ethmoidal n.
Levator palpebrae superioris
Superior rectus m.
Nasociliary n.
Zygomaticotemporal n.
Lacrimal n.
Ciliary ganglion
Communicating branch
of zygomaticotemporal
n. to lacrimal n.
Abducens n. (CN VI)
Lateral rectus m.
Short ciliary nn.
Zygomaticotemporal n.
Zygomaticofacial n.
Zygomatic n.
Infraorbital n.

Anterior
ethmoidal n.
Lateral branch
of supraorbital n.
Medial branch
of supraorbital n.
Supratrochlear n.
Superior oblique m.
Trochlea of superior oblique m.
Trochlear n. (CN IV)
Infratrochlear n.
Optic n. (CN II)
Long ciliary n.
Oculomotor n. (CN III)
Medial rectus m. (cut)
Inferior branch
of oculomotor n.
Inferior rectus m.
Inferior oblique m.

Lateral

Medial

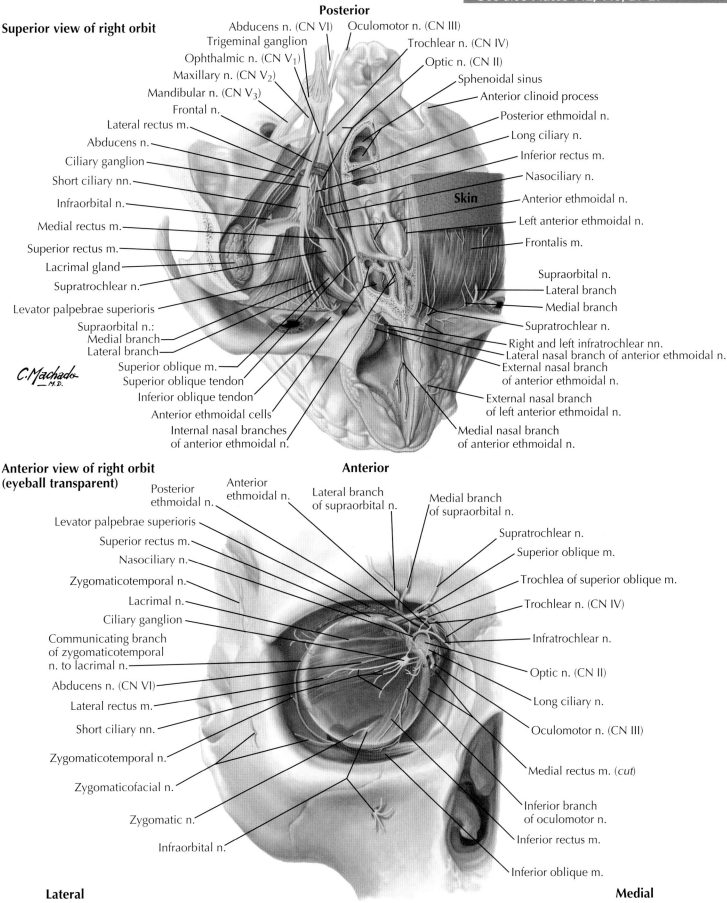

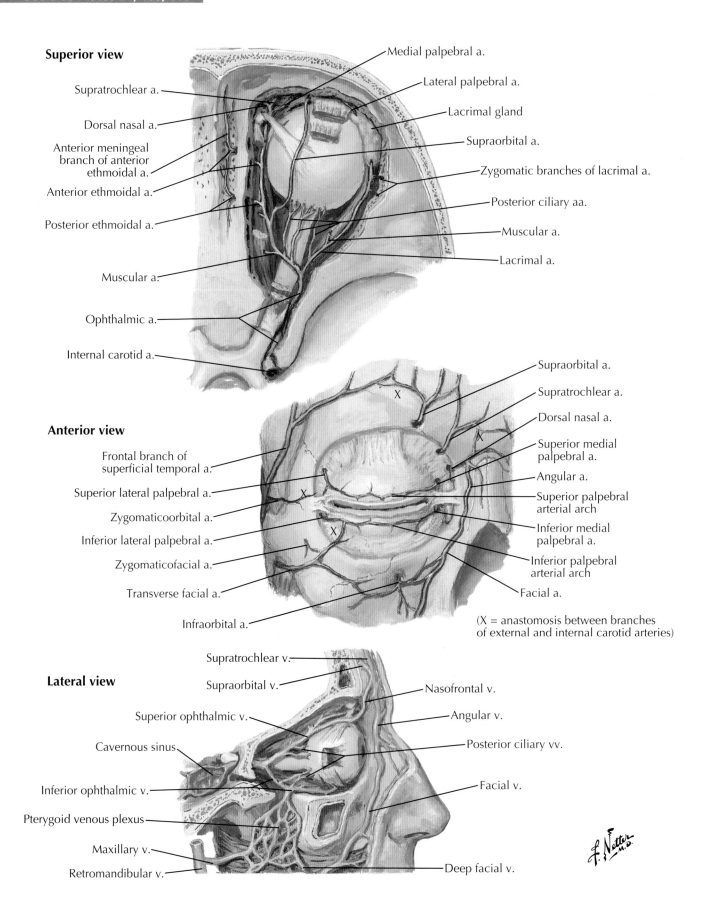

Superior view

Supratrochlear a.

Dorsal nasal a.

Anterior meningeal branch of anterior ethmoidal a.

Anterior ethmoidal a.

Posterior ethmoidal a.

Muscular a.

Ophthalmic a.

Internal carotid a.

Medial palpebral a.

Lateral palpebral a.

Lacrimal gland

Supraorbital a.

Zygomatic branches of lacrimal a.

Posterior ciliary aa.

Muscular a.

Lacrimal a.

Anterior view

Frontal branch of superficial temporal a.

Superior lateral palpebral a.

Zygomaticoorbital a.

Inferior lateral palpebral a.

Zygomaticofacial a.

Transverse facial a.

Infraorbital a.

Supraorbital a.

Supratrochlear a.

Dorsal nasal a.

Superior medial palpebral a.

Angular a.

Superior palpebral arterial arch

Inferior medial palpebral a.

Inferior palpebral arterial arch

Facial a.

(X = anastomosis between branches of external and internal carotid arteries)

Lateral view

Supratrochlear v.

Supraorbital v.

Superior ophthalmic v.

Cavernous sinus

Inferior ophthalmic v.

Pterygoid venous plexus

Maxillary v.

Retromandibular v.

Nasofrontal v.

Angular v.

Posterior ciliary vv.

Facial v.

Deep facial v.

Plate 115

Eye

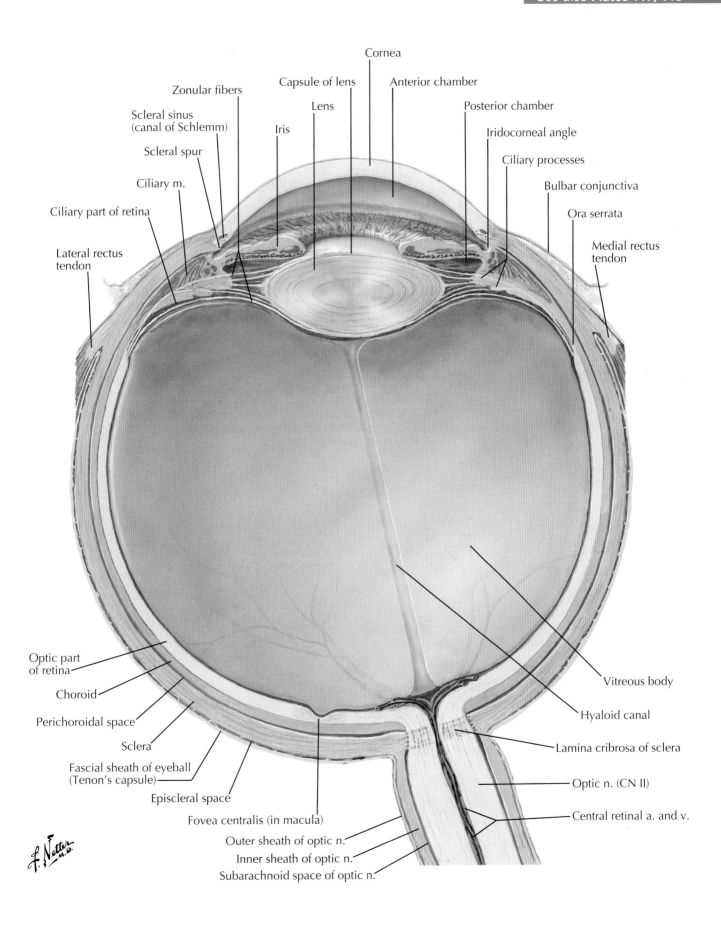

Zonular fibers

Scleral sinus
(canal of Schlemm)

Scleral spur

Ciliary m.

Ciliary part of retina

Lateral rectus
tendon

Capsule of lens

Lens

Iris

Cornea

Anterior chamber

Posterior chamber

Iridocorneal angle

Ciliary processes

Bulbar conjunctiva

Ora serrata

Medial rectus
tendon

Optic part
of retina

Choroid

Perichoroidal space

Sclera

Fascial sheath of eyeball
(Tenon's capsule)

Episcleral space

Fovea centralis (in macula)

Outer sheath of optic n.

Inner sheath of optic n.

Subarachnoid space of optic n.

Vitreous body

Hyaloid canal

Lamina cribrosa of sclera

Optic n. (CN II)

Central retinal a. and v.

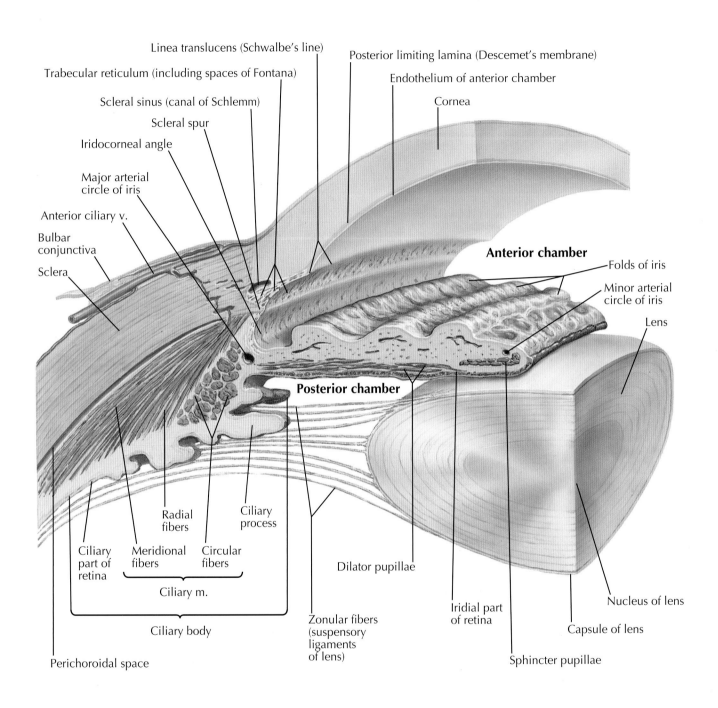

Linea translucens (Schwalbe's line)

Trabecular reticulum (including spaces of Fontana)

Posterior limiting lamina (Descemet's membrane)

Scleral sinus (canal of Schlemm)

Endothelium of anterior chamber

Scleral spur

Cornea

Iridocorneal angle

Major arterial circle of iris

Anterior ciliary v.

Bulbar conjunctiva

Anterior chamber

Sclera

Folds of iris

Minor arterial circle of iris

Lens

Posterior chamber

Radial fibers

Ciliary process

Ciliary part of retina

Meridional fibers

Circular fibers

Dilator pupillae

Nucleus of lens

Ciliary m.

Iridial part of retina

Ciliary body

Zonular fibers (suspensory ligaments of lens)

Capsule of lens

Perichoroidal space

Sphincter pupillae

Note: For clarity, only single plane of zonular fibers shown; actually, fibers surround entire circumference of lens.

Plate 117

Eye

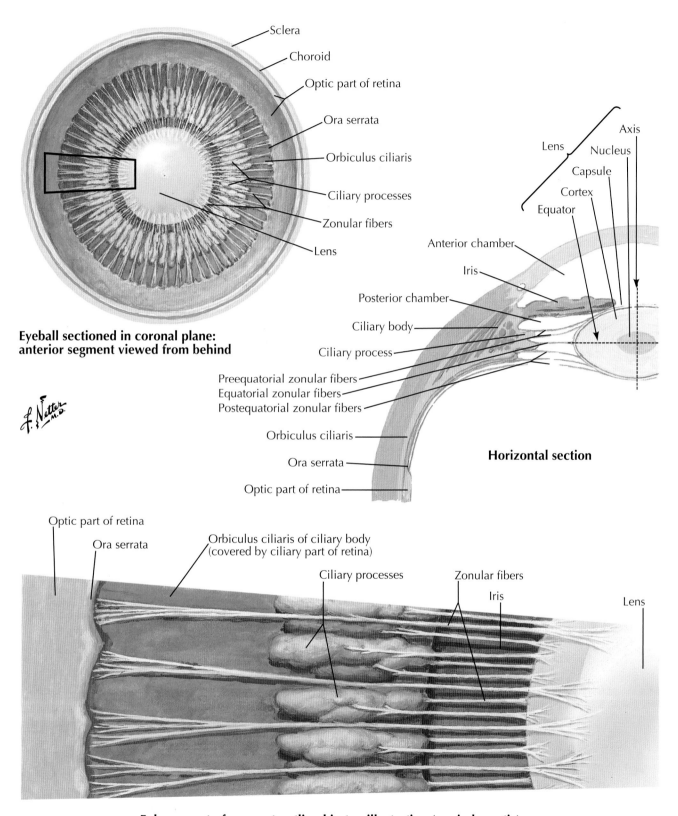

Sclera

Choroid

Optic part of retina

Ora serrata

Orbiculus ciliaris

Ciliary processes

Zonular fibers

Lens

**Eyeball sectioned in coronal plane:
anterior segment viewed from behind**

Axis

Lens

Nucleus

Capsule

Cortex

Equator

Anterior chamber

Iris

Posterior chamber

Ciliary body

Ciliary process

Preequatorial zonular fibers

Equatorial zonular fibers

Postequatorial zonular fibers

Orbiculus ciliaris

Ora serrata

Optic part of retina

Horizontal section

Optic part of retina

Ora serrata

Orbiculus ciliaris of ciliary body
(covered by ciliary part of retina)

Ciliary processes

Zonular fibers

Iris

Lens

Enlargement of segment outlined in top illustration (semischematic)

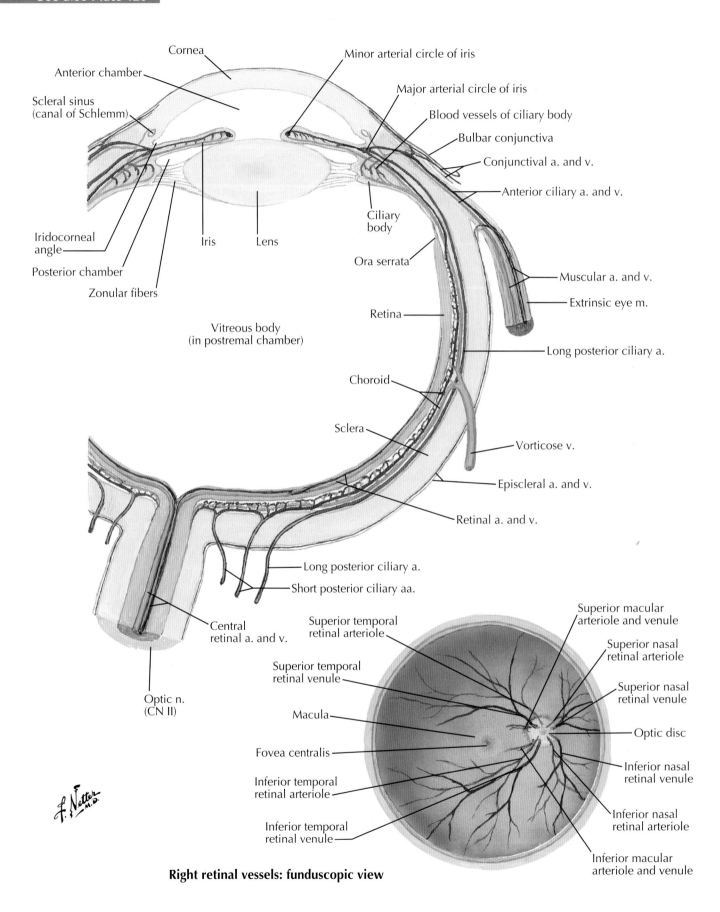

Right retinal vessels: funduscopic view

Plate 119

Eye

Vascular arrangements within the vascular layer of eyeball

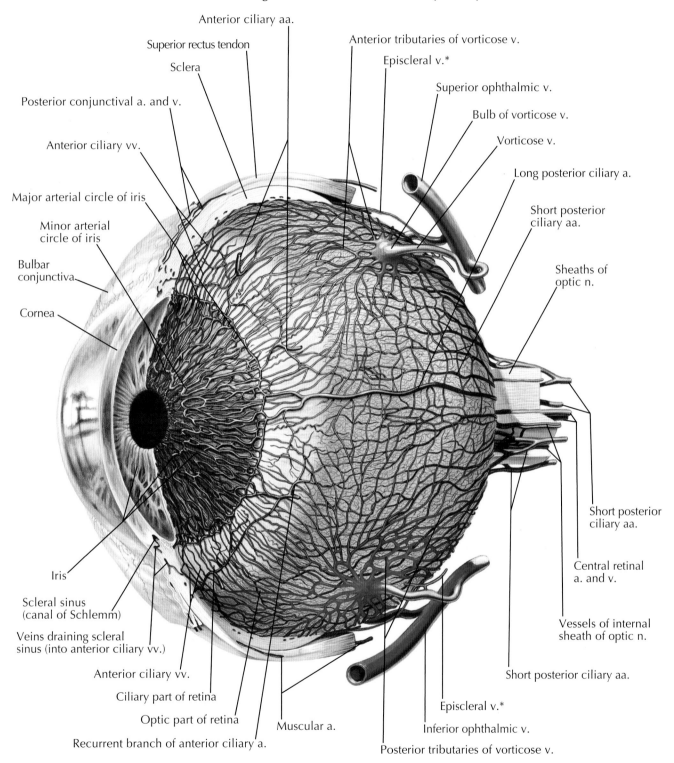

Anterior ciliary aa.

Superior rectus tendon

Sclera

Posterior conjunctival a. and v.

Anterior ciliary vv.

Major arterial circle of iris

Minor arterial circle of iris

Bulbar conjunctiva

Cornea

Iris

Scleral sinus (canal of Schlemm)

Veins draining scleral sinus (into anterior ciliary vv.)

Anterior ciliary vv.

Ciliary part of retina

Optic part of retina

Recurrent branch of anterior ciliary a.

Anterior tributaries of vorticose v.

Episcleral v.*

Superior ophthalmic v.

Bulb of vorticose v.

Vorticose v.

Long posterior ciliary a.

Short posterior ciliary aa.

Sheaths of optic n.

Short posterior ciliary aa.

Central retinal a. and v.

Vessels of internal sheath of optic n.

Short posterior ciliary aa.

Episcleral v.*

Inferior ophthalmic v.

Posterior tributaries of vorticose v.

Muscular a.

The episcleral veins are shown here anastomosing with the vorticose veins, which they do; however, they also drain into the anterior ciliary veins.

C. Machado
M.D.

Frontal section

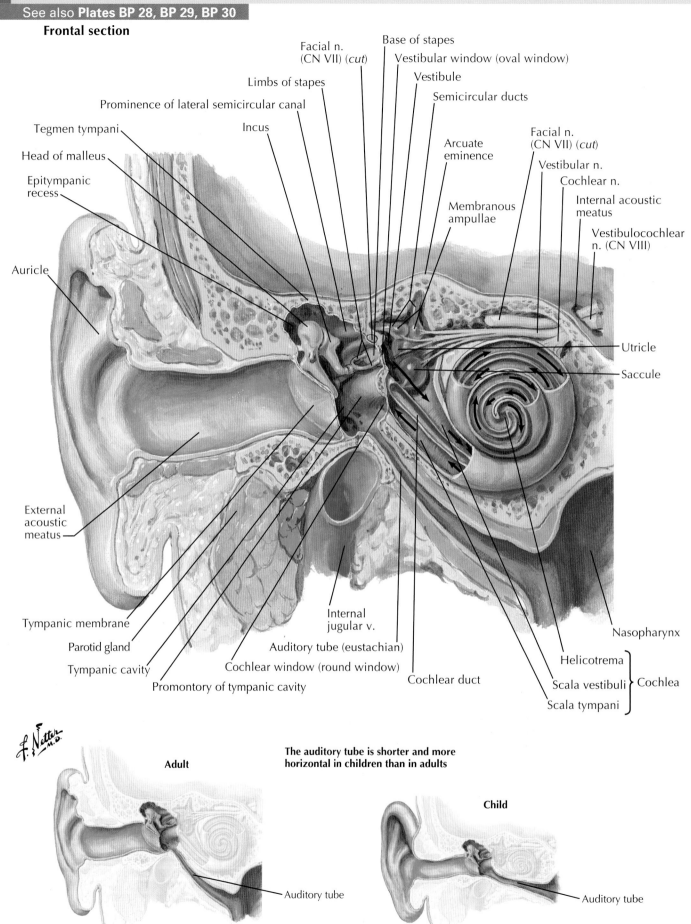

Facial n.
(CN VII) *(cut)*

Base of stapes

Vestibular window (oval window)

Vestibule

Limbs of stapes

Semicircular ducts

Prominence of lateral semicircular canal

Incus

Arcuate
eminence

Facial n.
(CN VII) *(cut)*

Tegmen tympani

Vestibular n.

Head of malleus

Cochlear n.

Membranous
ampullae

Internal acoustic
meatus

Epitympanic
recess

Vestibulocochlear
n. (CN VIII)

Auricle

Utricle

Saccule

External
acoustic
meatus

Nasopharynx

Tympanic membrane

Parotid gland

Tympanic cavity

Internal
jugular v.

Helicotrema

Promontory of tympanic cavity

Auditory tube (eustachian)

Scala vestibuli } Cochlea

Cochlear window (round window)

Scala tympani

Cochlear duct

**The auditory tube is shorter and more
horizontal in children than in adults**

Adult

Child

Auditory tube

Auditory tube

Plate 121

Ear

Right auricle

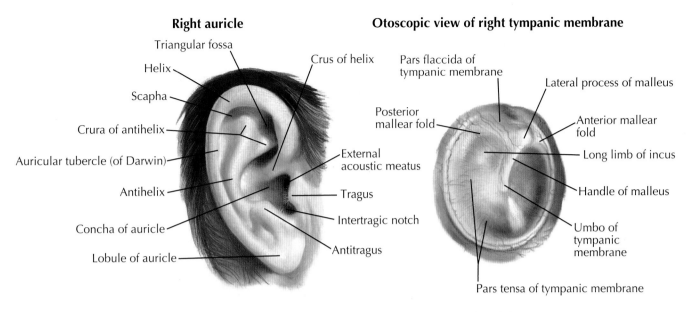

Triangular fossa

Helix

Scapha

Crura of antihelix

Auricular tubercle (of Darwin)

Antihelix

Concha of auricle

Lobule of auricle

Crus of helix

External acoustic meatus

Tragus

Intertragic notch

Antitragus

Otoscopic view of right tympanic membrane

Pars flaccida of tympanic membrane

Posterior mallear fold

Lateral process of malleus

Anterior mallear fold

Long limb of incus

Handle of malleus

Umbo of tympanic membrane

Pars tensa of tympanic membrane

Coronal oblique section of external acoustic meatus and middle ear

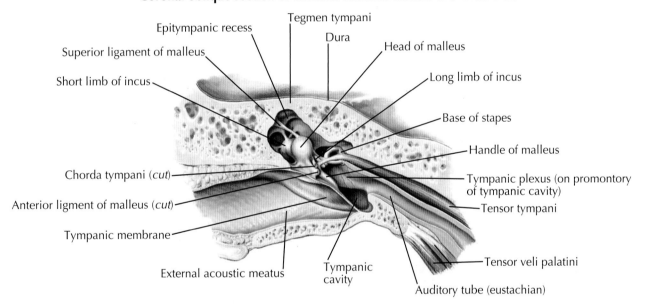

Epitympanic recess

Superior ligament of malleus

Short limb of incus

Chorda tympani (cut)

Anterior ligment of malleus (cut)

Tympanic membrane

External acoustic meatus

Tegmen tympani

Dura

Head of malleus

Long limb of incus

Base of stapes

Handle of malleus

Tympanic plexus (on promontory of tympanic cavity)

Tensor tympani

Tympanic cavity

Tensor veli palatini

Auditory tube (eustachian)

Right tympanic cavity after removal of tympanic membrane (lateral view)

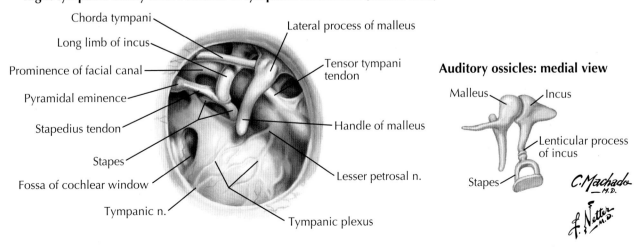

Chorda tympani

Long limb of incus

Prominence of facial canal

Pyramidal eminence

Stapedius tendon

Stapes

Fossa of cochlear window

Tympanic n.

Lateral process of malleus

Tensor tympani tendon

Handle of malleus

Lesser petrosal n.

Tympanic plexus

Auditory ossicles: medial view

Malleus

Incus

Lenticular process of incus

Stapes

C. Machado — M.D.

F. Netter M.D.

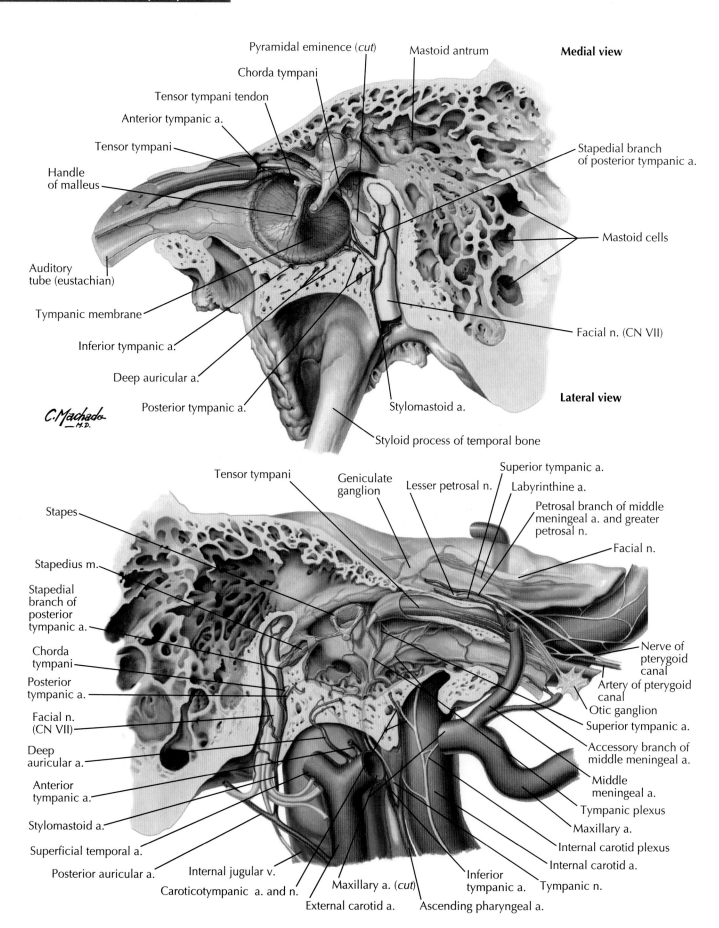

Pyramidal eminence *(cut)*

Mastoid antrum

Medial view

Chorda tympani

Tensor tympani tendon

Anterior tympanic a.

Tensor tympani

Stapedial branch of posterior tympanic a.

Handle of malleus

Mastoid cells

Auditory tube (eustachian)

Tympanic membrane

Inferior tympanic a.

Deep auricular a.

Facial n. (CN VII)

C. Machado —M.D.

Posterior tympanic a.

Lateral view

Stylomastoid a.

Styloid process of temporal bone

Tensor tympani

Geniculate ganglion

Lesser petrosal n.

Superior tympanic a.

Labyrinthine a.

Petrosal branch of middle meningeal a. and greater petrosal n.

Stapes

Facial n.

Stapedius m.

Stapedial branch of posterior tympanic a.

Chorda tympani

Nerve of pterygoid canal

Posterior tympanic a.

Artery of pterygoid canal

Otic ganglion

Facial n. (CN VII)

Superior tympanic a.

Deep auricular a.

Accessory branch of middle meningeal a.

Anterior tympanic a.

Middle meningeal a.

Stylomastoid a.

Tympanic plexus

Superficial temporal a.

Maxillary a.

Internal carotid plexus

Posterior auricular a.

Internal jugular v.

Internal carotid a.

Caroticotympanic a. and n.

Maxillary a. *(cut)*

Inferior tympanic a.

Tympanic n.

External carotid a.

Ascending pharyngeal a.

Plate 123

Ear

Right bony labyrinth (otic capsule), anterolateral view: surrounding cancellous bone removed

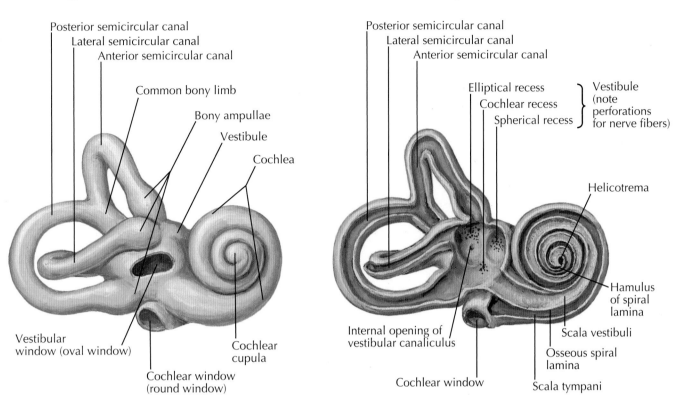

Posterior semicircular canal
Lateral semicircular canal
Anterior semicircular canal
Common bony limb
Bony ampullae
Vestibule
Cochlea
Vestibular window (oval window)
Cochlear cupula
Cochlear window (round window)

Dissected right bony labyrinth (otic capsule): membranous labyrinth removed

Posterior semicircular canal
Lateral semicircular canal
Anterior semicircular canal
Elliptical recess
Cochlear recess
Spherical recess
} Vestibule (note perforations for nerve fibers)
Helicotrema
Hamulus of spiral lamina
Internal opening of vestibular canaliculus
Scala vestibuli
Osseous spiral lamina
Cochlear window
Scala tympani

Right membranous labyrinth with nerves: medial view

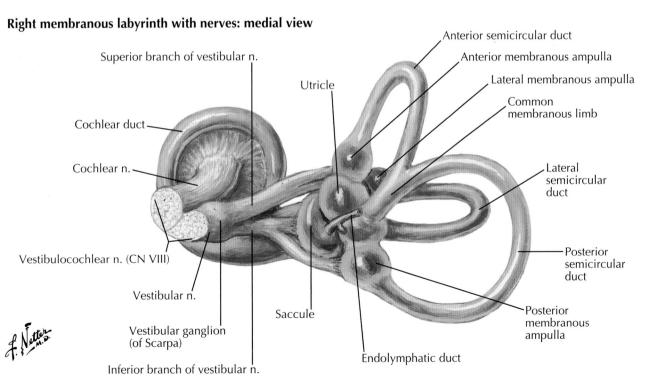

Superior branch of vestibular n.
Utricle
Anterior semicircular duct
Anterior membranous ampulla
Lateral membranous ampulla
Common membranous limb
Cochlear duct
Cochlear n.
Lateral semicircular duct
Vestibulocochlear n. (CN VIII)
Vestibular n.
Posterior semicircular duct
Vestibular ganglion (of Scarpa)
Saccule
Posterior membranous ampulla
Inferior branch of vestibular n.
Endolymphatic duct

Bony and membranous labyrinths: schema

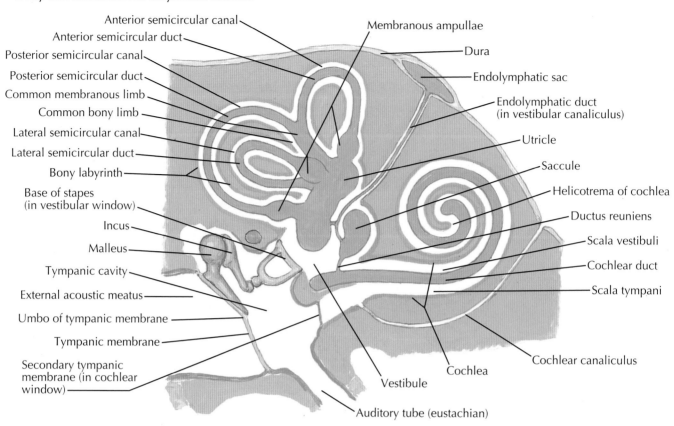

Anterior semicircular canal
Anterior semicircular duct
Posterior semicircular canal
Posterior semicircular duct
Common membranous limb
Common bony limb
Lateral semicircular canal
Lateral semicircular duct
Bony labyrinth
Base of stapes (in vestibular window)
Incus
Malleus
Tympanic cavity
External acoustic meatus
Umbo of tympanic membrane
Tympanic membrane
Secondary tympanic membrane (in cochlear window)

Membranous ampullae
Dura
Endolymphatic sac
Endolymphatic duct (in vestibular canaliculus)
Utricle
Saccule
Helicotrema of cochlea
Ductus reuniens
Scala vestibuli
Cochlear duct
Scala tympani
Cochlear canaliculus
Cochlea
Vestibule
Auditory tube (eustachian)

Section through turn of cochlea

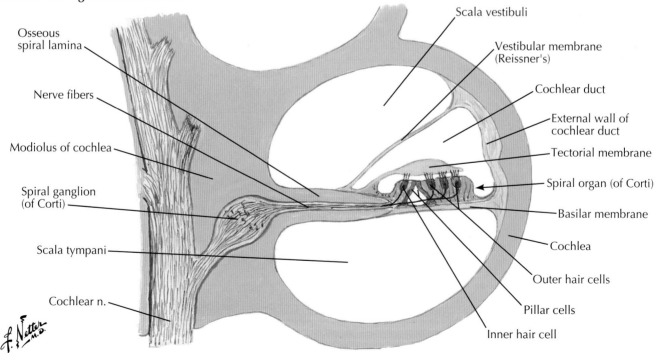

Osseous spiral lamina
Nerve fibers
Modiolus of cochlea
Spiral ganglion (of Corti)
Scala tympani
Cochlear n.

Scala vestibuli
Vestibular membrane (Reissner's)
Cochlear duct
External wall of cochlear duct
Tectorial membrane
Spiral organ (of Corti)
Basilar membrane
Cochlea
Outer hair cells
Pillar cells
Inner hair cell

Plate 125

Ear

Superior projection of right bony labyrinth on floor of cranium

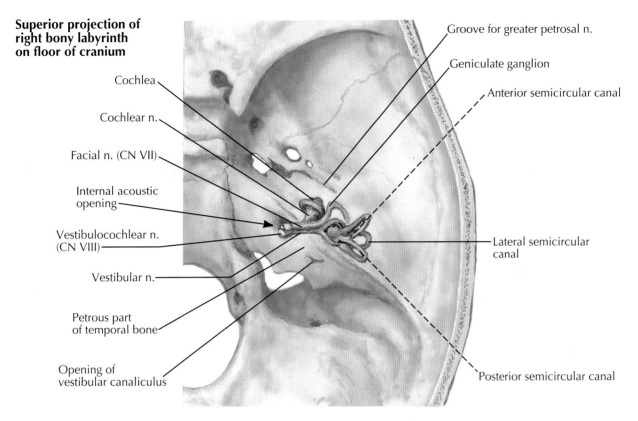

Cochlea

Cochlear n.

Facial n. (CN VII)

Internal acoustic opening

Vestibulocochlear n. (CN VIII)

Vestibular n.

Petrous part of temporal bone

Opening of vestibular canaliculus

Groove for greater petrosal n.

Geniculate ganglion

Anterior semicircular canal

Lateral semicircular canal

Posterior semicircular canal

Lateral projection of right membranous labyrinth

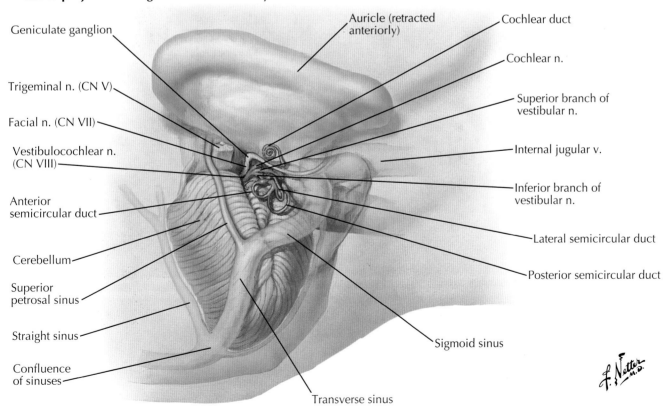

Geniculate ganglion

Trigeminal n. (CN V)

Facial n. (CN VII)

Vestibulocochlear n. (CN VIII)

Anterior semicircular duct

Cerebellum

Superior petrosal sinus

Straight sinus

Confluence of sinuses

Auricle (retracted anteriorly)

Cochlear duct

Cochlear n.

Superior branch of vestibular n.

Internal jugular v.

Inferior branch of vestibular n.

Lateral semicircular duct

Posterior semicircular duct

Sigmoid sinus

Transverse sinus

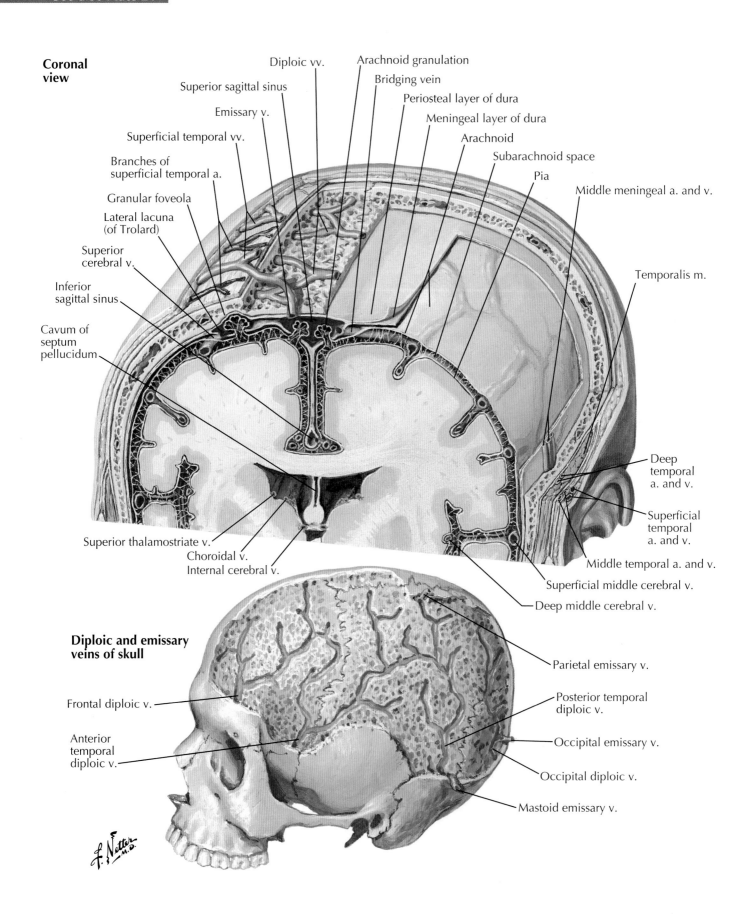

Coronal view

Diploic vv.

Superior sagittal sinus

Arachnoid granulation

Bridging vein

Emissary v.

Periosteal layer of dura

Meningeal layer of dura

Superficial temporal vv.

Arachnoid

Branches of superficial temporal a.

Subarachnoid space

Pia

Granular foveola

Middle meningeal a. and v.

Lateral lacuna (of Trolard)

Superior cerebral v.

Temporalis m.

Inferior sagittal sinus

Cavum of septum pellucidum

Deep temporal a. and v.

Superficial temporal a. and v.

Superior thalamostriate v.

Choroidal v.

Internal cerebral v.

Middle temporal a. and v.

Superficial middle cerebral v.

Deep middle cerebral v.

Diploic and emissary veins of skull

Frontal diploic v.

Anterior temporal diploic v.

Parietal emissary v.

Posterior temporal diploic v.

Occipital emissary v.

Occipital diploic v.

Mastoid emissary v.

Plate 127

Brain and Meninges

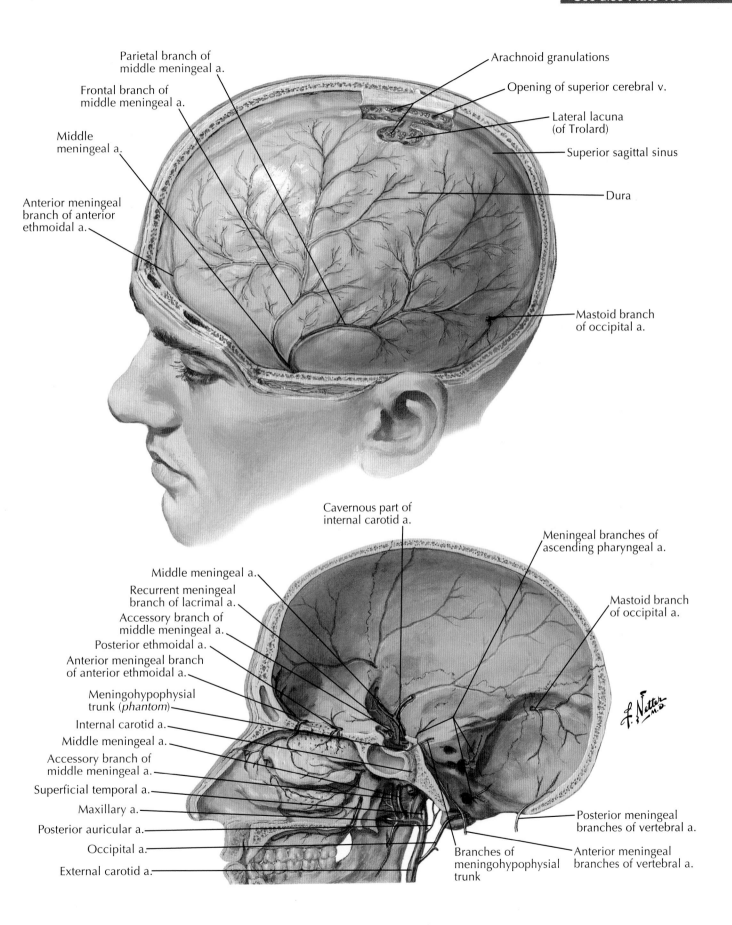

Parietal branch of middle meningeal a.

Frontal branch of middle meningeal a.

Middle meningeal a.

Anterior meningeal branch of anterior ethmoidal a.

Arachnoid granulations

Opening of superior cerebral v.

Lateral lacuna (of Trolard)

Superior sagittal sinus

Dura

Mastoid branch of occipital a.

Cavernous part of internal carotid a.

Meningeal branches of ascending pharyngeal a.

Middle meningeal a.

Recurrent meningeal branch of lacrimal a.

Accessory branch of middle meningeal a.

Posterior ethmoidal a.

Anterior meningeal branch of anterior ethmoidal a.

Meningohypophysial trunk (*phantom*)

Internal carotid a.

Middle meningeal a.

Accessory branch of middle meningeal a.

Superficial temporal a.

Maxillary a.

Posterior auricular a.

Occipital a.

External carotid a.

Mastoid branch of occipital a.

Posterior meningeal branches of vertebral a.

Branches of meningohypophysial trunk

Anterior meningeal branches of vertebral a.

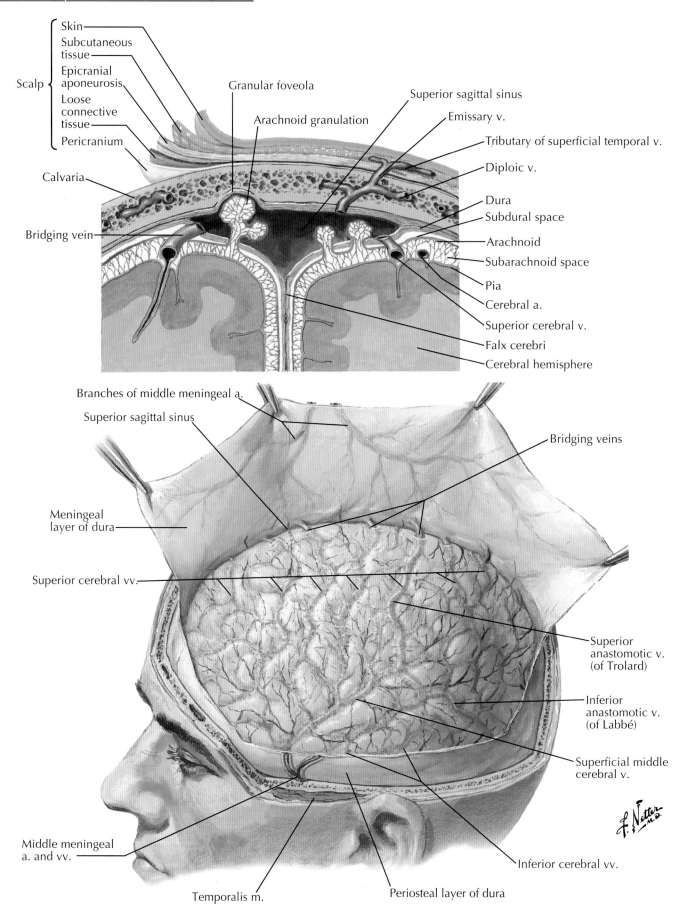

Skin

Subcutaneous tissue

Epicranial aponeurosis

Scalp

Loose connective tissue

Pericranium

Calvaria

Bridging vein

Granular foveola

Arachnoid granulation

Superior sagittal sinus

Emissary v.

Tributary of superficial temporal v.

Diploic v.

Dura

Subdural space

Arachnoid

Subarachnoid space

Pia

Cerebral a.

Superior cerebral v.

Falx cerebri

Cerebral hemisphere

Branches of middle meningeal a.

Superior sagittal sinus

Bridging veins

Meningeal layer of dura

Superior cerebral vv.

Superior anastomotic v. (of Trolard)

Inferior anastomotic v. (of Labbé)

Superficial middle cerebral v.

Middle meningeal a. and vv.

Inferior cerebral vv.

Temporalis m.

Periosteal layer of dura

Plate 129

Brain and Meninges

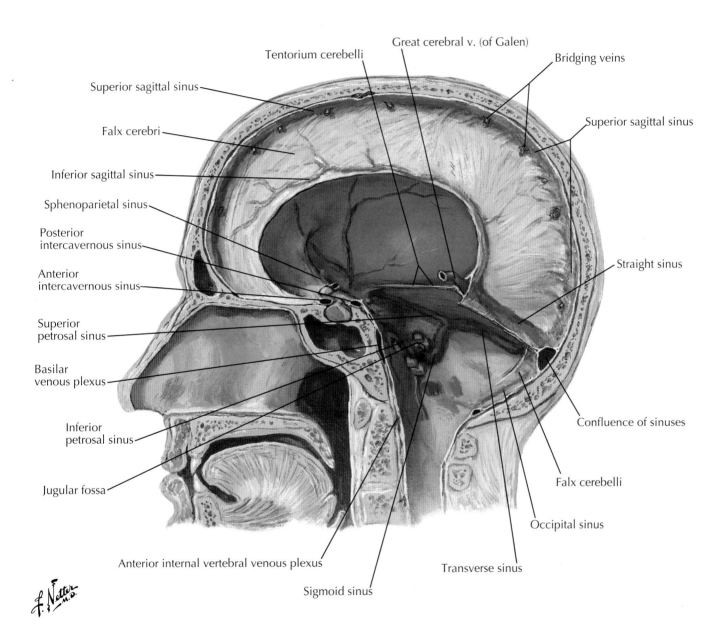

Tentorium cerebelli

Great cerebral v. (of Galen)

Bridging veins

Superior sagittal sinus

Superior sagittal sinus

Falx cerebri

Inferior sagittal sinus

Sphenoparietal sinus

Posterior intercavernous sinus

Anterior intercavernous sinus

Straight sinus

Superior petrosal sinus

Basilar venous plexus

Inferior petrosal sinus

Confluence of sinuses

Jugular fossa

Falx cerebelli

Occipital sinus

Anterior internal vertebral venous plexus

Transverse sinus

Sigmoid sinus

Skull sectioned horizontally: superior view

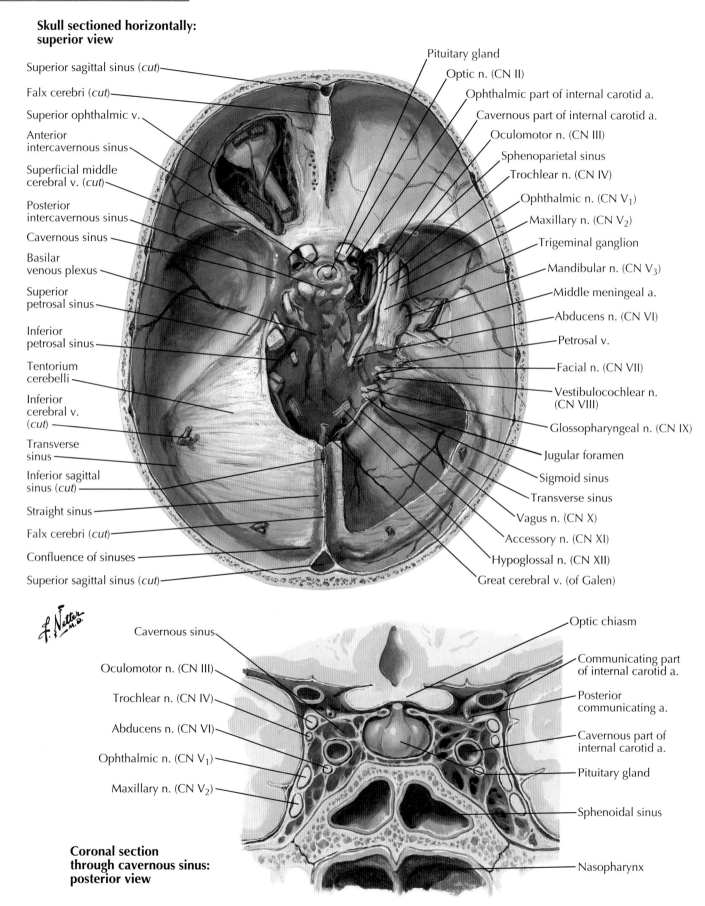

Superior sagittal sinus (*cut*)

Falx cerebri (*cut*)

Superior ophthalmic v.

Anterior intercavernous sinus

Superficial middle cerebral v. (*cut*)

Posterior intercavernous sinus

Cavernous sinus

Basilar venous plexus

Superior petrosal sinus

Inferior petrosal sinus

Tentorium cerebelli

Inferior cerebral v. (*cut*)

Transverse sinus

Inferior sagittal sinus (*cut*)

Straight sinus

Falx cerebri (*cut*)

Confluence of sinuses

Superior sagittal sinus (*cut*)

Pituitary gland

Optic n. (CN II)

Ophthalmic part of internal carotid a.

Cavernous part of internal carotid a.

Oculomotor n. (CN III)

Sphenoparietal sinus

Trochlear n. (CN IV)

Ophthalmic n. (CN V$_1$)

Maxillary n. (CN V$_2$)

Trigeminal ganglion

Mandibular n. (CN V$_3$)

Middle meningeal a.

Abducens n. (CN VI)

Petrosal v.

Facial n. (CN VII)

Vestibulocochlear n. (CN VIII)

Glossopharyngeal n. (CN IX)

Jugular foramen

Sigmoid sinus

Transverse sinus

Vagus n. (CN X)

Accessory n. (CN XI)

Hypoglossal n. (CN XII)

Great cerebral v. (of Galen)

Cavernous sinus

Oculomotor n. (CN III)

Trochlear n. (CN IV)

Abducens n. (CN VI)

Ophthalmic n. (CN V$_1$)

Maxillary n. (CN V$_2$)

Optic chiasm

Communicating part of internal carotid a.

Posterior communicating a.

Cavernous part of internal carotid a.

Pituitary gland

Sphenoidal sinus

Nasopharynx

Coronal section through cavernous sinus: posterior view

Plate 131

Brain and Meninges

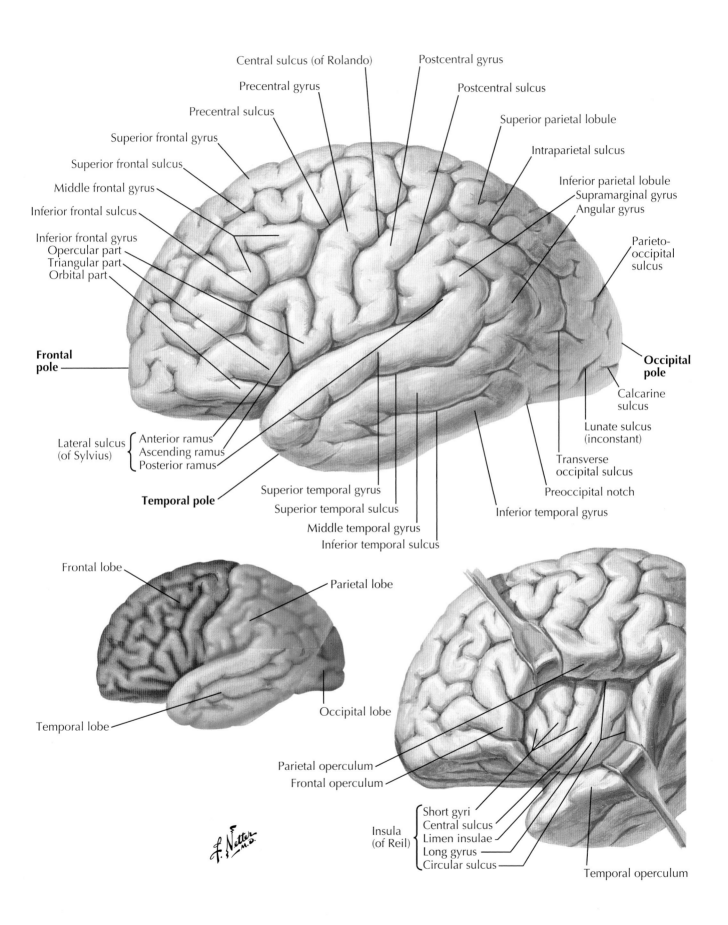

Central sulcus (of Rolando)
Postcentral gyrus
Precentral gyrus
Postcentral sulcus
Precentral sulcus
Superior parietal lobule
Superior frontal gyrus
Intraparietal sulcus
Superior frontal sulcus
Inferior parietal lobule
Middle frontal gyrus
Supramarginal gyrus
Inferior frontal sulcus
Angular gyrus
Inferior frontal gyrus
Opercular part
Parieto-occipital sulcus
Triangular part
Orbital part
Frontal pole
Occipital pole
Calcarine sulcus
Lunate sulcus (inconstant)
Lateral sulcus (of Sylvius) { Anterior ramus / Ascending ramus / Posterior ramus
Transverse occipital sulcus
Temporal pole
Preoccipital notch
Superior temporal gyrus
Inferior temporal gyrus
Superior temporal sulcus
Middle temporal gyrus
Inferior temporal sulcus

Frontal lobe
Parietal lobe
Occipital lobe
Temporal lobe

Parietal operculum
Frontal operculum

Insula (of Reil) { Short gyri / Central sulcus / Limen insulae / Long gyrus / Circular sulcus
Temporal operculum

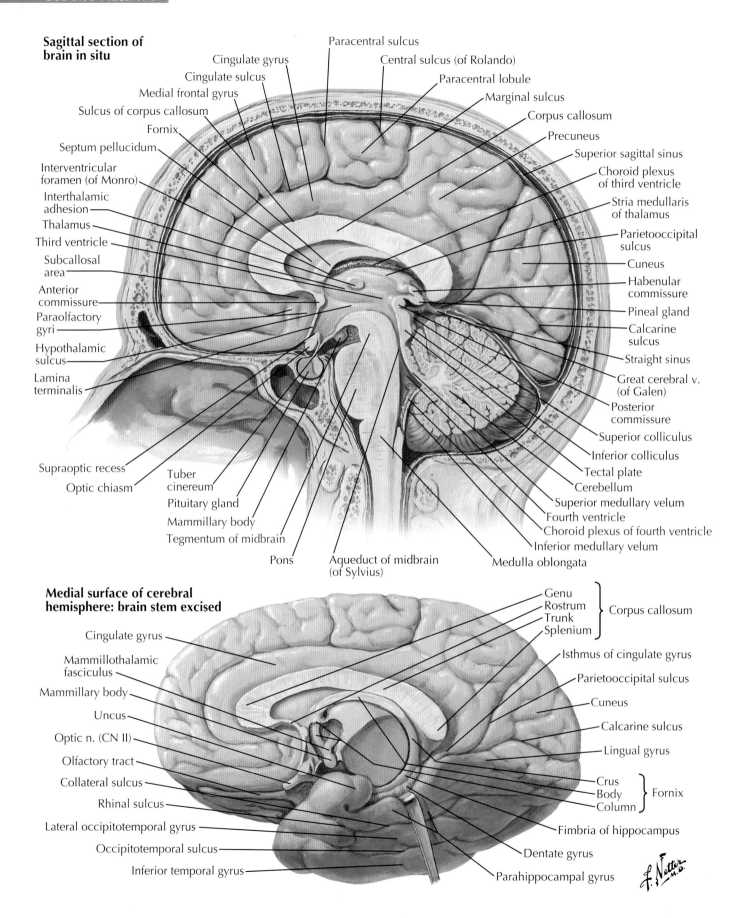

Sagittal section of brain in situ

Paracentral sulcus

Central sulcus (of Rolando)

Cingulate gyrus

Cingulate sulcus

Medial frontal gyrus

Sulcus of corpus callosum

Fornix

Septum pellucidum

Interventricular foramen (of Monro)

Interthalamic adhesion

Thalamus

Third ventricle

Subcallosal area

Anterior commissure

Paraolfactory gyri

Hypothalamic sulcus

Lamina terminalis

Paracentral lobule

Marginal sulcus

Corpus callosum

Precuneus

Superior sagittal sinus

Choroid plexus of third ventricle

Stria medullaris of thalamus

Parietooccipital sulcus

Cuneus

Habenular commissure

Pineal gland

Calcarine sulcus

Straight sinus

Great cerebral v. (of Galen)

Posterior commissure

Superior colliculus

Inferior colliculus

Tectal plate

Cerebellum

Superior medullary velum

Fourth ventricle

Choroid plexus of fourth ventricle

Inferior medullary velum

Medulla oblongata

Supraoptic recess

Optic chiasm

Tuber cinereum

Pituitary gland

Mammillary body

Tegmentum of midbrain

Pons

Aqueduct of midbrain (of Sylvius)

Medial surface of cerebral hemisphere: brain stem excised

Cingulate gyrus

Mammillothalamic fasciculus

Mammillary body

Uncus

Optic n. (CN II)

Olfactory tract

Collateral sulcus

Rhinal sulcus

Lateral occipitotemporal gyrus

Occipitotemporal sulcus

Inferior temporal gyrus

Genu

Rostrum   } Corpus callosum

Trunk

Splenium

Isthmus of cingulate gyrus

Parietooccipital sulcus

Cuneus

Calcarine sulcus

Lingual gyrus

Crus

Body   } Fornix

Column

Fimbria of hippocampus

Dentate gyrus

Parahippocampal gyrus

Plate 133

Brain and Meninges

Sectioned brain stem

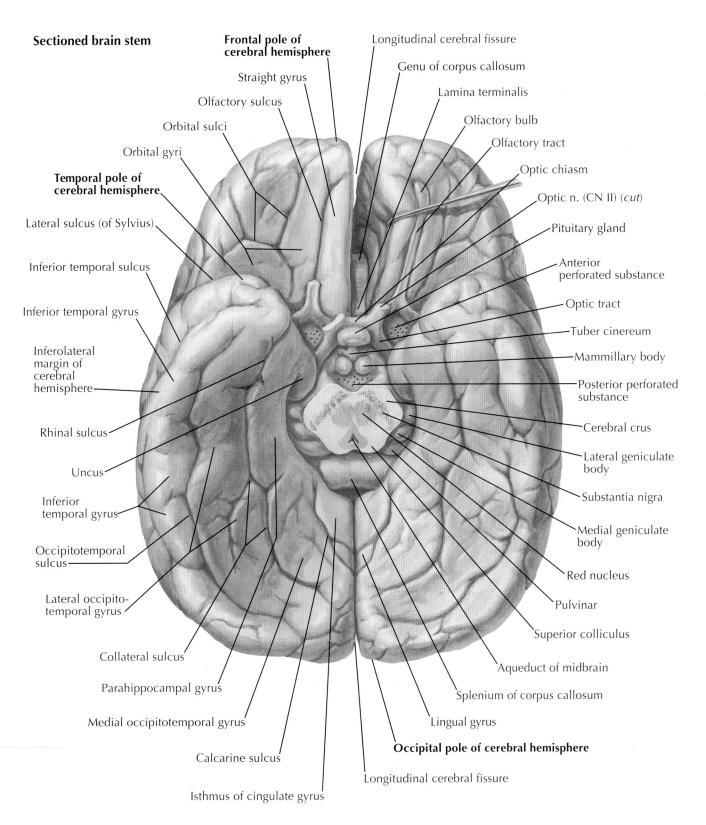

Frontal pole of cerebral hemisphere

Straight gyrus

Olfactory sulcus

Orbital sulci

Orbital gyri

Temporal pole of cerebral hemisphere

Lateral sulcus (of Sylvius)

Inferior temporal sulcus

Inferior temporal gyrus

Inferolateral margin of cerebral hemisphere

Rhinal sulcus

Uncus

Inferior temporal gyrus

Occipitotemporal sulcus

Lateral occipito-temporal gyrus

Collateral sulcus

Parahippocampal gyrus

Medial occipitotemporal gyrus

Calcarine sulcus

Isthmus of cingulate gyrus

Longitudinal cerebral fissure

Genu of corpus callosum

Lamina terminalis

Olfactory bulb

Olfactory tract

Optic chiasm

Optic n. (CN II) (cut)

Pituitary gland

Anterior perforated substance

Optic tract

Tuber cinereum

Mammillary body

Posterior perforated substance

Cerebral crus

Lateral geniculate body

Substantia nigra

Medial geniculate body

Red nucleus

Pulvinar

Superior colliculus

Aqueduct of midbrain

Splenium of corpus callosum

Lingual gyrus

Occipital pole of cerebral hemisphere

Longitudinal cerebral fissure

Left lateral phantom view

Right lateral ventricle

Frontal horn

Body

Temporal horn

Occipital horn

Lateral ventricle

Aqueduct of midbrain (of Sylvius)

Fourth ventricle

Lateral aperture of fourth ventricle (foramen of Luschka)

Lateral recess of fourth ventricle

Median aperture of fourth ventricle (foramen of Magendie)

Interventricular foramen (of Monro)

Third ventricle

Supraoptic recess

Infundibular recess

Pineal recess

Suprapineal recess

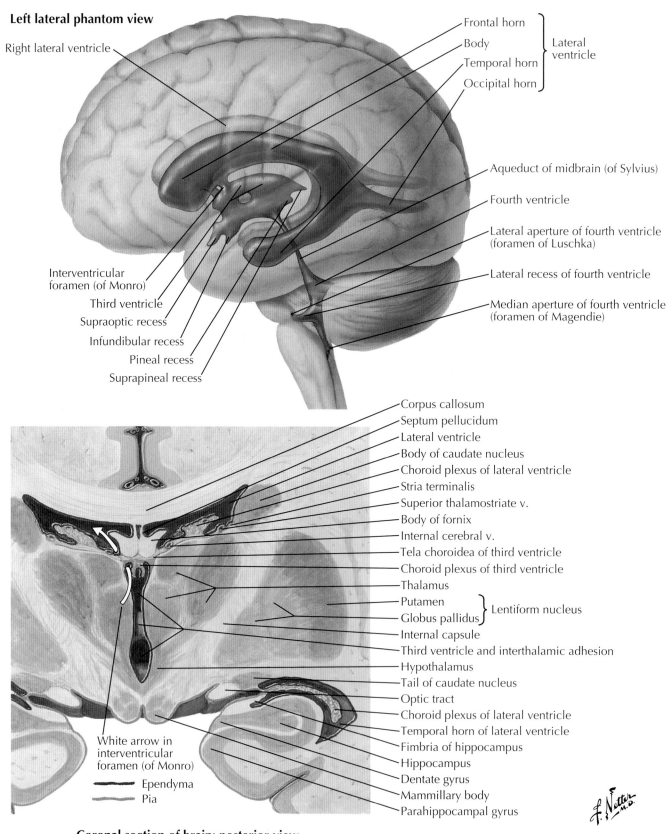

Corpus callosum
Septum pellucidum
Lateral ventricle
Body of caudate nucleus
Choroid plexus of lateral ventricle
Stria terminalis
Superior thalamostriate v.
Body of fornix
Internal cerebral v.
Tela choroidea of third ventricle
Choroid plexus of third ventricle
Thalamus
Putamen
Globus pallidus
Lentiform nucleus
Internal capsule
Third ventricle and interthalamic adhesion
Hypothalamus
Tail of caudate nucleus
Optic tract
Choroid plexus of lateral ventricle
Temporal horn of lateral ventricle
Fimbria of hippocampus
Hippocampus
Dentate gyrus
Mammillary body
Parahippocampal gyrus

White arrow in interventricular foramen (of Monro)

Ependyma
Pia

Coronal section of brain: posterior view

Plate 135

Brain and Meninges

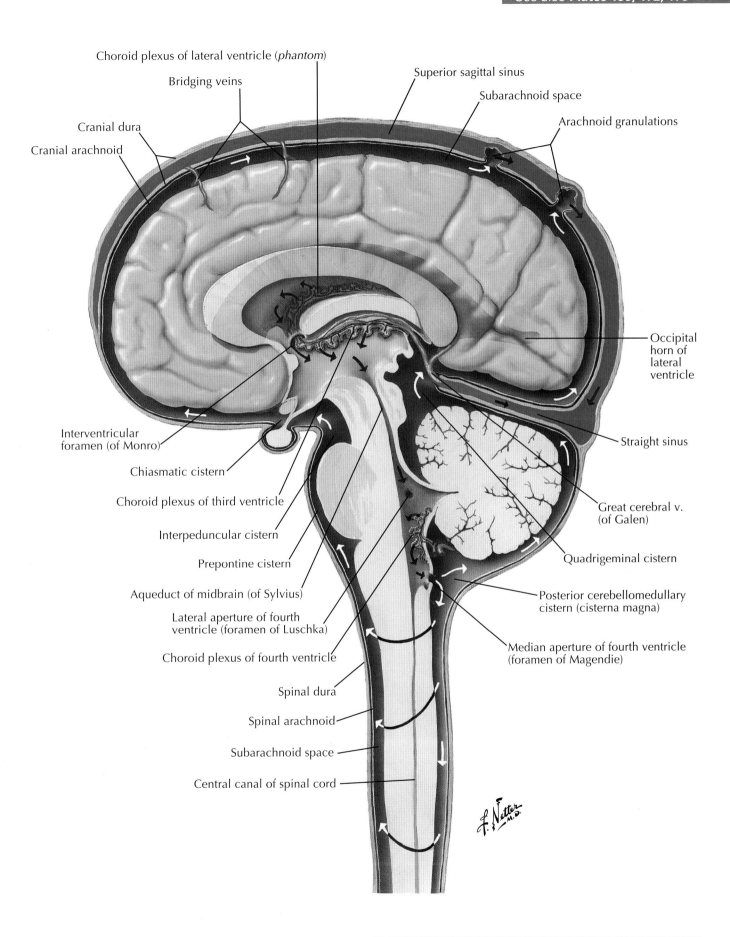

Choroid plexus of lateral ventricle (*phantom*)

Bridging veins

Cranial dura

Cranial arachnoid

Superior sagittal sinus

Subarachnoid space

Arachnoid granulations

Occipital horn of lateral ventricle

Interventricular foramen (of Monro)

Chiasmatic cistern

Choroid plexus of third ventricle

Interpeduncular cistern

Prepontine cistern

Aqueduct of midbrain (of Sylvius)

Lateral aperture of fourth ventricle (foramen of Luschka)

Choroid plexus of fourth ventricle

Spinal dura

Spinal arachnoid

Subarachnoid space

Central canal of spinal cord

Straight sinus

Great cerebral v. (of Galen)

Quadrigeminal cistern

Posterior cerebellomedullary cistern (cisterna magna)

Median aperture of fourth ventricle (foramen of Magendie)

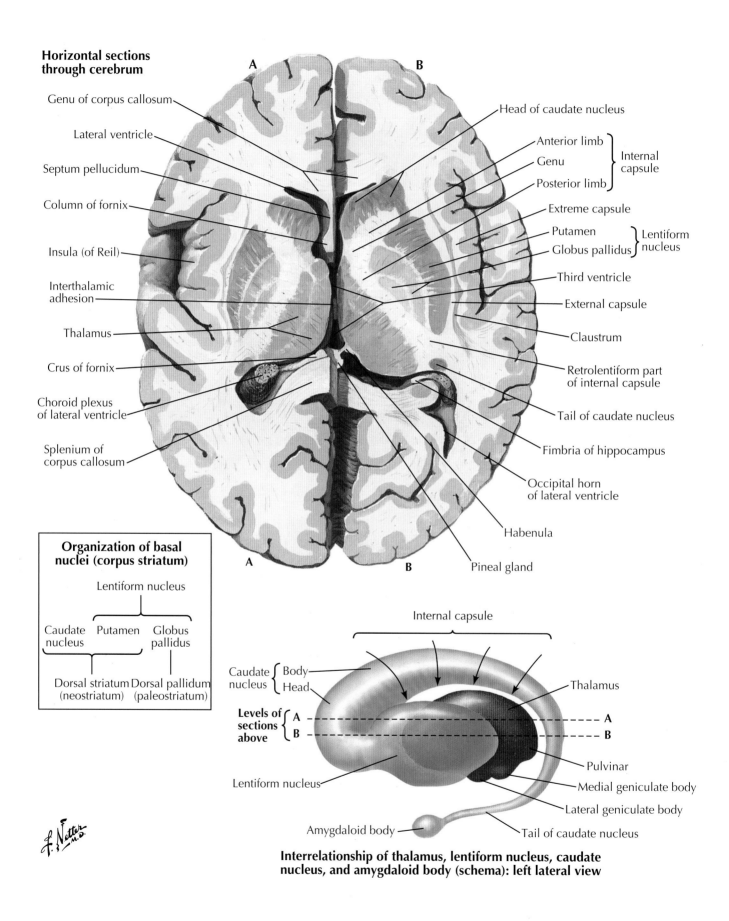

Horizontal sections through cerebrum

Genu of corpus callosum

Lateral ventricle

Septum pellucidum

Column of fornix

Insula (of Reil)

Interthalamic adhesion

Thalamus

Crus of fornix

Choroid plexus of lateral ventricle

Splenium of corpus callosum

A   B

Head of caudate nucleus

Anterior limb

Genu — Internal capsule

Posterior limb

Extreme capsule

Putamen — Lentiform nucleus

Globus pallidus

Third ventricle

External capsule

Claustrum

Retrolentiform part of internal capsule

Tail of caudate nucleus

Fimbria of hippocampus

Occipital horn of lateral ventricle

Habenula

Pineal gland

A   B

Organization of basal nuclei (corpus striatum)

Lentiform nucleus

| Caudate nucleus | Putamen | Globus pallidus |

Dorsal striatum (neostriatum)   Dorsal pallidum (paleostriatum)

Internal capsule

Caudate nucleus { Body / Head }

Thalamus

Levels of sections above { A / B }

A

B

Pulvinar

Lentiform nucleus

Amygdaloid body

Medial geniculate body

Lateral geniculate body

Tail of caudate nucleus

Interrelationship of thalamus, lentiform nucleus, caudate nucleus, and amygdaloid body (schema): left lateral view

Plate 137   **Brain and Meninges**

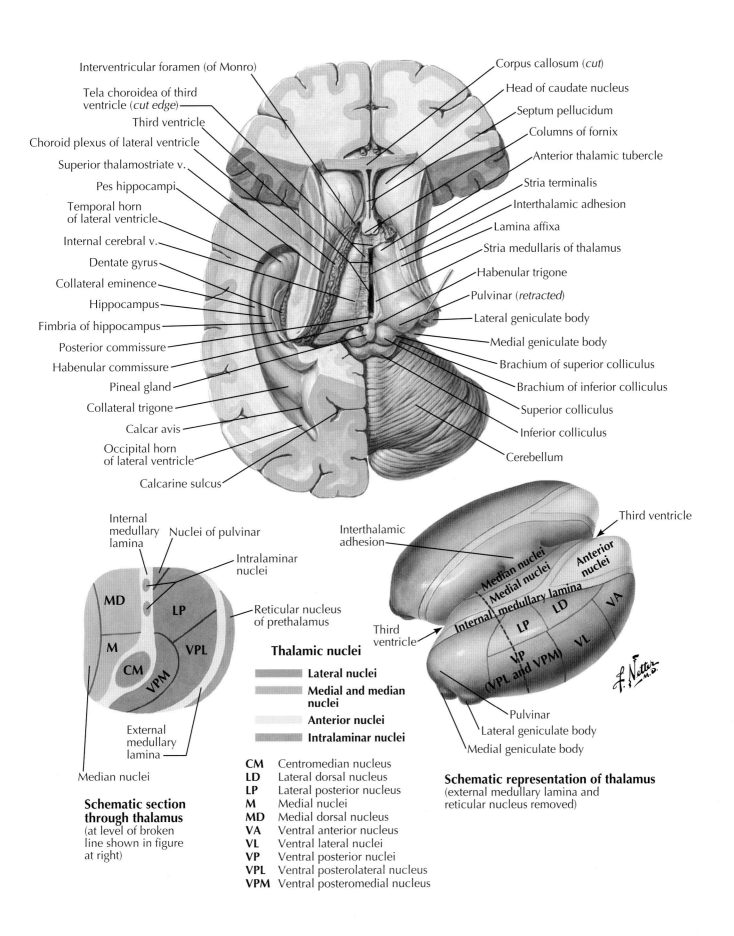

Interventricular foramen (of Monro)

Tela choroidea of third ventricle (*cut edge*)

Third ventricle

Choroid plexus of lateral ventricle

Superior thalamostriate v.

Pes hippocampi

Temporal horn of lateral ventricle

Internal cerebral v.

Dentate gyrus

Collateral eminence

Hippocampus

Fimbria of hippocampus

Posterior commissure

Habenular commissure

Pineal gland

Collateral trigone

Calcar avis

Occipital horn of lateral ventricle

Calcarine sulcus

Corpus callosum (*cut*)

Head of caudate nucleus

Septum pellucidum

Columns of fornix

Anterior thalamic tubercle

Stria terminalis

Interthalamic adhesion

Lamina affixa

Stria medullaris of thalamus

Habenular trigone

Pulvinar (*retracted*)

Lateral geniculate body

Medial geniculate body

Brachium of superior colliculus

Brachium of inferior colliculus

Superior colliculus

Inferior colliculus

Cerebellum

Internal medullary lamina

Nuclei of pulvinar

Intralaminar nuclei

MD

LP

M

VPL

CM

VPM

Reticular nucleus of prethalamus

Thalamic nuclei

Lateral nuclei

Medial and median nuclei

Anterior nuclei

Intralaminar nuclei

External medullary lamina

Median nuclei

Schematic section through thalamus
(at level of broken line shown in figure at right)

CM Centromedian nucleus
LD Lateral dorsal nucleus
LP Lateral posterior nucleus
M Medial nuclei
MD Medial dorsal nucleus
VA Ventral anterior nucleus
VL Ventral lateral nuclei
VP Ventral posterior nuclei
VPL Ventral posterolateral nucleus
VPM Ventral posteromedial nucleus

Interthalamic adhesion

Third ventricle

Median nuclei

Medial nuclei

Internal medullary lamina

Anterior nuclei

LD

LP

VA

VP (VPL and VPM)

VL

Pulvinar

Lateral geniculate body

Medial geniculate body

Third ventricle

Schematic representation of thalamus
(external medullary lamina and reticular nucleus removed)

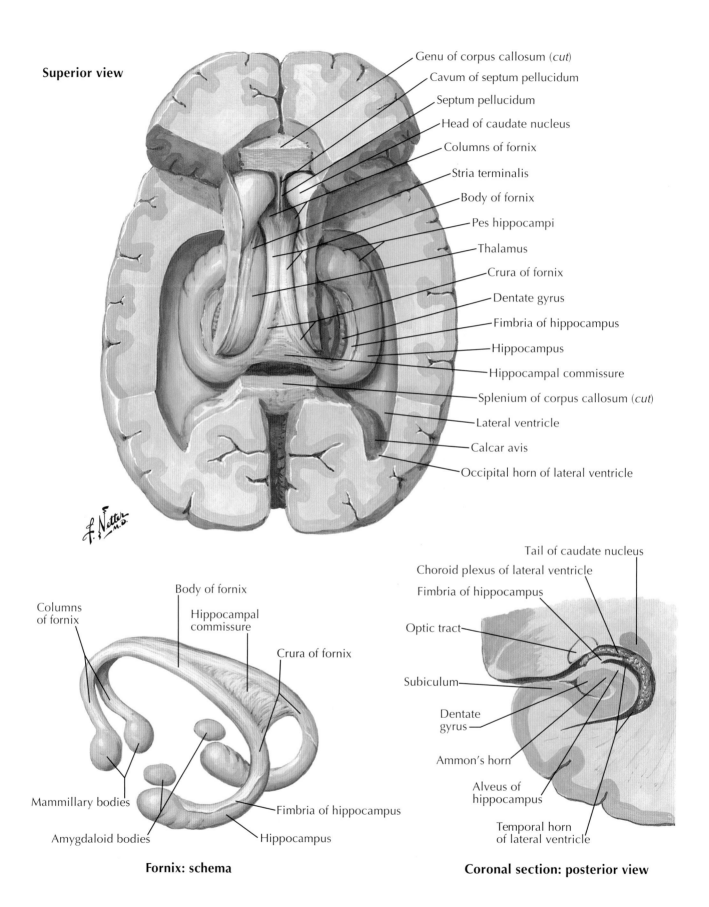

Superior view

Genu of corpus callosum (*cut*)

Cavum of septum pellucidum

Septum pellucidum

Head of caudate nucleus

Columns of fornix

Stria terminalis

Body of fornix

Pes hippocampi

Thalamus

Crura of fornix

Dentate gyrus

Fimbria of hippocampus

Hippocampus

Hippocampal commissure

Splenium of corpus callosum (*cut*)

Lateral ventricle

Calcar avis

Occipital horn of lateral ventricle

Columns of fornix

Body of fornix

Hippocampal commissure

Crura of fornix

Mammillary bodies

Amygdaloid bodies

Fimbria of hippocampus

Hippocampus

Fornix: schema

Tail of caudate nucleus

Choroid plexus of lateral ventricle

Fimbria of hippocampus

Optic tract

Subiculum

Dentate gyrus

Ammon's horn

Alveus of hippocampus

Temporal horn of lateral ventricle

Coronal section: posterior view

Plate 139   **Brain and Meninges**

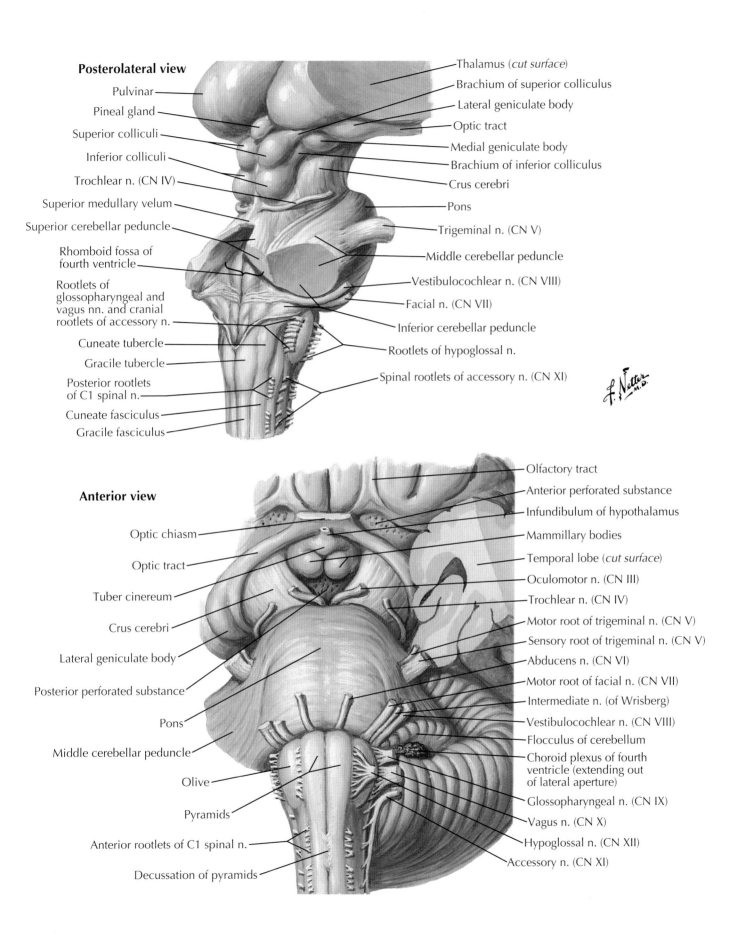

Posterolateral view

Pulvinar

Pineal gland

Superior colliculi

Inferior colliculi

Trochlear n. (CN IV)

Superior medullary velum

Superior cerebellar peduncle

Rhomboid fossa of fourth ventricle

Rootlets of glossopharyngeal and vagus nn. and cranial rootlets of accessory n.

Cuneate tubercle

Gracile tubercle

Posterior rootlets of C1 spinal n.

Cuneate fasciculus

Gracile fasciculus

Thalamus (*cut surface*)

Brachium of superior colliculus

Lateral geniculate body

Optic tract

Medial geniculate body

Brachium of inferior colliculus

Crus cerebri

Pons

Trigeminal n. (CN V)

Middle cerebellar peduncle

Vestibulocochlear n. (CN VIII)

Facial n. (CN VII)

Inferior cerebellar peduncle

Rootlets of hypoglossal n.

Spinal rootlets of accessory n. (CN XI)

Anterior view

Optic chiasm

Optic tract

Tuber cinereum

Crus cerebri

Lateral geniculate body

Posterior perforated substance

Pons

Middle cerebellar peduncle

Olive

Pyramids

Anterior rootlets of C1 spinal n.

Decussation of pyramids

Olfactory tract

Anterior perforated substance

Infundibulum of hypothalamus

Mammillary bodies

Temporal lobe (*cut surface*)

Oculomotor n. (CN III)

Trochlear n. (CN IV)

Motor root of trigeminal n. (CN V)

Sensory root of trigeminal n. (CN V)

Abducens n. (CN VI)

Motor root of facial n. (CN VII)

Intermediate n. (of Wrisberg)

Vestibulocochlear n. (CN VIII)

Flocculus of cerebellum

Choroid plexus of fourth ventricle (extending out of lateral aperture)

Glossopharyngeal n. (CN IX)

Vagus n. (CN X)

Hypoglossal n. (CN XII)

Accessory n. (CN XI)

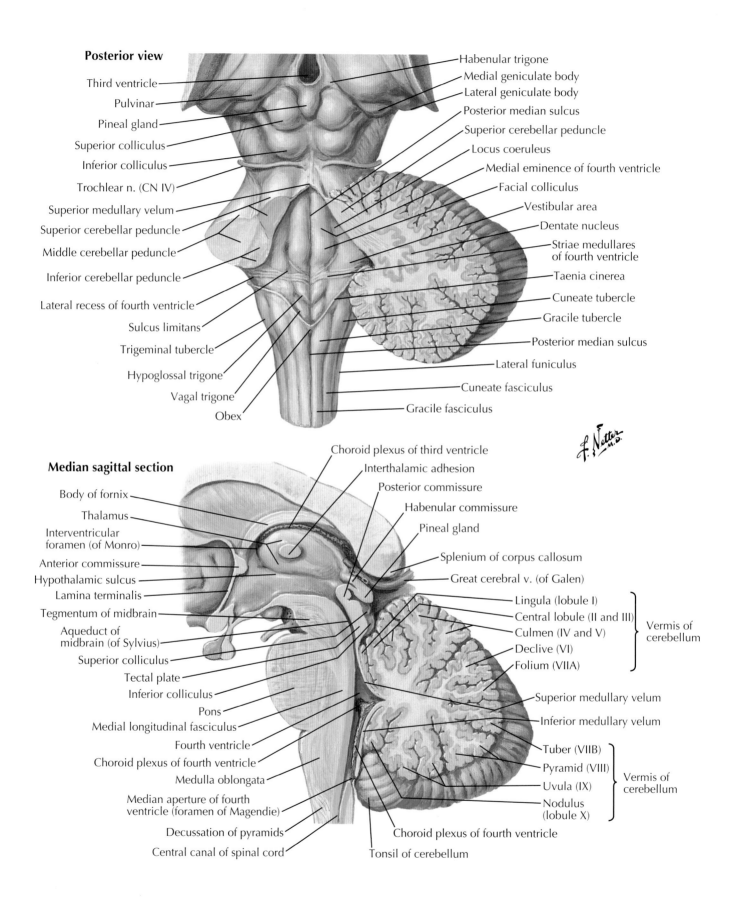

Posterior view

Third ventricle
Pulvinar
Pineal gland
Superior colliculus
Inferior colliculus
Trochlear n. (CN IV)
Superior medullary velum
Superior cerebellar peduncle
Middle cerebellar peduncle
Inferior cerebellar peduncle
Lateral recess of fourth ventricle
Sulcus limitans
Trigeminal tubercle
Hypoglossal trigone
Vagal trigone
Obex

Habenular trigone
Medial geniculate body
Lateral geniculate body
Posterior median sulcus
Superior cerebellar peduncle
Locus coeruleus
Medial eminence of fourth ventricle
Facial colliculus
Vestibular area
Dentate nucleus
Striae medullares of fourth ventricle
Taenia cinerea
Cuneate tubercle
Gracile tubercle
Posterior median sulcus
Lateral funiculus
Cuneate fasciculus
Gracile fasciculus

Median sagittal section

Body of fornix
Thalamus
Interventricular foramen (of Monro)
Anterior commissure
Hypothalamic sulcus
Lamina terminalis
Tegmentum of midbrain
Aqueduct of midbrain (of Sylvius)
Superior colliculus
Tectal plate
Inferior colliculus
Pons
Medial longitudinal fasciculus
Fourth ventricle
Choroid plexus of fourth ventricle
Medulla oblongata
Median aperture of fourth ventricle (foramen of Magendie)
Decussation of pyramids
Central canal of spinal cord

Choroid plexus of third ventricle
Interthalamic adhesion
Posterior commissure
Habenular commissure
Pineal gland
Splenium of corpus callosum
Great cerebral v. (of Galen)
Lingula (lobule I)
Central lobule (II and III)
Culmen (IV and V)
Declive (VI)
Folium (VIIA)
Vermis of cerebellum
Superior medullary velum
Inferior medullary velum
Tuber (VIIB)
Pyramid (VIII)
Uvula (IX)
Nodulus (lobule X)
Vermis of cerebellum
Choroid plexus of fourth ventricle
Tonsil of cerebellum

Plate 141

Brain and Meninges

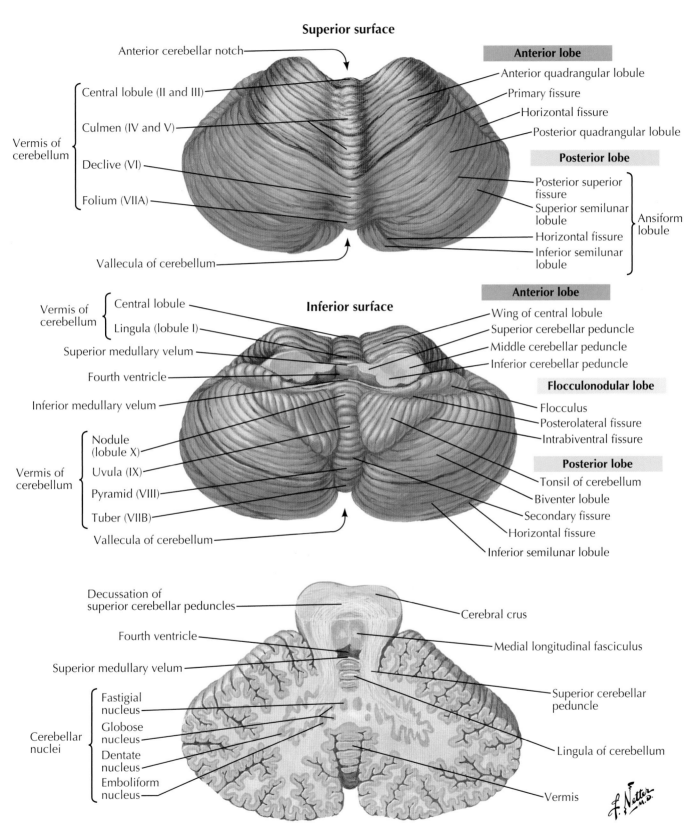

Superior surface

Anterior cerebellar notch

Central lobule (II and III)

Culmen (IV and V)

Vermis of cerebellum

Declive (VI)

Folium (VIIA)

Vallecula of cerebellum

Anterior lobe

Anterior quadrangular lobule

Primary fissure

Horizontal fissure

Posterior quadrangular lobule

Posterior lobe

Posterior superior fissure

Superior semilunar lobule

Horizontal fissure

Inferior semilunar lobule

Ansiform lobule

Inferior surface

Vermis of cerebellum

Central lobule

Lingula (lobule I)

Superior medullary velum

Fourth ventricle

Inferior medullary velum

Vermis of cerebellum

Nodule (lobule X)

Uvula (IX)

Pyramid (VIII)

Tuber (VIIB)

Vallecula of cerebellum

Anterior lobe

Wing of central lobule

Superior cerebellar peduncle

Middle cerebellar peduncle

Inferior cerebellar peduncle

Flocculonodular lobe

Flocculus

Posterolateral fissure

Intrabiventral fissure

Posterior lobe

Tonsil of cerebellum

Biventer lobule

Secondary fissure

Horizontal fissure

Inferior semilunar lobule

Decussation of superior cerebellar peduncles

Fourth ventricle

Superior medullary velum

Cerebellar nuclei

Fastigial nucleus

Globose nucleus

Dentate nucleus

Emboliform nucleus

Cerebral crus

Medial longitudinal fasciculus

Superior cerebellar peduncle

Lingula of cerebellum

Vermis

Section in plane of superior cerebellar peduncle

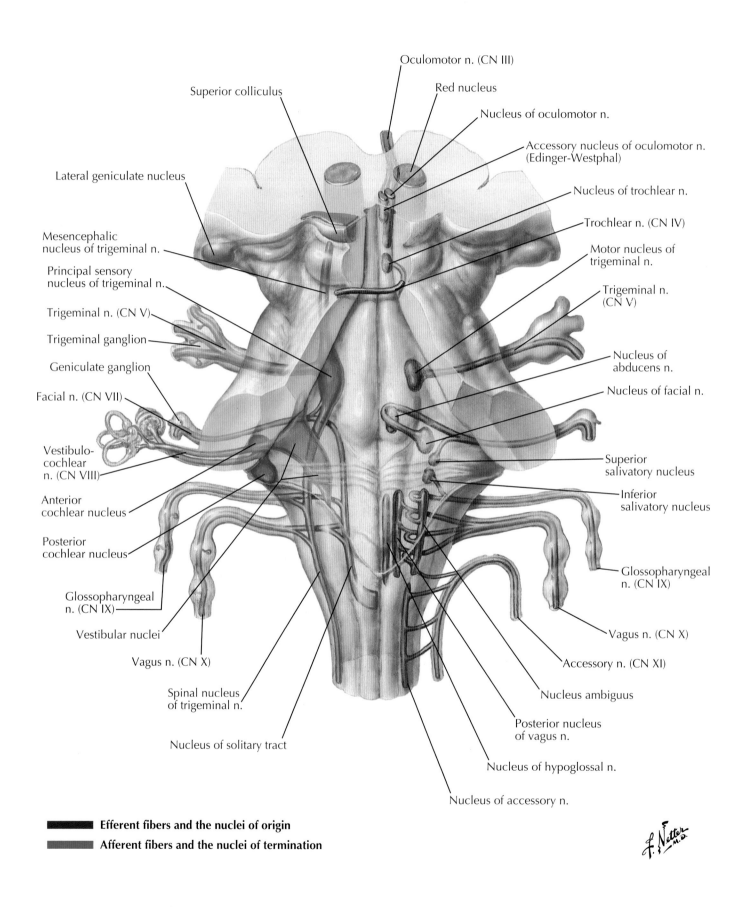

Oculomotor n. (CN III)

Superior colliculus

Red nucleus

Nucleus of oculomotor n.

Accessory nucleus of oculomotor n. (Edinger-Westphal)

Lateral geniculate nucleus

Nucleus of trochlear n.

Trochlear n. (CN IV)

Mesencephalic nucleus of trigeminal n.

Motor nucleus of trigeminal n.

Principal sensory nucleus of trigeminal n.

Trigeminal n. (CN V)

Trigeminal n. (CN V)

Trigeminal ganglion

Nucleus of abducens n.

Geniculate ganglion

Facial n. (CN VII)

Nucleus of facial n.

Vestibulo-cochlear n. (CN VIII)

Superior salivatory nucleus

Anterior cochlear nucleus

Inferior salivatory nucleus

Posterior cochlear nucleus

Glossopharyngeal n. (CN IX)

Glossopharyngeal n. (CN IX)

Vestibular nuclei

Vagus n. (CN X)

Vagus n. (CN X)

Accessory n. (CN XI)

Spinal nucleus of trigeminal n.

Nucleus ambiguus

Posterior nucleus of vagus n.

Nucleus of solitary tract

Nucleus of hypoglossal n.

Nucleus of accessory n.

■■■■ Efferent fibers and the nuclei of origin

■■■■ Afferent fibers and the nuclei of termination

Plate 143

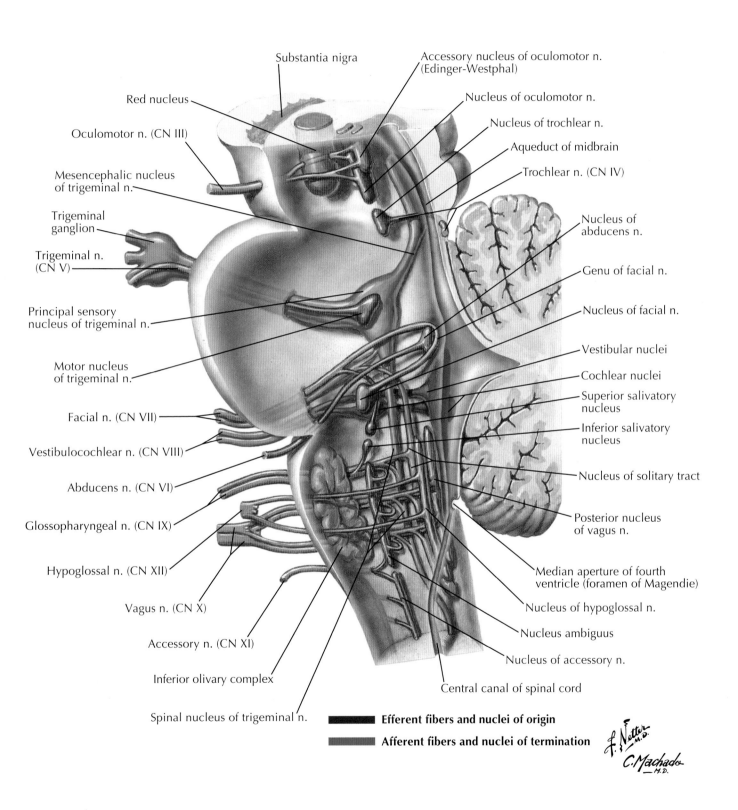

Substantia nigra

Accessory nucleus of oculomotor n. (Edinger-Westphal)

Red nucleus

Nucleus of oculomotor n.

Oculomotor n. (CN III)

Nucleus of trochlear n.

Aqueduct of midbrain

Mesencephalic nucleus of trigeminal n.

Trochlear n. (CN IV)

Trigeminal ganglion

Nucleus of abducens n.

Trigeminal n. (CN V)

Genu of facial n.

Principal sensory nucleus of trigeminal n.

Nucleus of facial n.

Vestibular nuclei

Cochlear nuclei

Motor nucleus of trigeminal n.

Superior salivatory nucleus

Facial n. (CN VII)

Inferior salivatory nucleus

Vestibulocochlear n. (CN VIII)

Nucleus of solitary tract

Abducens n. (CN VI)

Glossopharyngeal n. (CN IX)

Posterior nucleus of vagus n.

Hypoglossal n. (CN XII)

Median aperture of fourth ventricle (foramen of Magendie)

Vagus n. (CN X)

Nucleus of hypoglossal n.

Accessory n. (CN XI)

Nucleus ambiguus

Inferior olivary complex

Nucleus of accessory n.

Central canal of spinal cord

Spinal nucleus of trigeminal n.

Efferent fibers and nuclei of origin

Afferent fibers and nuclei of termination

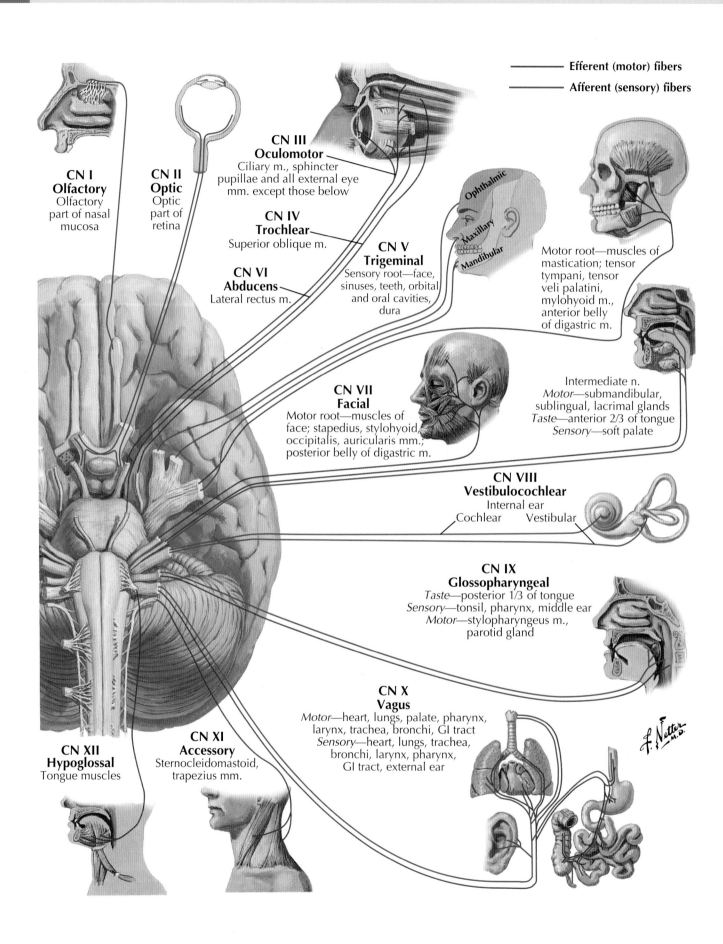

—— Efferent (motor) fibers
—— Afferent (sensory) fibers

CN I
Olfactory
Olfactory
part of nasal
mucosa

CN II
Optic
Optic
part of
retina

CN III
Oculomotor
Ciliary m., sphincter
pupillae and all external eye
mm. except those below

CN IV
Trochlear
Superior oblique m.

CN V
Trigeminal
Sensory root—face,
sinuses, teeth, orbital
and oral cavities,
dura

CN VI
Abducens
Lateral rectus m.

Ophthalmic
Maxillary
Mandibular

Motor root—muscles of
mastication; tensor
tympani, tensor
veli palatini,
mylohyoid m.,
anterior belly
of digastric m.

Intermediate n.
Motor—submandibular,
sublingual, lacrimal glands
Taste—anterior 2/3 of tongue
Sensory—soft palate

CN VII
Facial
Motor root—muscles of
face; stapedius, stylohyoid,
occipitalis, auricularis mm.;
posterior belly of digastric m.

CN VIII
Vestibulocochlear
Internal ear
Cochlear   Vestibular

CN IX
Glossopharyngeal
Taste—posterior 1/3 of tongue
Sensory—tonsil, pharynx, middle ear
Motor—stylopharyngeus m.,
parotid gland

CN X
Vagus
Motor—heart, lungs, palate, pharynx,
larynx, trachea, bronchi, GI tract
Sensory—heart, lungs, trachea,
bronchi, larynx, pharynx,
GI tract, external ear

CN XII
Hypoglossal
Tongue muscles

CN XI
Accessory
Sternocleidomastoid,
trapezius mm.

f. Netter
M.D.

Plate 145   **Cranial and Cervical Nerves**

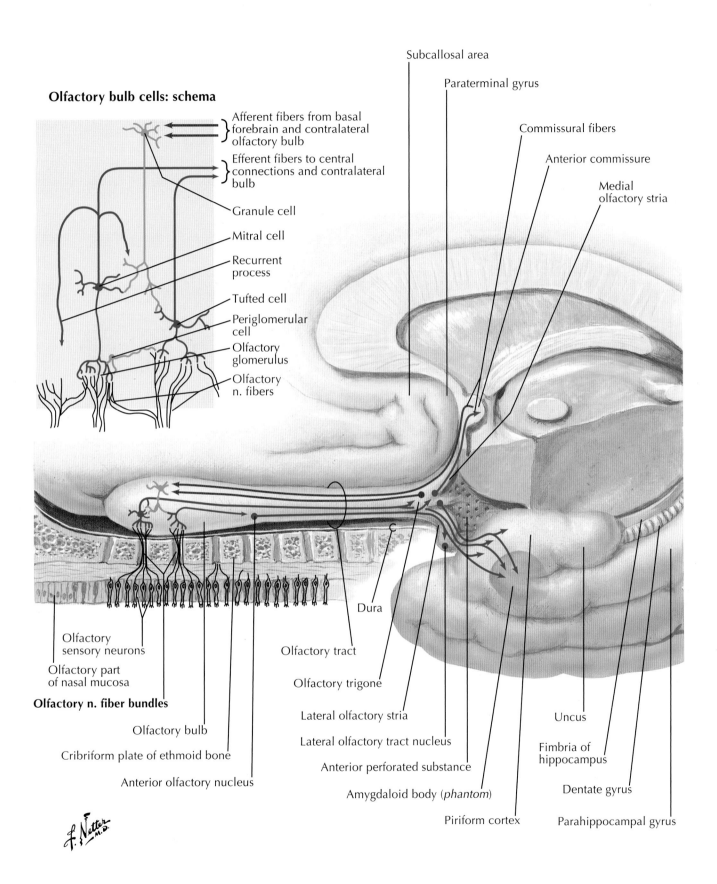

Olfactory bulb cells: schema

Afferent fibers from basal forebrain and contralateral olfactory bulb

Efferent fibers to central connections and contralateral bulb

Granule cell

Mitral cell

Recurrent process

Tufted cell

Periglomerular cell

Olfactory glomerulus

Olfactory n. fibers

Subcallosal area

Paraterminal gyrus

Commissural fibers

Anterior commissure

Medial olfactory stria

Olfactory sensory neurons

Olfactory part of nasal mucosa

Olfactory n. fiber bundles

Olfactory bulb

Cribriform plate of ethmoid bone

Anterior olfactory nucleus

Dura

Olfactory tract

Olfactory trigone

Lateral olfactory stria

Lateral olfactory tract nucleus

Anterior perforated substance

Amygdaloid body (*phantom*)

Piriform cortex

Uncus

Fimbria of hippocampus

Dentate gyrus

Parahippocampal gyrus

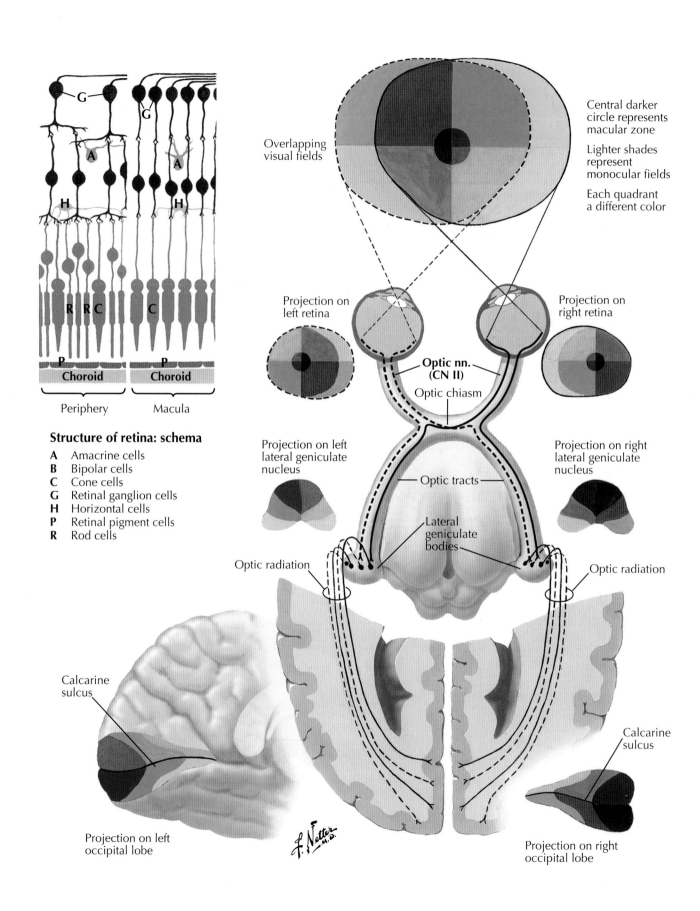

Structure of retina: schema

A   Amacrine cells
B   Bipolar cells
C   Cone cells
G   Retinal ganglion cells
H   Horizontal cells
P   Retinal pigment cells
R   Rod cells

Periphery   Macula

Choroid   Choroid

Overlapping visual fields

Central darker circle represents macular zone

Lighter shades represent monocular fields

Each quadrant a different color

Projection on left retina

Projection on right retina

Optic nn. (CN II)

Optic chiasm

Projection on left lateral geniculate nucleus

Projection on right lateral geniculate nucleus

Optic tracts

Lateral geniculate bodies

Optic radiation

Optic radiation

Calcarine sulcus

Calcarine sulcus

Projection on left occipital lobe

Projection on right occipital lobe

Plate 147

Cranial and Cervical Nerves

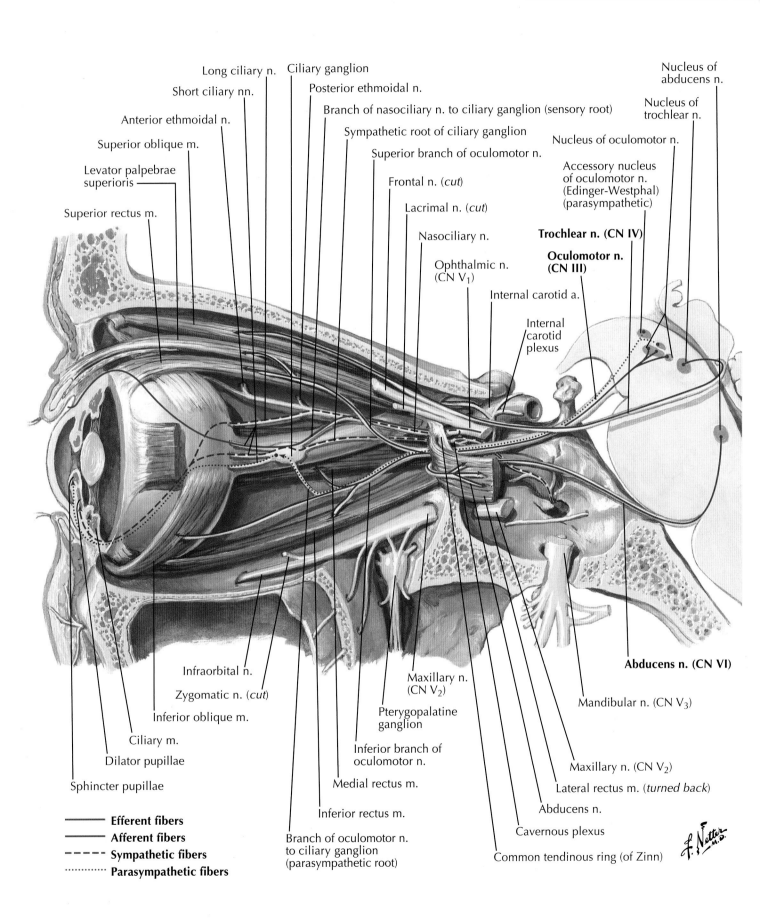

Long ciliary n.

Ciliary ganglion

Short ciliary nn.

Posterior ethmoidal n.

Anterior ethmoidal n.

Branch of nasociliary n. to ciliary ganglion (sensory root)

Superior oblique m.

Sympathetic root of ciliary ganglion

Levator palpebrae superioris

Superior branch of oculomotor n.

Frontal n. (*cut*)

Superior rectus m.

Lacrimal n. (*cut*)

Nasociliary n.

Ophthalmic n. (CN V$_1$)

Nucleus of abducens n.

Nucleus of trochlear n.

Nucleus of oculomotor n.

Accessory nucleus of oculomotor n. (Edinger-Westphal) (parasympathetic)

Trochlear n. (CN IV)

Oculomotor n. (CN III)

Internal carotid a.

Internal carotid plexus

Infraorbital n.

Zygomatic n. (*cut*)

Inferior oblique m.

Maxillary n. (CN V$_2$)

Ciliary m.

Pterygopalatine ganglion

Dilator pupillae

Abducens n. (CN VI)

Mandibular n. (CN V$_3$)

Sphincter pupillae

Inferior branch of oculomotor n.

Maxillary n. (CN V$_2$)

Medial rectus m.

Lateral rectus m. (*turned back*)

Inferior rectus m.

Abducens n.

Branch of oculomotor n. to ciliary ganglion (parasympathetic root)

Cavernous plexus

Common tendinous ring (of Zinn)

——— **Efferent fibers**
——— **Afferent fibers**
– – – **Sympathetic fibers**
·········· **Parasympathetic fibers**

Cranial and Cervical Nerves

Plate 148

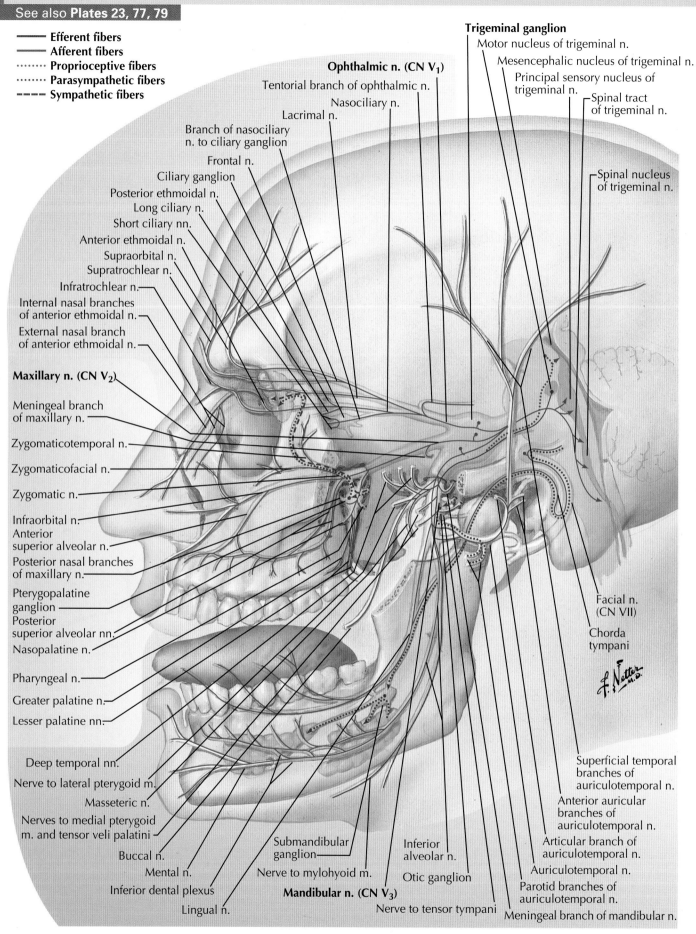

—— Efferent fibers
—— Afferent fibers
······· Proprioceptive fibers
······· Parasympathetic fibers
----- Sympathetic fibers

Trigeminal ganglion
Motor nucleus of trigeminal n.
Mesencephalic nucleus of trigeminal n.
Principal sensory nucleus of trigeminal n.
Spinal tract of trigeminal n.

Ophthalmic n. (CN V₁)
Tentorial branch of ophthalmic n.
Nasociliary n.
Lacrimal n.
Branch of nasociliary n. to ciliary ganglion
Frontal n.
Ciliary ganglion
Posterior ethmoidal n.
Long ciliary n.
Short ciliary nn.
Anterior ethmoidal n.
Supraorbital n.
Supratrochlear n.
Infratrochlear n.
Internal nasal branches of anterior ethmoidal n.
External nasal branch of anterior ethmoidal n.

Spinal nucleus of trigeminal n.

Maxillary n. (CN V₂)

Meningeal branch of maxillary n.
Zygomaticotemporal n.
Zygomaticofacial n.
Zygomatic n.
Infraorbital n.
Anterior superior alveolar n.
Posterior nasal branches of maxillary n.
Pterygopalatine ganglion
Posterior superior alveolar nn.
Nasopalatine n.
Pharyngeal n.
Greater palatine n.
Lesser palatine nn.

Facial n. (CN VII)
Chorda tympani

Deep temporal nn.
Nerve to lateral pterygoid m.
Masseteric n.
Nerves to medial pterygoid m. and tensor veli palatini
Buccal n.
Mental n.
Inferior dental plexus
Lingual n.

Submandibular ganglion
Nerve to mylohyoid m.
Mandibular n. (CN V₃)

Inferior alveolar n.
Otic ganglion
Nerve to tensor tympani

Superficial temporal branches of auriculotemporal n.
Anterior auricular branches of auriculotemporal n.
Articular branch of auriculotemporal n.
Auriculotemporal n.
Parotid branches of auriculotemporal n.
Meningeal branch of mandibular n.

Plate 149

Cranial and Cervical Nerves

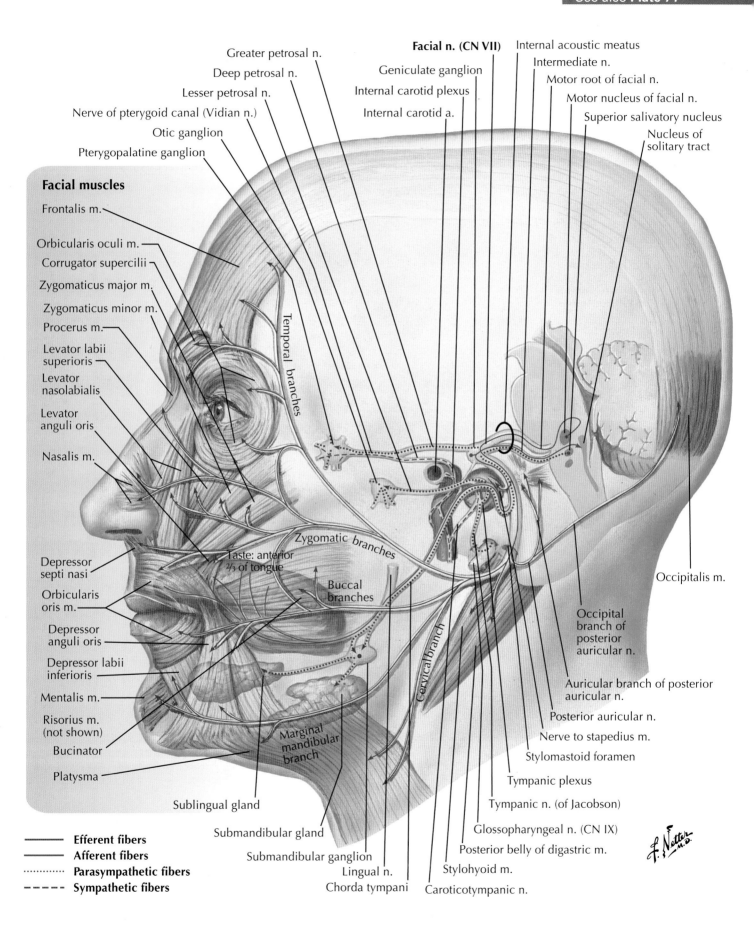

Greater petrosal n.
Deep petrosal n.
Lesser petrosal n.
Nerve of pterygoid canal (Vidian n.)
Otic ganglion
Pterygopalatine ganglion

Facial n. (CN VII)
Geniculate ganglion
Internal carotid plexus
Internal carotid a.

Internal acoustic meatus
Intermediate n.
Motor root of facial n.
Motor nucleus of facial n.
Superior salivatory nucleus
Nucleus of solitary tract

Facial muscles
Frontalis m.
Orbicularis oculi m.
Corrugator supercilii
Zygomaticus major m.
Zygomaticus minor m.
Procerus m.
Levator labii superioris
Levator nasolabialis
Levator anguli oris
Nasalis m.
Depressor septi nasi
Orbicularis oris m.
Depressor anguli oris
Depressor labii inferioris
Mentalis m.
Risorius m. (not shown)
Bucinator
Platysma

Temporal branches

Zygomatic branches

Taste: anterior ⅔ of tongue

Buccal branches

Cervical branch

Occipitalis m.
Occipital branch of posterior auricular n.
Auricular branch of posterior auricular n.
Posterior auricular n.
Nerve to stapedius m.
Stylomastoid foramen
Tympanic plexus
Tympanic n. (of Jacobson)
Glossopharyngeal n. (CN IX)
Posterior belly of digastric m.
Stylohyoid m.
Caroticotympanic n.

Marginal mandibular branch

Sublingual gland
Submandibular gland
Submandibular ganglion
Lingual n.
Chorda tympani

—— Efferent fibers
—— Afferent fibers
·········· Parasympathetic fibers
– – – – Sympathetic fibers

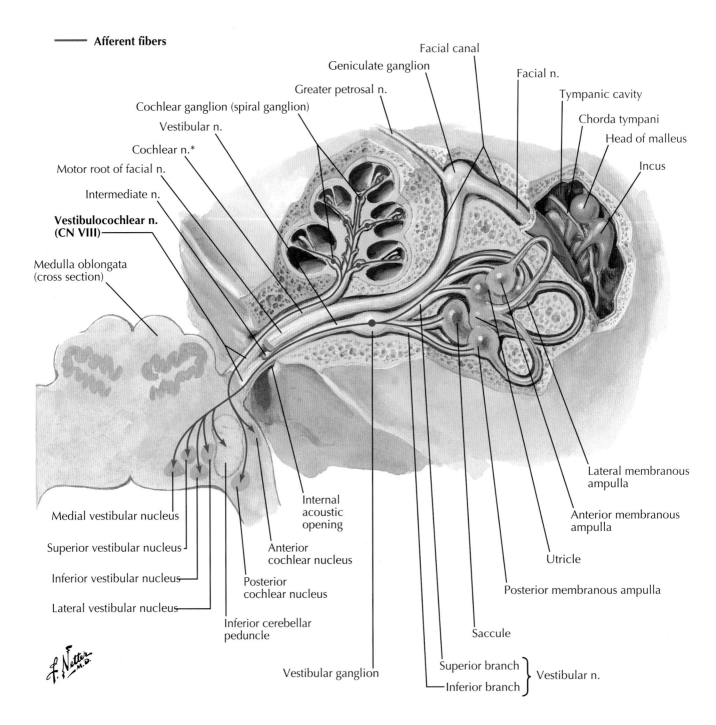

—— **Afferent fibers**

Facial canal

Geniculate ganglion

Greater petrosal n.

Facial n.

Tympanic cavity

Cochlear ganglion (spiral ganglion)

Chorda tympani

Vestibular n.

Head of malleus

Cochlear n.*

Incus

Motor root of facial n.

Intermediate n.

Vestibulocochlear n. (CN VIII)

Medulla oblongata (cross section)

Lateral membranous ampulla

Anterior membranous ampulla

Internal acoustic opening

Medial vestibular nucleus

Superior vestibular nucleus

Anterior cochlear nucleus

Utricle

Inferior vestibular nucleus

Posterior cochlear nucleus

Posterior membranous ampulla

Lateral vestibular nucleus

Inferior cerebellar peduncle

Saccule

Vestibular ganglion

Superior branch

Inferior branch

} Vestibular n.

Note: The cochlear nerve also contains efferent fibers to the sensory epithelium. These fibers are derived from the vestibular nerve while in the internal acoustic meatus.

Plate 151

Cranial and Cervical Nerves

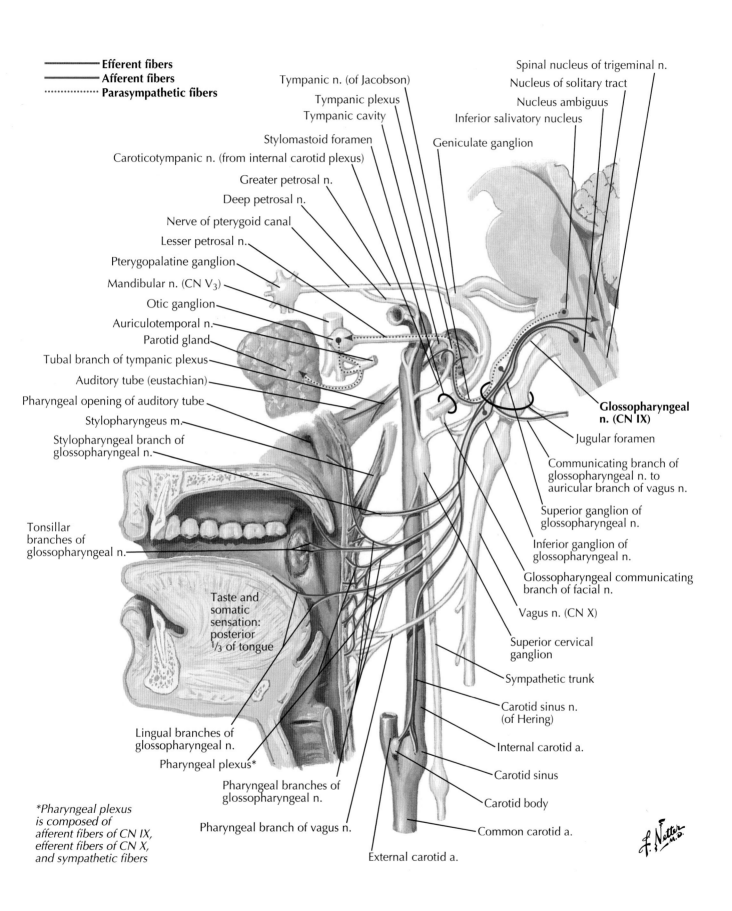

Efferent fibers
Afferent fibers
Parasympathetic fibers

Tympanic n. (of Jacobson)
Tympanic plexus
Tympanic cavity
Stylomastoid foramen
Caroticotympanic n. (from internal carotid plexus)
Greater petrosal n.
Deep petrosal n.
Nerve of pterygoid canal
Lesser petrosal n.
Pterygopalatine ganglion
Mandibular n. (CN V₃)
Otic ganglion
Auriculotemporal n.
Parotid gland
Tubal branch of tympanic plexus
Auditory tube (eustachian)
Pharyngeal opening of auditory tube
Stylopharyngeus m.
Stylopharyngeal branch of glossopharyngeal n.

Tonsillar branches of glossopharyngeal n.

Taste and somatic sensation: posterior ¹/₃ of tongue

Lingual branches of glossopharyngeal n.
Pharyngeal plexus*
Pharyngeal branches of glossopharyngeal n.
Pharyngeal branch of vagus n.

*Pharyngeal plexus is composed of afferent fibers of CN IX, efferent fibers of CN X, and sympathetic fibers

External carotid a.

Spinal nucleus of trigeminal n.
Nucleus of solitary tract
Nucleus ambiguus
Inferior salivatory nucleus
Geniculate ganglion

Glossopharyngeal n. (CN IX)
Jugular foramen
Communicating branch of glossopharyngeal n. to auricular branch of vagus n.
Superior ganglion of glossopharyngeal n.
Inferior ganglion of glossopharyngeal n.
Glossopharyngeal communicating branch of facial n.
Vagus n. (CN X)
Superior cervical ganglion
Sympathetic trunk
Carotid sinus n. (of Hering)
Internal carotid a.
Carotid sinus
Carotid body
Common carotid a.

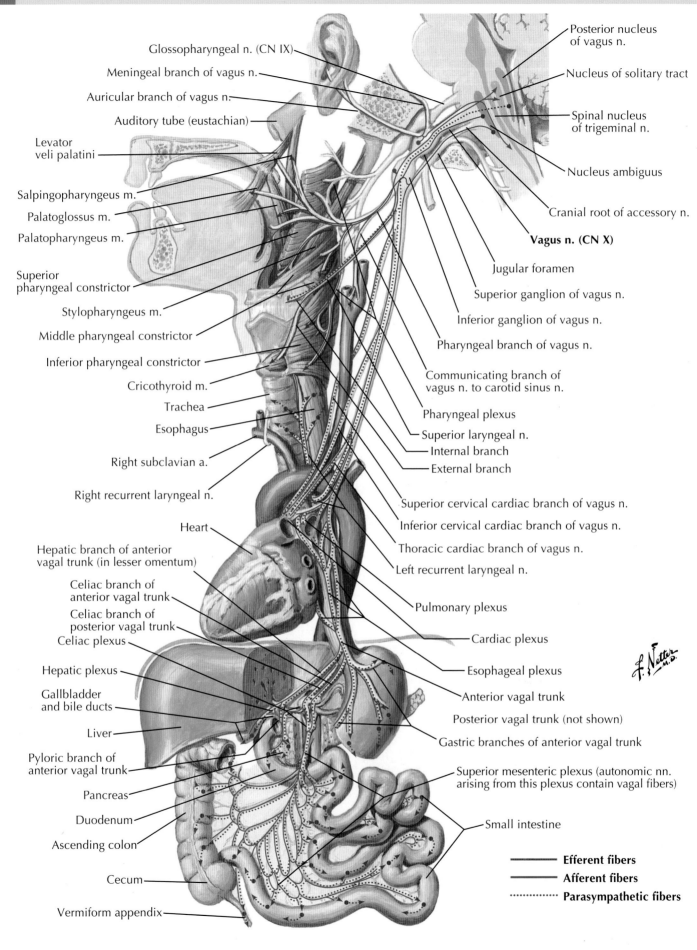

Glossopharyngeal n. (CN IX)

Meningeal branch of vagus n.

Auricular branch of vagus n.

Auditory tube (eustachian)

Levator veli palatini

Salpingopharyngeus m.

Palatoglossus m.

Palatopharyngeus m.

Superior pharyngeal constrictor

Stylopharyngeus m.

Middle pharyngeal constrictor

Inferior pharyngeal constrictor

Cricothyroid m.

Trachea

Esophagus

Right subclavian a.

Right recurrent laryngeal n.

Heart

Hepatic branch of anterior vagal trunk (in lesser omentum)

Celiac branch of anterior vagal trunk

Celiac branch of posterior vagal trunk

Celiac plexus

Hepatic plexus

Gallbladder and bile ducts

Liver

Pyloric branch of anterior vagal trunk

Pancreas

Duodenum

Ascending colon

Cecum

Vermiform appendix

Posterior nucleus of vagus n.

Nucleus of solitary tract

Spinal nucleus of trigeminal n.

Nucleus ambiguus

Cranial root of accessory n.

Vagus n. (CN X)

Jugular foramen

Superior ganglion of vagus n.

Inferior ganglion of vagus n.

Pharyngeal branch of vagus n.

Communicating branch of vagus n. to carotid sinus n.

Pharyngeal plexus

Superior laryngeal n.

Internal branch

External branch

Superior cervical cardiac branch of vagus n.

Inferior cervical cardiac branch of vagus n.

Thoracic cardiac branch of vagus n.

Left recurrent laryngeal n.

Pulmonary plexus

Cardiac plexus

Esophageal plexus

Anterior vagal trunk

Posterior vagal trunk (not shown)

Gastric branches of anterior vagal trunk

Superior mesenteric plexus (autonomic nn. arising from this plexus contain vagal fibers)

Small intestine

—————— **Efferent fibers**

—————— **Afferent fibers**

·············· **Parasympathetic fibers**

Plate 153

Cranial and Cervical Nerves

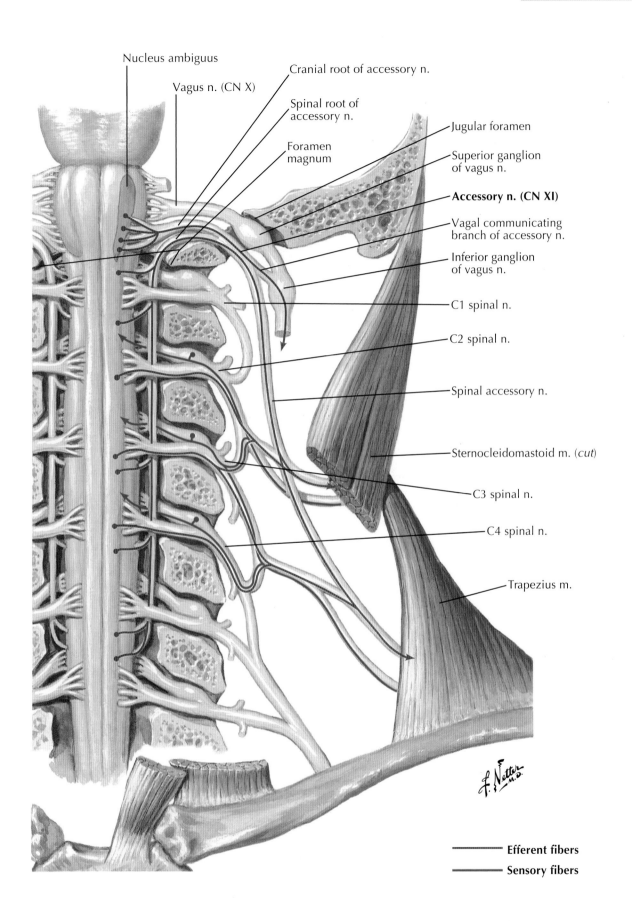

Nucleus ambiguus

Vagus n. (CN X)

Cranial root of accessory n.

Spinal root of accessory n.

Foramen magnum

Jugular foramen

Superior ganglion of vagus n.

Accessory n. (CN XI)

Vagal communicating branch of accessory n.

Inferior ganglion of vagus n.

C1 spinal n.

C2 spinal n.

Spinal accessory n.

Sternocleidomastoid m. (*cut*)

C3 spinal n.

C4 spinal n.

Trapezius m.

—— **Efferent fibers**

—— **Sensory fibers**

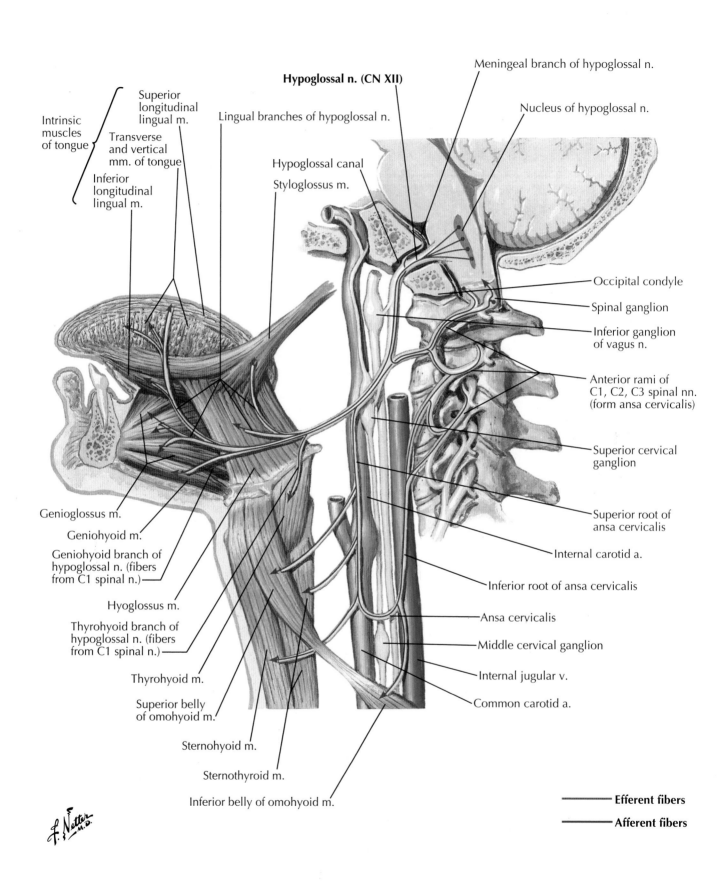

Superior
longitudinal
lingual m.

Intrinsic
muscles
of tongue

Transverse
and vertical
mm. of tongue

Inferior
longitudinal
lingual m.

Lingual branches of hypoglossal n.

Hypoglossal n. (CN XII)

Meningeal branch of hypoglossal n.

Nucleus of hypoglossal n.

Hypoglossal canal

Styloglossus m.

Occipital condyle

Spinal ganglion

Inferior ganglion
of vagus n.

Anterior rami of
C1, C2, C3 spinal nn.
(form ansa cervicalis)

Superior cervical
ganglion

Superior root of
ansa cervicalis

Internal carotid a.

Inferior root of ansa cervicalis

Ansa cervicalis

Middle cervical ganglion

Internal jugular v.

Common carotid a.

Genioglossus m.

Geniohyoid m.

Geniohyoid branch of
hypoglossal n. (fibers
from C1 spinal n.)

Hyoglossus m.

Thyrohyoid branch of
hypoglossal n. (fibers
from C1 spinal n.)

Thyrohyoid m.

Superior belly
of omohyoid m.

Sternohyoid m.

Sternothyroid m.

Inferior belly of omohyoid m.

———— **Efferent fibers**

———— **Afferent fibers**

Plate 155

Cranial and Cervical Nerves

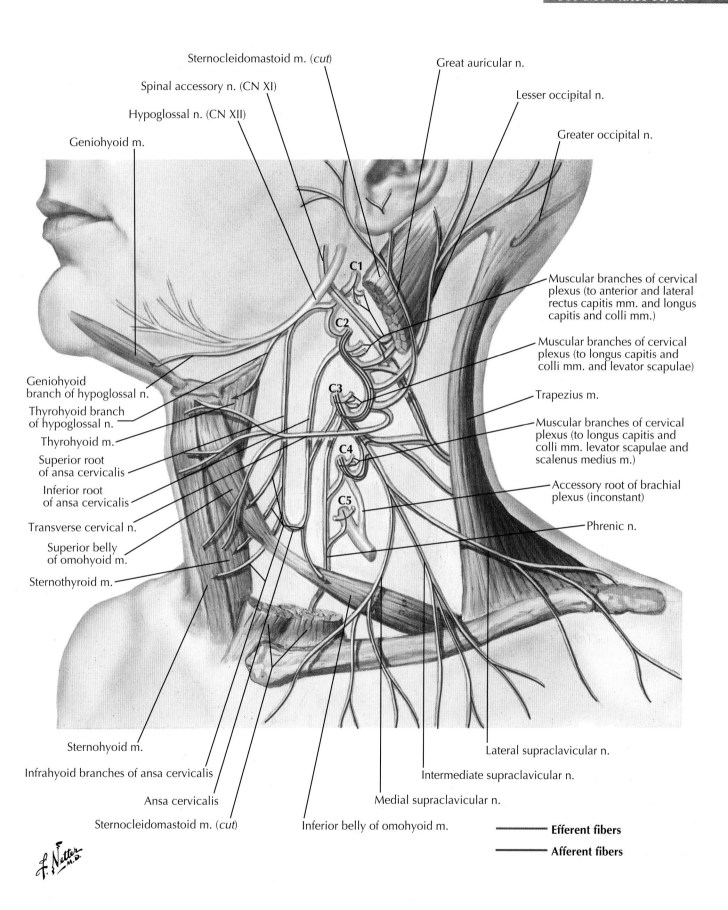

Sternocleidomastoid m. (*cut*)

Spinal accessory n. (CN XI)

Hypoglossal n. (CN XII)

Geniohyoid m.

Great auricular n.

Lesser occipital n.

Greater occipital n.

C1

C2

C3

C4

C5

Muscular branches of cervical plexus (to anterior and lateral rectus capitis mm. and longus capitis and colli mm.)

Muscular branches of cervical plexus (to longus capitis and colli mm. and levator scapulae)

Trapezius m.

Muscular branches of cervical plexus (to longus capitis and colli mm. levator scapulae and scalenus medius m.)

Accessory root of brachial plexus (inconstant)

Phrenic n.

Geniohyoid branch of hypoglossal n.

Thyrohyoid branch of hypoglossal n.

Thyrohyoid m.

Superior root of ansa cervicalis

Inferior root of ansa cervicalis

Transverse cervical n.

Superior belly of omohyoid m.

Sternothyroid m.

Sternohyoid m.

Infrahyoid branches of ansa cervicalis

Ansa cervicalis

Sternocleidomastoid m. (*cut*)

Inferior belly of omohyoid m.

Medial supraclavicular n.

Intermediate supraclavicular n.

Lateral supraclavicular n.

———— **Efferent fibers**

———— **Afferent fibers**

F. Netter M.D.

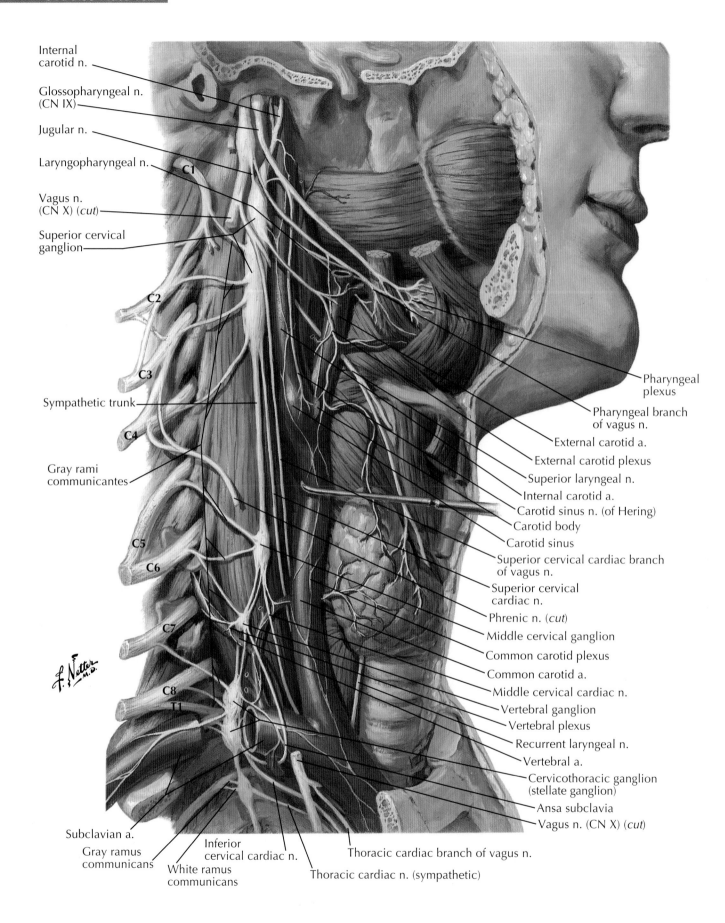

Internal carotid n.

Glossopharyngeal n. (CN IX)

Jugular n.

Laryngopharyngeal n.

Vagus n. (CN X) (cut)

Superior cervical ganglion

C1

C2

C3

Sympathetic trunk

C4

Gray rami communicantes

C5

C6

C7

C8

T1

Subclavian a.

Gray ramus communicans

Inferior cervical cardiac n.

White ramus communicans

Thoracic cardiac n. (sympathetic)

Thoracic cardiac branch of vagus n.

Pharyngeal plexus

Pharyngeal branch of vagus n.

External carotid a.

External carotid plexus

Superior laryngeal n.

Internal carotid a.

Carotid sinus n. (of Hering)

Carotid body

Carotid sinus

Superior cervical cardiac branch of vagus n.

Superior cervical cardiac n.

Phrenic n. (cut)

Middle cervical ganglion

Common carotid plexus

Common carotid a.

Middle cervical cardiac n.

Vertebral ganglion

Vertebral plexus

Recurrent laryngeal n.

Vertebral a.

Cervicothoracic ganglion (stellate ganglion)

Ansa subclavia

Vagus n. (CN X) (cut)

Plate 157

Cranial and Cervical Nerves

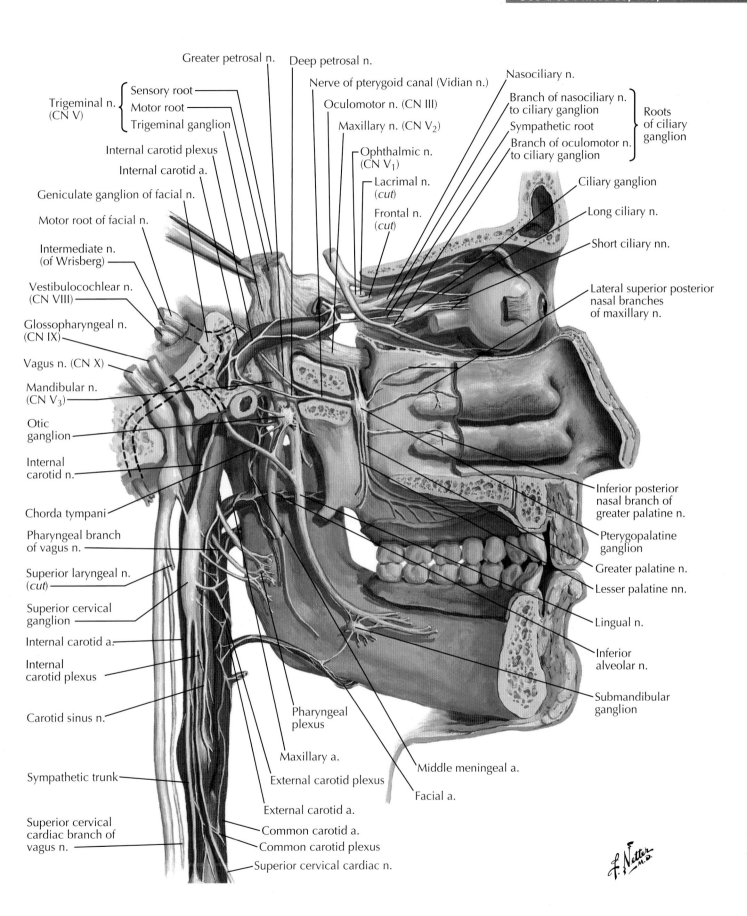

Greater petrosal n.

Deep petrosal n.

Nerve of pterygoid canal (Vidian n.)

Nasociliary n.

Trigeminal n. (CN V) { Sensory root — Motor root — Trigeminal ganglion }

Oculomotor n. (CN III)

Branch of nasociliary n. to ciliary ganglion }
Sympathetic root
Branch of oculomotor n. to ciliary ganglion } Roots of ciliary ganglion

Maxillary n. (CN V₂)

Internal carotid plexus

Internal carotid a.

Ophthalmic n. (CN V₁)

Ciliary ganglion

Lacrimal n. (cut)

Long ciliary n.

Geniculate ganglion of facial n.

Frontal n. (cut)

Short ciliary nn.

Motor root of facial n.

Intermediate n. (of Wrisberg)

Lateral superior posterior nasal branches of maxillary n.

Vestibulocochlear n. (CN VIII)

Glossopharyngeal n. (CN IX)

Vagus n. (CN X)

Mandibular n. (CN V₃)

Otic ganglion

Internal carotid n.

Inferior posterior nasal branch of greater palatine n.

Chorda tympani

Pterygopalatine ganglion

Pharyngeal branch of vagus n.

Greater palatine n.

Superior laryngeal n. (cut)

Lesser palatine nn.

Superior cervical ganglion

Lingual n.

Internal carotid a.

Internal carotid plexus

Inferior alveolar n.

Carotid sinus n.

Submandibular ganglion

Pharyngeal plexus

Sympathetic trunk

Maxillary a.

Middle meningeal a.

External carotid plexus

Facial a.

External carotid a.

Superior cervical cardiac branch of vagus n.

Common carotid a.

Common carotid plexus

Superior cervical cardiac n.

f. Netter M.D.

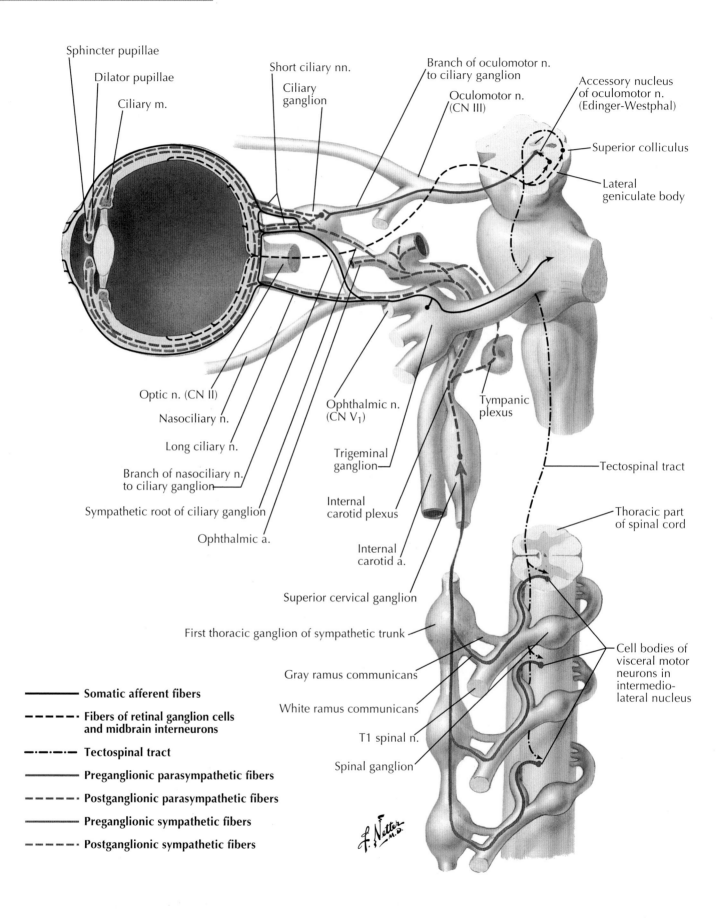

Sphincter pupillae

Dilator pupillae

Ciliary m.

Short ciliary nn.

Ciliary ganglion

Branch of oculomotor n. to ciliary ganglion

Oculomotor n. (CN III)

Accessory nucleus of oculomotor n. (Edinger-Westphal)

Superior colliculus

Lateral geniculate body

Optic n. (CN II)

Nasociliary n.

Long ciliary n.

Branch of nasociliary n. to ciliary ganglion

Sympathetic root of ciliary ganglion

Ophthalmic a.

Ophthalmic n. (CN V₁)

Trigeminal ganglion

Internal carotid plexus

Internal carotid a.

Superior cervical ganglion

First thoracic ganglion of sympathetic trunk

Gray ramus communicans

White ramus communicans

T1 spinal n.

Spinal ganglion

Tympanic plexus

Tectospinal tract

Thoracic part of spinal cord

Cell bodies of visceral motor neurons in intermediolateral nucleus

—————— Somatic afferent fibers

- - - - - Fibers of retinal ganglion cells and midbrain interneurons

—·—·— Tectospinal tract

—————— Preganglionic parasympathetic fibers

- - - - - Postganglionic parasympathetic fibers

—————— Preganglionic sympathetic fibers

- - - - - Postganglionic sympathetic fibers

Plate 159

Cranial and Cervical Nerves

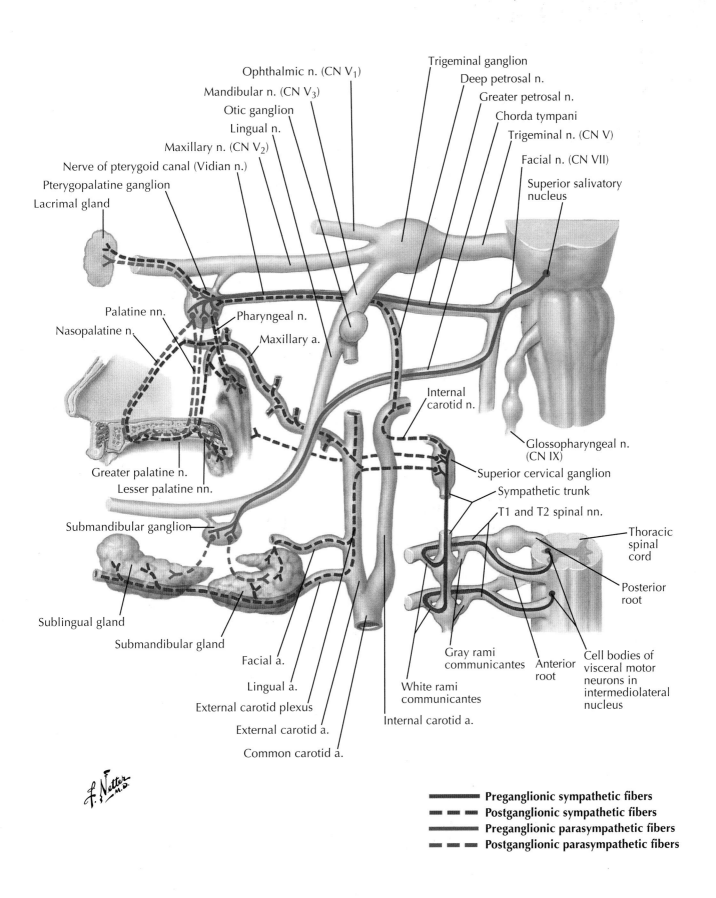

Ophthalmic n. (CN V$_1$)

Mandibular n. (CN V$_3$)

Otic ganglion

Lingual n.

Maxillary n. (CN V$_2$)

Nerve of pterygoid canal (Vidian n.)

Pterygopalatine ganglion

Lacrimal gland

Trigeminal ganglion

Deep petrosal n.

Greater petrosal n.

Chorda tympani

Trigeminal n. (CN V)

Facial n. (CN VII)

Superior salivatory nucleus

Palatine nn.

Nasopalatine n.

Pharyngeal n.

Maxillary a.

Greater palatine n.

Lesser palatine nn.

Internal carotid n.

Glossopharyngeal n. (CN IX)

Superior cervical ganglion

Sympathetic trunk

T1 and T2 spinal nn.

Submandibular ganglion

Thoracic spinal cord

Posterior root

Sublingual gland

Submandibular gland

Facial a.

Lingual a.

External carotid plexus

External carotid a.

Common carotid a.

Gray rami communicantes

Anterior root

White rami communicantes

Internal carotid a.

Cell bodies of visceral motor neurons in intermediolateral nucleus

━━━━━ **Preganglionic sympathetic fibers**

▬ ▬ ▬ **Postganglionic sympathetic fibers**

━━━━━ **Preganglionic parasympathetic fibers**

▬ ▬ ▬ **Postganglionic parasympathetic fibers**

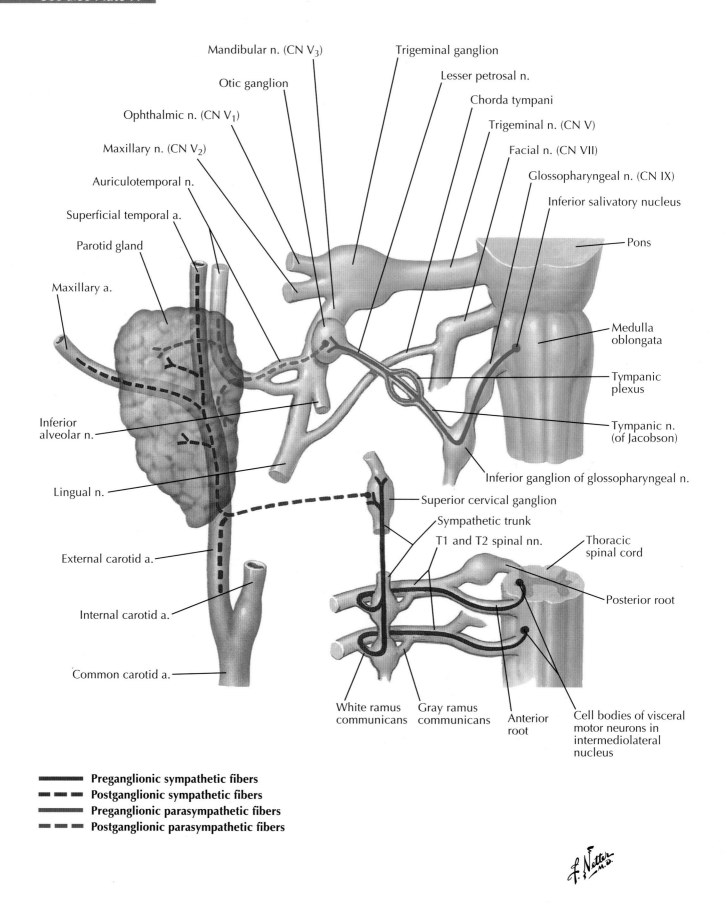

Mandibular n. (CN V₃)

Trigeminal ganglion

Otic ganglion

Lesser petrosal n.

Chorda tympani

Ophthalmic n. (CN V₁)

Trigeminal n. (CN V)

Maxillary n. (CN V₂)

Facial n. (CN VII)

Auriculotemporal n.

Glossopharyngeal n. (CN IX)

Superficial temporal a.

Inferior salivatory nucleus

Parotid gland

Pons

Maxillary a.

Medulla oblongata

Tympanic plexus

Inferior alveolar n.

Tympanic n. (of Jacobson)

Lingual n.

Inferior ganglion of glossopharyngeal n.

Superior cervical ganglion

Sympathetic trunk

T1 and T2 spinal nn.

Thoracic spinal cord

External carotid a.

Posterior root

Internal carotid a.

Common carotid a.

White ramus communicans

Gray ramus communicans

Anterior root

Cell bodies of visceral motor neurons in intermediolateral nucleus

━━━━ Preganglionic sympathetic fibers
▬ ▬ ▬ Postganglionic sympathetic fibers
━━━━ Preganglionic parasympathetic fibers
▬ ▬ ▬ Postganglionic parasympathetic fibers

Plate 161

Cranial and Cervical Nerves

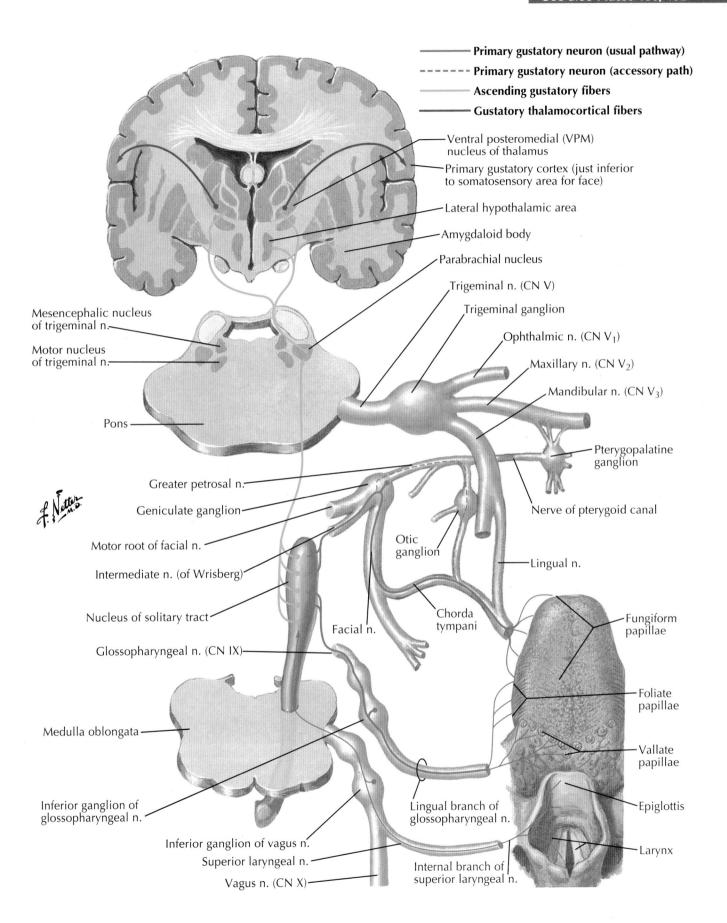

Primary gustatory neuron (usual pathway)
Primary gustatory neuron (accessory path)
Ascending gustatory fibers
Gustatory thalamocortical fibers

Ventral posteromedial (VPM) nucleus of thalamus

Primary gustatory cortex (just inferior to somatosensory area for face)

Lateral hypothalamic area

Amygdaloid body

Parabrachial nucleus

Trigeminal n. (CN V)

Trigeminal ganglion

Ophthalmic n. (CN V₁)

Maxillary n. (CN V₂)

Mandibular n. (CN V₃)

Mesencephalic nucleus of trigeminal n.

Motor nucleus of trigeminal n.

Pons

Pterygopalatine ganglion

Nerve of pterygoid canal

Greater petrosal n.

Geniculate ganglion

Motor root of facial n.

Intermediate n. (of Wrisberg)

Nucleus of solitary tract

Glossopharyngeal n. (CN IX)

Otic ganglion

Chorda tympani

Facial n.

Lingual n.

Fungiform papillae

Foliate papillae

Vallate papillae

Medulla oblongata

Inferior ganglion of glossopharyngeal n.

Lingual branch of glossopharyngeal n.

Epiglottis

Inferior ganglion of vagus n.

Superior laryngeal n.

Vagus n. (CN X)

Internal branch of superior laryngeal n.

Larynx

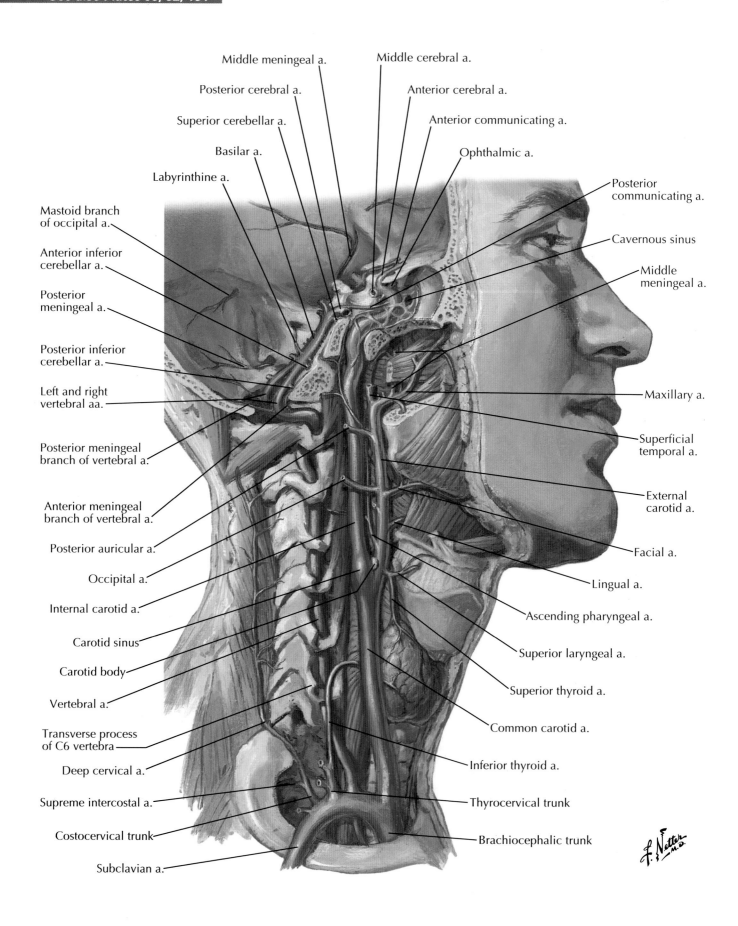

Middle meningeal a.

Middle cerebral a.

Posterior cerebral a.

Anterior cerebral a.

Superior cerebellar a.

Anterior communicating a.

Basilar a.

Ophthalmic a.

Labyrinthine a.

Posterior communicating a.

Mastoid branch of occipital a.

Cavernous sinus

Anterior inferior cerebellar a.

Middle meningeal a.

Posterior meningeal a.

Posterior inferior cerebellar a.

Left and right vertebral aa.

Maxillary a.

Posterior meningeal branch of vertebral a.

Superficial temporal a.

Anterior meningeal branch of vertebral a.

External carotid a.

Posterior auricular a.

Facial a.

Occipital a.

Lingual a.

Internal carotid a.

Ascending pharyngeal a.

Carotid sinus

Superior laryngeal a.

Carotid body

Superior thyroid a.

Vertebral a.

Common carotid a.

Transverse process of C6 vertebra

Inferior thyroid a.

Deep cervical a.

Supreme intercostal a.

Thyrocervical trunk

Costocervical trunk

Brachiocephalic trunk

Subclavian a.

Plate 163

Cerebral Vasculature

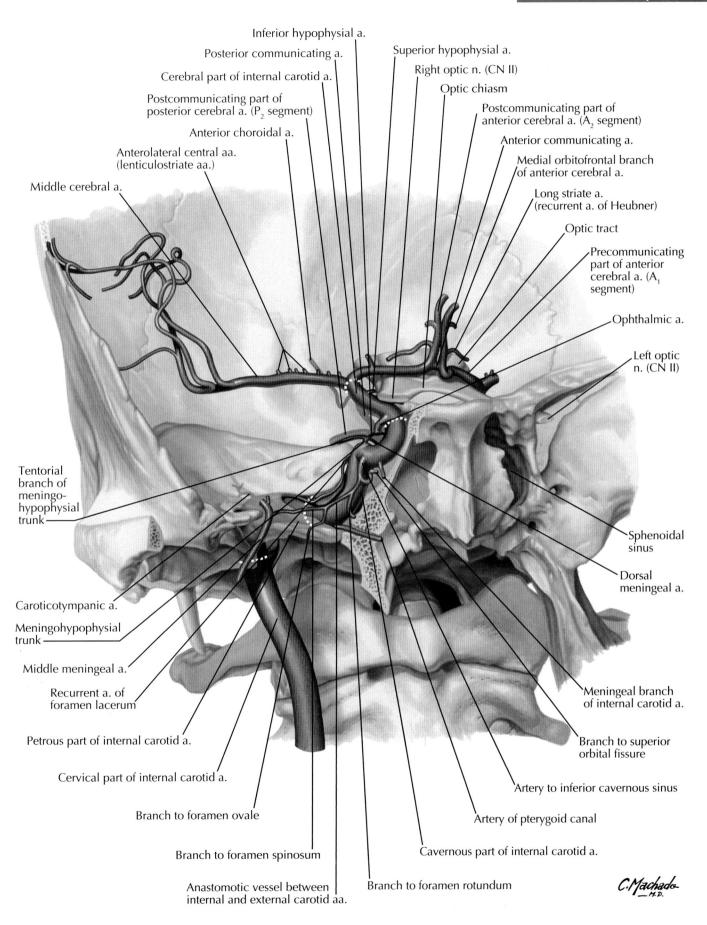

Inferior hypophysial a.

Posterior communicating a.

Cerebral part of internal carotid a.

Postcommunicating part of
posterior cerebral a. (P₂ segment)

Anterior choroidal a.

Anterolateral central aa.
(lenticulostriate aa.)

Middle cerebral a.

Superior hypophysial a.

Right optic n. (CN II)

Optic chiasm

Postcommunicating part of
anterior cerebral a. (A₂ segment)

Anterior communicating a.

Medial orbitofrontal branch
of anterior cerebral a.

Long striate a.
(recurrent a. of Heubner)

Optic tract

Precommunicating
part of anterior
cerebral a. (A₁
segment)

Ophthalmic a.

Left optic
n. (CN II)

Tentorial
branch of
meningo-
hypophysial
trunk

Caroticotympanic a.

Meningohypophysial
trunk

Middle meningeal a.

Recurrent a. of
foramen lacerum

Petrous part of internal carotid a.

Cervical part of internal carotid a.

Branch to foramen ovale

Branch to foramen spinosum

Anastomotic vessel between
internal and external carotid aa.

Branch to foramen rotundum

Cavernous part of internal carotid a.

Artery of pterygoid canal

Artery to inferior cavernous sinus

Branch to superior
orbital fissure

Meningeal branch
of internal carotid a.

Dorsal
meningeal a.

Sphenoidal
sinus

C. Machado
—M.D.

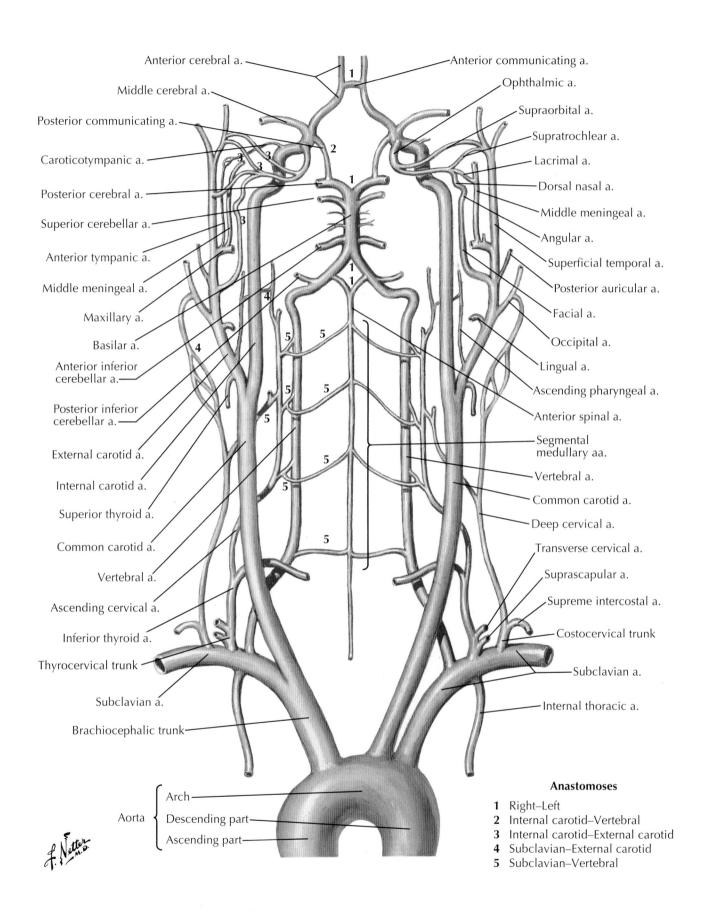

Anterior cerebral a.

Middle cerebral a.

Posterior communicating a.

Caroticotympanic a.

Posterior cerebral a.

Superior cerebellar a.

Anterior tympanic a.

Middle meningeal a.

Maxillary a.

Basilar a.

Anterior inferior
cerebellar a.

Posterior inferior
cerebellar a.

External carotid a.

Internal carotid a.

Superior thyroid a.

Common carotid a.

Vertebral a.

Ascending cervical a.

Inferior thyroid a.

Thyrocervical trunk

Subclavian a.

Brachiocephalic trunk

Anterior communicating a.

Ophthalmic a.

Supraorbital a.

Supratrochlear a.

Lacrimal a.

Dorsal nasal a.

Middle meningeal a.

Angular a.

Superficial temporal a.

Posterior auricular a.

Facial a.

Occipital a.

Lingual a.

Ascending pharyngeal a.

Anterior spinal a.

Segmental
medullary aa.

Vertebral a.

Common carotid a.

Deep cervical a.

Transverse cervical a.

Suprascapular a.

Supreme intercostal a.

Costocervical trunk

Subclavian a.

Internal thoracic a.

Aorta
Arch
Descending part
Ascending part

Anastomoses

1 Right–Left
2 Internal carotid–Vertebral
3 Internal carotid–External carotid
4 Subclavian–External carotid
5 Subclavian–Vertebral

Plate 165   **Cerebral Vasculature**

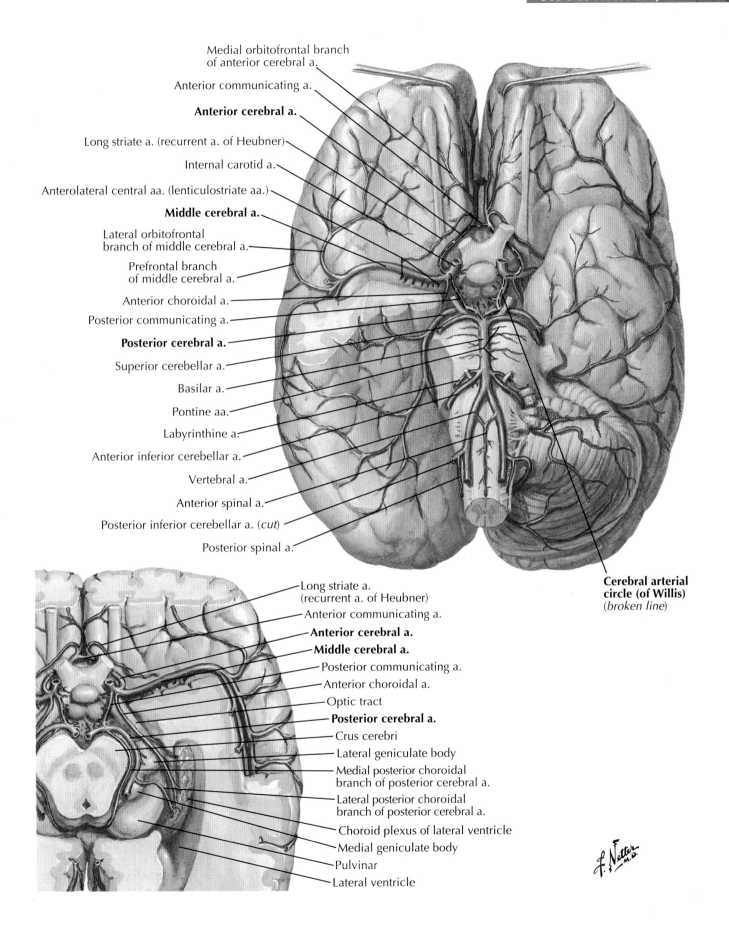

Medial orbitofrontal branch of anterior cerebral a.

Anterior communicating a.

Anterior cerebral a.

Long striate a. (recurrent a. of Heubner)

Internal carotid a.

Anterolateral central aa. (lenticulostriate aa.)

Middle cerebral a.

Lateral orbitofrontal branch of middle cerebral a.

Prefrontal branch of middle cerebral a.

Anterior choroidal a.

Posterior communicating a.

Posterior cerebral a.

Superior cerebellar a.

Basilar a.

Pontine aa.

Labyrinthine a.

Anterior inferior cerebellar a.

Vertebral a.

Anterior spinal a.

Posterior inferior cerebellar a. (*cut*)

Posterior spinal a.

Cerebral arterial circle (of Willis) (*broken line*)

Long striate a. (recurrent a. of Heubner)

Anterior communicating a.

Anterior cerebral a.

Middle cerebral a.

Posterior communicating a.

Anterior choroidal a.

Optic tract

Posterior cerebral a.

Crus cerebri

Lateral geniculate body

Medial posterior choroidal branch of posterior cerebral a.

Lateral posterior choroidal branch of posterior cerebral a.

Choroid plexus of lateral ventricle

Medial geniculate body

Pulvinar

Lateral ventricle

Vessels dissected out: inferior view

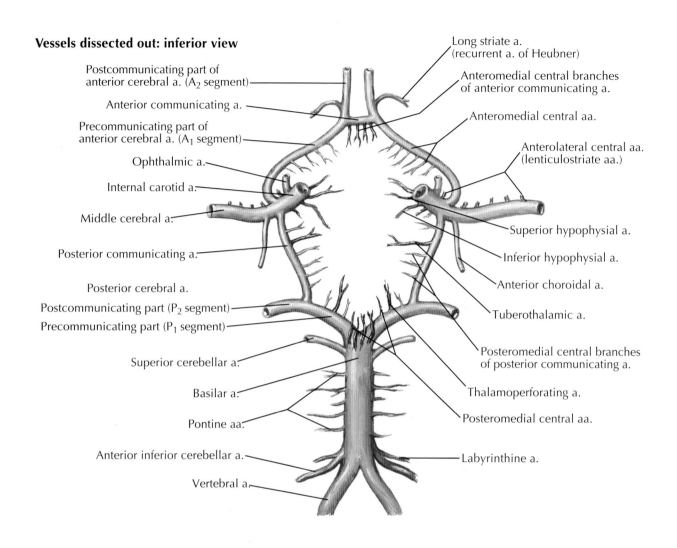

Postcommunicating part of anterior cerebral a. (A_2 segment)

Anterior communicating a.

Precommunicating part of anterior cerebral a. (A_1 segment)

Ophthalmic a.

Internal carotid a.

Middle cerebral a.

Posterior communicating a.

Posterior cerebral a.

Postcommunicating part (P_2 segment)

Precommunicating part (P_1 segment)

Superior cerebellar a.

Basilar a.

Pontine aa.

Anterior inferior cerebellar a.

Vertebral a.

Long striate a. (recurrent a. of Heubner)

Anteromedial central branches of anterior communicating a.

Anteromedial central aa.

Anterolateral central aa. (lenticulostriate aa.)

Superior hypophysial a.

Inferior hypophysial a.

Anterior choroidal a.

Tuberothalamic a.

Posteromedial central branches of posterior communicating a.

Thalamoperforating a.

Posteromedial central aa.

Labyrinthine a.

Vessels in situ: inferior view

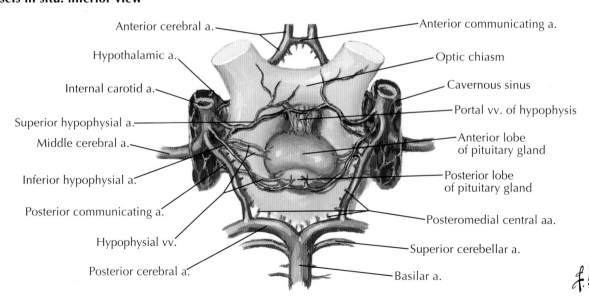

Anterior cerebral a.

Hypothalamic a.

Internal carotid a.

Superior hypophysial a.

Middle cerebral a.

Inferior hypophysial a.

Posterior communicating a.

Hypophysial vv.

Posterior cerebral a.

Anterior communicating a.

Optic chiasm

Cavernous sinus

Portal vv. of hypophysis

Anterior lobe of pituitary gland

Posterior lobe of pituitary gland

Posteromedial central aa.

Superior cerebellar a.

Basilar a.

Plate 167

Cerebral Vasculature

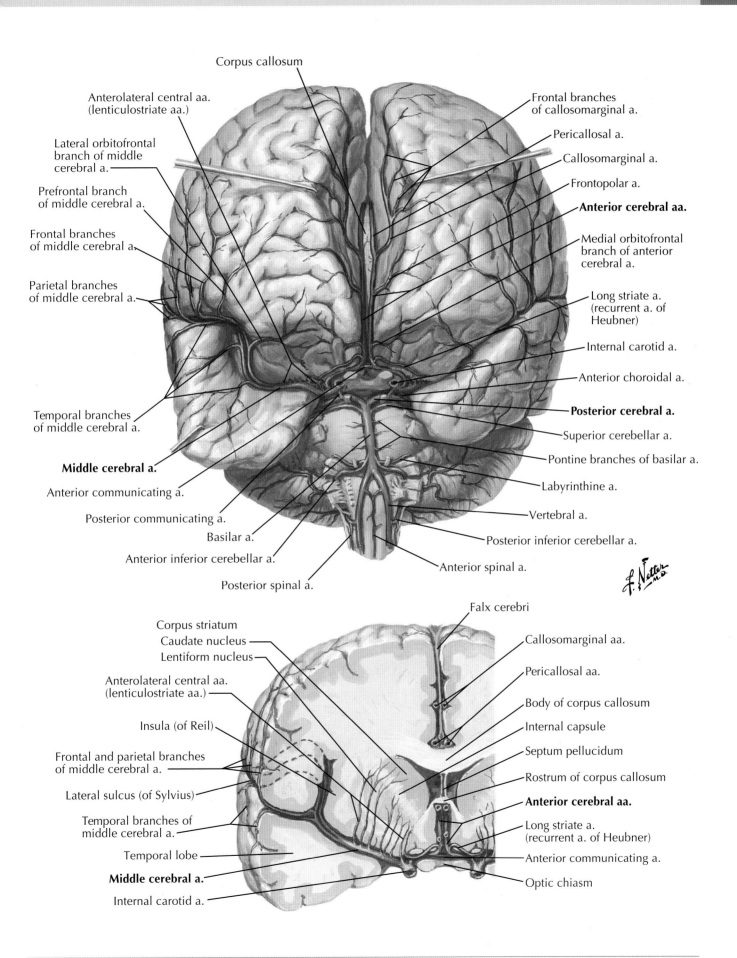

Corpus callosum

Anterolateral central aa. (lenticulostriate aa.)

Lateral orbitofrontal branch of middle cerebral a.

Prefrontal branch of middle cerebral a.

Frontal branches of middle cerebral a.

Parietal branches of middle cerebral a.

Temporal branches of middle cerebral a.

Middle cerebral a.

Anterior communicating a.

Posterior communicating a.

Basilar a.

Anterior inferior cerebellar a.

Posterior spinal a.

Frontal branches of callosomarginal a.

Pericallosal a.

Callosomarginal a.

Frontopolar a.

Anterior cerebral aa.

Medial orbitofrontal branch of anterior cerebral a.

Long striate a. (recurrent a. of Heubner)

Internal carotid a.

Anterior choroidal a.

Posterior cerebral a.

Superior cerebellar a.

Pontine branches of basilar a.

Labyrinthine a.

Vertebral a.

Posterior inferior cerebellar a.

Anterior spinal a.

Corpus striatum
Caudate nucleus
Lentiform nucleus

Anterolateral central aa. (lenticulostriate aa.)

Insula (of Reil)

Frontal and parietal branches of middle cerebral a.

Lateral sulcus (of Sylvius)

Temporal branches of middle cerebral a.

Temporal lobe

Middle cerebral a.

Internal carotid a.

Falx cerebri

Callosomarginal aa.

Pericallosal aa.

Body of corpus callosum

Internal capsule

Septum pellucidum

Rostrum of corpus callosum

Anterior cerebral aa.

Long striate a. (recurrent a. of Heubner)

Anterior communicating a.

Optic chiasm

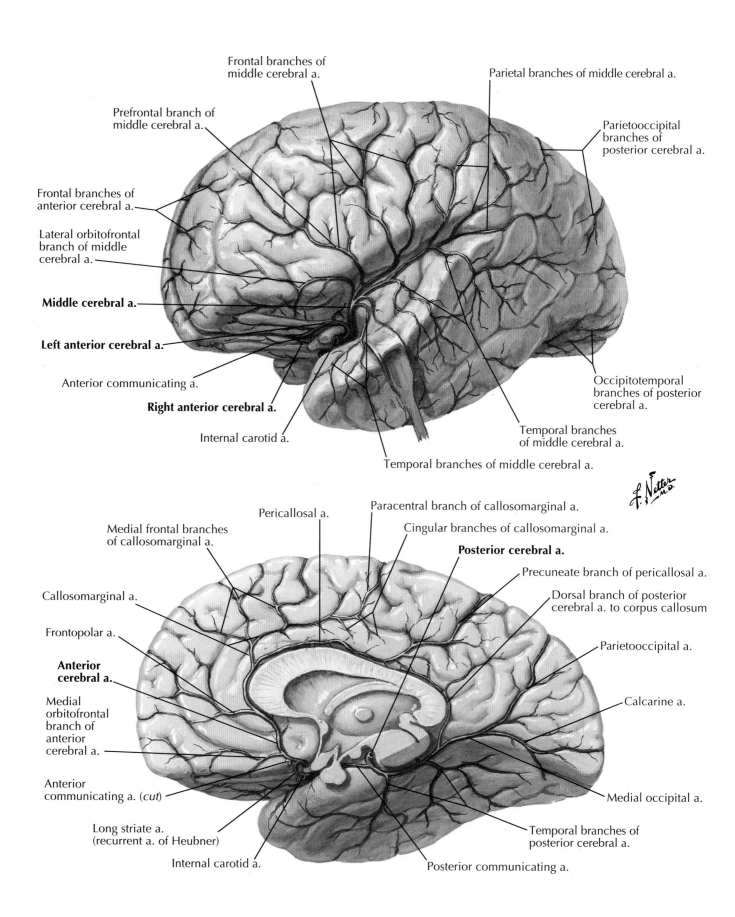

Frontal branches of
middle cerebral a.

Parietal branches of middle cerebral a.

Prefrontal branch of
middle cerebral a.

Parietooccipital
branches of
posterior cerebral a.

Frontal branches of
anterior cerebral a.

Lateral orbitofrontal
branch of middle
cerebral a.

Middle cerebral a.

Left anterior cerebral a.

Anterior communicating a.

Right anterior cerebral a.

Internal carotid a.

Occipitotemporal
branches of posterior
cerebral a.

Temporal branches
of middle cerebral a.

Temporal branches of middle cerebral a.

Pericallosal a.

Paracentral branch of callosomarginal a.

Medial frontal branches
of callosomarginal a.

Cingular branches of callosomarginal a.

Posterior cerebral a.

Precuneate branch of pericallosal a.

Callosomarginal a.

Dorsal branch of posterior
cerebral a. to corpus callosum

Frontopolar a.

Parietooccipital a.

**Anterior
cerebral a.**

Medial
orbitofrontal
branch of
anterior
cerebral a.

Calcarine a.

Anterior
communicating a. (*cut*)

Medial occipital a.

Long striate a.
(recurrent a. of Heubner)

Temporal branches of
posterior cerebral a.

Internal carotid a.

Posterior communicating a.

Plate 169

Cerebral Vasculature

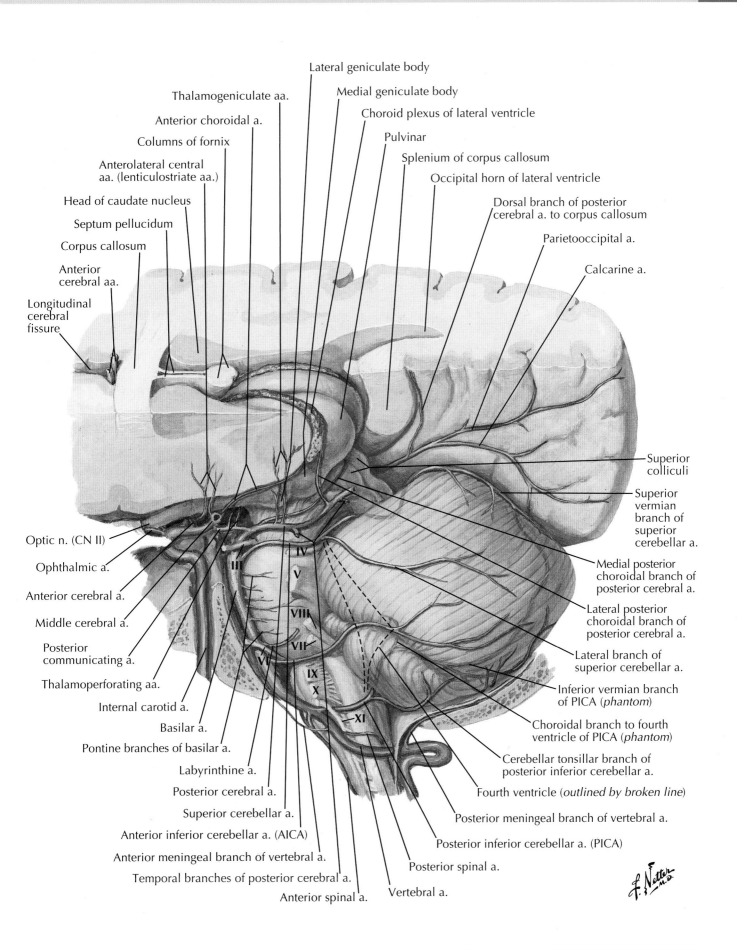

Lateral geniculate body

Thalamogeniculate aa.

Medial geniculate body

Anterior choroidal a.

Choroid plexus of lateral ventricle

Columns of fornix

Pulvinar

Anterolateral central
aa. (lenticulostriate aa.)

Splenium of corpus callosum

Occipital horn of lateral ventricle

Head of caudate nucleus

Dorsal branch of posterior
cerebral a. to corpus callosum

Septum pellucidum

Parietooccipital a.

Corpus callosum

Anterior
cerebral aa.

Calcarine a.

Longitudinal
cerebral
fissure

Superior
colliculi

Superior
vermian
branch of
superior
cerebellar a.

Optic n. (CN II)

Ophthalmic a.

Medial posterior
choroidal branch of
posterior cerebral a.

Anterior cerebral a.

Lateral posterior
choroidal branch of
posterior cerebral a.

Middle cerebral a.

Posterior
communicating a.

Lateral branch of
superior cerebellar a.

Thalamoperforating aa.

Inferior vermian branch
of PICA (*phantom*)

Internal carotid a.

Choroidal branch to fourth
ventricle of PICA (*phantom*)

Basilar a.

Pontine branches of basilar a.

Cerebellar tonsillar branch of
posterior inferior cerebellar a.

Labyrinthine a.

Fourth ventricle (*outlined by broken line*)

Posterior cerebral a.

Posterior meningeal branch of vertebral a.

Superior cerebellar a.

Anterior inferior cerebellar a. (AICA)

Posterior inferior cerebellar a. (PICA)

Anterior meningeal branch of vertebral a.

Temporal branches of posterior cerebral a.

Posterior spinal a.

Vertebral a.

Anterior spinal a.

III   IV   V   VIII   VII   VI   IX   X   XI

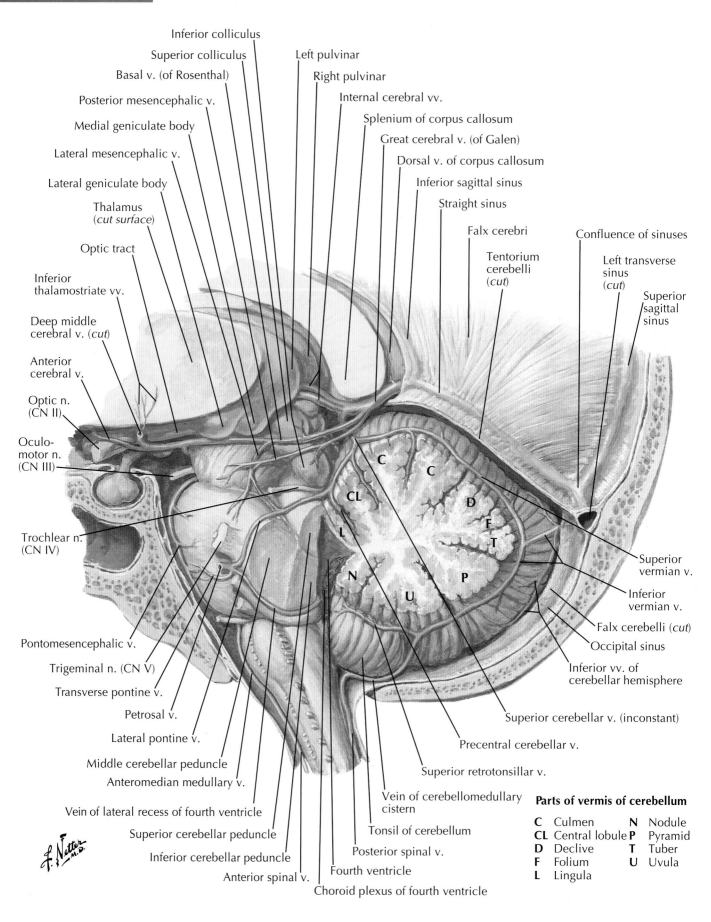

Inferior colliculus

Superior colliculus

Basal v. (of Rosenthal)

Posterior mesencephalic v.

Medial geniculate body

Lateral mesencephalic v.

Lateral geniculate body

Thalamus
(*cut surface*)

Optic tract

Inferior
thalamostriate vv.

Deep middle
cerebral v. (*cut*)

Anterior
cerebral v.

Optic n.
(CN II)

Oculo-
motor n.
(CN III)

Trochlear n.
(CN IV)

Pontomesencephalic v.

Trigeminal n. (CN V)

Transverse pontine v.

Petrosal v.

Lateral pontine v.

Middle cerebellar peduncle

Anteromedian medullary v.

Vein of lateral recess of fourth ventricle

Superior cerebellar peduncle

Inferior cerebellar peduncle

Anterior spinal v.

Left pulvinar

Right pulvinar

Internal cerebral vv.

Splenium of corpus callosum

Great cerebral v. (of Galen)

Dorsal v. of corpus callosum

Inferior sagittal sinus

Straight sinus

Falx cerebri

Tentorium
cerebelli
(*cut*)

Confluence of sinuses

Left transverse
sinus
(*cut*)

Superior
sagittal
sinus

Superior
vermian v.

Inferior
vermian v.

Falx cerebelli (*cut*)

Occipital sinus

Inferior vv. of
cerebellar hemisphere

Superior cerebellar v. (inconstant)

Precentral cerebellar v.

Superior retrotonsillar v.

Vein of cerebellomedullary
cistern

Tonsil of cerebellum

Posterior spinal v.

Fourth ventricle

Choroid plexus of fourth ventricle

Parts of vermis of cerebellum

C	Culmen	**N**	Nodule
CL	Central lobule	**P**	Pyramid
D	Declive	**T**	Tuber
F	Folium	**U**	Uvula
L	Lingula		

Plate 171

Cerebral Vasculature

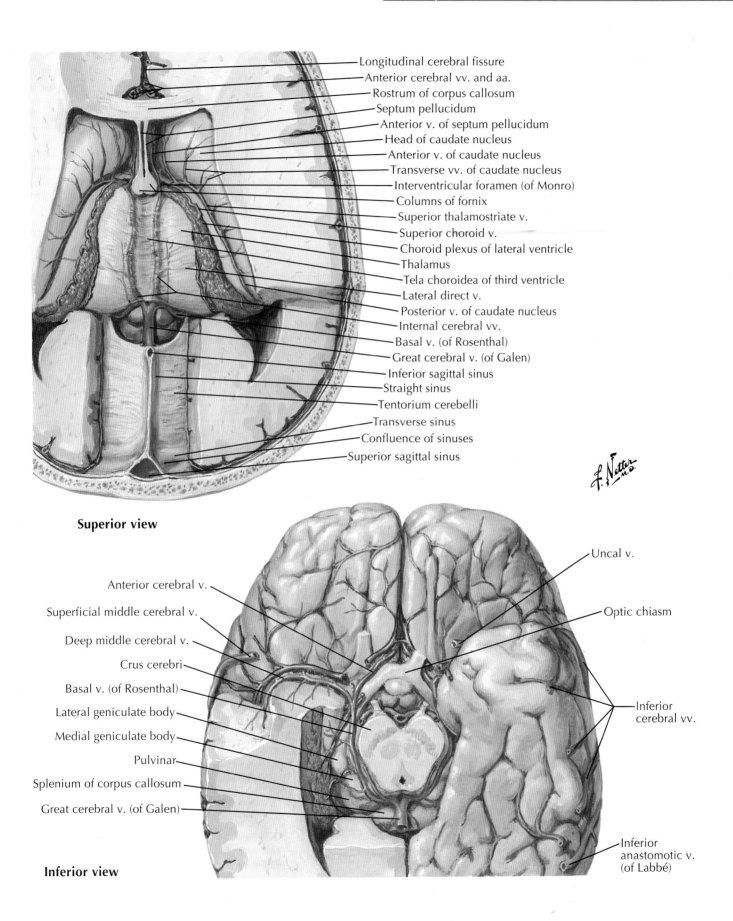

Longitudinal cerebral fissure
Anterior cerebral vv. and aa.
Rostrum of corpus callosum
Septum pellucidum
Anterior v. of septum pellucidum
Head of caudate nucleus
Anterior v. of caudate nucleus
Transverse vv. of caudate nucleus
Interventricular foramen (of Monro)
Columns of fornix
Superior thalamostriate v.
Superior choroid v.
Choroid plexus of lateral ventricle
Thalamus
Tela choroidea of third ventricle
Lateral direct v.
Posterior v. of caudate nucleus
Internal cerebral vv.
Basal v. (of Rosenthal)
Great cerebral v. (of Galen)
Inferior sagittal sinus
Straight sinus
Tentorium cerebelli
Transverse sinus
Confluence of sinuses
Superior sagittal sinus

Superior view

Uncal v.

Anterior cerebral v.

Superficial middle cerebral v.

Optic chiasm

Deep middle cerebral v.

Crus cerebri

Basal v. (of Rosenthal)

Inferior cerebral vv.

Lateral geniculate body

Medial geniculate body

Pulvinar

Splenium of corpus callosum

Great cerebral v. (of Galen)

Inferior anastomotic v. (of Labbé)

Inferior view

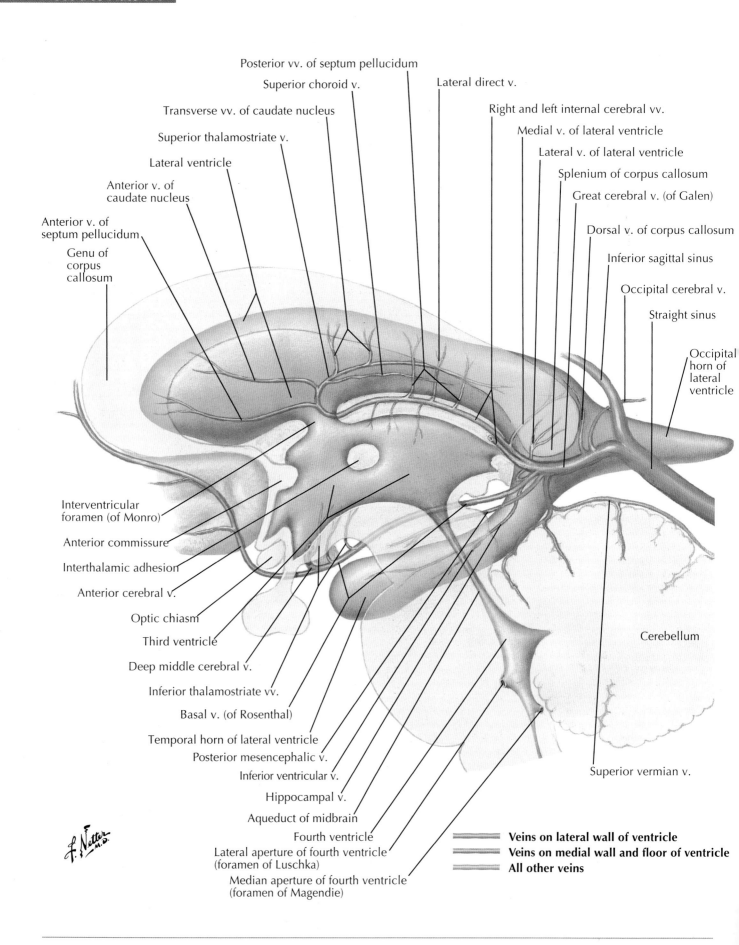

Posterior vv. of septum pellucidum

Superior choroid v.

Lateral direct v.

Transverse vv. of caudate nucleus

Right and left internal cerebral vv.

Superior thalamostriate v.

Medial v. of lateral ventricle

Lateral ventricle

Lateral v. of lateral ventricle

Anterior v. of caudate nucleus

Splenium of corpus callosum

Great cerebral v. (of Galen)

Anterior v. of septum pellucidum

Dorsal v. of corpus callosum

Genu of corpus callosum

Inferior sagittal sinus

Occipital cerebral v.

Straight sinus

Occipital horn of lateral ventricle

Interventricular foramen (of Monro)

Anterior commissure

Interthalamic adhesion

Anterior cerebral v.

Optic chiasm

Third ventricle

Cerebellum

Deep middle cerebral v.

Inferior thalamostriate vv.

Basal v. (of Rosenthal)

Temporal horn of lateral ventricle

Posterior mesencephalic v.

Inferior ventricular v.

Hippocampal v.

Superior vermian v.

Aqueduct of midbrain

Fourth ventricle

Lateral aperture of fourth ventricle (foramen of Luschka)

Median aperture of fourth ventricle (foramen of Magendie)

Veins on lateral wall of ventricle
Veins on medial wall and floor of ventricle
All other veins

Plate 173

Cerebral Vasculature

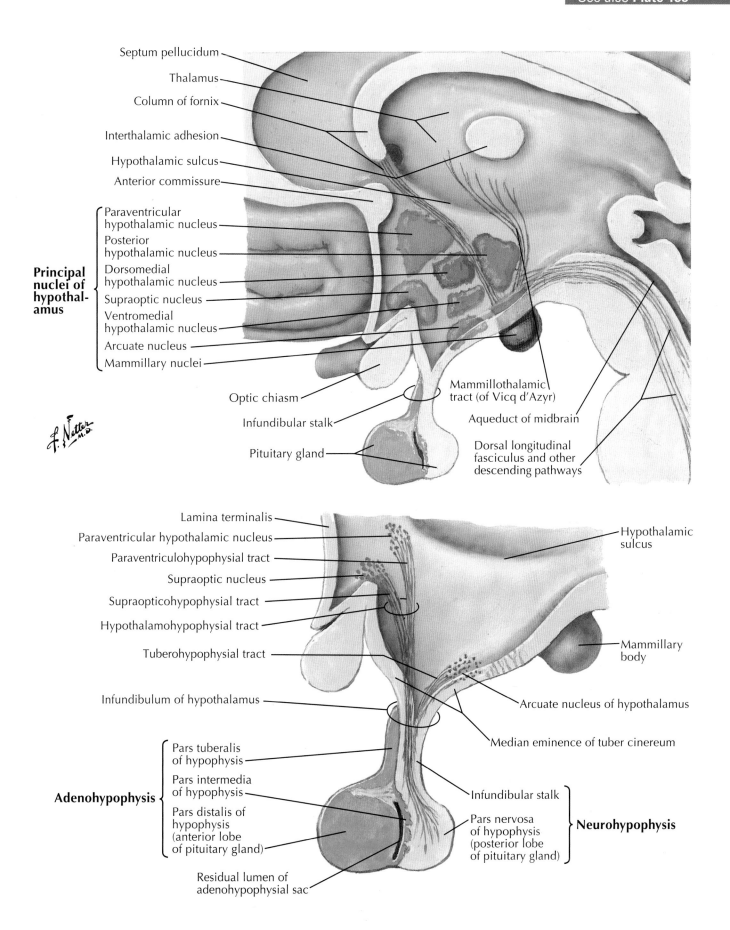

Septum pellucidum

Thalamus

Column of fornix

Interthalamic adhesion

Hypothalamic sulcus

Anterior commissure

Principal nuclei of hypothalamus
- Paraventricular hypothalamic nucleus
- Posterior hypothalamic nucleus
- Dorsomedial hypothalamic nucleus
- Supraoptic nucleus
- Ventromedial hypothalamic nucleus
- Arcuate nucleus
- Mammillary nuclei

Optic chiasm

Infundibular stalk

Pituitary gland

Mammillothalamic tract (of Vicq d'Azyr)

Aqueduct of midbrain

Dorsal longitudinal fasciculus and other descending pathways

Lamina terminalis

Paraventricular hypothalamic nucleus

Paraventriculohypophysial tract

Supraoptic nucleus

Supraopticohypophysial tract

Hypothalamohypophysial tract

Tuberohypophysial tract

Infundibulum of hypothalamus

Hypothalamic sulcus

Mammillary body

Arcuate nucleus of hypothalamus

Median eminence of tuber cinereum

Adenohypophysis
- Pars tuberalis of hypophysis
- Pars intermedia of hypophysis
- Pars distalis of hypophysis (anterior lobe of pituitary gland)

Residual lumen of adenohypophysial sac

Infundibular stalk

Pars nervosa of hypophysis (posterior lobe of pituitary gland) **Neurohypophysis**

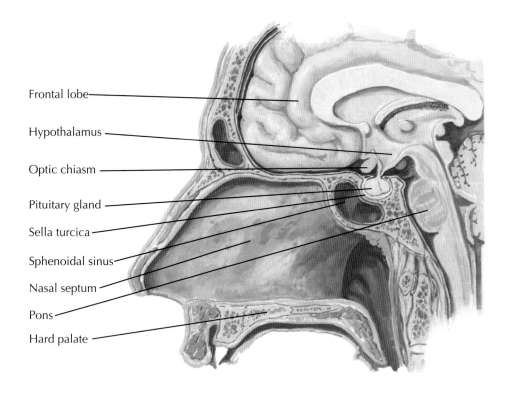

Frontal lobe

Hypothalamus

Optic chiasm

Pituitary gland

Sella turcica

Sphenoidal sinus

Nasal septum

Pons

Hard palate

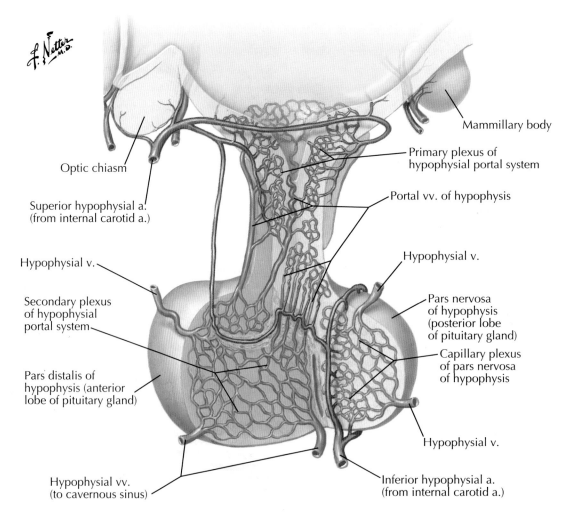

Mammillary body

Primary plexus of hypophysial portal system

Optic chiasm

Portal vv. of hypophysis

Superior hypophysial a. (from internal carotid a.)

Hypophysial v.

Hypophysial v.

Secondary plexus of hypophysial portal system

Pars nervosa of hypophysis (posterior lobe of pituitary gland)

Capillary plexus of pars nervosa of hypophysis

Pars distalis of hypophysis (anterior lobe of pituitary gland)

Hypophysial v.

Hypophysial vv. (to cavernous sinus)

Inferior hypophysial a. (from internal carotid a.)

Plate 175   *Cerebral Vasculature*

Magnetic resonance angiography (MRA) at level of cerebral arterial circle (3D time-of-flight image without contrast)

Postcommunicating part of anterior cerebral a. (A2 segment)

Middle cerebral a. (M2 segments)

Anterior communicating a.

Internal carotid a.

Superior cerebellar a.

Basilar a.

Precommunicating part of anterior cerebral a. (A1 segment)

Middle cerebral a. (M1 segment)

Posterior communicating a.

Posterior cerebral a.

Anterior inferior cerebellar a.

Magnetic resonance venography (MRV) (2D time-of-flight image without contrast)

Superior cerebral v.

Superior sagittal sinus

Internal cerebral v.

Great cerebral v. (of Galen)

Straight sinus

Confluence of sinuses

Transverse sinus

Sigmoid sinus

Internal jugular v.

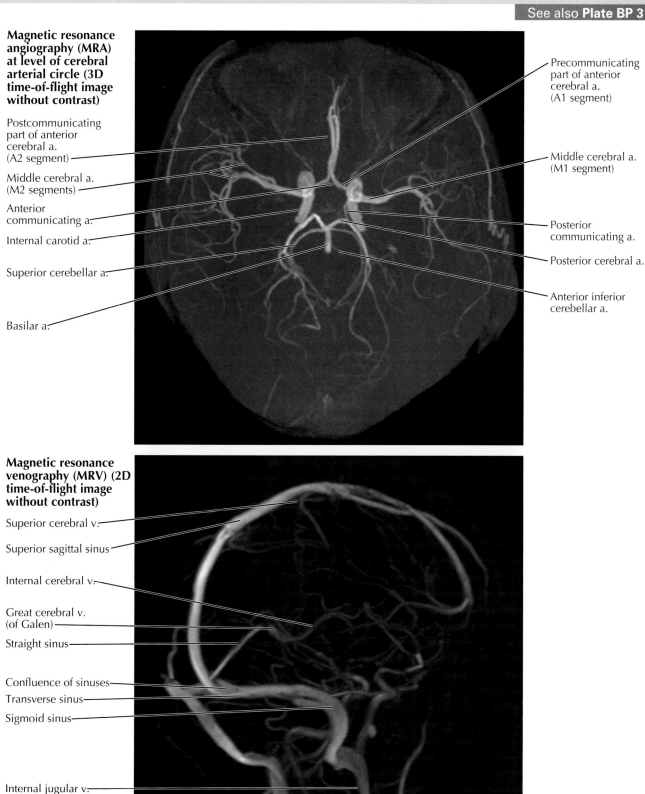

T1-weighted MRI, sagittal view

Body of corpus callosum
Genu of corpus callosum
Rostrum of corpus callosum
Mammillary body
Optic chiasm
Infundibular stalk
Crista galli
Pituitary gland
Sphenoidal sinus
Clivus
Pharyngeal tonsil
Hard palate
Soft palate
Tongue

Splenium of corpus callosum
Third ventricle
Aqueduct of midbrain
Tectum
Midbrain
Cerebellum
Fourth ventricle
Pons
Medulla oblongata
Anterior arch of atlas
Body of axis
Cervical part of spinal cord

T2-weighted MRI, axial views without contrast

Internal carotid a.
Basilar a.
Trigeminal n. (CN V)
Pons
Middle cerebellar peduncle
Fourth ventricle
Cerebellum

Eyeball
Ethmoidal cells
Sphenoidal sinus
Temporal lobe
Trigeminal cave

Gray matter (cerebral cortex)
White matter of telencephalon
Head of caudate nucleus
Anterior limb of internal capsule
External capsule
Putamen
Genu of internal capsule
Posterior limb of internal capsule
Thalamus

Longitudinal cerebral fissure
Genu of corpus callosum
Frontal horn of lateral ventricle
Interventricular foramen (of Monro)
Third ventricle
Occipital horn of lateral ventricle
Splenium of corpus callosum

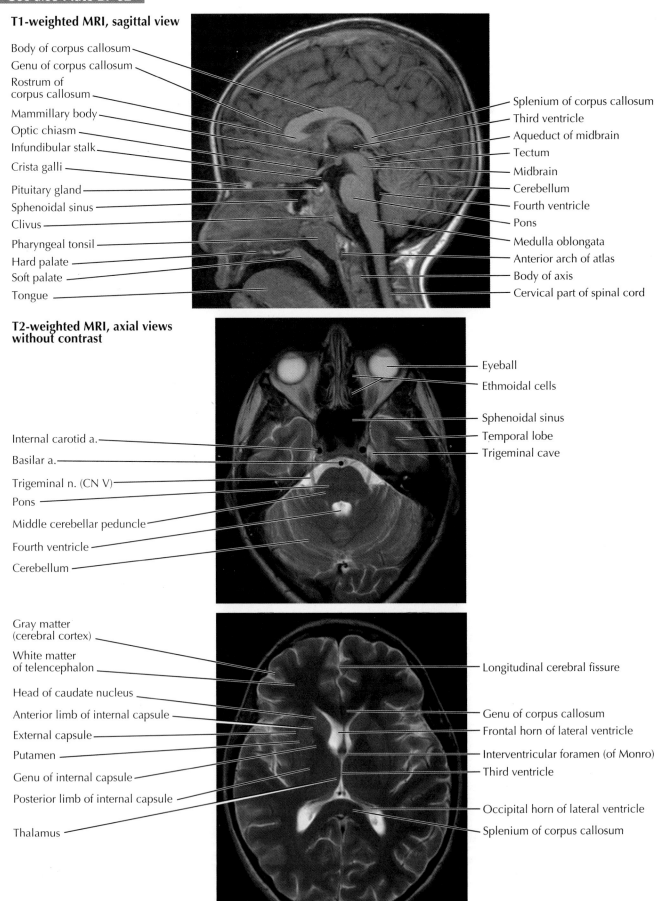

Plate 177

Regional Imaging

ANATOMICAL STRUCTURE	CLINICAL COMMENT	PLATE NUMBERS
Nervous System and Sense Organs		
Spinal accessory nerve (CN XI)	Lymph node biopsy in posterior cervical triangle can cause iatrogenic injury of CN XI	56
Cervical plexus	Cervical plexus blocks are performed for neck procedures	56, 57
Trigeminal nerve (CN V)	Branches of CN V are anesthetized for procedures on face or anterior scalp; compression of the nerve may result in painful condition known as trigeminal neuralgia	60, 149
Olfactory nerve (CN I)	One of most commonly injured cranial nerves; can be avulsed at cribriform plate following falls, resulting in anosmia	64, 146
Facial nerve (CN VII)	Idiopathic unilateral facial nerve palsy (Bell's palsy) can result in inability to fully close eye and result in desiccated cornea ipsilaterally	71
Recurrent laryngeal nerve	May be compressed or damaged by procedures in neck (e.g., thyroidectomy), aortic arch aneurysm, or lung cancer, producing hoarseness of voice; identified by (posterior) suspensory ligament of thyroid and/or inferior thyroid artery and/or tracheoesophageal groove	104, 105
Oculomotor (CN III), trochlear (CN IV), and abducens (CN VI) nerves	Cavernous sinus thrombosis can result in dysfunction of extraocular muscles caused by compression of one, two, or all three nerves; abducens nerve is most commonly affected	131
Superior colliculus and aqueduct of midbrain	Tumor of midbrain can result in compression of aqueduct of midbrain, with resultant hydrocephalus	136
Optic nerve (CN II)	Pituitary gland mass may cause compression at optic chiasm and resulting bitemporal hemianopsia; optic radiations (Meyer's loop) can be affected by temporal lobe tumors; early sign of ophthalmic artery aneurysm is visual loss due to compression of overlying optic nerve.	147, 167
Fovea centralis	Location of highest density of retinal cones, making this part of macula lutea the site of greatest visual acuity and color vision	116
Utricle and saccule	Location of calcium carbonate crystals known as otoconia; accumulation of otoconia in semicircular canals is most common cause of vertigo and is known as benign paroxysmal positional vertigo (BPPV)	124, 125
Lens	Degeneration and opacification, known as cataract, may lead to progressive vision loss	116
Orbital septum	Infections anterior to this structure are known as preseptal/periorbital cellulitis and are milder than infections extending posterior to it, known as orbital cellulitis	110
Anterior segment of eyeball	Ciliary body produces aqueous humor, which flows through iris into anterior chamber and drains through scleral sinus; anterior displacement of lens may obstruct flow of aqueous humor and cause elevated intraocular pressure, a painful, vision-threatening condition known as acute angle closure glaucoma	117, 119
Tympanic membrane	Visualized with otoscope; bulging indicates middle ear infection with effusion; rupture and otorrhea may occur in severe infection; tympanostomy tubes (T-tubes) may be placed in children with recurrent effusive infections	122
External acoustic meatus	May become infected or inflamed in children, a condition known as otitis externa; can be diagnosed with auricle (pinna) pull technique	121
Skeletal System		
Orbit	Most facial trauma involves orbit; traumatic fractures can occur in rim or walls; a "blowout" fracture involves inferior wall and may injure inferior rectus muscle and/or infraorbital nerve; rim fractures affect contours of orbital rim and occur in zygomaticomaxillary fractures	25, 69

Structures with High Clinical Significance   Table 2.1

ANATOMICAL STRUCTURE	CLINICAL COMMENT	PLATE NUMBERS
Skeletal System—Continued		
Pterion	Intersection of frontal, parietal, temporal, and sphenoid bones; thin, weak region of skull that is susceptible to fracture; frontal branch of middle meningeal artery lies immediately deep to this region and may be injured	27
Asterion	Landmark at posterior end of parietotemporal suture, used in lateral neurosurgical approaches to posterior cranial fossa	27
Temporomandibular joint	Temporomandibular joint disorders are common sources of pain and joint dysfunction; poor replacement results to date; dislocations/subluxations can be reduced via retromolar fossa	42
Cranial sutures	Premature fusion may result in skull deformity known as craniosynostosis; sagittal suture is most often affected	26, 35
Cervical vertebrae	Degenerative changes causing narrowing of intervertebral foramina may result in cervical radiculopathy; C1–C4 and C5–C7 vertebrae are common areas of pathology for children and adults, respectively; neck hyperextension from abrupt deceleration may cause bilateral pedicle fractures in axis (C2 vertebra) known as hangman's fracture; hyperflexion may cause anterior wedge vertebral fracture; axial loads may cause burst fracture; dens axis (odontoid) fractures may occur with either forceful extension or flexion; all fractures are classified as "stable" or "unstable" depending on whether structural integrity of cervical spine has been sufficiently disrupted to permit compression of spinal cord	43–45
Laryngeal cartilages	Thyroid and cricoid cartilages are palpable landmarks of anterior neck used for cricothyrotomy, tracheostomy, and cricoid pressure during airway intubation	103, 106
Hyoid bone	Palpable anterior neck landmark at C3 vertebral level that can be fractured during sporting activities, compromising swallowing and speech; fractures may also indicate strangulation	54
Auditory ossicles	Pathologic conditions involving ossicles (e.g., otosclerosis) can cause conductive hearing loss	121, 122
Muscular System		
Muscles of facial expression	Used to assess function of facial nerve (CN VII) during cranial nerve examination; may become weak or paralyzed with CN VII dysfunction (e.g., Bell's palsy); often targeted with botulinum toxin injections for aesthetics (wrinkles), headache, and teeth grinding	48, 150
Sternocleidomastoid muscle	Important landmark that divides neck into anterior and posterior cervical triangles; palpated to identify "nerve point of neck" for administration of anesthesia to cervical plexus and also used as landmark for insertion of central venous catheters; in children, abnormal shortening or fibrosis of sternocleidomastoid muscle results in head tilt, a condition known as torticollis	56
Sternocleidomastoid and trapezius muscles	Used to assess function of spinal accessory nerve (CN XI) during cranial nerve examination	56, 154
Muscles of mastication	Used to assess function of trigeminal nerve (CN V) during cranial nerve examination; masseter is involved in teeth grinding and associated headaches, which can be treated with botulinum toxin injection	72, 73
Levator veli palatini and musculus uvulae	Used to assess function of vagus nerve (CN X) during cranial nerve examination; contralateral deviation of uvula of soft palate during elevation indicates CN X dysfunction	85
Genioglossus muscle	Used to assess hypoglossal nerve (CN XII) function during cranial nerve examination; tongue deviates to side of lesion when protruded following CN XII injury	88, 155
Styloglossus muscle	Posterosuperior movement covers laryngeal inlet, protecting vocal cords and airway, which is especially crucial if epiglottis has been surgically removed because of malignancy	88, 95, 96

Table 2.2　　　　　　　　　　　　　　　　　　　　　　　　　　**Structures with High Clinical Significance**

ANATOMICAL STRUCTURE	CLINICAL COMMENT	PLATE NUMBERS
Stapedius muscle	Smallest skeletal muscle in the body, regulates movement of stapes bone to control amplitude of sound	123
Levator palpebrae superioris and superior tarsal muscle	Responsible for elevating eyelid; ptosis indicates pathologic change in oculomotor nerve (CN III) or sympathetic fibers (if only superior tarsal muscle is affected)	110, 112
Extraocular muscles	Used to assess function of oculomotor (CN III), trochlear (CN IV), and abducens (CN VI) nerves during cranial nerve examination; abnormalities in tone result in disconjugate eye movements, a condition known as strabismus	112, 114
Dilator pupillae	Important in assessment of sympathetic function in head; lack of dilation indicates interruption in sympathetic outflow (e.g., Horner's syndrome)	117, 148
Sphincter pupillae	Involved in pupillary light reflex and accommodation reflex	117, 148
Pterygomandibular raphe	Valuable intraoral landmark for inferior alveolar nerve blocks to anesthetize mandibular teeth	96, 97–99
Modiolus of angulus oris	A fibrous intersection of facial muscles, approximately 1 cm lateral to corner of mouth; valuable landmark when considering reconstruction of injured facial muscles associated with oral region	48
Retromolar fossa	Important clinical landmark for reducing dislocated temporomandibular joints; buccal and lingual nerves pass by the retromolar fossa and may be harmed in dental molar implant surgery	39
Cardiovascular System		
Right internal and external jugular veins	Examined to assess right atrial pressure, estimated as height of jugular pulsation above sternal angle (in centimeters) plus 5; right internal jugular vein is preferred because it is in line with superior vena cava	50
Internal jugular vein	Thrombosis may occur secondary to local extension of inflammation from severe pharyngitis, a condition known as Lemierre's syndrome	50
Internal jugular and subclavian veins	Used to obtain venous access via insertion of central venous catheter	50, 100
Inferior thyroid artery	At risk during thyroidectomy; must be preserved to maintain blood supply to parathyroid glands; identified by its redundant loop shape and is landmark to identify recurrent laryngeal nerve	57, 104
Common carotid artery	Palpate in neck to assess carotid pulse; bifurcation typically at C4 vertebral level	58, 165
Internal carotid artery	Common site for atherosclerosis, which may be treated with stent or endarterectomy for stroke prevention; carotid sinus is sensitive to changes in circulating blood volume and may be massaged to induce vagal reaction; internal carotid artery has no branches in neck; classified into seven parts: C1, cervical; C2, petrous; C3, lacerum; C4, cavernous; C5, clinoid; C6, ophthalmic; and C7, communicating	163, 164
Anterior ethmoidal, sphenopalatine, and facial arteries	Anastomosis site of branches of these vessels in nasal vestibule, known as Kiesselbach's plexus or Little's area, is common site of anterior nosebleeds (epistaxis); sphenopalatine artery injury causes posterior nosebleeds	65
Pterygoid venous plexus	Common route for spread of infection due to connections between face, orbit, and venous sinuses; valveless veins allow retrograde flow	100, 115
Ophthalmic artery	Primary source of blood to retina; blindness may occur if artery is occluded	115, 119
Arteries of scalp	Scalp lacerations bleed profusely owing to rich blood supply	24, 127
Superior cerebral veins	May be torn from their junction with superior sagittal sinus, producing subdural hematoma	129, 130, 136
Middle meningeal artery	Trauma to pterion can tear middle meningeal artery (frontal branch), often causing epidural hematoma	128

Structures with High Clinical Significance

Table 2.3

Structures with High* Clinical Significance

ANATOMICAL STRUCTURE	CLINICAL COMMENT	PLATE NUMBERS
Cardiovascular System—Continued		
Dural venous sinuses	Infections in head may spread to sinuses, causing dural venous sinus thrombosis; cavernous sinus is most common site	130, 131
Cavernous sinus	Fistula (anastomosis) between internal carotid artery and cavernous sinus may form, especially following trauma	131,167
Carotid sinus	Compressed during carotid sinus massage, which may result in bradycardia and/or hypertension; carotid sinus hypersensitivity, most common among older adults, may cause syncope	157
Cerebral arterial circle of Willis	Common site of aneurysms and important site of collateral cerebral circulation; aneurysmal rupture produces subarachnoid hemorrhage	166
Emissary veins	Valveless veins, which can convey infection extracranially to intracranially and allow alternative route for drainage when dural venous sinuses are obstructed	129
Lymph Vessels and Lymphoid Organs		
Thoracic duct	Thoracic duct may be injured during surgeries to neck and thorax due to multiple and frequent variants; injury in lower neck region is often at junction of left internal jugular and subclavian veins; increased risk during esophageal surgery and left central venous line placement	101, 260
Superior and inferior deep lateral cervical nodes	Palpated during neck examination to assess size and shape; if coalescent, malignancy must be excluded	101, 102
Palatine and pharyngeal tonsils	Palatine tonsils are commonly involved with viral and bacterial infection; exudative lesions combined with fever, lymphadenopathy, and lack of cough suggestive of streptococcal infection (strep throat); enlarged pharyngeal tonsils (adenoids) cause snoring and can cover auditory (eustachian) tube opening, increasing risk of middle ear infections	90
Respiratory System		
Epiglottis	Crucial landmark during endotracheal intubation; bacterial or viral infection (e.g., with *Haemophilus influenzae*) may cause epiglottitis, which presents with respiratory distress, sore throat, and hoarse voice	93–95
Paranasal sinuses (maxillary, ethmoid, frontal, and sphenoid)	Cavities in skull prone to mucosal inflammation due to bacterial or viral infection	67, 69
Nasal septum	Congenital or acquired deviation may lead to nasal obstruction, treated with septoplasty; perforations may occur with cocaine use or in granulomatosis with polyangiitis	37, 69
Digestive System		
Parotid gland	Swelling of gland due to infection (parotitis), such as from viruses or bacteria, may cause pain and compress branches of facial nerve, producing facial muscle weakness; external carotid, superficial temporal, and maxillary arteries also pass through the gland	70, 71
Endocrine System		
Thyroid gland	Enlargement is known as goiter; may be partially or completely removed in malignancy or hyperthyroidism; during examination should move relatively equally bilaterally with hyoid bone and laryngeal cartilages; malignancy may anchor portion of gland and cause asymmetrical movement	103

*Selections were based largely on clinical data and commonly discussed clinical correlations in macroscopic ("gross") anatomy courses.

Cranial nerves are traditionally described as tree-like structures that emerge from the brain and branch peripherally. This matches the direction that action potentials travel in efferent fibers in nerves. It must be remembered that action potentials travel in the opposite direction in afferent fibers within these nerves.

NERVE	ORIGIN	COURSE	BRANCHES	MOTOR	SENSORY
Olfactory nerve (CN I)	Olfactory bulb	Neurons of olfactory mucosa send approximately 20 axon bundles through cribriform plate to synapse on olfactory bulb neurons			SVA (smell): olfactory epithelium
Optic nerve (CN II)	Optic chiasm	Axons of retinal ganglion neurons exit orbit through optic canal to enter cranial cavity			SSA (vision): optic part of retina
Oculomotor nerve (CN III)	Interpeduncular fossa of midbrain	Exits midbrain into posterior cranial fossa and then middle cranial fossa; traverses cavernous sinus to enter orbit via superior orbital fissure	Superior and inferior branches	GSE: medial, superior, and inferior rectus muscles, inferior oblique muscle, and levator palpebrae superioris GVE: ciliary ganglion	
Trochlear nerve (CN IV)	Posterior surface of midbrain	Exits dorsal midbrain, coursing lateral to cerebral peduncle to anterior surface of brain stem; follows medial edge of tentorium cerebelli to enter middle cranial fossa; traverses cavernous sinus to enter orbit via superior orbital fissure		GSE: superior oblique muscle	
Trigeminal nerve (CN V)	Motor and sensory roots (arising from pons)	Exits anterolateral pons into posterior cranial fossa; large sensory and small motor roots enter middle cranial fossa; sensory root forms trigeminal ganglion, which gives rise to ophthalmic, maxillary, and mandibular nerves; motor root passes deep to ganglion and contributes only to mandibular nerve	Ophthalmic, maxillary, and mandibular nerves	SVE: see branches	GSA: see branches
Ophthalmic nerve (CN V$_1$)	Trigeminal nerve	Exits anterior border of trigeminal ganglion and traverses cavernous sinus to leave cranium through superior orbital fissure into orbit	Lacrimal, frontal, and nasociliary nerves, meningeal branch		GSA: forehead, upper eyelid, conjunctiva
Maxillary nerve (CN V$_2$)	Trigeminal nerve	Exits trigeminal ganglion, passes through foramen rotundum into pterygopalatine fossa, and enters orbit via inferior orbital fissure	Nasopalatine, pharyngeal, palatine, zygomatic, posterior superior alveolar, and infraorbital nerves, superior posterior nasal and meningeal branches		GSA: midface, nasal cavity, paranasal sinuses, palate, maxillary teeth
Mandibular nerve (CN V$_3$)	Trigeminal nerve	Exits trigeminal ganglion inferiorly, leaves cranium through foramen ovale and enters infratemporal fossa	Deep temporal, buccal, auriculotemporal, lingual, and inferior alveolar nerves, meningeal branch	SVE: muscles of mastication, mylohyoid and digastric (anterior belly) muscles, tensor tympani, tensor veli palatini	GSA: mandibular teeth, anterior tongue, floor of oral cavity, temporomandibular joint

Continued

NERVE	ORIGIN	COURSE	BRANCHES	MOTOR	SENSORY
Abducens nerve (CN VI)	Medullopontine sulcus (medial to CN VII)	Exits between pons and medulla near midline, pierces dura on clivus and grooves on petrous part of temporal bone to access middle cranial fossa; traverses cavernous sinus to enter orbit through superior orbital fissure		GSE: lateral rectus muscle	
Facial nerve (CN VII)	Motor root and intermediate nerve (sensory root)	Exits laterally between pons and medulla oblongata as a larger motor root and smaller intermediate nerve (carrying SVA, GVE, and GSA fibers); both roots traverse internal acoustic meatus to enter facial canal, in which a sharp turn (genu) occurs just before geniculate (facial) ganglion; facial nerve gives off several branches within facial canal, before it exits through stylomastoid foramen and forms terminal branches within parotid gland	Greater petrosal and posterior auricular nerves, chorda tympani, and temporal, zygomatic, buccal, marginal mandibular, and cervical branches	SVE: muscles of facial expression (including epicranial muscles and platysma), digastric (posterior belly), stylohyoid, and stapedius muscles GVE: pterygopalatine and submandibular ganglia	SVA (taste): anterior tongue and palate GSA: part of external ear
Vestibulocochlear nerve (CN VIII)	Medullopontine sulcus (lateral to CN VII)	Exits brain stem between pons and medulla oblongata, near inferior cerebellar peduncle, and divides into two branches as it traverses posterior cranial fossa	Vestibular and cochlear nerves		SSA: internal ear (see branches)
Cochlear nerve	Vestibulocochlear nerve	Contains processes of cochlear ganglion cells; passes through internal acoustic meatus into internal ear			SSA (hearing): spiral organ of cochlear duct
Vestibular nerve	Vestibulocochlear nerve	Contains processes of vestibular ganglion cells; passes through internal acoustic meatus into internal ear	Superior and inferior branches		SSA (equilibrium and motion): maculae and cristae ampullares of vestibular labyrinth
Glossopharyngeal nerve (CN IX)	Retro-olivary groove of medulla oblongata	Emerges from upper medulla between olive and inferior cerebral peduncle and exits cranium by traversing jugular foramen with vagus and accessory nerves; superior and inferior ganglia for afferent components of CN IX are located just inferior to jugular foramen; passes inferiorly, innervating and following stylopharyngeus muscle, ultimately sending a branch into posterior oral cavity that passes deep to hyoglossus muscle and a branch that enters pharynx by passing between superior and middle pharyngeal constrictors	Tympanic nerve	SVE: stylopharyngeus muscle GVE: otic ganglion	SVA (taste): posterior tongue GVA: carotid body and sinus GSA: posterior tongue, oropharynx, middle ear

Table 2.6

NERVE	ORIGIN	COURSE	BRANCHES	MOTOR	SENSORY
Vagus nerve (CN X)	Retro-olivary groove of medulla oblongata (between CN IX and cranial roots of CN XI)	Exits from retro-olivary groove of medulla oblongata to traverse jugular foramen with accessory and glossopharyngeal nerves; initially runs between internal carotid artery and internal jugular vein; courses of right and left vagus nerves differ, with right vagus nerve coursing inferiorly between subclavian artery and vein, at which point right recurrent laryngeal nerve ascends and vagus nerve descends along trachea and root of right lung, forming pulmonary and esophageal plexuses; left vagus nerve descends between left subclavian and common carotid arteries posteriorly to left brachiocephalic vein, and anteriorly to aortic arch, at which point left recurrent laryngeal nerve ascends posteriorly to aorta and rest of vagus nerve descends posteriorly to root of lung, forming pulmonary and esophageal plexuses; vagus nerve fibers continue into abdomen via esophageal plexuses and vagal trunks	Pharyngeal branch, superior and recurrent laryngeal nerves	GVE: thoracic and abdominal visceral ganglia SVE: see branches	GVA: aortic arch and bodies SVA (taste): epiglottis, oropharynx GSA: external ear (also see branches)
Pharyngeal branch of vagus nerve	Vagus nerve	Passes between carotid arteries superficial to middle pharyngeal constrictor		SVE: pharyngeal constrictors; palatoglossus, palatopharyngeus, and salpingopharyngeus muscles; and levator veli palatini	
Superior laryngeal nerve	Vagus nerve	Exits inferior ganglion, passing inferiorly and medially deep to internal carotid artery, dividing into external (motor) branch and internal (sensory) branch that pierces thyrohyoid membrane	Internal and external branches	SVE: cricothyroid muscle	GSA: superior larynx
Recurrent laryngeal nerve	Vagus nerve	On right side, arises anterior to subclavian artery and spirals to run superiorly and posteriorly to common carotid and inferior thyroid arteries, lateral to trachea; on left side, arises inferior to aortic arch, passing posterior to arch and lateral to ligamentum arteriosum to ascend superiorly lateral to trachea; both right and left nerves enter larynx at junction of esophagus and inferior pharyngeal constrictor		SVE: intrinsic muscles of larynx (except cricothyroid), striated muscle of esophagus	GSA: inferior larynx

Continued

NERVE	ORIGIN	COURSE	BRANCHES	MOTOR	SENSORY
Accessory nerve (CN XI)	Cranial and spinal roots	Spinal roots from C1–C5 segments of spinal cord ascend and enter cranium through foramen magnum, where they course with cranial root arising from retroolivary groove of medulla oblongata; spinal accessory nerve exits from cranium via jugular foramen; it passes inferiorly and posteriorly into upper third of sternocleidomastoid muscle and then inferiorly, crossing occipital triangle to enter trapezius muscle; vagal communicating branch joins vagus nerve at jugular foramen to innervate larynx, palate, and pharyngeal muscles	Spinal accessory nerve, vagal communicating branch	GSE: trapezius and sternocleidomastoid muscles	
Hypoglossal nerve (CN XII)	Hypoglossal rootlets and hypoglossal communicating branch of C1 spinal nerve	Rootlets from anterolateral sulcus of medulla oblongata exit cranium through hypoglossal canal, then pass inferiorly and anteriorly between vagus and spinal accessory nerves to inferior border of posterior belly of digastric muscle, eventually entering oral cavity by passing between mylohyoid and hyoglossus muscles	Lingual branches and branches conveying fibers from C1 spinal nerve (thyrohoid and geniohyoid branches and superior root of ansa cervicalis)	GSE: intrinsic muscles of tongue and genioglossus, hyoglossus, and styloglossus muscles C1: thyrohyoid, geniohyoid, and omohyoid (superior belly) muscles	

Table 2.8   **Cranial Nerves**

The roots of the cervical plexus are the anterior rami of C1–C4 spinal nerves.

NERVE	ORIGIN	COURSE	BRANCHES	MOTOR	CUTANEOUS SENSORY
Hypoglossal communicating branch	Anterior ramus of C1 spinal nerve	Emerges and briefly adheres to hypoglossal nerve; superior root of ansa cervicalis exits just posterior to greater horn of hyoid bone and descends along carotid sheath, where it joins inferior root at C4–C5 level	Superior root of ansa cervicalis, thyrohyoid and geniohyoid branches of hypoglossal nerve	Omohyoid (superior belly), thyrohyoid, and geniohyoid muscles	
Inferior root of ansa cervicalis	Anterior rami of C2–C3 spinal nerves	Descends along anterolateral carotid sheath, joining superior root at C4–C5 level	Infrahyoid branches	Omohyoid (inferior belly), sternohyoid, and sternothyroid muscles	
Muscular branches of cervical plexus	Anterior rami of C1–C4 spinal nerves	Three loops form along C1–C4 vertebrae that course laterally between levator scapulae and scalenus medius muscle, deep to sternocleidomastoid muscle		Rectus anterior and lateralis capitis muscles; longus capitis and colli muscles; scalenus anterior, medius and posterior muscles; levator scapulae	
Phrenic nerve	Anterior rami of C3–C5 spinal nerves	Descends on scalenus anterior muscle deep to inferior belly of omohyoid muscle and transverses cervical and suprascapular vessels; enters thorax between subclavian vein and artery, passing anterior to root of lung and along lateral border of pericardium to pierce diaphragm		Diaphragm	
Lesser occipital nerve	Anterior ramus of C2 spinal nerve	Formed in posterior cervical triangle deep to sternocleidomastoid muscle, ascending along its posterior border; perforates deep fascia at mastoid process, ascending posterior to ear			Temporal, auricular, and mastoid regions
Great auricular nerve	Anterior rami of C2–C3 spinal nerves	Formed in posterior cervical triangle deep to sternocleidomastoid muscle, ascending obliquely between that muscle and platysma	Anterior and posterior branches		Parotid, auricular, and mastoid regions
Transverse cervical nerve	Anterior rami of C2–C3 spinal nerves	Formed in posterior cervical triangle deep to sternocleidomastoid muscle, runs superficial to that muscle, passing deep to external jugular vein	Superior and inferior branches		Anterior and lateral regions of neck
Supraclavicular nerve	Anterior rami of C3–C4 spinal nerves	Formed in posterior cervical triangle deep to mid-third of sternocleidomastoid muscle, passes lateral to external jugular vein, and descends just inferior to clavicle	Medial, intermediate, and lateral supraclavicular nerves		Clavicular and infraclavicular regions

MUSCLE	MUSCLE GROUP	ORIGIN ATTACHMENT	INSERTION ATTACHMENT	INNERVATION	BLOOD SUPPLY	MAIN ACTIONS
Auricularis anterior m.	Facial expression (external ear)	Temporal fascia, epicranial aponeurosis	Anterior part of medial surface of helix of ear	Temporal branches of facial nerve	Posterior auricular and superficial temporal arteries	Elevates and draws auricle forward
Auricularis posterior m.	Facial expression (external ear)	Mastoid process	Inferior part of medial surface of auricle	Posterior auricular nerve (branch of facial nerve)	Posterior auricular and superficial temporal arteries	Retracts and elevates auricle
Auricularis superior m.	Facial expression (external ear)	Temporal fascia, epicranial aponeurosis	Superior part of medial surface of auricle	Temporal branches of facial nerve	Posterior auricular and superficial temporal arteries	Retracts and elevates auricle
Bucinator	Facial expression	Posterior portions of alveolar processes of maxilla and mandible, anterior border of pterygomandibular raphe	Modiolus of angle of mouth	Buccal branches of facial nerve	Facial and maxillary arteries	Compresses cheeks
Ciliary m.	Intrinsic eye (smooth muscle)	Scleral spur	Choroid	Short ciliary nerves (parasympathetic fibers from ciliary ganglion)	Ophthalmic artery	Constricts ciliary body and lens rounds up (accommodation)
Corrugator supercilii	Facial expression	Medial part of supra-orbital margin	Skin of medial half of eyebrow	Temporal branches of facial nerve	Superficial temporal artery	Draws eyebrows inferiorly and medially, produces vertical wrinkles of skin between eyebrows
Cricothyroid m.	Laryngeal	Arch of cricoid cartilage	Inferior border of lamina and inferior horn of thyroid cartilage	External branch of superior laryngeal nerve	Superior and inferior thyroid arteries	Lengthens and tenses vocal ligaments
Depressor anguli oris	Facial expression	Oblique line of mandible	Modiolus of angle of mouth	Marginal mandibular and buccal branches of facial nerve	Inferior labial artery	Depresses angle of mouth
Depressor labii inferioris	Facial expression	External surface of mandible between symphysis and mental foramen	Skin of lower lip	Marginal mandibular branches of facial nerve	Inferior labial artery	Depresses lower lip and draws it laterally
Depressor septi nasi	Facial expression	Incisive fossa of maxilla	Nasal septum and posterior part of ala of nose	Zygomatic and buccal branches of facial nerve	Superior labial artery	Narrows nostril, draws septum inferiorly
Digastric m.	Suprahyoid	*Anterior belly:* digastric fossa of mandible	Intermediate tendon attached to body of hyoid	*Anterior belly:* inferior alveolar nerve (branch of mandibular nerve)	*Anterior belly:* submental artery	Elevates hyoid bone and base of tongue, steadies hyoid bone, opens mouth by depressing mandible
		Posterior belly: mastoid notch of temporal bone		*Posterior belly:* digastric branch of facial nerve	*Posterior belly:* posterior auricular and occipital arteries	
Dilator pupillae	Intrinsic eye (smooth muscle)	Iridal part of retina	Pupillary margin of iris	Long ciliary nerves (sympathetic fibers from superior cervical ganglion)	Ophthalmic artery	Dilates pupil
Frontalis m.	Facial expression (epicranial)	Epicranial aponeurosis (at level of coronal suture)	Skin of frontal region, epicranial aponeurosis	Temporal branches of facial nerve	Superficial temporal artery	Horizontally wrinkles skin of forehead, elevates eyebrows
Genioglossus m.	Extrinsic tongue	Superior mental spine of mandible	Dorsum of tongue, hyoid bone	Hypoglossal nerve (CN XII)	Sublingual and submental arteries	Depresses and protrudes tongue
Geniohyoid m.	Suprahyoid	Inferior mental spine of mandible	Anterior surface of body of hyoid bone	Anterior ramus of C1 spinal nerve (via hypoglossal nerve)	Sublingual artery	Elevates hyoid bone and depresses mandible
Hyoglossus m.	Extrinsic tongue	Body and greater horn of hyoid bone	Margin and inferior surface of tongue	Hypoglossal nerve (CN XII)	Sublingual and submental arteries	Depresses and retracts tongue
Inferior longitudinal lingual m.	Intrinsic tongue	Inferior surface of tongue	Apex of tongue	Hypoglossal nerve (CN XII)	Lingual and facial arteries	Shortens tongue, turns tip and sides inferiorly

Table 2.10   **Muscles**

MUSCLE	MUSCLE GROUP	ORIGIN ATTACHMENT	INSERTION ATTACHMENT	INNERVATION	BLOOD SUPPLY	MAIN ACTIONS
Inferior oblique m.	Extraocular	Anterior floor of orbit (lateral to nasolacrimal canal)	Sclera lateral to corneoscleral junction	Oculomotor nerve (CN III)	Ophthalmic artery	Abducts, elevates, and laterally rotates eyeball
Inferior pharyngeal constrictor	Circular pharyngeal	Oblique line of thyroid cartilage, cricoid cartilage	Pharyngeal raphe	Vagus nerve (via pharyngeal plexus)	Ascending pharyngeal and superior thyroid arteries	Constricts wall of pharynx during swallowing
Inferior rectus m.	Extraocular	Common tendinous ring	Sclera inferior to corneoscleral junction	Oculomotor nerve (CN III)	Ophthalmic artery	Depresses, adducts, and laterally rotates eyeball
Lateral cricoarytenoid m.	Laryngeal	Arch of cricoid cartilage	Muscular process of arytenoid cartilage	Recurrent laryngeal nerve	Superior and inferior thyroid arteries	Adducts vocal folds
Lateral pterygoid m.	Mastication	*Superior head:* infratemporal surface of greater wing of sphenoid bone	Pterygoid fovea of mandible, capsule and articular disc of temporomandibular joint	Mandibular nerve (CN V$_3$)	Maxillary artery	*Bilaterally:* protrudes mandible
		Inferior head: lateral pterygoid plate				*Unilaterally and alternately:* produces side-to-side grinding
Lateral rectus m.	Extraocular	Common tendinous ring	Sclera lateral to corneoscleral junction	Abducens nerve (CN VI)	Ophthalmic artery	Abducts eyeball
Levator anguli oris	Facial expression	Canine fossa of maxilla	Modiolus of angle of mouth	Zygomatic and buccal branches of facial nerve	Superior labial artery	Elevates angle of mouth
Levator labii superioris	Facial expression	Maxilla superior to infraorbital foramen	Skin of upper lip	Zygomatic and buccal branches of facial nerve	Superior labial and angular arteries	Elevates upper lip, dilates nares
Levator nasolabialis (levator labii superioris alaeque nasi)	Facial expression	Superior part of frontal process of maxilla	Major alar cartilage, skin of nose, lateral part of upper lip	Zygomatic and buccal branches of facial nerve	Superior labial and angular arteries	Elevates upper lip and dilates nostril
Levator palpebrae superioris	Extraocular	Lesser wing of sphenoid bone (anterior to optic canal)	Superior tarsus	Oculomotor nerve (CN III)	Ophthalmic artery	Elevates upper eyelid
Levator veli palatini	Palatal	Petrous part of temporal bone, auditory tube	Palatine aponeurosis	Vagus nerve (via pharyngeal plexus)	Ascending and descending palatine arteries	Elevates soft palate during swallowing
Longus capitis m.	Anterior vertebral	Anterior tubercles of transverse processes of C3–C6 vertebrae	Inferior surface of basilar part of occipital bone	Anterior rami of C1–C3 spinal nerves	Ascending cervical, ascending pharyngeal and vertebral arteries	Flexes head
Longus colli m.	Anterior vertebral	*Vertical portion:* C5–T3 vertebrae	*Vertical portion:* C2–C4 vertebrae	Anterior rami of C2–C8 spinal nerves	Ascending pharyngeal, ascending cervical and vertebral arteries	*Bilaterally:* flex and assist in rotating cervical vertebrae and head
		Inferior oblique portion: T1–T3 vertebrae	*Inferior oblique portion:* anterior tubercles of transverse processes of C3–C6 vertebrae			*Unilaterally:* laterally flexes vertebral column
		Superior oblique portion: anterior tubercles of transverse processes of C3–C5 vertebrae	*Superior oblique portion:* anterior tubercle of atlas			
Masseter	Mastication	Zygomatic arch	Ramus of mandible, coronoid process	Mandibular nerve (CN V$_3$)	Transverse facial, masseteric, and facial arteries	Elevates and protrudes mandible; deep fibers retract it

Continued

MUSCLE	MUSCLE GROUP	ORIGIN ATTACHMENT	INSERTION ATTACHMENT	INNERVATION	BLOOD SUPPLY	MAIN ACTIONS
Medial pterygoid m.	Mastication	Medial surface of lateral pterygoid plate, pyramidal process of palatine bone, maxillary tuberosity	Medial surface of ramus and angle of mandible (inferior to inferior alveolar foramen)	Mandibular nerve (V_3)	Facial and maxillary arteries	Bilaterally: protrudes and elevates mandible. Unilaterally and alternately: produces side-to-side movements
Medial rectus m.	Extraocular	Common tendinous ring	Sclera medial to corneoscleral junction	Oculomotor nerve (CN III)	Ophthalmic artery	Adducts eyeball
Mentalis m.	Facial expression	Incisive fossa of mandible	Skin of chin	Marginal mandibular branch of facial nerve	Inferior labial artery	Elevates and protrudes lower lip
Middle pharyngeal constrictor	Circular pharyngeal	Stylohyoid ligament, horns of hyoid bone	Pharyngeal raphe	Vagus nerve (via pharyngeal plexus)	Ascending pharyngeal and ascending palatine arteries, tonsillar branches of facial artery, dorsal lingual branches of lingual artery	Constricts wall of pharynx during swallowing
Musculus uvulae	Palatal	Posterior nasal spine (of palatine bone), palatine aponeurosis	Mucosa of uvula of palate	Vagus nerve (via pharyngeal plexus)	Ascending and descending palatine arteries	Shortens, elevates, and retracts uvula of the palate
Mylohyoid m.	Suprahyoid	Mylohyoid line of mandible	Median raphe of mylohyoid, body of hyoid bone	Inferior alveolar nerve (branch of mandibular nerve)	Sublingual and submental arteries	Elevates hyoid bone, base of tongue, floor of mouth; depresses mandible
Nasalis m.	Facial expression	Canine eminence (superior and lateral to incisive fossa of maxilla)	Aponeurosis on nasal cartilages	Buccal branches of facial nerve	Superior labial artery, septal and lateral nasal branches of facial artery	Draws ala of nose toward nasal septum, compresses nostrils; alar part opens nostrils
Oblique arytenoid m.	Laryngeal	Arytenoid cartilage	Opposite arytenoid cartilage	Recurrent laryngeal nerve	Superior and inferior thyroid arteries	Closes intercartilaginous portion of rima glottidis
Occipitalis m.	Facial expression (epicranial)	Lateral two-thirds of superior nuchal line, mastoid process	Skin of occipital region, epicranial aponeurosis	Posterior auricular nerve (branch of facial nerve)	Posterior auricular and occipital arteries	Moves scalp backward
Omohyoid m.	Infrahyoid	*Inferior belly:* superior border of scapula, superior transverse scapular ligament *Superior belly:* body of hyoid bone	Intermediate tendon of omohyoid	*Inferior belly:* ansa cervicalis (C2–C3) *Superior belly:* ansa cervicalis (C1)	Lingual and superior thyroid arteries	Steadies and depresses hyoid bone
Orbicularis oculi m.	Facial expression	Medial orbital margin, medial palpebral ligament, lacrimal bone	Skin around orbit, lateral palpebral ligament, upper and lower eyelids	Temporal and zygomatic branches of facial nerve	Facial and superficial temporal arteries	Closes eyelids
Orbicularis oris m.	Facial expression	Maxilla and mandible, perioral skin and muscles	Modiolus of angle of mouth	Buccal and marginal mandibular branches of facial nerve	Inferior and superior labial arteries	Compression, contraction, and protrusion of lips
Palatoglossus m.	Palatal	Palatine aponeurosis	Margin of tongue	Vagus nerve (via pharyngeal plexus)	Ascending pharyngeal artery, palatine branches of facial and maxillary arteries	Elevates posterior tongue, depresses palate
Palatopharyngeus m.	Palatal	Hard palate, palatine aponeurosis	Lateral pharyngeal wall	Vagus nerve (via pharyngeal plexus)	Ascending and descending palatine arteries	Tenses soft palate; pulls walls of pharynx superiorly, anteriorly, and medially during swallowing

Table 2.12   **Muscles**

MUSCLE	MUSCLE GROUP	ORIGIN ATTACHMENT	INSERTION ATTACHMENT	INNERVATION	BLOOD SUPPLY	MAIN ACTIONS
Platysma	Facial expression (neck)	Skin of infraclavicular region	Mandible, muscles of lower lip	Cervical branch of facial nerve	Submental and suprascapular arteries	Tenses skin of neck
Posterior cricoarytenoid m.	Laryngeal	Posterior surface of lamina of cricoid cartilage	Muscular process of arytenoid cartilage	Recurrent laryngeal nerve	Superior and inferior thyroid arteries	Abducts vocal folds
Procerus m.	Facial expression	Fascia covering inferior part of nasal bone and superior part of lateral nasal cartilage	Skin medial and superior to eyebrow	Temporal and zygomatic branches of facial nerve	Angular artery, lateral nasal branches of facial artery	Draws down medial angle of eyebrows, produces transverse wrinkles over bridge of nose
Rectus anterior capitis m.	Anterior vertebral	Lateral mass of atlas	Basilar part of occipital bone	Anterior rami of C1–C2 spinal nerves	Vertebral and ascending pharyngeal arteries	Flexes head
Rectus lateralis capitis m.	Anterior vertebral	Superior surface of transverse process of atlas	Inferior surface of jugular process of occipital bone	Anterior rami of C1–C2 spinal nerves	Vertebral, occipital, and ascending pharyngeal arteries	Flexes head laterally to same side
Risorius m.	Facial expression	Fascia over masseter	Modiolus of angle of mouth	Buccal branches of facial nerve	Superior labial artery	Retracts angle of mouth
Salpingopharyngeus m.	Longitudinal pharyngeal	Auditory tube	Lateral pharyngeal wall	Vagus nerve (via pharyngeal plexus)	Ascending pharyngeal artery	Elevates pharynx and larynx during swallowing and speaking
Scalenus anterior m.	Lateral vertebral	Anterior tubercles of transverse processes of C3–C6 vertebrae	Scalene tubercle on 1st rib	Anterior rami of C5–C8 spinal nerves	Ascending cervical artery	Elevates 1st rib, bends neck
Scalenus medius m.	Lateral vertebral	Posterior tubercles of transverse processes of C2–C7 vertebrae	Superior surface of 1st rib (behind subclavian groove)	Anterior rami of C3–C7 spinal nerves	Ascending cervical artery	Elevates 1st rib, bends neck
Scalenus posterior m.	Lateral vertebral	Posterior tubercles of transverse processes of C4–C6 vertebrae	External surface of 2nd rib	Anterior rami of C5–C8 spinal nerves	Ascending cervical artery, superficial branch of transverse cervical artery	Elevates 2nd rib, bends neck
Sphincter pupillae	Intrinsic eye (smooth muscle)	Circular smooth muscle of iris that passes around pupil	Blends with dilator pupillae fibers	Short ciliary nerves (parasympathetic fibers from ciliary ganglion)	Ophthalmic artery	Constricts pupil
Stapedius m.	Middle ear	Pyramidal eminence of temporal bone (in tympanic cavity)	Stapes	Nerve to stapedius muscle (branch of facial nerve)	Posterior auricular, anterior tympanic, and middle meningeal arteries	Pulls stapes posteriorly to lessen oscillation of tympanic membrane
Sternocleidomastoid m.	Neck	*Sternal head:* anterior surface of manubrium stemi; *Clavicular head:* medial third of superior surface of clavicle	Lateral surface of mastoid process; lateral half of superior nuchal line of occipital bone	Spinal accessory nerve (CN XI)	Superior thyroid, occipital, suprascapular, and posterior auricular arteries	*Bilaterally:* flexes head, elevates thorax; *Unilaterally:* turns face toward opposite side
Sternohyoid m.	Infrahyoid	Posterior surface of manubrium sterni, posterior sternoclavicular ligament, sternal end of clavicle	Medial part of inferior border of body of hyoid bone	Ansa cervicalis (C1–C3)	Superior thyroid and lingual arteries	Depresses larynx and hyoid bone, steadies hyoid bone
Sternothyroid m.	Infrahyoid	Posterior surface of manubrium sterni, posterior border of 1st costal cartilage	Oblique line of thyroid cartilage	Ansa cervicalis (C1–C3)	Cricothyroid branch of superior thyroid artery	Depresses larynx and thyroid cartilage
Styloglossus m.	Extrinsic tongue	Styloid process of temporal bone, stylohyoid ligament	Margin and inferior surface of tongue	Hypoglossal nerve (CN XII)	Sublingual artery	Retracts tongue and draws it up for swallowing
Stylohyoid m.	Suprahyoid	Posterior border of styloid process of temporal bone	Body of hyoid bone (at junction with greater horn)	Stylohyoid branch of facial nerve	Facial and occipital arteries	Elevates and retracts hyoid bone

Continued

MUSCLE	MUSCLE GROUP	ORIGIN ATTACHMENT	INSERTION ATTACHMENT	INNERVATION	BLOOD SUPPLY	MAIN ACTIONS
Stylopharyngeus m.	Longitudinal pharyngeal	Medial surface of styloid process of temporal bone	Pharyngeal wall, posterior border of thyroid cartilage	Glossopharyngeal nerve (CN IX)	Ascending pharyngeal and ascending palatine arteries, tonsillar branches of facial artery, dorsal lingual branches of lingual artery	Elevates pharynx and larynx during swallowing and speaking
Superior longitudinal lingual m.	Intrinsic tongue	Submucosa of posterior part of dorsum of tongue	Apex of tongue; (unites with muscle of opposite side)	Hypoglossal nerve (CN XII)	Deep lingual and facial arteries	Shortens tongue, turns tip and sides superiorly
Superior oblique m.	Extraocular	Body of sphenoid bone (above optic canal)	Sclera superior to corneoscleral junction (after passing through trochlea)	Trochlear nerve (CN IV)	Ophthalmic artery	Abducts, depresses, and medially rotates eyeball
Superior pharyngeal constrictor	Circular pharyngeal	Pterygoid hamulus, pterygomandibular raphe, mylohyoid line of mandible	Pharyngeal raphe	Vagus nerve (via pharyngeal plexus)	Ascending pharyngeal and ascending palatine arteries, tonsillar branches of facial artery, dorsal branches of lingual artery	Constricts wall of pharynx during swallowing
Superior rectus m.	Extraocular	Common tendinous ring	Sclera superior to corneoscleral junction	Oculomotor nerve (CN III)	Ophthalmic artery	Elevates, adducts, and medially rotates eyeball
Temporalis m.	Mastication	Temporal fossa, deep layer of temporal fascia	Coronoid process, ramus of mandible	Deep temporal nerves (branches of mandibular nerve)	Superficial temporal and maxillary arteries	Elevates mandible; posterior fibers retract mandible
Tensor tympani	Middle ear	Cartilage of auditory tube	Handle of malleus	Mandibular nerve (CN V$_3$)	Superior tympanic artery	Tenses tympanic membrane by drawing it medially
Tensor veli palatini	Palatal	Scaphoid fossa and spine of sphenoid bone, auditory tube	Palatine aponeurosis	Mandibular nerve (CN V$_3$)	Ascending and descending palatine arteries	Tenses soft palate, opens auditory tube during swallowing and yawning
Thyroarytenoid m.	Laryngeal	Internal surface of thyroid cartilage	Muscular process of arytenoid cartilage	Recurrent laryngeal nerve	Superior and inferior thyroid arteries	Shortens and relaxes vocal cords, sphincter of vestibule
Thyrohyoid m.	Infrahyoid	Oblique line of thyroid cartilage	Inferior border of body and greater horn of hyoid bone	Anterior ramus of C1 spinal nerve (via hypoglossal nerve)	Superior thyroid artery	Depresses larynx and hyoid bone, elevates larynx when hyoid bone is fixed
Transverse arytenoid m.	Laryngeal	Arytenoid cartilage	Opposite arytenoid cartilage	Recurrent laryngeal nerve	Superior and inferior thyroid arteries	Closes intercartilaginous portion of rima glottidis
Transversus linguae m.	Intrinsic tongue	Septum of tongue	Dorsum and margin of tongue	Hypoglossal nerve (CN XII)	Deep lingual and facial arteries	Narrows and elongates tongue
Verticalis linguae m.	Intrinsic tongue	Mucosa of anterior part of dorsum of tongue	Inferior surface of tongue	Hypoglossal nerve (CN XII)	Deep lingual and facial arteries	Flattens and broadens tongue
Vocalis m.	Laryngeal	Vocal process of arytenoid cartilage	Vocal ligament	Recurrent laryngeal nerve	Superior and inferior thyroid arteries	Tenses anterior part of vocal ligament, relaxes posterior part of vocal ligament
Zygomaticus major m.	Facial expression	Zygomatic arch	Modiolus of angle of mouth	Zygomatic and buccal branches of facial nerve	Superior labial artery	Draws angle of mouth posteriorly and superiorly
Zygomaticus minor m.	Facial expression	Zygomatic arch	Modiolus of angle of mouth, upper lip	Zygomatic and buccal branches of facial nerve	Superior labial artery	Elevates upper lip

Variations in spinal nerve contributions to the innervation of muscles, their arterial supply, their attachments, and their actions are common themes in human anatomy. Therefore, expect differences between texts and realize that anatomical variation is normal.

Table 2.14

Muscles

BACK 3

ELECTRONIC BONUS PLATES

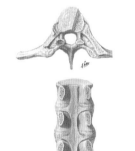

BP 33 Ligaments of Vertebral Column

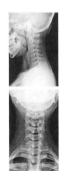

BP 34 Cervical Spine: Radiographs

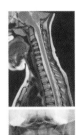

BP 35 Cervical Spine: MRI and Radiograph

BP 36 Thoracolumbar Spine: Lateral Radiograph

BP 37 Lumbar Vertebrae: Radiographs

BP 38 Lumbar Spine: MRIs

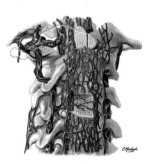

BP 39 Vertebral Veins: Detail Showing Venous Communications

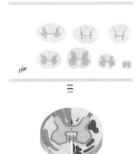

BP 40 Spinal Cord Cross Sections: Fiber Tracts

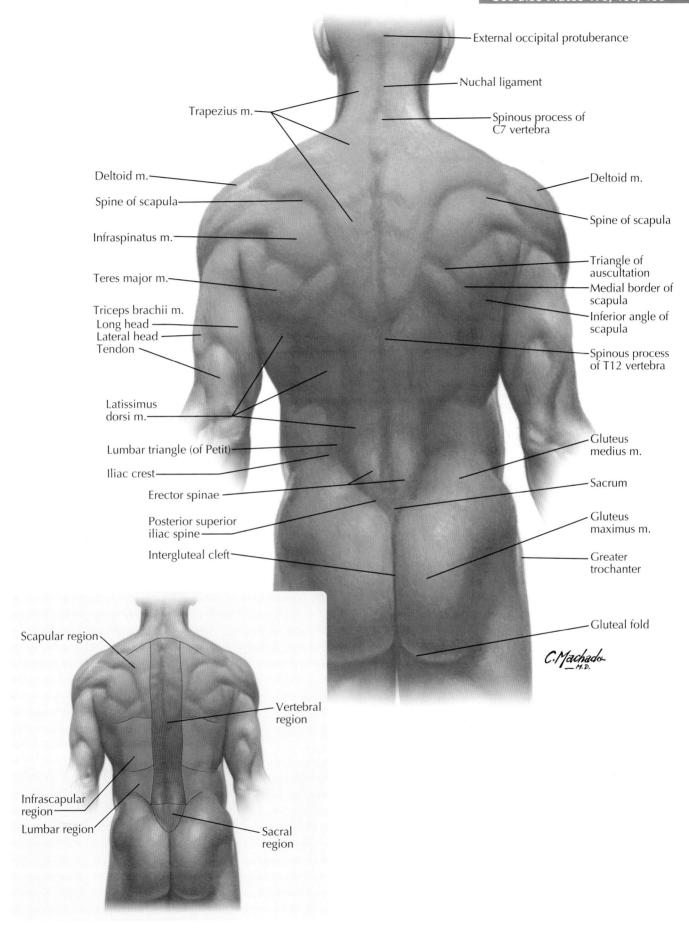

External occipital protuberance

Nuchal ligament

Trapezius m.

Spinous process of C7 vertebra

Deltoid m.

Deltoid m.

Spine of scapula

Spine of scapula

Infraspinatus m.

Triangle of auscultation

Teres major m.

Medial border of scapula

Triceps brachii m.
Long head
Lateral head
Tendon

Inferior angle of scapula

Spinous process of T12 vertebra

Latissimus dorsi m.

Lumbar triangle (of Petit)

Gluteus medius m.

Iliac crest

Sacrum

Erector spinae

Posterior superior iliac spine

Gluteus maximus m.

Intergluteal cleft

Greater trochanter

Gluteal fold

C. Machado
M.D.

Scapular region

Vertebral region

Infrascapular region

Lumbar region

Sacral region

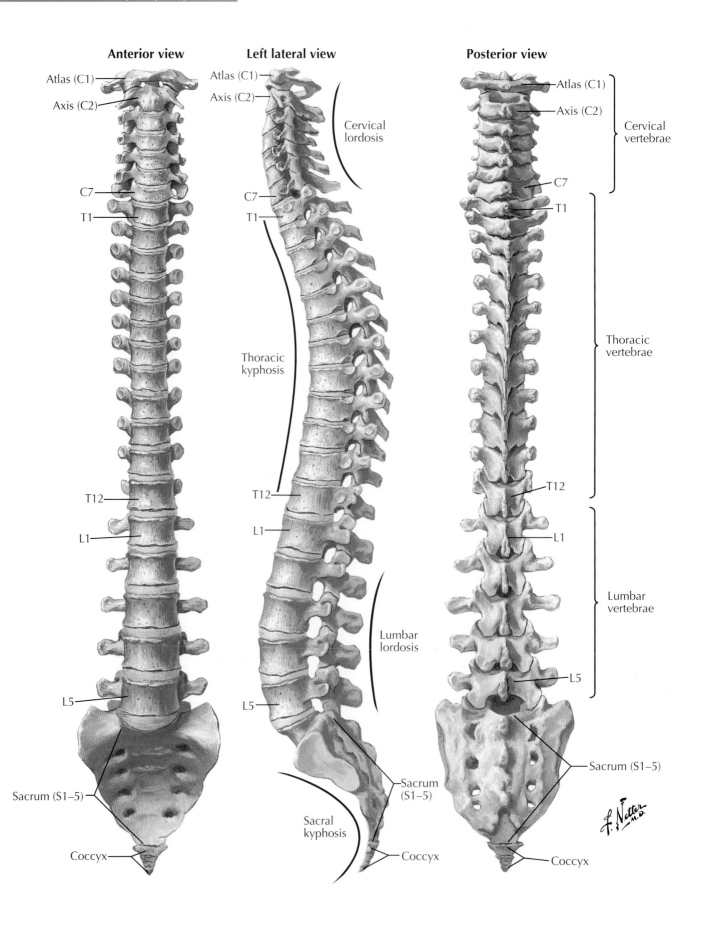

Anterior view

- Atlas (C1)
- Axis (C2)
- C7
- T1
- T12
- L1
- L5
- Sacrum (S1–5)
- Coccyx

Left lateral view

- Atlas (C1)
- Axis (C2)
- Cervical lordosis
- C7
- T1
- Thoracic kyphosis
- T12
- L1
- Lumbar lordosis
- L5
- Sacrum (S1–5)
- Sacral kyphosis
- Coccyx

Posterior view

- Atlas (C1)
- Axis (C2)
- Cervical vertebrae
- C7
- T1
- Thoracic vertebrae
- T12
- L1
- Lumbar vertebrae
- L5
- Sacrum (S1–5)
- Coccyx

Plate 179

Vertebral Column

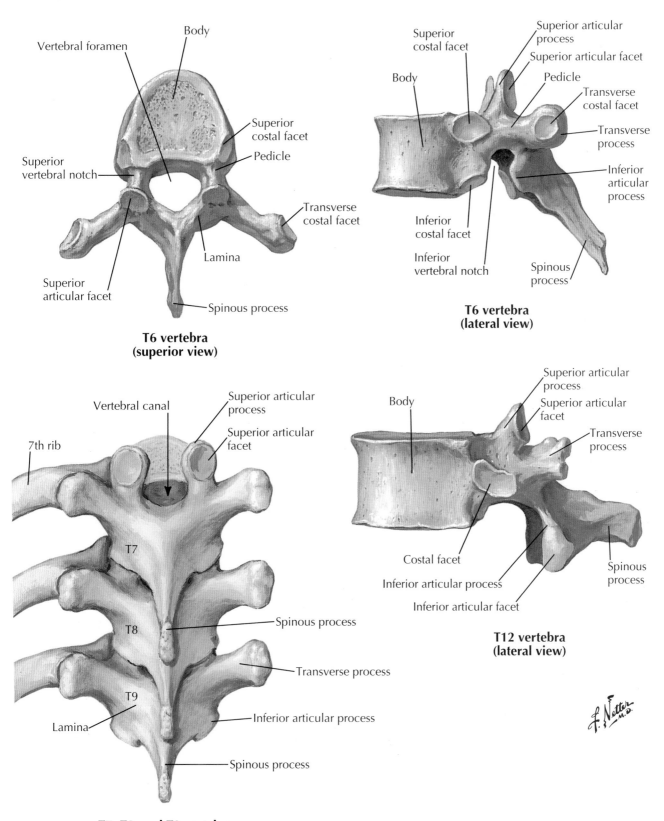

Vertebral foramen

Body

Superior costal facet

Pedicle

Superior vertebral notch

Superior articular facet

Lamina

Transverse costal facet

Spinous process

T6 vertebra (superior view)

Superior costal facet

Body

Superior articular process

Superior articular facet

Pedicle

Transverse costal facet

Transverse process

Inferior articular process

Inferior costal facet

Inferior vertebral notch

Spinous process

T6 vertebra (lateral view)

Vertebral canal

Superior articular process

Superior articular facet

7th rib

T7

T8

T9

Lamina

Spinous process

Transverse process

Inferior articular process

Spinous process

T7, T8, and T9 vertebrae (posterior view)

Body

Superior articular process

Superior articular facet

Transverse process

Costal facet

Inferior articular process

Inferior articular facet

Spinous process

T12 vertebra (lateral view)

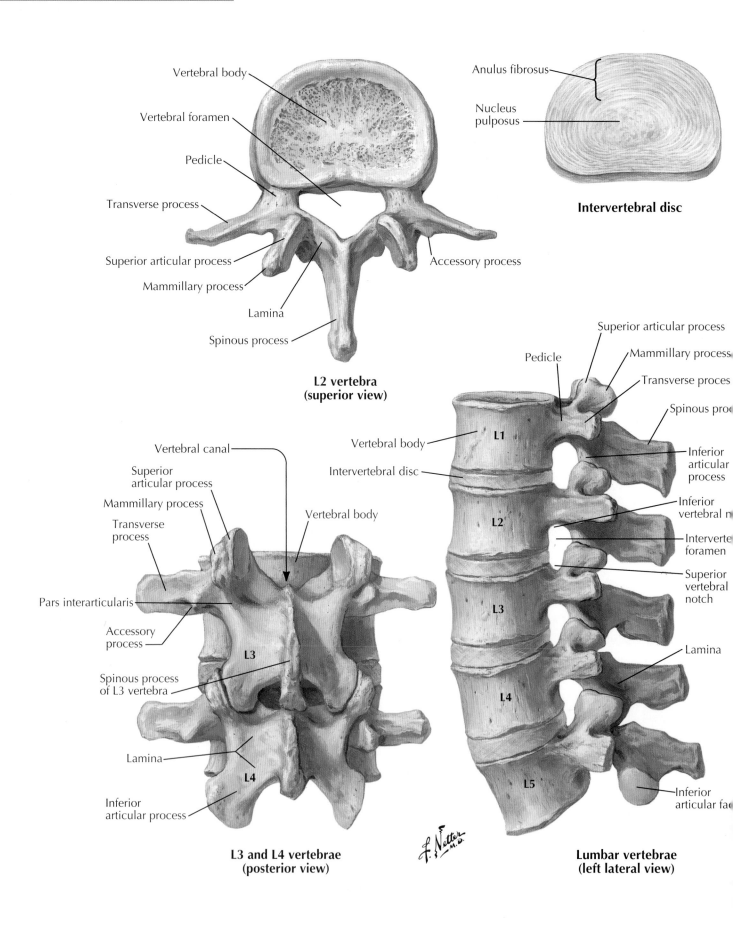

Vertebral body

Vertebral foramen

Pedicle

Transverse process

Superior articular process

Mammillary process

Lamina

Spinous process

**L2 vertebra
(superior view)**

Anulus fibrosus

Nucleus pulposus

Intervertebral disc

Vertebral canal

Superior articular process

Mammillary process

Transverse process

Pars interarticularis

Accessory process

Spinous process of L3 vertebra

Vertebral body

L3

L4

Lamina

Inferior articular process

**L3 and L4 vertebrae
(posterior view)**

Pedicle

Vertebral body

Intervertebral disc

L1

L2

L3

L4

L5

Superior articular process

Mammillary process

Transverse process

Spinous process

Inferior articular process

Inferior vertebral notch

Intervertebral foramen

Superior vertebral notch

Lamina

Inferior articular facet

**Lumbar vertebrae
(left lateral view)**

Plate 181

Vertebral Column

See also **Plates 179, 268, BP 36**

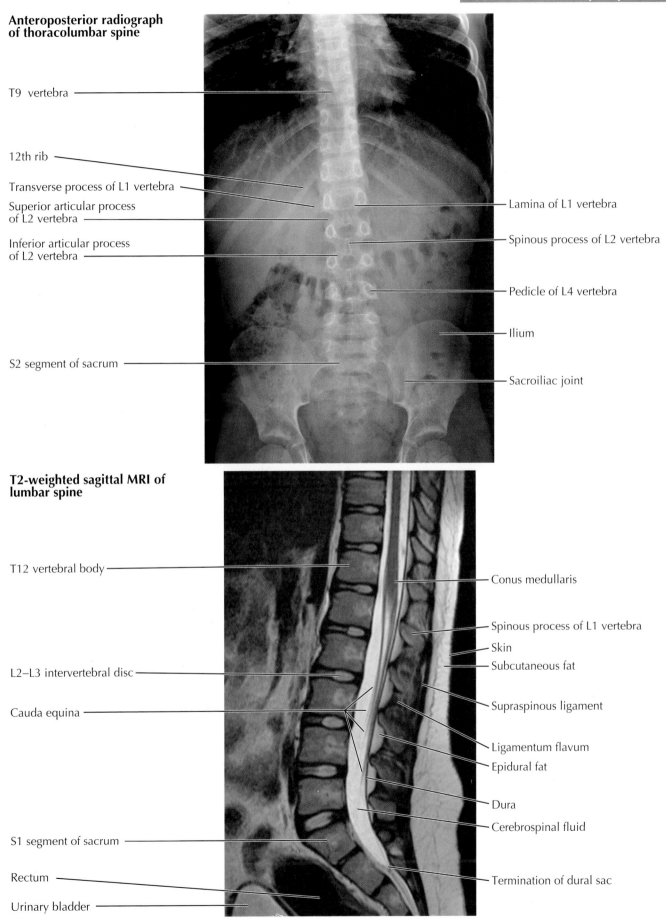

Anteroposterior radiograph of thoracolumbar spine

T9 vertebra

12th rib

Transverse process of L1 vertebra

Superior articular process of L2 vertebra

Inferior articular process of L2 vertebra

S2 segment of sacrum

Lamina of L1 vertebra

Spinous process of L2 vertebra

Pedicle of L4 vertebra

Ilium

Sacroiliac joint

T2-weighted sagittal MRI of lumbar spine

T12 vertebral body

L2–L3 intervertebral disc

Cauda equina

S1 segment of sacrum

Rectum

Urinary bladder

Conus medullaris

Spinous process of L1 vertebra

Skin

Subcutaneous fat

Supraspinous ligament

Ligamentum flavum

Epidural fat

Dura

Cerebrospinal fluid

Termination of dural sac

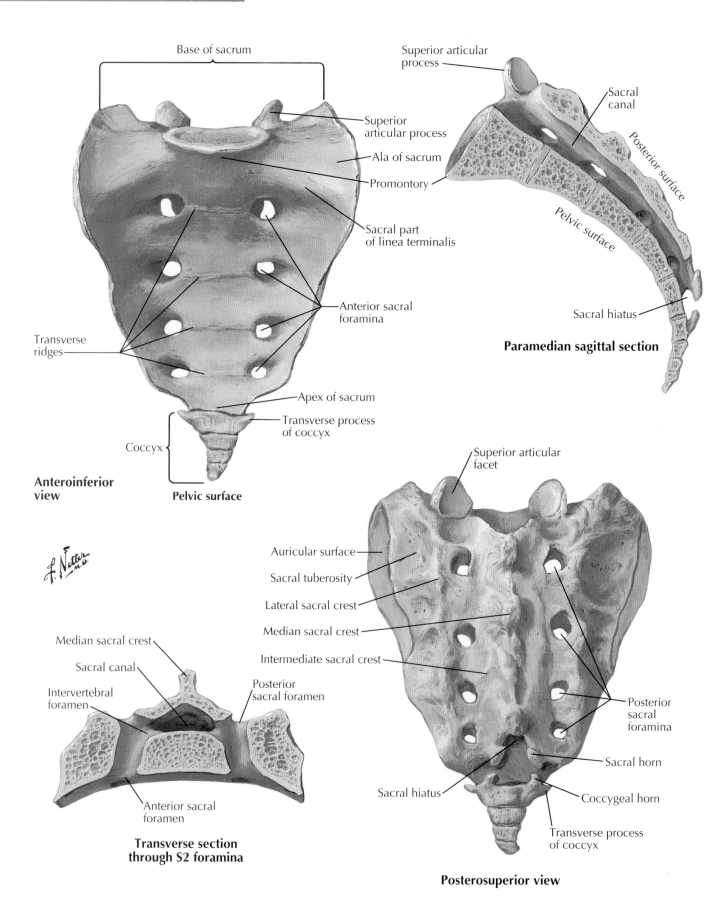

Base of sacrum

Superior articular process

Superior articular process

Ala of sacrum

Promontory

Sacral part of linea terminalis

Anterior sacral foramina

Transverse ridges

Apex of sacrum

Transverse process of coccyx

Coccyx

Anteroinferior view

Pelvic surface

Superior articular process

Sacral canal

Posterior surface

Pelvic surface

Sacral hiatus

Paramedian sagittal section

Median sacral crest

Sacral canal

Intervertebral foramen

Posterior sacral foramen

Anterior sacral foramen

Transverse section through S2 foramina

Superior articular facet

Auricular surface

Sacral tuberosity

Lateral sacral crest

Median sacral crest

Intermediate sacral crest

Posterior sacral foramina

Sacral horn

Coccygeal horn

Sacral hiatus

Transverse process of coccyx

Posterosuperior view

Plate 183

Vertebral Column

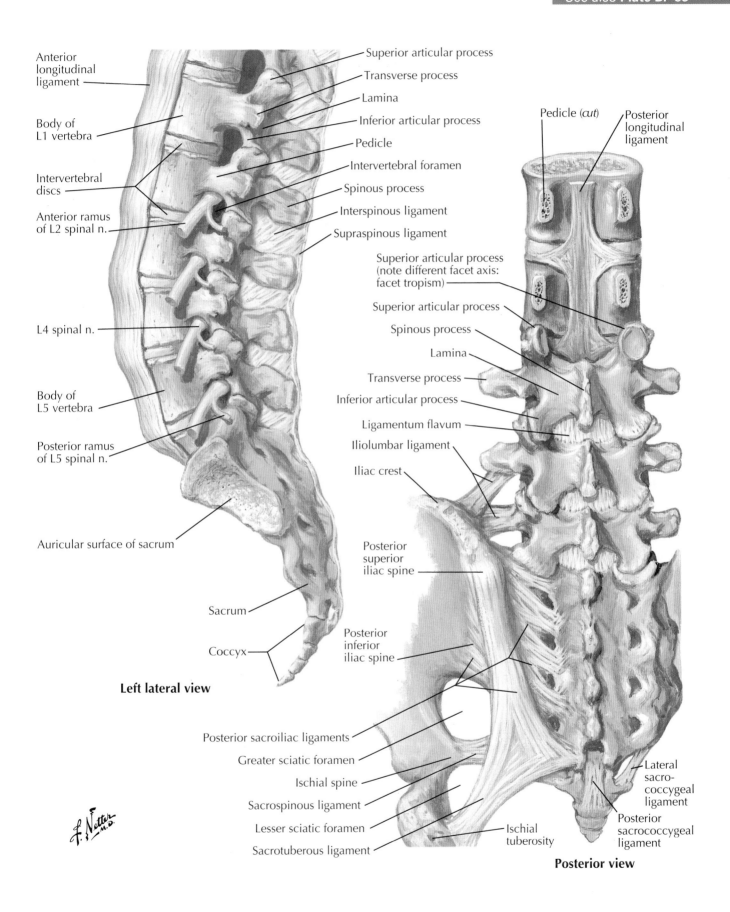

Anterior longitudinal ligament

Body of L1 vertebra

Intervertebral discs

Anterior ramus of L2 spinal n.

L4 spinal n.

Body of L5 vertebra

Posterior ramus of L5 spinal n.

Auricular surface of sacrum

Sacrum

Coccyx

Left lateral view

Superior articular process

Transverse process

Lamina

Inferior articular process

Pedicle

Intervertebral foramen

Spinous process

Interspinous ligament

Supraspinous ligament

Pedicle (*cut*)

Posterior longitudinal ligament

Superior articular process (note different facet axis: facet tropism)

Superior articular process

Spinous process

Lamina

Transverse process

Inferior articular process

Ligamentum flavum

Iliolumbar ligament

Iliac crest

Posterior superior iliac spine

Posterior inferior iliac spine

Posterior sacroiliac ligaments

Greater sciatic foramen

Ischial spine

Sacrospinous ligament

Lesser sciatic foramen

Sacrotuberous ligament

Ischial tuberosity

Lateral sacro-coccygeal ligament

Posterior sacrococcygeal ligament

Posterior view

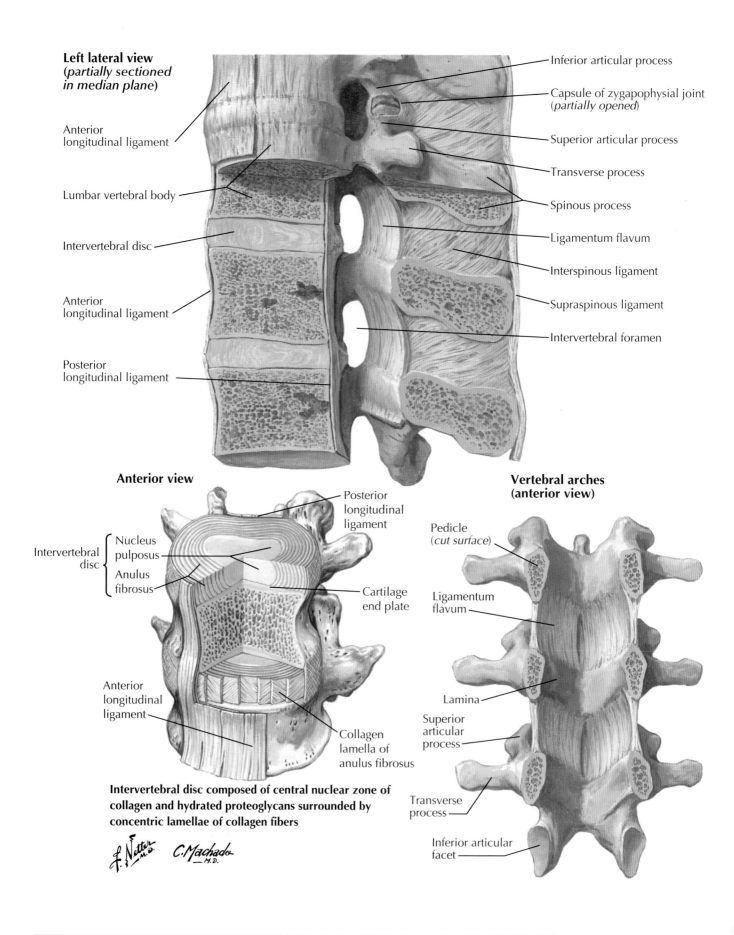

Left lateral view (*partially sectioned in median plane*)

Anterior longitudinal ligament

Lumbar vertebral body

Intervertebral disc

Anterior longitudinal ligament

Posterior longitudinal ligament

Inferior articular process

Capsule of zygapophysial joint (*partially opened*)

Superior articular process

Transverse process

Spinous process

Ligamentum flavum

Interspinous ligament

Supraspinous ligament

Intervertebral foramen

Anterior view

Intervertebral disc
  Nucleus pulposus
  Anulus fibrosus

Anterior longitudinal ligament

Posterior longitudinal ligament

Cartilage end plate

Collagen lamella of anulus fibrosus

Intervertebral disc composed of central nuclear zone of collagen and hydrated proteoglycans surrounded by concentric lamellae of collagen fibers

Vertebral arches (**anterior view**)

Pedicle (*cut surface*)

Ligamentum flavum

Lamina

Superior articular process

Transverse process

Inferior articular facet

Plate 185

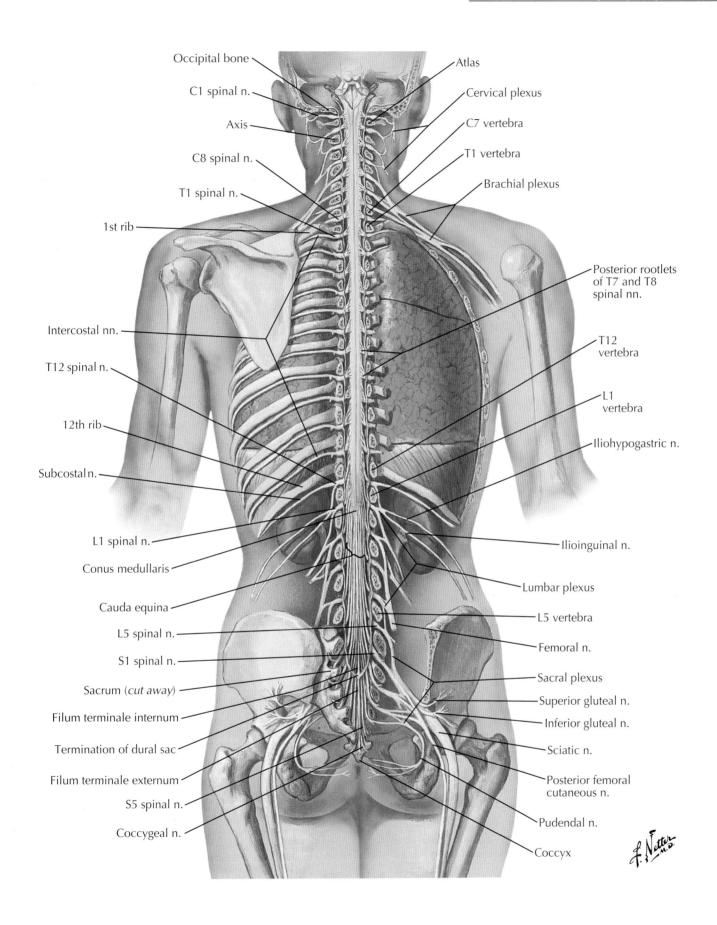

Occipital bone

C1 spinal n.

Axis

C8 spinal n.

T1 spinal n.

1st rib

Intercostal nn.

T12 spinal n.

12th rib

Subcostal n.

L1 spinal n.

Conus medullaris

Cauda equina

L5 spinal n.

S1 spinal n.

Sacrum (*cut away*)

Filum terminale internum

Termination of dural sac

Filum terminale externum

S5 spinal n.

Coccygeal n.

Atlas

Cervical plexus

C7 vertebra

T1 vertebra

Brachial plexus

Posterior rootlets of T7 and T8 spinal nn.

T12 vertebra

L1 vertebra

Iliohypogastric n.

Ilioinguinal n.

Lumbar plexus

L5 vertebra

Femoral n.

Sacral plexus

Superior gluteal n.

Inferior gluteal n.

Sciatic n.

Posterior femoral cutaneous n.

Pudendal n.

Coccyx

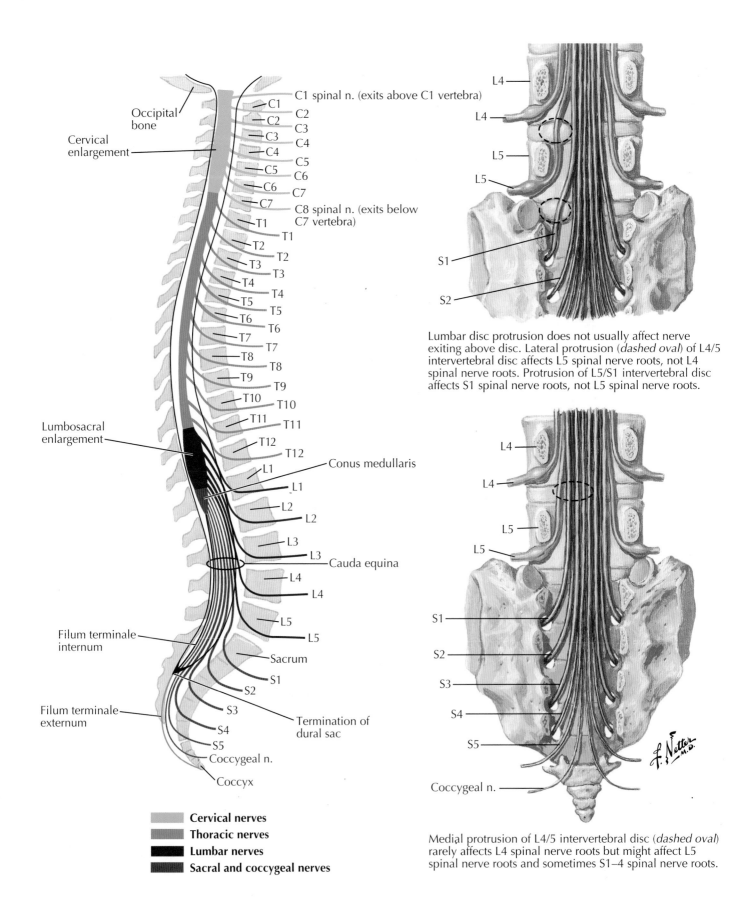

Occipital bone

Cervical enlargement

C1 spinal n. (exits above C1 vertebra)

C1
C2
C3
C4
C5
C6
C7

C1
C2
C3
C4
C5
C6
C7

C8 spinal n. (exits below C7 vertebra)

T1
T2
T3
T4
T5
T6
T7
T8
T9
T10
T11
T12

T1
T2
T3
T4
T5
T6
T7
T8
T9
T10
T11
T12

Lumbosacral enlargement

Conus medullaris

L1
L2
L3
L4
L5

L1
L2
L3
L4
L5

Cauda equina

Filum terminale internum

Sacrum

S1
S2
S3
S4
S5
Coccygeal n.

Termination of dural sac

Filum terminale externum

Coccyx

- **Cervical nerves**
- **Thoracic nerves**
- **Lumbar nerves**
- **Sacral and coccygeal nerves**

L4
L4
L5
L5
S1
S2

Lumbar disc protrusion does not usually affect nerve exiting above disc. Lateral protrusion (*dashed oval*) of L4/5 intervertebral disc affects L5 spinal nerve roots, not L4 spinal nerve roots. Protrusion of L5/S1 intervertebral disc affects S1 spinal nerve roots, not L5 spinal nerve roots.

L4
L4
L5
L5
S1
S2
S3
S4
S5
Coccygeal n.

Medial protrusion of L4/5 intervertebral disc (*dashed oval*) rarely affects L4 spinal nerve roots but might affect L5 spinal nerve roots and sometimes S1–4 spinal nerve roots.

Plate 187

Spinal Cord

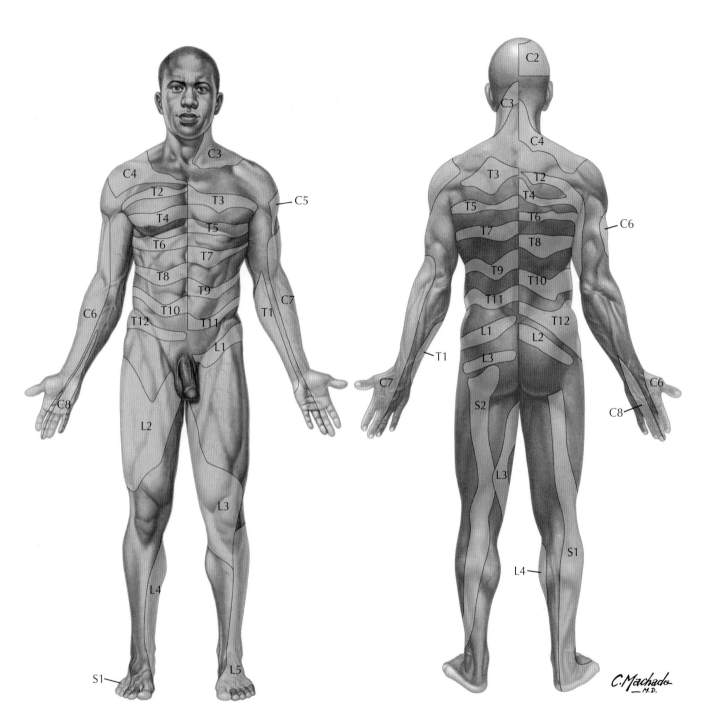

Levels of principal dermatomes

C4	Level of clavicles	**T10**	Level of umbilicus
C5, C6, C7	Lateral surfaces of upper limbs	**L1**	Inguinal region and proximal anterior thigh
C8, T1	Medial surfaces of upper limbs	**L1, L2, L3, L4**	Anteromedial lower limb and gluteal region
C6	Lateral digits	**L4, L5, S1**	Foot
C6, C7, C8	Hand	**L4**	Medial leg
C8	Medial digits	**L5, S1**	Posterolateral lower limb and dorsum of foot
T4	Level of nipples	**S1**	Lateral foot

Schematic based on Lee MW, McPhee RW, Stringer MD. An evidence-based approach to human dermatomes. Clin Anat. 2008; 21(5):363–373. doi: 10.1002/ca.20636. PMID: 18470936. Please note that these areas are not absolute and vary from person to person. S3, S4, S5, and Co supply the perineum but are not shown for reasons of clarity. Of note, the dermatomes are larger than illustrated as the figure is based on best evidence; gaps represent areas in which the data are inconclusive.

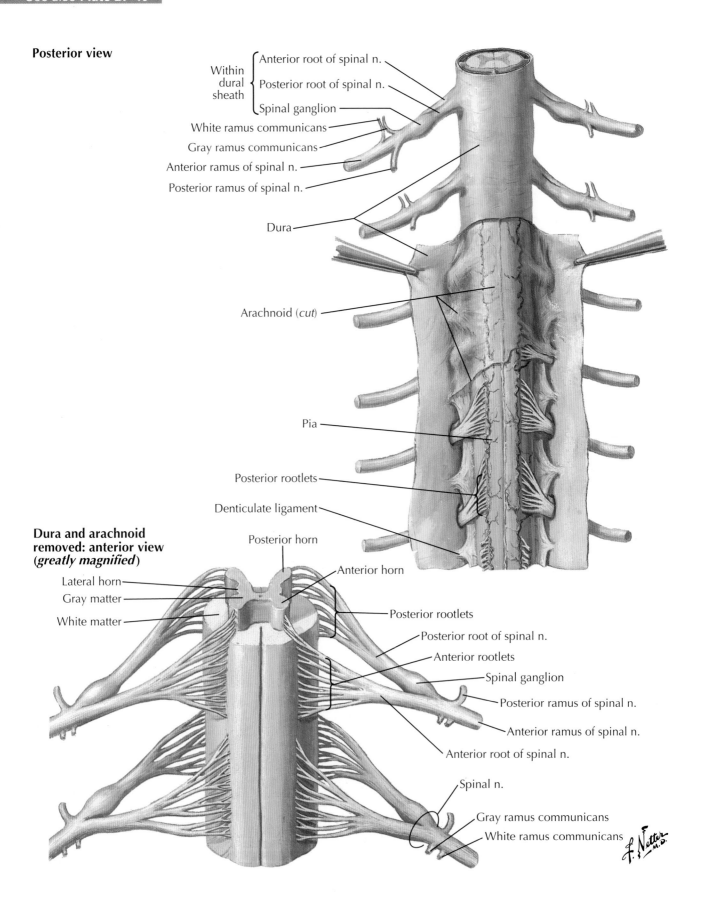

Posterior view

Within dural sheath {
Anterior root of spinal n.
Posterior root of spinal n.
Spinal ganglion
}

White ramus communicans
Gray ramus communicans
Anterior ramus of spinal n.
Posterior ramus of spinal n.

Dura

Arachnoid (*cut*)

Pia

Posterior rootlets

Denticulate ligament

Dura and arachnoid removed: anterior view (*greatly magnified*)

Lateral horn
Gray matter
White matter

Posterior horn

Anterior horn

Posterior rootlets

Posterior root of spinal n.

Anterior rootlets

Spinal ganglion

Posterior ramus of spinal n.

Anterior ramus of spinal n.

Anterior root of spinal n.

Spinal n.

Gray ramus communicans
White ramus communicans

Plate 189

Spinal Cord

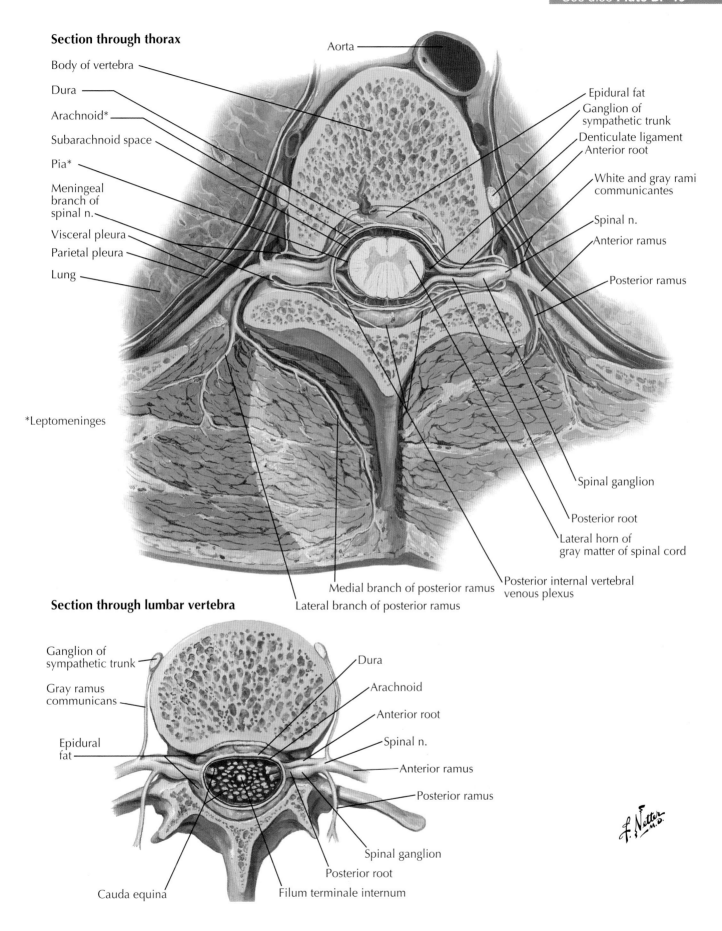

Section through thorax

Aorta

Body of vertebra

Dura

Arachnoid*

Subarachnoid space

Pia*

Meningeal branch of spinal n.

Visceral pleura

Parietal pleura

Lung

*Leptomeninges

Epidural fat

Ganglion of sympathetic trunk

Denticulate ligament

Anterior root

White and gray rami communicantes

Spinal n.

Anterior ramus

Posterior ramus

Spinal ganglion

Posterior root

Lateral horn of gray matter of spinal cord

Posterior internal vertebral venous plexus

Medial branch of posterior ramus

Lateral branch of posterior ramus

Section through lumbar vertebra

Ganglion of sympathetic trunk

Gray ramus communicans

Epidural fat

Cauda equina

Filum terminale internum

Posterior root

Spinal ganglion

Posterior ramus

Anterior ramus

Spinal n.

Anterior root

Arachnoid

Dura

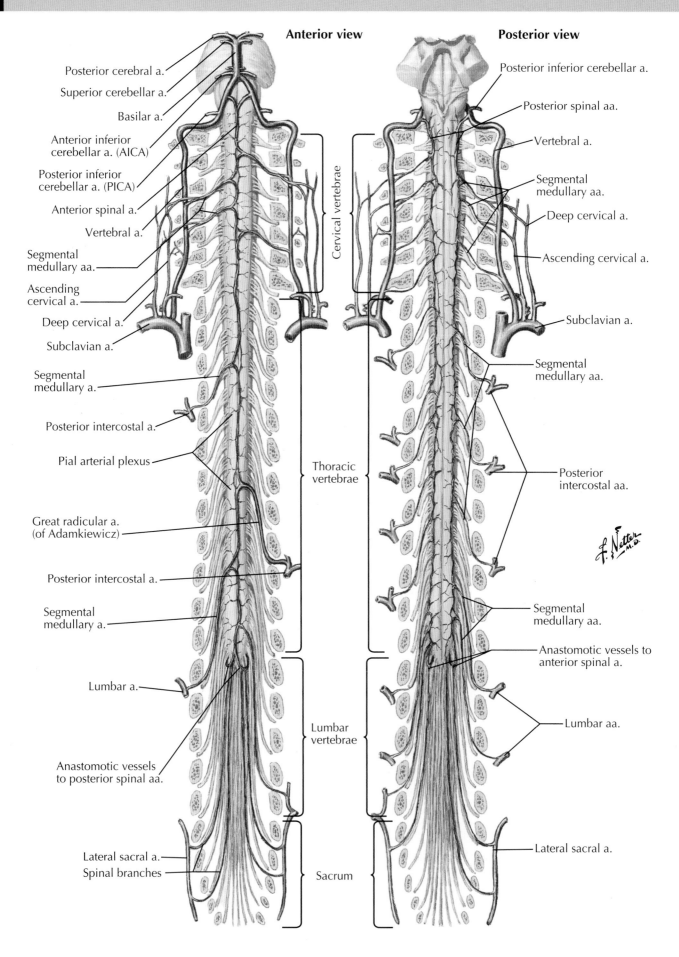

Anterior view

Posterior view

Posterior cerebral a.

Superior cerebellar a.

Basilar a.

Anterior inferior cerebellar a. (AICA)

Posterior inferior cerebellar a. (PICA)

Anterior spinal a.

Vertebral a.

Segmental medullary aa.

Ascending cervical a.

Deep cervical a.

Subclavian a.

Segmental medullary a.

Posterior intercostal a.

Pial arterial plexus

Great radicular a. (of Adamkiewicz)

Posterior intercostal a.

Segmental medullary a.

Lumbar a.

Anastomotic vessels to posterior spinal aa.

Lateral sacral a.

Spinal branches

Cervical vertebrae

Thoracic vertebrae

Lumbar vertebrae

Sacrum

Posterior inferior cerebellar a.

Posterior spinal aa.

Vertebral a.

Segmental medullary aa.

Deep cervical a.

Ascending cervical a.

Subclavian a.

Segmental medullary aa.

Posterior intercostal aa.

Segmental medullary aa.

Anastomotic vessels to anterior spinal a.

Lumbar aa.

Lateral sacral a.

Note: All spinal nerve roots have associated radicular or segmental medullary arteries. Most roots have radicular arteries (see Plate 192). Both types of arteries run along roots, but radicular arteries end before reaching anterior or posterior spinal arteries; larger segmental medullary arteries continue on to supply a segment of these arteries.

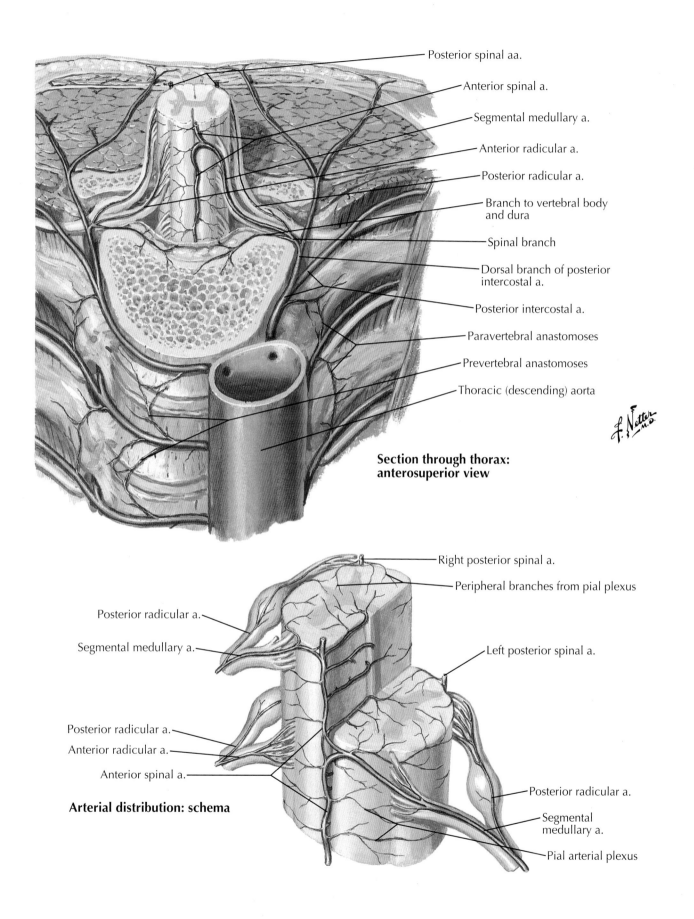

Posterior spinal aa.

Anterior spinal a.

Segmental medullary a.

Anterior radicular a.

Posterior radicular a.

Branch to vertebral body and dura

Spinal branch

Dorsal branch of posterior intercostal a.

Posterior intercostal a.

Paravertebral anastomoses

Prevertebral anastomoses

Thoracic (descending) aorta

Section through thorax: anterosuperior view

Right posterior spinal a.

Peripheral branches from pial plexus

Posterior radicular a.

Segmental medullary a.

Left posterior spinal a.

Posterior radicular a.

Anterior radicular a.

Anterior spinal a.

Posterior radicular a.

Segmental medullary a.

Pial arterial plexus

Arterial distribution: schema

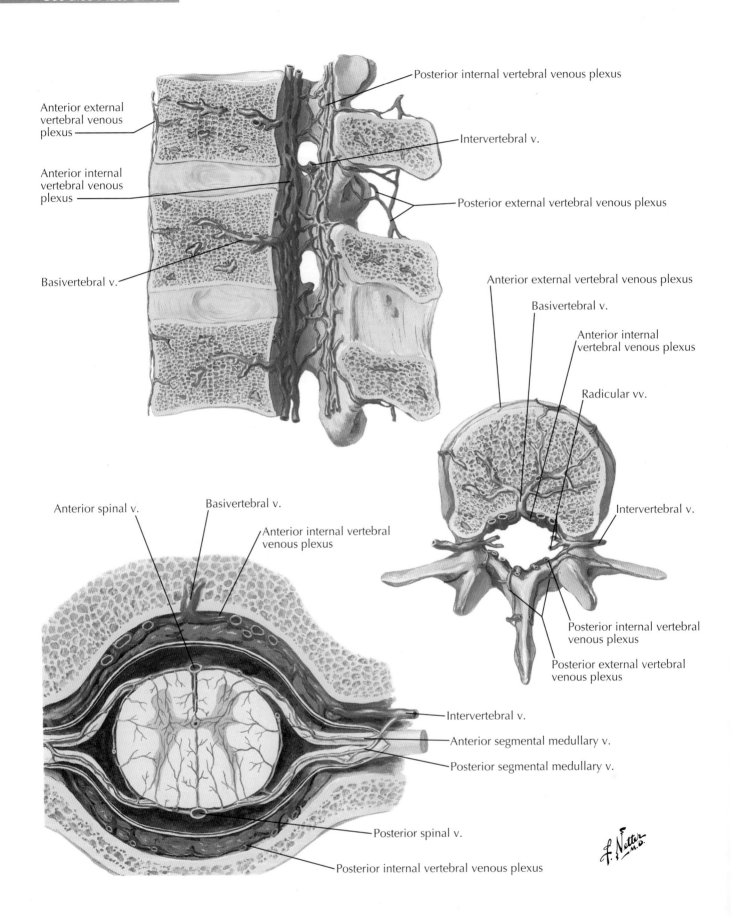

Posterior internal vertebral venous plexus

Anterior external vertebral venous plexus

Intervertebral v.

Anterior internal vertebral venous plexus

Posterior external vertebral venous plexus

Basivertebral v.

Anterior external vertebral venous plexus

Basivertebral v.

Anterior internal vertebral venous plexus

Radicular vv.

Intervertebral v.

Anterior spinal v.

Basivertebral v.

Anterior internal vertebral venous plexus

Posterior internal vertebral venous plexus

Posterior external vertebral venous plexus

Intervertebral v.

Anterior segmental medullary v.

Posterior segmental medullary v.

Posterior spinal v.

Posterior internal vertebral venous plexus

Plate 193

Spinal Cord

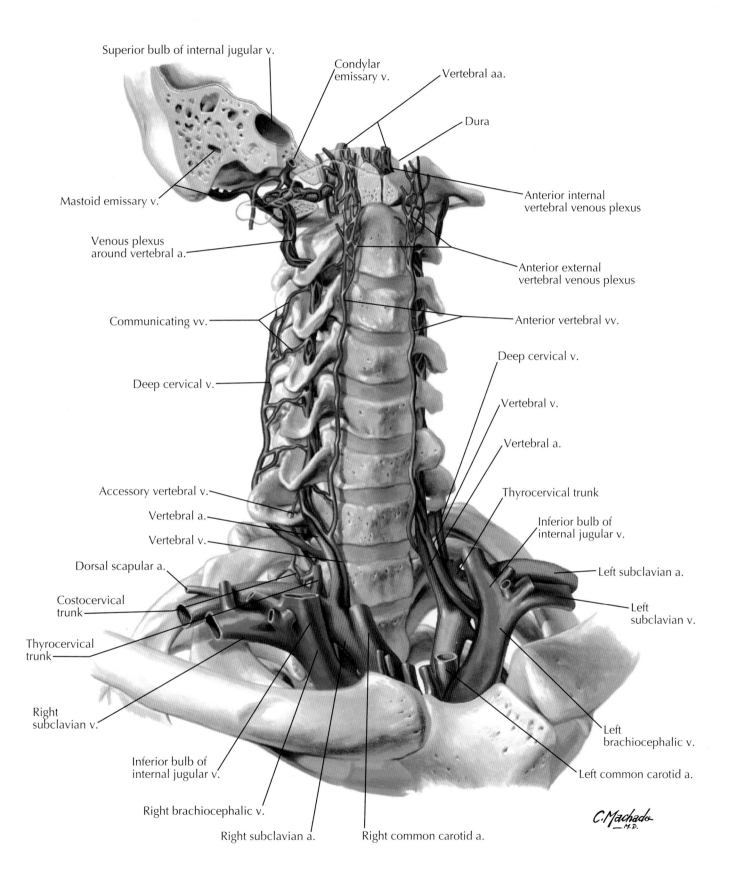

Superior bulb of internal jugular v.

Condylar emissary v.

Vertebral aa.

Dura

Mastoid emissary v.

Anterior internal vertebral venous plexus

Venous plexus around vertebral a.

Anterior external vertebral venous plexus

Communicating vv.

Anterior vertebral vv.

Deep cervical v.

Deep cervical v.

Vertebral v.

Vertebral a.

Accessory vertebral v.

Thyrocervical trunk

Vertebral a.

Inferior bulb of internal jugular v.

Vertebral v.

Left subclavian a.

Dorsal scapular a.

Costocervical trunk

Left subclavian v.

Thyrocervical trunk

Right subclavian v.

Left brachiocephalic v.

Inferior bulb of internal jugular v.

Left common carotid a.

Right brachiocephalic v.

Right subclavian a.

Right common carotid a.

C. Machado
_M.D.

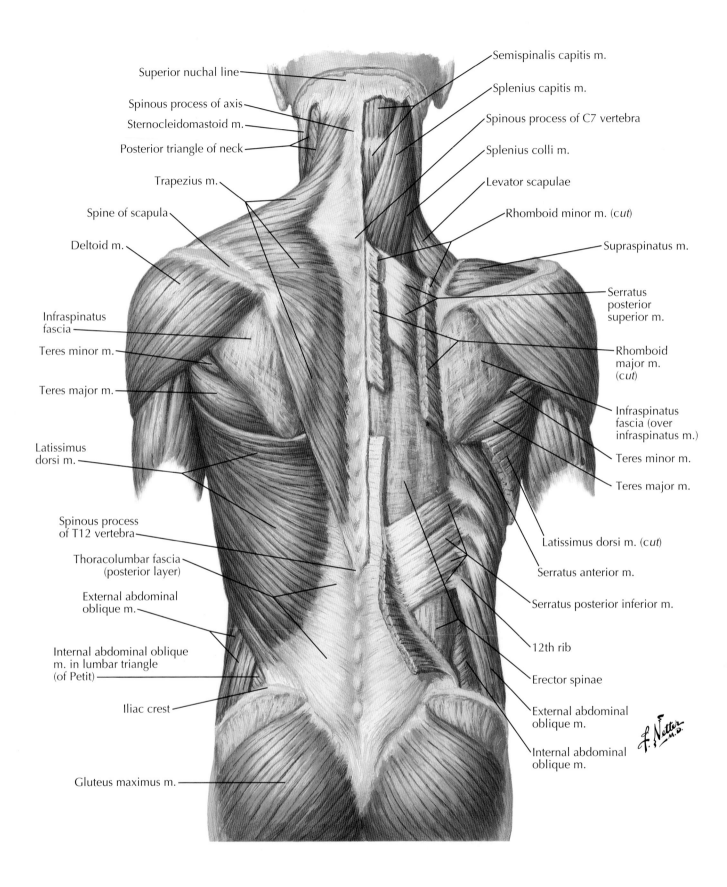

Superior nuchal line

Spinous process of axis

Sternocleidomastoid m.

Posterior triangle of neck

Trapezius m.

Spine of scapula

Deltoid m.

Infraspinatus fascia

Teres minor m.

Teres major m.

Latissimus dorsi m.

Spinous process of T12 vertebra

Thoracolumbar fascia (posterior layer)

External abdominal oblique m.

Internal abdominal oblique m. in lumbar triangle (of Petit)

Iliac crest

Gluteus maximus m.

Semispinalis capitis m.

Splenius capitis m.

Spinous process of C7 vertebra

Splenius colli m.

Levator scapulae

Rhomboid minor m. (*cut*)

Supraspinatus m.

Serratus posterior superior m.

Rhomboid major m. (*cut*)

Infraspinatus fascia (over infraspinatus m.)

Teres minor m.

Teres major m.

Latissimus dorsi m. (*cut*)

Serratus anterior m.

Serratus posterior inferior m.

12th rib

Erector spinae

External abdominal oblique m.

Internal abdominal oblique m.

Plate 195

Muscles and Nerves

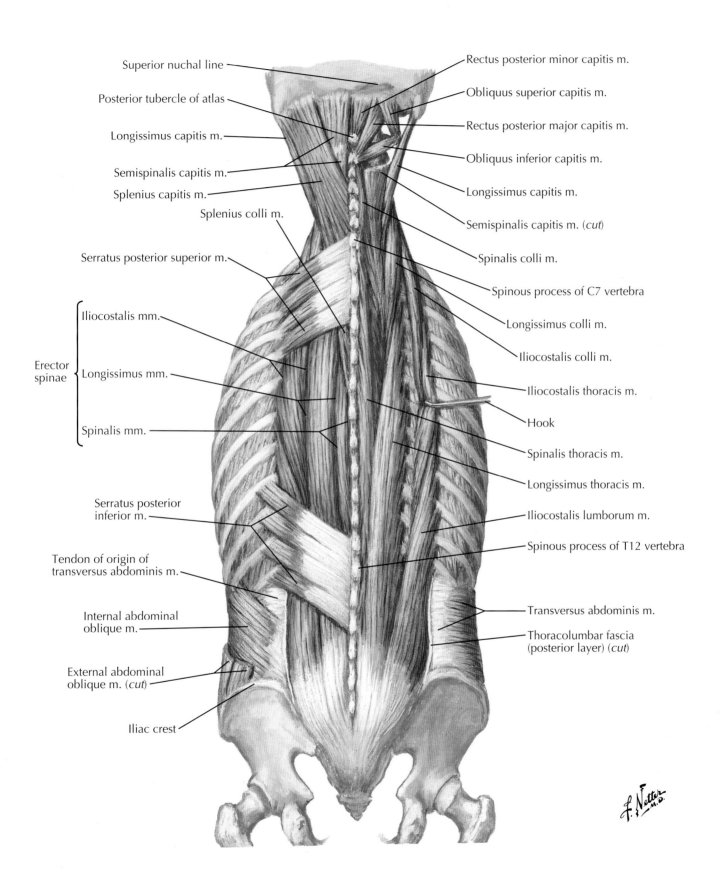

Superior nuchal line

Posterior tubercle of atlas

Longissimus capitis m.

Semispinalis capitis m.

Splenius capitis m.

Splenius colli m.

Serratus posterior superior m.

Iliocostalis mm.

Erector spinae { Longissimus mm.

Spinalis mm.

Serratus posterior inferior m.

Tendon of origin of transversus abdominis m.

Internal abdominal oblique m.

External abdominal oblique m. (*cut*)

Iliac crest

Rectus posterior minor capitis m.

Obliquus superior capitis m.

Rectus posterior major capitis m.

Obliquus inferior capitis m.

Longissimus capitis m.

Semispinalis capitis m. (*cut*)

Spinalis colli m.

Spinous process of C7 vertebra

Longissimus colli m.

Iliocostalis colli m.

Iliocostalis thoracis m.

Hook

Spinalis thoracis m.

Longissimus thoracis m.

Iliocostalis lumborum m.

Spinous process of T12 vertebra

Transversus abdominis m.

Thoracolumbar fascia (posterior layer) (*cut*)

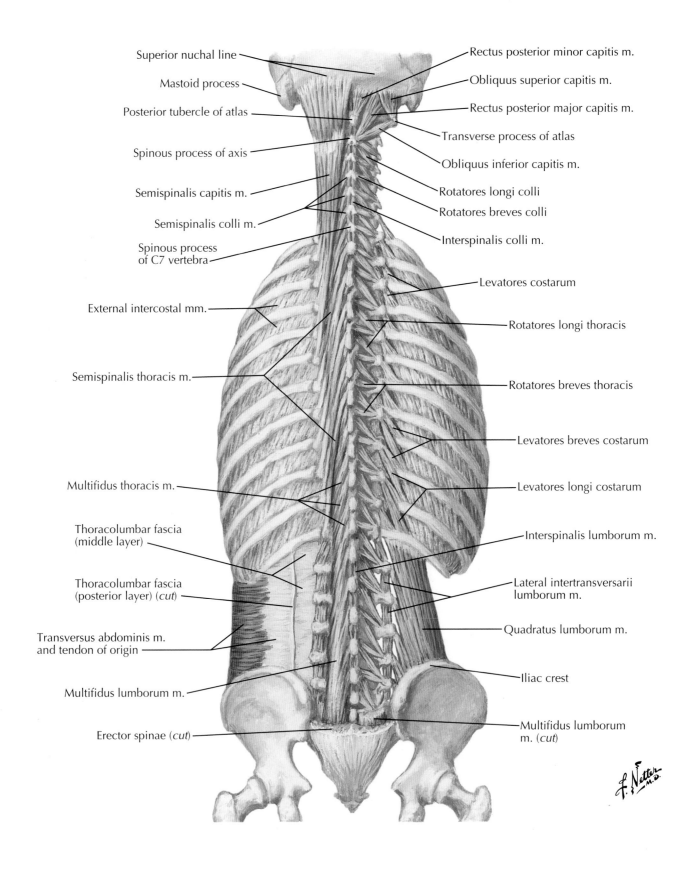

Superior nuchal line

Mastoid process

Posterior tubercle of atlas

Spinous process of axis

Semispinalis capitis m.

Semispinalis colli m.

Spinous process
of C7 vertebra

External intercostal mm.

Semispinalis thoracis m.

Multifidus thoracis m.

Thoracolumbar fascia
(middle layer)

Thoracolumbar fascia
(posterior layer) (*cut*)

Transversus abdominis m.
and tendon of origin

Multifidus lumborum m.

Erector spinae (*cut*)

Rectus posterior minor capitis m.

Obliquus superior capitis m.

Rectus posterior major capitis m.

Transverse process of atlas

Obliquus inferior capitis m.

Rotatores longi colli

Rotatores breves colli

Interspinalis colli m.

Levatores costarum

Rotatores longi thoracis

Rotatores breves thoracis

Levatores breves costarum

Levatores longi costarum

Interspinalis lumborum m.

Lateral intertransversarii
lumborum m.

Quadratus lumborum m.

Iliac crest

Multifidus lumborum
m. (*cut*)

Plate 197

Muscles and Nerves

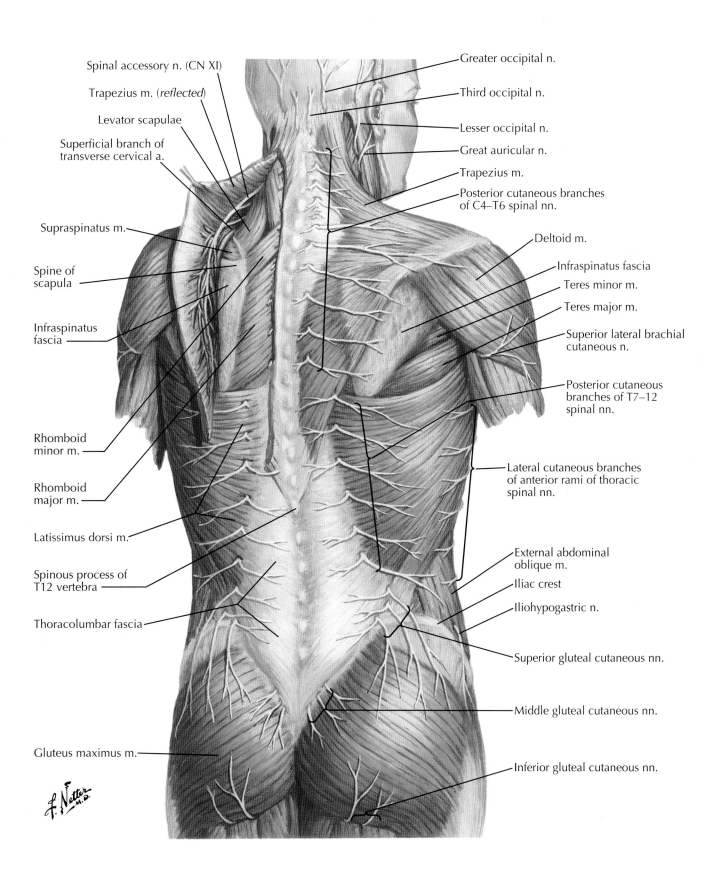

Spinal accessory n. (CN XI)

Trapezius m. (*reflected*)

Levator scapulae

Superficial branch of
transverse cervical a.

Supraspinatus m.

Spine of
scapula

Infraspinatus
fascia

Rhomboid
minor m.

Rhomboid
major m.

Latissimus dorsi m.

Spinous process of
T12 vertebra

Thoracolumbar fascia

Gluteus maximus m.

Greater occipital n.

Third occipital n.

Lesser occipital n.

Great auricular n.

Trapezius m.

Posterior cutaneous branches
of C4–T6 spinal nn.

Deltoid m.

Infraspinatus fascia

Teres minor m.

Teres major m.

Superior lateral brachial
cutaneous n.

Posterior cutaneous
branches of T7–12
spinal nn.

Lateral cutaneous branches
of anterior rami of thoracic
spinal nn.

External abdominal
oblique m.

Iliac crest

Iliohypogastric n.

Superior gluteal cutaneous nn.

Middle gluteal cutaneous nn.

Inferior gluteal cutaneous nn.

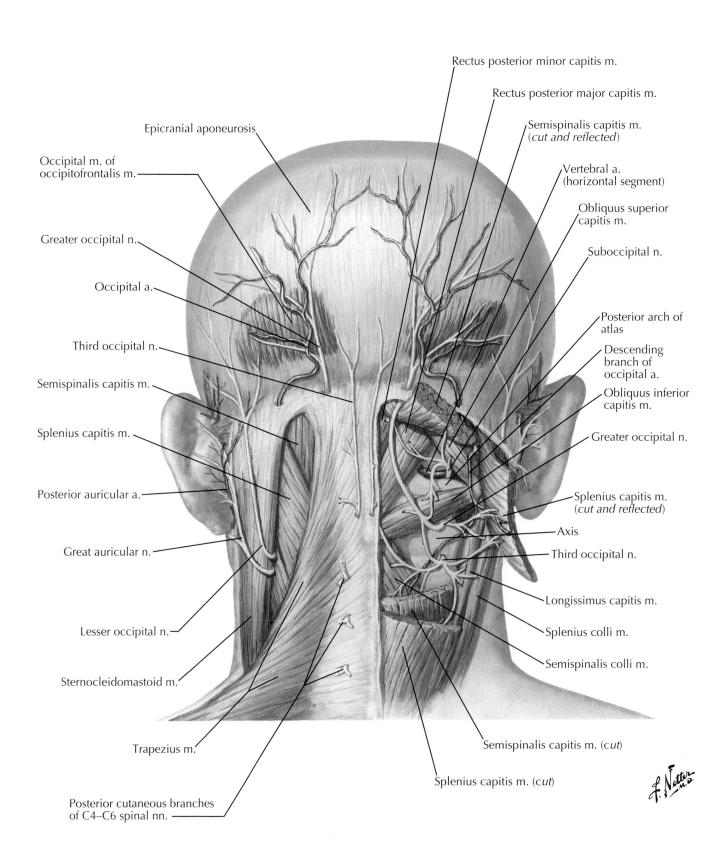

Rectus posterior minor capitis m.

Rectus posterior major capitis m.

Semispinalis capitis m.
(*cut and reflected*)

Vertebral a.
(horizontal segment)

Obliquus superior
capitis m.

Suboccipital n.

Posterior arch of
atlas

Descending
branch of
occipital a.

Obliquus inferior
capitis m.

Greater occipital n.

Splenius capitis m.
(*cut and reflected*)

Axis

Third occipital n.

Longissimus capitis m.

Splenius colli m.

Semispinalis colli m.

Semispinalis capitis m. (*cut*)

Splenius capitis m. (*cut*)

Epicranial aponeurosis

Occipital m. of
occipitofrontalis m.

Greater occipital n.

Occipital a.

Third occipital n.

Semispinalis capitis m.

Splenius capitis m.

Posterior auricular a.

Great auricular n.

Lesser occipital n.

Sternocleidomastoid m.

Trapezius m.

Posterior cutaneous branches
of C4–C6 spinal nn.

Plate 199

Muscles and Nerves

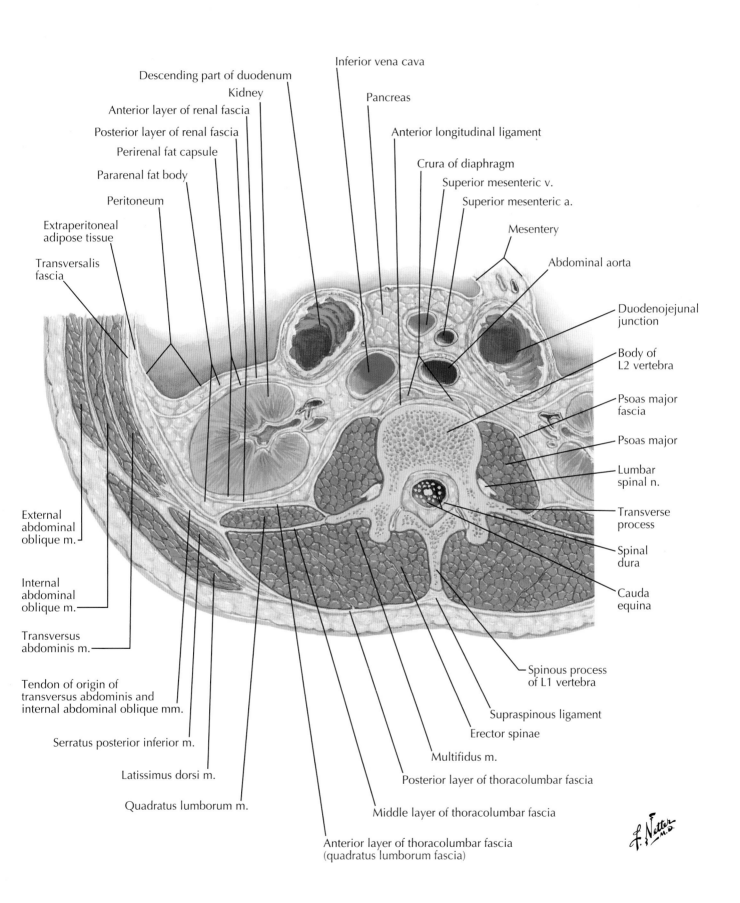

Inferior vena cava

Descending part of duodenum

Kidney

Pancreas

Anterior layer of renal fascia

Anterior longitudinal ligament

Posterior layer of renal fascia

Crura of diaphragm

Perirenal fat capsule

Superior mesenteric v.

Pararenal fat body

Superior mesenteric a.

Peritoneum

Mesentery

Extraperitoneal adipose tissue

Abdominal aorta

Transversalis fascia

Duodenojejunal junction

Body of L2 vertebra

Psoas major fascia

Psoas major

Lumbar spinal n.

Transverse process

External abdominal oblique m.

Spinal dura

Internal abdominal oblique m.

Cauda equina

Transversus abdominis m.

Tendon of origin of transversus abdominis and internal abdominal oblique mm.

Spinous process of L1 vertebra

Serratus posterior inferior m.

Supraspinous ligament

Latissimus dorsi m.

Erector spinae

Multifidus m.

Quadratus lumborum m.

Posterior layer of thoracolumbar fascia

Middle layer of thoracolumbar fascia

Anterior layer of thoracolumbar fascia (quadratus lumborum fascia)

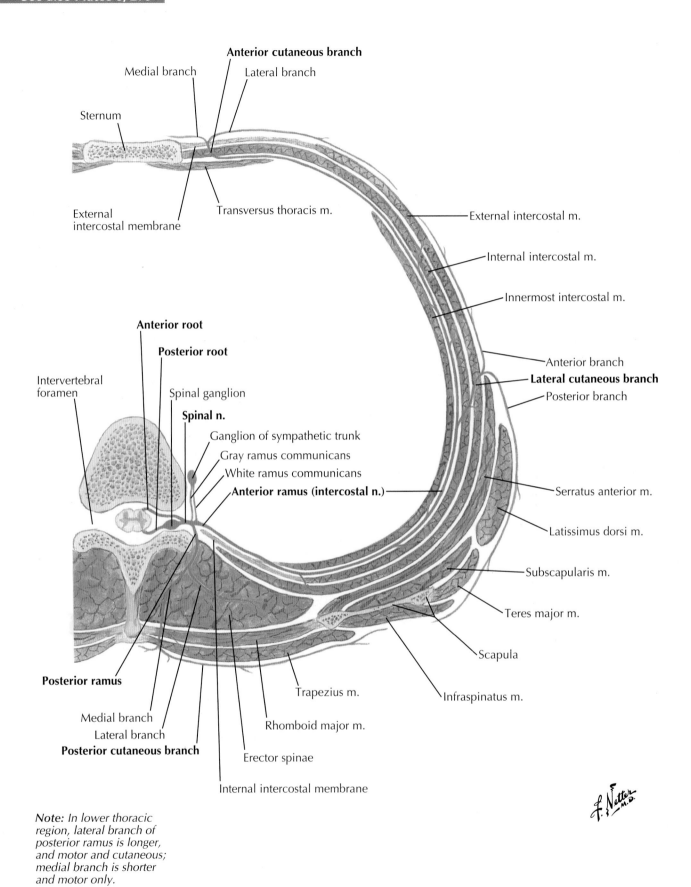

Anterior cutaneous branch

Medial branch

Lateral branch

Sternum

External intercostal membrane

Transversus thoracis m.

External intercostal m.

Internal intercostal m.

Innermost intercostal m.

Anterior root

Posterior root

Intervertebral foramen

Spinal ganglion

Spinal n.

Ganglion of sympathetic trunk

Gray ramus communicans

White ramus communicans

Anterior ramus (intercostal n.)

Anterior branch

Lateral cutaneous branch

Posterior branch

Serratus anterior m.

Latissimus dorsi m.

Subscapularis m.

Teres major m.

Scapula

Infraspinatus m.

Trapezius m.

Rhomboid major m.

Erector spinae

Internal intercostal membrane

Posterior ramus

Medial branch

Lateral branch

Posterior cutaneous branch

Note: In lower thoracic region, lateral branch of posterior ramus is longer, and motor and cutaneous; medial branch is shorter and motor only.

Plate 201

Cross-Sectional Anatomy

ANATOMICAL STRUCTURE	CLINICAL COMMENT	PLATE NUMBERS
Nervous System and Sense Organs		
Conus medullaris	Is inferior limit of spinal cord; can lie as inferior as L4 vertebra in neonates and as superior as T12 in adults (average is L1–L2); necessary to determine its location in procedures such as lumbar puncture; in adults, start at L2 or lower	186
Cauda equina	Lumbar and sacral nerve roots may be anesthetized with anesthesia injected into subarachnoid space (spinal block)	186, 187
Spinal meninges	Access to epidural and subarachnoid spaces is necessary for clinical procedures such as epidural anesthesia and lumbar puncture; meningitis is a life-threatening infection	182, 190
Skeletal System		
Spinous processes	Palpable landmarks used to assess spinal curvatures and determine location of spinal cord for procedures such as lumbar puncture and injection of spinal anesthesia	178, 195
Spinous process of C7 vertebra (vertebra prominens)	Most prominent spinous process in cervical region; often used to begin counting vertebrae	178, 180
Intervertebral disc	Age-related changes may produce herniation of nucleus pulposus, causing back pain; occurs most commonly in lower lumbar regions of vertebral column	181, 187
Lamina of vertebral arch	Surgically removed in laminectomy to gain access to vertebral canal and spinal cord	181
Intervertebral foramen	May become narrowed by age-related changes (e.g., osteophyte formation) or changes in intervertebral disc height, producing compression of its contents	181, 184, 185
Sacral hiatus	Provides access to epidural space for administration of caudal epidural anesthesia	183
Fifth lumbar vertebra	Spondylolysis is clinical condition in which vertebral body separates from part of its vertebral arch bearing inferior articulating process (defect is through pars interarticularis); if this occurs bilaterally, L5 vertebral body and transverse process may slide forward over sacrum, giving rise to spondylolisthesis	184
L5–S1 intervertebral disc	Most common level of intervertebral disc herniation, which may result in nerve compression and lower back pain associated with pain and weakness in ipsilateral lower limb (sciatica)	184, 187
Vertebral foramen	May be congenitally stenotic in cervical region, or narrowed by arthritic changes in lumbar vertebrae; can lead to back pain, sciatica, numbness or tingling, and weakness in lower limbs	180, 181
Muscular System		
Trapezius muscle	Responsible for holding scapula against thoracic wall against gravity; drooping of shoulder may indicate injury to spinal accessory nerve	195
Deep (or intrinsic) back muscles	Microscopic stretching or tearing of muscle fibers produces back strain, a common cause of low back pain	196, 197
Cardiovascular System		
Arteries of spinal cord	Narrowing or damage caused by atherosclerosis, vertebral fractures, or vertebral dislocations may cause ischemia of spinal cord	191
Vertebral venous plexuses	Mostly valveless veins along vertebral column allow retrograde flow and can act as conduits for metastasis of cancer cells to spine, lungs, and brain	194, BP 39

*Selections were based largely on clinical data and commonly discussed clinical correlations in macroscopic ("gross") anatomy courses.

MUSCLE	MUSCLE GROUP	SUPERIOR ATTACHMENT	INFERIOR ATTACHMENT	INNERVATION	BLOOD SUPPLY	MAIN ACTIONS
Iliocostalis mm.	Deep back (erector spinae)	*Colli muscle:* posterior tubercles of C4–C6 vertebrae	*Colli muscle:* angles of ribs 3–6	Posterior rami of spinal nerves	*Cervical portions:* occipital, deep cervical, and vertebral arteries	Extend and laterally bend vertebral column and head
		Thoracis muscle: angles of ribs 1–6	*Thoracis muscle:* angles of ribs 7–12		*Thoracic portions:* dorsal branches of posterior intercostal and subcostal arteries	
		Lumborum muscle: ribs 4–12, transverse processes of L1–L4 vertebrae	*Lumborum muscle:* sacrum (via erector spinae aponeurosis), iliac crest		*Lumbar portions:* dorsal branches of lumbar and lateral sacral arteries	
Interspinalis mm.	Deep back	*Colli muscle:* spinous processes of C2–C7 vertebrae	Spinous processes of vertebrae subjacent to vertebrae of muscle origin	Posterior rami of spinal nerves	*Cervical portions:* occipital, deep cervical, and vertebral arteries	Aid in extension of vertebral column
		Thoracis muscle: spinous processes of T1–T2 and T11–T12 vertebrae			*Thoracic portions:* dorsal branches of posterior intercostal and subcostal arteries	
		Lumborum muscle: spinous processes of L1–L4 vertebrae			*Lumbar portions:* dorsal branches of lumbar arteries	
Intertransversarii mm.	Deep back	*Medial muscles:* transverse processes of C1–C7 and T10–T12 vertebrae, mammillary processes of L1–L4 vertebrae	*Medial muscles:* transverse processes of subjacent cervical and thoracic vertebrae, mammillary processes of subjacent lumbar vertebrae	*Medial muscles:* posterior rami of spinal nerves	*Cervical portions:* occipital, deep cervical, and vertebral arteries	Assist in lateral flexion of vertebral column
					Thoracic portions: dorsal branches of posterior intercostal and subcostal arteries	
		Lateral and anterior muscles: transverse processes of C1–C7 and L1–L4 vertebrae	*Lateral and anterior muscles:* transverse processes of subjacent vertebrae	*Lateral and anterior muscles:* anterior rami of spinal nerves	*Lumbar portions:* dorsal branches of lumbar arteries	
Latissimus dorsi m.	Superficial back	Spinous processes of T7–T12 vertebrae, posterior layer of thoracolumbar fascia (and thus to L1–L5 vertebrae and iliac crest), ribs 10–12	Intertubercular sulcus of humerus	Thoracodorsal nerve	Thoracodorsal artery, dorsal perforating branches of 9th, 10th, and 11th posterior intercostal, subcostal, and first three lumbar arteries	Extends, adducts, and medially rotates humerus
Levator scapulae	Superficial back	Transverse processes of atlas and axis, posterior tubercles of transverse processes of C3–C4 vertebrae	Medial border of scapula (superior to spine)	Anterior rami of C3–C4 spinal nerves, dorsal scapular nerve	Dorsal scapular and transverse and ascending cervical arteries	Elevates scapula medially, inferiorly rotates glenoid fossa
Longissimus mm.	Deep back (erector spinae)	*Capitis muscle:* mastoid process	*Capitis muscle:* transverse processes of C4–T4 vertebrae	Posterior rami of spinal nerves	*Cervical portions:* occipital, deep cervical, and vertebral arteries	Extend and laterally bend vertebral column and head
		Colli muscle: transverse processes of C2–C6 vertebrae	*Colli muscle:* transverse processes of T1–T5 vertebrae		*Thoracic portions:* dorsal branches of posterior intercostal and subcostal arteries	
		Thoracis muscle: transverse processes of T1–T12 vertebrae, ribs 5–12, accessory and transverse processes of L1–L5 vertebrae	*Thoracis muscle:* spinous processes of L1–L5 vertebrae, posterior surface of sacrum, iliac tuberosity, posterior sacroiliac ligament		*Lumbar portions:* dorsal branches of lumbar and lateral sacral arteries	

Table 3.2   **Muscles**

MUSCLE	MUSCLE GROUP	SUPERIOR ATTACHMENT	INFERIOR ATTACHMENT	INNERVATION	BLOOD SUPPLY	MAIN ACTIONS
Multifidus mm.	Deep back (transversospinales)	Spinous processes of C2–L5 vertebrae (2–5 levels above muscle insertion)	*Colli muscle:* superior articular processes of C4–C7 vertebrae	Posterior rami of spinal nerves	*Cervical portions:* occipital, deep cervical, and vertebral arteries	Stabilize spine
			Thoracis muscle: transverse processes of T1–T12 vertebrae		*Thoracic portions:* dorsal branches of posterior intercostal and subcostal arteries	
			Lumborum muscle: mammillary processes of L1–L5 vertebrae, sacrum, iliac crest		*Lumbar portions:* dorsal branches of lumbar and lateral sacral arteries	
Obliquus inferior capitis m.	Suboccipital	Transverse process of atlas	Spinous process of axis	Suboccipital nerve	Vertebral artery and occipital arteries	Rotates atlas to turn face to same side
Obliquus superior capitis m.	Suboccipital	Lateral part of inferior nuchal line	Transverse process of atlas	Suboccipital nerve	Vertebral and occipital arteries	Extends and bends head laterally
Rectus posterior major capitis m.	Suboccipital	Middle part of inferior nuchal line	Spinous process of axis	Suboccipital nerve	Vertebral and occipital arteries	Extends and rotates head to same side
Rectus posterior minor capitis m.	Suboccipital	Medial part of inferior nuchal line	Posterior tubercle of atlas	Suboccipital nerve	Vertebral and occipital arteries	Extends head
Rhomboid major m.	Superficial back	Spinous processes of T2–T5 vertebrae	Medial border of scapula (inferior to spine of scapula)	Dorsal scapular nerve	Dorsal scapular artery *OR* deep branch of transverse cervical artery, dorsal perforating branches of upper five or six posterior intercostal arteries	Fixes scapula to thoracic wall and retracts and rotates it to depress glenoid fossa
Rhomboid minor m.	Superficial back	Nuchal ligament, spinous processes of C7 and T1 vertebrae	Medial border of scapula (at spine of scapula)	Dorsal scapular nerve	Dorsal scapular artery *OR* deep branch of transverse cervical artery, dorsal perforating branches of upper five or six posterior intercostal arteries	Fixes scapula to thoracic wall and retracts and rotates it to depress glenoid fossa
Rotatores	Deep back (transversospinales)	*Colli muscle:* spinous processes of cervical vertebrae	*Colli muscle:* superior articular processes of cervical vertebrae 1 or 2 levels below vertebra of muscle origin	Posterior rami of spinal nerves	*Cervical portions:* occipital, deep cervical, and vertebral arteries	Stabilize, extend, and rotate spine
		Thoracis muscle: spinous processes and laminae of T1–T11 vertebrae	*Thoracis muscle:* transverse processes of T2–T12 vertebrae (brevis muscles insert into adjacent vertebra; longus muscles into vertebra 2 levels down)		*Thoracic portions:* dorsal branches of posterior intercostal and subcostal arteries	
		Lumborum muscle: spinous processes of lumbar vertebrae	*Lumborum muscle:* mammillary processes of lumbar vertebrae 2 levels down		*Lumbar portions:* dorsal branches of lumbar arteries	

Muscles

MUSCLE	MUSCLE GROUP	SUPERIOR ATTACHMENT	INFERIOR ATTACHMENT	INNERVATION	BLOOD SUPPLY	MAIN ACTIONS
Semispinalis mm.	Deep back (transversospinales)	*Capitis muscle:* occipital bone (between superior and inferior nuchal lines)	*Capitis muscle:* superior articular processes of C4–C7 vertebrae, transverse processes of T1–T6 vertebrae	Posterior rami of spinal nerves	*Cervical portions:* occipital, deep cervical, and vertebral arteries	Extend head and neck and rotate them to opposite side
		Colli muscle: spinous processes of C2–C5 vertebrae	*Colli muscle:* transverse processes of T1–T6 vertebrae			
		Thoracis muscle: spinous processes of C6–T4 vertebrae	*Thoracis muscle:* transverse processes of T6–T10 vertebrae		*Thoracic portions:* dorsal branches of posterior intercostal arteries	
Serratus posterior inferior m.	Superficial back	Inferior aspect of ribs 9–12	Spinous processes of T11–L2 vertebrae	Anterior rami of T9–T12 spinal nerves	Posterior intercostal arteries	Depresses ribs
Serratus posterior superior m.	Superficial back	Nuchal ligament, spinous processes of C7–T3 vertebrae	Superior aspect of ribs 2–5	Anterior rami of T2–T5 spinal nerves	Posterior intercostal arteries	Elevates ribs
Spinalis mm.	Deep back (erector spinae)	*Capitis muscle:* external occipital protuberance	*Capitis muscle:* spinous processes of C7 and T1 vertebrae	Posterior rami of spinal nerves	*Cervical portions:* occipital, deep cervical, and vertebral arteries	Extend and laterally bend vertebral column and head
		Colli muscle: spinous processes of C2–C4 vertebrae	*Colli muscle:* spinous processes of C7–T2 vertebrae		*Thoracic portions:* dorsal branches of posterior intercostal and subcostal arteries	
		Thoracis muscle: spinous processes of T2–T8 vertebrae	*Thoracis muscle:* spinous processes of T11–L2 vertebrae		*Lumbar portions:* dorsal branches of lumbar and lateral sacral arteries	
Splenius capitis m.	Spinotransversales	Mastoid process of temporal bone, lateral one-third of superior nuchal line	Nuchal ligament, spinous processes of C7–T4 vertebrae	Posterior rami of C2–C3 spinal nerves	Occipital artery and deep cervical arteries	*Bilaterally:* extends head
						Unilaterally: laterally bends (flexes) and rotates face to same side
Splenius colli m.	Spinotransversales	Transverse processes of C1–C3 vertebrae	Spinous processes of T3–T6 vertebrae	Posterior rami of C4–C6 spinal nerves	Occipital and deep cervical arteries	*Bilaterally:* extends neck
						Unilaterally: laterally bends (flexes) and rotates neck toward same side
Trapezius m.	Superficial back	*Descending part:* superior nuchal line, external occipital protuberance, nuchal ligament	*Descending part:* lateral one-third of clavicle	Spinal accessory nerve (CN XI)	Transverse cervical and posterior intercostal arteries	Elevates, retracts, and rotates scapula; lower fibers depress scapula
		Transverse part: spinous processes of C7–T3 vertebrae	*Transverse part:* acromion			
		Ascending part: spinous processes of T4–T12 vertebrae	*Ascending part:* spine of scapula			

Variations in spinal nerve contributions to the innervation of muscles, their arterial supply, their attachments, and their actions are common themes in human anatomy. Therefore, expect differences between texts and realize that anatomical variation is normal.

Table 3.4   **Muscles**

THORAX 4

ELECTRONIC BONUS PLATES

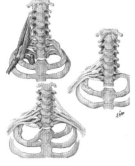

BP 41 Cervical Ribs and Related Variations

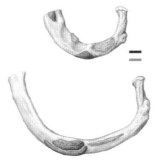

BP 42 Muscle Attachments of Ribs

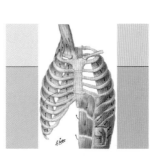

BP 43 Muscles of Respiration

BP 44 Intrapulmonary Airways: Schema

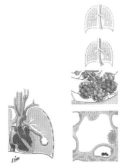

BP 45 Anatomy of Ventilation and Respiration

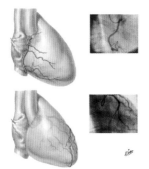

BP 46 Coronary Arteries: Right Anterolateral Views with Arteriograms

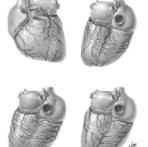

BP 47 Coronary Arteries and Cardiac Veins: Variations

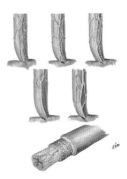

BP 48 Intrinsic Nerves and Variations in Nerves of Esophagus

ELECTRONIC BONUS PLATES—*cont'd*

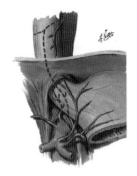

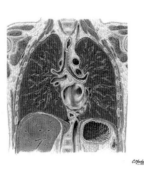

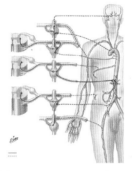

BP 49 Arteries of Esophagus: Variations

BP 50 Thorax: Coronal Section

BP 51 Thorax: Coronal CTs

BP 52 Innervation of Blood Vessels: Schema

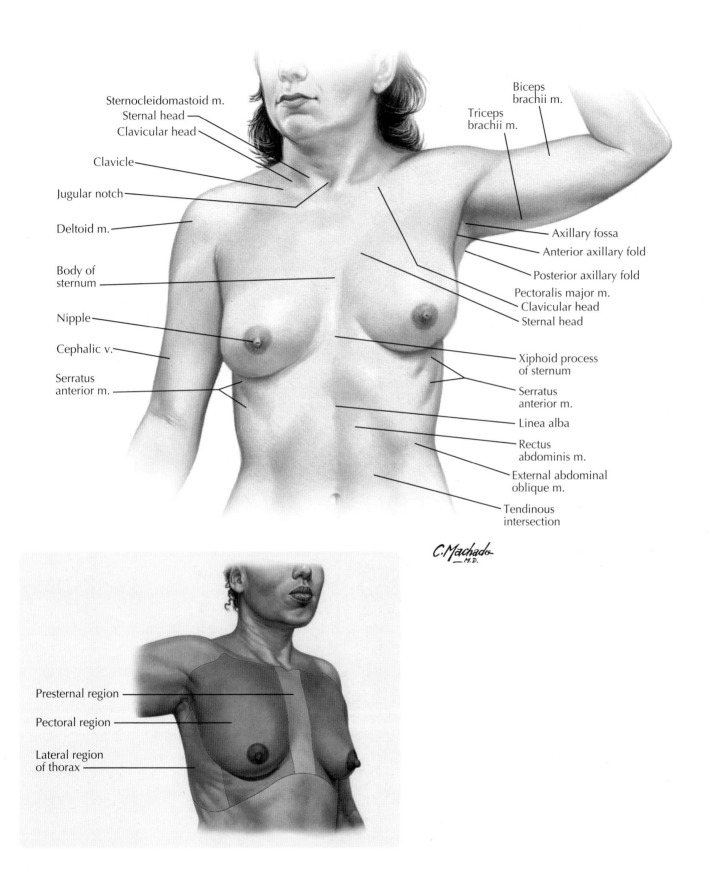

Sternocleidomastoid m.
Sternal head
Clavicular head

Clavicle

Jugular notch

Deltoid m.

Body of
sternum

Nipple

Cephalic v.

Serratus
anterior m.

Biceps
brachii m.

Triceps
brachii m.

Axillary fossa

Anterior axillary fold

Posterior axillary fold

Pectoralis major m.
Clavicular head
Sternal head

Xiphoid process
of sternum

Serratus
anterior m.

Linea alba

Rectus
abdominis m.

External abdominal
oblique m.

Tendinous
intersection

C. Machado
_M.D.

Presternal region

Pectoral region

Lateral region
of thorax

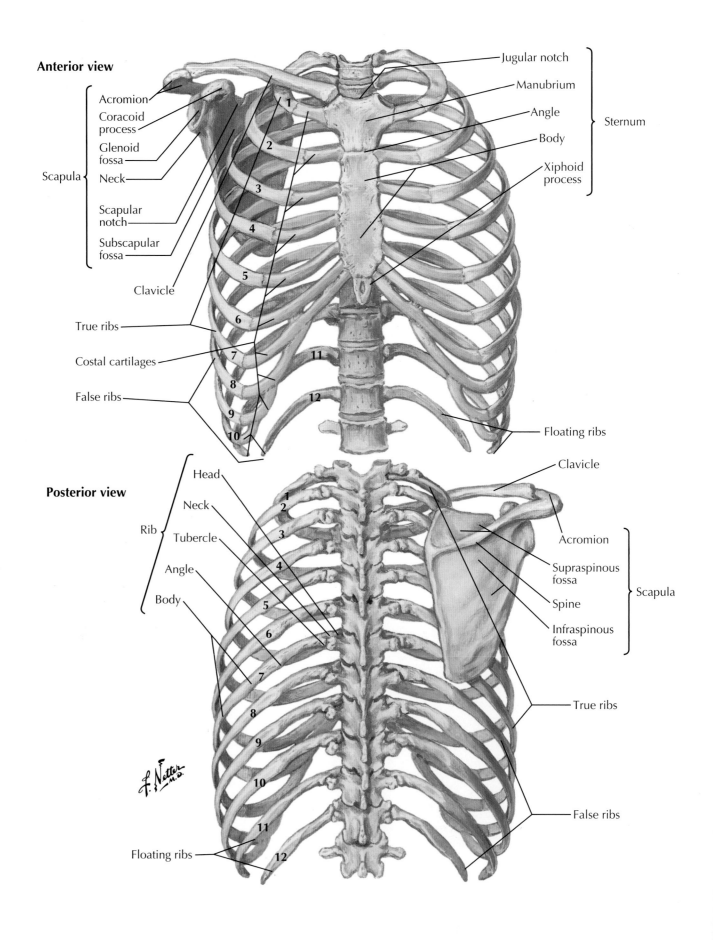

Anterior view

Scapula {
- Acromion
- Coracoid process
- Glenoid fossa
- Neck
- Scapular notch
- Subscapular fossa

Clavicle

True ribs

Costal cartilages

False ribs

Jugular notch

Manubrium

Angle — } Sternum

Body

Xiphoid process

1
2
3
4
5
6
7
8
9
10
11
12

Floating ribs

Posterior view

Rib {
- Head
- Neck
- Tubercle
- Angle
- Body

1
2
3
4
5
6
7
8
9
10
11
12

Floating ribs

Clavicle

Acromion

Supraspinous fossa

Spine — } Scapula

Infraspinous fossa

True ribs

False ribs

Plate 203   **Thoracic Skeleton**

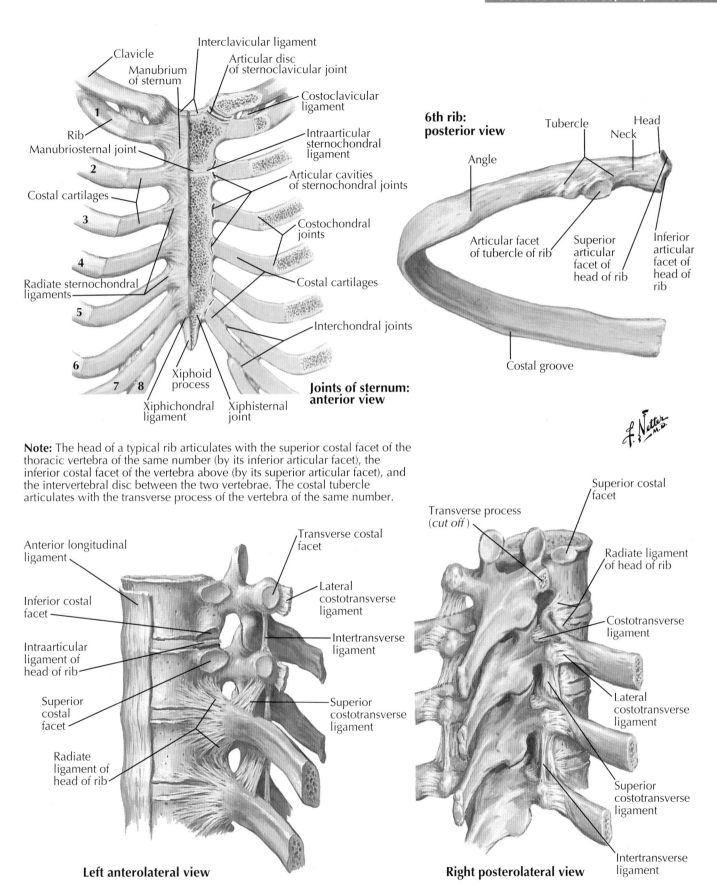

Clavicle

Interclavicular ligament

Manubrium of sternum

Articular disc of sternoclavicular joint

Costoclavicular ligament

Rib

Manubriosternal joint

Intraarticular sternochondral ligament

Costal cartilages

Articular cavities of sternochondral joints

Radiate sternochondral ligaments

Costochondral joints

Costal cartilages

Interchondral joints

Xiphoid process

Xiphichondral ligament

Xiphisternal joint

Joints of sternum: anterior view

6th rib: posterior view

Angle

Tubercle

Neck

Head

Articular facet of tubercle of rib

Superior articular facet of head of rib

Inferior articular facet of head of rib

Costal groove

Note: The head of a typical rib articulates with the superior costal facet of the thoracic vertebra of the same number (by its inferior articular facet), the inferior costal facet of the vertebra above (by its superior articular facet), and the intervertebral disc between the two vertebrae. The costal tubercle articulates with the transverse process of the vertebra of the same number.

Anterior longitudinal ligament

Transverse costal facet

Inferior costal facet

Lateral costotransverse ligament

Intraarticular ligament of head of rib

Intertransverse ligament

Superior costal facet

Superior costotransverse ligament

Radiate ligament of head of rib

Left anterolateral view

Transverse process (*cut off*)

Superior costal facet

Radiate ligament of head of rib

Costotransverse ligament

Lateral costotransverse ligament

Superior costotransverse ligament

Intertransverse ligament

Right posterolateral view

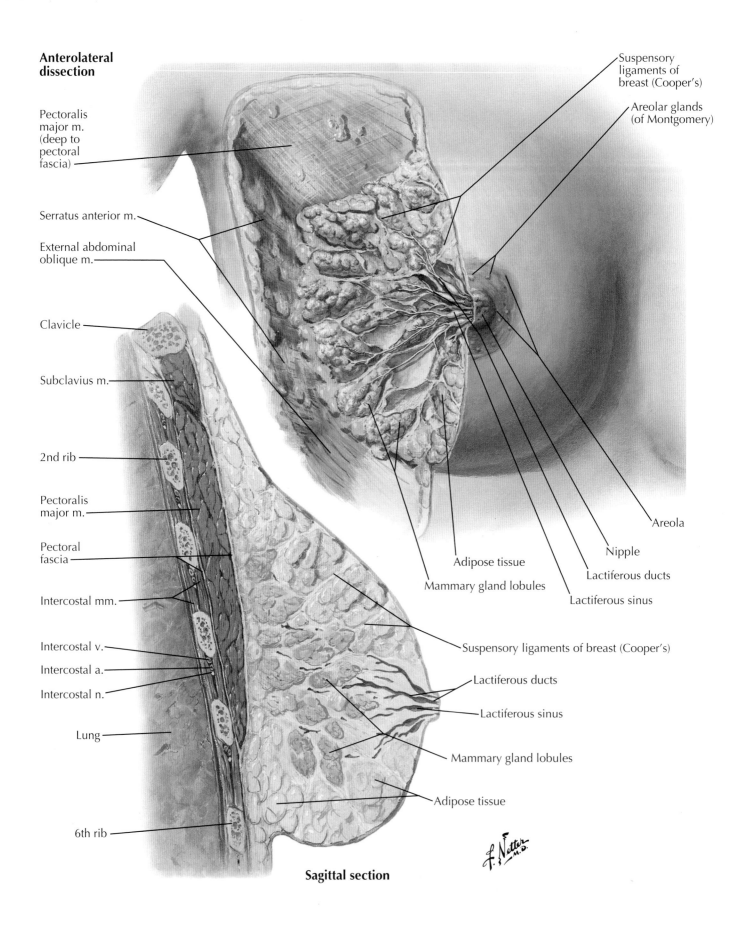

Anterolateral dissection

Pectoralis major m. (deep to pectoral fascia)

Serratus anterior m.

External abdominal oblique m.

Clavicle

Subclavius m.

2nd rib

Pectoralis major m.

Pectoral fascia

Intercostal mm.

Intercostal v.

Intercostal a.

Intercostal n.

Lung

6th rib

Suspensory ligaments of breast (Cooper's)

Areolar glands (of Montgomery)

Areola

Nipple

Lactiferous ducts

Lactiferous sinus

Adipose tissue

Mammary gland lobules

Suspensory ligaments of breast (Cooper's)

Lactiferous ducts

Lactiferous sinus

Mammary gland lobules

Adipose tissue

Sagittal section

Plate 205

Mammary Glands

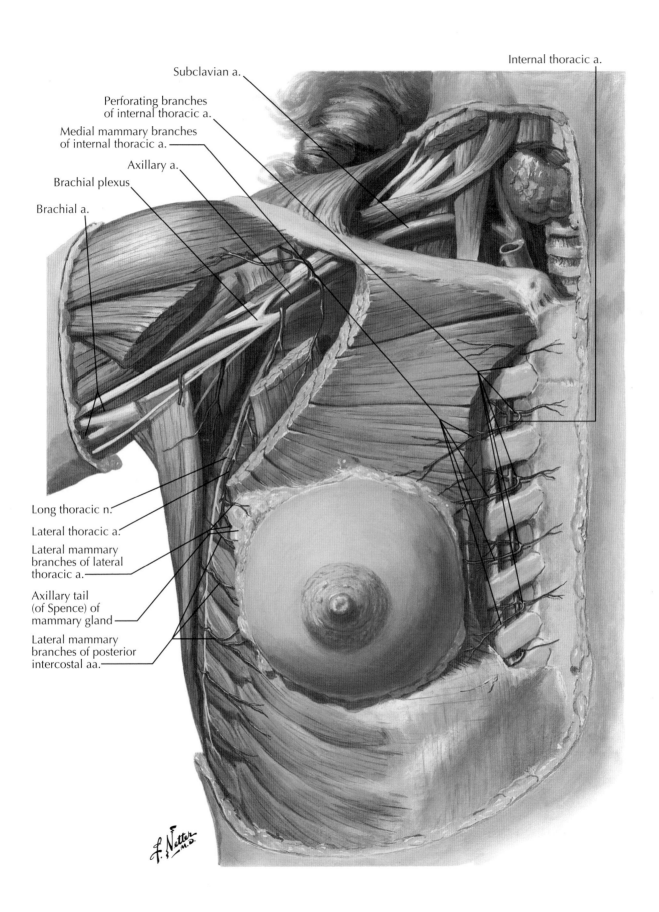

Internal thoracic a.

Subclavian a.

Perforating branches
of internal thoracic a.

Medial mammary branches
of internal thoracic a.

Axillary a.

Brachial plexus

Brachial a.

Long thoracic n.

Lateral thoracic a.

Lateral mammary
branches of lateral
thoracic a.

Axillary tail
(of Spence) of
mammary gland

Lateral mammary
branches of posterior
intercostal aa.

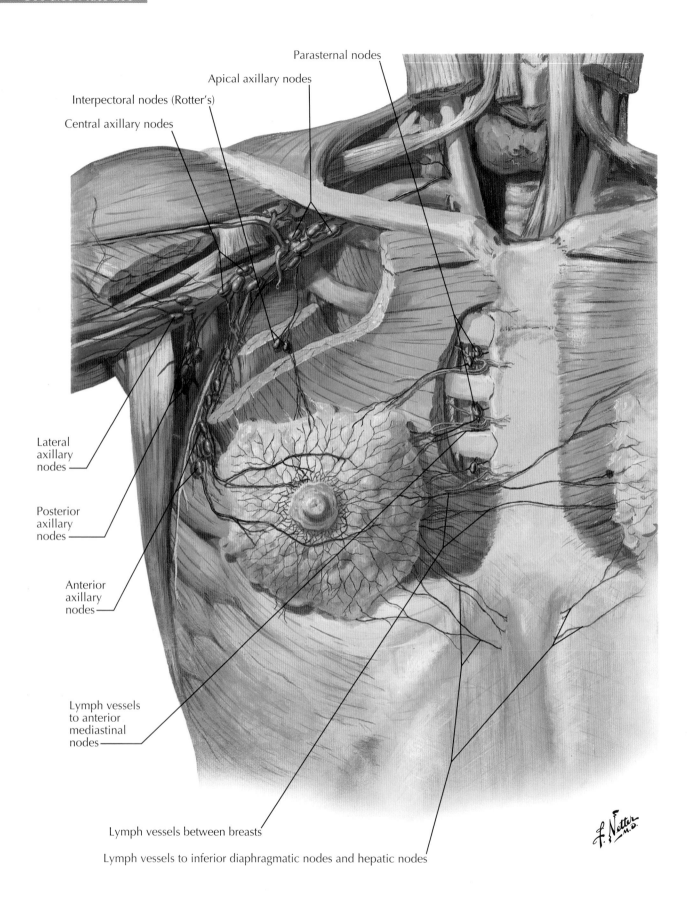

Parasternal nodes

Apical axillary nodes

Interpectoral nodes (Rotter's)

Central axillary nodes

Lateral
axillary
nodes

Posterior
axillary
nodes

Anterior
axillary
nodes

Lymph vessels
to anterior
mediastinal
nodes

Lymph vessels between breasts

Lymph vessels to inferior diaphragmatic nodes and hepatic nodes

Plate 207

Mammary Glands

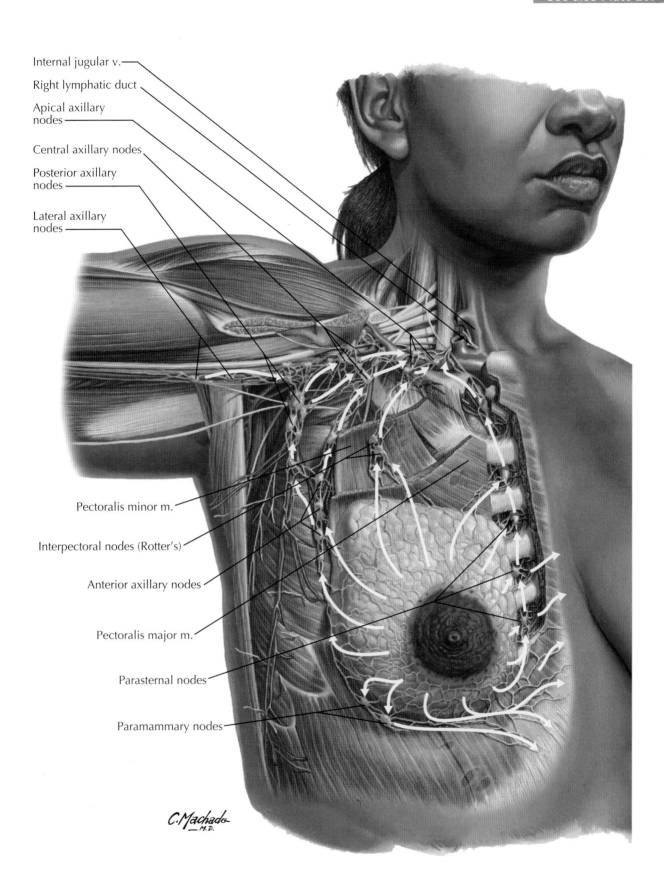

Internal jugular v.

Right lymphatic duct

Apical axillary nodes

Central axillary nodes

Posterior axillary nodes

Lateral axillary nodes

Pectoralis minor m.

Interpectoral nodes (Rotter's)

Anterior axillary nodes

Pectoralis major m.

Parasternal nodes

Paramammary nodes

C. Machado
—M.D.

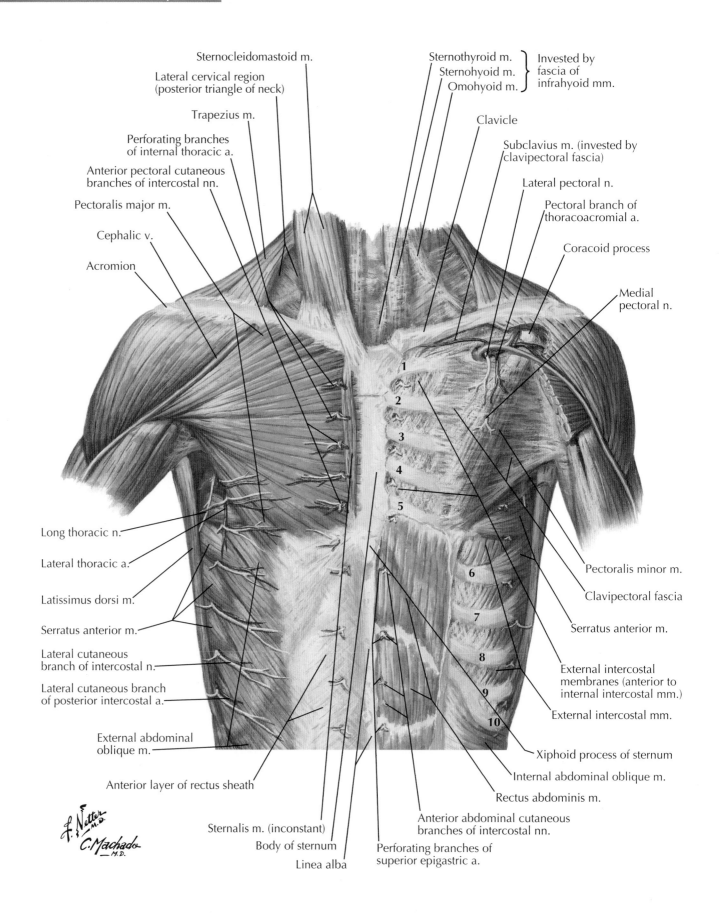

Sternocleidomastoid m.

Lateral cervical region
(posterior triangle of neck)

Trapezius m.

Perforating branches
of internal thoracic a.

Anterior pectoral cutaneous
branches of intercostal nn.

Pectoralis major m.

Cephalic v.

Acromion

Sternothyroid m. ⎫ Invested by
Sternohyoid m. ⎬ fascia of
Omohyoid m. ⎭ infrahyoid mm.

Clavicle

Subclavius m. (invested by
clavipectoral fascia)

Lateral pectoral n.

Pectoral branch of
thoracoacromial a.

Coracoid process

Medial
pectoral n.

Long thoracic n.

Lateral thoracic a.

Latissimus dorsi m.

Serratus anterior m.

Lateral cutaneous
branch of intercostal n.

Lateral cutaneous branch
of posterior intercostal a.

External abdominal
oblique m.

Anterior layer of rectus sheath

Sternalis m. (inconstant)

Body of sternum

Linea alba

Anterior abdominal cutaneous
branches of intercostal nn.

Perforating branches of
superior epigastric a.

Rectus abdominis m.

Internal abdominal oblique m.

Xiphoid process of sternum

External intercostal mm.

External intercostal
membranes (anterior to
internal intercostal mm.)

Serratus anterior m.

Clavipectoral fascia

Pectoralis minor m.

C. Machado
M.D.

Plate 209

Thoracic Wall and Diaphragm

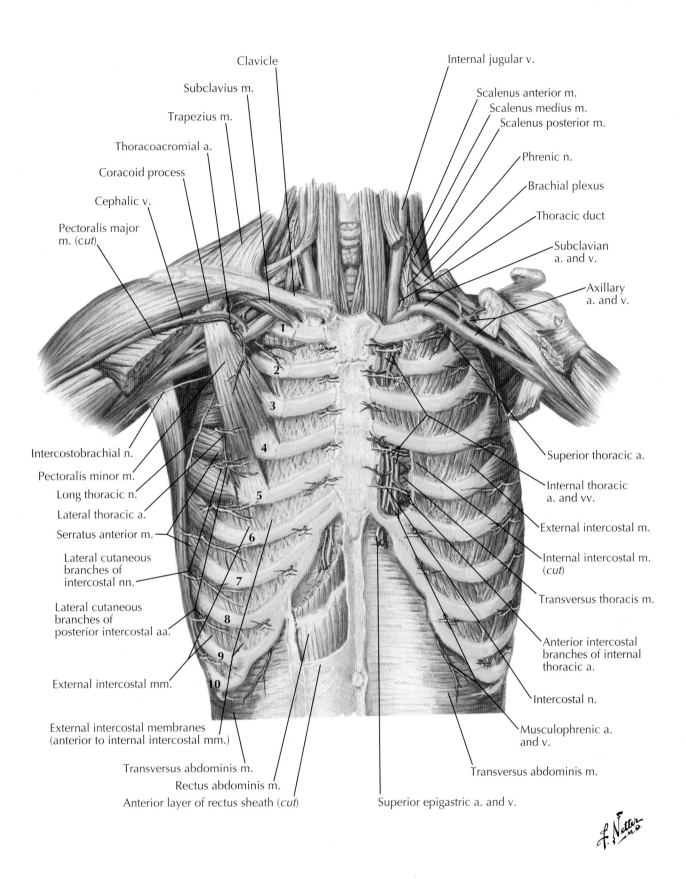

Clavicle

Subclavius m.

Trapezius m.

Thoracoacromial a.

Coracoid process

Cephalic v.

Pectoralis major m. (cut)

Internal jugular v.

Scalenus anterior m.

Scalenus medius m.

Scalenus posterior m.

Phrenic n.

Brachial plexus

Thoracic duct

Subclavian a. and v.

Axillary a. and v.

Intercostobrachial n.

Pectoralis minor m.

Long thoracic n.

Lateral thoracic a.

Serratus anterior m.

Lateral cutaneous branches of intercostal nn.

Lateral cutaneous branches of posterior intercostal aa.

External intercostal mm.

External intercostal membranes (anterior to internal intercostal mm.)

Transversus abdominis m.

Rectus abdominis m.

Anterior layer of rectus sheath (cut)

Superior thoracic a.

Internal thoracic a. and vv.

External intercostal m.

Internal intercostal m. (cut)

Transversus thoracis m.

Anterior intercostal branches of internal thoracic a.

Intercostal n.

Musculophrenic a. and v.

Transversus abdominis m.

Superior epigastric a. and v.

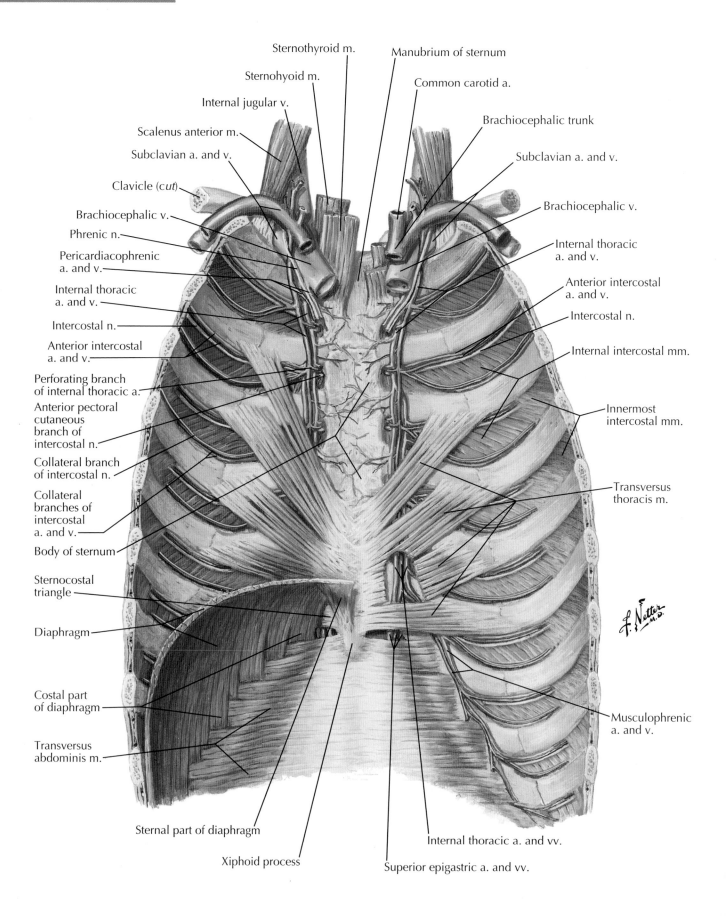

Sternothyroid m.

Sternohyoid m.

Internal jugular v.

Scalenus anterior m.

Subclavian a. and v.

Clavicle (*cut*)

Brachiocephalic v.

Phrenic n.

Pericardiacophrenic a. and v.

Internal thoracic a. and v.

Intercostal n.

Anterior intercostal a. and v.

Perforating branch of internal thoracic a.

Anterior pectoral cutaneous branch of intercostal n.

Collateral branch of intercostal n.

Collateral branches of intercostal a. and v.

Body of sternum

Sternocostal triangle

Diaphragm

Costal part of diaphragm

Transversus abdominis m.

Sternal part of diaphragm

Xiphoid process

Manubrium of sternum

Common carotid a.

Brachiocephalic trunk

Subclavian a. and v.

Brachiocephalic v.

Internal thoracic a. and v.

Anterior intercostal a. and v.

Intercostal n.

Internal intercostal mm.

Innermost intercostal mm.

Transversus thoracis m.

Musculophrenic a. and v.

Internal thoracic a. and vv.

Superior epigastric a. and vv.

Plate 211

Thoracic Wall and Diaphragm

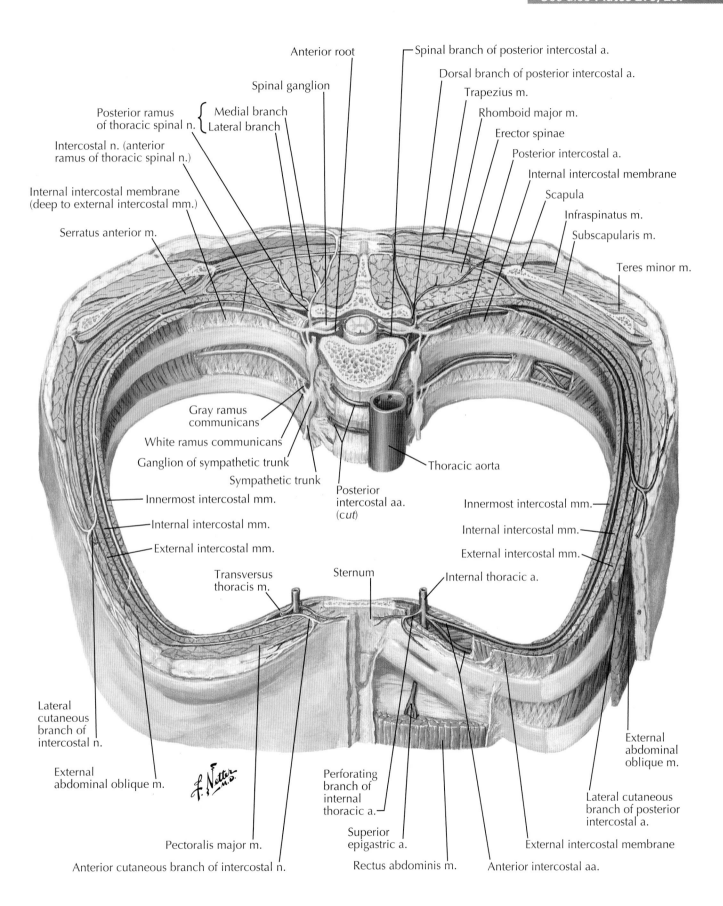

Anterior root

Spinal branch of posterior intercostal a.

Spinal ganglion

Dorsal branch of posterior intercostal a.

Trapezius m.

Posterior ramus of thoracic spinal n. { Medial branch / Lateral branch }

Rhomboid major m.

Erector spinae

Intercostal n. (anterior ramus of thoracic spinal n.)

Posterior intercostal a.

Internal intercostal membrane

Internal intercostal membrane (deep to external intercostal mm.)

Scapula

Serratus anterior m.

Infraspinatus m.

Subscapularis m.

Teres minor m.

Gray ramus communicans

White ramus communicans

Ganglion of sympathetic trunk

Sympathetic trunk

Thoracic aorta

Posterior intercostal aa. (cut)

Innermost intercostal mm.

Internal intercostal mm.

External intercostal mm.

Innermost intercostal mm.

Internal intercostal mm.

External intercostal mm.

Transversus thoracis m.

Sternum

Internal thoracic a.

Lateral cutaneous branch of intercostal n.

External abdominal oblique m.

Perforating branch of internal thoracic a.

Superior epigastric a.

Pectoralis major m.

Anterior cutaneous branch of intercostal n.

Rectus abdominis m.

Anterior intercostal aa.

External abdominal oblique m.

Lateral cutaneous branch of posterior intercostal a.

External intercostal membrane

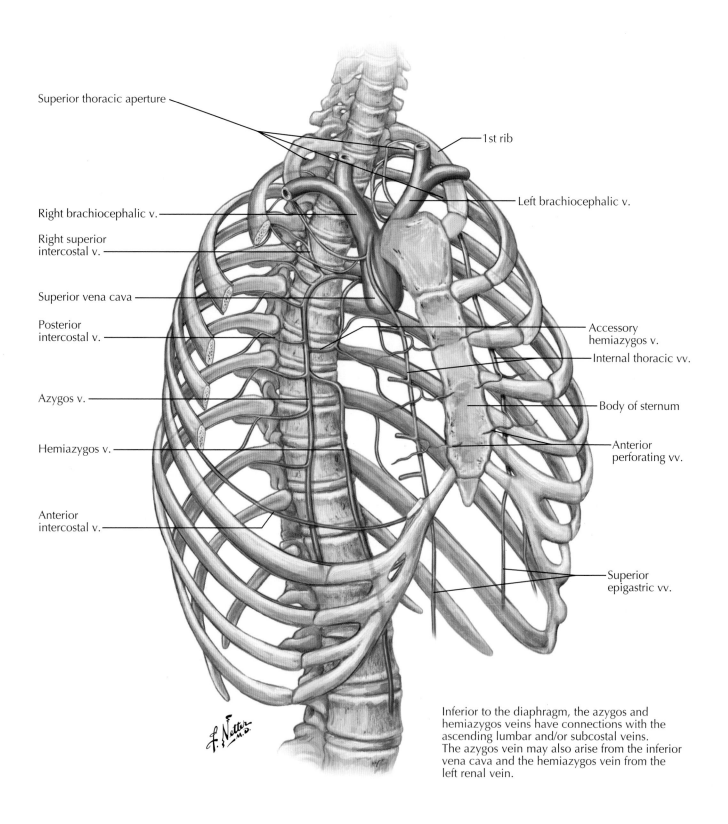

Superior thoracic aperture

1st rib

Right brachiocephalic v.

Left brachiocephalic v.

Right superior
intercostal v.

Superior vena cava

Posterior
intercostal v.

Accessory
hemiazygos v.

Internal thoracic vv.

Azygos v.

Body of sternum

Hemiazygos v.

Anterior
perforating vv.

Anterior
intercostal v.

Superior
epigastric vv.

Inferior to the diaphragm, the azygos and
hemiazygos veins have connections with the
ascending lumbar and/or subcostal veins.
The azygos vein may also arise from the inferior
vena cava and the hemiazygos vein from the
left renal vein.

Plate 213

Thoracic Wall and Diaphragm

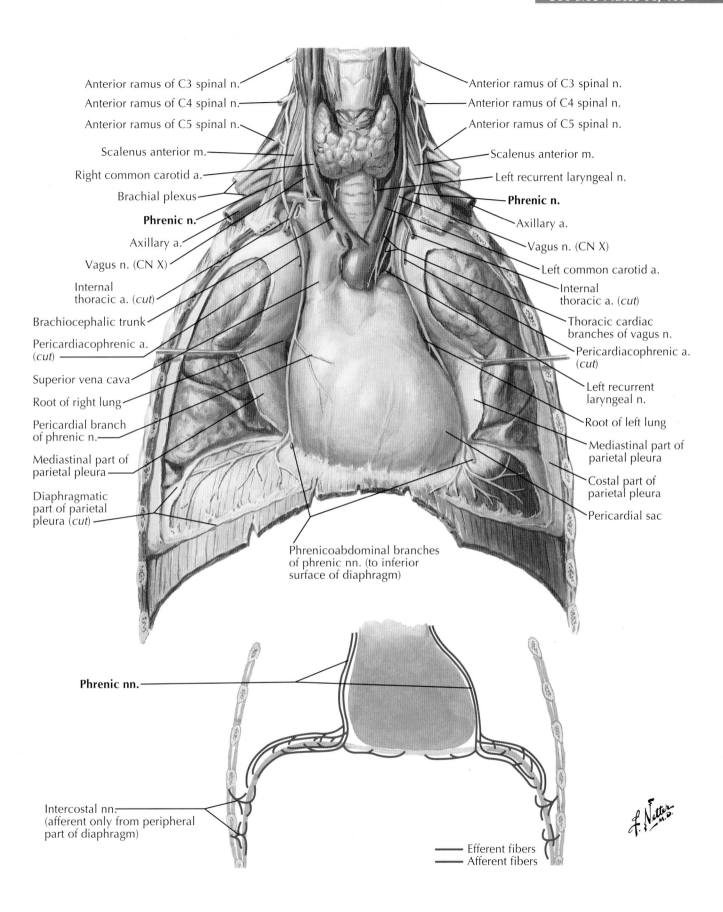

Anterior ramus of C3 spinal n.

Anterior ramus of C4 spinal n.

Anterior ramus of C5 spinal n.

Scalenus anterior m.

Right common carotid a.

Brachial plexus

Phrenic n.

Axillary a.

Vagus n. (CN X)

Internal thoracic a. (*cut*)

Brachiocephalic trunk

Pericardiacophrenic a. (*cut*)

Superior vena cava

Root of right lung

Pericardial branch of phrenic n.

Mediastinal part of parietal pleura

Diaphragmatic part of parietal pleura (*cut*)

Anterior ramus of C3 spinal n.

Anterior ramus of C4 spinal n.

Anterior ramus of C5 spinal n.

Scalenus anterior m.

Left recurrent laryngeal n.

Phrenic n.

Axillary a.

Vagus n. (CN X)

Left common carotid a.

Internal thoracic a. (*cut*)

Thoracic cardiac branches of vagus n.

Pericardiacophrenic a. (*cut*)

Left recurrent laryngeal n.

Root of left lung

Mediastinal part of parietal pleura

Costal part of parietal pleura

Pericardial sac

Phrenicoabdominal branches of phrenic nn. (to inferior surface of diaphragm)

Phrenic nn.

Intercostal nn. (afferent only from peripheral part of diaphragm)

Efferent fibers
Afferent fibers

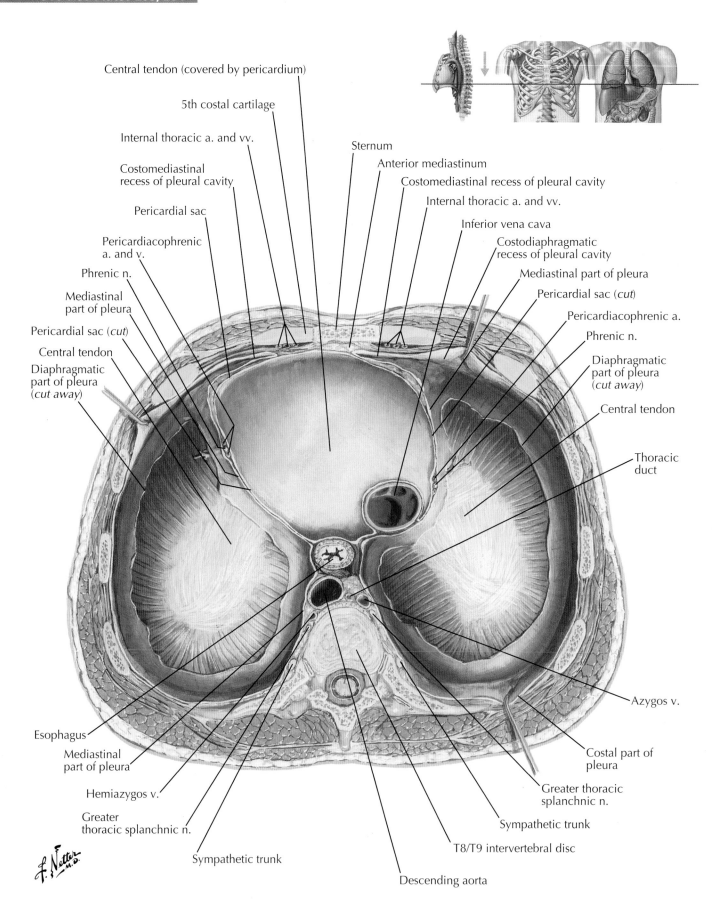

Central tendon (covered by pericardium)

5th costal cartilage

Internal thoracic a. and vv.

Costomediastinal
recess of pleural cavity

Pericardial sac

Pericardiacophrenic
a. and v.

Phrenic n.

Mediastinal
part of pleura

Pericardial sac (*cut*)

Central tendon

Diaphragmatic
part of pleura
(*cut away*)

Sternum

Anterior mediastinum

Costomediastinal recess of pleural cavity

Internal thoracic a. and vv.

Inferior vena cava

Costodiaphragmatic
recess of pleural cavity

Mediastinal part of pleura

Pericardial sac (*cut*)

Pericardiacophrenic a.

Phrenic n.

Diaphragmatic
part of pleura
(*cut away*)

Central tendon

Thoracic
duct

Azygos v.

Costal part of
pleura

Greater thoracic
splanchnic n.

Sympathetic trunk

T8/T9 intervertebral disc

Descending aorta

Sympathetic trunk

Greater
thoracic splanchnic n.

Hemiazygos v.

Mediastinal
part of pleura

Esophagus

Plate 215

Thoracic Wall and Diaphragm

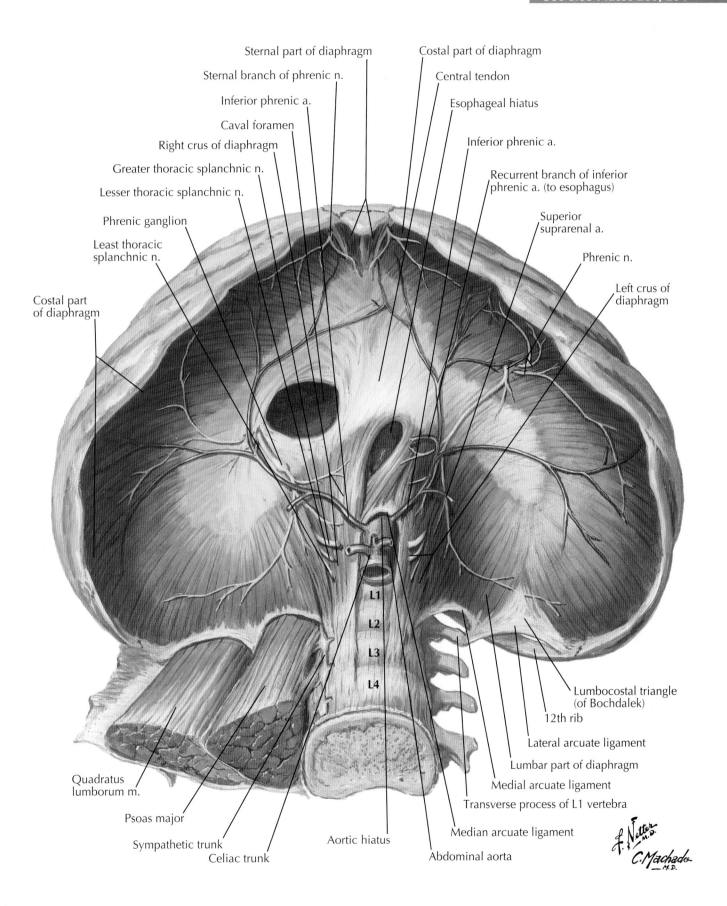

Sternal part of diaphragm

Sternal branch of phrenic n.

Inferior phrenic a.

Caval foramen

Right crus of diaphragm

Greater thoracic splanchnic n.

Lesser thoracic splanchnic n.

Phrenic ganglion

Least thoracic splanchnic n.

Costal part of diaphragm

Costal part of diaphragm

Central tendon

Esophageal hiatus

Inferior phrenic a.

Recurrent branch of inferior phrenic a. (to esophagus)

Superior suprarenal a.

Phrenic n.

Left crus of diaphragm

L1

L2

L3

L4

Lumbocostal triangle (of Bochdalek)

12th rib

Lateral arcuate ligament

Lumbar part of diaphragm

Medial arcuate ligament

Transverse process of L1 vertebra

Median arcuate ligament

Abdominal aorta

Quadratus lumborum m.

Psoas major

Sympathetic trunk

Celiac trunk

Aortic hiatus

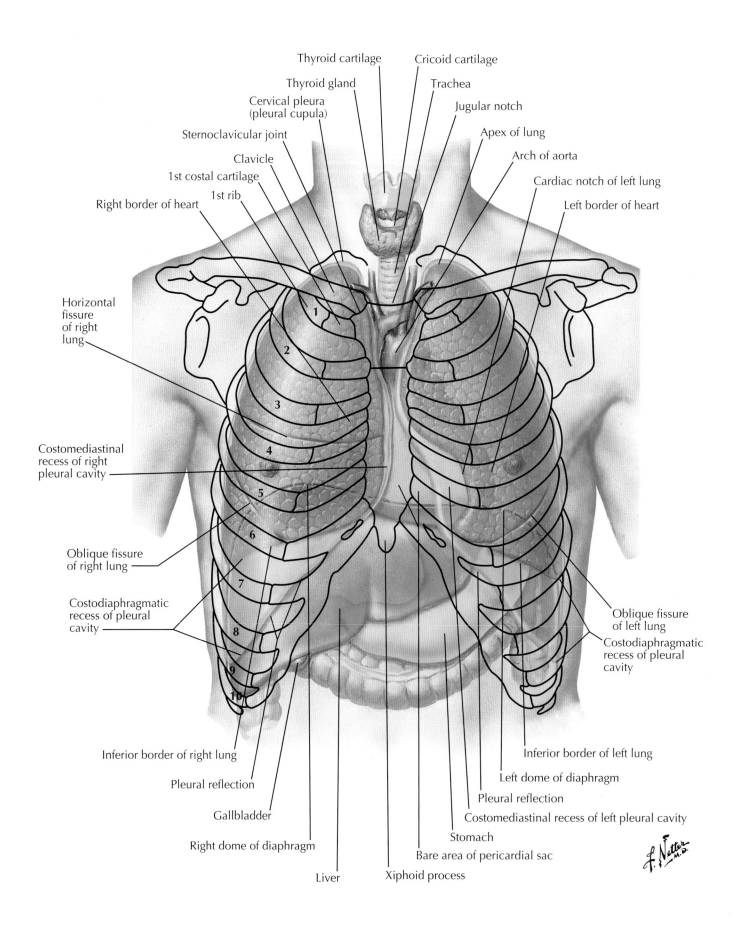

Thyroid cartilage

Cricoid cartilage

Thyroid gland

Trachea

Cervical pleura
(pleural cupula)

Jugular notch

Sternoclavicular joint

Apex of lung

Clavicle

Arch of aorta

1st costal cartilage

Cardiac notch of left lung

1st rib

Right border of heart

Left border of heart

Horizontal
fissure
of right
lung

Costomediastinal
recess of right
pleural cavity

Oblique fissure
of right lung

Oblique fissure
of left lung

Costodiaphragmatic
recess of pleural
cavity

Costodiaphragmatic
recess of pleural
cavity

Inferior border of right lung

Inferior border of left lung

Pleural reflection

Left dome of diaphragm

Gallbladder

Pleural reflection

Costomediastinal recess of left pleural cavity

Right dome of diaphragm

Stomach

Bare area of pericardial sac

Liver

Xiphoid process

Plate 217

Lungs, Trachea, and Bronchi

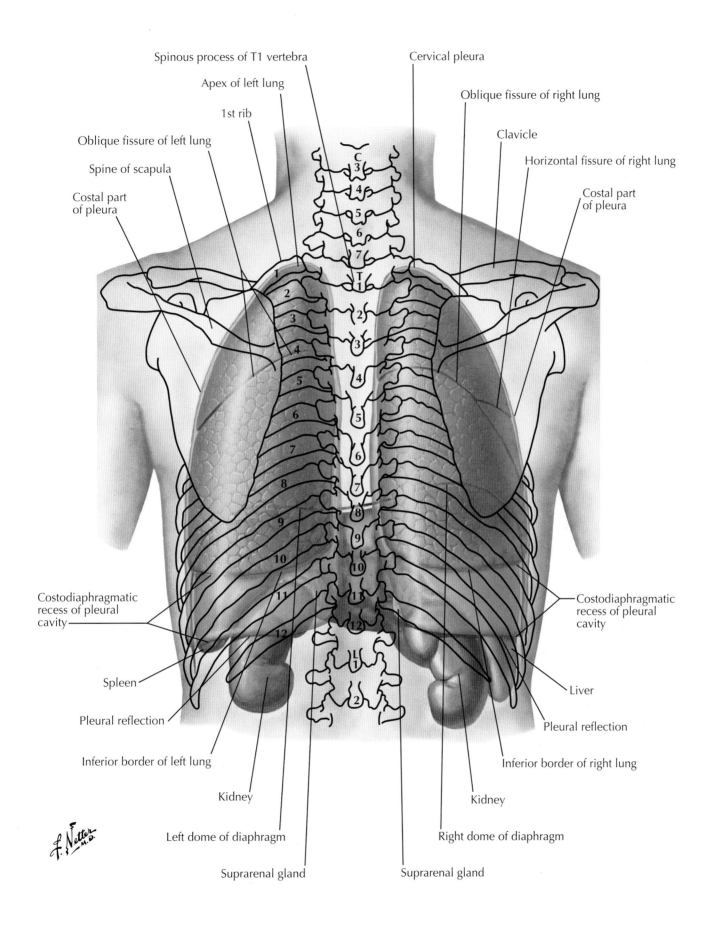

Spinous process of T1 vertebra

Apex of left lung

1st rib

Oblique fissure of left lung

Spine of scapula

Costal part of pleura

Cervical pleura

Oblique fissure of right lung

Clavicle

Horizontal fissure of right lung

Costal part of pleura

Costodiaphragmatic recess of pleural cavity

Costodiaphragmatic recess of pleural cavity

Spleen

Liver

Pleural reflection

Pleural reflection

Inferior border of left lung

Inferior border of right lung

Kidney

Kidney

Left dome of diaphragm

Right dome of diaphragm

Suprarenal gland

Suprarenal gland

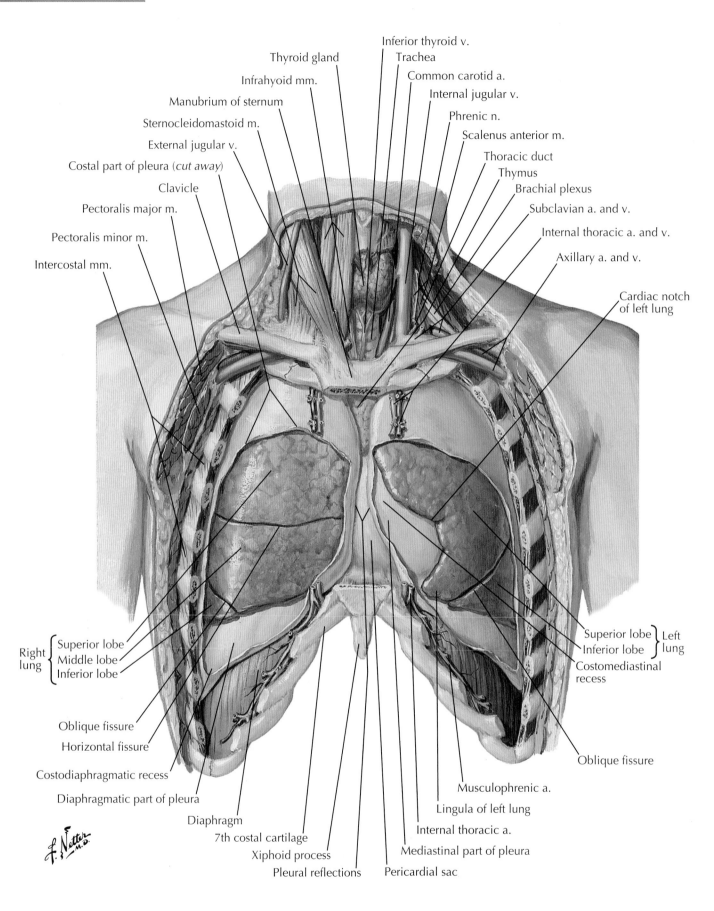

Inferior thyroid v.

Thyroid gland

Trachea

Infrahyoid mm.

Common carotid a.

Manubrium of sternum

Internal jugular v.

Sternocleidomastoid m.

Phrenic n.

External jugular v.

Scalenus anterior m.

Costal part of pleura (*cut away*)

Thoracic duct

Clavicle

Thymus

Pectoralis major m.

Brachial plexus

Pectoralis minor m.

Subclavian a. and v.

Intercostal mm.

Internal thoracic a. and v.

Axillary a. and v.

Cardiac notch of left lung

Right lung { Superior lobe / Middle lobe / Inferior lobe

Superior lobe } Left lung
Inferior lobe

Costomediastinal recess

Oblique fissure

Horizontal fissure

Oblique fissure

Costodiaphragmatic recess

Diaphragmatic part of pleura

Musculophrenic a.

Lingula of left lung

Diaphragm

Internal thoracic a.

7th costal cartilage

Xiphoid process

Mediastinal part of pleura

Pleural reflections

Pericardial sac

Plate 219

Lungs, Trachea, and Bronchi

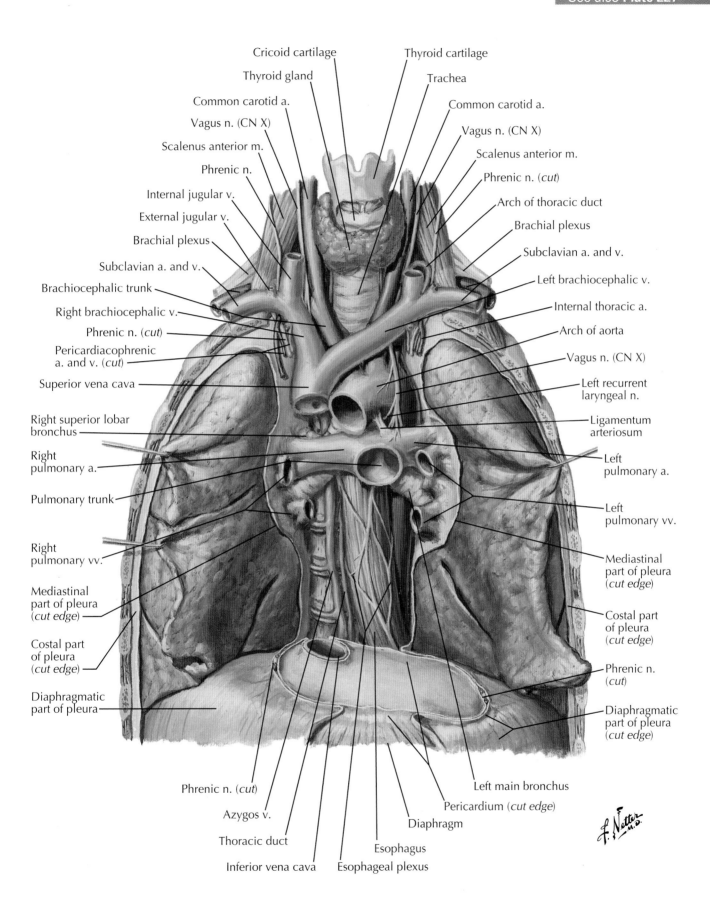

Cricoid cartilage

Thyroid cartilage

Thyroid gland

Trachea

Common carotid a.

Common carotid a.

Vagus n. (CN X)

Vagus n. (CN X)

Scalenus anterior m.

Scalenus anterior m.

Phrenic n.

Phrenic n. (*cut*)

Internal jugular v.

Arch of thoracic duct

External jugular v.

Brachial plexus

Brachial plexus

Subclavian a. and v.

Subclavian a. and v.

Left brachiocephalic v.

Brachiocephalic trunk

Internal thoracic a.

Right brachiocephalic v.

Arch of aorta

Phrenic n. (*cut*)

Vagus n. (CN X)

Pericardiacophrenic
a. and v. (*cut*)

Left recurrent
laryngeal n.

Superior vena cava

Ligamentum
arteriosum

Right superior lobar
bronchus

Right
pulmonary a.

Left
pulmonary a.

Pulmonary trunk

Left
pulmonary vv.

Right
pulmonary vv.

Mediastinal
part of pleura
(*cut edge*)

Mediastinal
part of pleura
(*cut edge*)

Costal part
of pleura
(*cut edge*)

Costal part
of pleura
(*cut edge*)

Phrenic n.
(*cut*)

Diaphragmatic
part of pleura

Diaphragmatic
part of pleura
(*cut edge*)

Phrenic n. (*cut*)

Left main bronchus

Azygos v.

Pericardium (*cut edge*)

Thoracic duct

Diaphragm

Esophagus

Inferior vena cava

Esophageal plexus

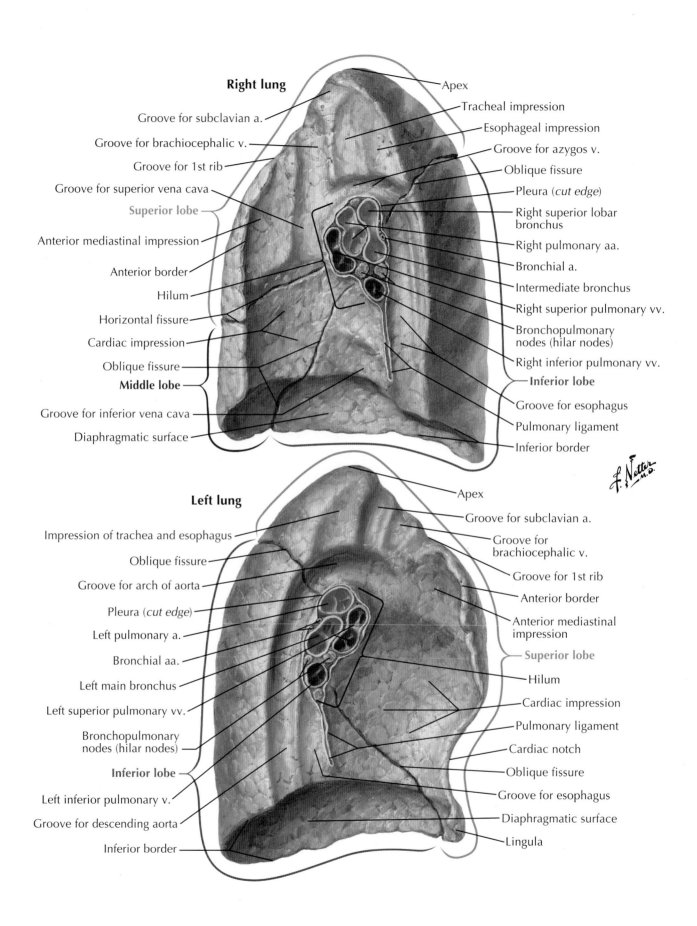

Right lung

Groove for subclavian a.
Groove for brachiocephalic v.
Groove for 1st rib
Groove for superior vena cava
Superior lobe
Anterior mediastinal impression
Anterior border
Hilum
Horizontal fissure
Cardiac impression
Oblique fissure
Middle lobe
Groove for inferior vena cava
Diaphragmatic surface

Apex
Tracheal impression
Esophageal impression
Groove for azygos v.
Oblique fissure
Pleura (*cut edge*)
Right superior lobar bronchus
Right pulmonary aa.
Bronchial a.
Intermediate bronchus
Right superior pulmonary vv.
Bronchopulmonary nodes (hilar nodes)
Right inferior pulmonary vv.
Inferior lobe
Groove for esophagus
Pulmonary ligament
Inferior border

Left lung

Impression of trachea and esophagus
Oblique fissure
Groove for arch of aorta
Pleura (*cut edge*)
Left pulmonary a.
Bronchial aa.
Left main bronchus
Left superior pulmonary vv.
Bronchopulmonary nodes (hilar nodes)
Inferior lobe
Left inferior pulmonary v.
Groove for descending aorta
Inferior border

Apex
Groove for subclavian a.
Groove for brachiocephalic v.
Groove for 1st rib
Anterior border
Anterior mediastinal impression
Superior lobe
Hilum
Cardiac impression
Pulmonary ligament
Cardiac notch
Oblique fissure
Groove for esophagus
Diaphragmatic surface
Lingula

Plate 221   **Lungs, Trachea, and Bronchi**

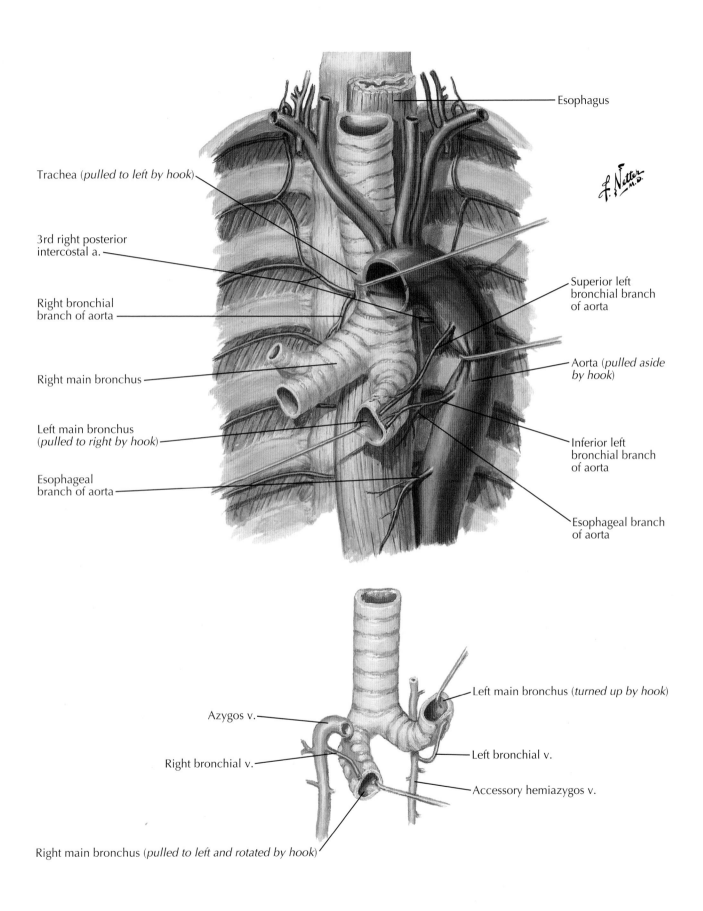

Esophagus

Trachea (*pulled to left by hook*)

3rd right posterior intercostal a.

Right bronchial branch of aorta

Right main bronchus

Left main bronchus (*pulled to right by hook*)

Esophageal branch of aorta

Superior left bronchial branch of aorta

Aorta (*pulled aside by hook*)

Inferior left bronchial branch of aorta

Esophageal branch of aorta

Left main bronchus (*turned up by hook*)

Azygos v.

Right bronchial v.

Left bronchial v.

Accessory hemiazygos v.

Right main bronchus (*pulled to left and rotated by hook*)

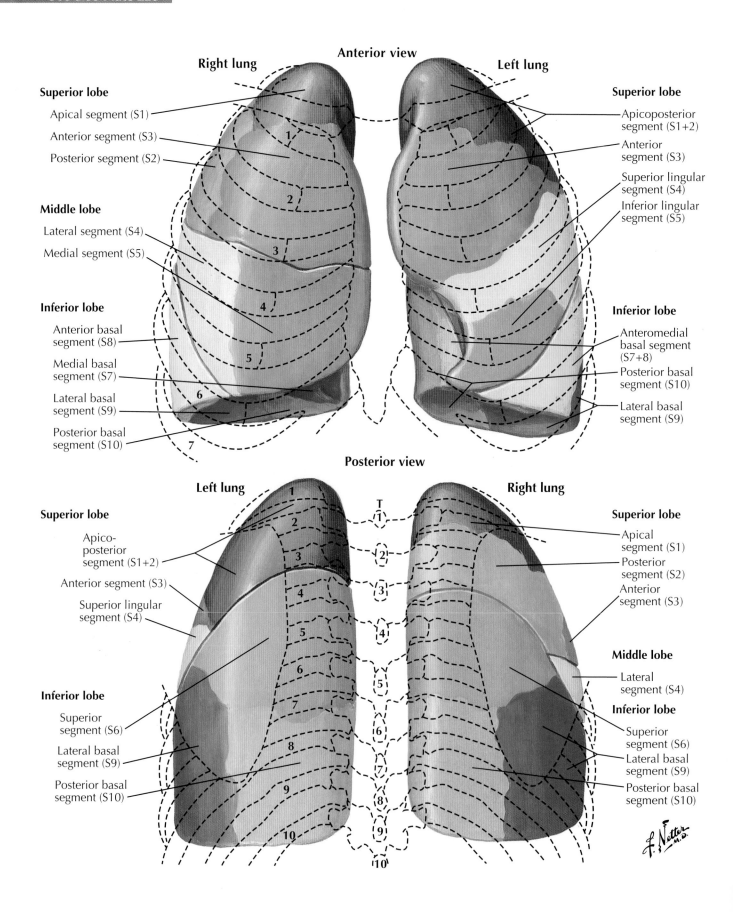

Anterior view

Right lung

Left lung

Superior lobe
- Apical segment (S1)
- Anterior segment (S3)
- Posterior segment (S2)

Middle lobe
- Lateral segment (S4)
- Medial segment (S5)

Inferior lobe
- Anterior basal segment (S8)
- Medial basal segment (S7)
- Lateral basal segment (S9)
- Posterior basal segment (S10)

Superior lobe
- Apicoposterior segment (S1+2)
- Anterior segment (S3)
- Superior lingular segment (S4)
- Inferior lingular segment (S5)

Inferior lobe
- Anteromedial basal segment (S7+8)
- Posterior basal segment (S10)
- Lateral basal segment (S9)

Posterior view

Left lung

Right lung

Superior lobe
- Apico-posterior segment (S1+2)
- Anterior segment (S3)
- Superior lingular segment (S4)

Inferior lobe
- Superior segment (S6)
- Lateral basal segment (S9)
- Posterior basal segment (S10)

Superior lobe
- Apical segment (S1)
- Posterior segment (S2)
- Anterior segment (S3)

Middle lobe
- Lateral segment (S4)

Inferior lobe
- Superior segment (S6)
- Lateral basal segment (S9)
- Posterior basal segment (S10)

Plate 223

Lungs, Trachea, and Bronchi

Lateral views

Right lung

Superior lobe

Apical segment (S1)

Posterior segment (S2)

Anterior segment (S3)

Middle lobe

Lateral segment (S4)

Medial segment (S5)

Inferior lobe

Superior segment (S6)

Anterior basal segment (S8)

Lateral basal segment (S9)

Left lung

Superior lobe

Apico-posterior segment (S1+2)

Anterior segment (S3)

Superior lingular segment (S4)

Inferior lingular segment (S5)

Inferior lobe

Superior segment (S6)

Anteromedial basal segment (S7+8)

Lateral basal segment (S9)

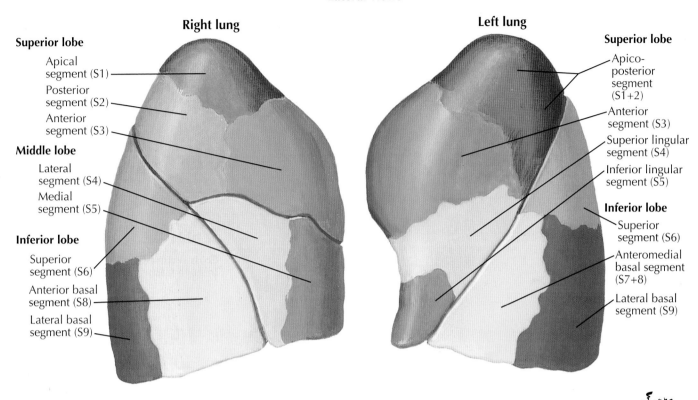

Medial views

Right lung

Superior lobe

Apical segment (S1)

Posterior segment (S2)

Anterior segment (S3)

Middle lobe

Medial segment (S5)

Inferior lobe

Superior segment (S6)

Medial basal segment (S7)

Anterior basal segment (S8)

Lateral basal segment (S9)

Posterior basal segment (S10)

Left lung

Superior lobe

Apico-posterior segment (S1+2)

Anterior segment (S3)

Superior lingular segment (S4)

Inferior lingular segment (S5)

Inferior lobe

Superior segment (S6)

Anteromedial basal segment (S7+8)

Lateral basal segment (S9)

Posterior basal segment (S10)

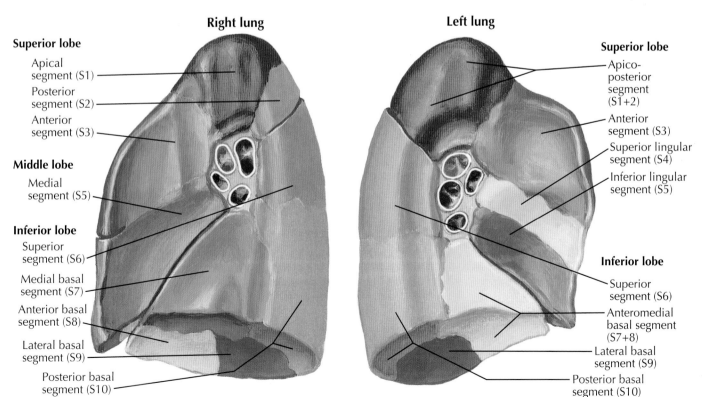

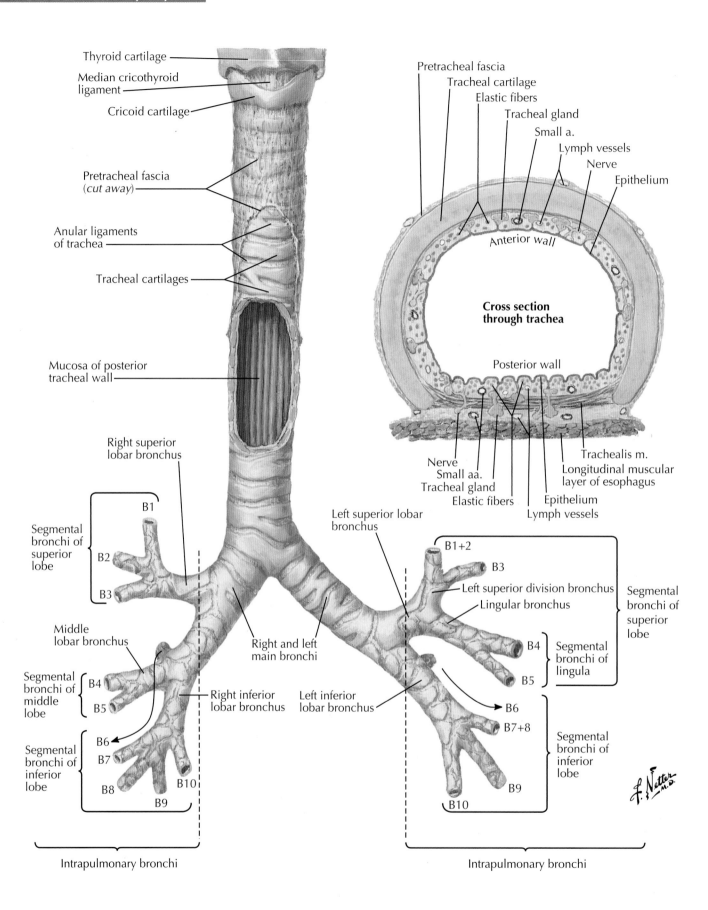

Thyroid cartilage

Median cricothyroid ligament

Cricoid cartilage

Pretracheal fascia (*cut away*)

Anular ligaments of trachea

Tracheal cartilages

Mucosa of posterior tracheal wall

Right superior lobar bronchus

B1

Segmental bronchi of superior lobe

B2

B3

Middle lobar bronchus

Segmental bronchi of middle lobe

B4

B5

Segmental bronchi of inferior lobe

B6

B7

B8

B9

B10

Right inferior lobar bronchus

Right and left main bronchi

Left inferior lobar bronchus

Intrapulmonary bronchi

Pretracheal fascia

Tracheal cartilage

Elastic fibers

Tracheal gland

Small a.

Lymph vessels

Nerve

Epithelium

Anterior wall

Cross section through trachea

Posterior wall

Nerve

Small aa.

Tracheal gland

Elastic fibers

Epithelium

Lymph vessels

Trachealis m.

Longitudinal muscular layer of esophagus

Left superior lobar bronchus

B1+2

B3

Left superior division bronchus

Lingular bronchus

Segmental bronchi of superior lobe

B4

B5

Segmental bronchi of lingula

B6

B7+8

Segmental bronchi of inferior lobe

B9

B10

Intrapulmonary bronchi

Plate 225

Lungs, Trachea, and Bronchi

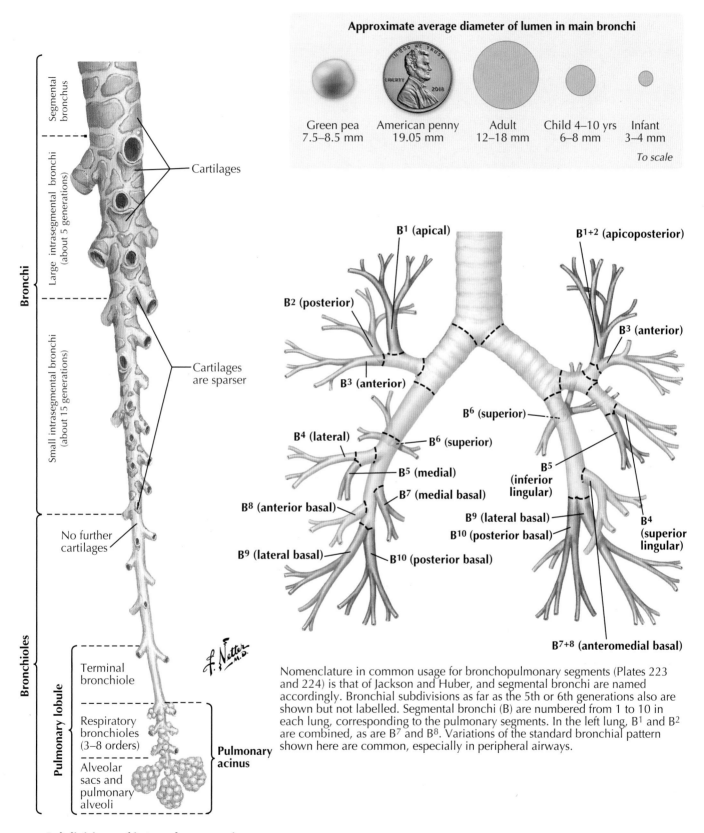

Approximate average diameter of lumen in main bronchi

| Green pea 7.5–8.5 mm | American penny 19.05 mm | Adult 12–18 mm | Child 4–10 yrs 6–8 mm | Infant 3–4 mm |

To scale

Bronchi

Segmental bronchus

Large intrasegmental bronchi (about 5 generations)

Cartilages

Small intrasegmental bronchi (about 15 generations)

Cartilages are sparser

Bronchioles

No further cartilages

Pulmonary lobule

Terminal bronchiole

Respiratory bronchioles (3–8 orders)

Pulmonary acinus

Alveolar sacs and pulmonary alveoli

Subdivisions of intrapulmonary airways

B¹ (apical)

B¹⁺² (apicoposterior)

B² (posterior)

B³ (anterior)

B³ (anterior)

B⁶ (superior)

B⁴ (lateral)

B⁶ (superior)

B⁵ (medial)

B⁵ (inferior lingular)

B⁷ (medial basal)

B⁸ (anterior basal)

B⁹ (lateral basal)

B¹⁰ (posterior basal)

B⁴ (superior lingular)

B⁹ (lateral basal)

B¹⁰ (posterior basal)

B⁷⁺⁸ (anteromedial basal)

Nomenclature in common usage for bronchopulmonary segments (Plates 223 and 224) is that of Jackson and Huber, and segmental bronchi are named accordingly. Bronchial subdivisions as far as the 5th or 6th generations also are shown but not labelled. Segmental bronchi (B) are numbered from 1 to 10 in each lung, corresponding to the pulmonary segments. In the left lung, B¹ and B² are combined, as are B⁷ and B⁸. Variations of the standard bronchial pattern shown here are common, especially in peripheral airways.

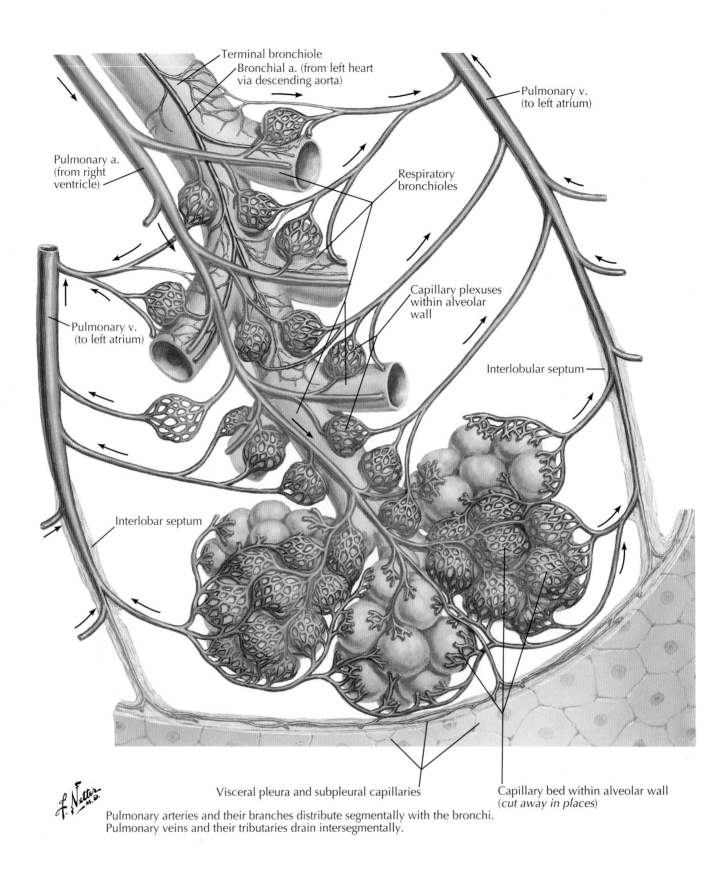

Terminal bronchiole

Bronchial a. (from left heart via descending aorta)

Pulmonary v. (to left atrium)

Pulmonary a. (from right ventricle)

Respiratory bronchioles

Capillary plexuses within alveolar wall

Pulmonary v. (to left atrium)

Interlobular septum

Interlobar septum

Visceral pleura and subpleural capillaries

Capillary bed within alveolar wall (*cut away in places*)

Pulmonary arteries and their branches distribute segmentally with the bronchi.
Pulmonary veins and their tributaries drain intersegmentally.

Plate 227

Lungs, Trachea, and Bronchi

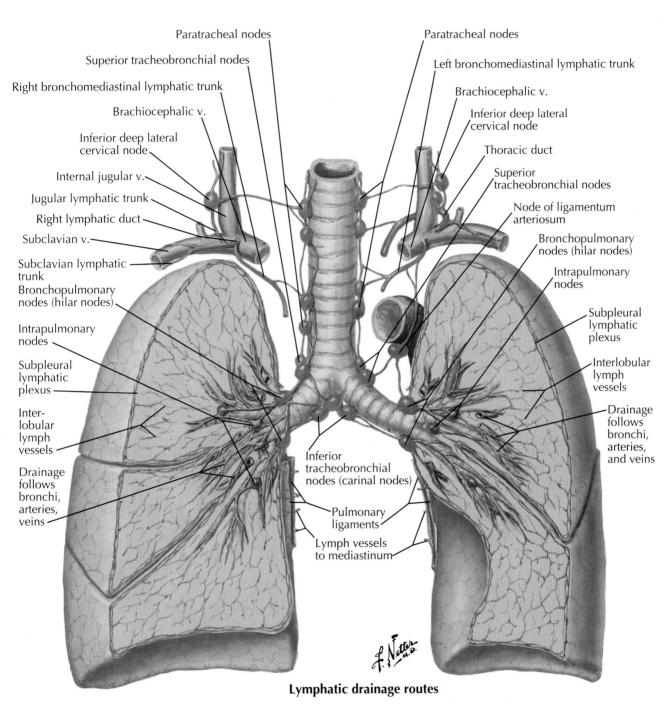

Paratracheal nodes

Superior tracheobronchial nodes

Right bronchomediastinal lymphatic trunk

Brachiocephalic v.

Inferior deep lateral cervical node

Internal jugular v.

Jugular lymphatic trunk

Right lymphatic duct

Subclavian v.

Subclavian lymphatic trunk

Bronchopulmonary nodes (hilar nodes)

Intrapulmonary nodes

Subpleural lymphatic plexus

Inter-lobular lymph vessels

Drainage follows bronchi, arteries, veins

Paratracheal nodes

Left bronchomediastinal lymphatic trunk

Brachiocephalic v.

Inferior deep lateral cervical node

Thoracic duct

Superior tracheobronchial nodes

Node of ligamentum arteriosum

Bronchopulmonary nodes (hilar nodes)

Intrapulmonary nodes

Subpleural lymphatic plexus

Interlobular lymph vessels

Drainage follows bronchi, arteries, and veins

Inferior tracheobronchial nodes (carinal nodes)

Pulmonary ligaments

Lymph vessels to mediastinum

Lymphatic drainage routes

Right lung: All lobes drain to intrapulmonary and bronchopulmonary nodes, then to inferior tracheobronchial nodes, right superior tracheobronchial nodes, and right paratracheal nodes on the way to the brachiocephalic vein via the right bronchomediastinal and jugular trunks.

Left lung: The superior lobe drains to intrapulmonary and bronchopulmonary nodes, then to inferior tracheobronchial nodes, left superior tracheobronchial nodes, left paratracheal nodes and the node of the ligamentum arteriosum on the way to the brachiocephalic vein via the left bronchomediastinal trunk and thoracic duct. The intrapulmonary and bronchopulmonary nodes of the left lung also drain to right superior tracheo-bronchial nodes, where the lymph follows the same route as lymph from the right lung.

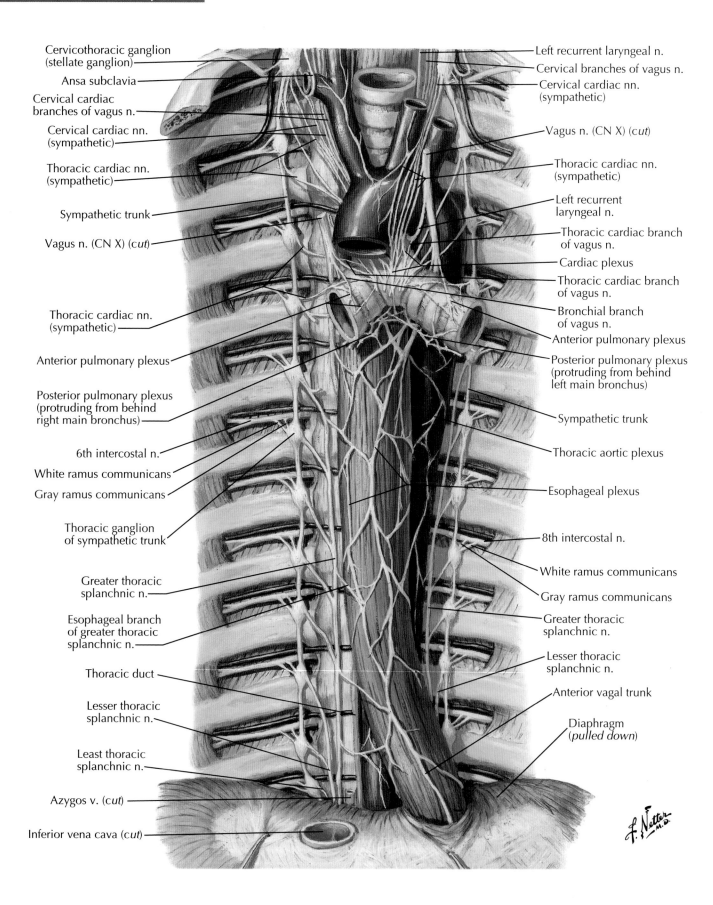

Cervicothoracic ganglion (stellate ganglion)

Ansa subclavia

Cervical cardiac branches of vagus n.

Cervical cardiac nn. (sympathetic)

Thoracic cardiac nn. (sympathetic)

Sympathetic trunk

Vagus n. (CN X) (cut)

Thoracic cardiac nn. (sympathetic)

Anterior pulmonary plexus

Posterior pulmonary plexus (protruding from behind right main bronchus)

6th intercostal n.

White ramus communicans

Gray ramus communicans

Thoracic ganglion of sympathetic trunk

Greater thoracic splanchnic n.

Esophageal branch of greater thoracic splanchnic n.

Thoracic duct

Lesser thoracic splanchnic n.

Least thoracic splanchnic n.

Azygos v. (cut)

Inferior vena cava (cut)

Left recurrent laryngeal n.

Cervical branches of vagus n.

Cervical cardiac nn. (sympathetic)

Vagus n. (CN X) (cut)

Thoracic cardiac nn. (sympathetic)

Left recurrent laryngeal n.

Thoracic cardiac branch of vagus n.

Cardiac plexus

Thoracic cardiac branch of vagus n.

Bronchial branch of vagus n.

Anterior pulmonary plexus

Posterior pulmonary plexus (protruding from behind left main bronchus)

Sympathetic trunk

Thoracic aortic plexus

Esophageal plexus

8th intercostal n.

White ramus communicans

Gray ramus communicans

Greater thoracic splanchnic n.

Lesser thoracic splanchnic n.

Anterior vagal trunk

Diaphragm (pulled down)

Plate 229

Lungs, Trachea, and Bronchi

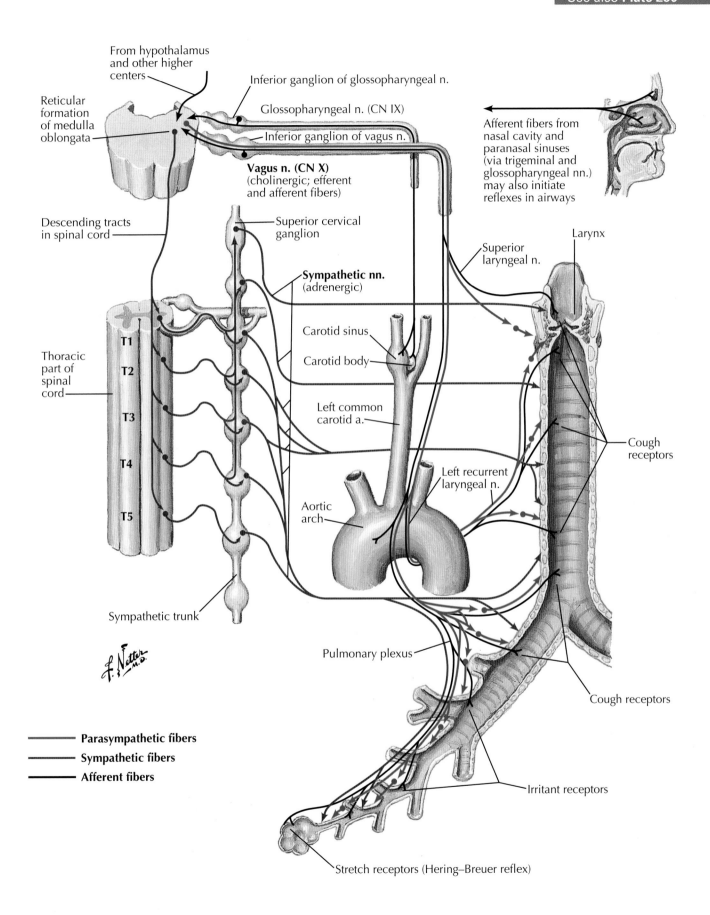

From hypothalamus and other higher centers

Inferior ganglion of glossopharyngeal n.

Glossopharyngeal n. (CN IX)

Reticular formation of medulla oblongata

Inferior ganglion of vagus n.

Vagus n. (CN X) (cholinergic; efferent and afferent fibers)

Afferent fibers from nasal cavity and paranasal sinuses (via trigeminal and glossopharyngeal nn.) may also initiate reflexes in airways

Descending tracts in spinal cord

Superior cervical ganglion

Larynx

Superior laryngeal n.

Sympathetic nn. (adrenergic)

Thoracic part of spinal cord

T1

T2

Carotid sinus

Carotid body

T3

Left common carotid a.

Cough receptors

T4

Left recurrent laryngeal n.

T5

Aortic arch

Sympathetic trunk

Pulmonary plexus

Cough receptors

Irritant receptors

Parasympathetic fibers

.... Sympathetic fibers

▬▬ Afferent fibers

Stretch receptors (Hering–Breuer reflex)

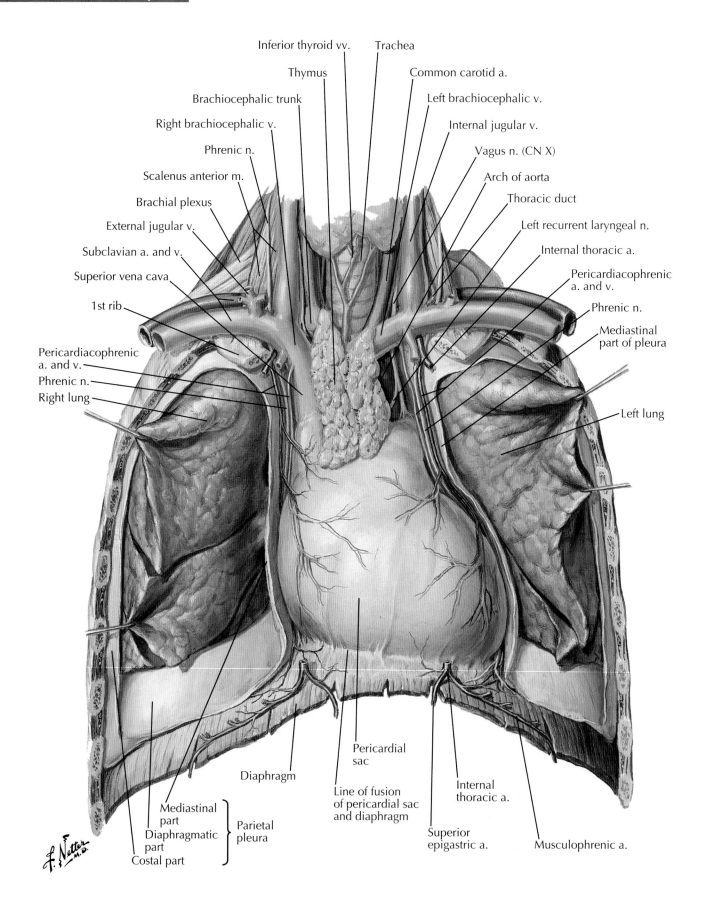

Inferior thyroid vv.

Trachea

Thymus

Common carotid a.

Brachiocephalic trunk

Left brachiocephalic v.

Right brachiocephalic v.

Internal jugular v.

Phrenic n.

Vagus n. (CN X)

Scalenus anterior m.

Arch of aorta

Brachial plexus

Thoracic duct

External jugular v.

Left recurrent laryngeal n.

Subclavian a. and v.

Internal thoracic a.

Superior vena cava

Pericardiacophrenic
a. and v.

1st rib

Phrenic n.

Pericardiacophrenic
a. and v.

Mediastinal
part of pleura

Phrenic n.

Right lung

Left lung

Pericardial
sac

Diaphragm

Line of fusion
of pericardial sac
and diaphragm

Internal
thoracic a.

Mediastinal
part

Diaphragmatic
part

Parietal
pleura

Superior
epigastric a.

Musculophrenic a.

Costal part

Plate 231

Heart

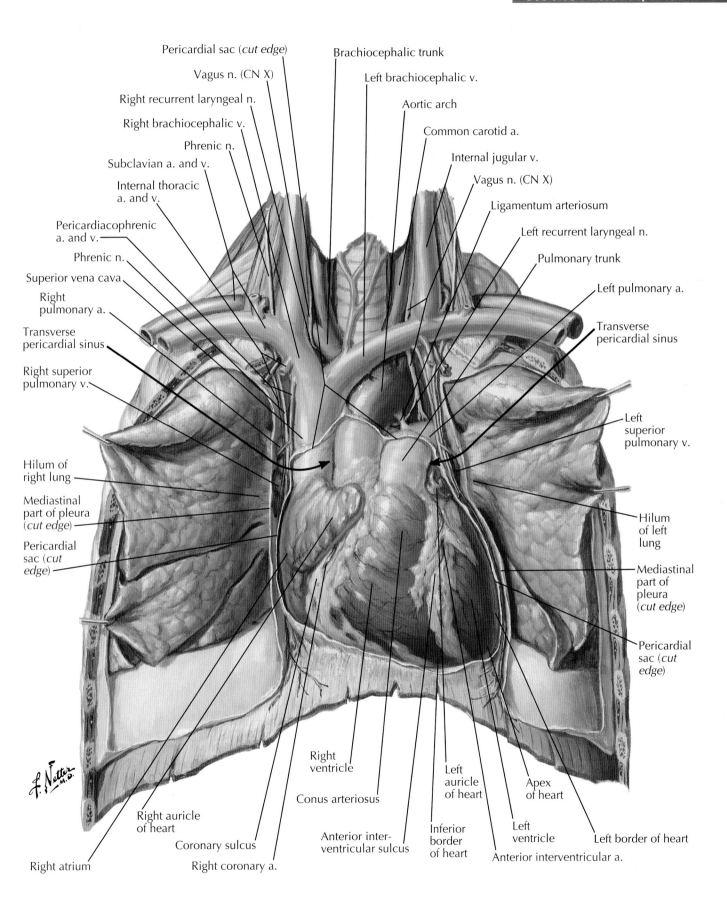

Pericardial sac (*cut edge*)

Vagus n. (CN X)

Right recurrent laryngeal n.

Right brachiocephalic v.

Phrenic n.

Subclavian a. and v.

Internal thoracic a. and v.

Pericardiacophrenic a. and v.

Phrenic n.

Superior vena cava

Right pulmonary a.

Transverse pericardial sinus

Right superior pulmonary v.

Hilum of right lung

Mediastinal part of pleura (*cut edge*)

Pericardial sac (*cut edge*)

Brachiocephalic trunk

Left brachiocephalic v.

Aortic arch

Common carotid a.

Internal jugular v.

Vagus n. (CN X)

Ligamentum arteriosum

Left recurrent laryngeal n.

Pulmonary trunk

Left pulmonary a.

Transverse pericardial sinus

Left superior pulmonary v.

Hilum of left lung

Mediastinal part of pleura (*cut edge*)

Pericardial sac (*cut edge*)

Right ventricle

Conus arteriosus

Right auricle of heart

Coronary sulcus

Right coronary a.

Right atrium

Anterior interventricular sulcus

Left auricle of heart

Inferior border of heart

Apex of heart

Left ventricle

Left border of heart

Anterior interventricular a.

Heart

Plate 232

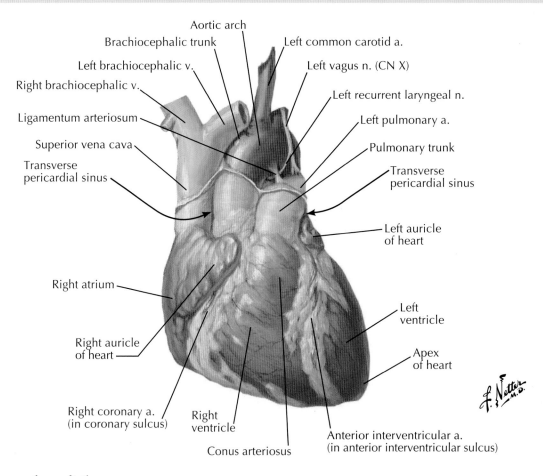

Aortic arch

Brachiocephalic trunk

Left common carotid a.

Left brachiocephalic v.

Left vagus n. (CN X)

Right brachiocephalic v.

Left recurrent laryngeal n.

Ligamentum arteriosum

Left pulmonary a.

Superior vena cava

Pulmonary trunk

Transverse pericardial sinus

Transverse pericardial sinus

Left auricle of heart

Right atrium

Left ventricle

Right auricle of heart

Apex of heart

Right coronary a. (in coronary sulcus)

Right ventricle

Conus arteriosus

Anterior interventricular a. (in anterior interventricular sulcus)

Precordial areas of auscultation:
One listens to the closing of a heart valve downstream from the heart valve, that is, in the right and left ventricles for the tricuspid and mitral valves, respectively, and over the pulmonary trunk and ascending aorta for the pulmonic and aortic valves, respectively.

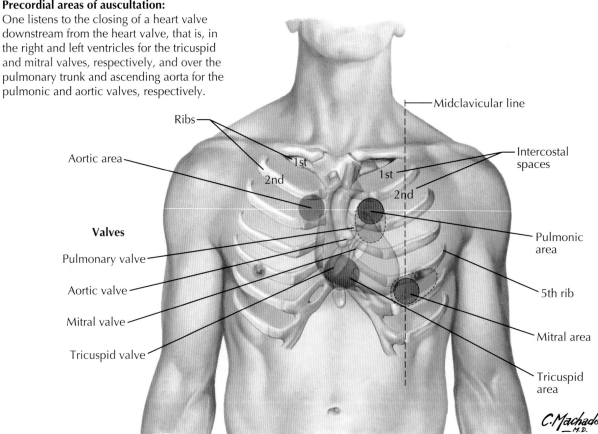

Ribs

Midclavicular line

Aortic area

1st

2nd

1st

Intercostal spaces

2nd

Valves

Pulmonic area

Pulmonary valve

Aortic valve

5th rib

Mitral valve

Mitral area

Tricuspid valve

Tricuspid area

Plate 233   **Heart**

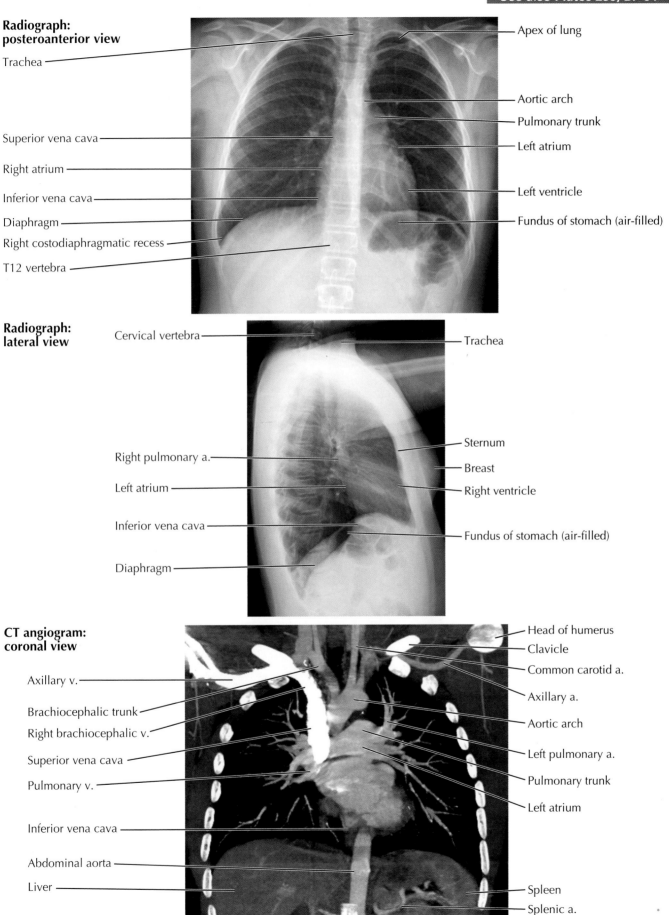

**Radiograph:
posteroanterior view**

Trachea

Superior vena cava

Right atrium

Inferior vena cava

Diaphragm

Right costodiaphragmatic recess

T12 vertebra

Apex of lung

Aortic arch

Pulmonary trunk

Left atrium

Left ventricle

Fundus of stomach (air-filled)

**Radiograph:
lateral view**

Cervical vertebra

Trachea

Right pulmonary a.

Left atrium

Inferior vena cava

Diaphragm

Sternum

Breast

Right ventricle

Fundus of stomach (air-filled)

**CT angiogram:
coronal view**

Axillary v.

Brachiocephalic trunk

Right brachiocephalic v.

Superior vena cava

Pulmonary v.

Inferior vena cava

Abdominal aorta

Liver

Head of humerus

Clavicle

Common carotid a.

Axillary a.

Aortic arch

Left pulmonary a.

Pulmonary trunk

Left atrium

Spleen

Splenic a.

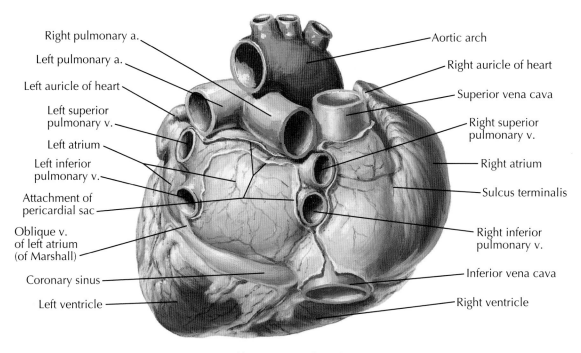

Right pulmonary a.

Left pulmonary a.

Left auricle of heart

Left superior pulmonary v.

Left atrium

Left inferior pulmonary v.

Attachment of pericardial sac

Oblique v. of left atrium (of Marshall)

Coronary sinus

Left ventricle

Aortic arch

Right auricle of heart

Superior vena cava

Right superior pulmonary v.

Right atrium

Sulcus terminalis

Right inferior pulmonary v.

Inferior vena cava

Right ventricle

Base of heart: posterior view

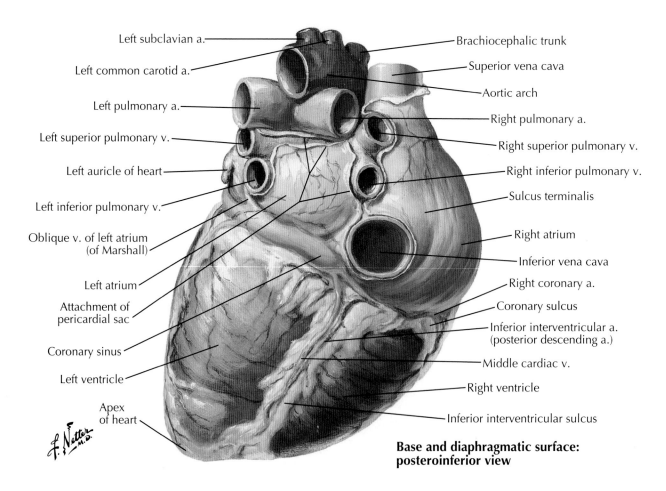

Left subclavian a.

Left common carotid a.

Left pulmonary a.

Left superior pulmonary v.

Left auricle of heart

Left inferior pulmonary v.

Oblique v. of left atrium (of Marshall)

Left atrium

Attachment of pericardial sac

Coronary sinus

Left ventricle

Apex of heart

Brachiocephalic trunk

Superior vena cava

Aortic arch

Right pulmonary a.

Right superior pulmonary v.

Right inferior pulmonary v.

Sulcus terminalis

Right atrium

Inferior vena cava

Right coronary a.

Coronary sulcus

Inferior interventricular a. (posterior descending a.)

Middle cardiac v.

Right ventricle

Inferior interventricular sulcus

Base and diaphragmatic surface: posteroinferior view

Plate 235   **Heart**

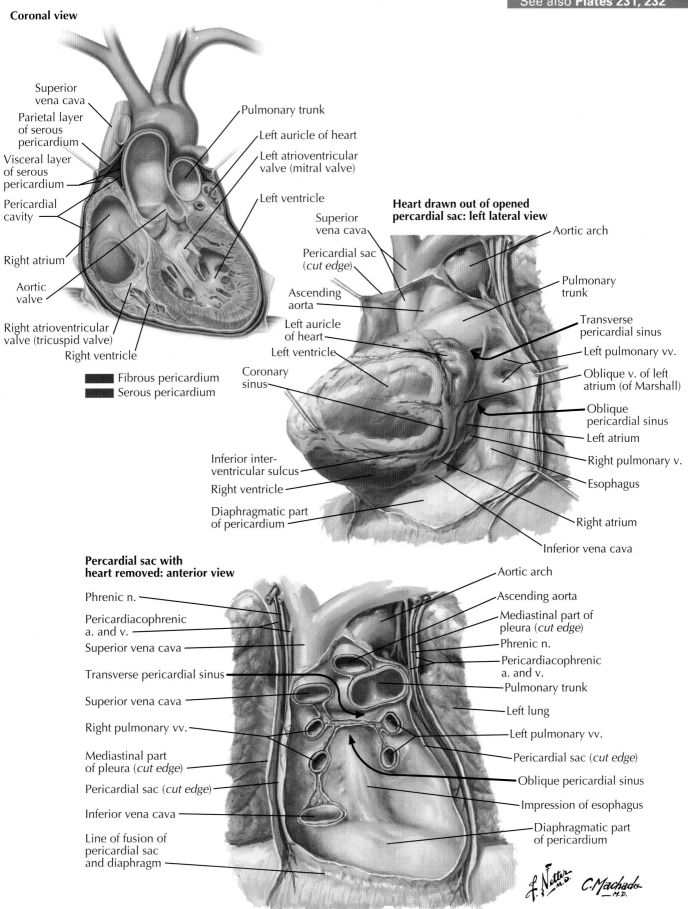

Coronal view

Superior vena cava

Parietal layer of serous pericardium

Visceral layer of serous pericardium

Pericardial cavity

Right atrium

Aortic valve

Right atrioventricular valve (tricuspid valve)

Right ventricle

Pulmonary trunk

Left auricle of heart

Left atrioventricular valve (mitral valve)

Left ventricle

■ Fibrous pericardium
■ Serous pericardium

Heart drawn out of opened percardial sac: left lateral view

Superior vena cava

Pericardial sac (cut edge)

Ascending aorta

Left auricle of heart

Left ventricle

Coronary sinus

Inferior inter-ventricular sulcus

Right ventricle

Diaphragmatic part of pericardium

Aortic arch

Pulmonary trunk

Transverse pericardial sinus

Left pulmonary vv.

Oblique v. of left atrium (of Marshall)

Oblique pericardial sinus

Left atrium

Right pulmonary v.

Esophagus

Right atrium

Inferior vena cava

Percardial sac with heart removed: anterior view

Phrenic n.

Pericardiacophrenic a. and v.

Superior vena cava

Transverse pericardial sinus

Superior vena cava

Right pulmonary vv.

Mediastinal part of pleura (cut edge)

Pericardial sac (cut edge)

Inferior vena cava

Line of fusion of pericardial sac and diaphragm

Aortic arch

Ascending aorta

Mediastinal part of pleura (cut edge)

Phrenic n.

Pericardiacophrenic a. and v.

Pulmonary trunk

Left lung

Left pulmonary vv.

Pericardial sac (cut edge)

Oblique pericardial sinus

Impression of esophagus

Diaphragmatic part of pericardium

J. Netter M.D.   *C. Machado M.D.*

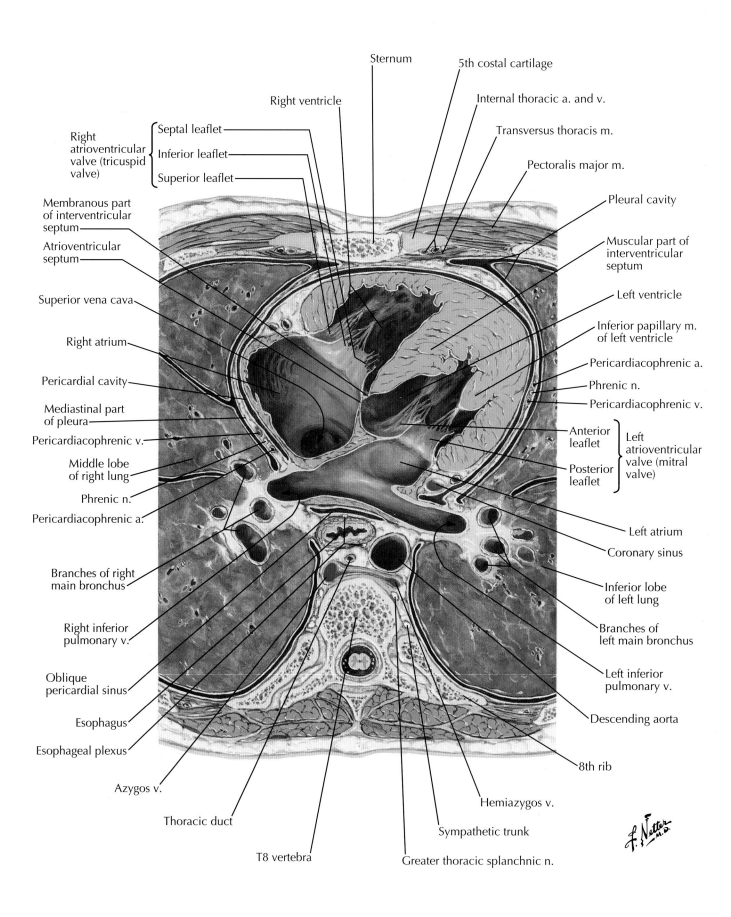

Sternum

5th costal cartilage

Right ventricle

Internal thoracic a. and v.

Transversus thoracis m.

Pectoralis major m.

Right atrioventricular valve (tricuspid valve)
- Septal leaflet
- Inferior leaflet
- Superior leaflet

Pleural cavity

Membranous part of interventricular septum

Atrioventricular septum

Muscular part of interventricular septum

Superior vena cava

Left ventricle

Inferior papillary m. of left ventricle

Right atrium

Pericardial cavity

Pericardiacophrenic a.

Phrenic n.

Pericardiacophrenic v.

Mediastinal part of pleura

Pericardiacophrenic v.

Left atrioventricular valve (mitral valve)
- Anterior leaflet
- Posterior leaflet

Middle lobe of right lung

Phrenic n.

Pericardiacophrenic a.

Left atrium

Coronary sinus

Inferior lobe of left lung

Branches of right main bronchus

Branches of left main bronchus

Right inferior pulmonary v.

Oblique pericardial sinus

Left inferior pulmonary v.

Esophagus

Descending aorta

Esophageal plexus

8th rib

Azygos v.

Hemiazygos v.

Thoracic duct

Sympathetic trunk

T8 vertebra

Greater thoracic splanchnic n.

Plate 237

Heart

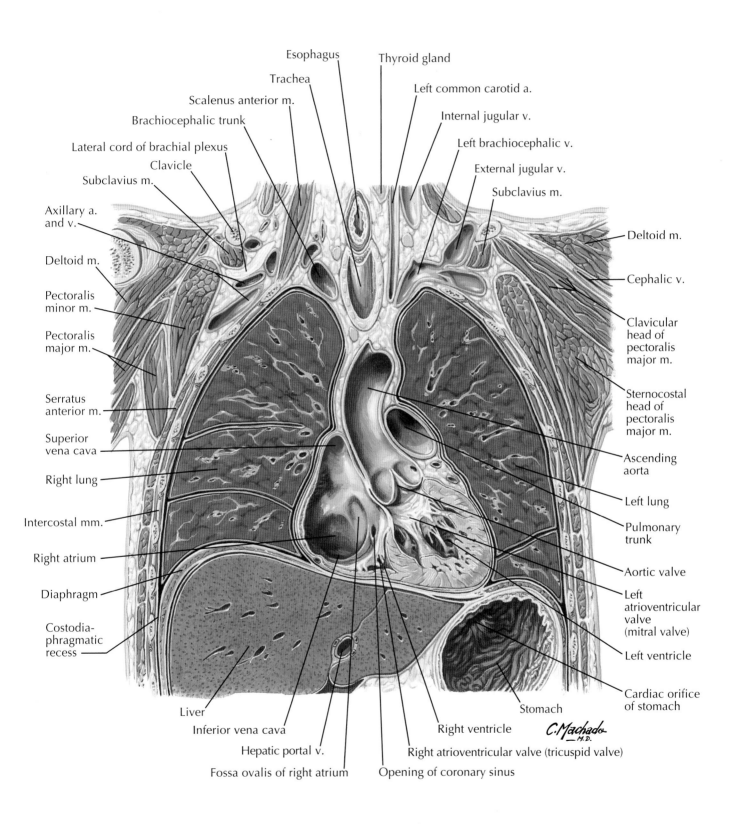

Esophagus

Thyroid gland

Trachea

Left common carotid a.

Scalenus anterior m.

Internal jugular v.

Brachiocephalic trunk

Left brachiocephalic v.

Lateral cord of brachial plexus

External jugular v.

Clavicle

Subclavius m.

Subclavius m.

Axillary a. and v.

Deltoid m.

Deltoid m.

Cephalic v.

Pectoralis minor m.

Clavicular head of pectoralis major m.

Pectoralis major m.

Serratus anterior m.

Sternocostal head of pectoralis major m.

Superior vena cava

Ascending aorta

Right lung

Left lung

Intercostal mm.

Pulmonary trunk

Right atrium

Aortic valve

Diaphragm

Left atrioventricular valve (mitral valve)

Costodiaphragmatic recess

Left ventricle

Cardiac orifice of stomach

Liver

Stomach

Inferior vena cava

Right ventricle

Hepatic portal v.

Right atrioventricular valve (tricuspid valve)

Fossa ovalis of right atrium

Opening of coronary sinus

C. Machado
_M.D._

Sternocostal surface

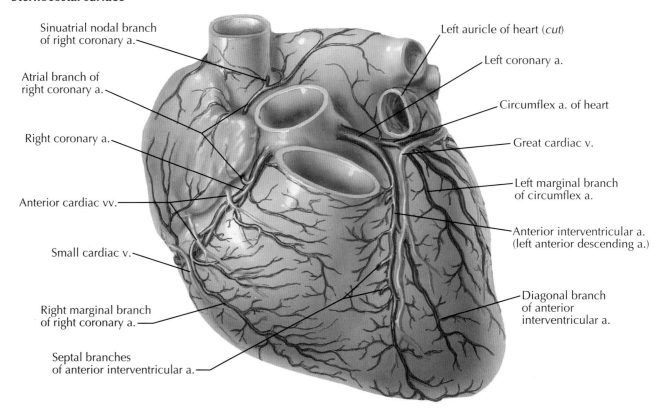

Sinuatrial nodal branch of right coronary a.

Atrial branch of right coronary a.

Right coronary a.

Anterior cardiac vv.

Small cardiac v.

Right marginal branch of right coronary a.

Septal branches of anterior interventricular a.

Left auricle of heart (*cut*)

Left coronary a.

Circumflex a. of heart

Great cardiac v.

Left marginal branch of circumflex a.

Anterior interventricular a. (left anterior descending a.)

Diagonal branch of anterior interventricular a.

Diaphragmatic surface

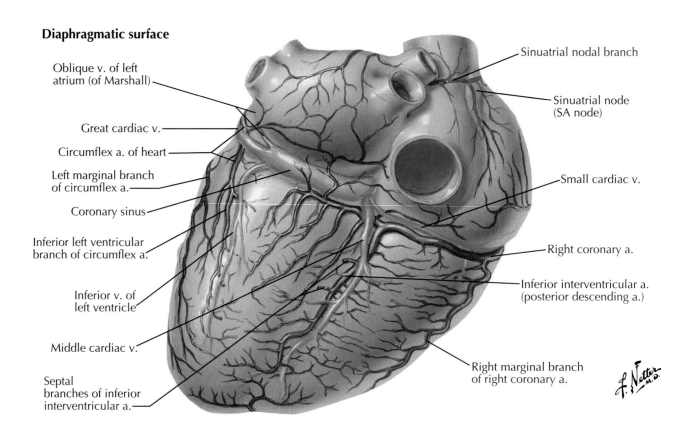

Oblique v. of left atrium (of Marshall)

Great cardiac v.

Circumflex a. of heart

Left marginal branch of circumflex a.

Coronary sinus

Inferior left ventricular branch of circumflex a.

Inferior v. of left ventricle

Middle cardiac v.

Septal branches of inferior interventricular a.

Sinuatrial nodal branch

Sinuatrial node (SA node)

Small cardiac v.

Right coronary a.

Inferior interventricular a. (posterior descending a.)

Right marginal branch of right coronary a.

Plate 239

Heart

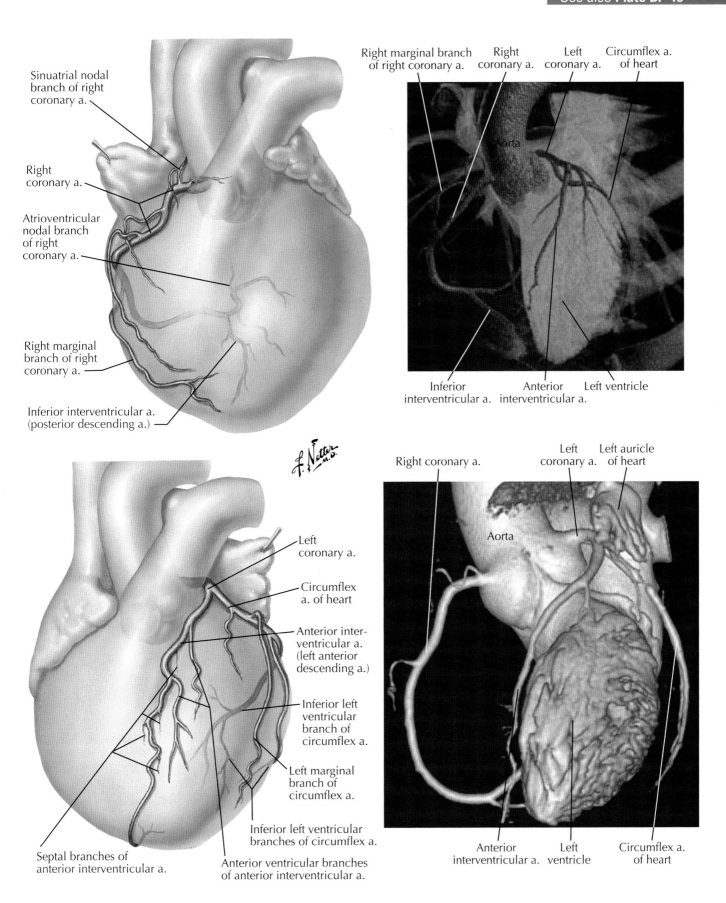

Sinuatrial nodal branch of right coronary a.

Right coronary a.

Atrioventricular nodal branch of right coronary a.

Right marginal branch of right coronary a.

Inferior interventricular a. (posterior descending a.)

Right marginal branch of right coronary a.

Right coronary a.

Left coronary a.

Circumflex a. of heart

Aorta

Inferior interventricular a.

Anterior interventricular a.

Left ventricle

Left coronary a.

Circumflex a. of heart

Anterior inter-ventricular a. (left anterior descending a.)

Inferior left ventricular branch of circumflex a.

Left marginal branch of circumflex a.

Inferior left ventricular branches of circumflex a.

Septal branches of anterior interventricular a.

Anterior ventricular branches of anterior interventricular a.

Right coronary a.

Left coronary a.

Left auricle of heart

Aorta

Anterior interventricular a.

Left ventricle

Circumflex a. of heart

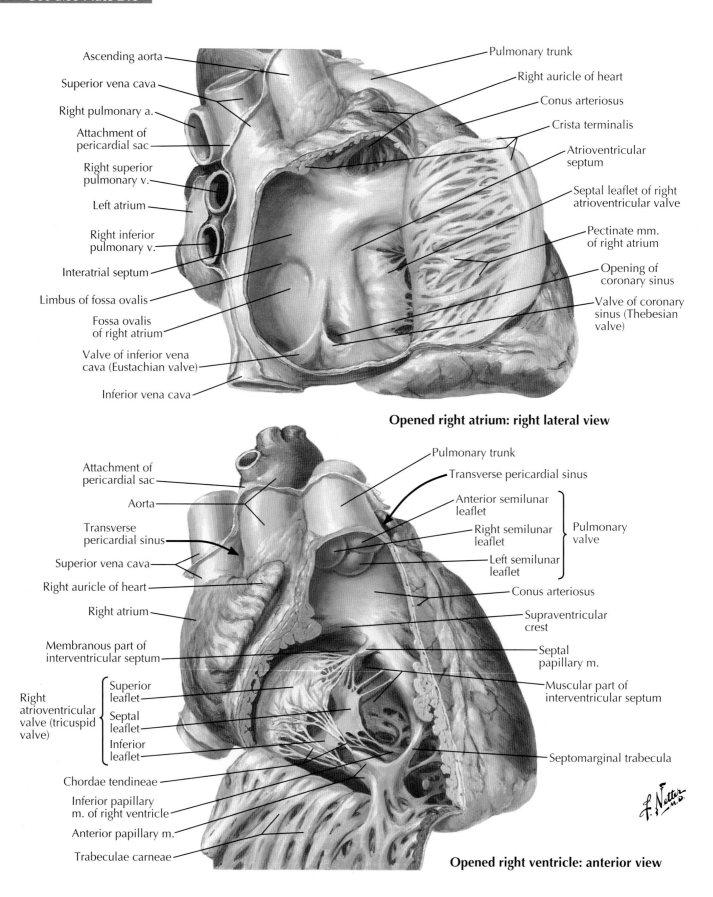

Ascending aorta

Superior vena cava

Right pulmonary a.

Attachment of pericardial sac

Right superior pulmonary v.

Left atrium

Right inferior pulmonary v.

Interatrial septum

Limbus of fossa ovalis

Fossa ovalis of right atrium

Valve of inferior vena cava (Eustachian valve)

Inferior vena cava

Pulmonary trunk

Right auricle of heart

Conus arteriosus

Crista terminalis

Atrioventricular septum

Septal leaflet of right atrioventricular valve

Pectinate mm. of right atrium

Opening of coronary sinus

Valve of coronary sinus (Thebesian valve)

Opened right atrium: right lateral view

Attachment of pericardial sac

Aorta

Transverse pericardial sinus

Superior vena cava

Right auricle of heart

Right atrium

Membranous part of interventricular septum

Right atrioventricular valve (tricuspid valve)
- Superior leaflet
- Septal leaflet
- Inferior leaflet

Chordae tendineae

Inferior papillary m. of right ventricle

Anterior papillary m.

Trabeculae carneae

Pulmonary trunk

Transverse pericardial sinus

Anterior semilunar leaflet

Right semilunar leaflet

Left semilunar leaflet

Pulmonary valve

Conus arteriosus

Supraventricular crest

Septal papillary m.

Muscular part of interventricular septum

Septomarginal trabecula

Opened right ventricle: anterior view

Plate 241

Heart

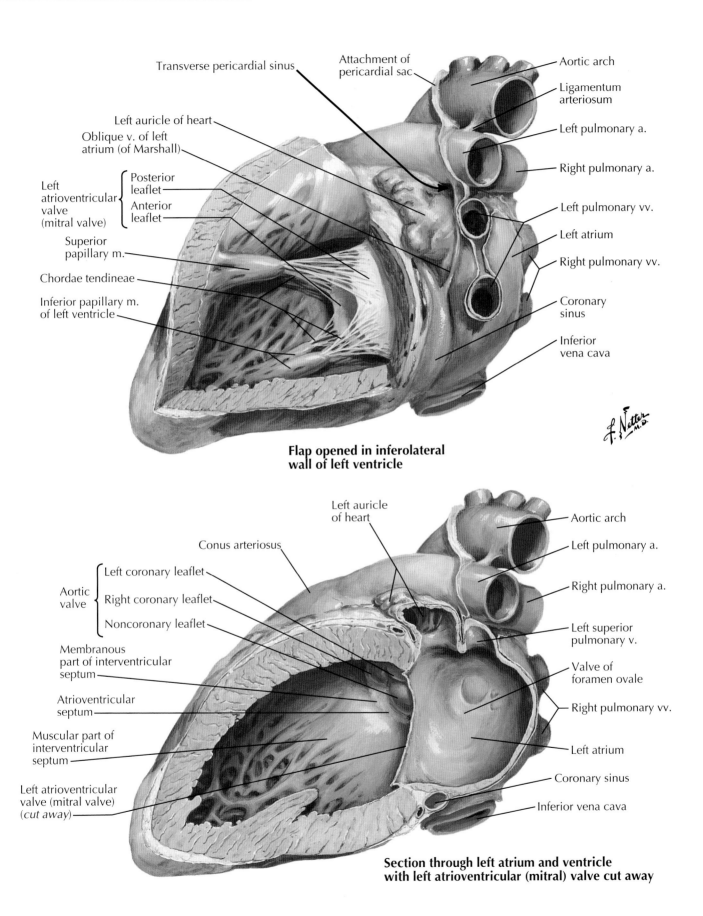

Transverse pericardial sinus

Attachment of pericardial sac

Aortic arch

Ligamentum arteriosum

Left auricle of heart

Oblique v. of left atrium (of Marshall)

Left pulmonary a.

Right pulmonary a.

Left atrioventricular valve (mitral valve) { Posterior leaflet / Anterior leaflet }

Left pulmonary vv.

Left atrium

Right pulmonary vv.

Superior papillary m.

Chordae tendineae

Coronary sinus

Inferior papillary m. of left ventricle

Inferior vena cava

Flap opened in inferolateral wall of left ventricle

Left auricle of heart

Conus arteriosus

Aortic arch

Left pulmonary a.

Right pulmonary a.

Aortic valve { Left coronary leaflet / Right coronary leaflet / Noncoronary leaflet }

Membranous part of interventricular septum

Atrioventricular septum

Muscular part of interventricular septum

Left superior pulmonary v.

Valve of foramen ovale

Right pulmonary vv.

Left atrium

Left atrioventricular valve (mitral valve) (*cut away*)

Coronary sinus

Inferior vena cava

Section through left atrium and ventricle with left atrioventricular (mitral) valve cut away

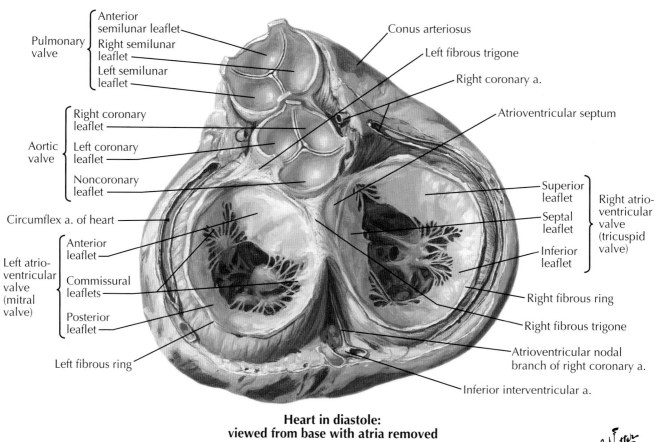

Pulmonary valve
- Anterior semilunar leaflet
- Right semilunar leaflet
- Left semilunar leaflet

Aortic valve
- Right coronary leaflet
- Left coronary leaflet
- Noncoronary leaflet

Circumflex a. of heart

Left atrioventricular valve (mitral valve)
- Anterior leaflet
- Commissural leaflets
- Posterior leaflet

Left fibrous ring

Conus arteriosus

Left fibrous trigone

Right coronary a.

Atrioventricular septum

Superior leaflet
Septal leaflet
Inferior leaflet

Right atrioventricular valve (tricuspid valve)

Right fibrous ring

Right fibrous trigone

Atrioventricular nodal branch of right coronary a.

Inferior interventricular a.

Heart in diastole:
viewed from base with atria removed

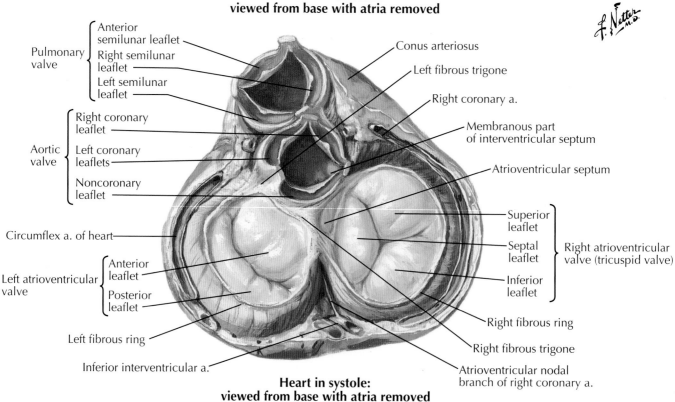

Pulmonary valve
- Anterior semilunar leaflet
- Right semilunar leaflet
- Left semilunar leaflet

Aortic valve
- Right coronary leaflet
- Left coronary leaflets
- Noncoronary leaflet

Circumflex a. of heart

Left atrioventricular valve
- Anterior leaflet
- Posterior leaflet

Left fibrous ring

Inferior interventricular a.

Conus arteriosus

Left fibrous trigone

Right coronary a.

Membranous part of interventricular septum

Atrioventricular septum

Superior leaflet
Septal leaflet
Inferior leaflet

Right atrioventricular valve (tricuspid valve)

Right fibrous ring

Right fibrous trigone

Atrioventricular nodal branch of right coronary a.

Heart in systole:
viewed from base with atria removed

Plate 243

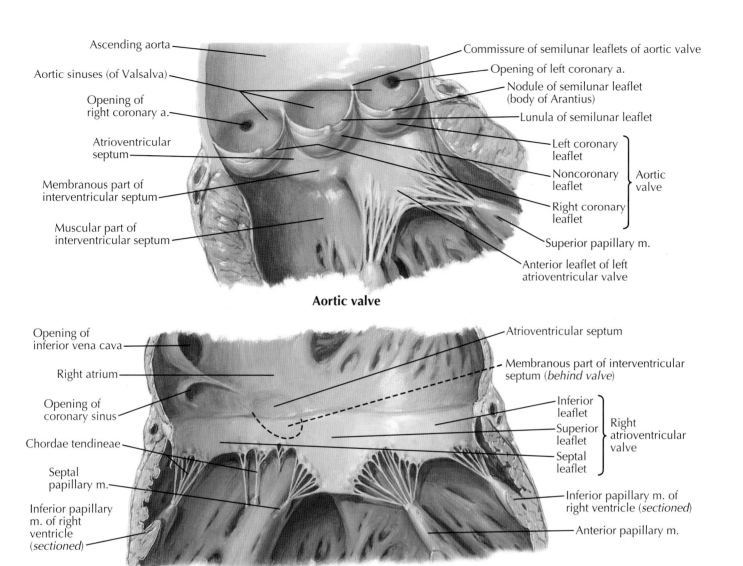

Ascending aorta

Aortic sinuses (of Valsalva)

Opening of right coronary a.

Atrioventricular septum

Membranous part of interventricular septum

Muscular part of interventricular septum

Commissure of semilunar leaflets of aortic valve

Opening of left coronary a.

Nodule of semilunar leaflet (body of Arantius)

Lunula of semilunar leaflet

Left coronary leaflet

Noncoronary leaflet

Right coronary leaflet

Aortic valve

Superior papillary m.

Anterior leaflet of left atrioventricular valve

Aortic valve

Opening of inferior vena cava

Right atrium

Opening of coronary sinus

Chordae tendineae

Septal papillary m.

Inferior papillary m. of right ventricle (*sectioned*)

Atrioventricular septum

Membranous part of interventricular septum (*behind valve*)

Inferior leaflet

Superior leaflet

Septal leaflet

Right atrioventricular valve

Inferior papillary m. of right ventricle (*sectioned*)

Anterior papillary m.

Right atrioventricular valve (tricuspid valve)

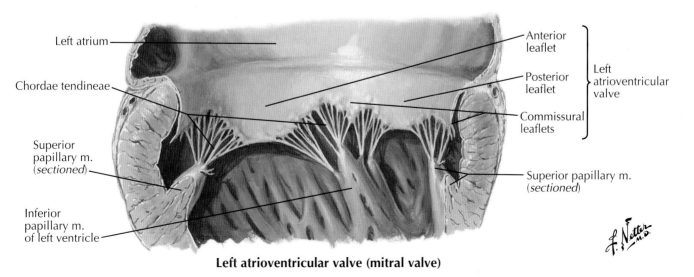

Left atrium

Chordae tendineae

Superior papillary m. (*sectioned*)

Inferior papillary m. of left ventricle

Anterior leaflet

Posterior leaflet

Left atrioventricular valve

Commissural leaflets

Superior papillary m. (*sectioned*)

Left atrioventricular valve (mitral valve)

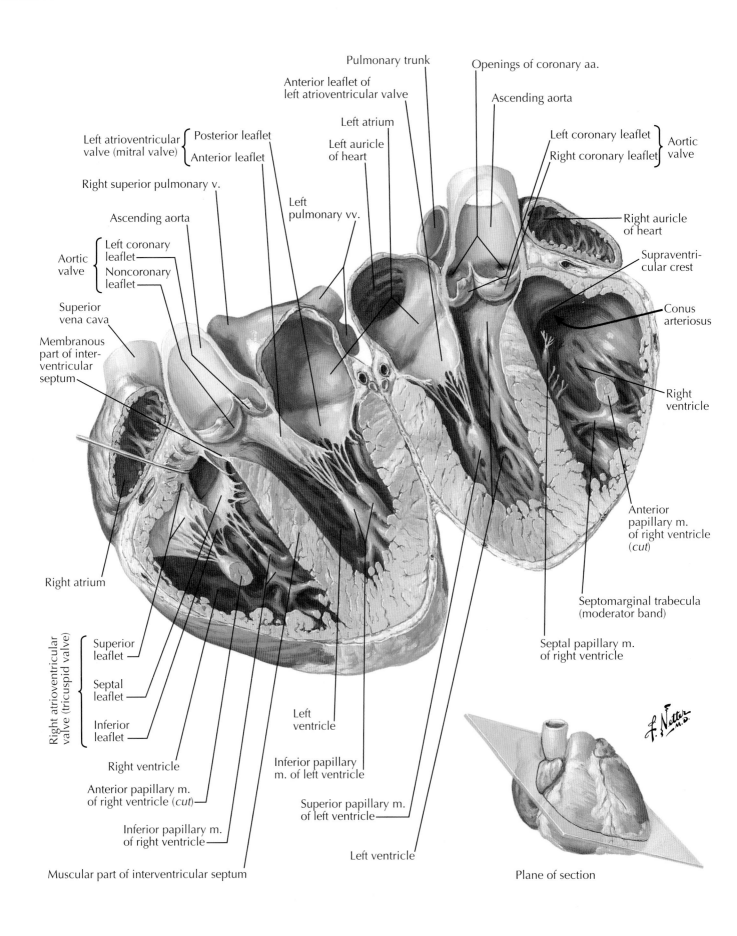

Pulmonary trunk

Openings of coronary aa.

Anterior leaflet of
left atrioventricular valve

Ascending aorta

Left atrium

Left coronary leaflet ⎫
Right coronary leaflet ⎭ Aortic valve

Left auricle
of heart

Left atrioventricular ⎧ Posterior leaflet
valve (mitral valve) ⎩ Anterior leaflet

Right superior pulmonary v.

Left
pulmonary vv.

Right auricle
of heart

Ascending aorta

Supraventri-
cular crest

Aortic ⎧ Left coronary
valve ⎨ leaflet
⎩ Noncoronary
leaflet

Conus
arteriosus

Superior
vena cava

Right
ventricle

Membranous
part of inter-
ventricular
septum

Anterior
papillary m.
of right ventricle
(cut)

Right atrium

Septomarginal trabecula
(moderator band)

Septal papillary m.
of right ventricle

Right atrioventricular valve (tricuspid valve) ⎧ Superior
leaflet

Septal
leaflet

Inferior
leaflet

Left
ventricle

Right ventricle

Inferior papillary
m. of left ventricle

Anterior papillary m.
of right ventricle (cut)

Superior papillary m.
of left ventricle

Inferior papillary m.
of right ventricle

Left ventricle

Plane of section

Muscular part of interventricular septum

Plate 245

Heart

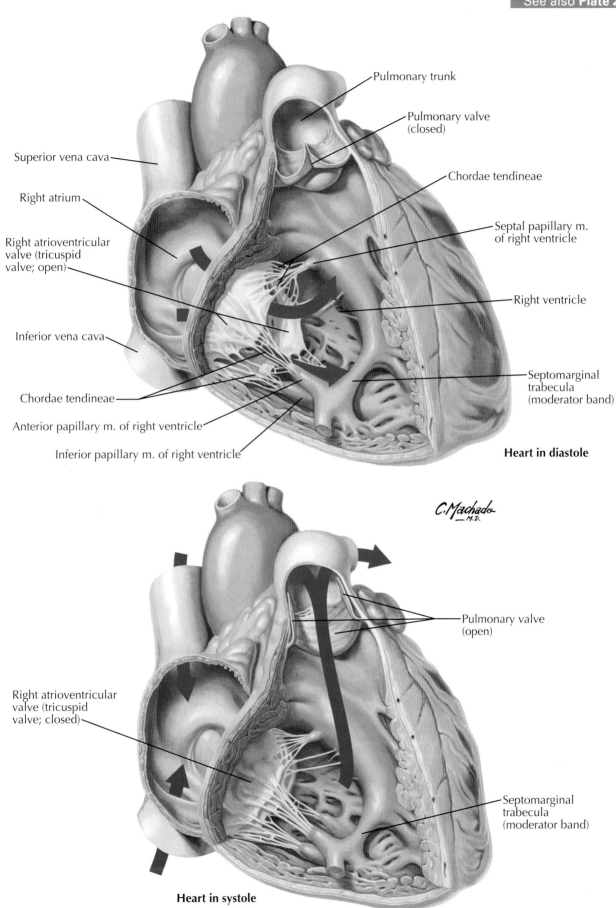

Pulmonary trunk

Pulmonary valve
(closed)

Chordae tendineae

Superior vena cava

Septal papillary m.
of right ventricle

Right atrium

Right atrioventricular
valve (tricuspid
valve; open)

Right ventricle

Inferior vena cava

Septomarginal
trabecula
(moderator band)

Chordae tendineae

Anterior papillary m. of right ventricle

Inferior papillary m. of right ventricle

Heart in diastole

C.Machado
_M.D._

Pulmonary valve
(open)

Right atrioventricular
valve (tricuspid
valve; closed)

Septomarginal
trabecula
(moderator band)

Heart in systole

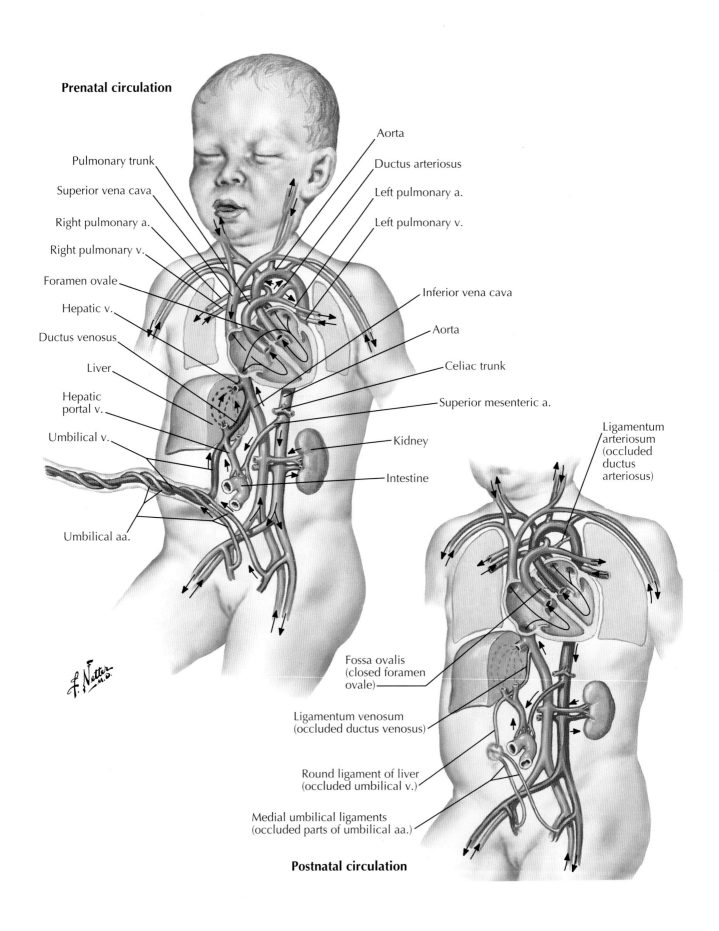

Prenatal circulation

Pulmonary trunk

Superior vena cava

Right pulmonary a.

Right pulmonary v.

Foramen ovale

Hepatic v.

Ductus venosus

Liver

Hepatic
portal v.

Umbilical v.

Umbilical aa.

Aorta

Ductus arteriosus

Left pulmonary a.

Left pulmonary v.

Inferior vena cava

Aorta

Celiac trunk

Superior mesenteric a.

Kidney

Intestine

Ligamentum
arteriosum
(occluded
ductus
arteriosus)

Fossa ovalis
(closed foramen
ovale)

Ligamentum venosum
(occluded ductus venosus)

Round ligament of liver
(occluded umbilical v.)

Medial umbilical ligaments
(occluded parts of umbilical aa.)

Postnatal circulation

Plate 247   **Heart**

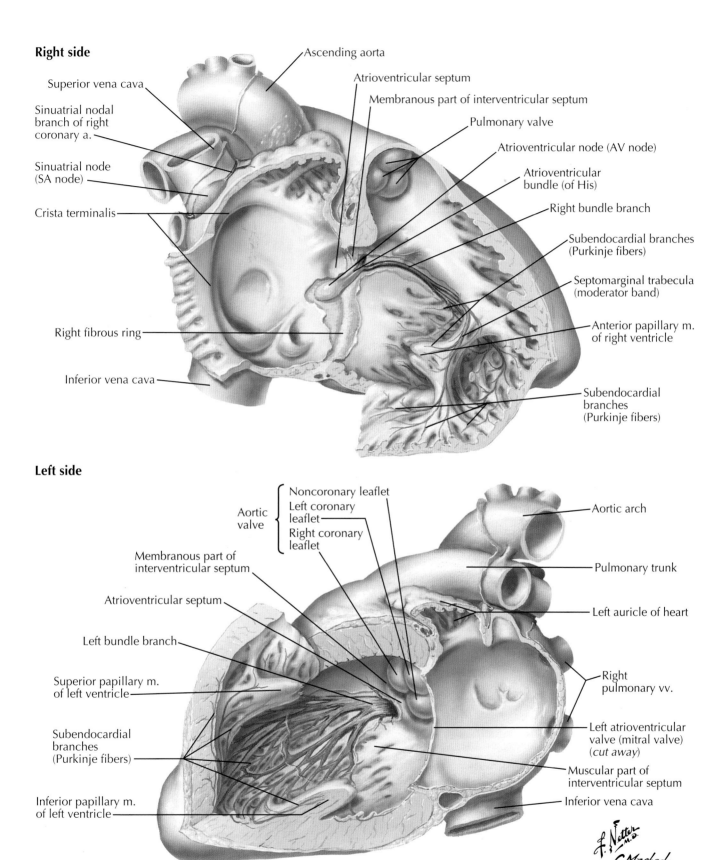

Right side

Superior vena cava

Sinuatrial nodal branch of right coronary a.

Sinuatrial node (SA node)

Crista terminalis

Right fibrous ring

Inferior vena cava

Ascending aorta

Atrioventricular septum

Membranous part of interventricular septum

Pulmonary valve

Atrioventricular node (AV node)

Atrioventricular bundle (of His)

Right bundle branch

Subendocardial branches (Purkinje fibers)

Septomarginal trabecula (moderator band)

Anterior papillary m. of right ventricle

Subendocardial branches (Purkinje fibers)

Left side

Aortic valve {
Noncoronary leaflet
Left coronary leaflet
Right coronary leaflet
}

Membranous part of interventricular septum

Atrioventricular septum

Left bundle branch

Superior papillary m. of left ventricle

Subendocardial branches (Purkinje fibers)

Inferior papillary m. of left ventricle

Aortic arch

Pulmonary trunk

Left auricle of heart

Right pulmonary vv.

Left atrioventricular valve (mitral valve) (*cut away*)

Muscular part of interventricular septum

Inferior vena cava

Heart

Plate 248

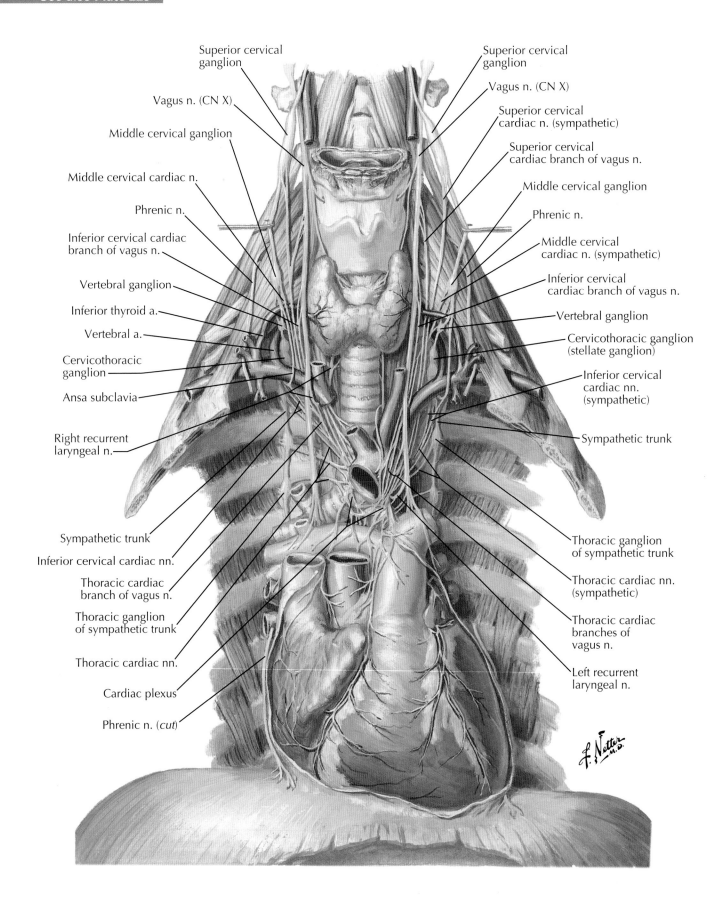

Superior cervical ganglion

Vagus n. (CN X)

Middle cervical ganglion

Middle cervical cardiac n.

Phrenic n.

Inferior cervical cardiac branch of vagus n.

Vertebral ganglion

Inferior thyroid a.

Vertebral a.

Cervicothoracic ganglion

Ansa subclavia

Right recurrent laryngeal n.

Sympathetic trunk

Inferior cervical cardiac nn.

Thoracic cardiac branch of vagus n.

Thoracic ganglion of sympathetic trunk

Thoracic cardiac nn.

Cardiac plexus

Phrenic n. (*cut*)

Superior cervical ganglion

Vagus n. (CN X)

Superior cervical cardiac n. (sympathetic)

Superior cervical cardiac branch of vagus n.

Middle cervical ganglion

Phrenic n.

Middle cervical cardiac n. (sympathetic)

Inferior cervical cardiac branch of vagus n.

Vertebral ganglion

Cervicothoracic ganglion (stellate ganglion)

Inferior cervical cardiac nn. (sympathetic)

Sympathetic trunk

Thoracic ganglion of sympathetic trunk

Thoracic cardiac nn. (sympathetic)

Thoracic cardiac branches of vagus n.

Left recurrent laryngeal n.

Plate 249

Heart

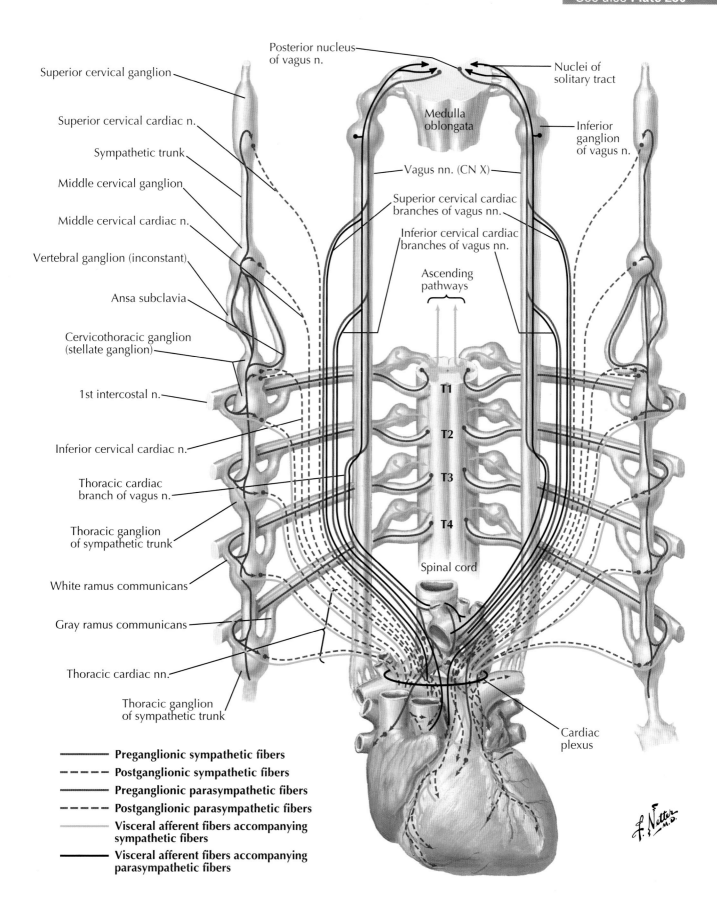

Posterior nucleus of vagus n.

Nuclei of solitary tract

Superior cervical ganglion

Superior cervical cardiac n.

Sympathetic trunk

Middle cervical ganglion

Middle cervical cardiac n.

Vertebral ganglion (inconstant)

Ansa subclavia

Cervicothoracic ganglion (stellate ganglion)

1st intercostal n.

Inferior cervical cardiac n.

Thoracic cardiac branch of vagus n.

Thoracic ganglion of sympathetic trunk

White ramus communicans

Gray ramus communicans

Thoracic cardiac nn.

Thoracic ganglion of sympathetic trunk

Medulla oblongata

Inferior ganglion of vagus n.

Vagus nn. (CN X)

Superior cervical cardiac branches of vagus nn.

Inferior cervical cardiac branches of vagus nn.

Ascending pathways

T1

T2

T3

T4

Spinal cord

Cardiac plexus

——————— **Preganglionic sympathetic fibers**

– – – – – **Postganglionic sympathetic fibers**

·········· **Preganglionic parasympathetic fibers**

– – – – – **Postganglionic parasympathetic fibers**

——————— **Visceral afferent fibers accompanying sympathetic fibers**

——————— **Visceral afferent fibers accompanying parasympathetic fibers**

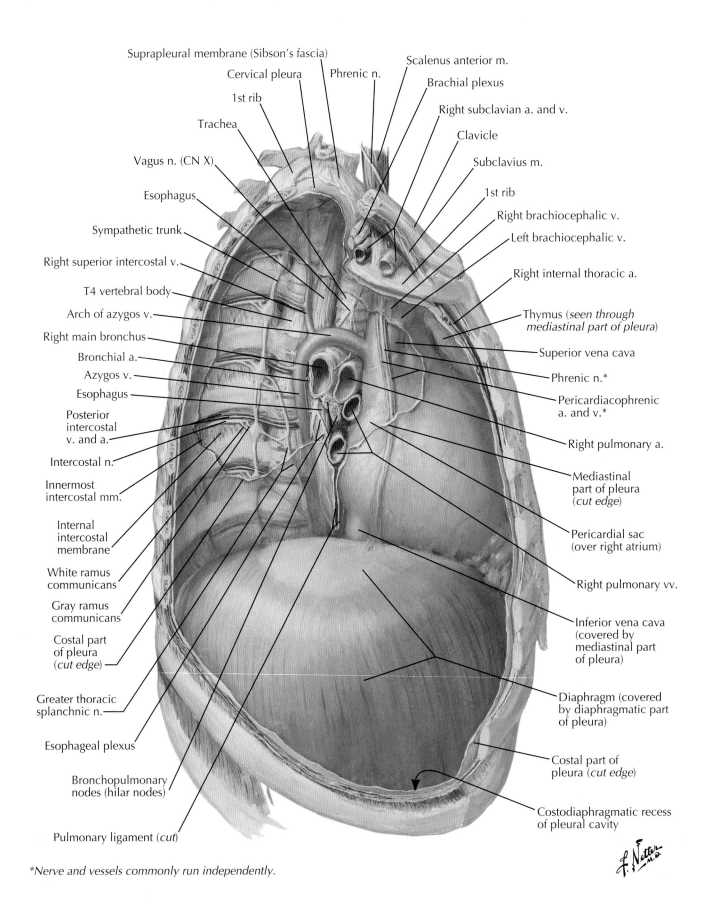

Suprapleural membrane (Sibson's fascia)

Cervical pleura

Phrenic n.

Scalenus anterior m.

Brachial plexus

1st rib

Right subclavian a. and v.

Trachea

Clavicle

Vagus n. (CN X)

Subclavius m.

Esophagus

1st rib

Sympathetic trunk

Right brachiocephalic v.

Left brachiocephalic v.

Right superior intercostal v.

Right internal thoracic a.

T4 vertebral body

Arch of azygos v.

Thymus (*seen through mediastinal part of pleura*)

Right main bronchus

Superior vena cava

Bronchial a.

Phrenic n.*

Azygos v.

Pericardiacophrenic a. and v.*

Esophagus

Posterior intercostal v. and a.

Right pulmonary a.

Intercostal n.

Mediastinal part of pleura (*cut edge*)

Innermost intercostal mm.

Internal intercostal membrane

Pericardial sac (over right atrium)

White ramus communicans

Right pulmonary vv.

Gray ramus communicans

Costal part of pleura (*cut edge*)

Inferior vena cava (covered by mediastinal part of pleura)

Greater thoracic splanchnic n.

Esophageal plexus

Diaphragm (covered by diaphragmatic part of pleura)

Bronchopulmonary nodes (hilar nodes)

Costal part of pleura (*cut edge*)

Pulmonary ligament (*cut*)

Costodiaphragmatic recess of pleural cavity

Nerve and vessels commonly run independently.

Plate 251

Mediastinum

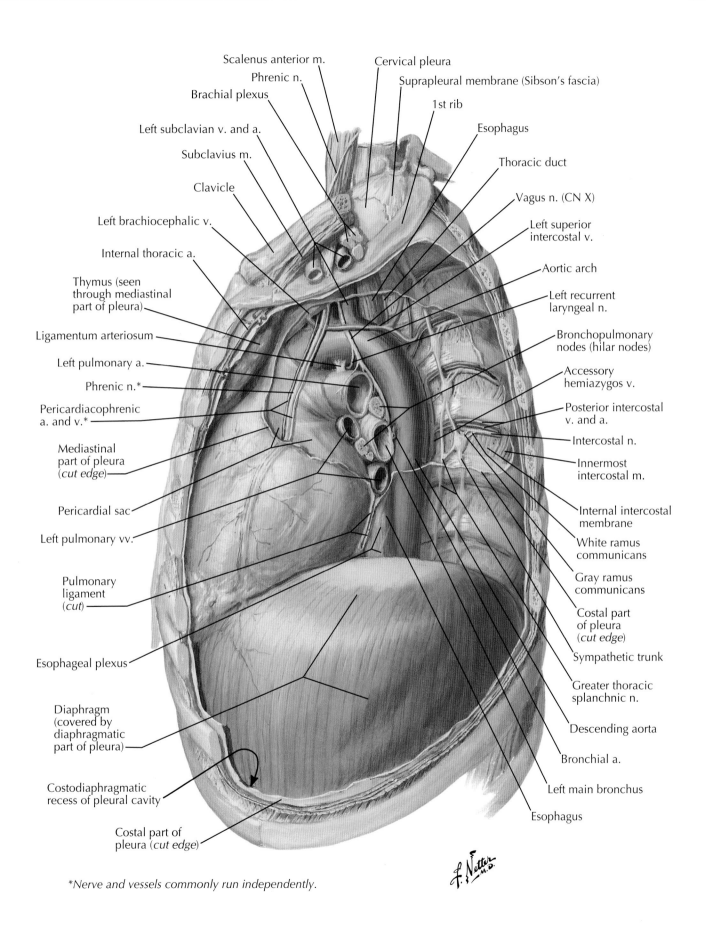

Scalenus anterior m.

Phrenic n.

Brachial plexus

Left subclavian v. and a.

Subclavius m.

Clavicle

Left brachiocephalic v.

Internal thoracic a.

Thymus (seen through mediastinal part of pleura)

Ligamentum arteriosum

Left pulmonary a.

Phrenic n.*

Pericardiacophrenic a. and v.*

Mediastinal part of pleura (*cut edge*)

Pericardial sac

Left pulmonary vv.

Pulmonary ligament (*cut*)

Esophageal plexus

Diaphragm (covered by diaphragmatic part of pleura)

Costodiaphragmatic recess of pleural cavity

Costal part of pleura (*cut edge*)

Cervical pleura

Suprapleural membrane (Sibson's fascia)

1st rib

Esophagus

Thoracic duct

Vagus n. (CN X)

Left superior intercostal v.

Aortic arch

Left recurrent laryngeal n.

Bronchopulmonary nodes (hilar nodes)

Accessory hemiazygos v.

Posterior intercostal v. and a.

Intercostal n.

Innermost intercostal m.

Internal intercostal membrane

White ramus communicans

Gray ramus communicans

Costal part of pleura (*cut edge*)

Sympathetic trunk

Greater thoracic splanchnic n.

Descending aorta

Bronchial a.

Left main bronchus

Esophagus

Nerve and vessels commonly run independently.

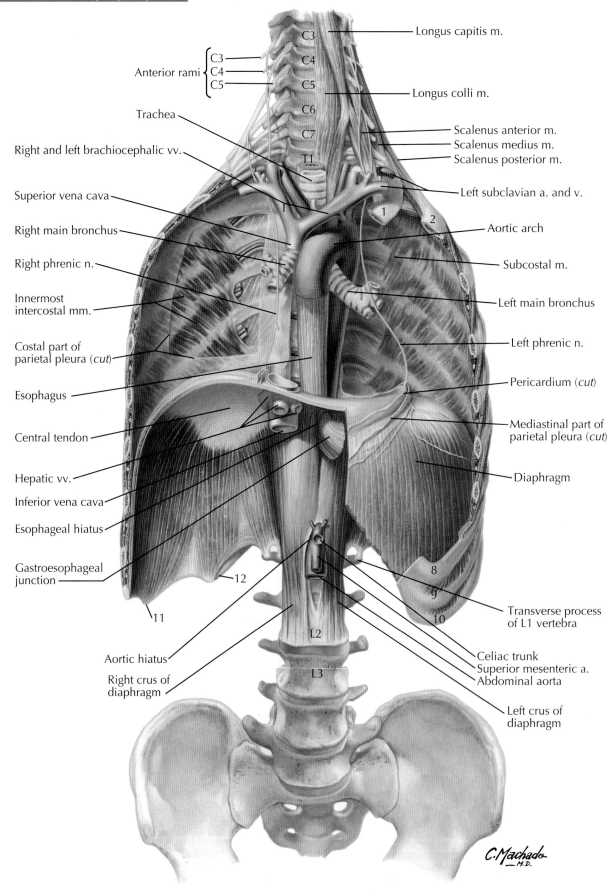

Longus capitis m.

C3

Anterior rami { C3 C4 C5

C4

C5

C6

C7

T1

Trachea

Right and left brachiocephalic vv.

Superior vena cava

Right main bronchus

Right phrenic n.

Innermost intercostal mm.

Costal part of parietal pleura (*cut*)

Esophagus

Central tendon

Hepatic vv.

Inferior vena cava

Esophageal hiatus

Gastroesophageal junction

12

11

Aortic hiatus

Right crus of diaphragm

L2

L3

Longus colli m.

Scalenus anterior m.
Scalenus medius m.
Scalenus posterior m.

Left subclavian a. and v.

1   2

Aortic arch

Subcostal m.

Left main bronchus

Left phrenic n.

Pericardium (*cut*)

Mediastinal part of parietal pleura (*cut*)

Diaphragm

8

9

10

Transverse process of L1 vertebra

Celiac trunk
Superior mesenteric a.
Abdominal aorta

Left crus of diaphragm

C. Machado
—M.D.

Plate 253

Mediastinum

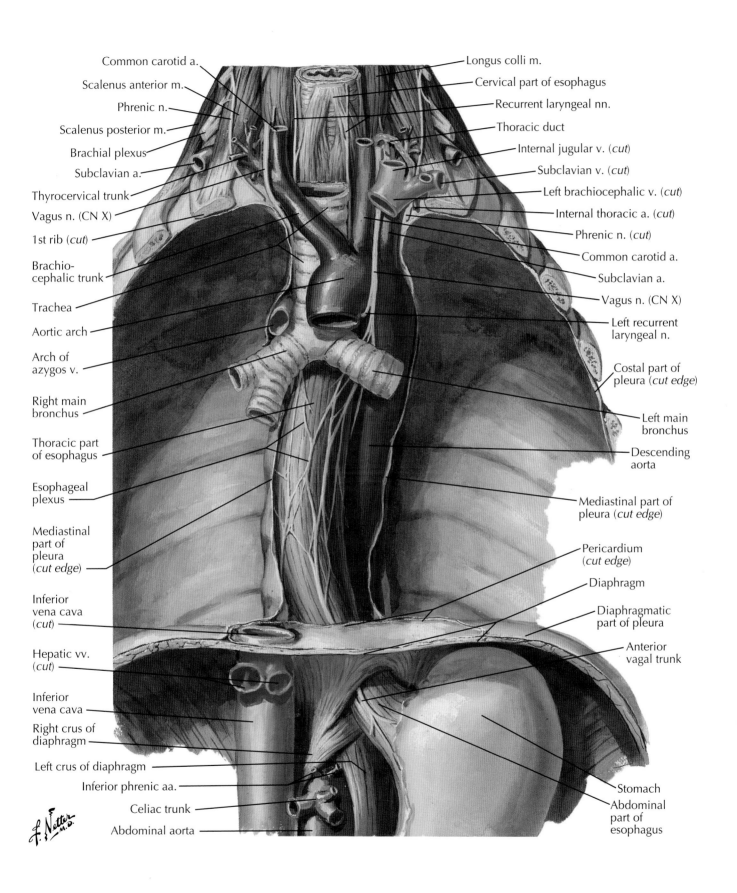

Common carotid a.

Scalenus anterior m.

Phrenic n.

Scalenus posterior m.

Brachial plexus

Subclavian a.

Thyrocervical trunk

Vagus n. (CN X)

1st rib (cut)

Brachio-
cephalic trunk

Trachea

Aortic arch

Arch of
azygos v.

Right main
bronchus

Thoracic part
of esophagus

Esophageal
plexus

Mediastinal
part of
pleura
(cut edge)

Inferior
vena cava
(cut)

Hepatic vv.
(cut)

Inferior
vena cava

Right crus of
diaphragm

Left crus of diaphragm

Inferior phrenic aa.

Celiac trunk

Abdominal aorta

Longus colli m.

Cervical part of esophagus

Recurrent laryngeal nn.

Thoracic duct

Internal jugular v. (cut)

Subclavian v. (cut)

Left brachiocephalic v. (cut)

Internal thoracic a. (cut)

Phrenic n. (cut)

Common carotid a.

Subclavian a.

Vagus n. (CN X)

Left recurrent
laryngeal n.

Costal part of
pleura (cut edge)

Left main
bronchus

Descending
aorta

Mediastinal part of
pleura (cut edge)

Pericardium
(cut edge)

Diaphragm

Diaphragmatic
part of pleura

Anterior
vagal trunk

Stomach

Abdominal
part of
esophagus

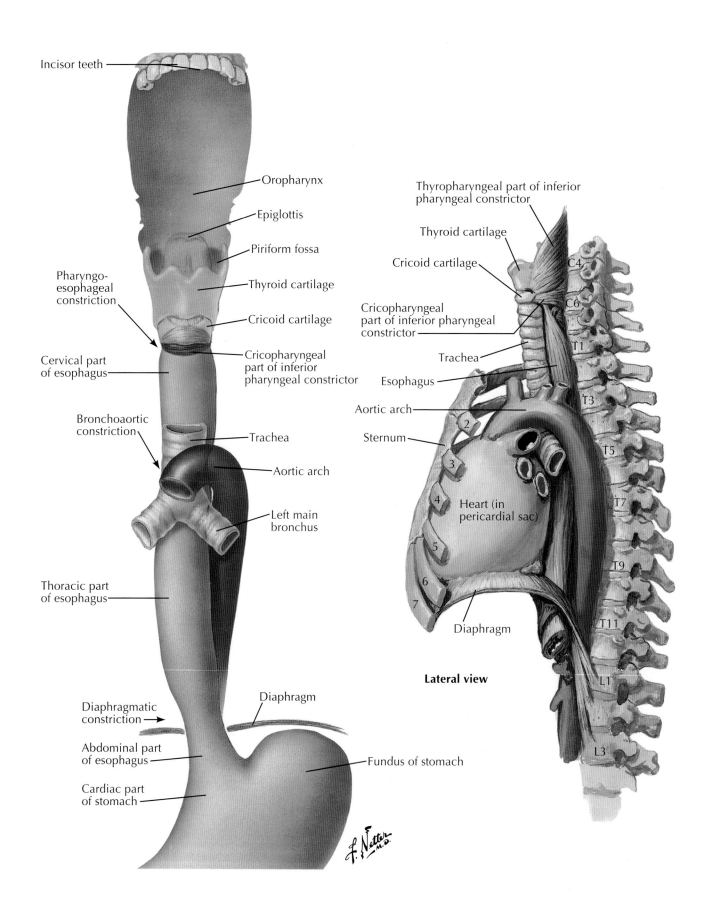

Incisor teeth

Oropharynx

Epiglottis

Piriform fossa

Pharyngo-
esophageal
constriction

Thyroid cartilage

Cricoid cartilage

Cricopharyngeal
part of inferior
pharyngeal constrictor

Cervical part
of esophagus

Bronchoaortic
constriction

Trachea

Aortic arch

Left main
bronchus

Thoracic part
of esophagus

Diaphragmatic
constriction →

Diaphragm

Abdominal part
of esophagus

Cardiac part
of stomach

Fundus of stomach

Thyropharyngeal part of inferior
pharyngeal constrictor

Thyroid cartilage

Cricoid cartilage

Cricopharyngeal
part of inferior pharyngeal
constrictor

Trachea

Esophagus

Aortic arch

Sternum

Heart (in
pericardial sac)

Diaphragm

Lateral view

C4

C6

T1

T3

T5

T7

T9

T11

L1

L3

2

3

4

5

6

7

Plate 255   **Mediastinum**

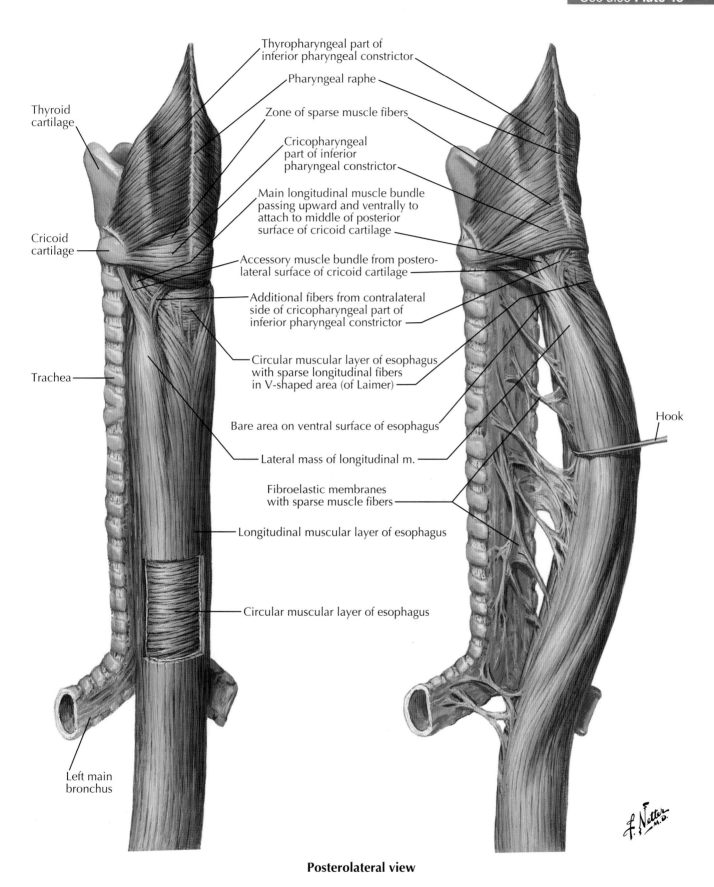

Thyropharyngeal part of inferior pharyngeal constrictor

Pharyngeal raphe

Zone of sparse muscle fibers

Cricopharyngeal part of inferior pharyngeal constrictor

Main longitudinal muscle bundle passing upward and ventrally to attach to middle of posterior surface of cricoid cartilage

Accessory muscle bundle from postero-lateral surface of cricoid cartilage

Additional fibers from contralateral side of cricopharyngeal part of inferior pharyngeal constrictor

Circular muscular layer of esophagus with sparse longitudinal fibers in V-shaped area (of Laimer)

Bare area on ventral surface of esophagus

Lateral mass of longitudinal m.

Fibroelastic membranes with sparse muscle fibers

Longitudinal muscular layer of esophagus

Circular muscular layer of esophagus

Thyroid cartilage

Cricoid cartilage

Trachea

Left main bronchus

Hook

Posterolateral view

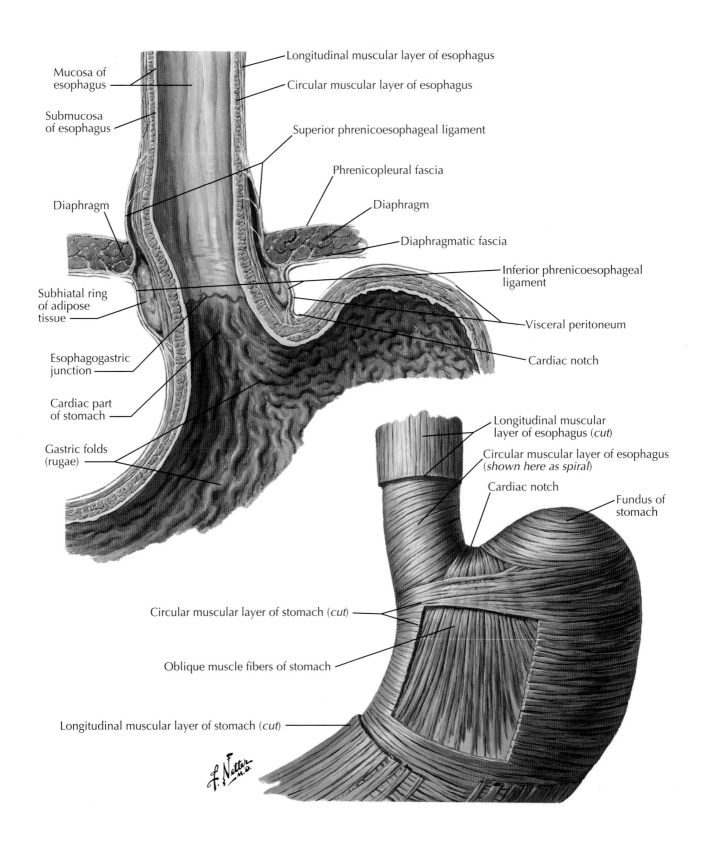

Mucosa of esophagus

Submucosa of esophagus

Diaphragm

Subhiatal ring of adipose tissue

Esophagogastric junction

Cardiac part of stomach

Gastric folds (rugae)

Longitudinal muscular layer of esophagus

Circular muscular layer of esophagus

Superior phrenicoesophageal ligament

Phrenicopleural fascia

Diaphragm

Diaphragmatic fascia

Inferior phrenicoesophageal ligament

Visceral peritoneum

Cardiac notch

Longitudinal muscular layer of esophagus (*cut*)

Circular muscular layer of esophagus (*shown here as spiral*)

Cardiac notch

Fundus of stomach

Circular muscular layer of stomach (*cut*)

Oblique muscle fibers of stomach

Longitudinal muscular layer of stomach (*cut*)

Plate 257

Mediastinum

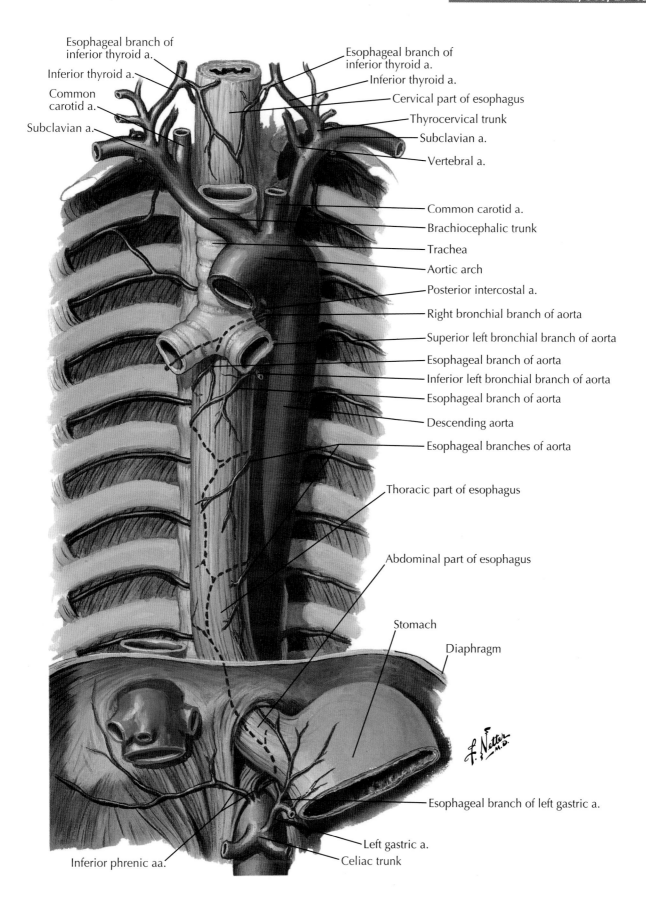

Esophageal branch of inferior thyroid a.

Inferior thyroid a.

Common carotid a.

Subclavian a.

Esophageal branch of inferior thyroid a.

Inferior thyroid a.

Cervical part of esophagus

Thyrocervical trunk

Subclavian a.

Vertebral a.

Common carotid a.

Brachiocephalic trunk

Trachea

Aortic arch

Posterior intercostal a.

Right bronchial branch of aorta

Superior left bronchial branch of aorta

Esophageal branch of aorta

Inferior left bronchial branch of aorta

Esophageal branch of aorta

Descending aorta

Esophageal branches of aorta

Thoracic part of esophagus

Abdominal part of esophagus

Stomach

Diaphragm

Esophageal branch of left gastric a.

Left gastric a.

Celiac trunk

Inferior phrenic aa.

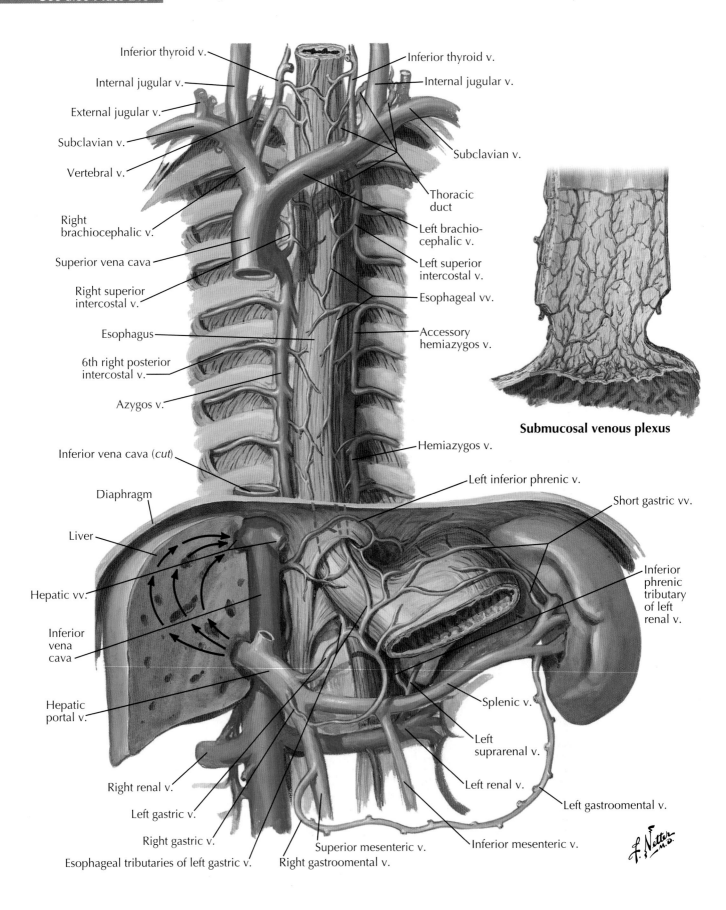

Inferior thyroid v.

Internal jugular v.

External jugular v.

Subclavian v.

Vertebral v.

Right brachiocephalic v.

Superior vena cava

Right superior intercostal v.

Esophagus

6th right posterior intercostal v.

Azygos v.

Inferior vena cava (*cut*)

Diaphragm

Liver

Hepatic vv.

Inferior vena cava

Hepatic portal v.

Right renal v.

Left gastric v.

Right gastric v.

Esophageal tributaries of left gastric v.

Inferior thyroid v.

Internal jugular v.

Subclavian v.

Thoracic duct

Left brachio-cephalic v.

Left superior intercostal v.

Esophageal vv.

Accessory hemiazygos v.

Hemiazygos v.

Left inferior phrenic v.

Short gastric vv.

Inferior phrenic tributary of left renal v.

Splenic v.

Left suprarenal v.

Left renal v.

Left gastroomental v.

Superior mesenteric v.

Right gastroomental v.

Inferior mesenteric v.

Submucosal venous plexus

Plate 259

Mediastinum

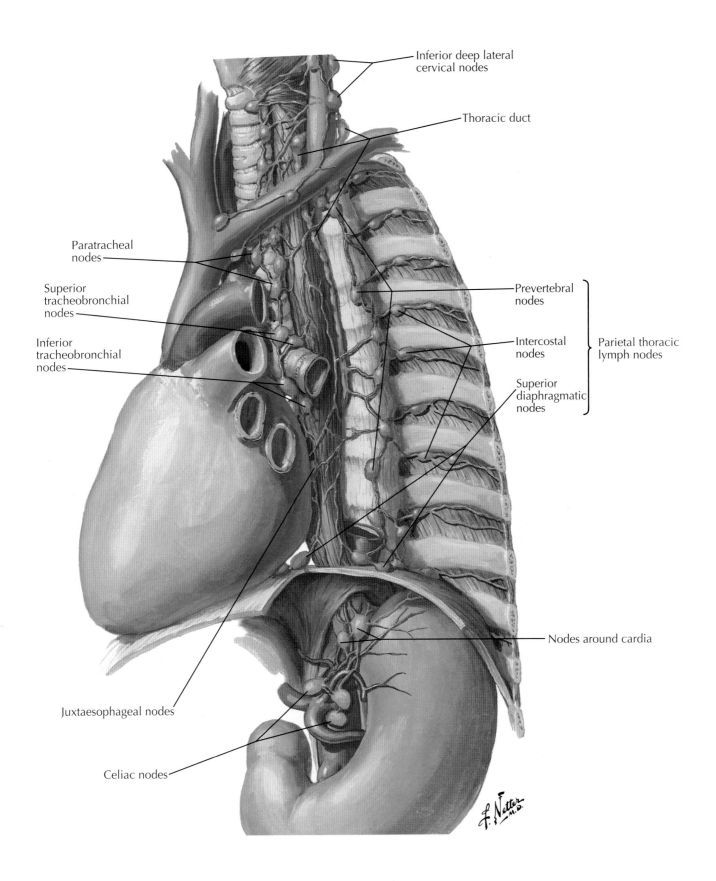

Inferior deep lateral cervical nodes

Thoracic duct

Paratracheal nodes

Superior tracheobronchial nodes

Inferior tracheobronchial nodes

Prevertebral nodes

Intercostal nodes

Parietal thoracic lymph nodes

Superior diaphragmatic nodes

Nodes around cardia

Juxtaesophageal nodes

Celiac nodes

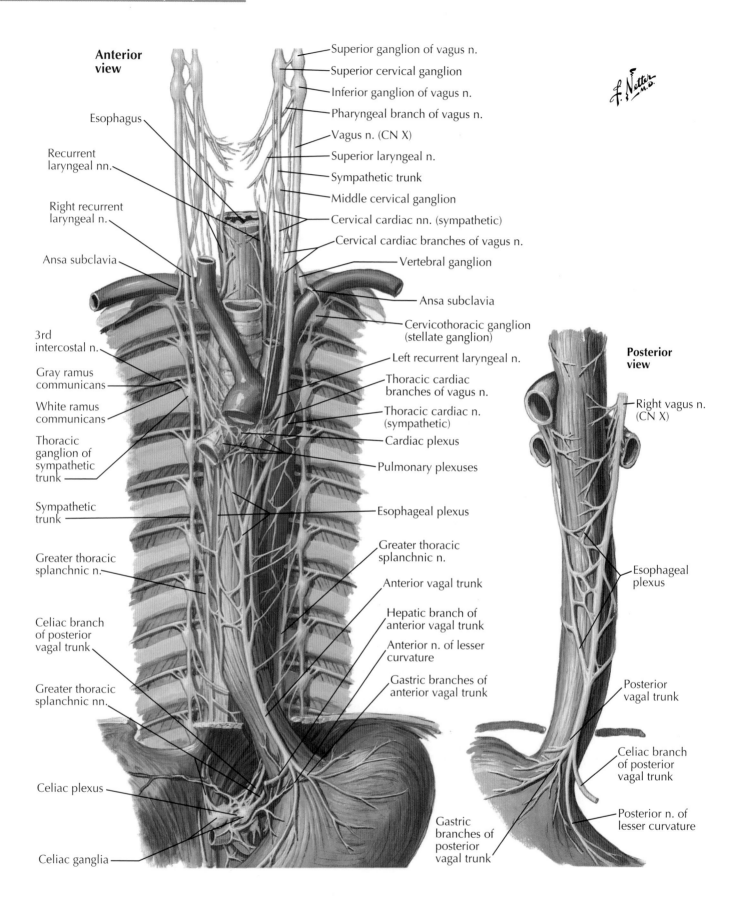

Anterior view

Esophagus

Recurrent laryngeal nn.

Right recurrent laryngeal n.

Ansa subclavia

3rd intercostal n.

Gray ramus communicans

White ramus communicans

Thoracic ganglion of sympathetic trunk

Sympathetic trunk

Greater thoracic splanchnic n.

Celiac branch of posterior vagal trunk

Greater thoracic splanchnic nn.

Celiac plexus

Celiac ganglia

Superior ganglion of vagus n.

Superior cervical ganglion

Inferior ganglion of vagus n.

Pharyngeal branch of vagus n.

Vagus n. (CN X)

Superior laryngeal n.

Sympathetic trunk

Middle cervical ganglion

Cervical cardiac nn. (sympathetic)

Cervical cardiac branches of vagus n.

Vertebral ganglion

Ansa subclavia

Cervicothoracic ganglion (stellate ganglion)

Left recurrent laryngeal n.

Thoracic cardiac branches of vagus n.

Thoracic cardiac n. (sympathetic)

Cardiac plexus

Pulmonary plexuses

Esophageal plexus

Greater thoracic splanchnic n.

Anterior vagal trunk

Hepatic branch of anterior vagal trunk

Anterior n. of lesser curvature

Gastric branches of anterior vagal trunk

Posterior view

Right vagus n. (CN X)

Esophageal plexus

Posterior vagal trunk

Celiac branch of posterior vagal trunk

Posterior n. of lesser curvature

Gastric branches of posterior vagal trunk

Plate 261

Mediastinum

Axial CT images of the thorax from superior (A) to inferior (C)

A

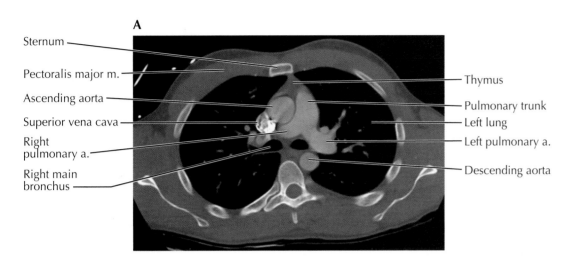

Sternum
Pectoralis major m.
Ascending aorta
Superior vena cava
Right pulmonary a.
Right main bronchus

Thymus
Pulmonary trunk
Left lung
Left pulmonary a.
Descending aorta

B

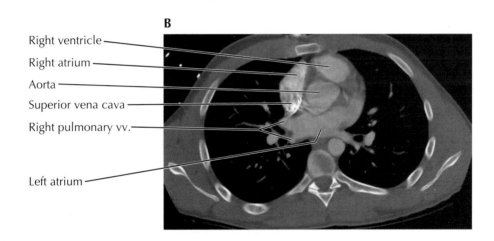

Right ventricle
Right atrium
Aorta
Superior vena cava
Right pulmonary vv.

Left atrium

C

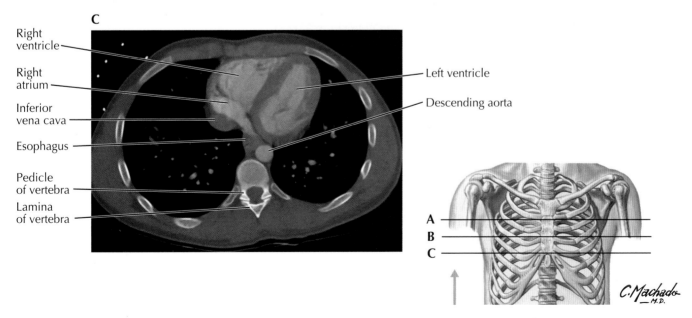

Right ventricle
Right atrium
Inferior vena cava
Esophagus
Pedicle of vertebra
Lamina of vertebra

Left ventricle
Descending aorta

A
B
C

C. Machado
— M.D.

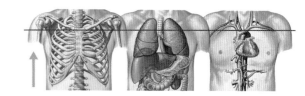

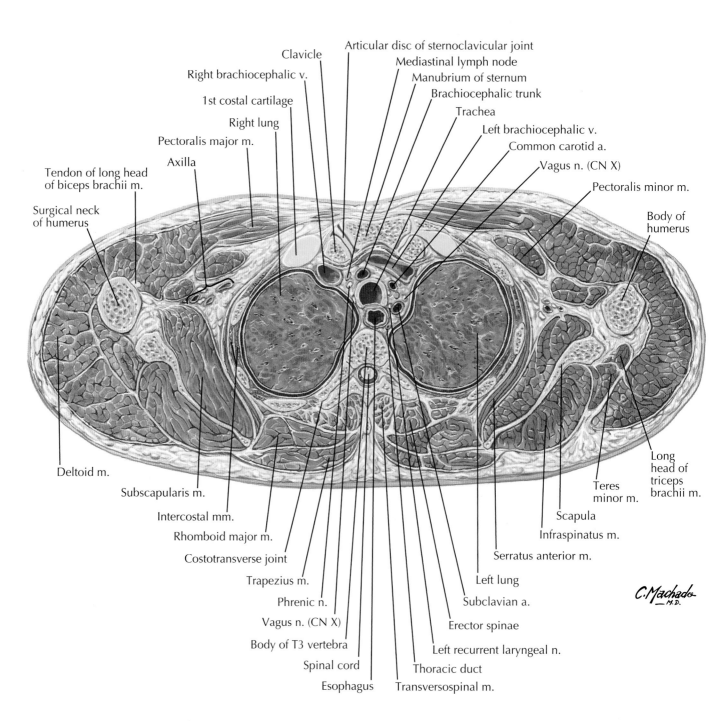

Clavicle

Articular disc of sternoclavicular joint

Right brachiocephalic v.

Mediastinal lymph node

Manubrium of sternum

1st costal cartilage

Brachiocephalic trunk

Right lung

Trachea

Pectoralis major m.

Left brachiocephalic v.

Axilla

Common carotid a.

Tendon of long head
of biceps brachii m.

Vagus n. (CN X)

Pectoralis minor m.

Surgical neck
of humerus

Body of
humerus

Deltoid m.

Long
head
of
triceps
brachii m.

Subscapularis m.

Teres
minor m.

Intercostal mm.

Scapula

Rhomboid major m.

Infraspinatus m.

Costotransverse joint

Serratus anterior m.

Trapezius m.

Left lung

Phrenic n.

Subclavian a.

Vagus n. (CN X)

Erector spinae

Body of T3 vertebra

Left recurrent laryngeal n.

Spinal cord

Thoracic duct

Esophagus

Transversospinal m.

C. Machado
M.D.

Plate 263

Cross-Sectional Anatomy

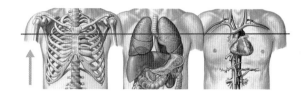

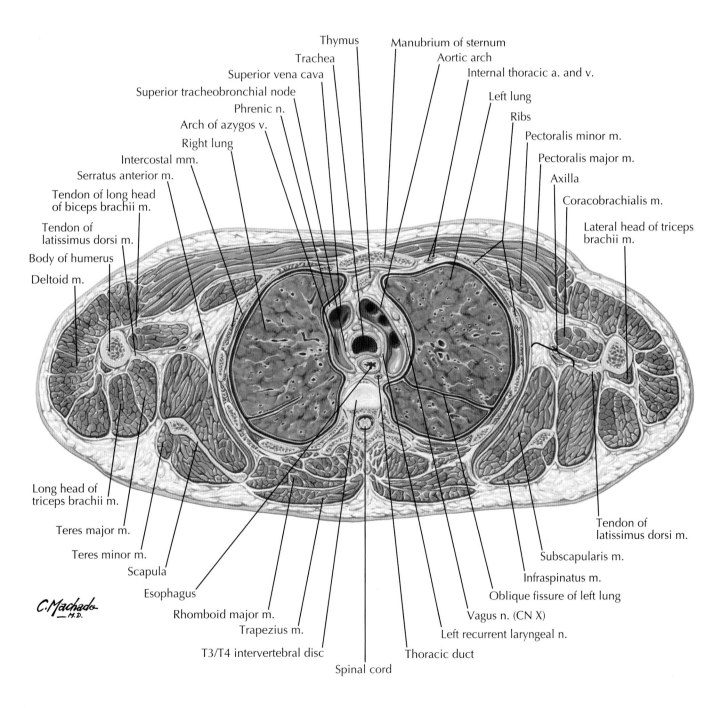

Thymus

Trachea

Superior vena cava

Superior tracheobronchial node

Phrenic n.

Arch of azygos v.

Right lung

Intercostal mm.

Serratus anterior m.

Tendon of long head of biceps brachii m.

Tendon of latissimus dorsi m.

Body of humerus

Deltoid m.

Manubrium of sternum

Aortic arch

Internal thoracic a. and v.

Left lung

Ribs

Pectoralis minor m.

Pectoralis major m.

Axilla

Coracobrachialis m.

Lateral head of triceps brachii m.

Long head of triceps brachii m.

Teres major m.

Teres minor m.

Scapula

Esophagus

Rhomboid major m.

Trapezius m.

T3/T4 intervertebral disc

Spinal cord

Thoracic duct

Left recurrent laryngeal n.

Vagus n. (CN X)

Oblique fissure of left lung

Infraspinatus m.

Subscapularis m.

Tendon of latissimus dorsi m.

C.Machado
_M.D._

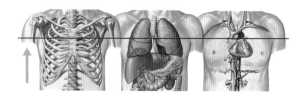

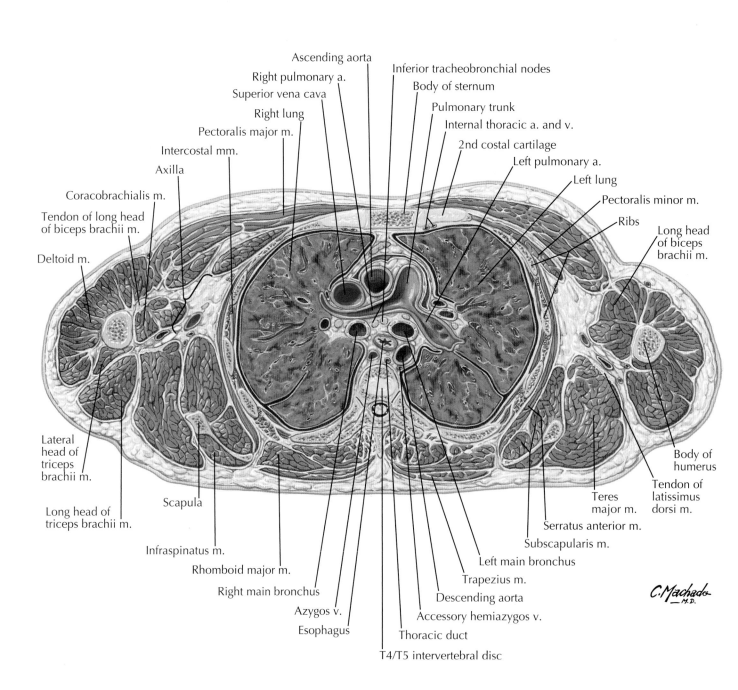

Ascending aorta

Right pulmonary a.

Superior vena cava

Right lung

Pectoralis major m.

Intercostal mm.

Axilla

Coracobrachialis m.

Tendon of long head of biceps brachii m.

Deltoid m.

Inferior tracheobronchial nodes

Body of sternum

Pulmonary trunk

Internal thoracic a. and v.

2nd costal cartilage

Left pulmonary a.

Left lung

Pectoralis minor m.

Ribs

Long head of biceps brachii m.

Lateral head of triceps brachii m.

Long head of triceps brachii m.

Scapula

Infraspinatus m.

Rhomboid major m.

Right main bronchus

Azygos v.

Esophagus

Left main bronchus

Trapezius m.

Descending aorta

Accessory hemiazygos v.

Thoracic duct

T4/T5 intervertebral disc

Subscapularis m.

Serratus anterior m.

Teres major m.

Tendon of latissimus dorsi m.

Body of humerus

C. Machado M.D.

Plate 265

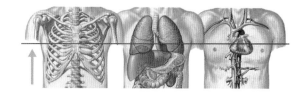

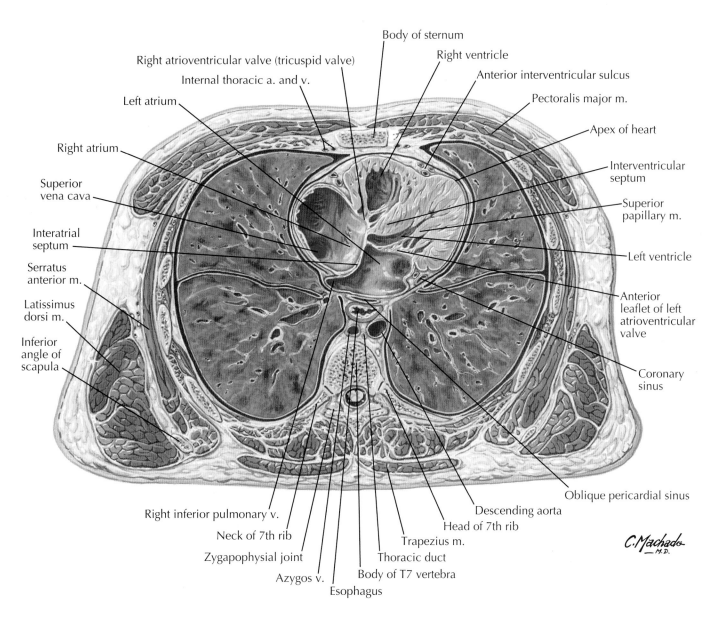

Body of sternum

Right atrioventricular valve (tricuspid valve)

Right ventricle

Internal thoracic a. and v.

Anterior interventricular sulcus

Left atrium

Pectoralis major m.

Right atrium

Apex of heart

Superior vena cava

Interventricular septum

Interatrial septum

Superior papillary m.

Serratus anterior m.

Left ventricle

Latissimus dorsi m.

Anterior leaflet of left atrioventricular valve

Inferior angle of scapula

Coronary sinus

Oblique pericardial sinus

Right inferior pulmonary v.

Descending aorta

Neck of 7th rib

Head of 7th rib

Zygapophysial joint

Trapezius m.

Azygos v.

Thoracic duct

Body of T7 vertebra

Esophagus

ANATOMICAL STRUCTURE	CLINICAL COMMENT	PLATE NUMBERS
Nervous System and Sense Organs		
Long thoracic nerve	May be damaged during chest tube placement or mastectomy, resulting in winged scapula (denervation of serratus anterior muscle)	206, 209
Intercostal nerve	Site of local anesthetic nerve block for procedures such as thoracostomy or to alleviate pain caused by herpes zoster (shingles)	211, 212
Spinal (posterior root) ganglion	Can house dormant varicella zoster virus, which, when activated, can result in herpes zoster (shingles)	212
Phrenic nerve	Surgical injury to phrenic nerve may cause ipsilateral paralysis of diaphragm; diaphragmatic irritation may manifest as shoulder pain because of referral to C3–C5 spinal levels	214, 216, 232
Recurrent laryngeal nerve	The left nerve takes a circuitous path around aorta and may rarely become compressed by large thoracic aortic aneurysm or enlarged left atrium, producing hoarse voice (Ortner's syndrome); more often, these branches are affected by malignancies, such as those of thyroid gland	252, 254
Thoracic cardiac nerves (sympathetic)	Pain of myocardial ischemia referred to upper thoracic dermatomes; may be perceived as somatic pain in thorax and medial upper limb	250
Skeletal System		
Ribs	Rib fractures may cause respiratory dysfunction, predispose to pneumonia, and injure underlying structures (e.g., liver and spleen); severe fractures may breach pleural space and cause pneumothorax; flail chest occurs when multiple fractures in adjacent ribs create "floating" area of thorax with paradoxical motion during inspiration	203
Costochondral and sternochondral (sternocostal) joints	Frequent sites of pain and tenderness after thoracic wall injury or excessive weightlifting (costochondritis); generally reproducible with palpation of joints	204
Clavicles	Common sites of fracture, often after falling onto outstretched limb or onto shoulder; fractures typically occur in middle third; supraclavicular nerve block relieves pain associated with fracture	203
Sternal angle (of Louis)	Surface landmark for counting ribs (2nd pair of ribs articulate between manubrium and body of sternum) and intercostal spaces; divides superior from inferior mediastinum, and marks transition from aortic arch to descending thoracic aorta	203
Superior thoracic aperture	Compression of neurovascular structures (inferior trunk of brachial plexus and great vessels) that traverse superior thoracic aperture may produce thoracic outlet syndrome	213
Intercostal spaces	Relationship of intercostal neurovascular bundle to ribs is crucial when placing chest tube to relieve pneumothorax or hemothorax, or needles to anesthetize nerves; tubes should be inserted along superior margin of ribs to avoid these bundles; of note, however, intercostal nerve may have collateral branch running along superior border of lower rib, which can result in pain	210, 238
Median cricothyroid ligament	Also known as cricothyroid membrane; site of cricothyrotomy, an emergency procedure to establish surgical airway	53, 103
Muscular System		
Diaphragm	Widening of esophageal hiatus at T8 vertebral level or congenital defect allows for protrusion of stomach into thorax (hiatal hernia), which increases incidence of gastroesophageal reflux	216, 257

Table 4.1　　　　　　　　　　　　　　　　　　　　　　**Structures with High Clinical Significance**

ANATOMICAL STRUCTURE	CLINICAL COMMENT	PLATE NUMBERS
Cardiovascular System		
Internal thoracic artery	Commonly used for coronary artery bypass grafts, most often for anterior interventricular artery (left anterior descending artery)	211, 212
Pulmonary arteries	Thromboemboli, most often from pelvic and femoral sources, may obstruct pulmonary arteries (pulmonary embolus), leading to hypoxemia, hemodynamic compromise, and pulmonary infarction	220, 227
Pericardium	Pericardial space can contain small amounts of physiologic fluid (15–50 mL); effusion may compromise heart function (cardiac tamponade); pericardial sac can expand and become quite large with slow, progressive fluid accumulation	231, 236
Coronary arteries	Fixed atherosclerotic disease may cause myocardial ischemia to manifest as thoracic pain; rupture and thrombosis of atherosclerotic plaque is main cause of acute myocardial infarction; severity and outcomes depend on amount of myocardium that vessel subtends, with proximal lesions of large vessels being most morbid	239
Pulmonary veins	Atrial fibrillation, a common arrhythmia, is believed to originate from within pulmonary veins; electrical ablation of this arrhythmia creates rings of fibrosis around pulmonary veins as they enter left atrium, thereby preventing propagation of electrical signals into heart	242, 247
Foramen ovale	Provides channel for interatrial flow (right-to-left shunt) during fetal development; remains patent in approximately one in four adults and may provide route for venous microthromboses to enter left side of heart and cause ischemic stroke	242
Interventricular septum	Ventricular septal defect is common congenital cardiac defect, most often involving membranous portion of septum; myocardial infarction in territory of anterior interventricular artery (left anterior descending artery), especially if not promptly treated, may generate sufficient ischemia of septum to cause perforation	242, 245
Heart valves	Valvular disease (e.g., aortic stenosis, mitral insufficiency) is common, especially among older adults, and may cause progressive heart failure	243
Aortic valve	1% of population has bicuspid aortic valve (i.e., containing two rather than three leaflets), which predisposes to aortic stenosis and insufficiency and is also associated with aortic aneurysm	243, 244
Sinuatrial node	Primary cardiac pacemaker, which generates action potentials that propagate through cardiac conduction system; aging, infiltrative diseases, and heart surgery may cause sinus node dysfunction, resulting in bradycardia	248
Atrioventricular node	Conducts action potentials from atria to ventricles; intrinsic refractory period prevents rapid atrial rhythms from causing equivalent tachycardia of ventricles; dysfunction secondary to fibrosis, medications, cardiac surgery may result in heart block; when block is complete, atria and ventricles have independent rhythms	248
Ligamentum arteriosum	Remnant of ductus arteriosus, which connects pulmonary and systemic circulations during fetal development; lack of ductus closure after birth may cause exertional dyspnea, pulmonary vascular disease, or heart failure; acts as landmark to identify left recurrent laryngeal nerve looping inferior to ligament	232, 247
Thoracic aorta	Lies naturally to left of vertebral column in thorax and commences at T4 vertebral level, where aortic arch terminates; congenital coarctation (narrowing) of aorta may cause hypertension in children and young adults; significant difference in blood pressure between upper and lower extremities is suggestive	258
Thoracic aorta	Aneurysm (enlargement) may occur secondary to age and atherosclerotic risk factors (such as tobacco abuse and hypertension), connective tissue disorders, in association with bicuspid aortic valve, or from infection (e.g., syphilis). Large aneurysm can rupture or dissect; the latter occurs when a tear in intimal layer allows blood to propagate into a false lumen between intima and media	258

Structures with High Clinical Significance

Table 4.2

ANATOMICAL STRUCTURE	CLINICAL COMMENT	PLATE NUMBERS
Cardiovascular System—Continued		
Azygos vein	Drains posterior thorax and provides important collateral channel between inferior vena cava and superior vena cava	259
Lymph Vessels and Lymphoid Organs		
Lymph vessels of breast	Metastatic spread of cancer cells from mammary gland to axilla and thorax via lymphatics draining breast	208
Axillary nodes	Primary nodes that receive lymphatic drainage from upper limb, thoracic wall, and breast; commonly enlarged in patients with breast cancer	207, 208
Respiratory System		
Pleura	Air or gas (spontaneous or traumatic) can leak into pleural space between visceral pleura and parietal pleura and compress lung, a condition known as pneumothorax; if severe enough to compromise venous return to heart, resulting in hypotension and dyspnea, condition is known as tension pneumothorax	217–220
Cervical pleura	Extends into neck superior to anterior aspect of angled 1st rib or to level of 1st rib posteriorly; it may therefore be punctured during neck procedures, producing pneumothorax	69, 217
Tracheal bifurcation	Important landmark when assessing position of endotracheal tube, which should terminate 4–5 cm above; often at T4–T5 vertebral level; right main bronchus is shorter, more vertical, and wider; therefore, aspirated objects more often enter right side; left main bronchus lies over esophagus and posterior to left atrium of heart	225
Apex of lung	Pancoast tumor (bronchogenic carcinoma of apex of lung) may compress sympathetic trunk, resulting in Horner's syndrome (ipsilateral miosis, ptosis, anhidrosis, facial flushing); apex is susceptible to pneumothorax from needles introduced in lower neck region	217, 251
Reproductive System		
Mammary gland	Breast cancer is most common malignancy in women; most common type originates in lactiferous duct and can either be localized within duct (ductal carcinoma in situ) or invasive into adjoining tissues (invasive ductal carcinoma)	205

*Selections were based largely on clinical data and commonly discussed clinical correlations in macroscopic ("gross") anatomy courses.

MUSCLE	MUSCLE GROUP	ORIGIN ATTACHMENT	INSERTION ATTACHMENT	INNERVATION	BLOOD SUPPLY	MAIN ACTIONS
Diaphragm	Diaphragm	Xiphoid process, costal cartilages 7–12, L1–L3 vertebrae	Central tendon	Phrenic nerve	Pericardiacophrenic, musculophrenic, posterior intercostal, and superior and inferior phrenic arteries	Draws central tendon down and forward during inspiration
External intercostal mm.	Thoracic wall	Lower borders of ribs	Upper borders of subjacent ribs	Intercostal nerves	Posterior intercostal arteries, supreme intercostal artery, internal thoracic artery, musculophrenic artery	Support intercostal spaces in inspiration and expiration, elevate ribs in inspiration
Innermost intercostal mm.	Thoracic wall	Lower borders of ribs	Upper borders of subjacent ribs	Intercostal nerves	Posterior intercostal arteries, supreme intercostal artery, internal thoracic artery, musculophrenic artery	Prevent pushing out or drawing in of intercostal spaces in inspiration and expiration, lower ribs in forced expiration
Internal intercostal mm.	Thoracic wall	Costal grooves, lower borders of costal cartilages	Upper borders of subjacent ribs	Intercostal nerves	Posterior intercostal arteries, supreme intercostal artery, internal thoracic artery, musculophrenic artery	Prevent pushing out or drawing in of intercostal spaces in inspiration and expiration, lower ribs in forced expiration
Levatores costarum	Thoracic wall	Transverse processes of C7–T11 vertebrae	Subjacent ribs (between tubercle and angle)	Posterior rami of thoracic spinal nerves	Posterior intercostal arteries	Elevate ribs
Pectoralis major m.	Pectoral region	*Clavicular part:* sternal half of clavicle *Sternal part:* anterior surface of sternum, cartilages of true ribs *Abdominal part:* aponeurosis of external abdominal oblique muscle	Lateral lip of intertubercular sulcus of humerus	Medial and lateral pectoral nerves	Pectoral branch of thoracoacromial artery, internal thoracic artery	Flexes, adducts, and medially rotates arm
Pectoralis minor m.	Pectoral region	External surface of upper margin of ribs 3–5	Coracoid process of scapula	Medial and lateral pectoral nerves	Pectoral branch of thoracoacromial artery, superior and lateral thoracic arteries	Lowers lateral angle of scapula and protracts scapula
Serratus anterior m.	Shoulder	Lateral surfaces of ribs 1–9	Costal surface of medial border of scapula	Long thoracic nerve	Lateral thoracic artery	Protracts and rotates scapula and holds it against thoracic wall
Subclavius m.	Shoulder	Upper border of 1st rib and costal cartilage	Inferior surface of middle one-third of clavicle	Subclavian nerve	Clavicular branch of thoracoacromial artery	Anchors and depresses clavicle
Subcostal mm.	Thoracic wall	Internal surfaces of lower ribs near their angles	Superior borders of ribs that are the 2nd or 3rd below the muscle origin	Intercostal nerves	Posterior intercostal arteries, musculophrenic artery	Depress ribs
Transversus thoracis m.	Thoracic wall	Internal surfaces of costal cartilages 2–6	Posterior surface of lower sternum	Intercostal nerves	Internal thoracic artery	Depresses ribs and costal cartilages

Variations in spinal nerve contributions to the innervation of muscles, their arterial supply, their attachments, and their actions are common themes in human anatomy. Therefore, expect differences between texts and realize that anatomical variation is normal.

ABDOMEN 5

ELECTRONIC BONUS PLATES

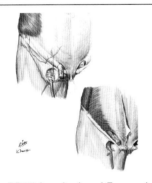

BP 53 Inguinal and Femoral Regions

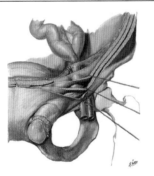

BP 54 Indirect Inguinal Hernia

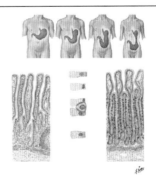

BP 55 Variations in Position and Contour of Stomach in Relation to Body Habitus

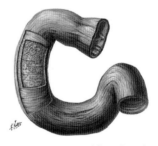

BP 56 Layers of Duodenal Wall

BP 57 CT and MRCP Showing Vermiform Appendix, Gallbladder, and Ducts; Nerve Branches on Hepatic Artery

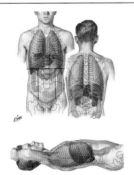

BP 58 Topography of Liver

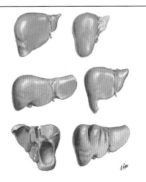

BP 59 Variations in Form of Liver

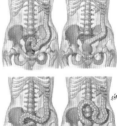

BP 60 Sigmoid Colon: Variations in Position

ELECTRONIC BONUS PLATES—*cont'd*

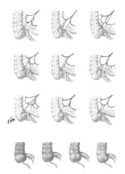

BP 61 Variations in Arterial Supply to Cecum and Posterior Peritoneal Attachment of Cecum

BP 62 Variations in Pancreatic Duct

BP 63 Variations in Cystic, Hepatic, and Pancreatic Ducts

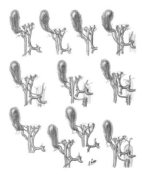

BP 64 Variations in Cystic Arteries

BP 65 Variations in Hepatic Arteries

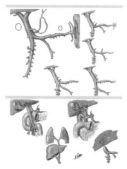

BP 66 Variations and Anomalies of Hepatic Portal Vein

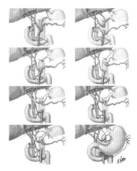

BP 67 Variations in Celiac Trunk

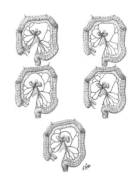

BP 68 Variations in Colic Arteries

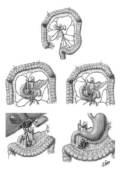

BP 69 Variations in Colic Arteries (continued)

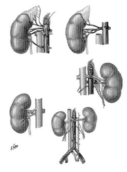

BP 70 Variations in Renal Artery and Vein

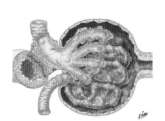

BP 71 Histology of Renal Corpuscle

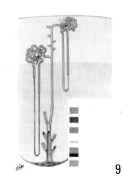

BP 72 Nephron and Collecting Tubule: Schema

9

ELECTRONIC BONUS PLATES—*cont'd*

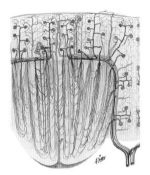

BP 73 Blood Vessels in Parenchyma of Kidney: Schema

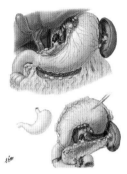

BP 74 Lymph Vessels and Lymph Nodes of Stomach

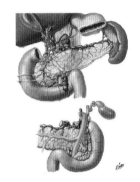

BP 75 Lymph Vessels and Lymph Nodes of Pancreas

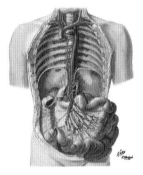

BP 76 Lymph Vessels and Lymph Nodes of Small Intestine

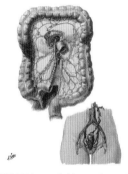

BP 77 Lymph Vessels and Lymph Nodes of Large Intestine

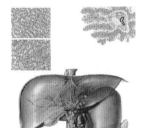

BP 78 Lymph Vessels and Lymph Nodes of Liver

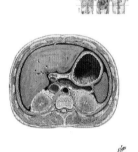

BP 79 Schematic Cross Section of Abdomen at T12 Vertebral Level

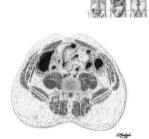

BP 80 Transverse Section of Abdomen: L5 Vertebral Level, Near Transtubercular Plane

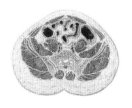

BP 81 Transverse Section of Abdomen: S1 Vertebral Level, Anterior Superior Iliac Spine

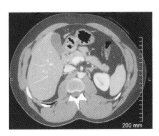

BP 82 Axial CT Image of Upper Abdomen

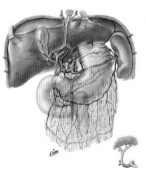

BP 83 Arterial Variations and Collateral Supply of Liver and Gallbladder

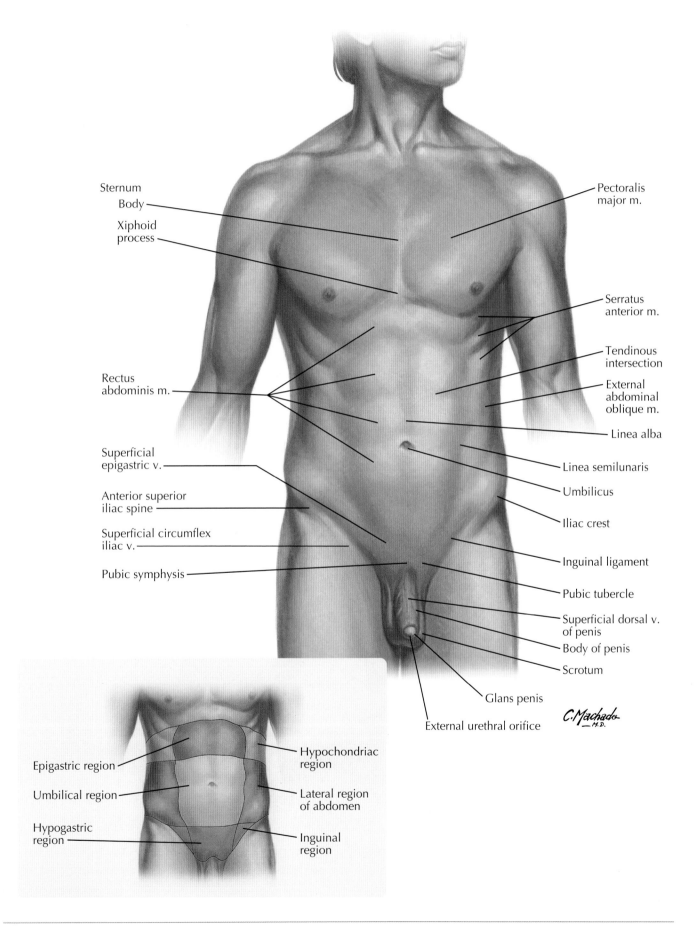

Sternum
Body
Xiphoid process

Pectoralis major m.

Serratus anterior m.

Tendinous intersection

External abdominal oblique m.

Rectus abdominis m.

Linea alba

Linea semilunaris

Superficial epigastric v.

Umbilicus

Anterior superior iliac spine

Iliac crest

Superficial circumflex iliac v.

Inguinal ligament

Pubic symphysis

Pubic tubercle

Superficial dorsal v. of penis

Body of penis

Scrotum

Glans penis

External urethral orifice

C. Machado
_M.D._

Epigastric region

Hypochondriac region

Umbilical region

Lateral region of abdomen

Hypogastric region

Inguinal region

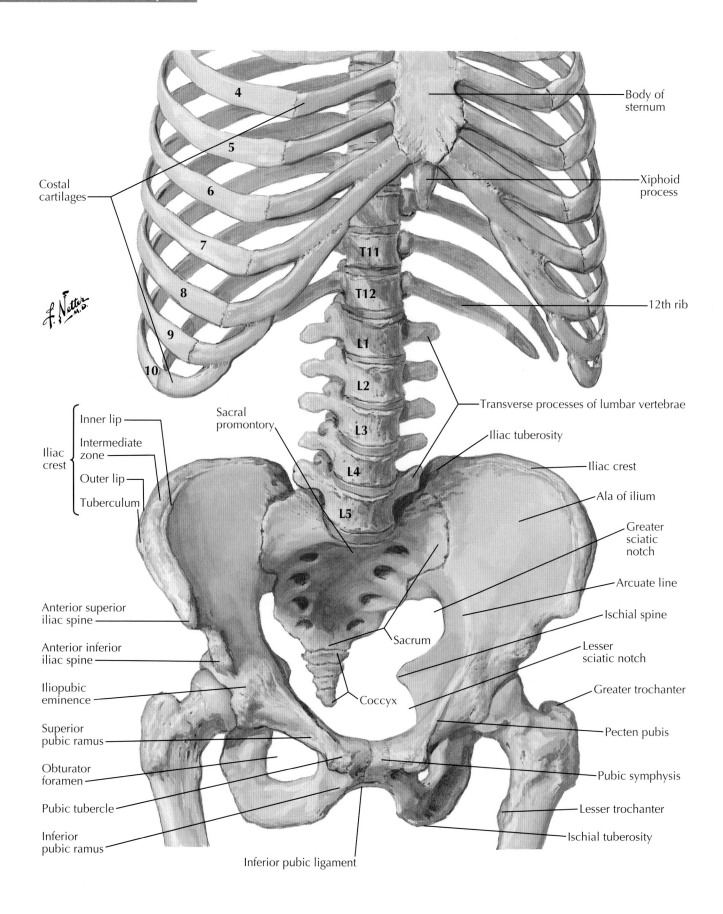

Costal cartilages

Body of sternum

Xiphoid process

12th rib

Transverse processes of lumbar vertebrae

Iliac tuberosity

Iliac crest

Ala of ilium

Greater sciatic notch

Arcuate line

Ischial spine

Lesser sciatic notch

Greater trochanter

Pecten pubis

Pubic symphysis

Lesser trochanter

Ischial tuberosity

Inner lip
Intermediate zone
Outer lip
Tuberculum

Iliac crest

Sacral promontory

Anterior superior iliac spine

Anterior inferior iliac spine

Iliopubic eminence

Superior pubic ramus

Obturator foramen

Pubic tubercle

Inferior pubic ramus

Sacrum

Coccyx

Inferior pubic ligament

T11
T12
L1
L2
L3
L4
L5

4
5
6
7
8
9
10

Plate 268

Abdominal Wall

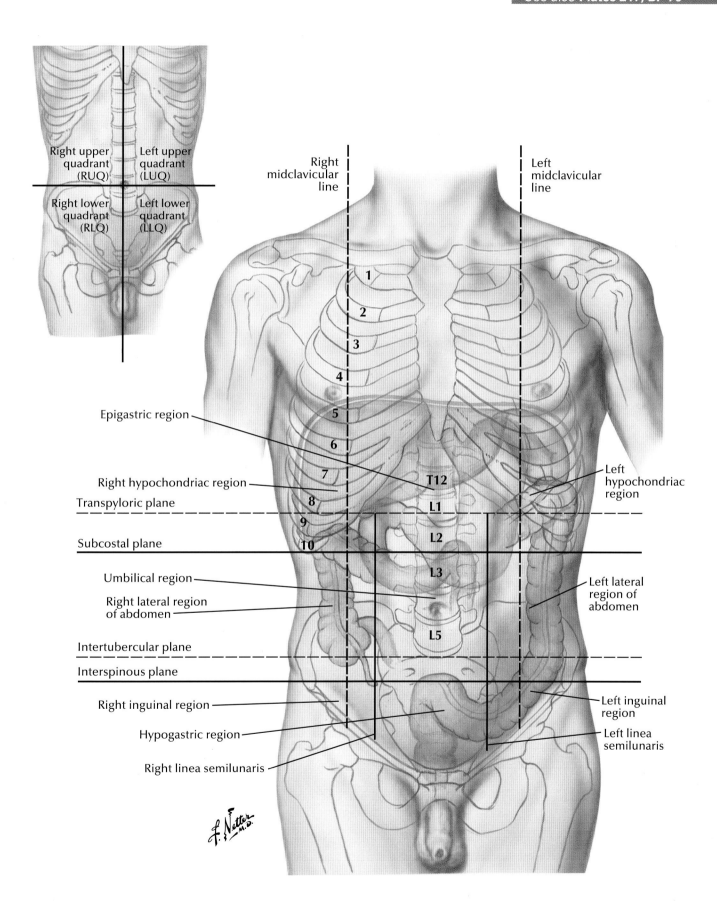

Right upper quadrant (RUQ)   Left upper quadrant (LUQ)

Right lower quadrant (RLQ)   Left lower quadrant (LLQ)

Right midclavicular line

Left midclavicular line

1
2
3
4
5
6
7
8
9
10

T12
L1
L2
L3
L5

Epigastric region

Right hypochondriac region

Transpyloric plane

Subcostal plane

Umbilical region

Right lateral region of abdomen

Intertubercular plane

Interspinous plane

Right inguinal region

Hypogastric region

Right linea semilunaris

Left hypochondriac region

Left lateral region of abdomen

Left inguinal region

Left linea semilunaris

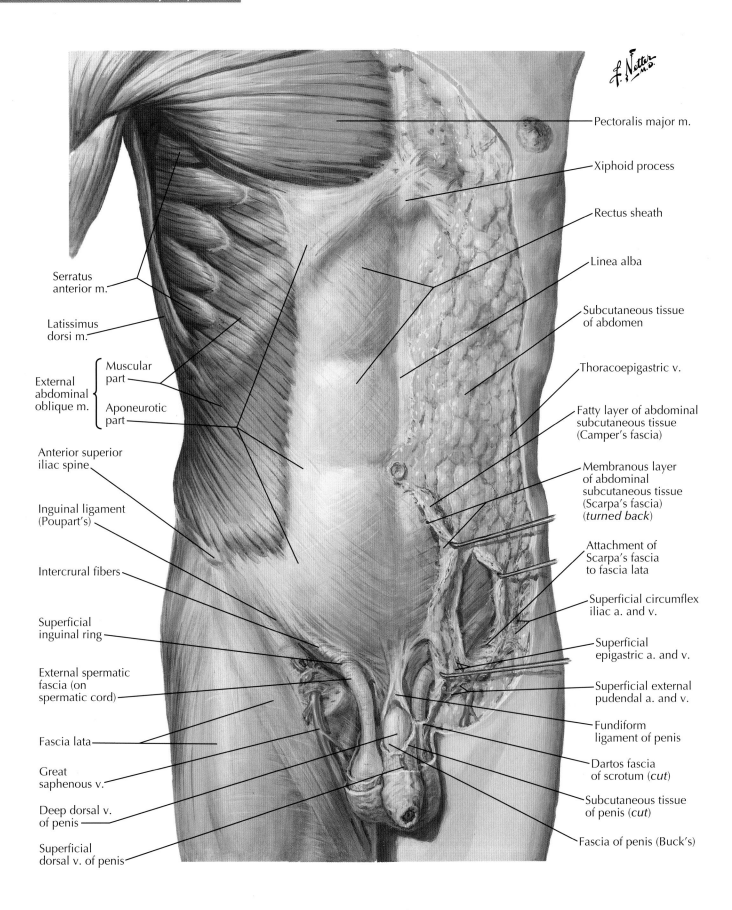

Serratus anterior m.

Latissimus dorsi m.

External abdominal oblique m. { Muscular part / Aponeurotic part

Anterior superior iliac spine

Inguinal ligament (Poupart's)

Intercrural fibers

Superficial inguinal ring

External spermatic fascia (on spermatic cord)

Fascia lata

Great saphenous v.

Deep dorsal v. of penis

Superficial dorsal v. of penis

Pectoralis major m.

Xiphoid process

Rectus sheath

Linea alba

Subcutaneous tissue of abdomen

Thoracoepigastric v.

Fatty layer of abdominal subcutaneous tissue (Camper's fascia)

Membranous layer of abdominal subcutaneous tissue (Scarpa's fascia) (*turned back*)

Attachment of Scarpa's fascia to fascia lata

Superficial circumflex iliac a. and v.

Superficial epigastric a. and v.

Superficial external pudendal a. and v.

Fundiform ligament of penis

Dartos fascia of scrotum (*cut*)

Subcutaneous tissue of penis (*cut*)

Fascia of penis (Buck's)

Plate 270

Abdominal Wall

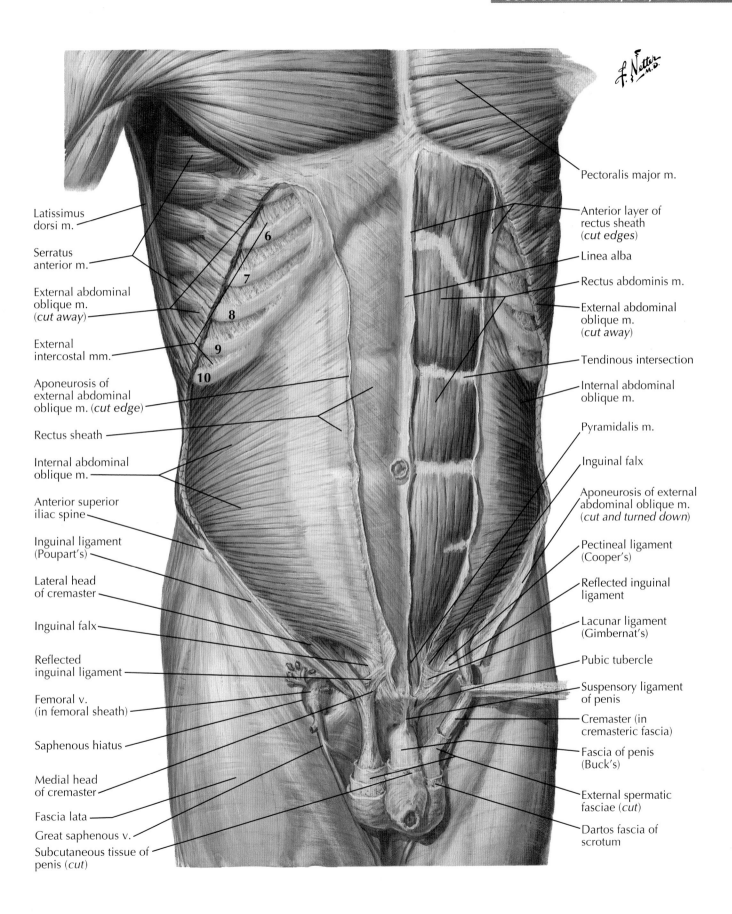

Latissimus dorsi m.

Serratus anterior m.

External abdominal oblique m. (*cut away*)

External intercostal mm.

Aponeurosis of external abdominal oblique m. (*cut edge*)

Rectus sheath

Internal abdominal oblique m.

Anterior superior iliac spine

Inguinal ligament (Poupart's)

Lateral head of cremaster

Inguinal falx

Reflected inguinal ligament

Femoral v. (in femoral sheath)

Saphenous hiatus

Medial head of cremaster

Fascia lata

Great saphenous v.

Subcutaneous tissue of penis (*cut*)

6
7
8
9
10

Pectoralis major m.

Anterior layer of rectus sheath (*cut edges*)

Linea alba

Rectus abdominis m.

External abdominal oblique m. (*cut away*)

Tendinous intersection

Internal abdominal oblique m.

Pyramidalis m.

Inguinal falx

Aponeurosis of external abdominal oblique m. (*cut and turned down*)

Pectineal ligament (Cooper's)

Reflected inguinal ligament

Lacunar ligament (Gimbernat's)

Pubic tubercle

Suspensory ligament of penis

Cremaster (in cremasteric fascia)

Fascia of penis (Buck's)

External spermatic fasciae (*cut*)

Dartos fascia of scrotum

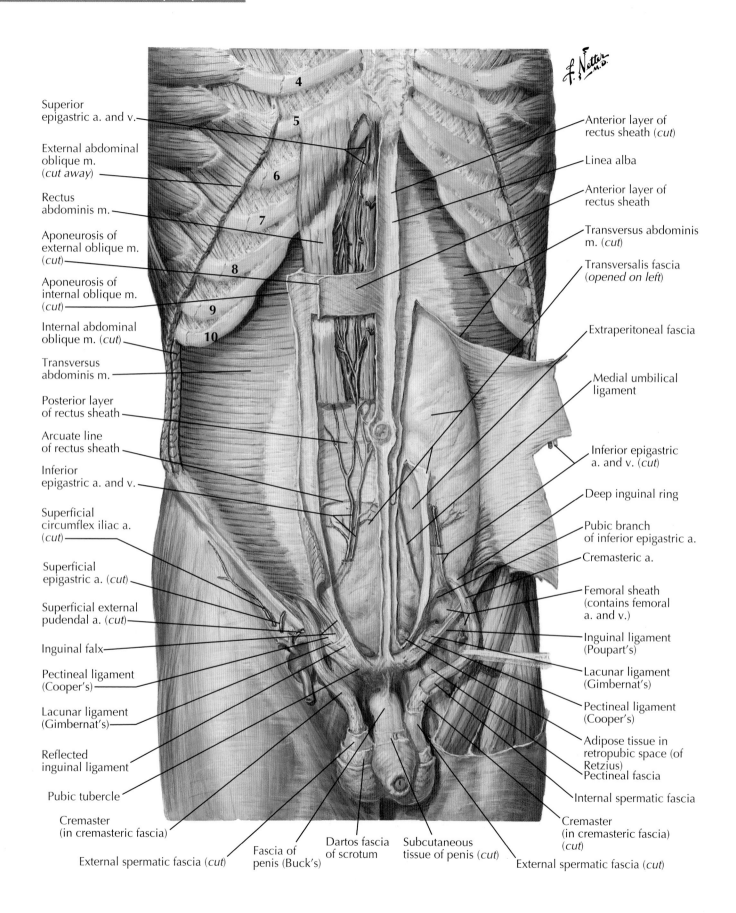

Superior epigastric a. and v.

External abdominal oblique m. (cut away)

Rectus abdominis m.

Aponeurosis of external oblique m. (cut)

Aponeurosis of internal oblique m. (cut)

Internal abdominal oblique m. (cut)

Transversus abdominis m.

Posterior layer of rectus sheath

Arcuate line of rectus sheath

Inferior epigastric a. and v.

Superficial circumflex iliac a. (cut)

Superficial epigastric a. (cut)

Superficial external pudendal a. (cut)

Inguinal falx

Pectineal ligament (Cooper's)

Lacunar ligament (Gimbernat's)

Reflected inguinal ligament

Pubic tubercle

Cremaster (in cremasteric fascia)

External spermatic fascia (cut)

Fascia of penis (Buck's)

Dartos fascia of scrotum

Subcutaneous tissue of penis (cut)

External spermatic fascia (cut)

Anterior layer of rectus sheath (cut)

Linea alba

Anterior layer of rectus sheath

Transversus abdominis m. (cut)

Transversalis fascia (opened on left)

Extraperitoneal fascia

Medial umbilical ligament

Inferior epigastric a. and v. (cut)

Deep inguinal ring

Pubic branch of inferior epigastric a.

Cremasteric a.

Femoral sheath (contains femoral a. and v.)

Inguinal ligament (Poupart's)

Lacunar ligament (Gimbernat's)

Pectineal ligament (Cooper's)

Adipose tissue in retropubic space (of Retzius)

Pectineal fascia

Internal spermatic fascia

Cremaster (in cremasteric fascia) (cut)

Plate 272

Abdominal Wall

Section superior to arcuate line of rectus sheath

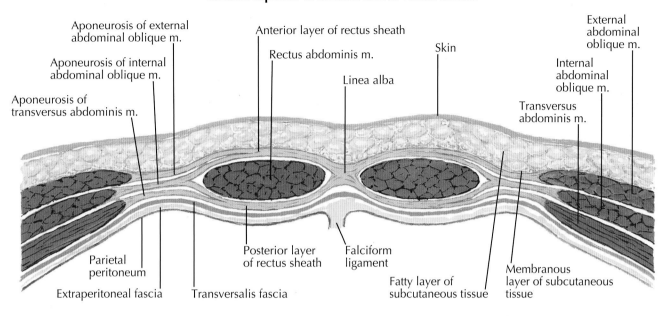

Aponeurosis of external abdominal oblique m.

Aponeurosis of internal abdominal oblique m.

Aponeurosis of transversus abdominis m.

Anterior layer of rectus sheath

Rectus abdominis m.

Linea alba

Skin

External abdominal oblique m.

Internal abdominal oblique m.

Transversus abdominis m.

Parietal peritoneum

Extraperitoneal fascia

Transversalis fascia

Posterior layer of rectus sheath

Falciform ligament

Fatty layer of subcutaneous tissue

Membranous layer of subcutaneous tissue

Aponeurosis of internal abdominal oblique muscle splits to form anterior and posterior layers of rectus sheath. Aponeurosis of external abdominal oblique muscle joins anterior layer; aponeurosis of transversus abdominis muscle joins posterior layer. Anterior and posterior layers of rectus sheath unite medially to form linea alba.

Section inferior to arcuate line of rectus sheath

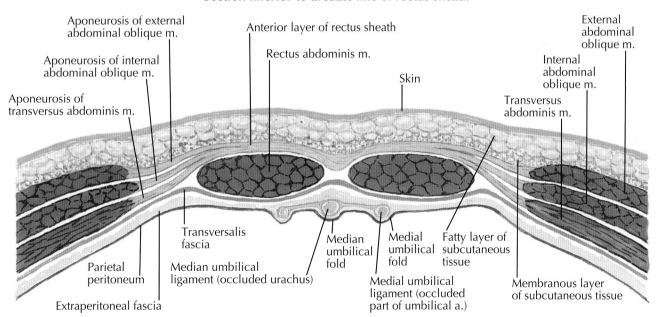

Aponeurosis of external abdominal oblique m.

Aponeurosis of internal abdominal oblique m.

Aponeurosis of transversus abdominis m.

Anterior layer of rectus sheath

Rectus abdominis m.

Skin

External abdominal oblique m.

Internal abdominal oblique m.

Transversus abdominis m.

Parietal peritoneum

Extraperitoneal fascia

Transversalis fascia

Median umbilical ligament (occluded urachus)

Median umbilical fold

Medial umbilical fold

Medial umbilical ligament (occluded part of umbilical a.)

Fatty layer of subcutaneous tissue

Membranous layer of subcutaneous tissue

Aponeurosis of internal abdominal oblique muscle does not split at this level but passes completely anterior to rectus abdominis muscle and is fused there with both the aponeurosis of external abdominal oblique muscle and that of transversus abdominis muscle. Thus, the posterior layer of the rectus sheath is absent below arcuate line, leaving only transversalis fascia.

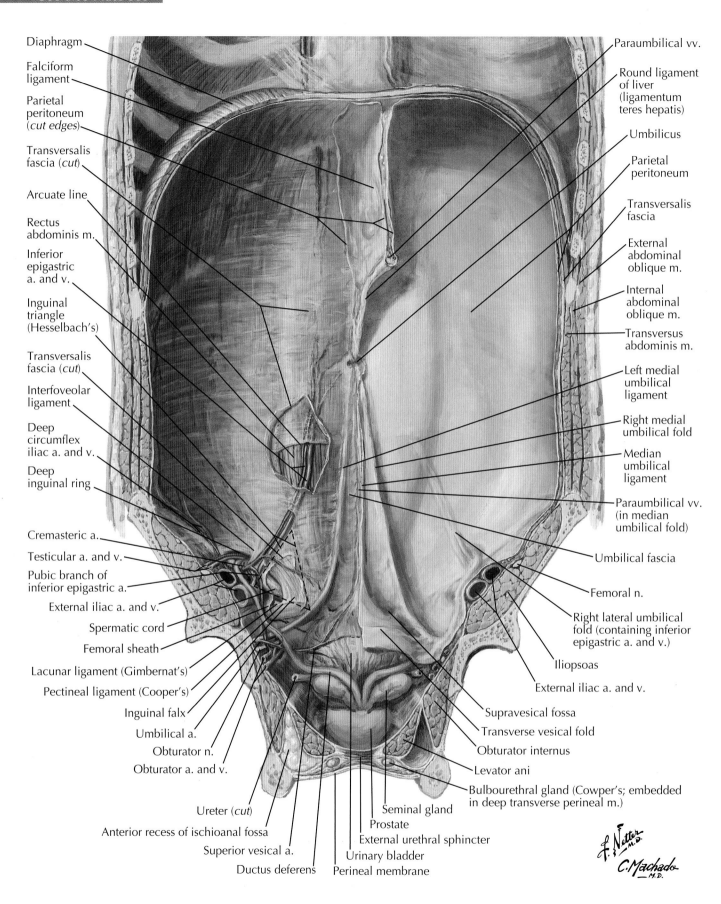

Diaphragm

Falciform ligament

Parietal peritoneum (*cut edges*)

Transversalis fascia (*cut*)

Arcuate line

Rectus abdominis m.

Inferior epigastric a. and v.

Inguinal triangle (Hesselbach's)

Transversalis fascia (*cut*)

Interfoveolar ligament

Deep circumflex iliac a. and v.

Deep inguinal ring

Cremasteric a.

Testicular a. and v.

Pubic branch of inferior epigastric a.

External iliac a. and v.

Spermatic cord

Femoral sheath

Lacunar ligament (Gimbernat's)

Pectineal ligament (Cooper's)

Inguinal falx

Umbilical a.

Obturator n.

Obturator a. and v.

Ureter (*cut*)

Anterior recess of ischioanal fossa

Superior vesical a.

Ductus deferens

Paraumbilical vv.

Round ligament of liver (ligamentum teres hepatis)

Umbilicus

Parietal peritoneum

Transversalis fascia

External abdominal oblique m.

Internal abdominal oblique m.

Transversus abdominis m.

Left medial umbilical ligament

Right medial umbilical fold

Median umbilical ligament

Paraumbilical vv. (in median umbilical fold)

Umbilical fascia

Femoral n.

Right lateral umbilical fold (containing inferior epigastric a. and v.)

Iliopsoas

External iliac a. and v.

Supravesical fossa

Transverse vesical fold

Obturator internus

Levator ani

Bulbourethral gland (Cowper's; embedded in deep transverse perineal m.)

Seminal gland

Prostate

External urethral sphincter

Urinary bladder

Perineal membrane

Plate 274

Abdominal Wall

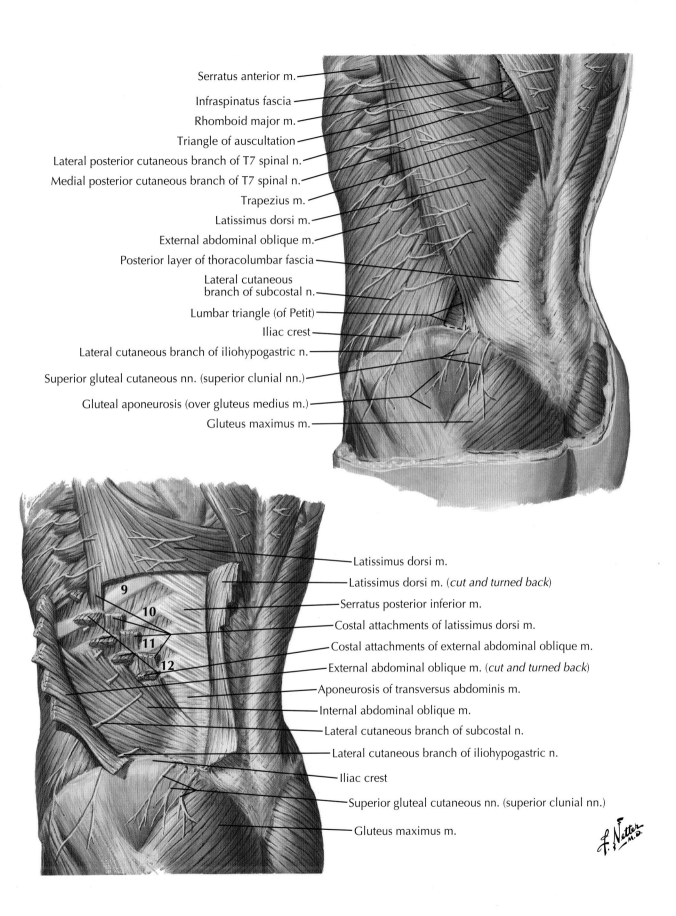

Serratus anterior m.

Infraspinatus fascia

Rhomboid major m.

Triangle of auscultation

Lateral posterior cutaneous branch of T7 spinal n.

Medial posterior cutaneous branch of T7 spinal n.

Trapezius m.

Latissimus dorsi m.

External abdominal oblique m.

Posterior layer of thoracolumbar fascia

Lateral cutaneous branch of subcostal n.

Lumbar triangle (of Petit)

Iliac crest

Lateral cutaneous branch of iliohypogastric n.

Superior gluteal cutaneous nn. (superior clunial nn.)

Gluteal aponeurosis (over gluteus medius m.)

Gluteus maximus m.

9

10

11

12

Latissimus dorsi m.

Latissimus dorsi m. (*cut and turned back*)

Serratus posterior inferior m.

Costal attachments of latissimus dorsi m.

Costal attachments of external abdominal oblique m.

External abdominal oblique m. (*cut and turned back*)

Aponeurosis of transversus abdominis m.

Internal abdominal oblique m.

Lateral cutaneous branch of subcostal n.

Lateral cutaneous branch of iliohypogastric n.

Iliac crest

Superior gluteal cutaneous nn. (superior clunial nn.)

Gluteus maximus m.

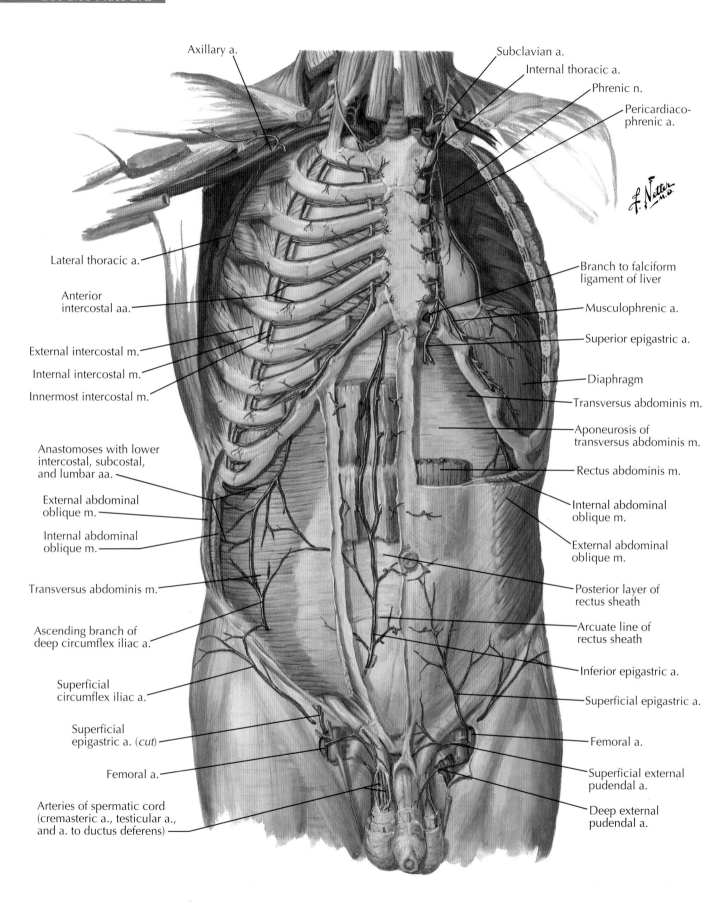

Axillary a.

Subclavian a.

Internal thoracic a.

Phrenic n.

Pericardiaco-
phrenic a.

Lateral thoracic a.

Anterior
intercostal aa.

External intercostal m.

Internal intercostal m.

Innermost intercostal m.

Branch to falciform
ligament of liver

Musculophrenic a.

Superior epigastric a.

Diaphragm

Transversus abdominis m.

Aponeurosis of
transversus abdominis m.

Rectus abdominis m.

Internal abdominal
oblique m.

External abdominal
oblique m.

Anastomoses with lower
intercostal, subcostal,
and lumbar aa.

External abdominal
oblique m.

Internal abdominal
oblique m.

Transversus abdominis m.

Ascending branch of
deep circumflex iliac a.

Superficial
circumflex iliac a.

Superficial
epigastric a. (cut)

Femoral a.

Arteries of spermatic cord
(cremasteric a., testicular a.,
and a. to ductus deferens)

Posterior layer of
rectus sheath

Arcuate line of
rectus sheath

Inferior epigastric a.

Superficial epigastric a.

Femoral a.

Superficial external
pudendal a.

Deep external
pudendal a.

Plate 276

Abdominal Wall

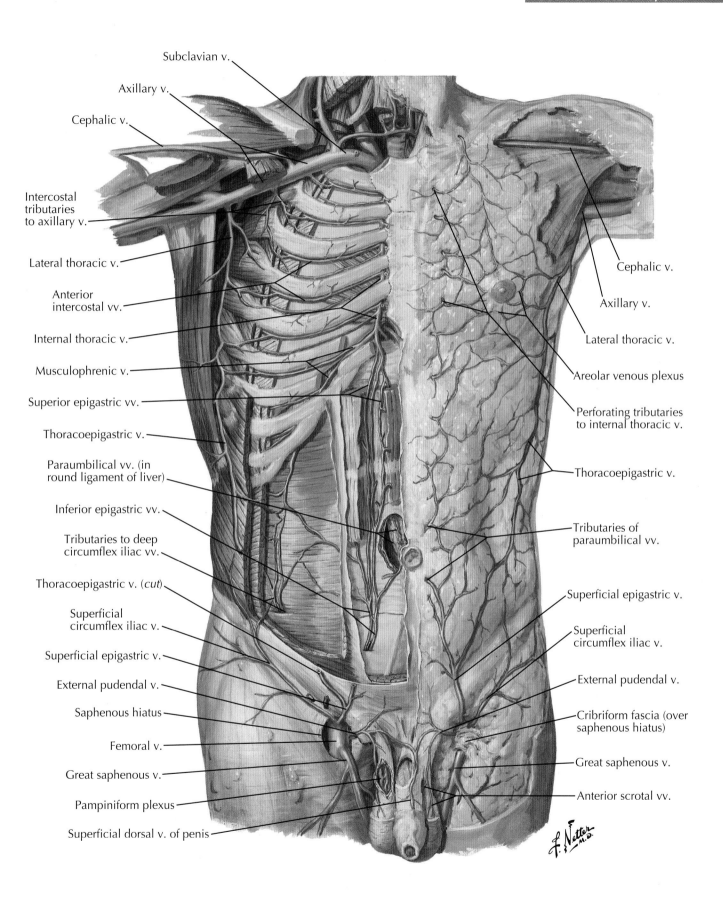

Subclavian v.

Axillary v.

Cephalic v.

Intercostal tributaries to axillary v.

Lateral thoracic v.

Anterior intercostal vv.

Internal thoracic v.

Musculophrenic v.

Superior epigastric vv.

Thoracoepigastric v.

Paraumbilical vv. (in round ligament of liver)

Inferior epigastric vv.

Tributaries to deep circumflex iliac vv.

Thoracoepigastric v. (cut)

Superficial circumflex iliac v.

Superficial epigastric v.

External pudendal v.

Saphenous hiatus

Femoral v.

Great saphenous v.

Pampiniform plexus

Superficial dorsal v. of penis

Cephalic v.

Axillary v.

Lateral thoracic v.

Areolar venous plexus

Perforating tributaries to internal thoracic v.

Thoracoepigastric v.

Tributaries of paraumbilical vv.

Superficial epigastric v.

Superficial circumflex iliac v.

External pudendal v.

Cribriform fascia (over saphenous hiatus)

Great saphenous v.

Anterior scrotal vv.

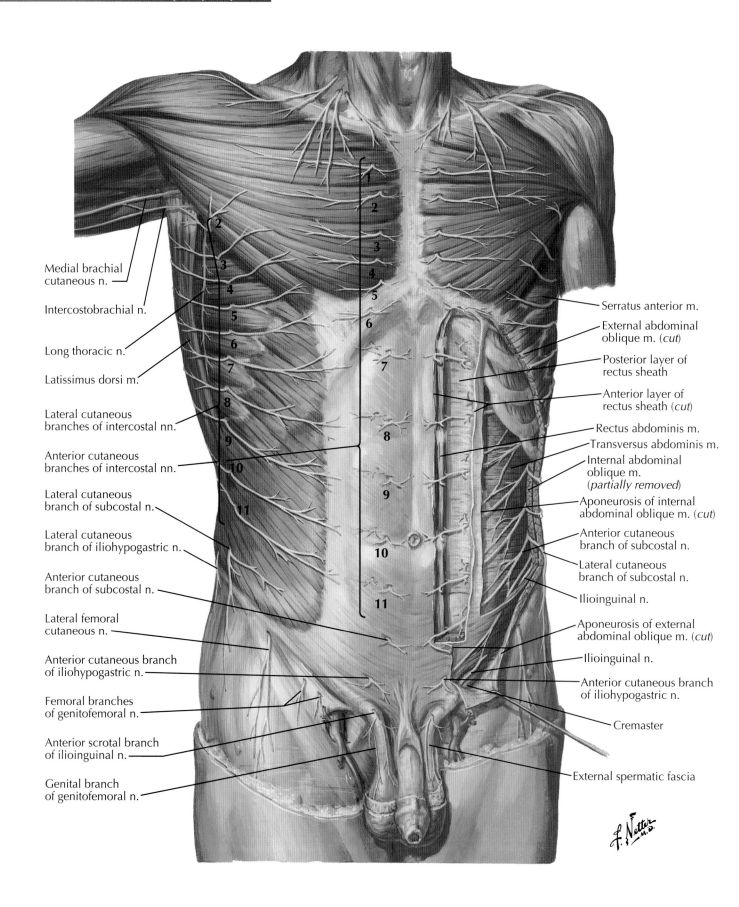

Medial brachial cutaneous n.

Intercostobrachial n.

Long thoracic n.

Latissimus dorsi m.

Lateral cutaneous branches of intercostal nn.

Anterior cutaneous branches of intercostal nn.

Lateral cutaneous branch of subcostal n.

Lateral cutaneous branch of iliohypogastric n.

Anterior cutaneous branch of subcostal n.

Lateral femoral cutaneous n.

Anterior cutaneous branch of iliohypogastric n.

Femoral branches of genitofemoral n.

Anterior scrotal branch of ilioinguinal n.

Genital branch of genitofemoral n.

Serratus anterior m.

External abdominal oblique m. (*cut*)

Posterior layer of rectus sheath

Anterior layer of rectus sheath (*cut*)

Rectus abdominis m.

Transversus abdominis m.

Internal abdominal oblique m. (*partially removed*)

Aponeurosis of internal abdominal oblique m. (*cut*)

Anterior cutaneous branch of subcostal n.

Lateral cutaneous branch of subcostal n.

Ilioinguinal n.

Aponeurosis of external abdominal oblique m. (*cut*)

Ilioinguinal n.

Anterior cutaneous branch of iliohypogastric n.

Cremaster

External spermatic fascia

Plate 278

Abdominal Wall

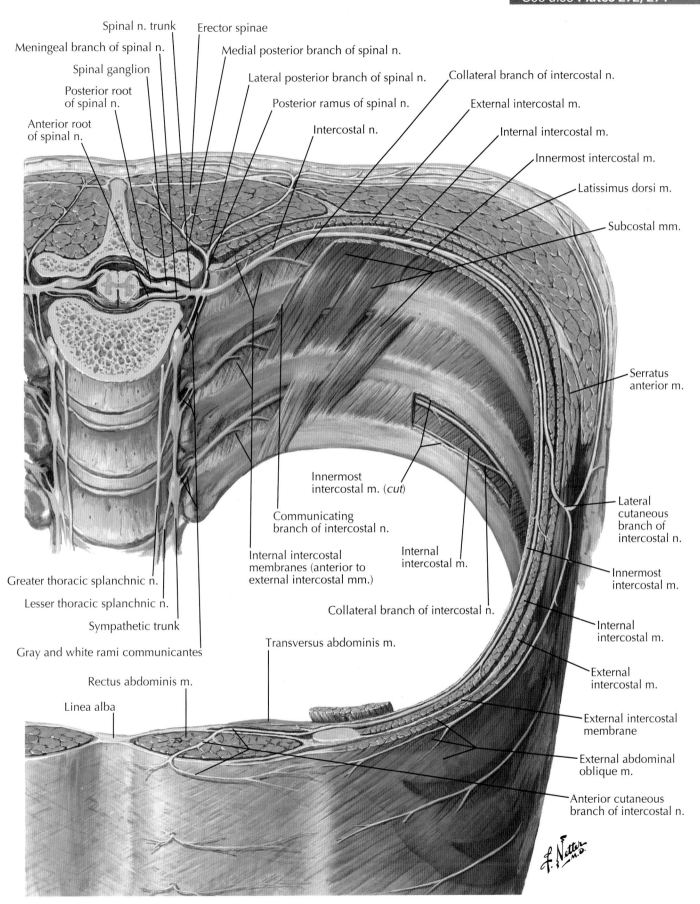

Spinal n. trunk
Erector spinae
Meningeal branch of spinal n.
Medial posterior branch of spinal n.
Spinal ganglion
Lateral posterior branch of spinal n.
Posterior root of spinal n.
Posterior ramus of spinal n.
Anterior root of spinal n.
Intercostal n.
Collateral branch of intercostal n.
External intercostal m.
Internal intercostal m.
Innermost intercostal m.
Latissimus dorsi m.
Subcostal mm.
Serratus anterior m.
Innermost intercostal m. (*cut*)
Communicating branch of intercostal n.
Internal intercostal membranes (anterior to external intercostal mm.)
Internal intercostal m.
Lateral cutaneous branch of intercostal n.
Innermost intercostal m.
Internal intercostal m.
Greater thoracic splanchnic n.
Lesser thoracic splanchnic n.
Collateral branch of intercostal n.
External intercostal m.
Sympathetic trunk
Gray and white rami communicantes
Transversus abdominis m.
External intercostal membrane
Rectus abdominis m.
External abdominal oblique m.
Linea alba
Anterior cutaneous branch of intercostal n.

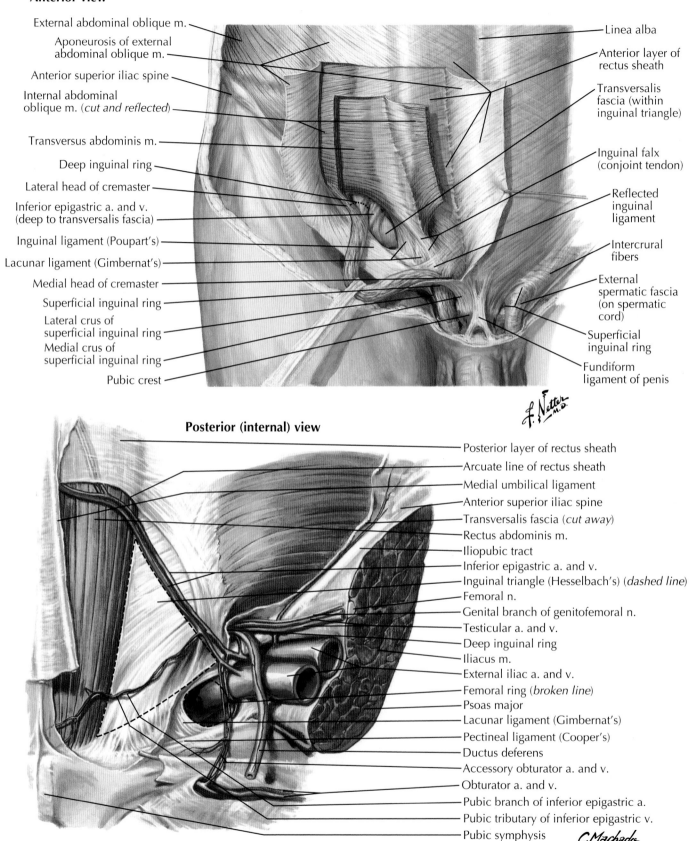

Anterior view

External abdominal oblique m.

Aponeurosis of external abdominal oblique m.

Anterior superior iliac spine

Internal abdominal oblique m. (*cut and reflected*)

Transversus abdominis m.

Deep inguinal ring

Lateral head of cremaster

Inferior epigastric a. and v. (deep to transversalis fascia)

Inguinal ligament (Poupart's)

Lacunar ligament (Gimbernat's)

Medial head of cremaster

Superficial inguinal ring

Lateral crus of superficial inguinal ring

Medial crus of superficial inguinal ring

Pubic crest

Linea alba

Anterior layer of rectus sheath

Transversalis fascia (within inguinal triangle)

Inguinal falx (conjoint tendon)

Reflected inguinal ligament

Intercrural fibers

External spermatic fascia (on spermatic cord)

Superficial inguinal ring

Fundiform ligament of penis

Posterior (internal) view

Posterior layer of rectus sheath

Arcuate line of rectus sheath

Medial umbilical ligament

Anterior superior iliac spine

Transversalis fascia (*cut away*)

Rectus abdominis m.

Iliopubic tract

Inferior epigastric a. and v.

Inguinal triangle (Hesselbach's) (*dashed line*)

Femoral n.

Genital branch of genitofemoral n.

Testicular a. and v.

Deep inguinal ring

Iliacus m.

External iliac a. and v.

Femoral ring (*broken line*)

Psoas major

Lacunar ligament (Gimbernat's)

Pectineal ligament (Cooper's)

Ductus deferens

Accessory obturator a. and v.

Obturator a. and v.

Pubic branch of inferior epigastric a.

Pubic tributary of inferior epigastric v.

Pubic symphysis

Plate 280

Abdominal Wall

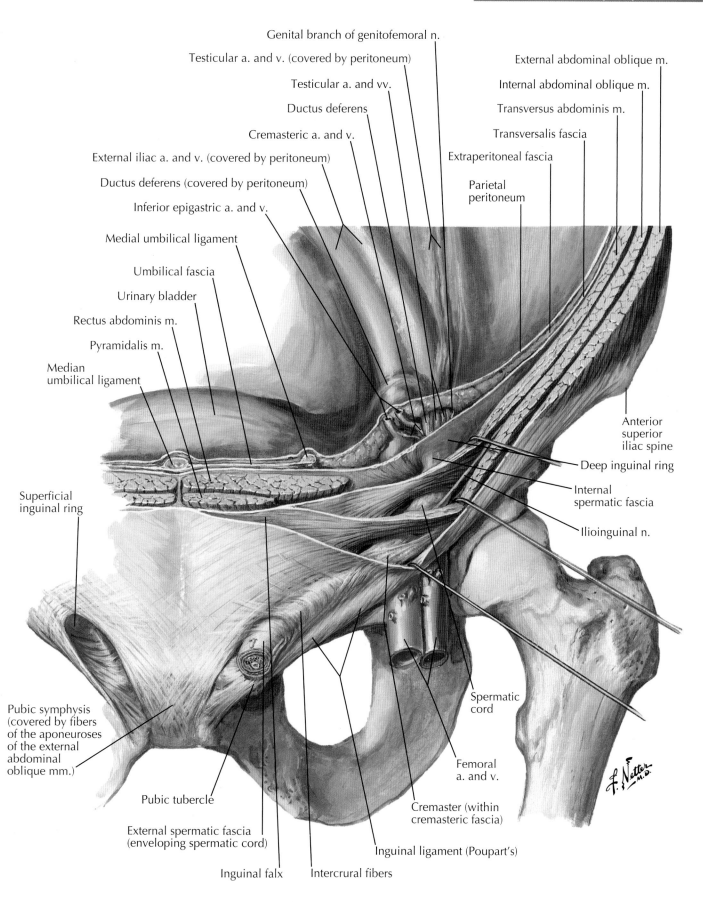

Genital branch of genitofemoral n.

Testicular a. and v. (covered by peritoneum)

External abdominal oblique m.

Testicular a. and vv.

Internal abdominal oblique m.

Ductus deferens

Transversus abdominis m.

Cremasteric a. and v.

Transversalis fascia

External iliac a. and v. (covered by peritoneum)

Extraperitoneal fascia

Ductus deferens (covered by peritoneum)

Parietal peritoneum

Inferior epigastric a. and v.

Medial umbilical ligament

Umbilical fascia

Urinary bladder

Rectus abdominis m.

Pyramidalis m.

Median umbilical ligament

Anterior superior iliac spine

Deep inguinal ring

Internal spermatic fascia

Superficial inguinal ring

Ilioinguinal n.

Spermatic cord

Pubic symphysis (covered by fibers of the aponeuroses of the external abdominal oblique mm.)

Femoral a. and v.

Pubic tubercle

Cremaster (within cremasteric fascia)

External spermatic fascia (enveloping spermatic cord)

Inguinal ligament (Poupart's)

Inguinal falx

Intercrural fibers

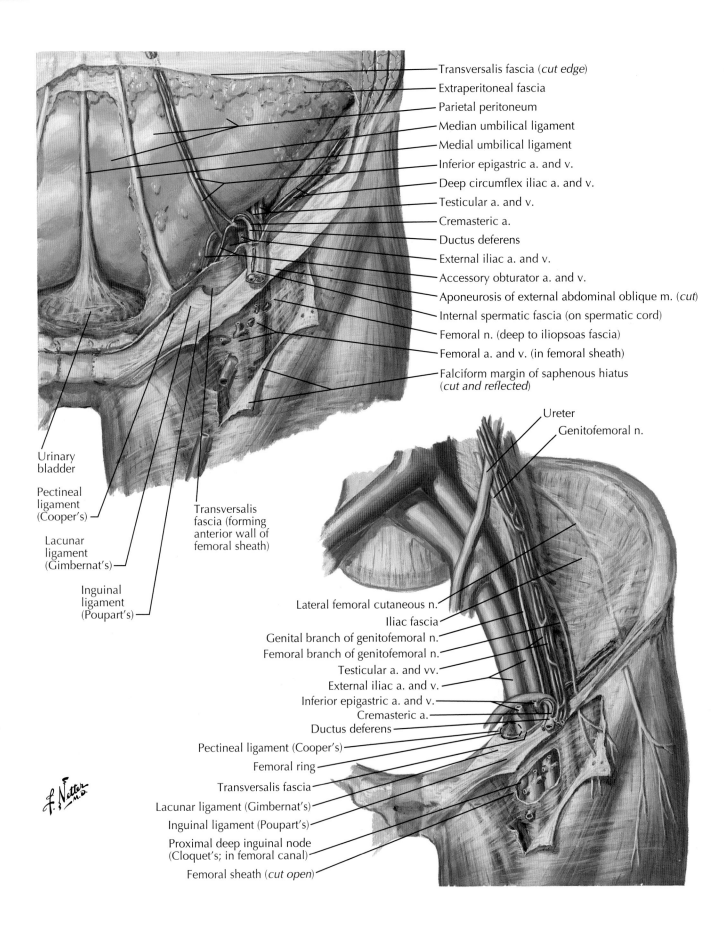

Transversalis fascia (*cut edge*)

Extraperitoneal fascia

Parietal peritoneum

Median umbilical ligament

Medial umbilical ligament

Inferior epigastric a. and v.

Deep circumflex iliac a. and v.

Testicular a. and v.

Cremasteric a.

Ductus deferens

External iliac a. and v.

Accessory obturator a. and v.

Aponeurosis of external abdominal oblique m. (*cut*)

Internal spermatic fascia (on spermatic cord)

Femoral n. (deep to iliopsoas fascia)

Femoral a. and v. (in femoral sheath)

Falciform margin of saphenous hiatus
(*cut and reflected*)

Urinary
bladder

Pectineal
ligament
(Cooper's)

Lacunar
ligament
(Gimbernat's)

Inguinal
ligament
(Poupart's)

Transversalis
fascia (forming
anterior wall of
femoral sheath)

Ureter

Genitofemoral n.

Lateral femoral cutaneous n.

Iliac fascia

Genital branch of genitofemoral n.

Femoral branch of genitofemoral n.

Testicular a. and vv.

External iliac a. and v.

Inferior epigastric a. and v.

Cremasteric a.

Ductus deferens

Pectineal ligament (Cooper's)

Femoral ring

Transversalis fascia

Lacunar ligament (Gimbernat's)

Inguinal ligament (Poupart's)

Proximal deep inguinal node
(Cloquet's; in femoral canal)

Femoral sheath (*cut open*)

Plate 282

Abdominal Wall

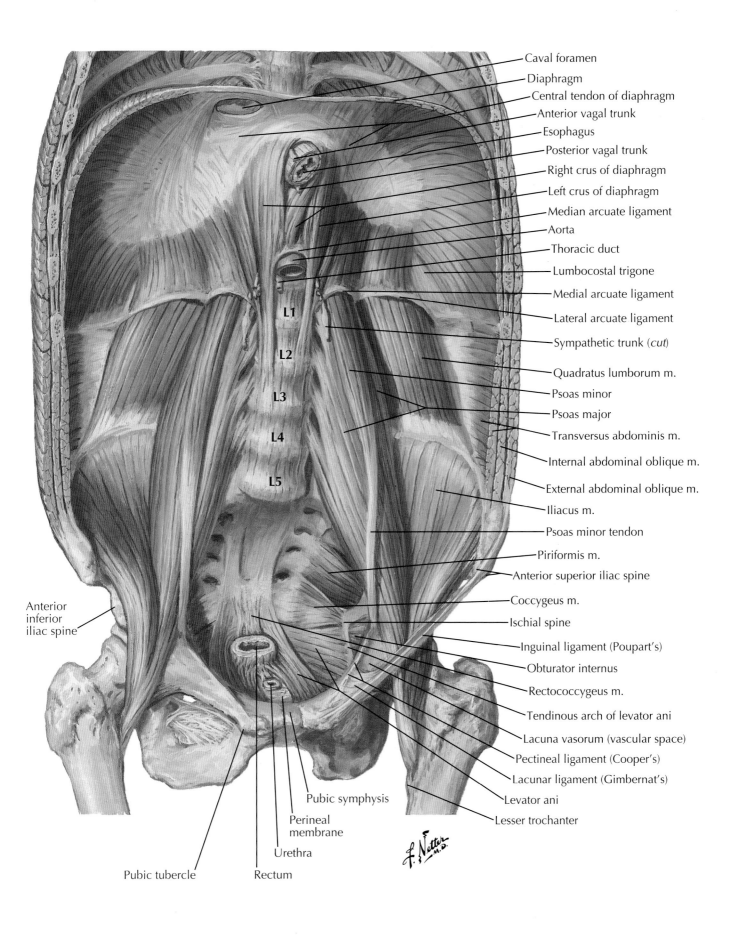

Caval foramen
Diaphragm
Central tendon of diaphragm
Anterior vagal trunk
Esophagus
Posterior vagal trunk
Right crus of diaphragm
Left crus of diaphragm
Median arcuate ligament
Aorta
Thoracic duct
Lumbocostal trigone
Medial arcuate ligament
Lateral arcuate ligament
Sympathetic trunk (cut)
Quadratus lumborum m.
Psoas minor
Psoas major
Transversus abdominis m.
Internal abdominal oblique m.
External abdominal oblique m.
Iliacus m.
Psoas minor tendon
Piriformis m.
Anterior superior iliac spine
Coccygeus m.
Ischial spine
Inguinal ligament (Poupart's)
Obturator internus
Rectococcygeus m.
Tendinous arch of levator ani
Lacuna vasorum (vascular space)
Pectineal ligament (Cooper's)
Lacunar ligament (Gimbernat's)
Levator ani
Lesser trochanter

L1
L2
L3
L4
L5

Anterior inferior iliac spine

Pubic symphysis
Perineal membrane
Urethra
Pubic tubercle
Rectum

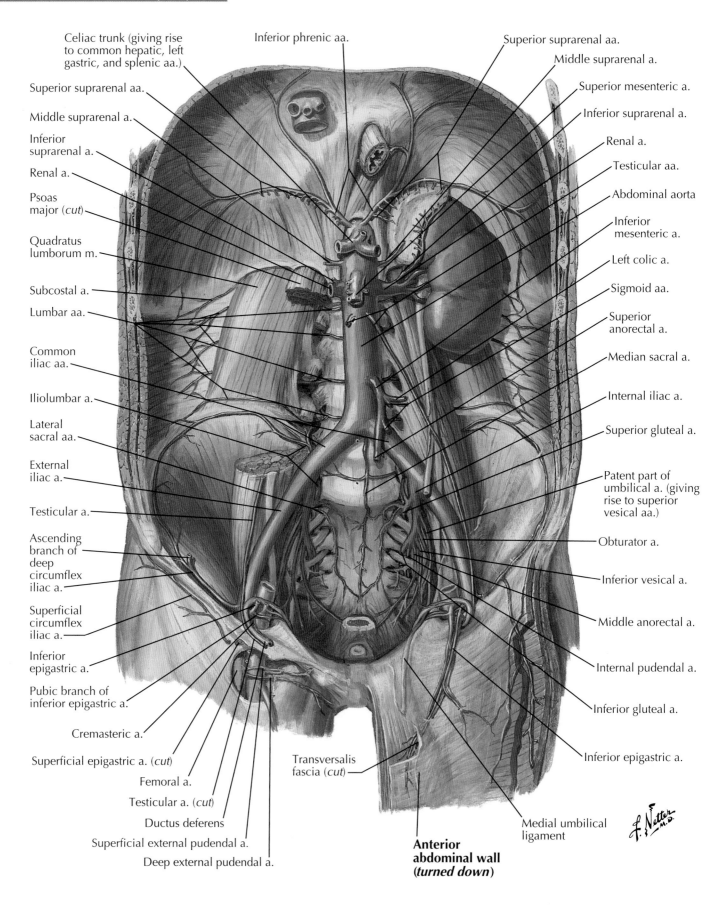

Celiac trunk (giving rise to common hepatic, left gastric, and splenic aa.)

Superior suprarenal aa.

Middle suprarenal a.

Inferior suprarenal a.

Renal a.

Psoas major (*cut*)

Quadratus lumborum m.

Subcostal a.

Lumbar aa.

Common iliac aa.

Iliolumbar a.

Lateral sacral aa.

External iliac a.

Testicular a.

Ascending branch of deep circumflex iliac a.

Superficial circumflex iliac a.

Inferior epigastric a.

Pubic branch of inferior epigastric a.

Cremasteric a.

Superficial epigastric a. (*cut*)

Femoral a.

Testicular a. (*cut*)

Ductus deferens

Superficial external pudendal a.

Deep external pudendal a.

Inferior phrenic aa.

Superior suprarenal aa.

Middle suprarenal a.

Superior mesenteric a.

Inferior suprarenal a.

Renal a.

Testicular aa.

Abdominal aorta

Inferior mesenteric a.

Left colic a.

Sigmoid aa.

Superior anorectal a.

Median sacral a.

Internal iliac a.

Superior gluteal a.

Patent part of umbilical a. (giving rise to superior vesical aa.)

Obturator a.

Inferior vesical a.

Middle anorectal a.

Internal pudendal a.

Inferior gluteal a.

Inferior epigastric a.

Medial umbilical ligament

Transversalis fascia (*cut*)

Anterior abdominal wall (*turned down*)

Plate 284   **Abdominal Wall**

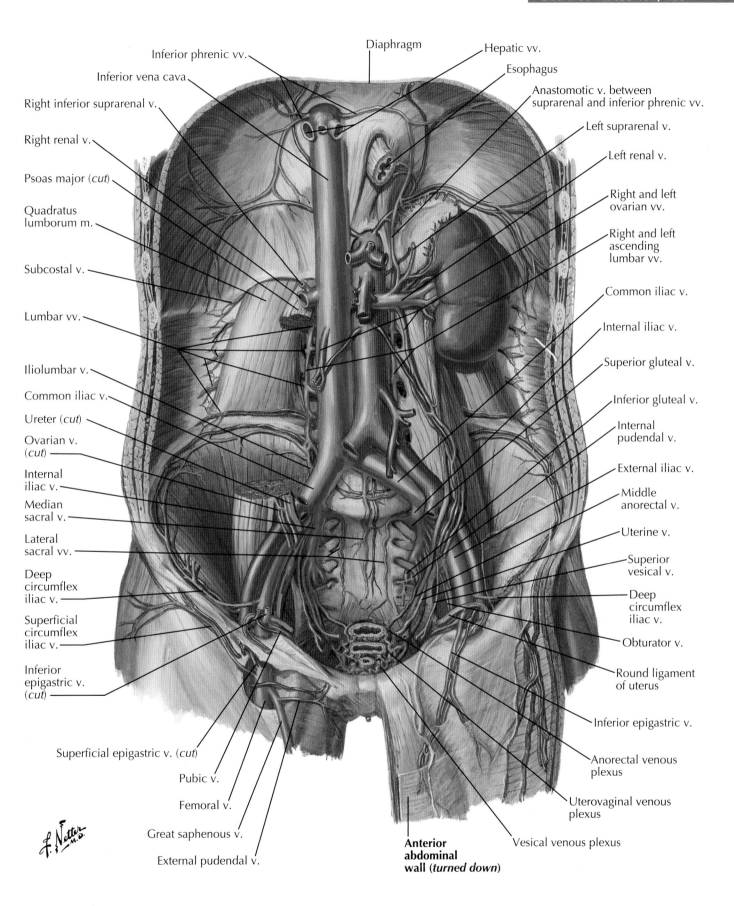

Inferior phrenic vv.

Inferior vena cava

Right inferior suprarenal v.

Right renal v.

Psoas major (*cut*)

Quadratus lumborum m.

Subcostal v.

Lumbar vv.

Iliolumbar v.

Common iliac v.

Ureter (*cut*)

Ovarian v. (*cut*)

Internal iliac v.

Median sacral v.

Lateral sacral vv.

Deep circumflex iliac v.

Superficial circumflex iliac v.

Inferior epigastric v. (*cut*)

Superficial epigastric v. (*cut*)

Pubic v.

Femoral v.

Great saphenous v.

External pudendal v.

Diaphragm

Hepatic vv.

Esophagus

Anastomotic v. between suprarenal and inferior phrenic vv.

Left suprarenal v.

Left renal v.

Right and left ovarian vv.

Right and left ascending lumbar vv.

Common iliac v.

Internal iliac v.

Superior gluteal v.

Inferior gluteal v.

Internal pudendal v.

External iliac v.

Middle anorectal v.

Uterine v.

Superior vesical v.

Deep circumflex iliac v.

Obturator v.

Round ligament of uterus

Inferior epigastric v.

Anorectal venous plexus

Uterovaginal venous plexus

Vesical venous plexus

Anterior abdominal wall (turned down)

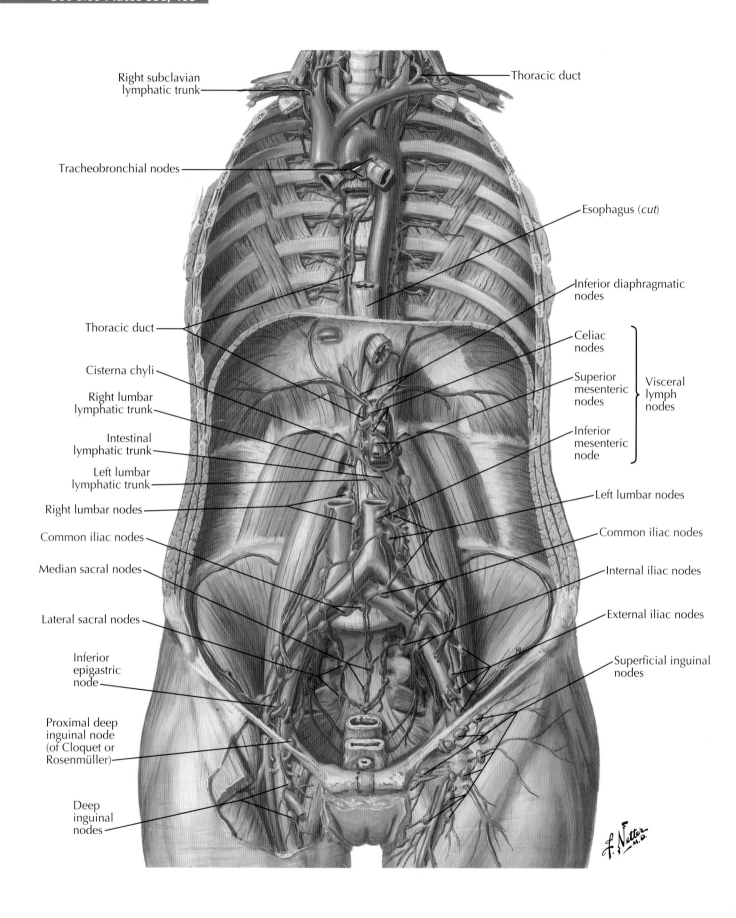

Right subclavian lymphatic trunk

Thoracic duct

Tracheobronchial nodes

Esophagus (*cut*)

Inferior diaphragmatic nodes

Thoracic duct

Celiac nodes

Cisterna chyli

Superior mesenteric nodes

Visceral lymph nodes

Right lumbar lymphatic trunk

Intestinal lymphatic trunk

Inferior mesenteric node

Left lumbar lymphatic trunk

Left lumbar nodes

Right lumbar nodes

Common iliac nodes

Common iliac nodes

Median sacral nodes

Internal iliac nodes

Lateral sacral nodes

External iliac nodes

Inferior epigastric node

Superficial inguinal nodes

Proximal deep inguinal node (of Cloquet or Rosenmüller)

Deep inguinal nodes

Plate 286

Abdominal Wall

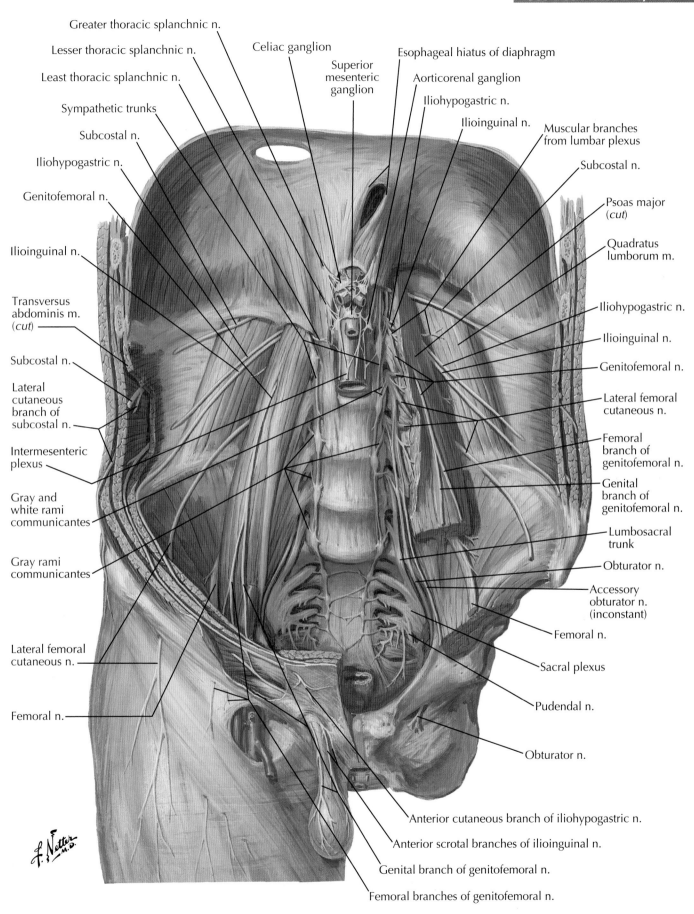

Greater thoracic splanchnic n.

Lesser thoracic splanchnic n.

Least thoracic splanchnic n.

Sympathetic trunks

Subcostal n.

Iliohypogastric n.

Genitofemoral n.

Ilioinguinal n.

Transversus abdominis m. (*cut*)

Subcostal n.

Lateral cutaneous branch of subcostal n.

Intermesenteric plexus

Gray and white rami communicantes

Gray rami communicantes

Lateral femoral cutaneous n.

Femoral n.

Celiac ganglion

Superior mesenteric ganglion

Esophageal hiatus of diaphragm

Aorticorenal ganglion

Iliohypogastric n.

Ilioinguinal n.

Muscular branches from lumbar plexus

Subcostal n.

Psoas major (*cut*)

Quadratus lumborum m.

Iliohypogastric n.

Ilioinguinal n.

Genitofemoral n.

Lateral femoral cutaneous n.

Femoral branch of genitofemoral n.

Genital branch of genitofemoral n.

Lumbosacral trunk

Obturator n.

Accessory obturator n. (inconstant)

Femoral n.

Sacral plexus

Pudendal n.

Obturator n.

Anterior cutaneous branch of iliohypogastric n.

Anterior scrotal branches of ilioinguinal n.

Genital branch of genitofemoral n.

Femoral branches of genitofemoral n.

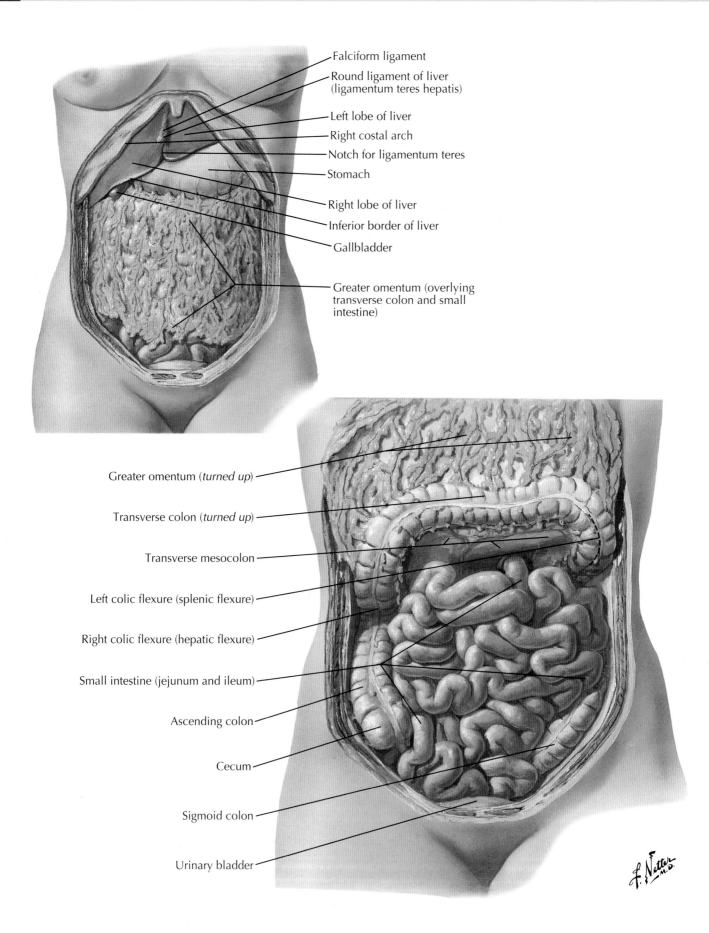

Falciform ligament

Round ligament of liver
(ligamentum teres hepatis)

Left lobe of liver

Right costal arch

Notch for ligamentum teres

Stomach

Right lobe of liver

Inferior border of liver

Gallbladder

Greater omentum (overlying
transverse colon and small
intestine)

Greater omentum (*turned up*)

Transverse colon (*turned up*)

Transverse mesocolon

Left colic flexure (splenic flexure)

Right colic flexure (hepatic flexure)

Small intestine (jejunum and ileum)

Ascending colon

Cecum

Sigmoid colon

Urinary bladder

Plate 288

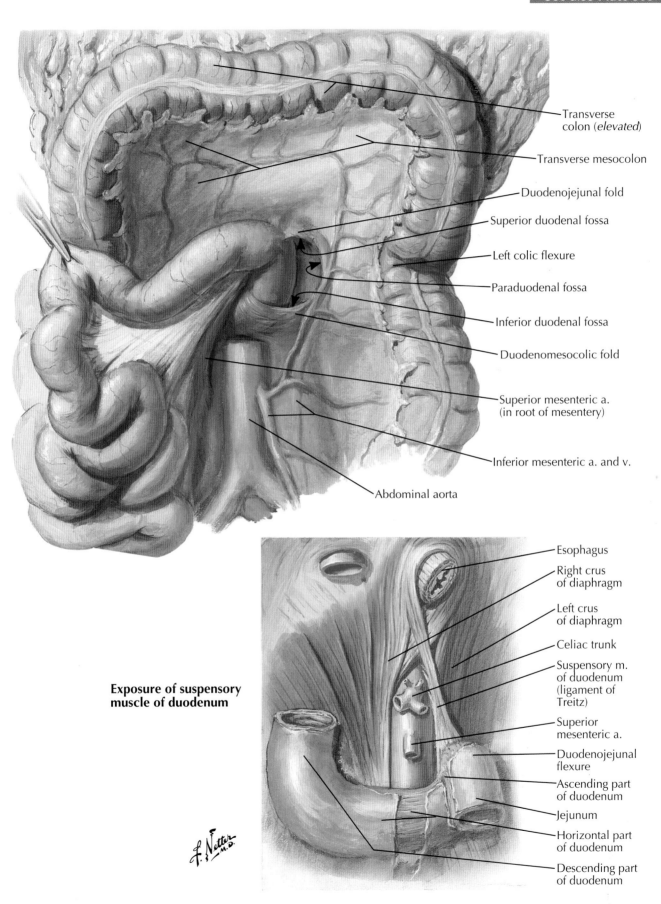

Transverse
colon (*elevated*)

Transverse mesocolon

Duodenojejunal fold

Superior duodenal fossa

Left colic flexure

Paraduodenal fossa

Inferior duodenal fossa

Duodenomesocolic fold

Superior mesenteric a.
(in root of mesentery)

Inferior mesenteric a. and v.

Abdominal aorta

Esophagus

Right crus
of diaphragm

Left crus
of diaphragm

Celiac trunk

Suspensory m.
of duodenum
(ligament of
Treitz)

Superior
mesenteric a.

Duodenojejunal
flexure

Ascending part
of duodenum

Jejunum

Horizontal part
of duodenum

Descending part
of duodenum

**Exposure of suspensory
muscle of duodenum**

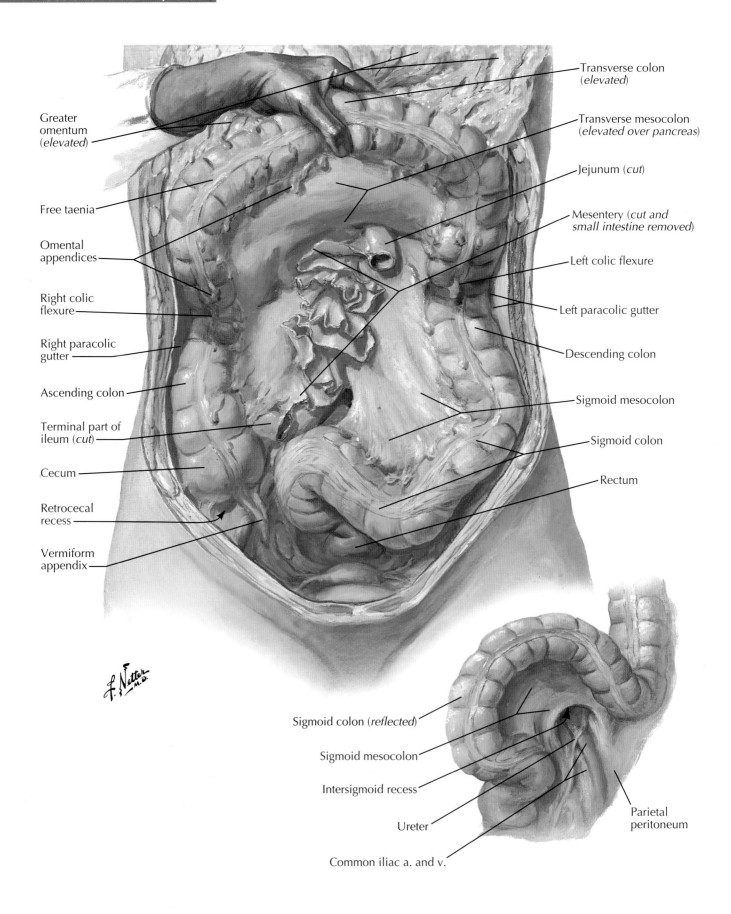

Transverse colon (*elevated*)

Transverse mesocolon (*elevated over pancreas*)

Jejunum (*cut*)

Mesentery (*cut and small intestine removed*)

Left colic flexure

Left paracolic gutter

Descending colon

Sigmoid mesocolon

Sigmoid colon

Rectum

Greater omentum (*elevated*)

Free taenia

Omental appendices

Right colic flexure

Right paracolic gutter

Ascending colon

Terminal part of ileum (*cut*)

Cecum

Retrocecal recess

Vermiform appendix

Sigmoid colon (*reflected*)

Sigmoid mesocolon

Intersigmoid recess

Ureter

Common iliac a. and v.

Parietal peritoneum

Plate 290

Peritoneal Cavity

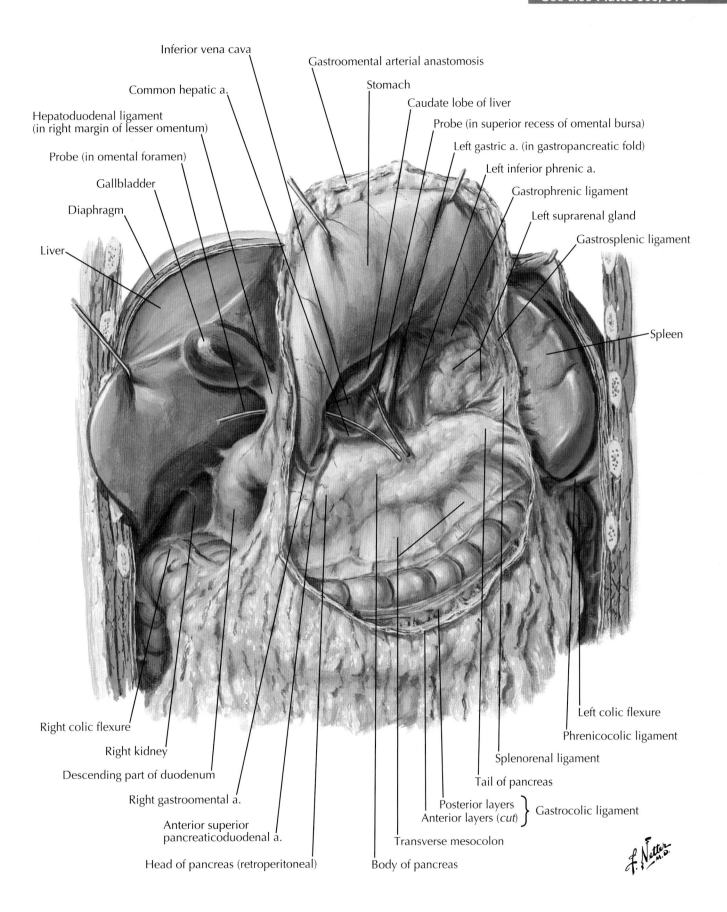

Inferior vena cava

Gastroomental arterial anastomosis

Common hepatic a.

Stomach

Caudate lobe of liver

Hepatoduodenal ligament
(in right margin of lesser omentum)

Probe (in superior recess of omental bursa)

Left gastric a. (in gastropancreatic fold)

Probe (in omental foramen)

Left inferior phrenic a.

Gallbladder

Gastrophrenic ligament

Diaphragm

Left suprarenal gland

Liver

Gastrosplenic ligament

Spleen

Right colic flexure

Left colic flexure

Right kidney

Phrenicocolic ligament

Descending part of duodenum

Splenorenal ligament

Right gastroomental a.

Tail of pancreas

Anterior superior
pancreaticoduodenal a.

Posterior layers
Anterior layers (cut) } Gastrocolic ligament

Head of pancreas (retroperitoneal)

Transverse mesocolon

Body of pancreas

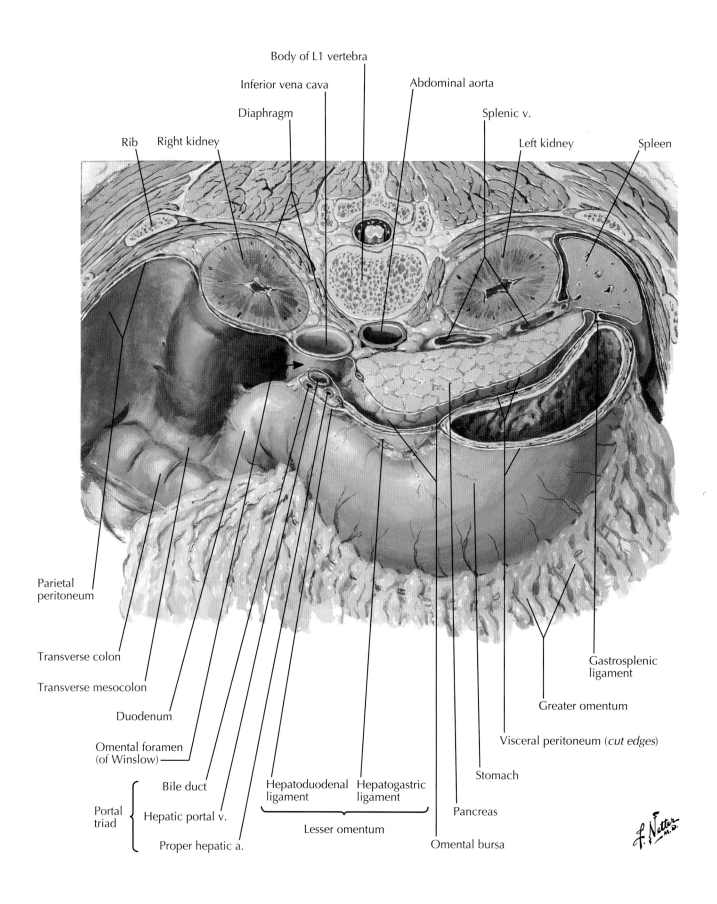

Body of L1 vertebra

Inferior vena cava

Abdominal aorta

Diaphragm

Splenic v.

Rib   Right kidney

Left kidney

Spleen

Parietal
peritoneum

Transverse colon

Transverse mesocolon

Duodenum

Omental foramen
(of Winslow)

Portal
triad

Bile duct

Hepatic portal v.

Proper hepatic a.

Hepatoduodenal
ligament

Hepatogastric
ligament

Lesser omentum

Stomach

Pancreas

Omental bursa

Visceral peritoneum (*cut edges*)

Greater omentum

Gastrosplenic
ligament

Plate 292

Peritoneal Cavity

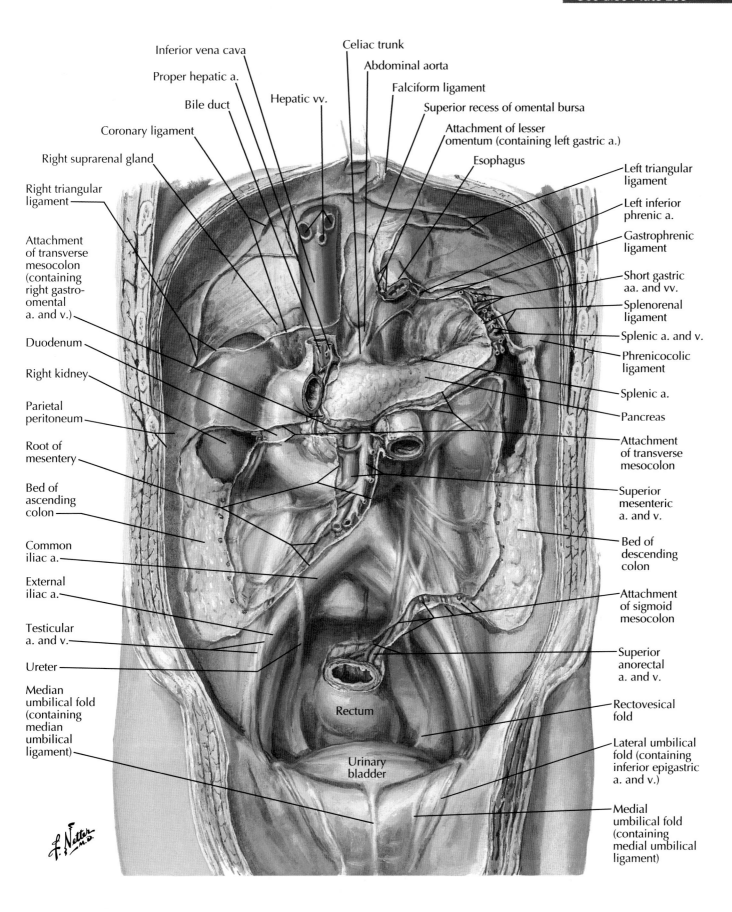

Inferior vena cava

Proper hepatic a.

Bile duct

Coronary ligament

Right suprarenal gland

Right triangular ligament

Attachment of transverse mesocolon (containing right gastro-omental a. and v.)

Duodenum

Right kidney

Parietal peritoneum

Root of mesentery

Bed of ascending colon

Common iliac a.

External iliac a.

Testicular a. and v.

Ureter

Median umbilical fold (containing median umbilical ligament)

Celiac trunk

Abdominal aorta

Falciform ligament

Superior recess of omental bursa

Hepatic vv.

Attachment of lesser omentum (containing left gastric a.)

Esophagus

Left triangular ligament

Left inferior phrenic a.

Gastrophrenic ligament

Short gastric aa. and vv.

Splenorenal ligament

Splenic a. and v.

Phrenicocolic ligament

Splenic a.

Pancreas

Attachment of transverse mesocolon

Superior mesenteric a. and v.

Bed of descending colon

Attachment of sigmoid mesocolon

Superior anorectal a. and v.

Rectovesical fold

Rectum

Lateral umbilical fold (containing inferior epigastric a. and v.)

Urinary bladder

Medial umbilical fold (containing medial umbilical ligament)

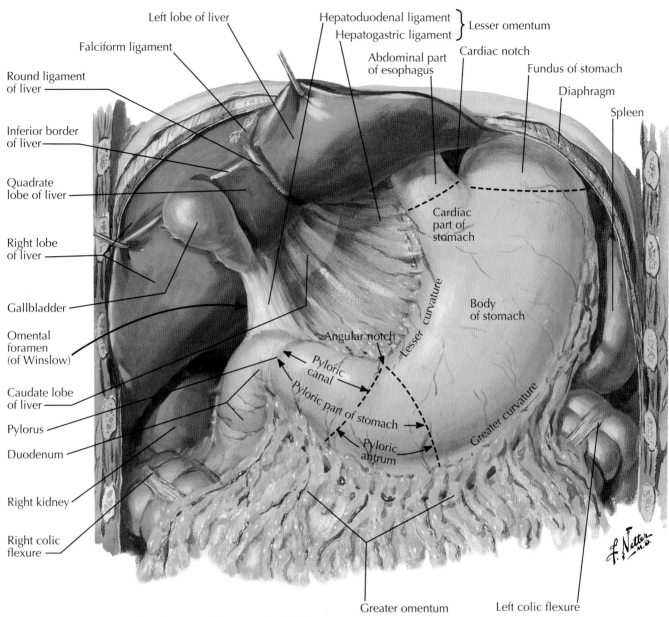

Left lobe of liver

Falciform ligament

Round ligament of liver

Inferior border of liver

Quadrate lobe of liver

Right lobe of liver

Gallbladder

Omental foramen (of Winslow)

Caudate lobe of liver

Pylorus

Duodenum

Right kidney

Right colic flexure

Hepatoduodenal ligament
Hepatogastric ligament } Lesser omentum

Abdominal part of esophagus

Cardiac notch

Fundus of stomach

Diaphragm

Spleen

Cardiac part of stomach

Lesser curvature

Body of stomach

Angular notch

Pyloric canal

Pyloric part of stomach

Pyloric antrum

Greater curvature

Greater omentum

Left colic flexure

Transverse gray-scale ultrasound image of midabdomen

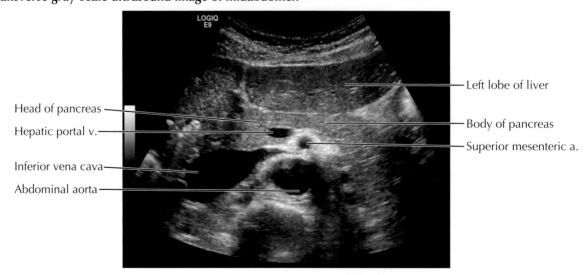

Head of pancreas

Hepatic portal v.

Inferior vena cava

Abdominal aorta

Left lobe of liver

Body of pancreas

Superior mesenteric a.

Plate 294

Stomach and Intestines

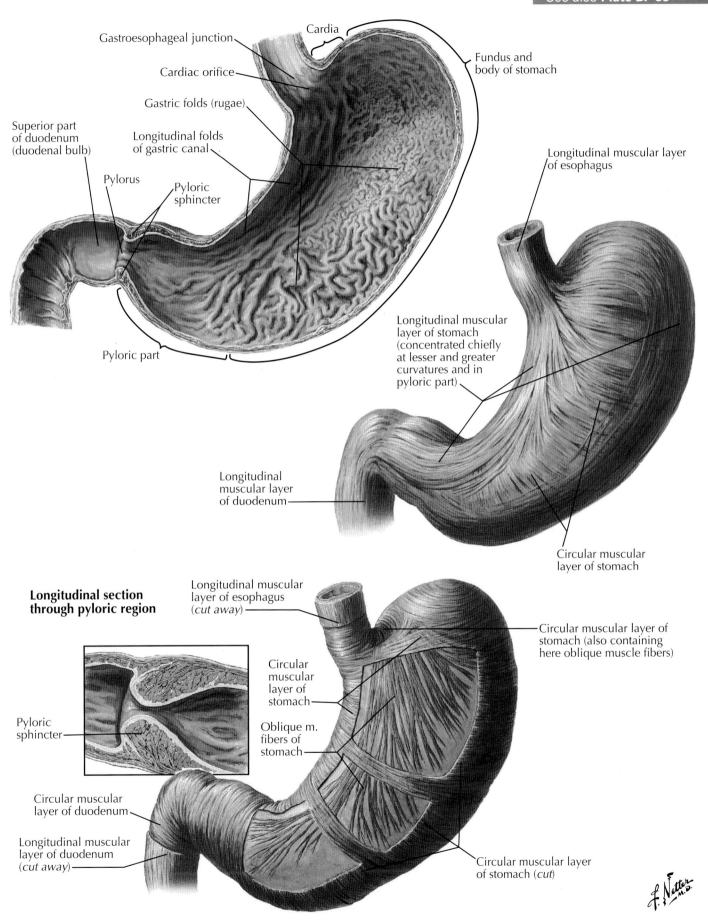

Gastroesophageal junction

Cardia

Cardiac orifice

Fundus and body of stomach

Gastric folds (rugae)

Superior part of duodenum (duodenal bulb)

Longitudinal folds of gastric canal

Pylorus

Pyloric sphincter

Longitudinal muscular layer of esophagus

Pyloric part

Longitudinal muscular layer of stomach (concentrated chiefly at lesser and greater curvatures and in pyloric part)

Longitudinal muscular layer of duodenum

Circular muscular layer of stomach

Longitudinal section through pyloric region

Longitudinal muscular layer of esophagus (*cut away*)

Circular muscular layer of stomach (also containing here oblique muscle fibers)

Pyloric sphincter

Circular muscular layer of stomach

Oblique m. fibers of stomach

Circular muscular layer of duodenum

Longitudinal muscular layer of duodenum (*cut away*)

Circular muscular layer of stomach (*cut*)

f. Netter.

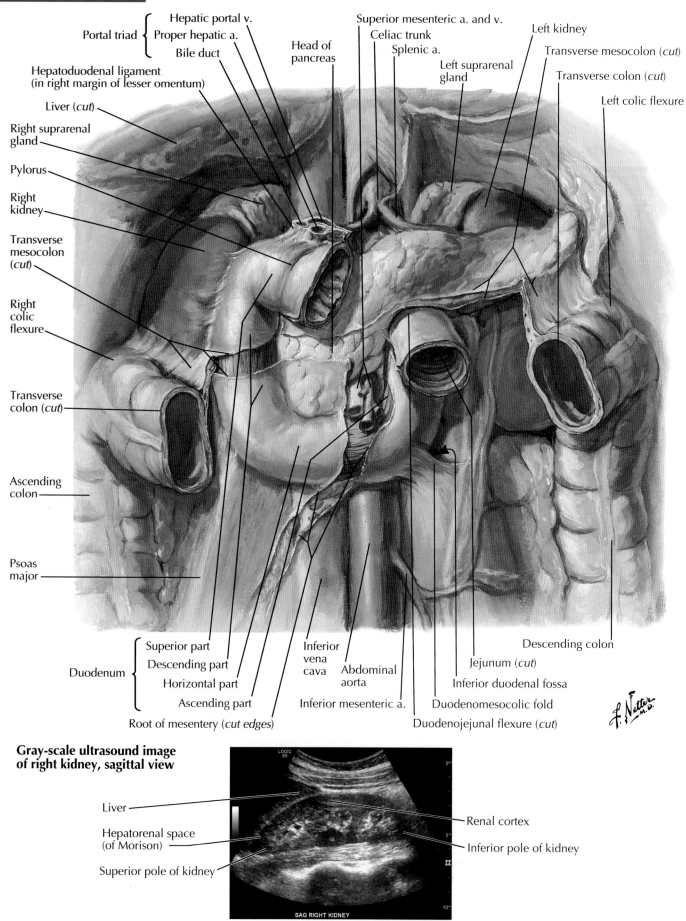

Portal triad { Hepatic portal v.
Proper hepatic a.
Bile duct

Hepatoduodenal ligament
(in right margin of lesser omentum)

Liver (*cut*)

Right suprarenal
gland

Pylorus

Right
kidney

Transverse
mesocolon
(*cut*)

Right
colic
flexure

Transverse
colon (*cut*)

Ascending
colon

Psoas
major

Head of
pancreas

Superior mesenteric a. and v.

Celiac trunk

Splenic a.

Left kidney

Left suprarenal
gland

Transverse mesocolon (*cut*)

Transverse colon (*cut*)

Left colic flexure

Duodenum { Superior part
Descending part
Horizontal part
Ascending part

Root of mesentery (*cut edges*)

Inferior
vena
cava

Abdominal
aorta

Inferior mesenteric a.

Jejunum (*cut*)

Inferior duodenal fossa

Duodenomesocolic fold

Duodenojejunal flexure (*cut*)

Descending colon

**Gray-scale ultrasound image
of right kidney, sagittal view**

Liver

Hepatorenal space
(of Morison)

Superior pole of kidney

Renal cortex

Inferior pole of kidney

Plate 296

Stomach and Intestines

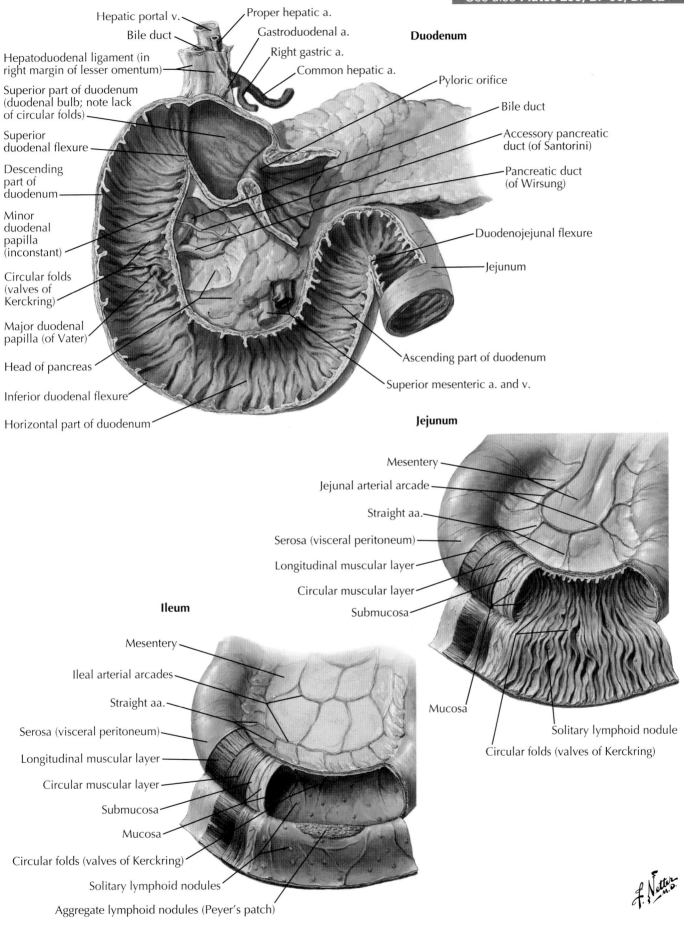

Duodenum

Hepatic portal v.

Bile duct

Hepatoduodenal ligament (in right margin of lesser omentum)

Superior part of duodenum (duodenal bulb; note lack of circular folds)

Superior duodenal flexure

Descending part of duodenum

Minor duodenal papilla (inconstant)

Circular folds (valves of Kerckring)

Major duodenal papilla (of Vater)

Head of pancreas

Inferior duodenal flexure

Horizontal part of duodenum

Proper hepatic a.

Gastroduodenal a.

Right gastric a.

Common hepatic a.

Pyloric orifice

Bile duct

Accessory pancreatic duct (of Santorini)

Pancreatic duct (of Wirsung)

Duodenojejunal flexure

Jejunum

Ascending part of duodenum

Superior mesenteric a. and v.

Jejunum

Mesentery

Jejunal arterial arcade

Straight aa.

Serosa (visceral peritoneum)

Longitudinal muscular layer

Circular muscular layer

Submucosa

Mucosa

Solitary lymphoid nodule

Circular folds (valves of Kerckring)

Ileum

Mesentery

Ileal arterial arcades

Straight aa.

Serosa (visceral peritoneum)

Longitudinal muscular layer

Circular muscular layer

Submucosa

Mucosa

Circular folds (valves of Kerckring)

Solitary lymphoid nodules

Aggregate lymphoid nodules (Peyer's patch)

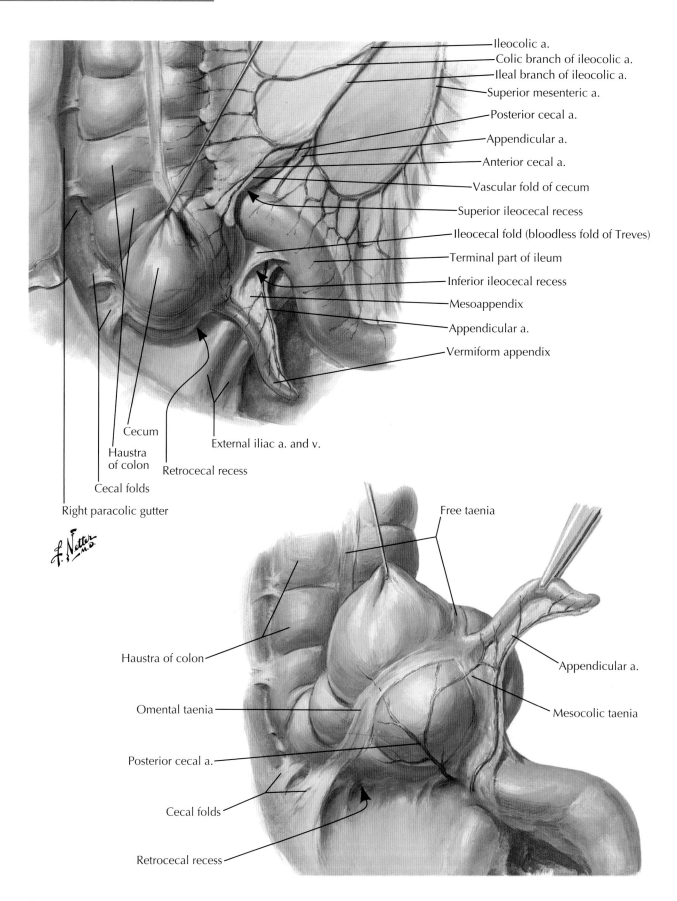

Ileocolic a.
Colic branch of ileocolic a.
Ileal branch of ileocolic a.
Superior mesenteric a.
Posterior cecal a.
Appendicular a.
Anterior cecal a.
Vascular fold of cecum
Superior ileocecal recess
Ileocecal fold (bloodless fold of Treves)
Terminal part of ileum
Inferior ileocecal recess
Mesoappendix
Appendicular a.
Vermiform appendix

Cecum
Haustra of colon
External iliac a. and v.
Retrocecal recess
Cecal folds
Right paracolic gutter

Free taenia
Appendicular a.
Mesocolic taenia
Haustra of colon
Omental taenia
Posterior cecal a.
Cecal folds
Retrocecal recess

Plate 298

Stomach and Intestines

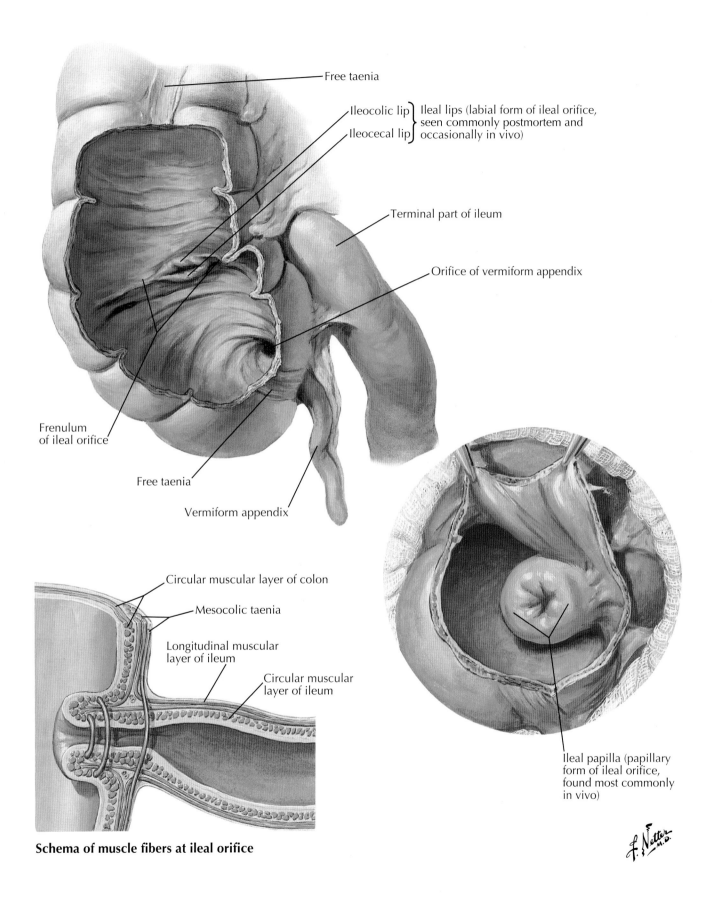

Free taenia

Ileocolic lip ⎫
⎬ Ileal lips (labial form of ileal orifice,
Ileocecal lip ⎭ seen commonly postmortem and
occasionally in vivo)

Terminal part of ileum

Orifice of vermiform appendix

Frenulum
of ileal orifice

Free taenia

Vermiform appendix

Circular muscular layer of colon

Mesocolic taenia

Longitudinal muscular
layer of ileum

Circular muscular
layer of ileum

Ileal papilla (papillary
form of ileal orifice,
found most commonly
in vivo)

Schema of muscle fibers at ileal orifice

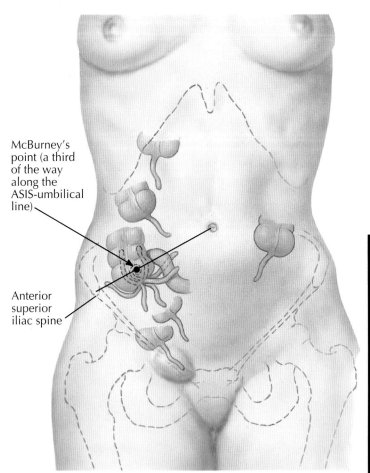

McBurney's point (a third of the way along the ASIS-umbilical line)

Anterior superior iliac spine

Variations in position of vermiform appendix

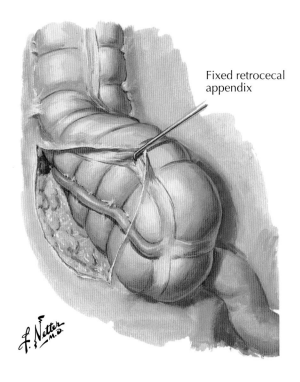

Fixed retrocecal appendix

Coronal CT image with oral and intravenous contrast

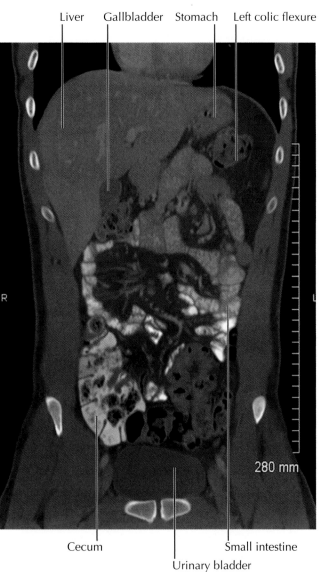

Liver Gallbladder Stomach Left colic flexure

R

L

280 mm

Cecum

Urinary bladder

Small intestine

Plate 300

Stomach and Intestines

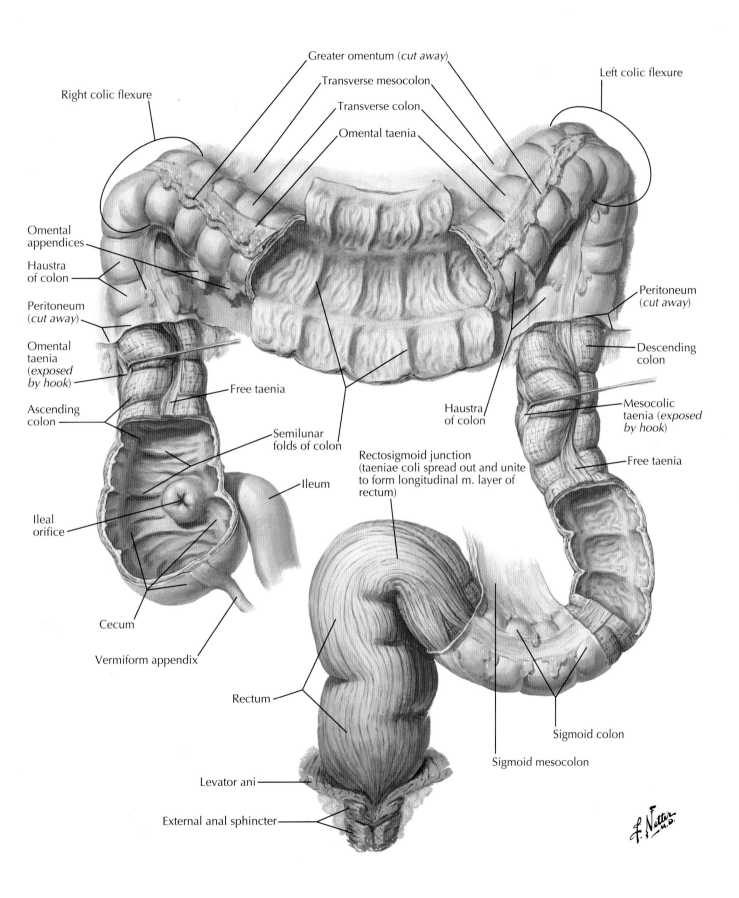

Right colic flexure

Greater omentum (*cut away*)

Transverse mesocolon

Transverse colon

Omental taenia

Left colic flexure

Omental appendices

Haustra of colon

Peritoneum (*cut away*)

Omental taenia (*exposed by hook*)

Ascending colon

Free taenia

Semilunar folds of colon

Ileal orifice

Ileum

Cecum

Vermiform appendix

Rectum

Levator ani

External anal sphincter

Peritoneum (*cut away*)

Descending colon

Haustra of colon

Mesocolic taenia (*exposed by hook*)

Free taenia

Rectosigmoid junction (taeniae coli spread out and unite to form longitudinal m. layer of rectum)

Sigmoid colon

Sigmoid mesocolon

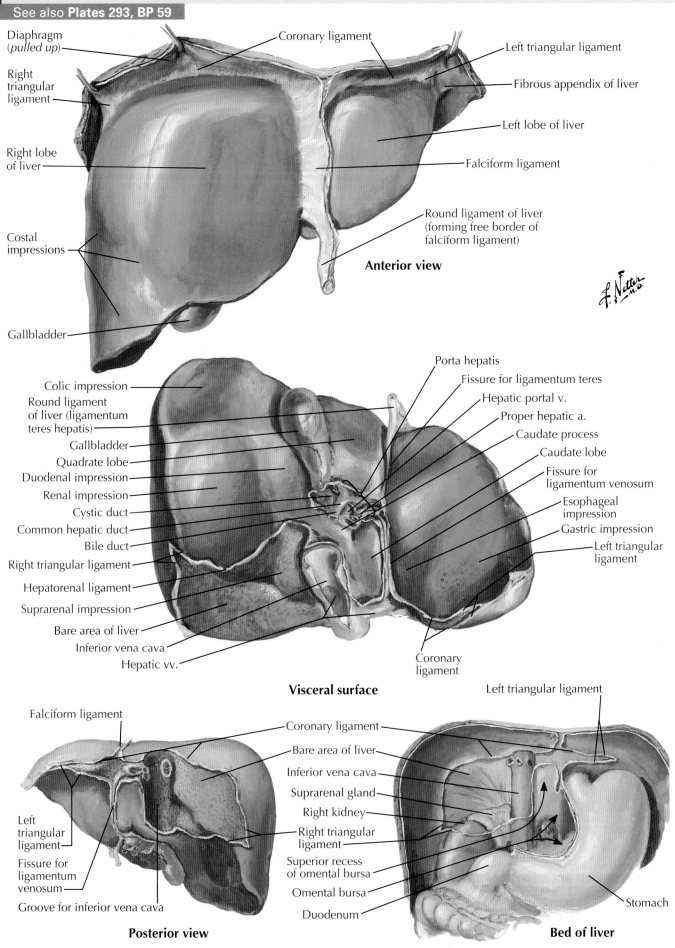

Diaphragm (*pulled up*)

Coronary ligament

Left triangular ligament

Right triangular ligament

Fibrous appendix of liver

Left lobe of liver

Right lobe of liver

Falciform ligament

Costal impressions

Round ligament of liver (forming free border of falciform ligament)

Anterior view

Gallbladder

Porta hepatis

Colic impression

Fissure for ligamentum teres

Round ligament of liver (ligamentum teres hepatis)

Hepatic portal v.

Proper hepatic a.

Gallbladder

Caudate process

Quadrate lobe

Caudate lobe

Duodenal impression

Fissure for ligamentum venosum

Renal impression

Esophageal impression

Cystic duct

Common hepatic duct

Gastric impression

Bile duct

Left triangular ligament

Right triangular ligament

Hepatorenal ligament

Suprarenal impression

Bare area of liver

Inferior vena cava

Hepatic vv.

Coronary ligament

Visceral surface

Left triangular ligament

Falciform ligament

Coronary ligament

Bare area of liver

Inferior vena cava

Suprarenal gland

Right kidney

Left triangular ligament

Right triangular ligament

Fissure for ligamentum venosum

Superior recess of omental bursa

Omental bursa

Groove for inferior vena cava

Duodenum

Stomach

Posterior view

Bed of liver

Plate 302

Liver, Gallbladder, Pancreas, and Spleen

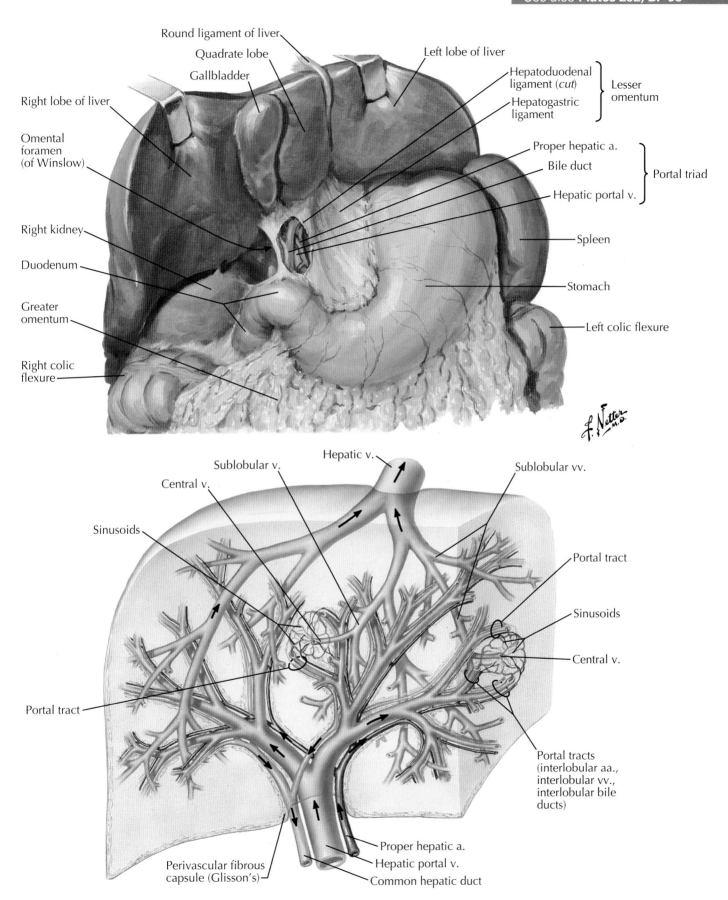

Round ligament of liver

Quadrate lobe

Gallbladder

Left lobe of liver

Hepatoduodenal ligament (*cut*)

Hepatogastric ligament

Lesser omentum

Right lobe of liver

Omental foramen (of Winslow)

Proper hepatic a.

Bile duct

Hepatic portal v.

Portal triad

Right kidney

Spleen

Duodenum

Stomach

Greater omentum

Left colic flexure

Right colic flexure

Hepatic v.

Sublobular v.

Sublobular vv.

Central v.

Sinusoids

Portal tract

Sinusoids

Central v.

Portal tract

Portal tracts (interlobular aa., interlobular vv., interlobular bile ducts)

Proper hepatic a.

Hepatic portal v.

Perivascular fibrous capsule (Glisson's)

Common hepatic duct

Liver, Gallbladder, Pancreas, and Spleen

Plate 303

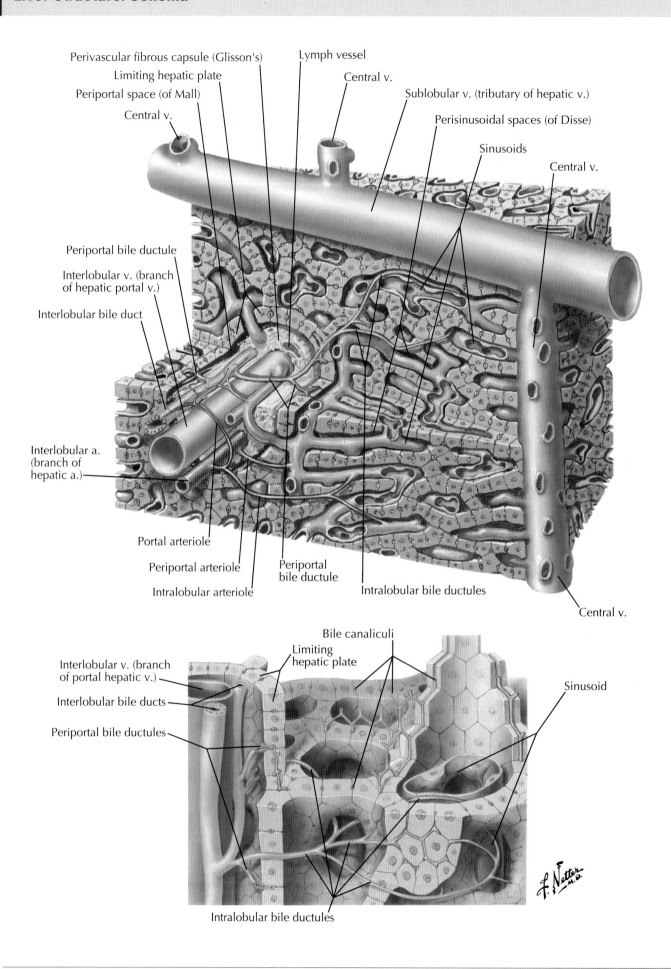

Perivascular fibrous capsule (Glisson's)

Limiting hepatic plate

Periportal space (of Mall)

Central v.

Lymph vessel

Central v.

Sublobular v. (tributary of hepatic v.)

Perisinusoidal spaces (of Disse)

Sinusoids

Central v.

Periportal bile ductule

Interlobular v. (branch of hepatic portal v.)

Interlobular bile duct

Interlobular a. (branch of hepatic a.)

Portal arteriole

Periportal arteriole

Intralobular arteriole

Periportal bile ductule

Intralobular bile ductules

Central v.

Bile canaliculi

Limiting hepatic plate

Interlobular v. (branch of portal hepatic v.)

Interlobular bile ducts

Periportal bile ductules

Sinusoid

Intralobular bile ductules

Plate 304

Liver, Gallbladder, Pancreas, and Spleen

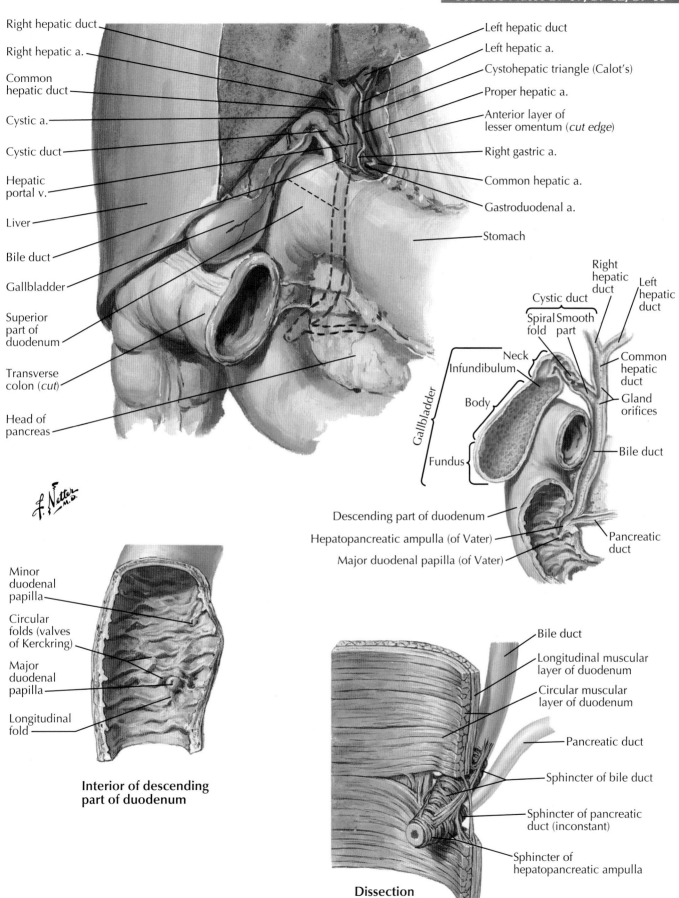

Right hepatic duct

Right hepatic a.

Common hepatic duct

Cystic a.

Cystic duct

Hepatic portal v.

Liver

Bile duct

Gallbladder

Superior part of duodenum

Transverse colon (cut)

Head of pancreas

Left hepatic duct

Left hepatic a.

Cystohepatic triangle (Calot's)

Proper hepatic a.

Anterior layer of lesser omentum (cut edge)

Right gastric a.

Common hepatic a.

Gastroduodenal a.

Stomach

Right hepatic duct

Left hepatic duct

Cystic duct

Spiral fold

Smooth part

Neck

Infundibulum

Body

Common hepatic duct

Gland orifices

Bile duct

Gallbladder

Fundus

Descending part of duodenum

Hepatopancreatic ampulla (of Vater)

Major duodenal papilla (of Vater)

Pancreatic duct

Minor duodenal papilla

Circular folds (valves of Kerckring)

Major duodenal papilla

Longitudinal fold

Interior of descending part of duodenum

Bile duct

Longitudinal muscular layer of duodenum

Circular muscular layer of duodenum

Pancreatic duct

Sphincter of bile duct

Sphincter of pancreatic duct (inconstant)

Sphincter of hepatopancreatic ampulla

Dissection

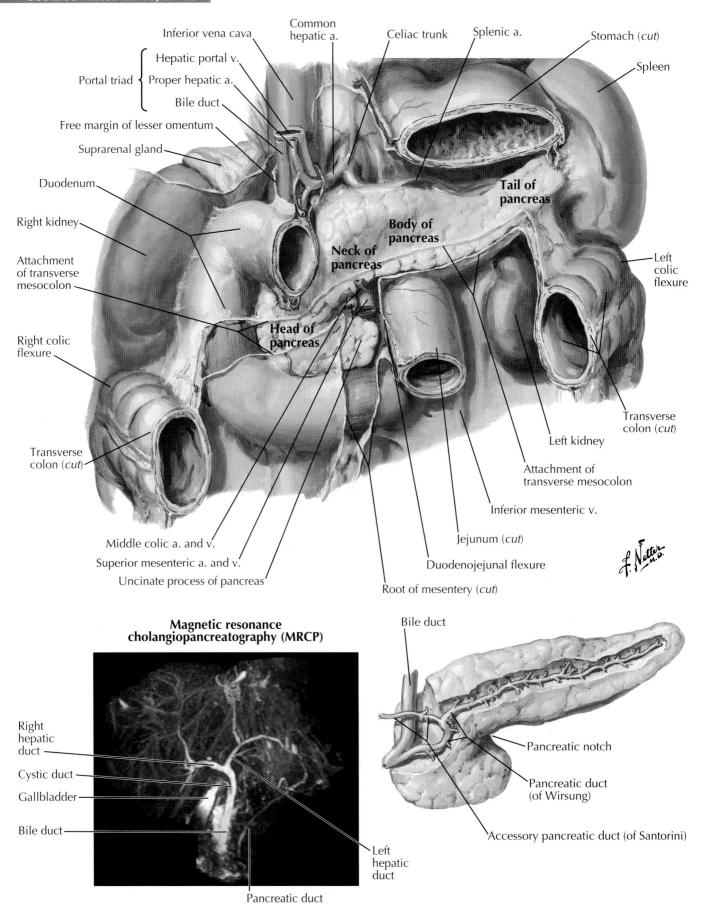

Inferior vena cava

Common hepatic a.

Celiac trunk

Splenic a.

Stomach (*cut*)

Spleen

Portal triad
- Hepatic portal v.
- Proper hepatic a.
- Bile duct

Free margin of lesser omentum

Suprarenal gland

Duodenum

Right kidney

Attachment of transverse mesocolon

Right colic flexure

Tail of pancreas

Body of pancreas

Neck of pancreas

Left colic flexure

Head of pancreas

Left kidney

Transverse colon (*cut*)

Attachment of transverse mesocolon

Inferior mesenteric v.

Middle colic a. and v.

Superior mesenteric a. and v.

Uncinate process of pancreas

Jejunum (*cut*)

Duodenojejunal flexure

Root of mesentery (*cut*)

Magnetic resonance cholangiopancreatography (MRCP)

Bile duct

Right hepatic duct

Cystic duct

Gallbladder

Bile duct

Pancreatic notch

Pancreatic duct (of Wirsung)

Accessory pancreatic duct (of Santorini)

Left hepatic duct

Pancreatic duct

Plate 306

Liver, Gallbladder, Pancreas, and Spleen

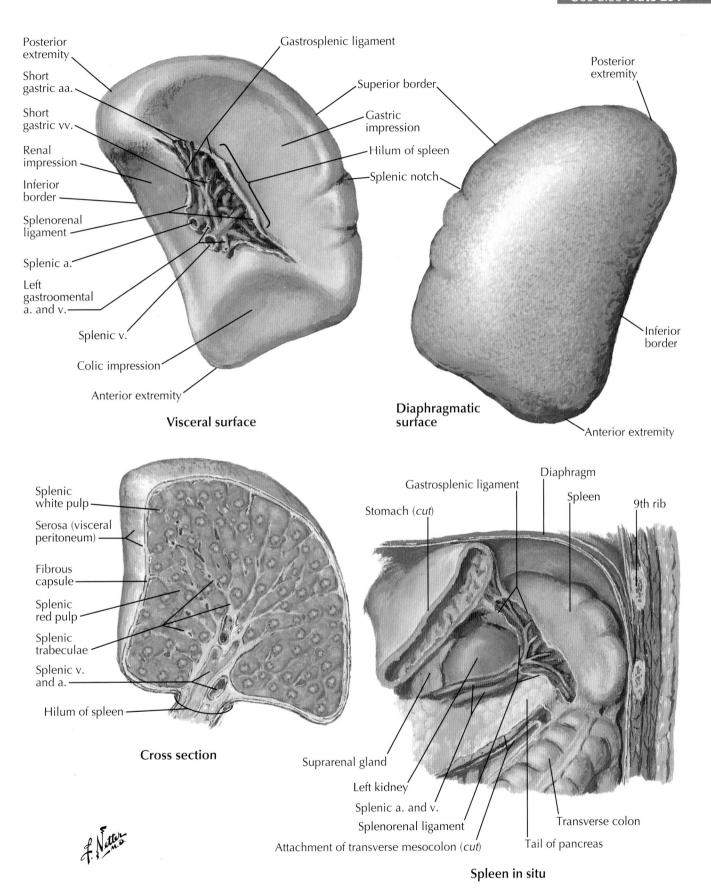

Posterior extremity

Gastrosplenic ligament

Short gastric aa.

Short gastric vv.

Renal impression

Inferior border

Splenorenal ligament

Splenic a.

Left gastroomental a. and v.

Splenic v.

Colic impression

Anterior extremity

Superior border

Gastric impression

Hilum of spleen

Splenic notch

Visceral surface

Posterior extremity

Inferior border

Anterior extremity

Diaphragmatic surface

Splenic white pulp

Serosa (visceral peritoneum)

Fibrous capsule

Splenic red pulp

Splenic trabeculae

Splenic v. and a.

Hilum of spleen

Cross section

Gastrosplenic ligament

Stomach (*cut*)

Diaphragm

Spleen

9th rib

Suprarenal gland

Left kidney

Splenic a. and v.

Splenorenal ligament

Attachment of transverse mesocolon (*cut*)

Transverse colon

Tail of pancreas

Spleen in situ

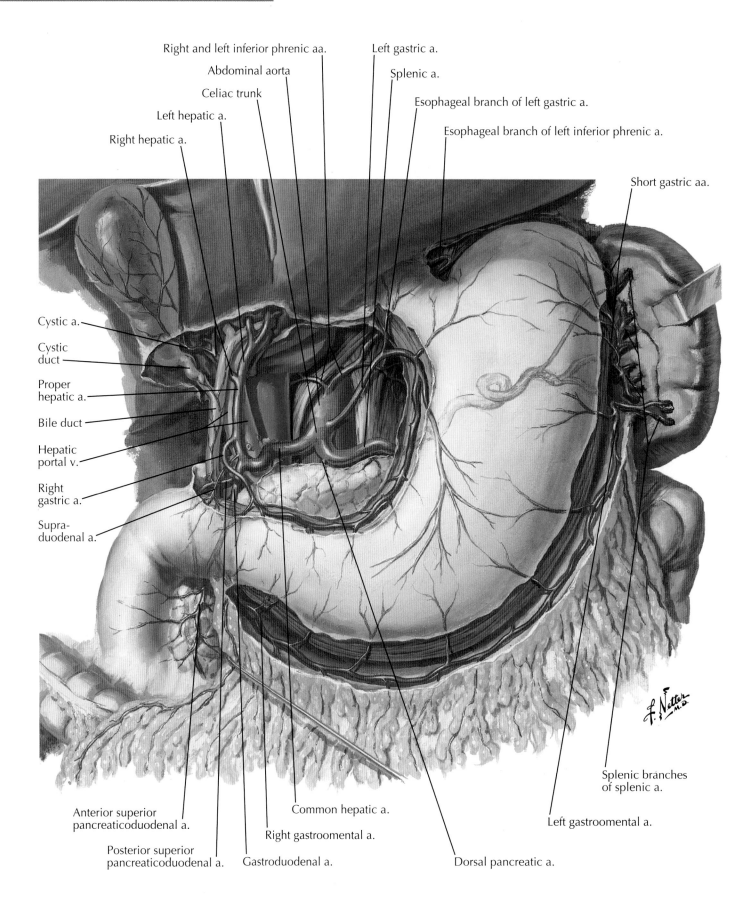

Right and left inferior phrenic aa.

Abdominal aorta

Celiac trunk

Left hepatic a.

Right hepatic a.

Left gastric a.

Splenic a.

Esophageal branch of left gastric a.

Esophageal branch of left inferior phrenic a.

Short gastric aa.

Cystic a.

Cystic duct

Proper hepatic a.

Bile duct

Hepatic portal v.

Right gastric a.

Supra-duodenal a.

Splenic branches of splenic a.

Left gastroomental a.

Anterior superior pancreaticoduodenal a.

Posterior superior pancreaticoduodenal a.

Gastroduodenal a.

Right gastroomental a.

Common hepatic a.

Dorsal pancreatic a.

Plate 308

Visceral Vasculature

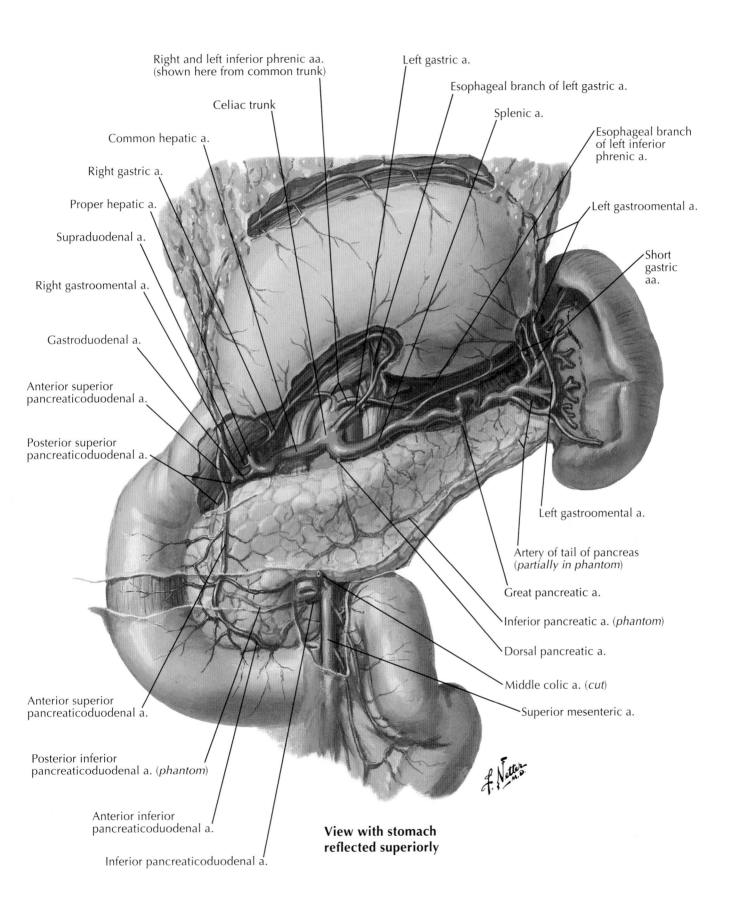

Right and left inferior phrenic aa.
(shown here from common trunk)

Left gastric a.

Esophageal branch of left gastric a.

Celiac trunk

Splenic a.

Common hepatic a.

Esophageal branch
of left inferior
phrenic a.

Right gastric a.

Proper hepatic a.

Left gastroomental a.

Supraduodenal a.

Short
gastric
aa.

Right gastroomental a.

Gastroduodenal a.

Anterior superior
pancreaticoduodenal a.

Posterior superior
pancreaticoduodenal a.

Left gastroomental a.

Artery of tail of pancreas
(*partially in phantom*)

Great pancreatic a.

Inferior pancreatic a. (*phantom*)

Dorsal pancreatic a.

Anterior superior
pancreaticoduodenal a.

Middle colic a. (*cut*)

Superior mesenteric a.

Posterior inferior
pancreaticoduodenal a. (*phantom*)

Anterior inferior
pancreaticoduodenal a.

**View with stomach
reflected superiorly**

Inferior pancreaticoduodenal a.

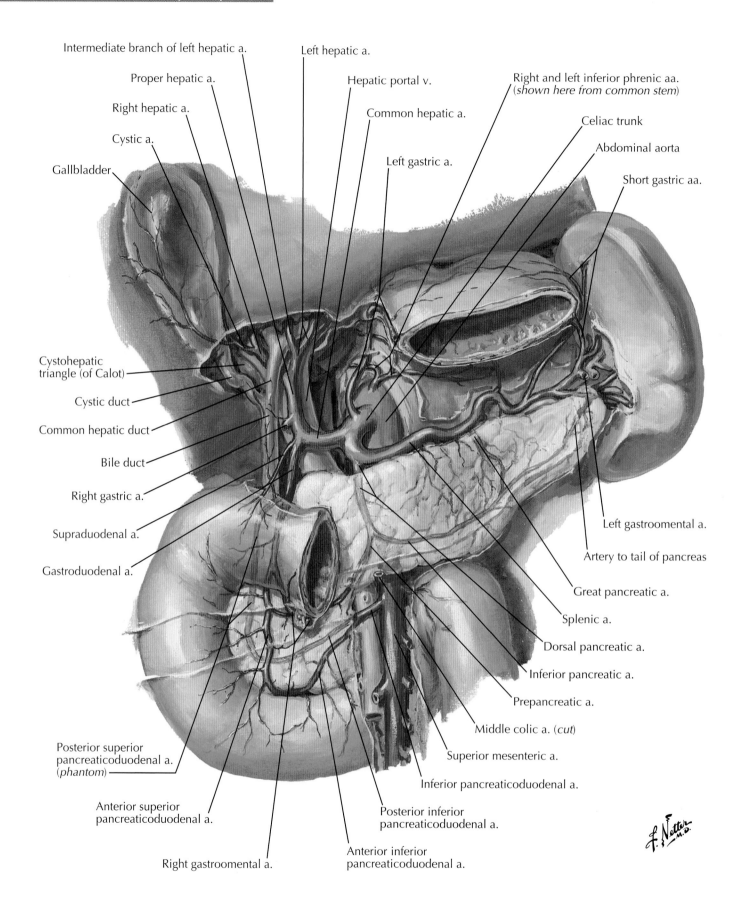

Intermediate branch of left hepatic a.

Proper hepatic a.

Right hepatic a.

Cystic a.

Gallbladder

Left hepatic a.

Hepatic portal v.

Common hepatic a.

Left gastric a.

Right and left inferior phrenic aa.
(*shown here from common stem*)

Celiac trunk

Abdominal aorta

Short gastric aa.

Cystohepatic triangle (of Calot)

Cystic duct

Common hepatic duct

Bile duct

Right gastric a.

Supraduodenal a.

Gastroduodenal a.

Left gastroomental a.

Artery to tail of pancreas

Great pancreatic a.

Splenic a.

Dorsal pancreatic a.

Inferior pancreatic a.

Prepancreatic a.

Middle colic a. (*cut*)

Superior mesenteric a.

Inferior pancreaticoduodenal a.

Posterior superior pancreaticoduodenal a. (*phantom*)

Anterior superior pancreaticoduodenal a.

Right gastroomental a.

Anterior inferior pancreaticoduodenal a.

Posterior inferior pancreaticoduodenal a.

Plate 310

Visceral Vasculature

3D volume-rendered CT image with intravenous contrast enhancement

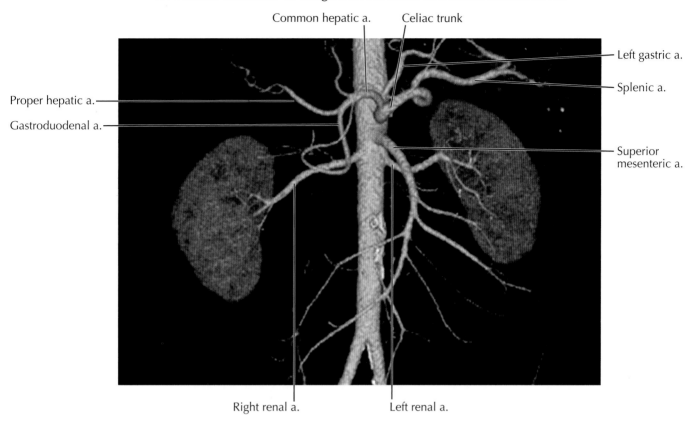

Common hepatic a.   Celiac trunk

Left gastric a.

Splenic a.

Proper hepatic a.

Gastroduodenal a.

Superior mesenteric a.

Right renal a.   Left renal a.

Selective digital subtraction angiogram, celiac trunk

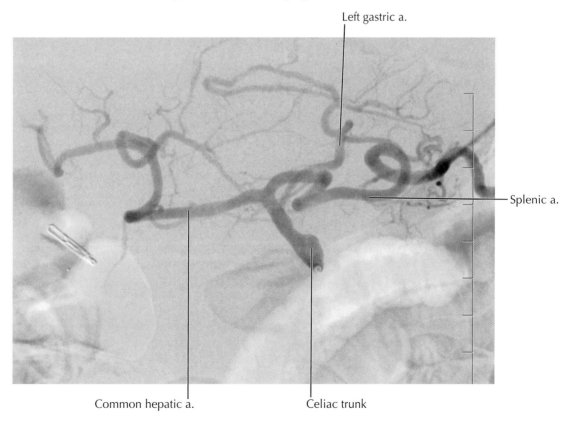

Left gastric a.

Splenic a.

Common hepatic a.   Celiac trunk

Duodenum and head of pancreas reflected to left

Bile duct

Gastroduodenal a.

Posterior superior
pancreaticoduodenal a.

Anterior superior
pancreaticoduodenal a.
(*phantom*)

Superior mesenteric a.

Inferior pancreaticoduodenal a.

Posterior inferior
pancreaticoduodenal a.

Anterior inferior
pancreaticoduodenal a.
(*partially in phantom*)

Inferior vena cava

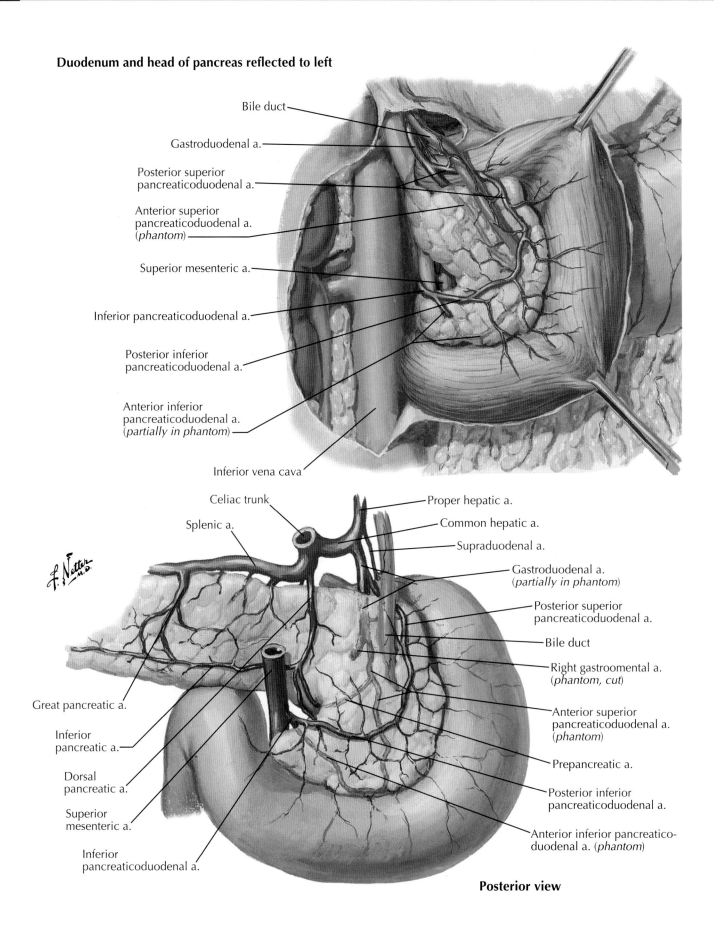

Celiac trunk

Splenic a.

Great pancreatic a.

Inferior
pancreatic a.

Dorsal
pancreatic a.

Superior
mesenteric a.

Inferior
pancreaticoduodenal a.

Proper hepatic a.

Common hepatic a.

Supraduodenal a.

Gastroduodenal a.
(*partially in phantom*)

Posterior superior
pancreaticoduodenal a.

Bile duct

Right gastroomental a.
(*phantom, cut*)

Anterior superior
pancreaticoduodenal a.
(*phantom*)

Prepancreatic a.

Posterior inferior
pancreaticoduodenal a.

Anterior inferior pancreaticoduodenal a. (*phantom*)

Posterior view

Plate 312

Visceral Vasculature

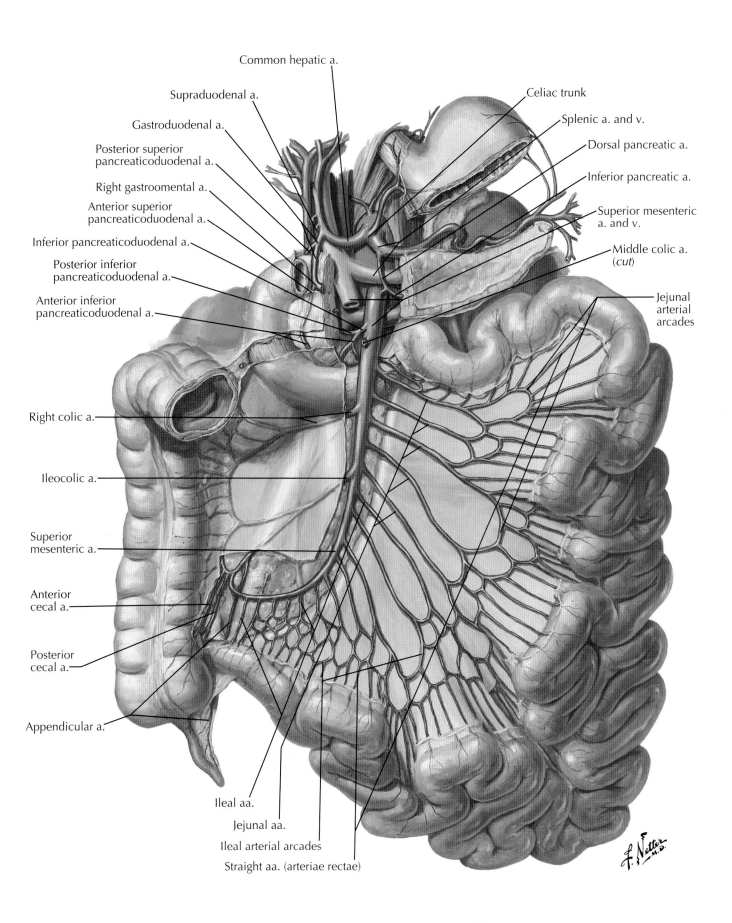

Common hepatic a.

Supraduodenal a.

Gastroduodenal a.

Posterior superior
pancreaticoduodenal a.

Right gastroomental a.

Anterior superior
pancreaticoduodenal a.

Inferior pancreaticoduodenal a.

Posterior inferior
pancreaticoduodenal a.

Anterior inferior
pancreaticoduodenal a.

Celiac trunk

Splenic a. and v.

Dorsal pancreatic a.

Inferior pancreatic a.

Superior mesenteric
a. and v.

Middle colic a.
(cut)

Jejunal
arterial
arcades

Right colic a.

Ileocolic a.

Superior
mesenteric a.

Anterior
cecal a.

Posterior
cecal a.

Appendicular a.

Ileal aa.

Jejunal aa.

Ileal arterial arcades

Straight aa. (arteriae rectae)

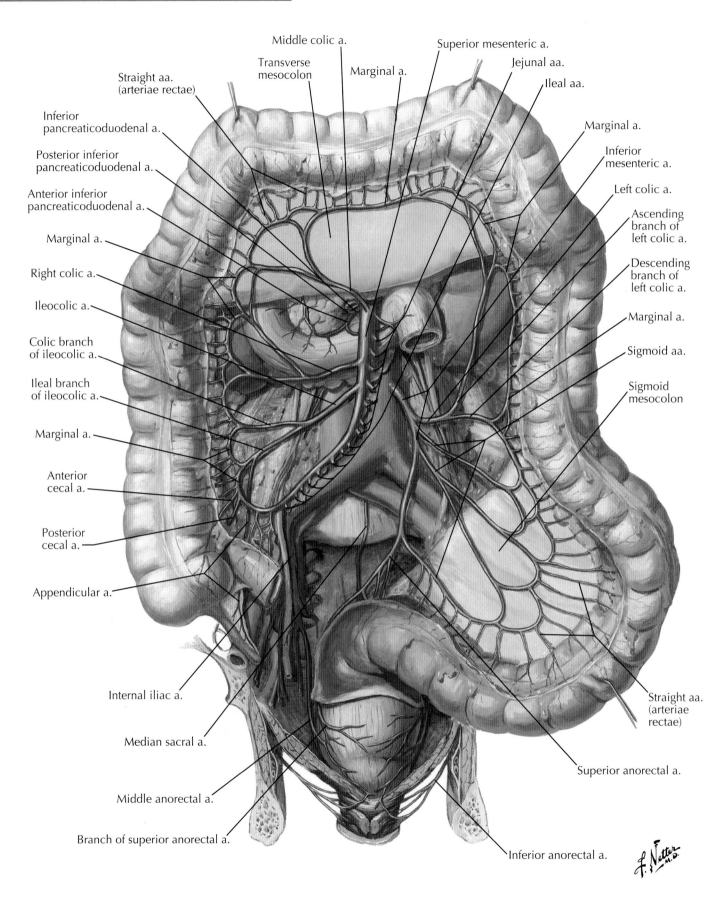

Middle colic a.

Transverse mesocolon

Marginal a.

Superior mesenteric a.

Jejunal aa.

Ileal aa.

Straight aa. (arteriae rectae)

Inferior pancreaticoduodenal a.

Posterior inferior pancreaticoduodenal a.

Anterior inferior pancreaticoduodenal a.

Marginal a.

Right colic a.

Ileocolic a.

Colic branch of ileocolic a.

Ileal branch of ileocolic a.

Marginal a.

Anterior cecal a.

Posterior cecal a.

Appendicular a.

Internal iliac a.

Median sacral a.

Middle anorectal a.

Branch of superior anorectal a.

Marginal a.

Inferior mesenteric a.

Left colic a.

Ascending branch of left colic a.

Descending branch of left colic a.

Marginal a.

Sigmoid aa.

Sigmoid mesocolon

Straight aa. (arteriae rectae)

Superior anorectal a.

Inferior anorectal a.

Plate 314

Visceral Vasculature

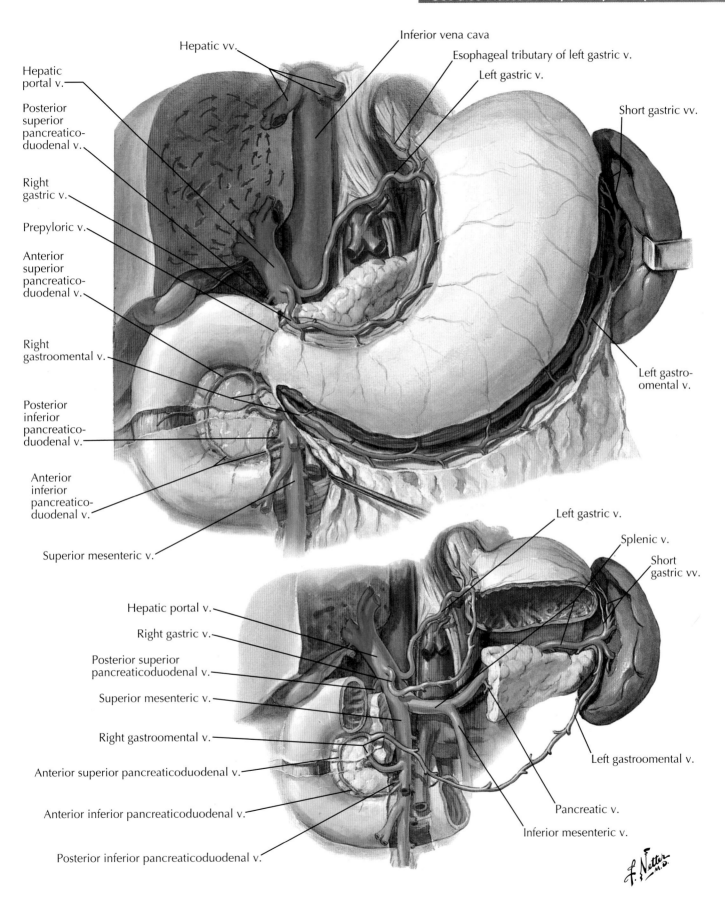

Hepatic vv.

Inferior vena cava

Esophageal tributary of left gastric v.

Left gastric v.

Hepatic portal v.

Short gastric vv.

Posterior superior pancreatico-duodenal v.

Right gastric v.

Prepyloric v.

Anterior superior pancreatico-duodenal v.

Right gastroomental v.

Posterior inferior pancreatico-duodenal v.

Anterior inferior pancreatico-duodenal v.

Superior mesenteric v.

Left gastro-omental v.

Hepatic portal v.

Right gastric v.

Posterior superior pancreaticoduodenal v.

Superior mesenteric v.

Right gastroomental v.

Anterior superior pancreaticoduodenal v.

Anterior inferior pancreaticoduodenal v.

Posterior inferior pancreaticoduodenal v.

Left gastric v.

Splenic v.

Short gastric vv.

Left gastroomental v.

Pancreatic v.

Inferior mesenteric v.

f. Netter
M.D.

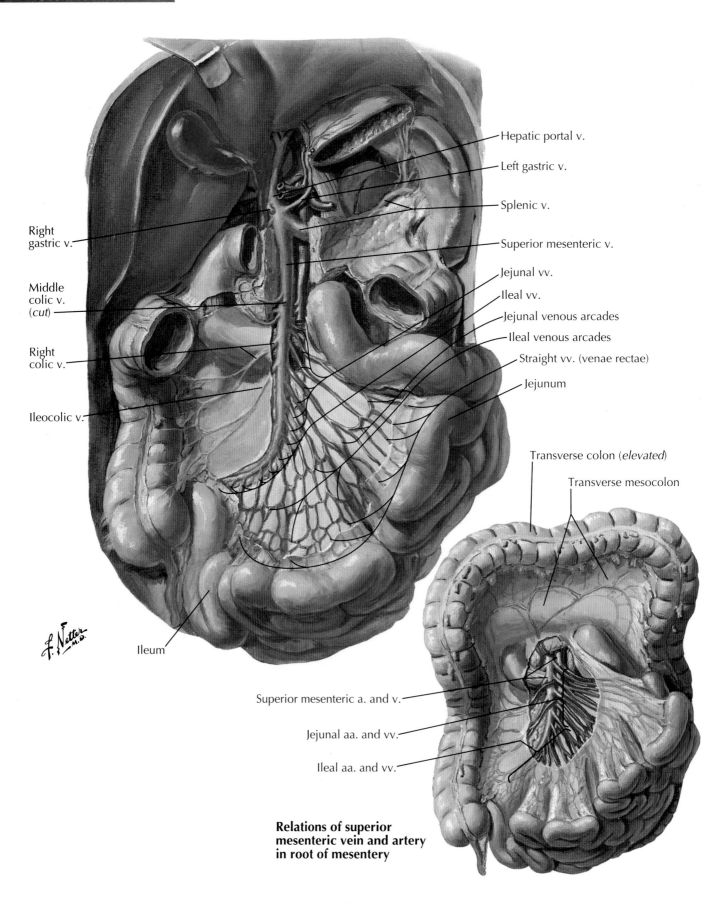

Hepatic portal v.

Left gastric v.

Splenic v.

Superior mesenteric v.

Jejunal vv.

Ileal vv.

Jejunal venous arcades

Ileal venous arcades

Straight vv. (venae rectae)

Jejunum

Right
gastric v.

Middle
colic v.
(cut)

Right
colic v.

Ileocolic v.

Transverse colon (elevated)

Transverse mesocolon

Ileum

Superior mesenteric a. and v.

Jejunal aa. and vv.

Ileal aa. and vv.

**Relations of superior
mesenteric vein and artery
in root of mesentery**

Plate 316

Visceral Vasculature

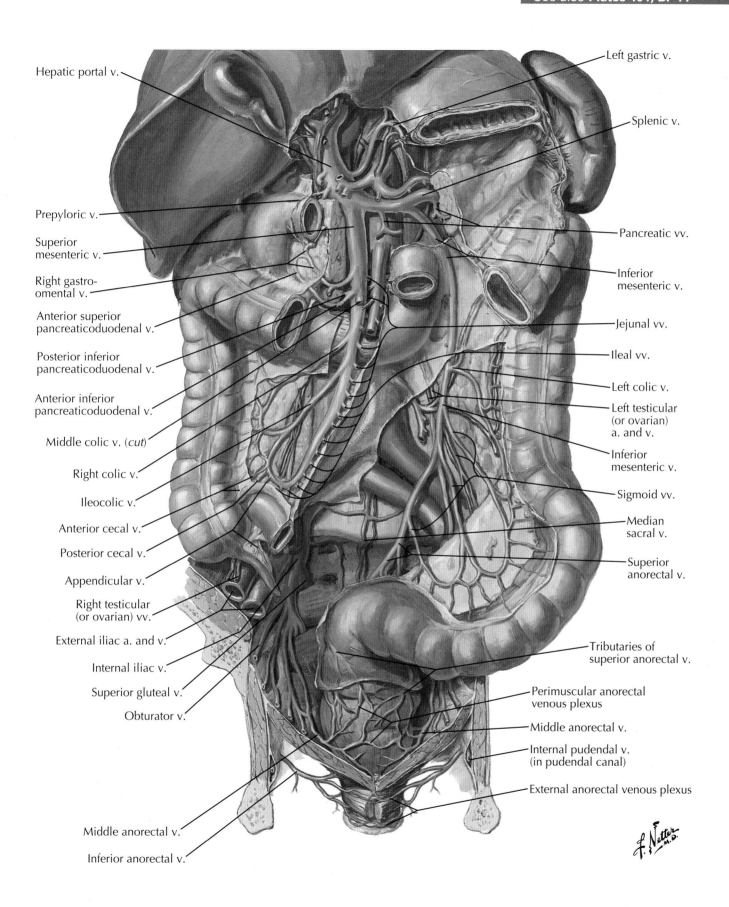

Hepatic portal v.

Prepyloric v.

Superior
mesenteric v.

Right gastro-
omental v.

Anterior superior
pancreaticoduodenal v.

Posterior inferior
pancreaticoduodenal v.

Anterior inferior
pancreaticoduodenal v.

Middle colic v. (*cut*)

Right colic v.

Ileocolic v.

Anterior cecal v.

Posterior cecal v.

Appendicular v.

Right testicular
(or ovarian) vv.

External iliac a. and v.

Internal iliac v.

Superior gluteal v.

Obturator v.

Middle anorectal v.

Inferior anorectal v.

Left gastric v.

Splenic v.

Pancreatic vv.

Inferior
mesenteric v.

Jejunal vv.

Ileal vv.

Left colic v.

Left testicular
(or ovarian)
a. and v.

Inferior
mesenteric v.

Sigmoid vv.

Median
sacral v.

Superior
anorectal v.

Tributaries of
superior anorectal v.

Perimuscular anorectal
venous plexus

Middle anorectal v.

Internal pudendal v.
(in pudendal canal)

External anorectal venous plexus

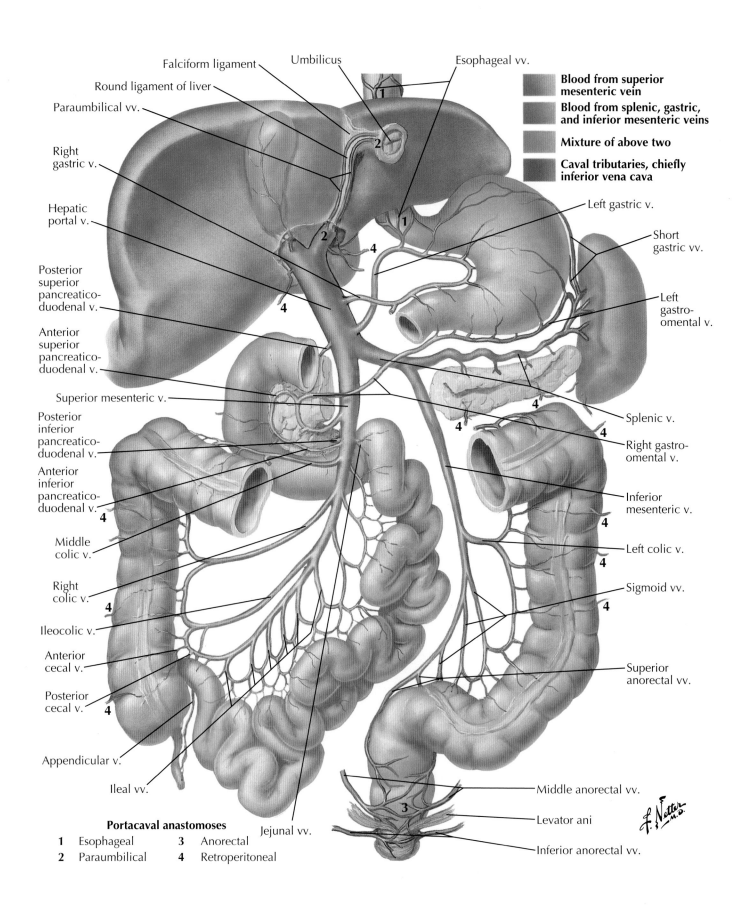

Falciform ligament

Round ligament of liver

Paraumbilical vv.

Umbilicus

Esophageal vv.

Right gastric v.

Hepatic portal v.

Posterior superior pancreatico-duodenal v.

Anterior superior pancreatico-duodenal v.

Superior mesenteric v.

Posterior inferior pancreatico-duodenal v.

Anterior inferior pancreatico-duodenal v.

Middle colic v.

Right colic v.

Ileocolic v.

Anterior cecal v.

Posterior cecal v.

Appendicular v.

Ileal vv.

Jejunal vv.

Blood from superior mesenteric vein

Blood from splenic, gastric, and inferior mesenteric veins

Mixture of above two

Caval tributaries, chiefly inferior vena cava

Left gastric v.

Short gastric vv.

Left gastro-omental v.

Splenic v.

Right gastro-omental v.

Inferior mesenteric v.

Left colic v.

Sigmoid vv.

Superior anorectal vv.

Middle anorectal vv.

Levator ani

Inferior anorectal vv.

Portacaval anastomoses

1	Esophageal	3	Anorectal
2	Paraumbilical	4	Retroperitoneal

Plate 318

Visceral Vasculature

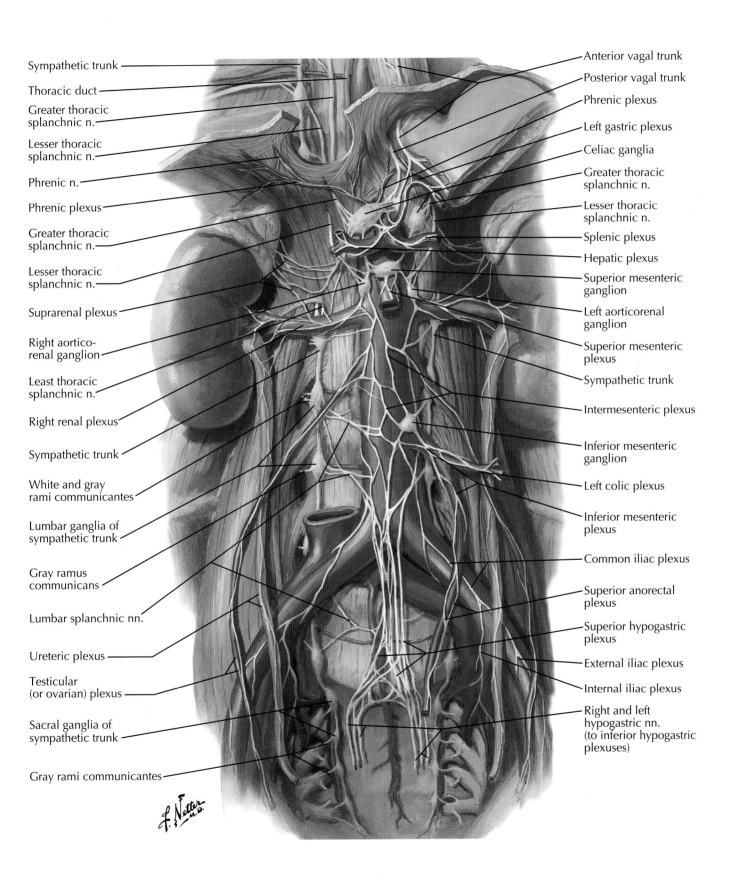

Sympathetic trunk

Thoracic duct

Greater thoracic splanchnic n.

Lesser thoracic splanchnic n.

Phrenic n.

Phrenic plexus

Greater thoracic splanchnic n.

Lesser thoracic splanchnic n.

Suprarenal plexus

Right aortico-renal ganglion

Least thoracic splanchnic n.

Right renal plexus

Sympathetic trunk

White and gray rami communicantes

Lumbar ganglia of sympathetic trunk

Gray ramus communicans

Lumbar splanchnic nn.

Ureteric plexus

Testicular (or ovarian) plexus

Sacral ganglia of sympathetic trunk

Gray rami communicantes

Anterior vagal trunk

Posterior vagal trunk

Phrenic plexus

Left gastric plexus

Celiac ganglia

Greater thoracic splanchnic n.

Lesser thoracic splanchnic n.

Splenic plexus

Hepatic plexus

Superior mesenteric ganglion

Left aorticorenal ganglion

Superior mesenteric plexus

Sympathetic trunk

Intermesenteric plexus

Inferior mesenteric ganglion

Left colic plexus

Inferior mesenteric plexus

Common iliac plexus

Superior anorectal plexus

Superior hypogastric plexus

External iliac plexus

Internal iliac plexus

Right and left hypogastric nn. (to inferior hypogastric plexuses)

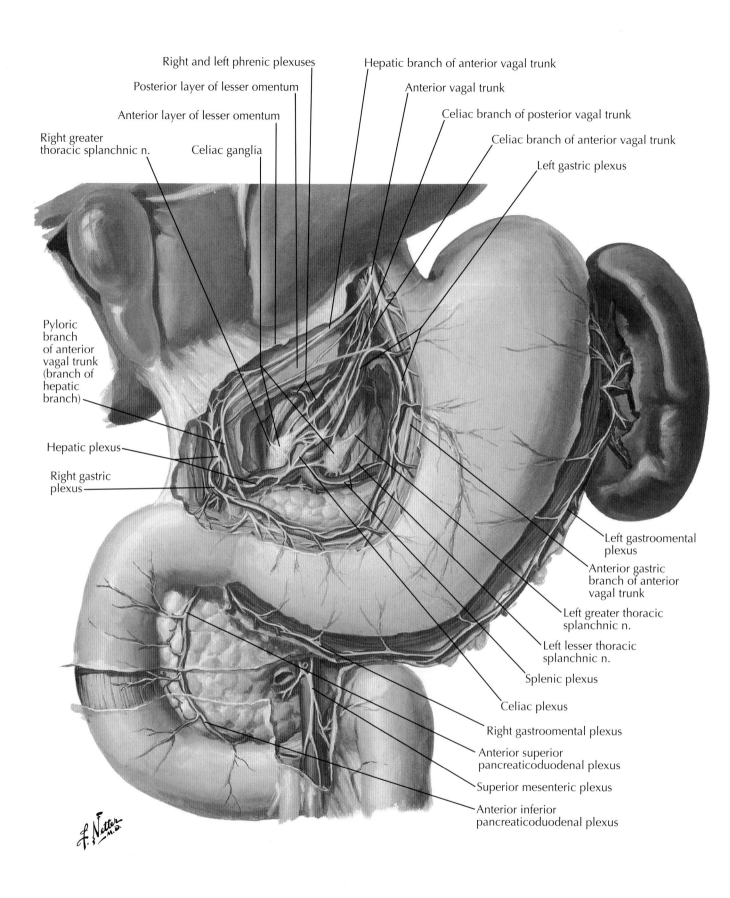

Right and left phrenic plexuses

Posterior layer of lesser omentum

Anterior layer of lesser omentum

Right greater thoracic splanchnic n.

Celiac ganglia

Hepatic branch of anterior vagal trunk

Anterior vagal trunk

Celiac branch of posterior vagal trunk

Celiac branch of anterior vagal trunk

Left gastric plexus

Pyloric branch of anterior vagal trunk (branch of hepatic branch)

Hepatic plexus

Right gastric plexus

Left gastroomental plexus

Anterior gastric branch of anterior vagal trunk

Left greater thoracic splanchnic n.

Left lesser thoracic splanchnic n.

Splenic plexus

Celiac plexus

Right gastroomental plexus

Anterior superior pancreaticoduodenal plexus

Superior mesenteric plexus

Anterior inferior pancreaticoduodenal plexus

Plate 320

Visceral Nerves and Plexuses

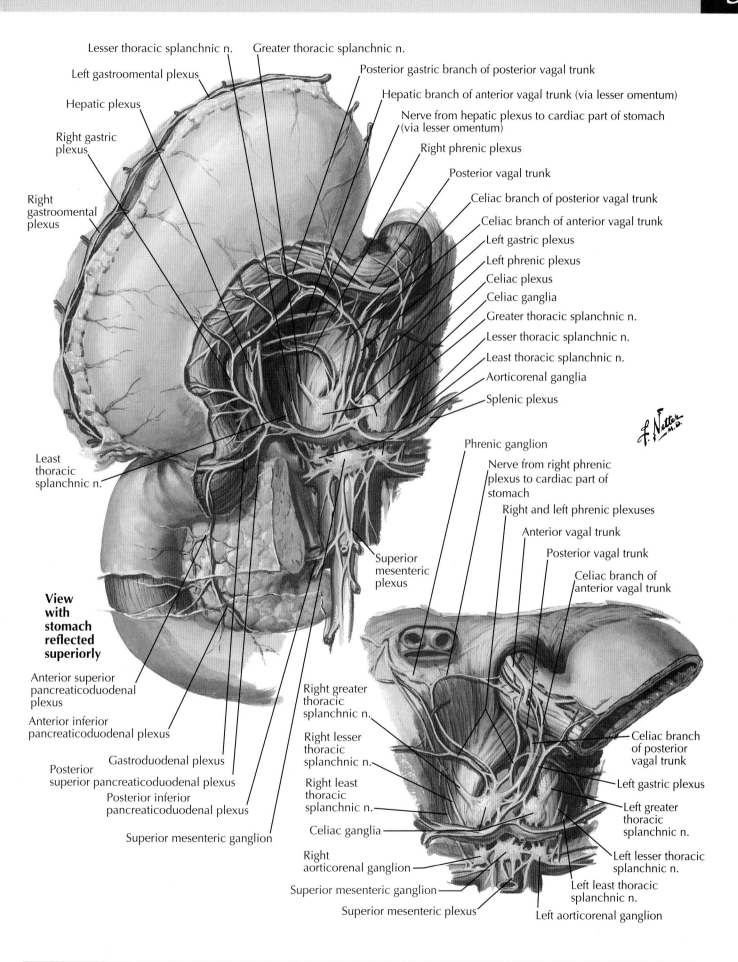

Lesser thoracic splanchnic n.

Greater thoracic splanchnic n.

Left gastroomental plexus

Posterior gastric branch of posterior vagal trunk

Hepatic plexus

Hepatic branch of anterior vagal trunk (via lesser omentum)

Right gastric plexus

Nerve from hepatic plexus to cardiac part of stomach (via lesser omentum)

Right phrenic plexus

Right gastroomental plexus

Posterior vagal trunk

Celiac branch of posterior vagal trunk

Celiac branch of anterior vagal trunk

Left gastric plexus

Left phrenic plexus

Celiac plexus

Celiac ganglia

Greater thoracic splanchnic n.

Lesser thoracic splanchnic n.

Least thoracic splanchnic n.

Aorticorenal ganglia

Splenic plexus

Least thoracic splanchnic n.

Phrenic ganglion

Nerve from right phrenic plexus to cardiac part of stomach

Right and left phrenic plexuses

Anterior vagal trunk

Posterior vagal trunk

Celiac branch of anterior vagal trunk

Superior mesenteric plexus

View with stomach reflected superiorly

Anterior superior pancreaticoduodenal plexus

Anterior inferior pancreaticoduodenal plexus

Gastroduodenal plexus

Posterior superior pancreaticoduodenal plexus

Posterior inferior pancreaticoduodenal plexus

Superior mesenteric ganglion

Right greater thoracic splanchnic n.

Right lesser thoracic splanchnic n.

Right least thoracic splanchnic n.

Celiac ganglia

Right aorticorenal ganglion

Superior mesenteric ganglion

Superior mesenteric plexus

Celiac branch of posterior vagal trunk

Left gastric plexus

Left greater thoracic splanchnic n.

Left lesser thoracic splanchnic n.

Left least thoracic splanchnic n.

Left aorticorenal ganglion

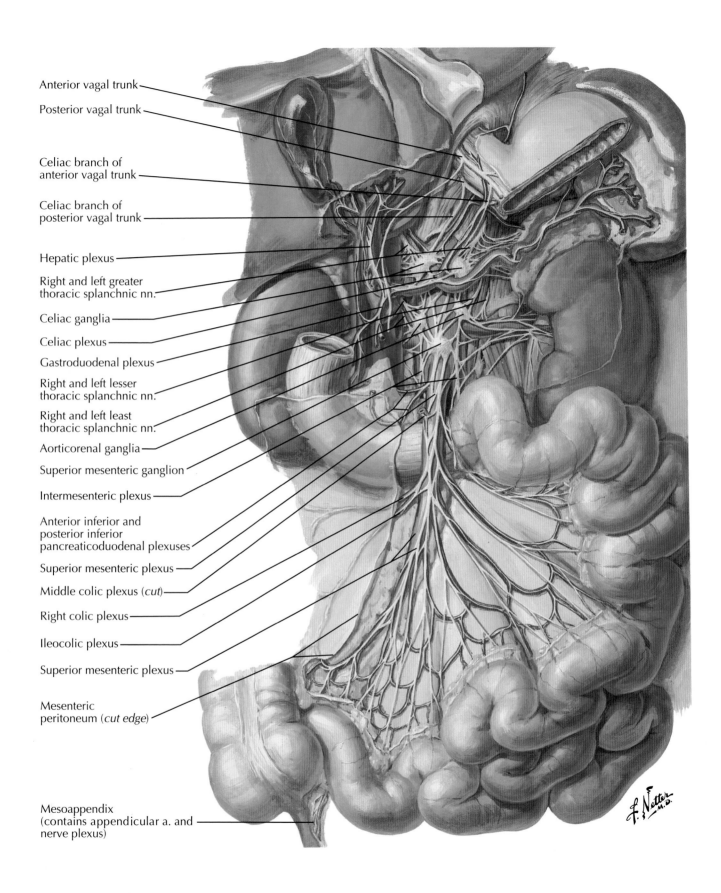

Anterior vagal trunk

Posterior vagal trunk

Celiac branch of anterior vagal trunk

Celiac branch of posterior vagal trunk

Hepatic plexus

Right and left greater thoracic splanchnic nn.

Celiac ganglia

Celiac plexus

Gastroduodenal plexus

Right and left lesser thoracic splanchnic nn.

Right and left least thoracic splanchnic nn.

Aorticorenal ganglia

Superior mesenteric ganglion

Intermesenteric plexus

Anterior inferior and posterior inferior pancreaticoduodenal plexuses

Superior mesenteric plexus

Middle colic plexus (*cut*)

Right colic plexus

Ileocolic plexus

Superior mesenteric plexus

Mesenteric peritoneum (*cut edge*)

Mesoappendix (contains appendicular a. and nerve plexus)

Plate 322

Visceral Nerves and Plexuses

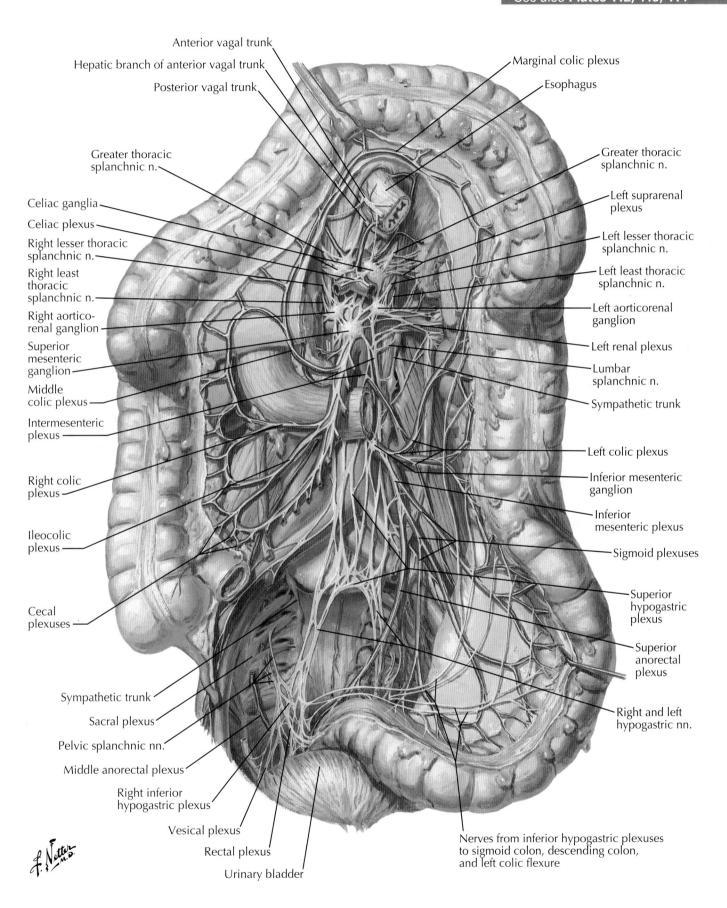

Anterior vagal trunk

Hepatic branch of anterior vagal trunk

Posterior vagal trunk

Greater thoracic splanchnic n.

Celiac ganglia

Celiac plexus

Right lesser thoracic splanchnic n.

Right least thoracic splanchnic n.

Right aortico-renal ganglion

Superior mesenteric ganglion

Middle colic plexus

Intermesenteric plexus

Right colic plexus

Ileocolic plexus

Cecal plexuses

Sympathetic trunk

Sacral plexus

Pelvic splanchnic nn.

Middle anorectal plexus

Right inferior hypogastric plexus

Vesical plexus

Rectal plexus

Urinary bladder

Marginal colic plexus

Esophagus

Greater thoracic splanchnic n.

Left suprarenal plexus

Left lesser thoracic splanchnic n.

Left least thoracic splanchnic n.

Left aorticorenal ganglion

Left renal plexus

Lumbar splanchnic n.

Sympathetic trunk

Left colic plexus

Inferior mesenteric ganglion

Inferior mesenteric plexus

Sigmoid plexuses

Superior hypogastric plexus

Superior anorectal plexus

Right and left hypogastric nn.

Nerves from inferior hypogastric plexuses to sigmoid colon, descending colon, and left colic flexure

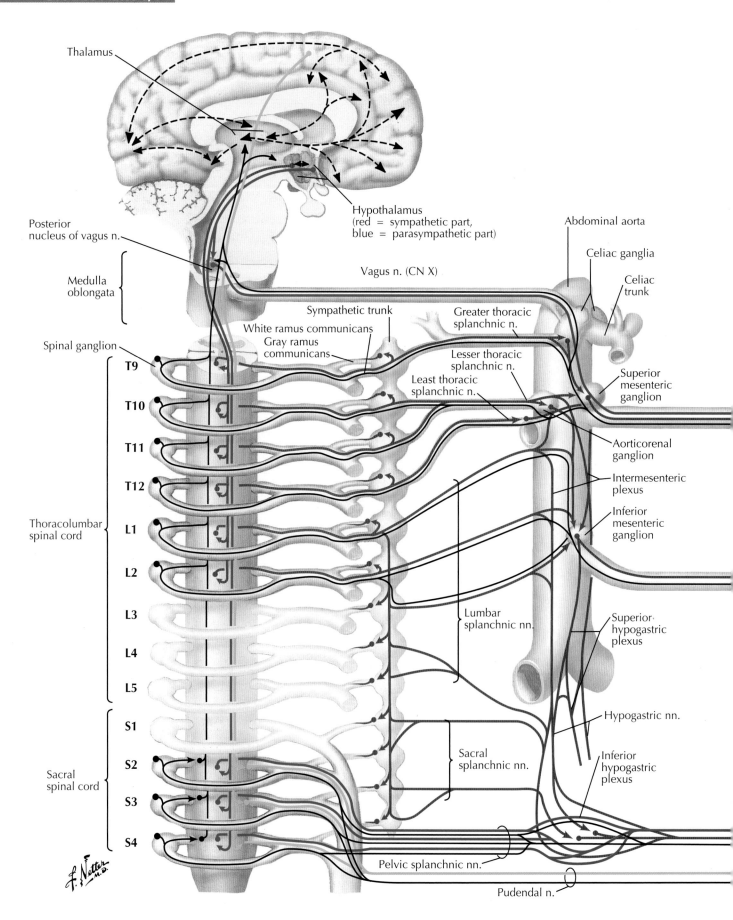

Thalamus

Hypothalamus
(red = sympathetic part,
blue = parasympathetic part)

Posterior
nucleus of vagus n.

Medulla
oblongata

Vagus n. (CN X)

Abdominal aorta

Celiac ganglia

Celiac
trunk

Spinal ganglion

Sympathetic trunk

Greater thoracic
splanchnic n.

White ramus communicans

Gray ramus
communicans

Lesser thoracic
splanchnic n.

Least thoracic
splanchnic n.

Superior
mesenteric
ganglion

T9

T10

T11

T12

L1

L2

L3

L4

L5

Thoracolumbar
spinal cord

Aorticorenal
ganglion

Intermesenteric
plexus

Inferior
mesenteric
ganglion

Lumbar
splanchnic nn.

Superior
hypogastric
plexus

S1

S2

S3

S4

Sacral
spinal cord

Hypogastric nn.

Sacral
splanchnic nn.

Inferior
hypogastric
plexus

Pelvic splanchnic nn.

Pudendal n.

Sympathetic fibers ▬▬▬▬
Parasympathetic fibers ▬▬▬
Somatic efferent fibers ▬▬▬
Afferents and CNS connections ▬▬▬
Indefinite paths ▬ ▬ ▬ ▬

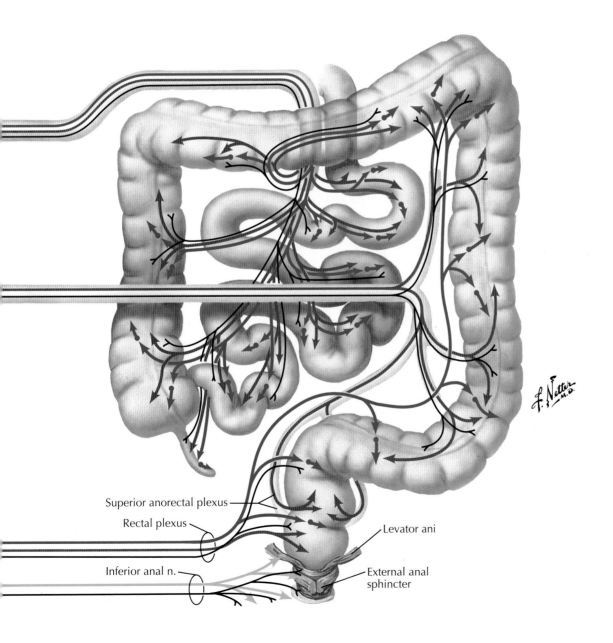

Superior anorectal plexus

Rectal plexus

Inferior anal n.

Levator ani

External anal sphincter

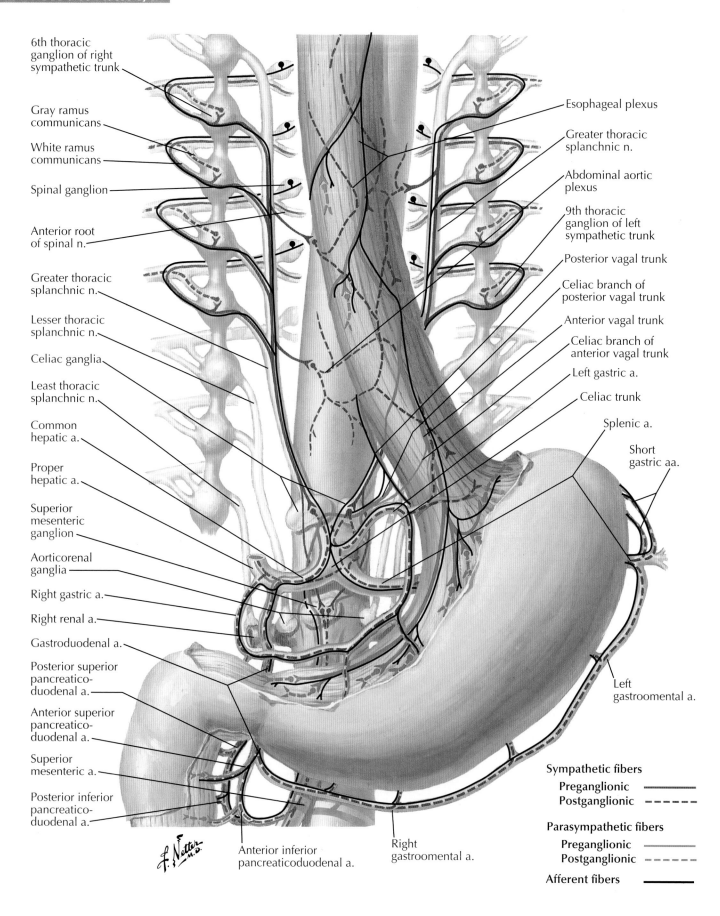

6th thoracic ganglion of right sympathetic trunk

Gray ramus communicans

White ramus communicans

Spinal ganglion

Anterior root of spinal n.

Greater thoracic splanchnic n.

Lesser thoracic splanchnic n.

Celiac ganglia

Least thoracic splanchnic n.

Common hepatic a.

Proper hepatic a.

Superior mesenteric ganglion

Aorticorenal ganglia

Right gastric a.

Right renal a.

Gastroduodenal a.

Posterior superior pancreatico-duodenal a.

Anterior superior pancreatico-duodenal a.

Superior mesenteric a.

Posterior inferior pancreatico-duodenal a.

Esophageal plexus

Greater thoracic splanchnic n.

Abdominal aortic plexus

9th thoracic ganglion of left sympathetic trunk

Posterior vagal trunk

Celiac branch of posterior vagal trunk

Anterior vagal trunk

Celiac branch of anterior vagal trunk

Left gastric a.

Celiac trunk

Splenic a.

Short gastric aa.

Left gastroomental a.

Anterior inferior pancreaticoduodenal a.

Right gastroomental a.

Sympathetic fibers

Preganglionic ———

Postganglionic – – – –

Parasympathetic fibers

Preganglionic ———

Postganglionic – – – –

Afferent fibers ▬▬▬

Plate 325

Visceral Nerves and Plexuses

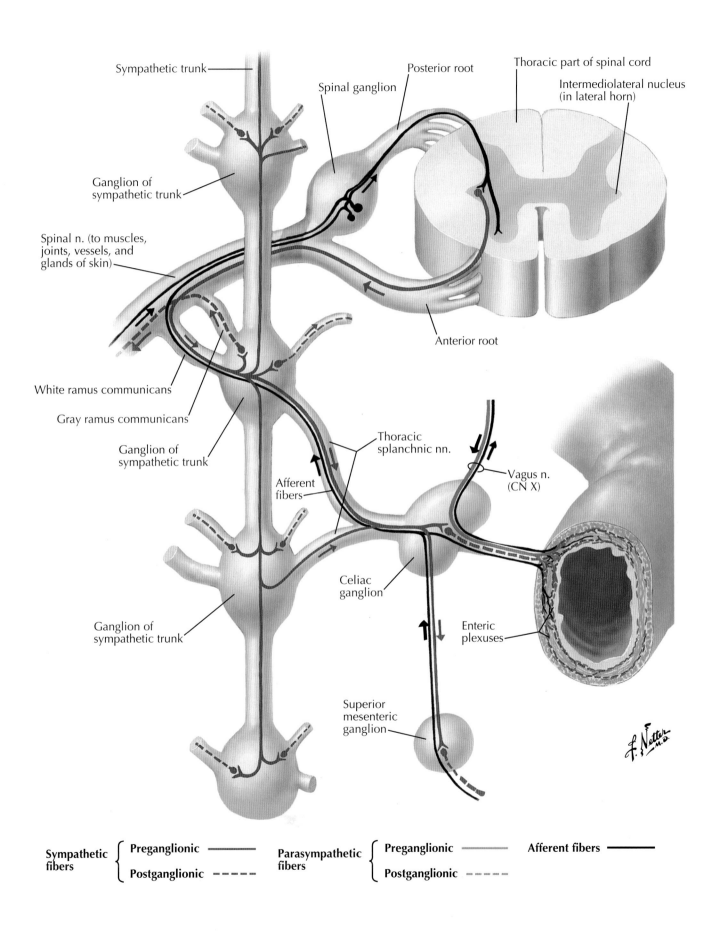

Sympathetic trunk

Spinal ganglion

Posterior root

Thoracic part of spinal cord

Intermediolateral nucleus (in lateral horn)

Ganglion of sympathetic trunk

Spinal n. (to muscles, joints, vessels, and glands of skin)

Anterior root

White ramus communicans

Gray ramus communicans

Ganglion of sympathetic trunk

Thoracic splanchnic nn.

Afferent fibers

Vagus n. (CN X)

Celiac ganglion

Enteric plexuses

Ganglion of sympathetic trunk

Superior mesenteric ganglion

Sympathetic fibers	Preganglionic ————	Parasympathetic fibers	Preganglionic ————	Afferent fibers ————
	Postganglionic - - - - -		Postganglionic - - - - -	

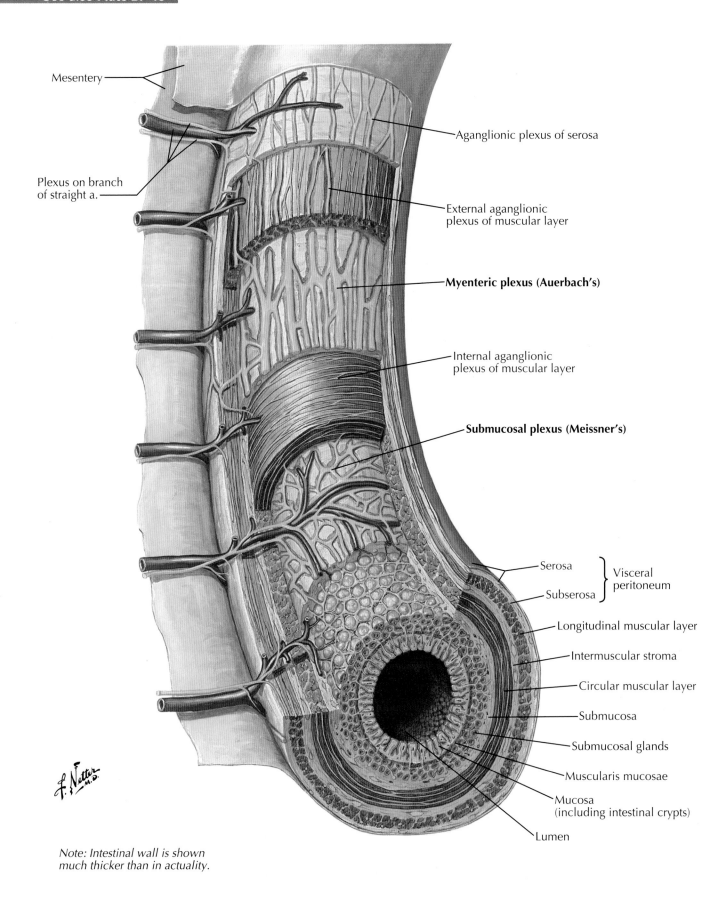

Mesentery

Plexus on branch
of straight a.

Aganglionic plexus of serosa

External aganglionic
plexus of muscular layer

Myenteric plexus (Auerbach's)

Internal aganglionic
plexus of muscular layer

Submucosal plexus (Meissner's)

Serosa ⎫
⎬ Visceral
Subserosa ⎭ peritoneum

Longitudinal muscular layer

Intermuscular stroma

Circular muscular layer

Submucosa

Submucosal glands

Muscularis mucosae

Mucosa
(including intestinal crypts)

Lumen

*Note: Intestinal wall is shown
much thicker than in actuality.*

Plate 327

Visceral Nerves and Plexuses

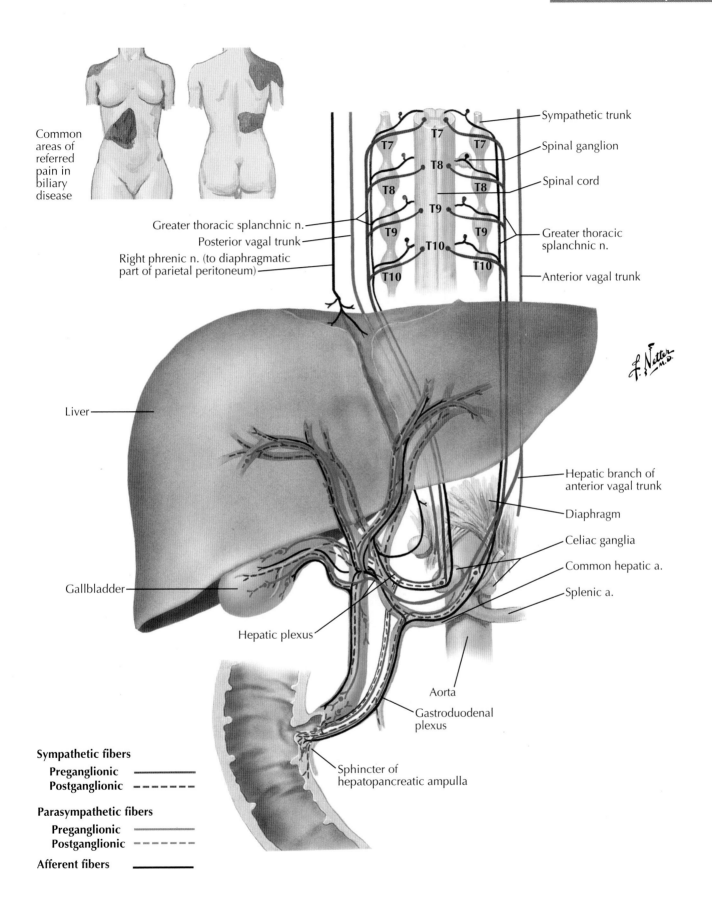

Common areas of referred pain in biliary disease

Sympathetic trunk

Spinal ganglion

Spinal cord

Greater thoracic splanchnic n.

Anterior vagal trunk

T7

T8

T9

T10

Greater thoracic splanchnic n.

Posterior vagal trunk

Right phrenic n. (to diaphragmatic part of parietal peritoneum)

Liver

Hepatic branch of anterior vagal trunk

Diaphragm

Celiac ganglia

Common hepatic a.

Splenic a.

Gallbladder

Hepatic plexus

Aorta

Gastroduodenal plexus

Sphincter of hepatopancreatic ampulla

Sympathetic fibers
 Preganglionic ———
 Postganglionic – – – –

Parasympathetic fibers
 Preganglionic ———
 Postganglionic – – – –

Afferent fibers ———

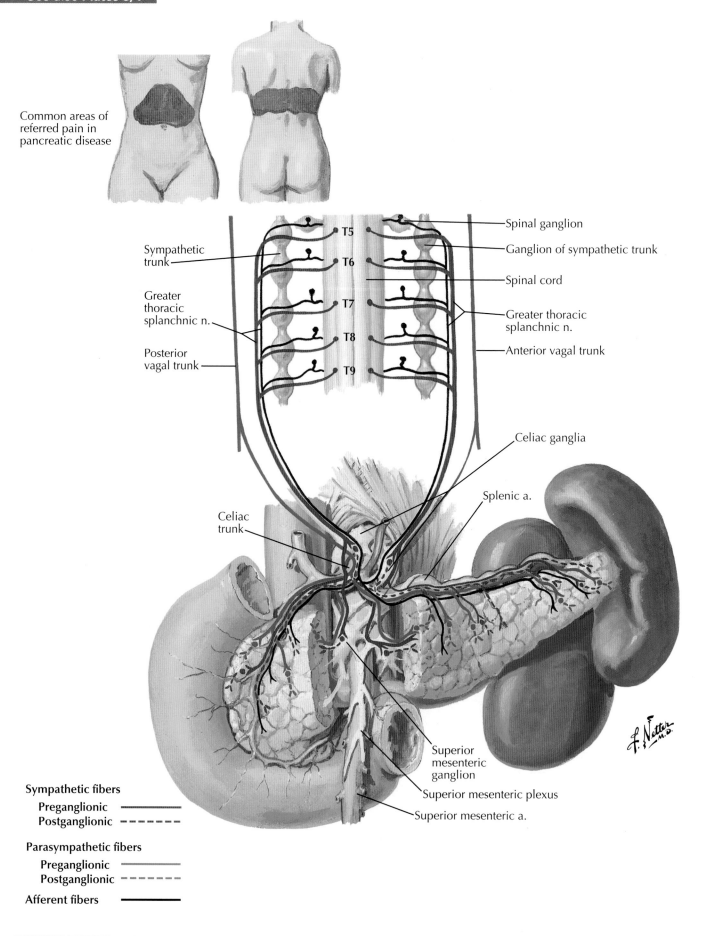

Common areas of referred pain in pancreatic disease

Spinal ganglion

T5

Sympathetic trunk

Ganglion of sympathetic trunk

T6

Spinal cord

Greater thoracic splanchnic n.

T7

Greater thoracic splanchnic n.

T8

Posterior vagal trunk

T9

Anterior vagal trunk

Celiac ganglia

Splenic a.

Celiac trunk

Superior mesenteric ganglion

Superior mesenteric plexus

Superior mesenteric a.

Sympathetic fibers

Preganglionic ——————

Postganglionic - - - - - -

Parasympathetic fibers

Preganglionic ——————

Postganglionic - - - - - -

Afferent fibers ——————

Plate 329

Visceral Nerves and Plexuses

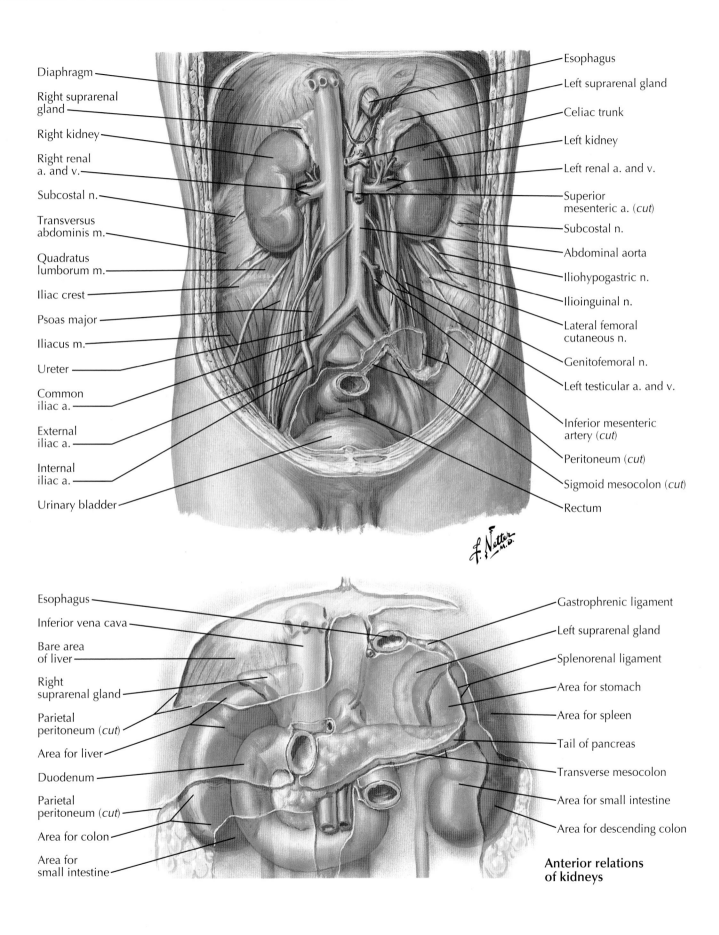

Diaphragm

Right suprarenal gland

Right kidney

Right renal a. and v.

Subcostal n.

Transversus abdominis m.

Quadratus lumborum m.

Iliac crest

Psoas major

Iliacus m.

Ureter

Common iliac a.

External iliac a.

Internal iliac a.

Urinary bladder

Esophagus

Left suprarenal gland

Celiac trunk

Left kidney

Left renal a. and v.

Superior mesenteric a. (*cut*)

Subcostal n.

Abdominal aorta

Iliohypogastric n.

Ilioinguinal n.

Lateral femoral cutaneous n.

Genitofemoral n.

Left testicular a. and v.

Inferior mesenteric artery (*cut*)

Peritoneum (*cut*)

Sigmoid mesocolon (*cut*)

Rectum

Esophagus

Inferior vena cava

Bare area of liver

Right suprarenal gland

Parietal peritoneum (*cut*)

Area for liver

Duodenum

Parietal peritoneum (*cut*)

Area for colon

Area for small intestine

Gastrophrenic ligament

Left suprarenal gland

Splenorenal ligament

Area for stomach

Area for spleen

Tail of pancreas

Transverse mesocolon

Area for small intestine

Area for descending colon

Anterior relations of kidneys

Kidneys and Suprarenal Glands

Plate 330

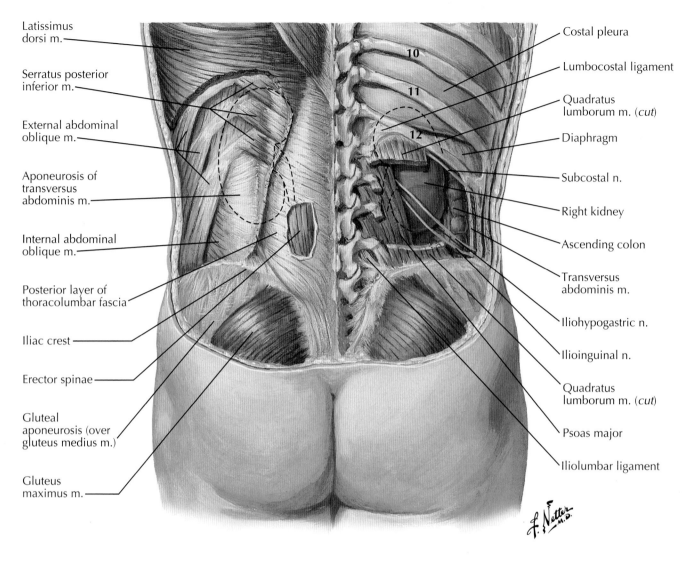

Latissimus dorsi m.

Serratus posterior inferior m.

External abdominal oblique m.

Aponeurosis of transversus abdominis m.

Internal abdominal oblique m.

Posterior layer of thoracolumbar fascia

Iliac crest

Erector spinae

Gluteal aponeurosis (over gluteus medius m.)

Gluteus maximus m.

Costal pleura

Lumbocostal ligament

Quadratus lumborum m. (cut)

Diaphragm

Subcostal n.

Right kidney

Ascending colon

Transversus abdominis m.

Iliohypogastric n.

Ilioinguinal n.

Quadratus lumborum m. (cut)

Psoas major

Iliolumbar ligament

10

11

12

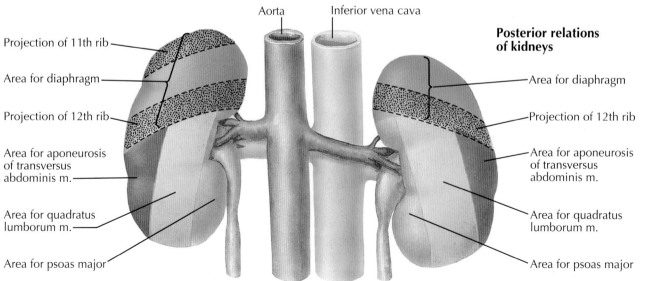

Aorta

Inferior vena cava

Posterior relations of kidneys

Projection of 11th rib

Area for diaphragm

Projection of 12th rib

Area for aponeurosis of transversus abdominis m.

Area for quadratus lumborum m.

Area for psoas major

Area for diaphragm

Projection of 12th rib

Area for aponeurosis of transversus abdominis m.

Area for quadratus lumborum m.

Area for psoas major

Plate 331

Kidneys and Suprarenal Glands

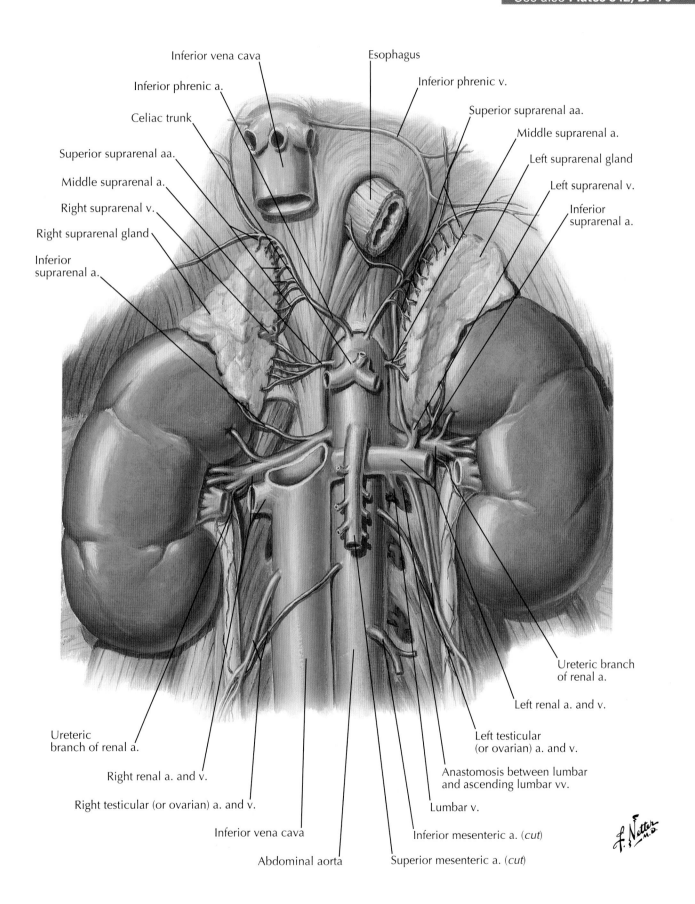

Inferior vena cava

Inferior phrenic a.

Celiac trunk

Superior suprarenal aa.

Middle suprarenal a.

Right suprarenal v.

Right suprarenal gland

Inferior suprarenal a.

Esophagus

Inferior phrenic v.

Superior suprarenal aa.

Middle suprarenal a.

Left suprarenal gland

Left suprarenal v.

Inferior suprarenal a.

Ureteric branch of renal a.

Left renal a. and v.

Left testicular (or ovarian) a. and v.

Anastomosis between lumbar and ascending lumbar vv.

Lumbar v.

Inferior mesenteric a. (*cut*)

Superior mesenteric a. (*cut*)

Ureteric branch of renal a.

Right renal a. and v.

Right testicular (or ovarian) a. and v.

Inferior vena cava

Abdominal aorta

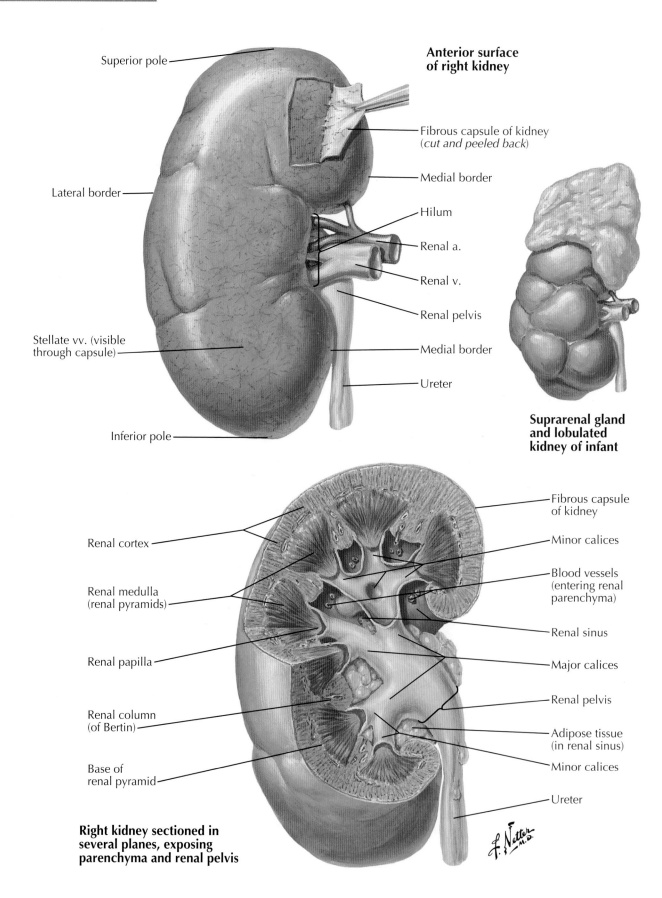

Superior pole

Anterior surface of right kidney

Fibrous capsule of kidney (*cut and peeled back*)

Medial border

Hilum

Renal a.

Renal v.

Renal pelvis

Lateral border

Medial border

Stellate vv. (visible through capsule)

Ureter

Inferior pole

Suprarenal gland and lobulated kidney of infant

Renal cortex

Fibrous capsule of kidney

Minor calices

Renal medulla (renal pyramids)

Blood vessels (entering renal parenchyma)

Renal sinus

Renal papilla

Major calices

Renal column (of Bertin)

Renal pelvis

Adipose tissue (in renal sinus)

Minor calices

Base of renal pyramid

Ureter

Right kidney sectioned in several planes, exposing parenchyma and renal pelvis

Plate 333

Kidneys and Suprarenal Glands

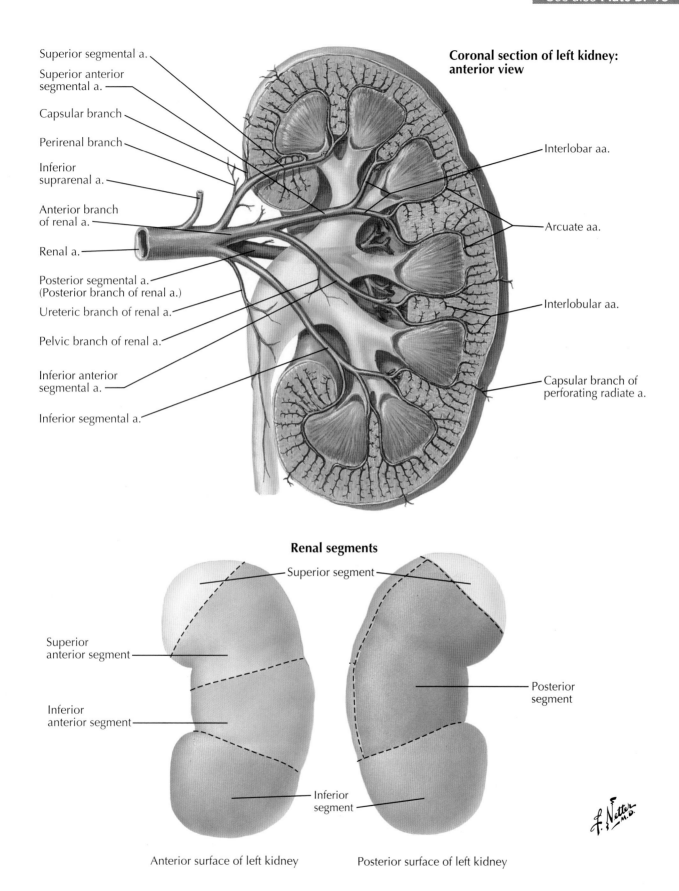

Superior segmental a.

Superior anterior segmental a.

Capsular branch

Perirenal branch

Inferior suprarenal a.

Anterior branch of renal a.

Renal a.

Posterior segmental a. (Posterior branch of renal a.)

Ureteric branch of renal a.

Pelvic branch of renal a.

Inferior anterior segmental a.

Inferior segmental a.

Coronal section of left kidney: anterior view

Interlobar aa.

Arcuate aa.

Interlobular aa.

Capsular branch of perforating radiate a.

Renal segments

Superior segment

Superior anterior segment

Inferior anterior segment

Posterior segment

Inferior segment

Anterior surface of left kidney

Posterior surface of left kidney

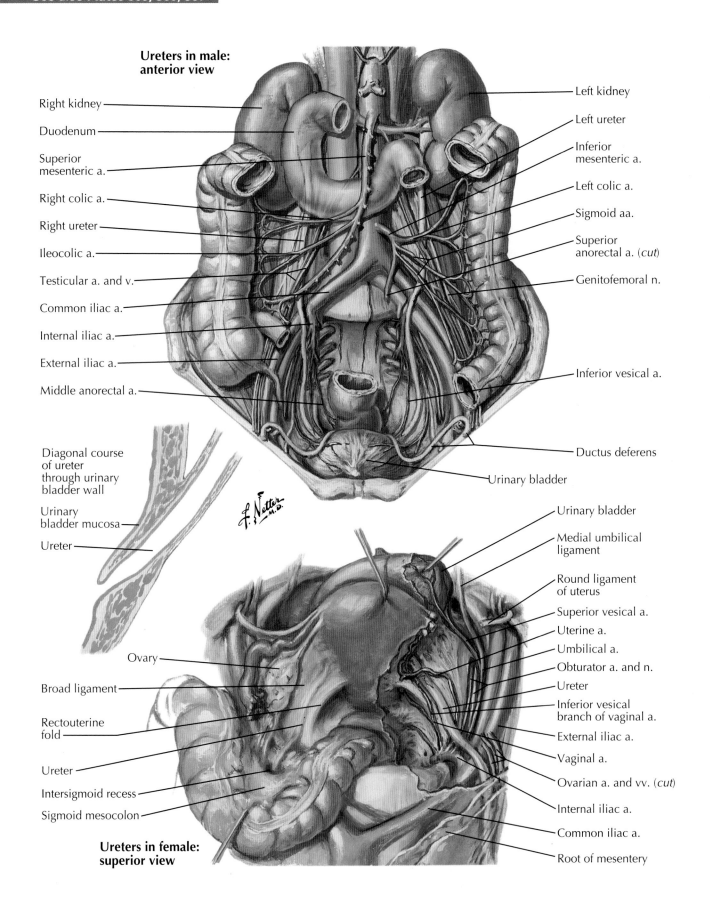

Ureters in male: anterior view

Right kidney

Duodenum

Superior mesenteric a.

Right colic a.

Right ureter

Ileocolic a.

Testicular a. and v.

Common iliac a.

Internal iliac a.

External iliac a.

Middle anorectal a.

Left kidney

Left ureter

Inferior mesenteric a.

Left colic a.

Sigmoid aa.

Superior anorectal a. (*cut*)

Genitofemoral n.

Inferior vesical a.

Ductus deferens

Urinary bladder

Diagonal course of ureter through urinary bladder wall

Urinary bladder mucosa

Ureter

Ovary

Broad ligament

Rectouterine fold

Ureter

Intersigmoid recess

Sigmoid mesocolon

Ureters in female: superior view

Urinary bladder

Medial umbilical ligament

Round ligament of uterus

Superior vesical a.

Uterine a.

Umbilical a.

Obturator a. and n.

Ureter

Inferior vesical branch of vaginal a.

External iliac a.

Vaginal a.

Ovarian a. and vv. (*cut*)

Internal iliac a.

Common iliac a.

Root of mesentery

Plate 335

Kidneys and Suprarenal Glands

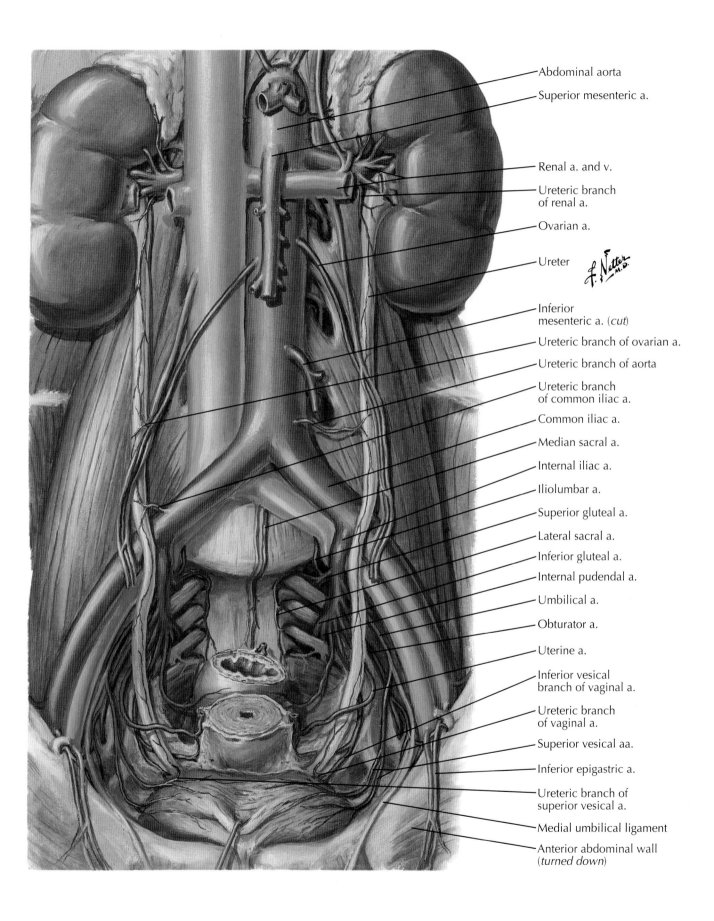

Abdominal aorta

Superior mesenteric a.

Renal a. and v.

Ureteric branch
of renal a.

Ovarian a.

Ureter

Inferior
mesenteric a. (*cut*)

Ureteric branch of ovarian a.

Ureteric branch of aorta

Ureteric branch
of common iliac a.

Common iliac a.

Median sacral a.

Internal iliac a.

Iliolumbar a.

Superior gluteal a.

Lateral sacral a.

Inferior gluteal a.

Internal pudendal a.

Umbilical a.

Obturator a.

Uterine a.

Inferior vesical
branch of vaginal a.

Ureteric branch
of vaginal a.

Superior vesical aa.

Inferior epigastric a.

Ureteric branch of
superior vesical a.

Medial umbilical ligament

Anterior abdominal wall
(*turned down*)

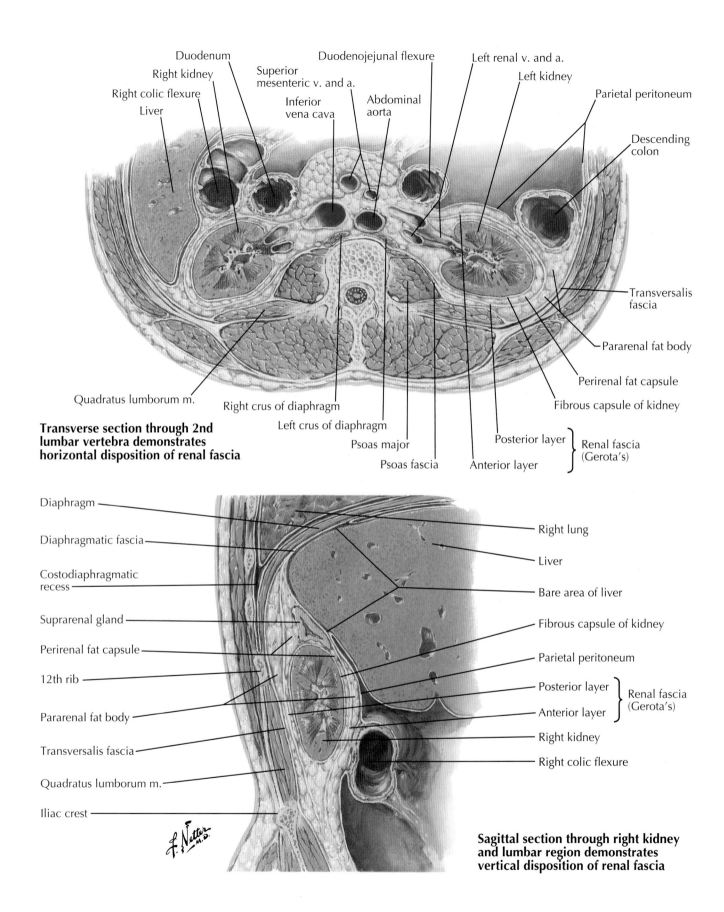

Duodenum
Right kidney
Right colic flexure
Liver
Superior mesenteric v. and a.
Duodenojejunal flexure
Left renal v. and a.
Left kidney
Parietal peritoneum
Inferior vena cava
Abdominal aorta
Descending colon
Transversalis fascia
Pararenal fat body
Perirenal fat capsule
Fibrous capsule of kidney
Quadratus lumborum m.
Right crus of diaphragm
Left crus of diaphragm
Psoas major
Psoas fascia
Posterior layer
Anterior layer
Renal fascia (Gerota's)

Transverse section through 2nd lumbar vertebra demonstrates horizontal disposition of renal fascia

Diaphragm
Diaphragmatic fascia
Costodiaphragmatic recess
Suprarenal gland
Perirenal fat capsule
12th rib
Pararenal fat body
Transversalis fascia
Quadratus lumborum m.
Iliac crest
Right lung
Liver
Bare area of liver
Fibrous capsule of kidney
Parietal peritoneum
Posterior layer
Anterior layer
Renal fascia (Gerota's)
Right kidney
Right colic flexure

Sagittal section through right kidney and lumbar region demonstrates vertical disposition of renal fascia

Plate 337

Kidneys and Suprarenal Glands

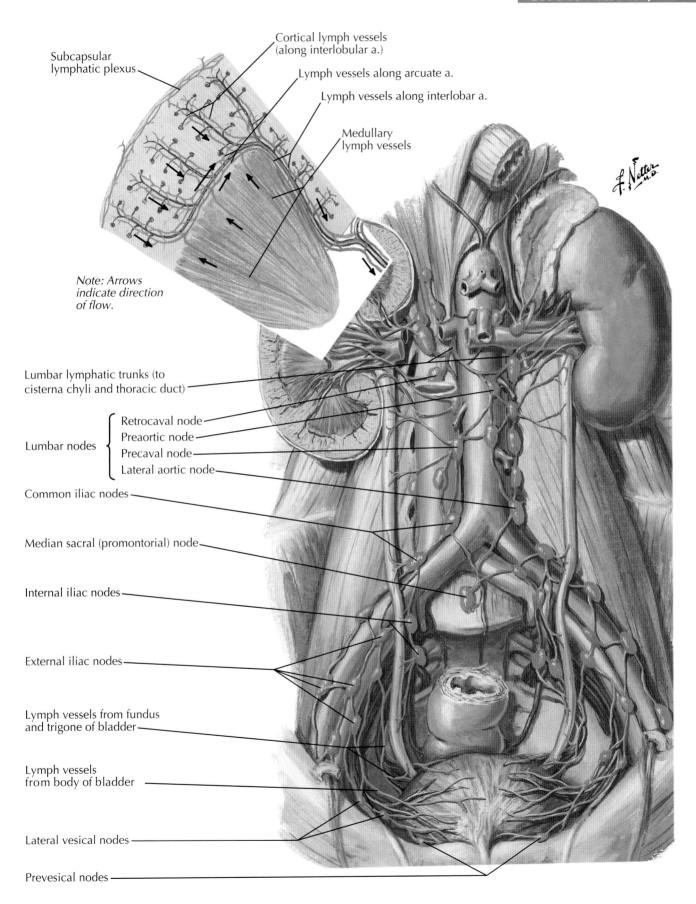

Subcapsular lymphatic plexus

Cortical lymph vessels (along interlobular a.)

Lymph vessels along arcuate a.

Lymph vessels along interlobar a.

Medullary lymph vessels

Note: Arrows indicate direction of flow.

Lumbar lymphatic trunks (to cisterna chyli and thoracic duct)

Lumbar nodes
{ Retrocaval node
Preaortic node
Precaval node
Lateral aortic node

Common iliac nodes

Median sacral (promontorial) node

Internal iliac nodes

External iliac nodes

Lymph vessels from fundus and trigone of bladder

Lymph vessels from body of bladder

Lateral vesical nodes

Prevesical nodes

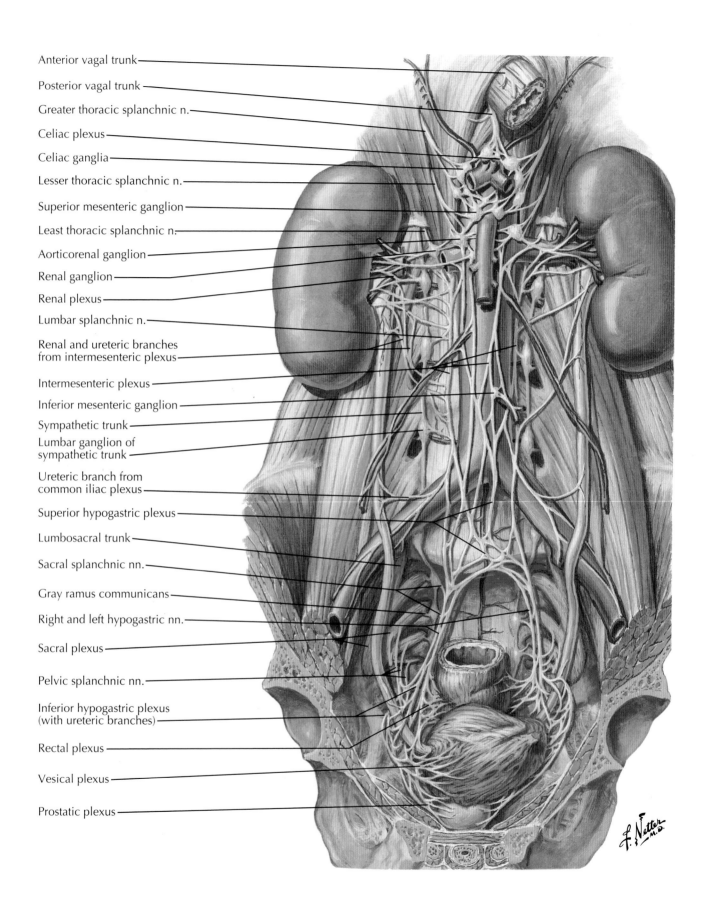

Anterior vagal trunk

Posterior vagal trunk

Greater thoracic splanchnic n.

Celiac plexus

Celiac ganglia

Lesser thoracic splanchnic n.

Superior mesenteric ganglion

Least thoracic splanchnic n.

Aorticorenal ganglion

Renal ganglion

Renal plexus

Lumbar splanchnic n.

Renal and ureteric branches from intermesenteric plexus

Intermesenteric plexus

Inferior mesenteric ganglion

Sympathetic trunk

Lumbar ganglion of sympathetic trunk

Ureteric branch from common iliac plexus

Superior hypogastric plexus

Lumbosacral trunk

Sacral splanchnic nn.

Gray ramus communicans

Right and left hypogastric nn.

Sacral plexus

Pelvic splanchnic nn.

Inferior hypogastric plexus (with ureteric branches)

Rectal plexus

Vesical plexus

Prostatic plexus

Plate 339

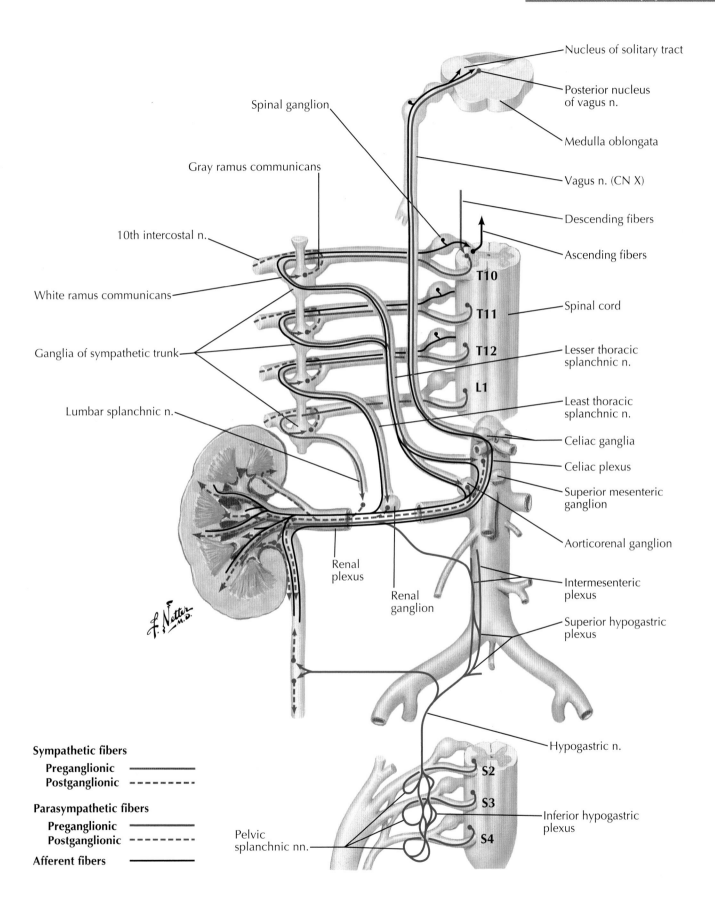

Nucleus of solitary tract

Spinal ganglion

Posterior nucleus of vagus n.

Gray ramus communicans

Medulla oblongata

10th intercostal n.

Vagus n. (CN X)

Descending fibers

Ascending fibers

White ramus communicans

T10

Spinal cord

T11

Ganglia of sympathetic trunk

T12

Lesser thoracic splanchnic n.

L1

Lumbar splanchnic n.

Least thoracic splanchnic n.

Celiac ganglia

Celiac plexus

Superior mesenteric ganglion

Aorticorenal ganglion

Renal plexus

Intermesenteric plexus

Renal ganglion

Superior hypogastric plexus

Hypogastric n.

Sympathetic fibers
  Preganglionic ————
  Postganglionic - - - - -

Parasympathetic fibers
  Preganglionic ————
  Postganglionic - - - - -

Afferent fibers ————

S2

S3

Inferior hypogastric plexus

S4

Pelvic splanchnic nn.

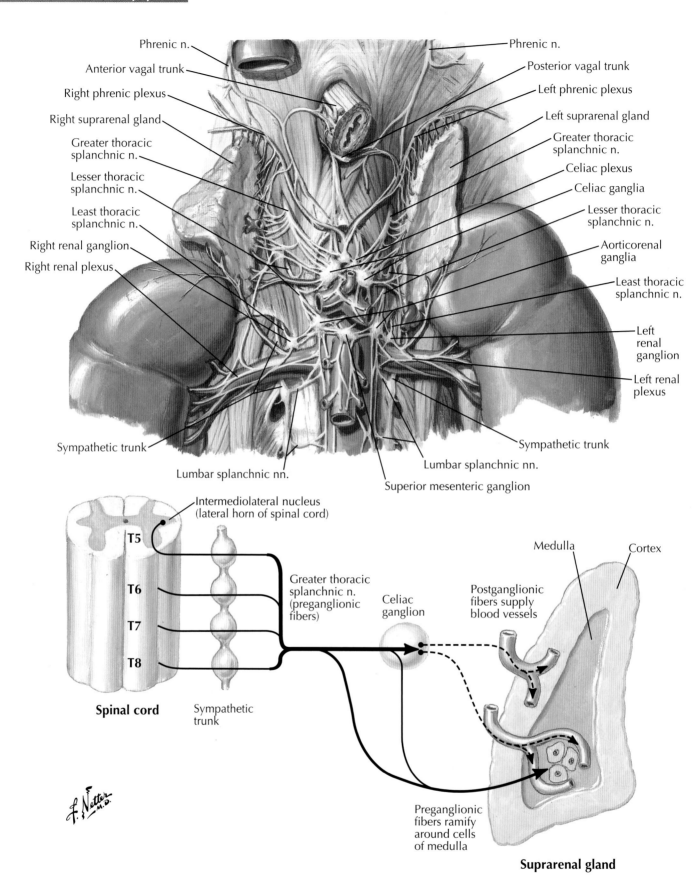

Phrenic n.

Anterior vagal trunk

Right phrenic plexus

Right suprarenal gland

Greater thoracic splanchnic n.

Lesser thoracic splanchnic n.

Least thoracic splanchnic n.

Right renal ganglion

Right renal plexus

Sympathetic trunk

Lumbar splanchnic nn.

Phrenic n.

Posterior vagal trunk

Left phrenic plexus

Left suprarenal gland

Greater thoracic splanchnic n.

Celiac plexus

Celiac ganglia

Lesser thoracic splanchnic n.

Aorticorenal ganglia

Least thoracic splanchnic n.

Left renal ganglion

Left renal plexus

Sympathetic trunk

Lumbar splanchnic nn.

Superior mesenteric ganglion

Intermediolateral nucleus (lateral horn of spinal cord)

T5

T6

T7

T8

Spinal cord

Sympathetic trunk

Greater thoracic splanchnic n. (preganglionic fibers)

Celiac ganglion

Postganglionic fibers supply blood vessels

Medulla

Cortex

Preganglionic fibers ramify around cells of medulla

Suprarenal gland

Plate 341

Kidneys and Suprarenal Glands

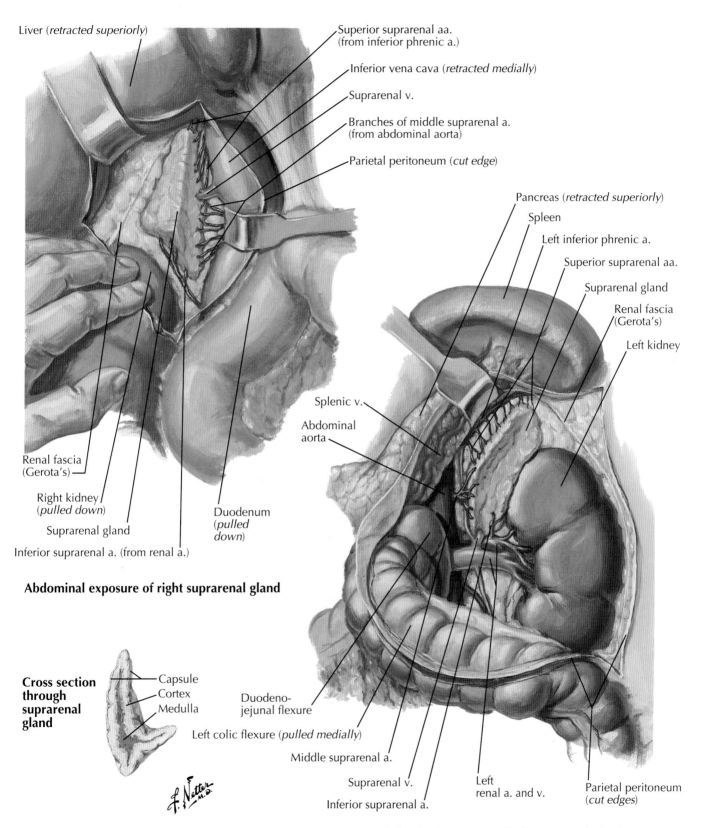

Liver (*retracted superiorly*)

Superior suprarenal aa. (from inferior phrenic a.)

Inferior vena cava (*retracted medially*)

Suprarenal v.

Branches of middle suprarenal a. (from abdominal aorta)

Parietal peritoneum (*cut edge*)

Pancreas (*retracted superiorly*)

Spleen

Left inferior phrenic a.

Superior suprarenal aa.

Suprarenal gland

Renal fascia (Gerota's)

Left kidney

Splenic v.

Abdominal aorta

Renal fascia (Gerota's)

Right kidney (*pulled down*)

Suprarenal gland

Inferior suprarenal a. (from renal a.)

Duodenum (*pulled down*)

Abdominal exposure of right suprarenal gland

Cross section through suprarenal gland

Capsule

Cortex

Medulla

Duodeno-jejunal flexure

Left colic flexure (*pulled medially*)

Middle suprarenal a.

Suprarenal v.

Inferior suprarenal a.

Left renal a. and v.

Parietal peritoneum (*cut edges*)

Abdominal exposure of left suprarenal gland

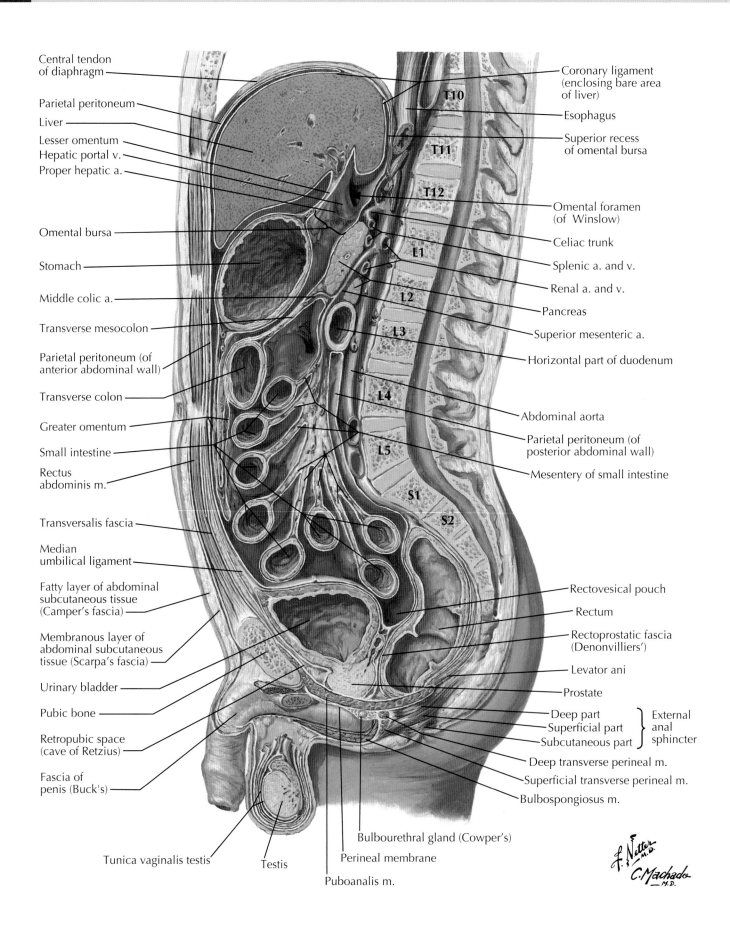

Central tendon of diaphragm

Parietal peritoneum

Liver

Lesser omentum

Hepatic portal v.

Proper hepatic a.

Omental bursa

Stomach

Middle colic a.

Transverse mesocolon

Parietal peritoneum (of anterior abdominal wall)

Transverse colon

Greater omentum

Small intestine

Rectus abdominis m.

Transversalis fascia

Median umbilical ligament

Fatty layer of abdominal subcutaneous tissue (Camper's fascia)

Membranous layer of abdominal subcutaneous tissue (Scarpa's fascia)

Urinary bladder

Pubic bone

Retropubic space (cave of Retzius)

Fascia of penis (Buck's)

Tunica vaginalis testis

Testis

Perineal membrane

Puboanalis m.

Bulbourethral gland (Cowper's)

Coronary ligament (enclosing bare area of liver)

Esophagus

Superior recess of omental bursa

Omental foramen (of Winslow)

Celiac trunk

Splenic a. and v.

Renal a. and v.

Pancreas

Superior mesenteric a.

Horizontal part of duodenum

Abdominal aorta

Parietal peritoneum (of posterior abdominal wall)

Mesentery of small intestine

Rectovesical pouch

Rectum

Rectoprostatic fascia (Denonvilliers')

Levator ani

Prostate

Deep part

Superficial part

Subcutaneous part

External anal sphincter

Deep transverse perineal m.

Superficial transverse perineal m.

Bulbospongiosus m.

T10

T11

T12

L1

L2

L3

L4

L5

S1

S2

Plate 343

Kidneys and Suprarenal Glands

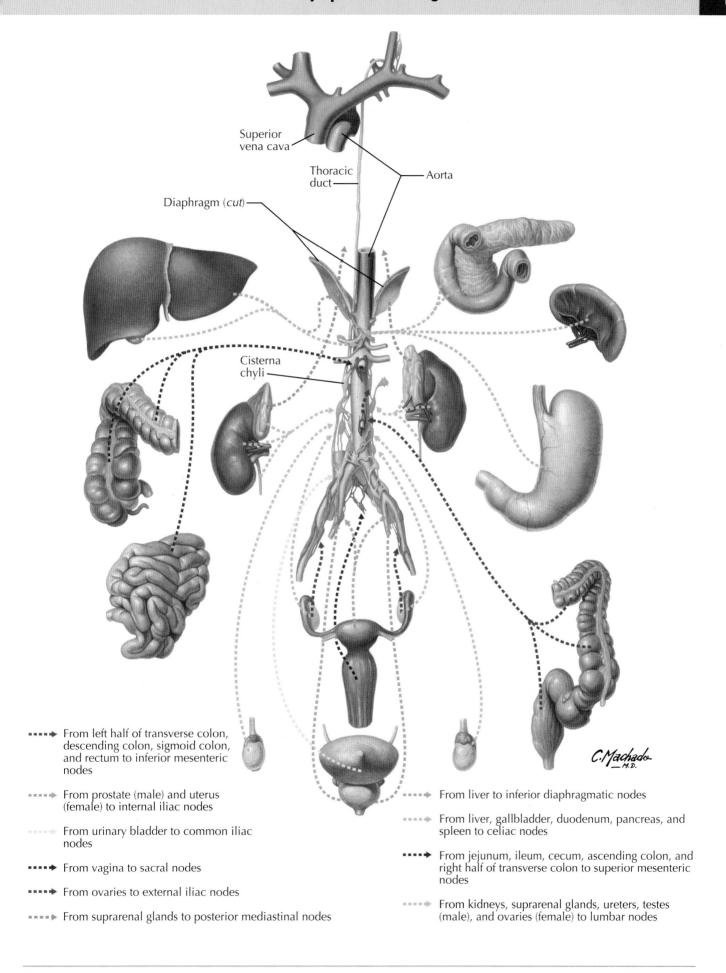

Superior vena cava

Thoracic duct

Aorta

Diaphragm (*cut*)

Cisterna chyli

---- From left half of transverse colon, descending colon, sigmoid colon, and rectum to inferior mesenteric nodes

---- From prostate (male) and uterus (female) to internal iliac nodes

---- From urinary bladder to common iliac nodes

---- From vagina to sacral nodes

---- From ovaries to external iliac nodes

---- From suprarenal glands to posterior mediastinal nodes

---- From liver to inferior diaphragmatic nodes

---- From liver, gallbladder, duodenum, pancreas, and spleen to celiac nodes

---- From jejunum, ileum, cecum, ascending colon, and right half of transverse colon to superior mesenteric nodes

---- From kidneys, suprarenal glands, ureters, testes (male), and ovaries (female) to lumbar nodes

C. Machado
_M.D.

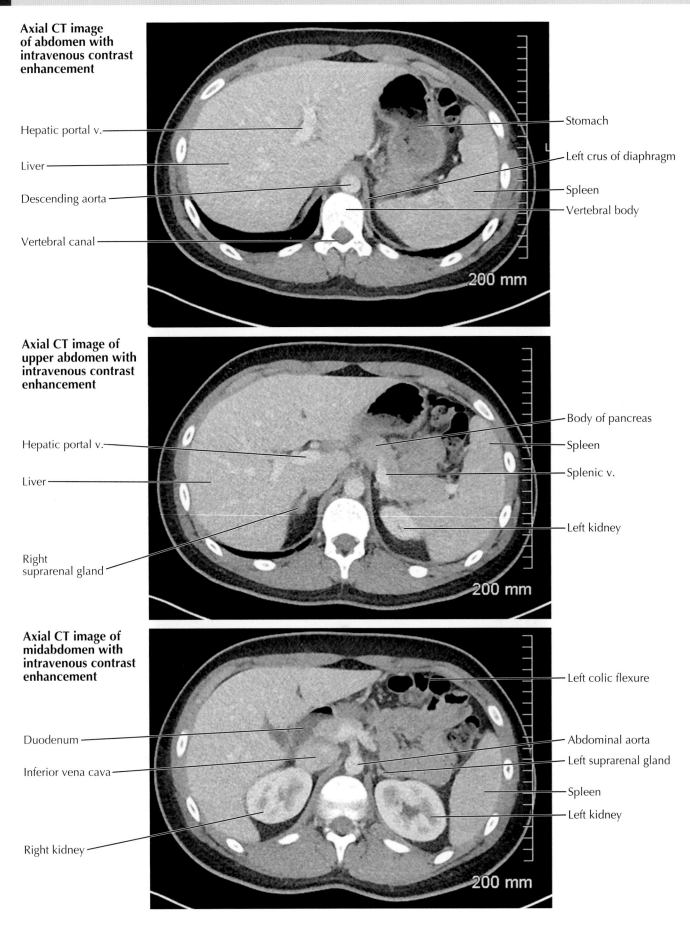

Axial CT image of abdomen with intravenous contrast enhancement

Hepatic portal v.

Liver

Descending aorta

Vertebral canal

Stomach

Left crus of diaphragm

Spleen

Vertebral body

Axial CT image of upper abdomen with intravenous contrast enhancement

Hepatic portal v.

Liver

Right suprarenal gland

Body of pancreas

Spleen

Splenic v.

Left kidney

Axial CT image of midabdomen with intravenous contrast enhancement

Duodenum

Inferior vena cava

Right kidney

Left colic flexure

Abdominal aorta

Left suprarenal gland

Spleen

Left kidney

Plate 345

Regional Imaging

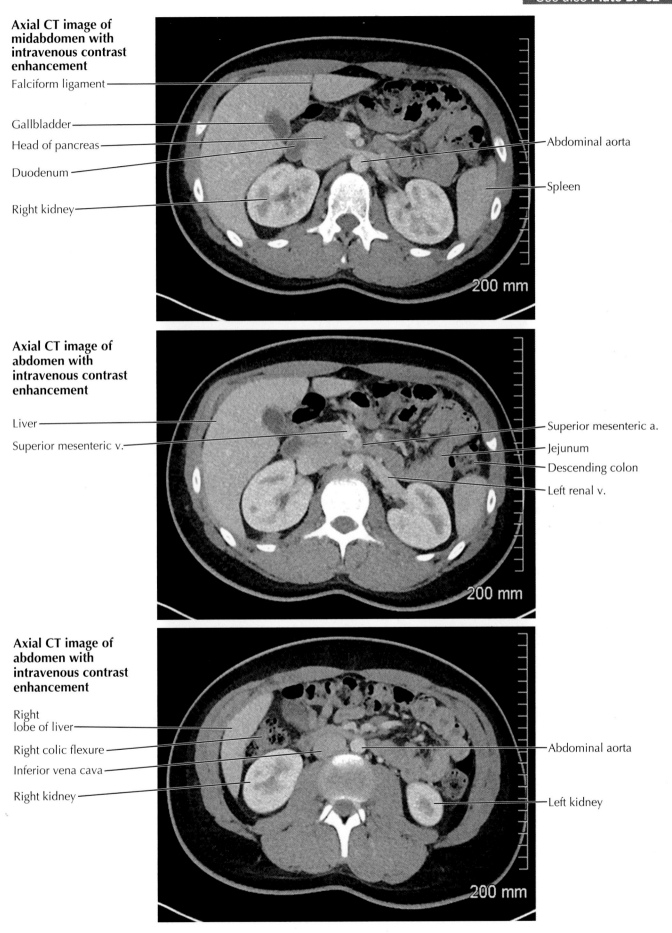

Axial CT image of midabdomen with intravenous contrast enhancement

Falciform ligament

Gallbladder

Head of pancreas

Duodenum

Right kidney

Abdominal aorta

Spleen

Axial CT image of abdomen with intravenous contrast enhancement

Liver

Superior mesenteric v.

Superior mesenteric a.

Jejunum

Descending colon

Left renal v.

Axial CT image of abdomen with intravenous contrast enhancement

Right lobe of liver

Right colic flexure

Inferior vena cava

Right kidney

Abdominal aorta

Left kidney

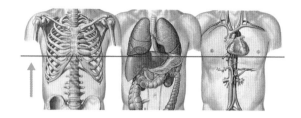

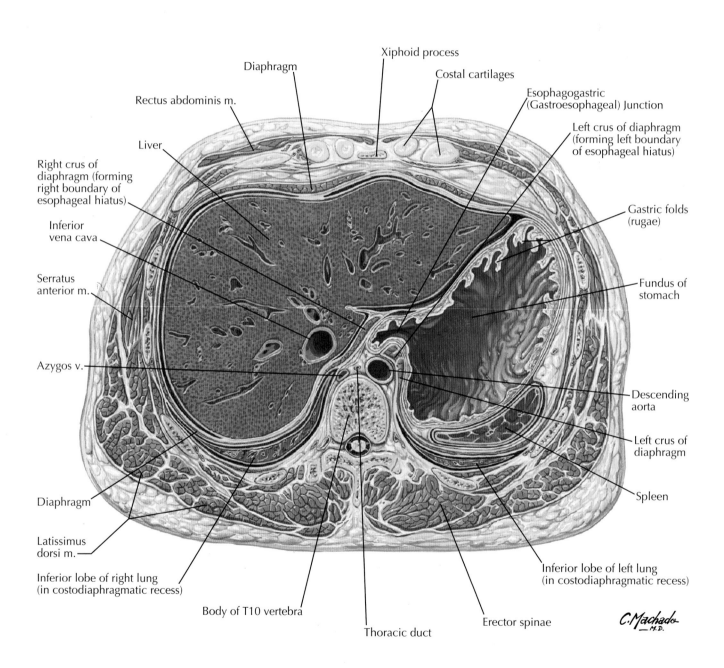

Diaphragm

Xiphoid process

Costal cartilages

Rectus abdominis m.

Esophagogastric (Gastroesophageal) Junction

Liver

Left crus of diaphragm (forming left boundary of esophageal hiatus)

Right crus of diaphragm (forming right boundary of esophageal hiatus)

Gastric folds (rugae)

Inferior vena cava

Serratus anterior m.

Fundus of stomach

Azygos v.

Descending aorta

Left crus of diaphragm

Diaphragm

Spleen

Latissimus dorsi m.

Inferior lobe of right lung (in costodiaphragmatic recess)

Inferior lobe of left lung (in costodiaphragmatic recess)

Body of T10 vertebra

Thoracic duct

Erector spinae

C. Machado
_M.D.

Plate 347

Cross-Sectional Anatomy

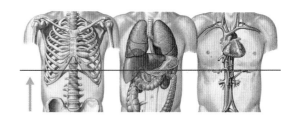

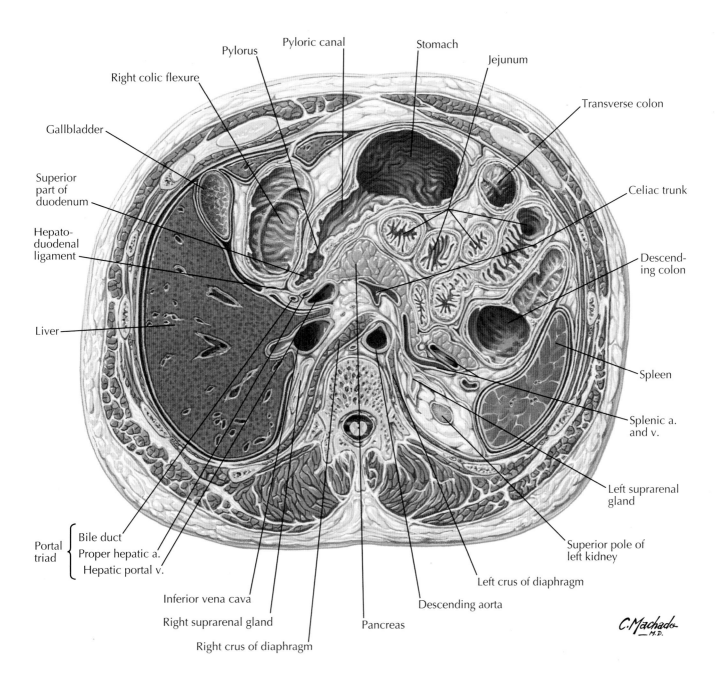

Pyloric canal

Pylorus

Stomach

Jejunum

Right colic flexure

Transverse colon

Gallbladder

Celiac trunk

Superior
part of
duodenum

Descending colon

Hepato-
duodenal
ligament

Liver

Spleen

Splenic a.
and v.

Left suprarenal
gland

Portal
triad {
Bile duct
Proper hepatic a.
Hepatic portal v.

Superior pole of
left kidney

Inferior vena cava

Left crus of diaphragm

Right suprarenal gland

Descending aorta

Right crus of diaphragm

Pancreas

C. Machado
— M.D.

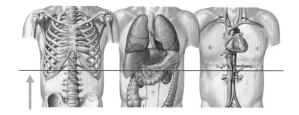

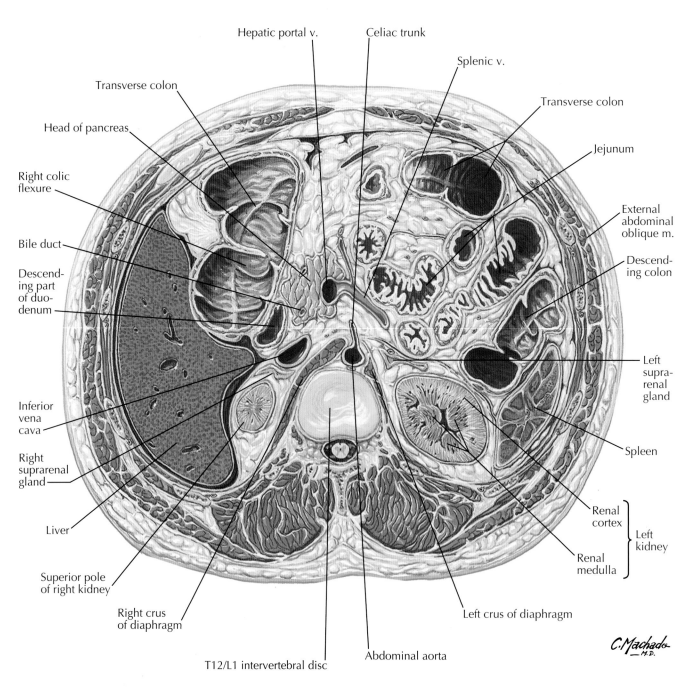

Hepatic portal v.

Celiac trunk

Splenic v.

Transverse colon

Transverse colon

Head of pancreas

Jejunum

Right colic flexure

External abdominal oblique m.

Bile duct

Descending colon

Descending part of duodenum

Left suprarenal gland

Inferior vena cava

Right suprarenal gland

Spleen

Liver

Renal cortex

Left kidney

Superior pole of right kidney

Renal medulla

Right crus of diaphragm

Left crus of diaphragm

T12/L1 intervertebral disc

Abdominal aorta

C. Machado
M.D.

Plate 349

Cross-Sectional Anatomy

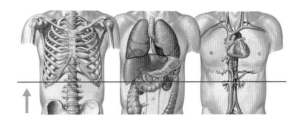

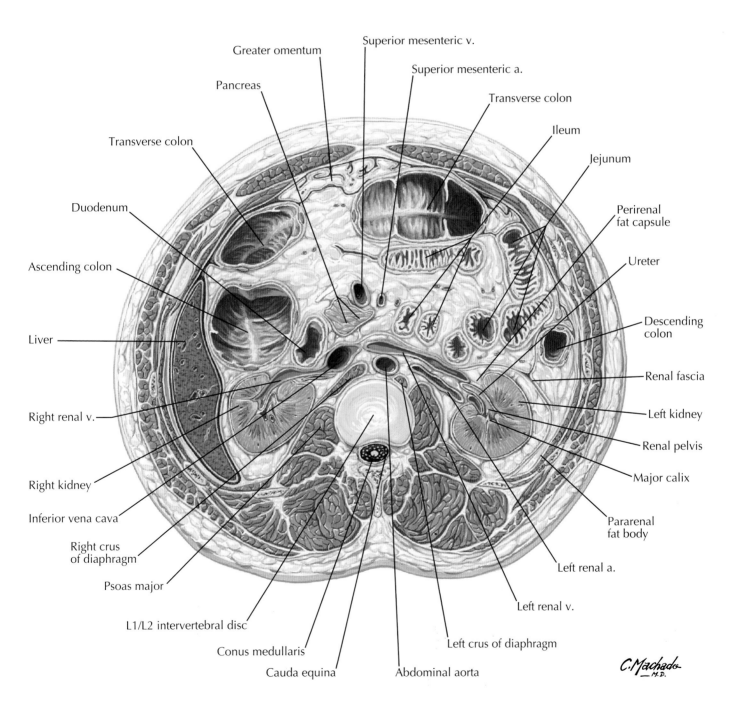

Superior mesenteric v.

Greater omentum

Superior mesenteric a.

Pancreas

Transverse colon

Ileum

Transverse colon

Jejunum

Duodenum

Perirenal
fat capsule

Ascending colon

Ureter

Descending
colon

Liver

Renal fascia

Right renal v.

Left kidney

Renal pelvis

Right kidney

Major calix

Inferior vena cava

Pararenal
fat body

Right crus
of diaphragm

Left renal a.

Psoas major

L1/L2 intervertebral disc

Left renal v.

Conus medullaris

Left crus of diaphragm

Cauda equina

Abdominal aorta

C. Machado
— M.D.

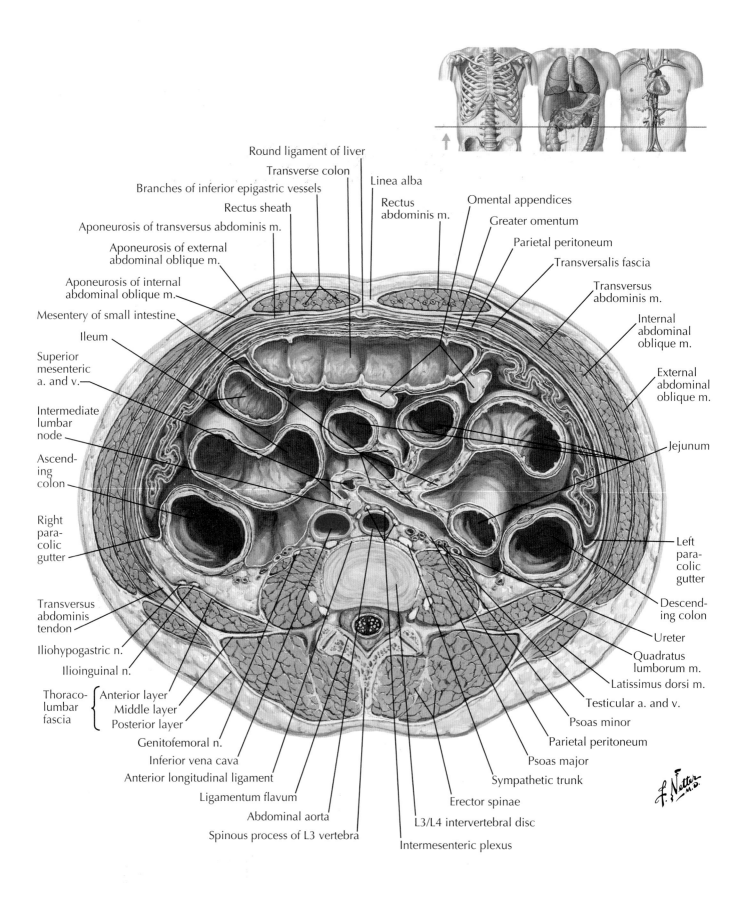

Round ligament of liver

Transverse colon

Branches of inferior epigastric vessels

Linea alba

Rectus sheath

Rectus abdominis m.

Omental appendices

Aponeurosis of transversus abdominis m.

Greater omentum

Aponeurosis of external abdominal oblique m.

Parietal peritoneum

Transversalis fascia

Aponeurosis of internal abdominal oblique m.

Transversus abdominis m.

Mesentery of small intestine

Internal abdominal oblique m.

Ileum

Superior mesenteric a. and v.

External abdominal oblique m.

Intermediate lumbar node

Ascending colon

Jejunum

Right para-colic gutter

Left para-colic gutter

Descending colon

Transversus abdominis tendon

Ureter

Iliohypogastric n.

Quadratus lumborum m.

Ilioinguinal n.

Latissimus dorsi m.

Thoraco-lumbar fascia { Anterior layer

Testicular a. and v.

Middle layer

Posterior layer

Psoas minor

Genitofemoral n.

Parietal peritoneum

Inferior vena cava

Psoas major

Anterior longitudinal ligament

Sympathetic trunk

Ligamentum flavum

Erector spinae

Abdominal aorta

L3/L4 intervertebral disc

Spinous process of L3 vertebra

Intermesenteric plexus

Plate 351

Cross-Sectional Anatomy

ANATOMICAL STRUCTURE	CLINICAL COMMENT	PLATE NUMBERS
Nervous System and Sense Organs		
Ilioinguinal and genitofemoral nerves	Mediate cremasteric reflex; femoral branch of genitofemoral nerve provides cutaneous innervation to skin over femoral triangle	287
Intercostal, subcostal, and iliohypogastric nerves	Convey well-localized pain sensations from abdominal wall and parietal peritoneum; pain in dermatomal distribution indicates problem with spinal nerves (e.g., herpes zoster infection)	278
Celiac ganglion	Some patients with medically intractable pain from chronic pancreatitis or advanced pancreatic malignancy undergo celiac ganglion block; located typically at upper or lower L1 vertebral level	320, 329
Thoracic and lumbar splanchnic nerves (sympathetic)	Convey pain sensations from abdominal viscera that are often referred to other sites; quadrant in which pain is located and site of radiation provide clues to source of pain	324, 325
Iliohypogastric nerve	Nephrectomy through quadratus lumborum muscle can damage iliohypogastric nerve, with resultant anesthesia superior to the pubic symphysis	331
Skeletal System		
Xiphoid process and pubic symphysis	Palpable landmarks used to locate transpyloric plane (of Addison; L1 plane), located halfway between these structures; plane may contain pylorus of stomach, horizontal portion of duodenum, head and neck of pancreas, superior mesenteric artery, and hilum of spleen	267
Anterior superior iliac spine (ASIS)	Palpable landmark used to locate McBurney's point; tenderness over McBurney's point is indication of appendicitis	268, 300
Muscular System		
Linea alba	Site used for abdominal wall incisions because this location provides access to many organs during exploratory surgery and is unlikely to have significant vessels crossing it	267, 272
Inguinal ligament	Surface landmark from ASIS to pubic tubercle that marks division between abdomen and lower limb; formed from external abdominal oblique aponeurosis	267, 271
Inguinal (Hesselbach's) triangle	Important region on interior surface of anterior abdominal wall, bounded by inferior epigastric vessels, inguinal ligament, and rectus abdominis muscle, through which abdominal contents may herniate to produce direct inguinal hernias	274, 280
Deep inguinal ring	Slit-like opening in transversalis fascia just above midpoint of inguinal ligament and lateral to inferior epigastric artery, through which abdominal contents may herniate to produce indirect inguinal hernias	280, 281
Superficial inguinal ring	Triangular opening in external abdominal oblique aponeurosis superior and lateral to pubic tubercle and medial to inferior epigastric artery, through which abdominal contents may herniate to produce indirect inguinal hernias	270, 280, 281
Femoral ring	Superior opening of femoral canal bounded by medial part of inguinal ligament, femoral vein, and lacunar ligament; abdominal contents may herniate through femoral ring into upper femoral triangle situated inferolateral to pubic tubercle, producing femoral hernia	282
Esophageal hiatus	Widening of opening through diaphragm allows stomach to protrude into mediastinum, which can increase incidence of gastroesophageal reflux disease (GERD)	287, 289
Rectus abdominis muscle	Separation (abdominal diastasis) commonly caused from multiple pregnancies, abdominal surgeries, and excessive weight gain; bleeding of inferior epigastric artery may cause blood to accumulate in rectus abdominis muscle (rectus sheath hematoma), which may be mistaken for acute abdominal pathologies, such as appendicitis	271
Cardiovascular System		
Paraumbilical veins	May become dilated in patients with portal hypertension and in late-term pregnancy	277, 318

Structures with High Clinical Significance   **Table 5.1**

ANATOMICAL STRUCTURE	CLINICAL COMMENT	PLATE NUMBERS
Cardiovascular System—Continued		
Cystic artery	Ligated during cholecystectomy; can have multiple origins; found classically in cystohepatic triangle (of Calot)	310
Superior mesenteric artery	May compress horizontal (third) part of duodenum in thin patient or patient who has recently lost a large amount of weight; can shear or tear in sudden deceleration injuries	310, 313
Intestinal arteries	Areas without significant collateral circulation between major vessels (watershed areas) are at risk for ischemia, which can occur secondary to atherosclerosis or thromboembolism of mesenteric arteries	313, 314
Marginal artery	Marginal artery connects right, middle, and left colic arteries, providing important anastomosis for collateral circulation	314
Esophageal veins	May become dilated in portal hypertension, resulting in esophageal varices; variceal hemorrhage can be life-threatening and often requires emergent endoscopic intervention	315, 318
Hepatic portal vein	Increased resistance to blood flow through liver (e.g., due to cirrhosis) may produce portal hypertension and dilation of tributaries of hepatic portal vein; blood may return to heart at sites of portosystemic anastomosis	317, 318
Superior anorectal (rectal) vein	Anastomoses with systemic middle and inferior anorectal veins may become dilated in portal hypertension	317, 318
Abdominal aorta	Common site for aneurysm in abdomen, especially inferior to renal arteries; assessed routinely with ultrasound to exclude aneurysms	336
Renal artery	Stenosis may occur secondary to atherosclerosis or fibromuscular dysplasia, resulting in difficult-to-control hypertension; can be affected in abdominal aortic aneurysms	334, 336
Lymph Vessels and Lymphoid Organs		
Spleen	May be ruptured by fracture of ribs 10 to 12; enlargement (splenomegaly) may occur in cirrhosis, viral infections, and hematologic malignancies; if palpable, the spleen is enlarged	291, 307
Digestive System		
Liver	Palpable inferior to right costal margin; hepatomegaly may occur in conditions such as hepatitis, heart failure, and infiltrative diseases; shrunken, nodular appearance of liver on imaging indicates cirrhosis; common site of metastasis	269, 288, 294
Gastroesophageal junction	Transient relaxations or decreased tone of lower esophageal sphincter can cause gastroesophageal reflux disease (GERD), a common cause of epigastric pain; common site of esophageal cancer	257, 295, 347
Stomach and duodenum	Primary sites of peptic ulcers; nonsteroidal anti-inflammatory drug (NSAID) overuse and/or *Helicobacter pylori* infection may cause ulceration	292, 294, 295
Pylorus	Infantile hypertrophic pyloric stenosis causes postprandial projectile vomiting among newborns	295
Major duodenal papilla (of Vater)	Catheterized and injected with contrast during endoscopic retrograde cholangiopancreatography (ERCP), a common diagnostic procedure; sphincter of hepatopancreatic ampulla (sphincter of Oddi) dysfunction may obstruct biliary flow through major duodenal papilla, resulting in right upper quadrant pain	297, 305
Vermiform appendix	Prone to inflammation and rupture (appendicitis); may have retrocecal position, in which case appendicitis causes inflammation of adjacent psoas fascia and atypical location of pain	298, 300
Colon	Common site of diverticula and malignancies; colonoscopy is performed to screen for colon cancer	301

Table 5.2   **Structures with High Clinical Significance**

ANATOMICAL STRUCTURE	CLINICAL COMMENT	PLATE NUMBERS
Gallbladder	May become inflamed (cholecystitis) and cause pain secondary to gallstones blocking cystic duct	302, 305, 308, 328
Bile duct	Gallstones may become impacted in bile duct (choledocholithiasis), resulting in hepatitis and jaundice; some cases may be complicated by pancreatitis and/or infection of obstructed biliary ducts (cholangitis); small stones can be problematic, whereas large stones usually remain in gallbladder and are more typically asymptomatic; ERCP is performed to locate and relieve obstruction	302, 303, 305
Umbilicus	Remnant of umbilical cord insertion into fetal umbilical vessels; landmark for locating trans-umbilical plane, which is used to divide abdomen into quadrants; marks position of T10 dermatome; used to locate McBurney's point; common site for hernias in abdominal wall	267, 269
Pancreas	Lies primarily retroperitoneal and deep to stomach; thus, inflamed pancreas may be compressed by stomach and cause intense pain referred to the back; inflamed pancreas, most often caused by biliary obstruction or alcohol abuse, may cause severe life-threatening complications; cancer of head/neck of pancreas can compress biliary tree	292, 306, 329
Urinary System		
Kidney	Right kidney is lower or more inferior than left kidney due to position inferior to liver; renal arteries are generally located at L2 vertebral level and may be involved in abdominal aortic aneurysms; renal calculi (kidney stones) cause significant pain when impacted in ureter	333
Renal pelvis	May become dilated secondary to ureteral or urinary bladder outlet obstruction, a condition known as hydronephrosis, which is readily visualized using ultrasound	333
Endocrine System		
Suprarenal gland	Produces hormones in its cortex (e.g., cortisol, aldosterone) and medulla (e.g., epinephrine, norepinephrine); small masses are frequently incidentally discovered on axial abdominal imaging; need for further workup depends in part on size	330

*Selections were based largely on clinical data and commonly discussed clinical correlations in macroscopic ("gross") anatomy courses.

MUSCLE	MUSCLE GROUP	ORIGIN ATTACHMENT	INSERTION ATTACHMENT	INNERVATION	BLOOD SUPPLY	MAIN ACTIONS
External abdominal oblique m.	Anterior abdominal wall	External surfaces of ribs 5–12	Linea alba, pubic tubercle, anterior half of iliac crest	Anterior rami of T7–T12 spinal nerves	Superior and inferior epigastric, and lumbar arteries	Compresses and supports abdominal viscera, flexes and rotates trunk
Internal abdominal oblique m.	Anterior abdominal wall	Thoracolumbar fascia, anterior two-thirds of iliac crest	Inferior borders of ribs 10–12, linea alba (via rectus sheath), pubic crest and pecten pubis (via inguinal falx)	Anterior rami of T7–L1 spinal nerves	Superior and inferior epigastric, deep circumflex iliac, and lumbar arteries	Compresses and supports abdominal viscera, flexes and rotates trunk
Pyramidalis m.	Anterior abdominal wall	Body of pubis and pubic symphysis (anterior to rectus abdominis)	Linea alba (inferior to umbilicus)	Anterior ramus of T12 spinal nerve via subcostal or iliohypogastric nerves	Inferior epigastric artery	Tenses linea alba
Quadratus lumborum m.	Posterior abdominal wall	Medial half of inferior border of 12th rib, tips of transverse processes of L1–L4 vertebrae	Internal lip of iliac crest, iliolumbar ligament	Anterior rami of T12–L1 spinal nerves	Iliolumbar artery	Extends and laterally flexes vertebral column, fixes 12th rib during inspiration
Rectus abdominis m.	Anterior abdominal wall	Pubic crest, pubic symphysis	Costal cartilages 5–7, xiphoid process	Anterior rami of T7–T12 spinal nerves	Superior and inferior epigastric arteries	Flexes trunk, compresses abdominal viscera
Transversus abdominis m.	Anterior abdominal wall	Internal surfaces of costal cartilages 7–12, thoracolumbar fascia, iliac crest	Linea alba (via rectus sheath), pubic crest and pecten pubis (via inguinal falx)	Anterior rami of T7–L1 spinal nerves	Deep circumflex iliac, inferior epigastric, and lumbar arteries	Compresses and supports abdominal viscera

Variations in spinal nerve contributions to the innervation of muscles, their arterial supply, their attachments, and their actions are common themes in human anatomy. Therefore, expect differences between texts and realize that anatomical variation is normal.

Table 5.4   **Muscles**

PELVIS 6

ELECTRONIC BONUS PLATES

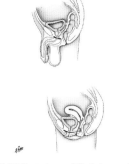

BP 84 Fasciae of Pelvis and Perineum: Male and Female

BP 85 Cystourethrograms: Male and Female

BP 86 Female Urethra

BP 87 Genetics of Reproduction

ELECTRONIC BONUS PLATES—*cont'd*

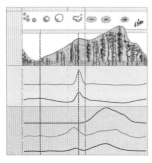

BP 88 Menstrual Cycle

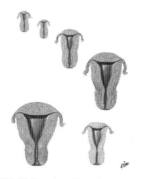

BP 89 Uterine Development

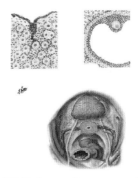

BP 90 Ovary, Oocytes, and Follicles

BP 91 Variations in Hymen

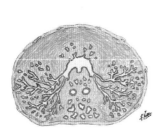

BP 92 Cross Section of Pelvis Through Prostate

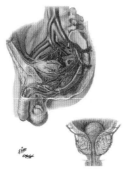

BP 93 Arteries and Veins of Pelvis: Male (Featuring Prostate)

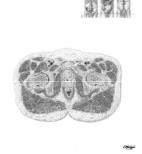

BP 94 Cross Section of Pelvis Through Vesicoprostatic Junction

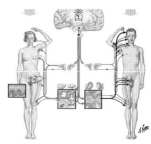

BP 95 Endocrine Glands, Hormones, and Puberty

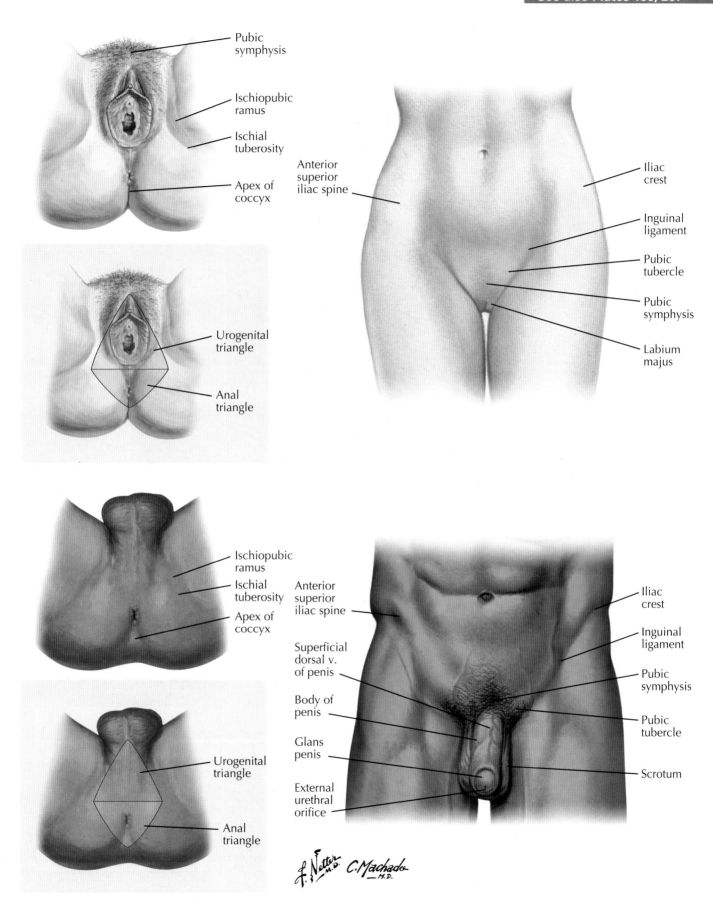

Pubic symphysis

Ischiopubic ramus

Ischial tuberosity

Apex of coccyx

Anterior superior iliac spine

Iliac crest

Inguinal ligament

Pubic tubercle

Pubic symphysis

Labium majus

Urogenital triangle

Anal triangle

Ischiopubic ramus

Ischial tuberosity

Apex of coccyx

Anterior superior iliac spine

Superficial dorsal v. of penis

Body of penis

Glans penis

External urethral orifice

Iliac crest

Inguinal ligament

Pubic symphysis

Pubic tubercle

Scrotum

Urogenital triangle

Anal triangle

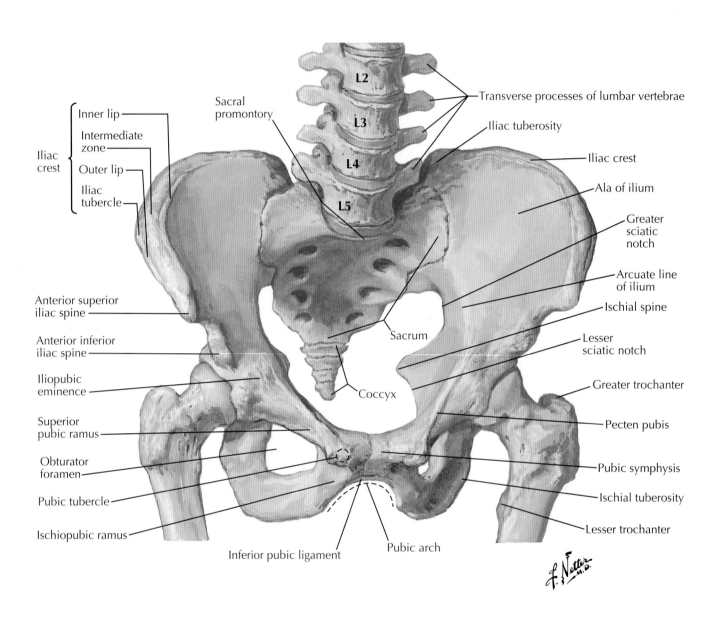

Iliac crest
- Inner lip
- Intermediate zone
- Outer lip
- Iliac tubercle

Sacral promontory

L2

L3

L4

L5

Transverse processes of lumbar vertebrae

Iliac tuberosity

Iliac crest

Ala of ilium

Greater sciatic notch

Arcuate line of ilium

Ischial spine

Lesser sciatic notch

Greater trochanter

Pecten pubis

Pubic symphysis

Ischial tuberosity

Lesser trochanter

Anterior superior iliac spine

Anterior inferior iliac spine

Iliopubic eminence

Superior pubic ramus

Obturator foramen

Pubic tubercle

Ischiopubic ramus

Sacrum

Coccyx

Inferior pubic ligament

Pubic arch

Plate 353

Bony Pelvis

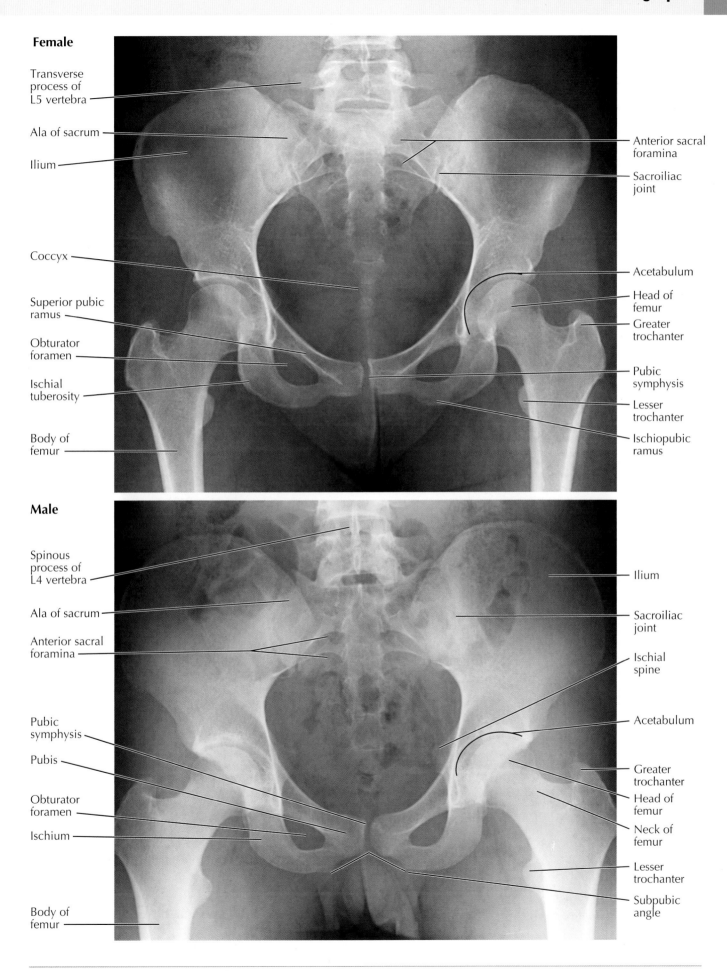

Female

Transverse process of L5 vertebra

Ala of sacrum

Ilium

Anterior sacral foramina

Sacroiliac joint

Coccyx

Acetabulum

Head of femur

Superior pubic ramus

Greater trochanter

Obturator foramen

Pubic symphysis

Ischial tuberosity

Lesser trochanter

Ischiopubic ramus

Body of femur

Male

Spinous process of L4 vertebra

Ala of sacrum

Ilium

Sacroiliac joint

Anterior sacral foramina

Ischial spine

Acetabulum

Pubic symphysis

Pubis

Greater trochanter

Head of femur

Obturator foramen

Ischium

Neck of femur

Lesser trochanter

Subpubic angle

Body of femur

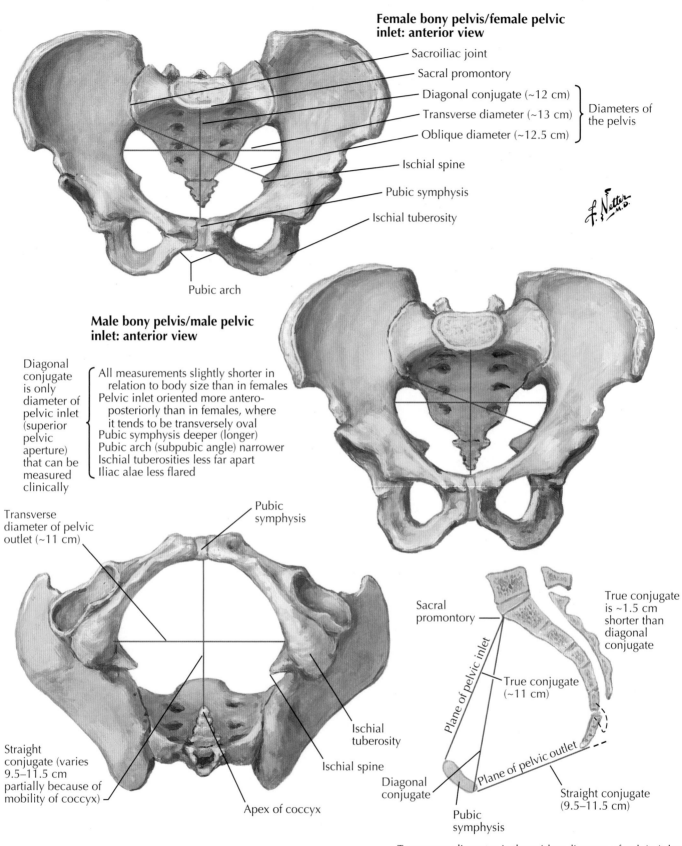

Female bony pelvis/female pelvic inlet: anterior view

- Sacroiliac joint
- Sacral promontory
- Diagonal conjugate (~12 cm)
- Transverse diameter (~13 cm) } Diameters of the pelvis
- Oblique diameter (~12.5 cm)
- Ischial spine
- Pubic symphysis
- Ischial tuberosity
- Pubic arch

Male bony pelvis/male pelvic inlet: anterior view

Diagonal conjugate is only diameter of pelvic inlet (superior pelvic aperture) that can be measured clinically

- All measurements slightly shorter in relation to body size than in females
- Pelvic inlet oriented more antero-posteriorly than in females, where it tends to be transversely oval
- Pubic symphysis deeper (longer)
- Pubic arch (subpubic angle) narrower
- Ischial tuberosities less far apart
- Iliac alae less flared

Transverse diameter of pelvic outlet (~11 cm)

Pubic symphysis

Straight conjugate (varies 9.5–11.5 cm partially because of mobility of coccyx)

Ischial tuberosity

Ischial spine

Apex of coccyx

Female bony pelvis/female pelvic outlet: inferior view

Sacral promontory

True conjugate is ~1.5 cm shorter than diagonal conjugate

Plane of pelvic inlet

True conjugate (~11 cm)

Diagonal conjugate

Plane of pelvic outlet

Straight conjugate (9.5–11.5 cm)

Pubic symphysis

Transverse diameter is the widest distance of pelvic inlet

Female: sagittal section

Plate 355

Bony Pelvis

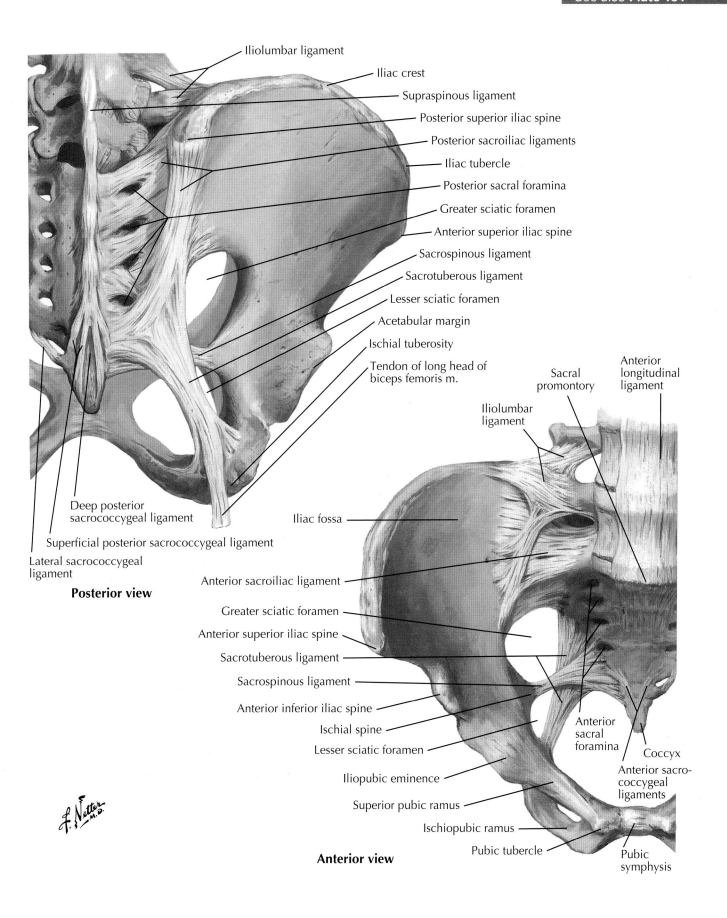

Iliolumbar ligament

Iliac crest

Supraspinous ligament

Posterior superior iliac spine

Posterior sacroiliac ligaments

Iliac tubercle

Posterior sacral foramina

Greater sciatic foramen

Anterior superior iliac spine

Sacrospinous ligament

Sacrotuberous ligament

Lesser sciatic foramen

Acetabular margin

Ischial tuberosity

Tendon of long head of biceps femoris m.

Sacral promontory

Anterior longitudinal ligament

Iliolumbar ligament

Deep posterior sacrococcygeal ligament

Superficial posterior sacrococcygeal ligament

Lateral sacrococcygeal ligament

Posterior view

Iliac fossa

Anterior sacroiliac ligament

Greater sciatic foramen

Anterior superior iliac spine

Sacrotuberous ligament

Sacrospinous ligament

Anterior inferior iliac spine

Ischial spine

Lesser sciatic foramen

Iliopubic eminence

Superior pubic ramus

Ischiopubic ramus

Pubic tubercle

Anterior sacral foramina

Coccyx

Anterior sacro-coccygeal ligaments

Pubic symphysis

Anterior view

Bony Pelvis

Plate 356

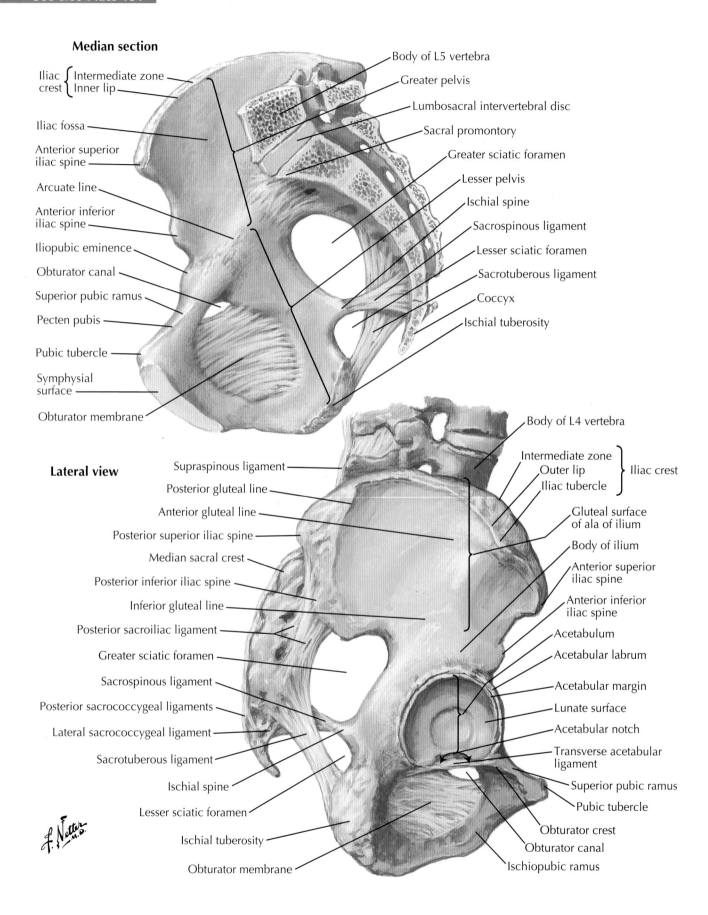

Median section

Iliac crest { Intermediate zone / Inner lip
Iliac fossa
Anterior superior iliac spine
Arcuate line
Anterior inferior iliac spine
Iliopubic eminence
Obturator canal
Superior pubic ramus
Pecten pubis
Pubic tubercle
Symphysial surface
Obturator membrane

Body of L5 vertebra
Greater pelvis
Lumbosacral intervertebral disc
Sacral promontory
Greater sciatic foramen
Lesser pelvis
Ischial spine
Sacrospinous ligament
Lesser sciatic foramen
Sacrotuberous ligament
Coccyx
Ischial tuberosity

Lateral view

Supraspinous ligament
Posterior gluteal line
Anterior gluteal line
Posterior superior iliac spine
Median sacral crest
Posterior inferior iliac spine
Inferior gluteal line
Posterior sacroiliac ligament
Greater sciatic foramen
Sacrospinous ligament
Posterior sacrococcygeal ligaments
Lateral sacrococcygeal ligament
Sacrotuberous ligament
Ischial spine
Lesser sciatic foramen
Ischial tuberosity
Obturator membrane

Body of L4 vertebra
Intermediate zone / Outer lip / Iliac tubercle } Iliac crest
Gluteal surface of ala of ilium
Body of ilium
Anterior superior iliac spine
Anterior inferior iliac spine
Acetabulum
Acetabular labrum
Acetabular margin
Lunate surface
Acetabular notch
Transverse acetabular ligament
Superior pubic ramus
Pubic tubercle
Obturator crest
Obturator canal
Ischiopubic ramus

Plate 357   **Bony Pelvis**

Medial view

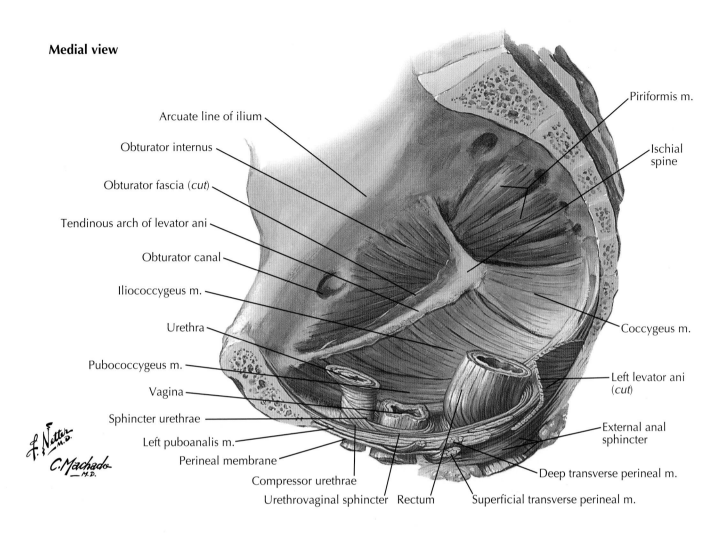

Arcuate line of ilium

Obturator internus

Obturator fascia (*cut*)

Tendinous arch of levator ani

Obturator canal

Iliococcygeus m.

Urethra

Pubococcygeus m.

Vagina

Sphincter urethrae

Left puboanalis m.

Perineal membrane

Compressor urethrae

Urethrovaginal sphincter   Rectum

Piriformis m.

Ischial spine

Coccygeus m.

Left levator ani (*cut*)

External anal sphincter

Deep transverse perineal m.

Superficial transverse perineal m.

Lateral view

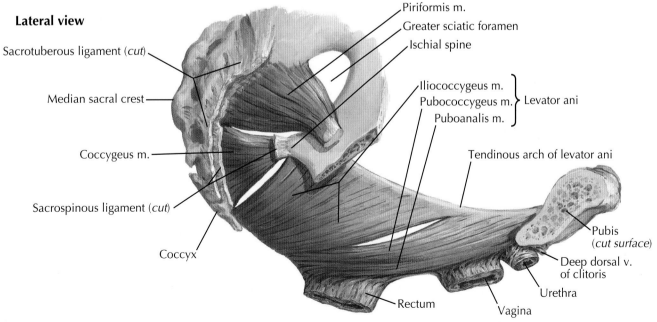

Sacrotuberous ligament (*cut*)

Median sacral crest

Coccygeus m.

Sacrospinous ligament (*cut*)

Coccyx

Piriformis m.

Greater sciatic foramen

Ischial spine

Iliococcygeus m.
Pubococcygeus m. } Levator ani
Puboanalis m.

Tendinous arch of levator ani

Pubis (*cut surface*)

Deep dorsal v. of clitoris

Urethra

Vagina

Rectum

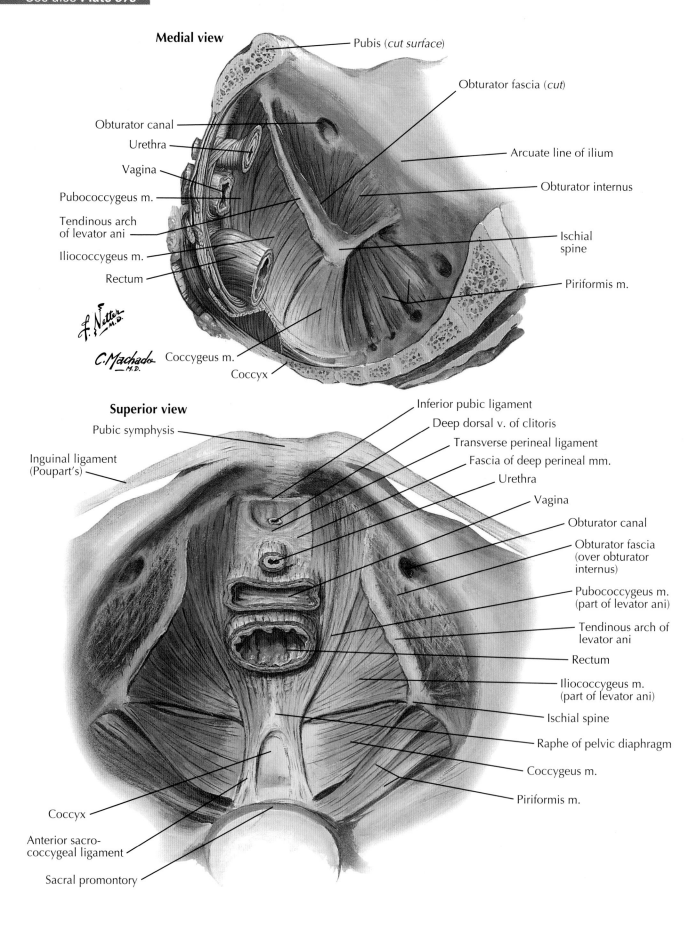

Medial view

Pubis (*cut surface*)

Obturator canal

Urethra

Vagina

Pubococcygeus m.

Tendinous arch
of levator ani

Iliococcygeus m.

Rectum

Coccygeus m.

Coccyx

Obturator fascia (*cut*)

Arcuate line of ilium

Obturator internus

Ischial
spine

Piriformis m.

Superior view

Pubic symphysis

Inguinal ligament
(Poupart's)

Inferior pubic ligament

Deep dorsal v. of clitoris

Transverse perineal ligament

Fascia of deep perineal mm.

Urethra

Vagina

Obturator canal

Obturator fascia
(over obturator
internus)

Pubococcygeus m.
(part of levator ani)

Tendinous arch of
levator ani

Rectum

Iliococcygeus m.
(part of levator ani)

Ischial spine

Raphe of pelvic diaphragm

Coccygeus m.

Piriformis m.

Coccyx

Anterior sacro-
coccygeal ligament

Sacral promontory

Plate 359

Pelvic Diaphragm and Viscera

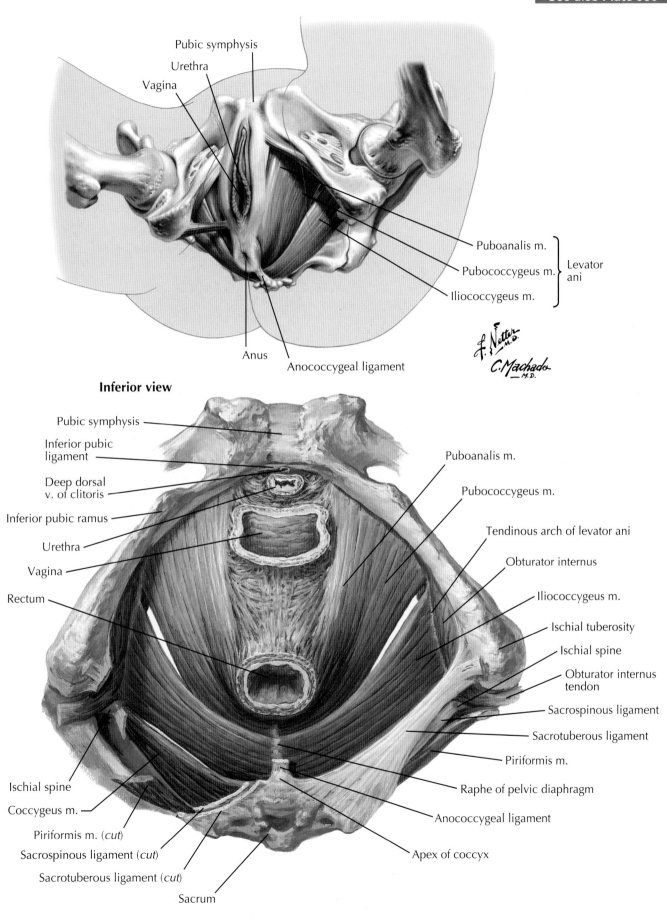

Pubic symphysis

Urethra

Vagina

Puboanalis m.

Pubococcygeus m. — Levator ani

Iliococcygeus m.

Anus

Anococcygeal ligament

Inferior view

Pubic symphysis

Inferior pubic ligament

Deep dorsal v. of clitoris

Inferior pubic ramus

Urethra

Vagina

Rectum

Ischial spine

Coccygeus m.

Piriformis m. (*cut*)

Sacrospinous ligament (*cut*)

Sacrotuberous ligament (*cut*)

Sacrum

Puboanalis m.

Pubococcygeus m.

Tendinous arch of levator ani

Obturator internus

Iliococcygeus m.

Ischial tuberosity

Ischial spine

Obturator internus tendon

Sacrospinous ligament

Sacrotuberous ligament

Piriformis m.

Raphe of pelvic diaphragm

Anococcygeal ligament

Apex of coccyx

Pelvic Diaphragm and Viscera

Superior view
(*viscera removed*)

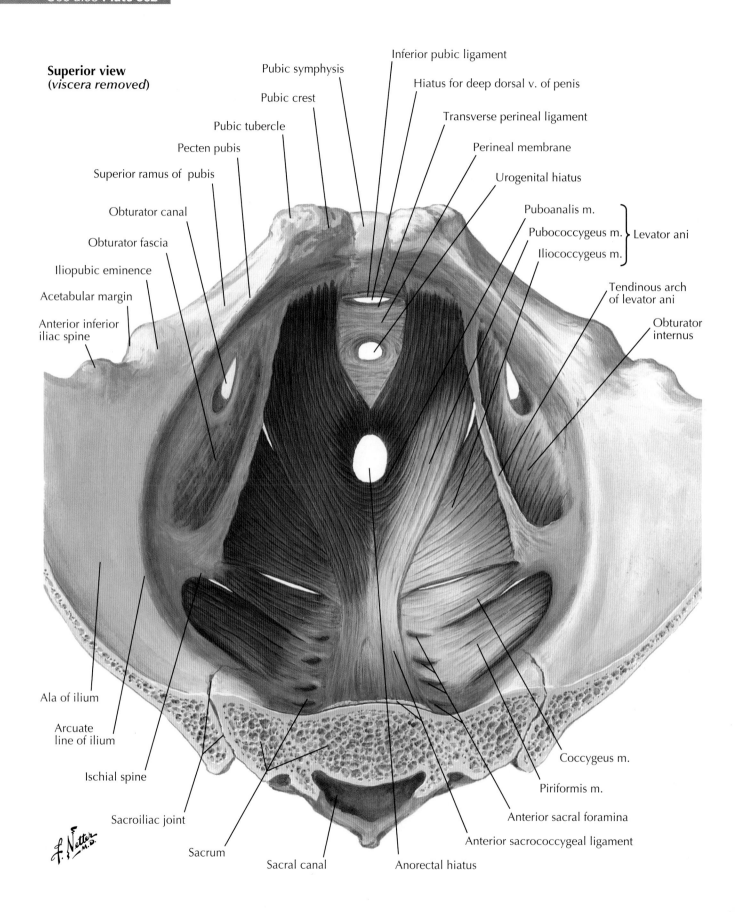

Pubic symphysis

Pubic crest

Pubic tubercle

Pecten pubis

Superior ramus of pubis

Obturator canal

Obturator fascia

Iliopubic eminence

Acetabular margin

Anterior inferior
iliac spine

Inferior pubic ligament

Hiatus for deep dorsal v. of penis

Transverse perineal ligament

Perineal membrane

Urogenital hiatus

Puboanalis m.

Pubococcygeus m. — Levator ani

Iliococcygeus m.

Tendinous arch
of levator ani

Obturator
internus

Ala of ilium

Arcuate
line of ilium

Ischial spine

Sacroiliac joint

Sacrum

Sacral canal

Anorectal hiatus

Anterior sacrococcygeal ligament

Anterior sacral foramina

Piriformis m.

Coccygeus m.

Plate 361

Pelvic Diaphragm and Viscera

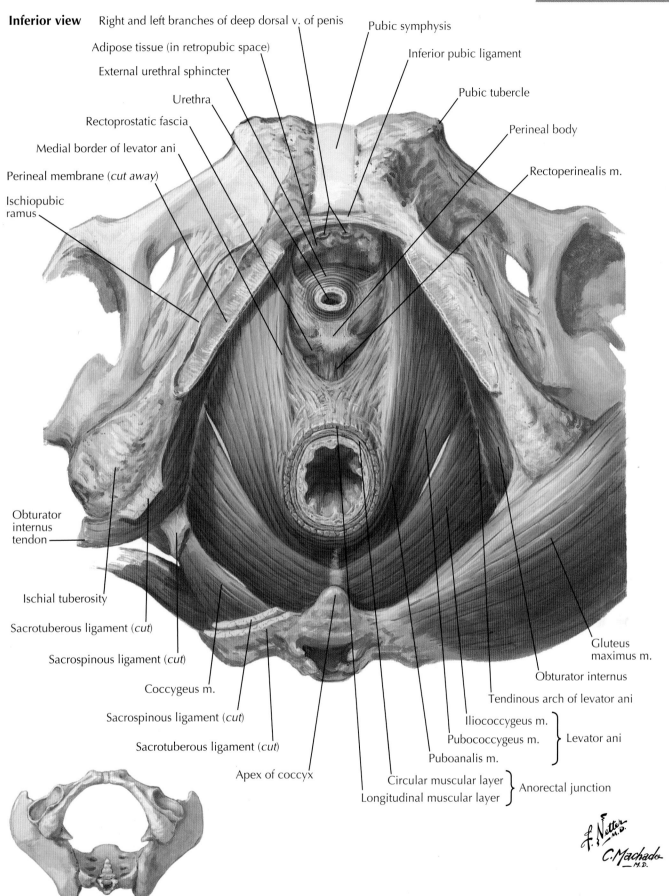

Inferior view

Right and left branches of deep dorsal v. of penis

Adipose tissue (*in retropubic space*)

External urethral sphincter

Urethra

Rectoprostatic fascia

Medial border of levator ani

Perineal membrane (*cut away*)

Ischiopubic ramus

Obturator internus tendon

Ischial tuberosity

Sacrotuberous ligament (*cut*)

Sacrospinous ligament (*cut*)

Coccygeus m.

Sacrospinous ligament (*cut*)

Sacrotuberous ligament (*cut*)

Apex of coccyx

Pubic symphysis

Inferior pubic ligament

Pubic tubercle

Perineal body

Rectoperinealis m.

Gluteus maximus m.

Obturator internus

Tendinous arch of levator ani

Iliococcygeus m. ⎫
Pubococcygeus m. ⎬ Levator ani
Puboanalis m. ⎭

Circular muscular layer ⎫ Anorectal junction
Longitudinal muscular layer ⎭

Superior view

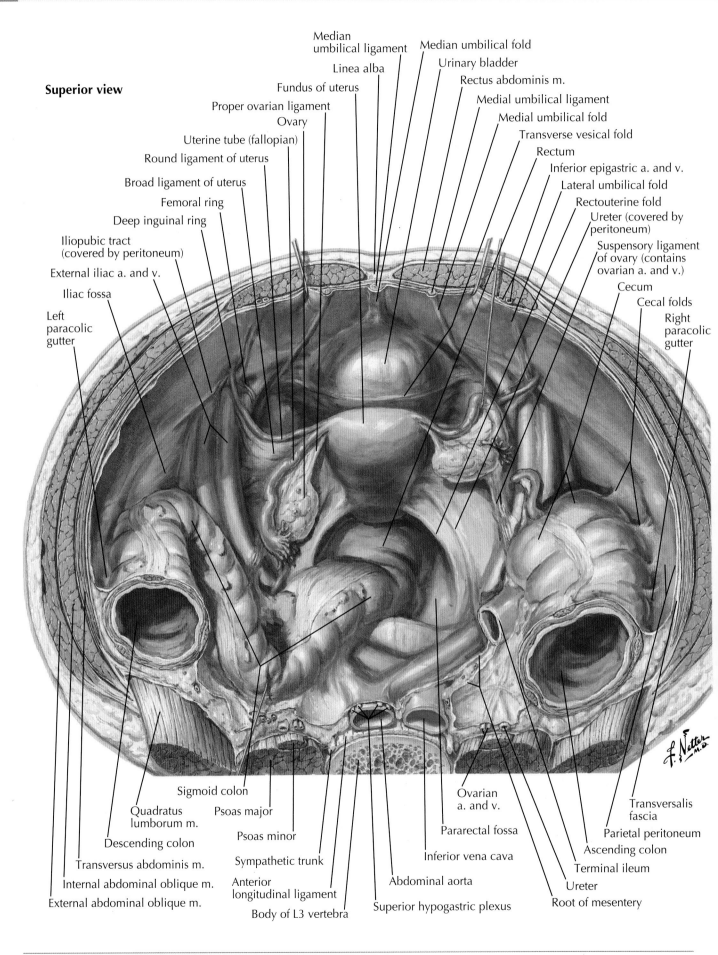

Median umbilical ligament

Linea alba

Fundus of uterus

Proper ovarian ligament

Ovary

Uterine tube (fallopian)

Round ligament of uterus

Broad ligament of uterus

Femoral ring

Deep inguinal ring

Iliopubic tract (covered by peritoneum)

External iliac a. and v.

Iliac fossa

Left paracolic gutter

Median umbilical fold

Urinary bladder

Rectus abdominis m.

Medial umbilical ligament

Medial umbilical fold

Transverse vesical fold

Rectum

Inferior epigastric a. and v.

Lateral umbilical fold

Rectouterine fold

Ureter (covered by peritoneum)

Suspensory ligament of ovary (contains ovarian a. and v.)

Cecum

Cecal folds

Right paracolic gutter

Sigmoid colon

Quadratus lumborum m.

Descending colon

Transversus abdominis m.

Internal abdominal oblique m.

External abdominal oblique m.

Psoas major

Psoas minor

Sympathetic trunk

Anterior longitudinal ligament

Body of L3 vertebra

Ovarian a. and v.

Pararectal fossa

Inferior vena cava

Abdominal aorta

Superior hypogastric plexus

Transversalis fascia

Parietal peritoneum

Ascending colon

Terminal ileum

Ureter

Root of mesentery

Plate 363

Pelvic Diaphragm and Viscera

Paramedian (sagittal) dissection

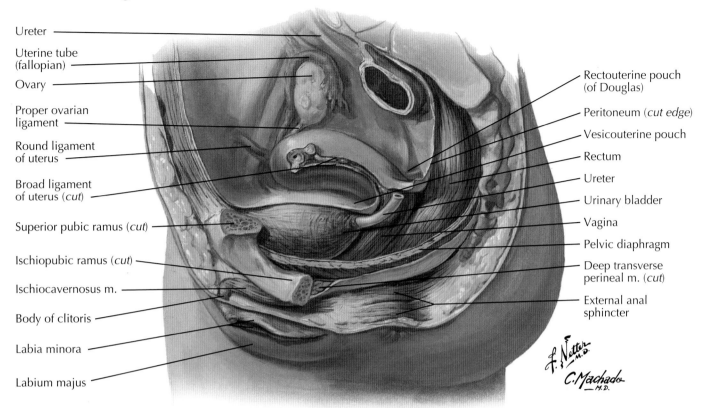

Ureter

Uterine tube (fallopian)

Ovary

Proper ovarian ligament

Round ligament of uterus

Broad ligament of uterus (*cut*)

Superior pubic ramus (*cut*)

Ischiopubic ramus (*cut*)

Ischiocavernosus m.

Body of clitoris

Labia minora

Labium majus

Rectouterine pouch (of Douglas)

Peritoneum (*cut edge*)

Vesicouterine pouch

Rectum

Ureter

Urinary bladder

Vagina

Pelvic diaphragm

Deep transverse perineal m. (*cut*)

External anal sphincter

Median (sagittal) section

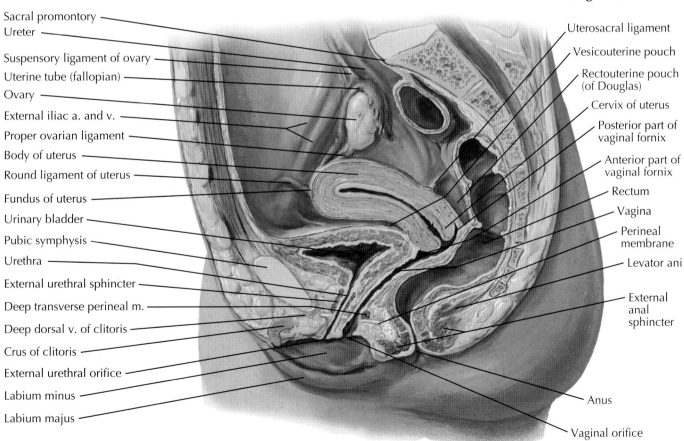

Sacral promontory

Ureter

Suspensory ligament of ovary

Uterine tube (fallopian)

Ovary

External iliac a. and v.

Proper ovarian ligament

Body of uterus

Round ligament of uterus

Fundus of uterus

Urinary bladder

Pubic symphysis

Urethra

External urethral sphincter

Deep transverse perineal m.

Deep dorsal v. of clitoris

Crus of clitoris

External urethral orifice

Labium minus

Labium majus

Uterosacral ligament

Vesicouterine pouch

Rectouterine pouch (of Douglas)

Cervix of uterus

Posterior part of vaginal fornix

Anterior part of vaginal fornix

Rectum

Vagina

Perineal membrane

Levator ani

External anal sphincter

Anus

Vaginal orifice

Superior view with peritoneum intact

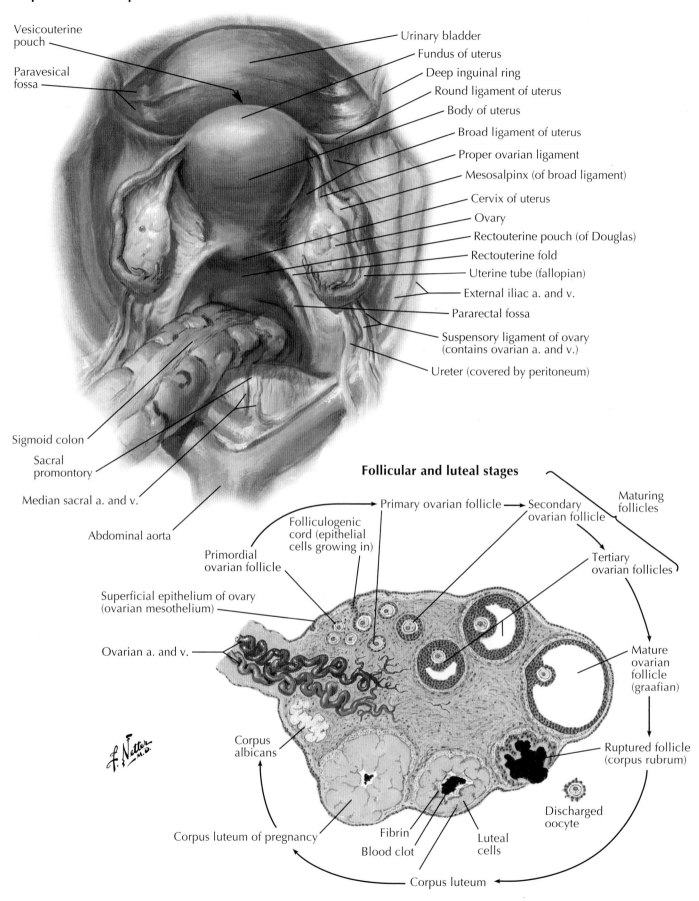

Vesicouterine pouch

Paravesical fossa

Urinary bladder

Fundus of uterus

Deep inguinal ring

Round ligament of uterus

Body of uterus

Broad ligament of uterus

Proper ovarian ligament

Mesosalpinx (of broad ligament)

Cervix of uterus

Ovary

Rectouterine pouch (of Douglas)

Rectouterine fold

Uterine tube (fallopian)

External iliac a. and v.

Pararectal fossa

Suspensory ligament of ovary (contains ovarian a. and v.)

Ureter (covered by peritoneum)

Sigmoid colon

Sacral promontory

Median sacral a. and v.

Abdominal aorta

Follicular and luteal stages

Primordial ovarian follicle

Folliculogenic cord (epithelial cells growing in)

Primary ovarian follicle

Secondary ovarian follicle

Maturing follicles

Tertiary ovarian follicles

Superficial epithelium of ovary (ovarian mesothelium)

Ovarian a. and v.

Mature ovarian follicle (graafian)

Corpus albicans

Ruptured follicle (corpus rubrum)

Discharged oocyte

Corpus luteum of pregnancy

Fibrin

Blood clot

Luteal cells

Corpus luteum

Plate 365

Pelvic Diaphragm and Viscera

Female: superior view (peritoneum and loose connective tissue removed)

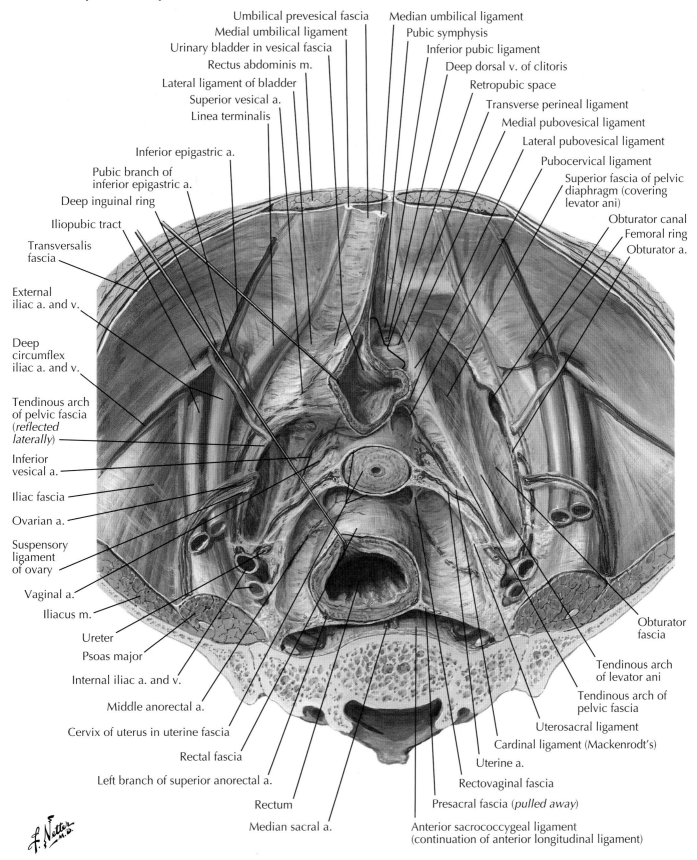

Umbilical prevesical fascia
Medial umbilical ligament
Urinary bladder in vesical fascia
Rectus abdominis m.
Lateral ligament of bladder
Superior vesical a.
Linea terminalis

Median umbilical ligament
Pubic symphysis
Inferior pubic ligament
Deep dorsal v. of clitoris
Retropubic space
Transverse perineal ligament
Medial pubovesical ligament
Lateral pubovesical ligament
Pubocervical ligament
Superior fascia of pelvic diaphragm (covering levator ani)
Obturator canal
Femoral ring
Obturator a.

Inferior epigastric a.
Pubic branch of inferior epigastric a.
Deep inguinal ring
Iliopubic tract
Transversalis fascia
External iliac a. and v.
Deep circumflex iliac a. and v.
Tendinous arch of pelvic fascia (*reflected laterally*)
Inferior vesical a.
Iliac fascia
Ovarian a.
Suspensory ligament of ovary
Vaginal a.
Iliacus m.
Ureter
Psoas major
Internal iliac a. and v.
Middle anorectal a.
Cervix of uterus in uterine fascia
Rectal fascia
Left branch of superior anorectal a.
Rectum
Median sacral a.

Obturator fascia
Tendinous arch of levator ani
Tendinous arch of pelvic fascia
Uterosacral ligament
Cardinal ligament (Mackenrodt's)
Uterine a.
Rectovaginal fascia
Presacral fascia (*pulled away*)
Anterior sacrococcygeal ligament (continuation of anterior longitudinal ligament)

Superior view

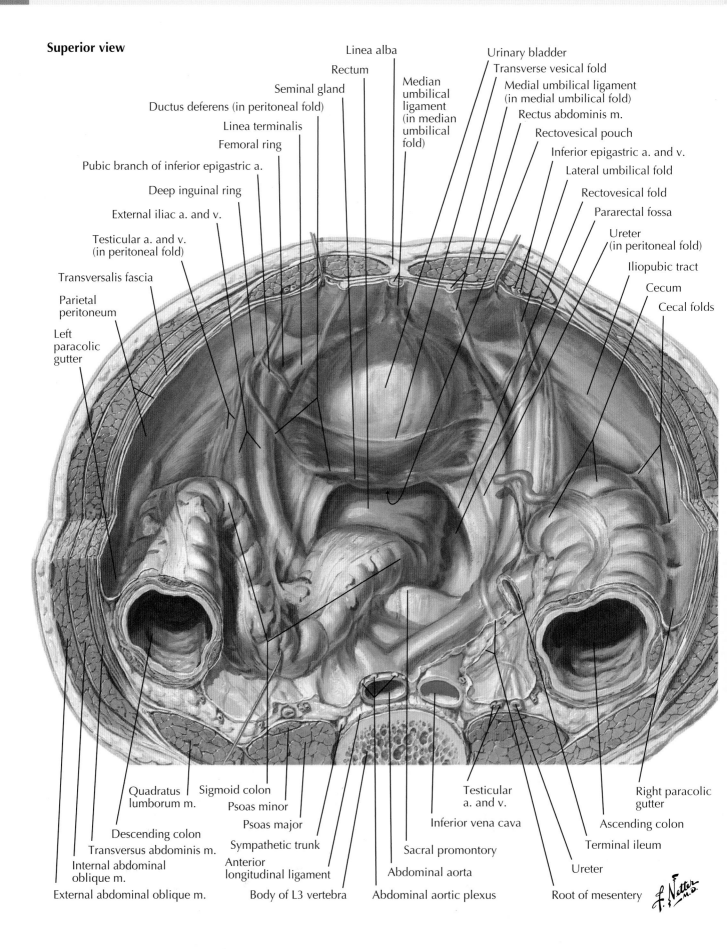

Linea alba

Rectum

Seminal gland

Ductus deferens (in peritoneal fold)

Linea terminalis

Femoral ring

Pubic branch of inferior epigastric a.

Deep inguinal ring

External iliac a. and v.

Testicular a. and v. (in peritoneal fold)

Transversalis fascia

Parietal peritoneum

Left paracolic gutter

Urinary bladder

Transverse vesical fold

Median umbilical ligament (in median umbilical fold)

Medial umbilical ligament (in medial umbilical fold)

Rectus abdominis m.

Rectovesical pouch

Inferior epigastric a. and v.

Lateral umbilical fold

Rectovesical fold

Pararectal fossa

Ureter (in peritoneal fold)

Iliopubic tract

Cecum

Cecal folds

Quadratus lumborum m.

Sigmoid colon

Psoas minor

Psoas major

Sympathetic trunk

Anterior longitudinal ligament

Body of L3 vertebra

Descending colon

Transversus abdominis m.

Internal abdominal oblique m.

External abdominal oblique m.

Testicular a. and v.

Inferior vena cava

Sacral promontory

Abdominal aorta

Abdominal aortic plexus

Right paracolic gutter

Ascending colon

Terminal ileum

Ureter

Root of mesentery

Plate 367

Pelvic Diaphragm and Viscera

Parasagittal view

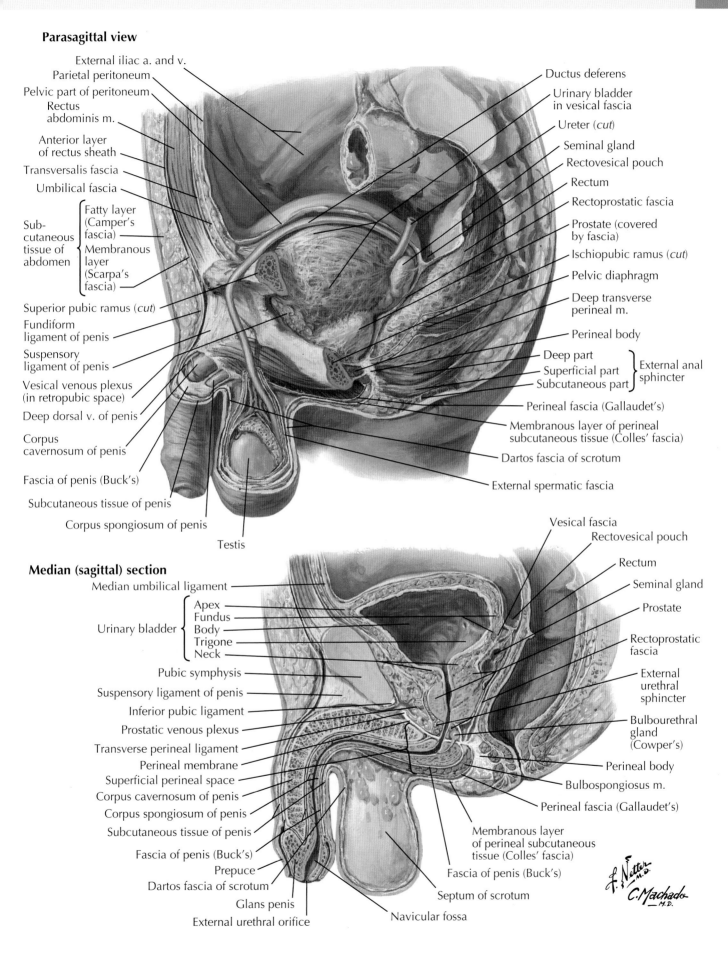

External iliac a. and v.
Parietal peritoneum
Pelvic part of peritoneum
Rectus abdominis m.
Anterior layer of rectus sheath
Transversalis fascia
Umbilical fascia
Subcutaneous tissue of abdomen
{ Fatty layer (Camper's fascia)
Membranous layer (Scarpa's fascia) }
Superior pubic ramus (*cut*)
Fundiform ligament of penis
Suspensory ligament of penis
Vesical venous plexus (in retropubic space)
Deep dorsal v. of penis
Corpus cavernosum of penis
Fascia of penis (Buck's)
Subcutaneous tissue of penis
Corpus spongiosum of penis
Testis

Ductus deferens
Urinary bladder in vesical fascia
Ureter (*cut*)
Seminal gland
Rectovesical pouch
Rectum
Rectoprostatic fascia
Prostate (covered by fascia)
Ischiopubic ramus (*cut*)
Pelvic diaphragm
Deep transverse perineal m.
Perineal body
Deep part
Superficial part
Subcutaneous part
} External anal sphincter
Perineal fascia (Gallaudet's)
Membranous layer of perineal subcutaneous tissue (Colles' fascia)
Dartos fascia of scrotum
External spermatic fascia

Median (sagittal) section

Median umbilical ligament
Urinary bladder {
Apex
Fundus
Body
Trigone
Neck }
Pubic symphysis
Suspensory ligament of penis
Inferior pubic ligament
Prostatic venous plexus
Transverse perineal ligament
Perineal membrane
Superficial perineal space
Corpus cavernosum of penis
Corpus spongiosum of penis
Subcutaneous tissue of penis
Fascia of penis (Buck's)
Prepuce
Dartos fascia of scrotum
Glans penis
External urethral orifice

Vesical fascia
Rectovesical pouch
Rectum
Seminal gland
Prostate
Rectoprostatic fascia
External urethral sphincter
Bulbourethral gland (Cowper's)
Perineal body
Bulbospongiosus m.
Perineal fascia (Gallaudet's)
Membranous layer of perineal subcutaneous tissue (Colles' fascia)
Fascia of penis (Buck's)
Septum of scrotum
Navicular fossa

F. Netter M.D.
C. Machado M.D.

Female: median section

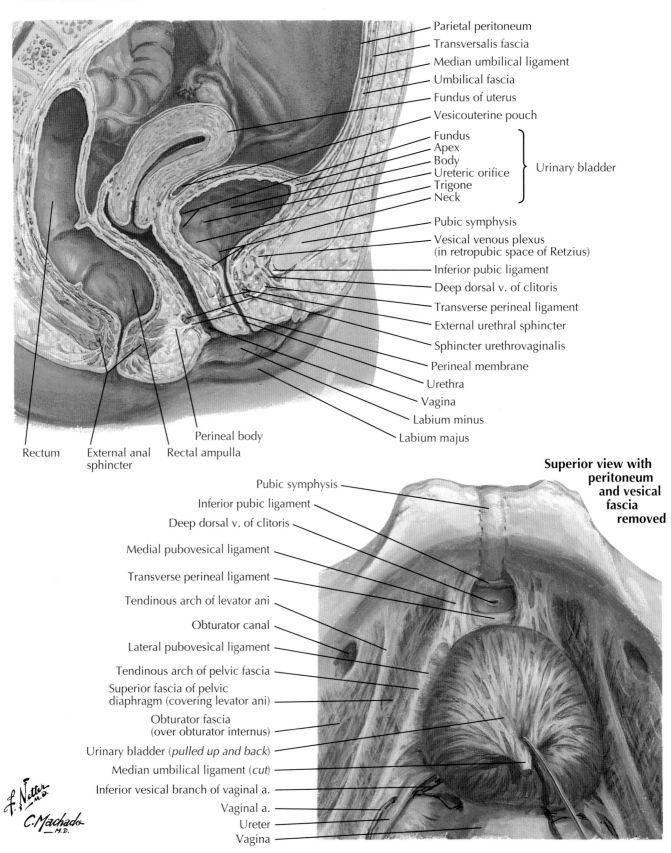

Parietal peritoneum
Transversalis fascia
Median umbilical ligament
Umbilical fascia
Fundus of uterus
Vesicouterine pouch
Fundus
Apex
Body
Ureteric orifice — Urinary bladder
Trigone
Neck
Pubic symphysis
Vesical venous plexus
(in retropubic space of Retzius)
Inferior pubic ligament
Deep dorsal v. of clitoris
Transverse perineal ligament
External urethral sphincter
Sphincter urethrovaginalis
Perineal membrane
Urethra
Vagina
Labium minus
Labium majus

Rectum
External anal sphincter
Rectal ampulla
Perineal body

Superior view with peritoneum and vesical fascia removed

Pubic symphysis
Inferior pubic ligament
Deep dorsal v. of clitoris
Medial pubovesical ligament
Transverse perineal ligament
Tendinous arch of levator ani
Obturator canal
Lateral pubovesical ligament
Tendinous arch of pelvic fascia
Superior fascia of pelvic diaphragm (covering levator ani)
Obturator fascia (over obturator internus)
Urinary bladder (*pulled up and back*)
Median umbilical ligament (*cut*)
Inferior vesical branch of vaginal a.
Vaginal a.
Ureter
Vagina

Plate 369   **Urinary Bladder**

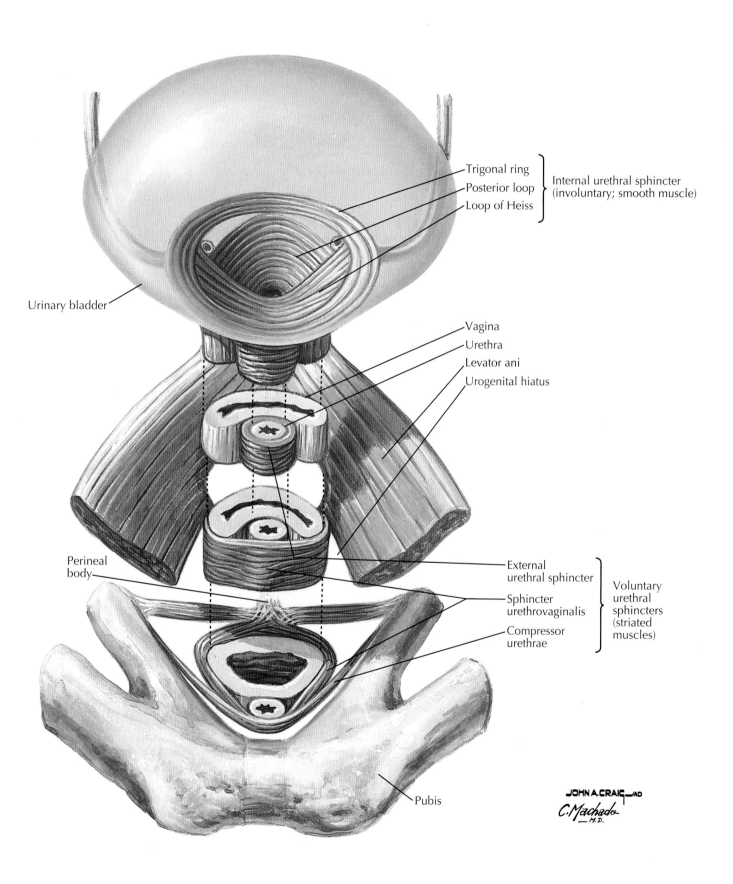

Trigonal ring
Posterior loop
Loop of Heiss
} Internal urethral sphincter (involuntary; smooth muscle)

Urinary bladder

Vagina
Urethra
Levator ani
Urogenital hiatus

Perineal body

External urethral sphincter
Sphincter urethrovaginalis
Compressor urethrae
} Voluntary urethral sphincters (striated muscles)

Pubis

JOHN A. CRAIG—AD
C. Machado
—M.D.

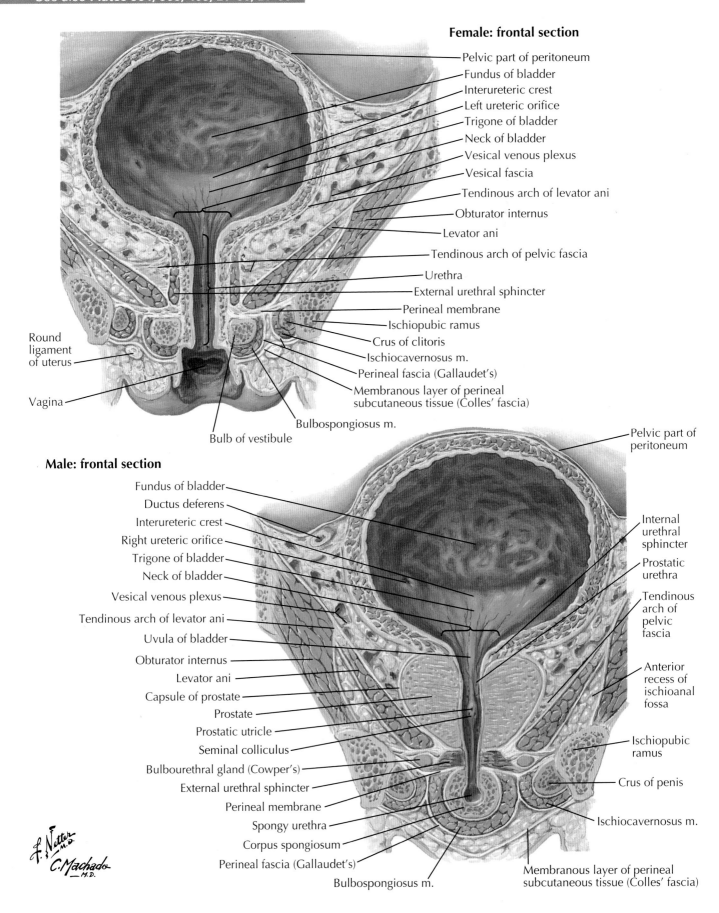

Female: frontal section

- Pelvic part of peritoneum
- Fundus of bladder
- Interureteric crest
- Left ureteric orifice
- Trigone of bladder
- Neck of bladder
- Vesical venous plexus
- Vesical fascia
- Tendinous arch of levator ani
- Obturator internus
- Levator ani
- Tendinous arch of pelvic fascia
- Urethra
- External urethral sphincter
- Perineal membrane
- Ischiopubic ramus
- Crus of clitoris
- Ischiocavernosus m.
- Perineal fascia (Gallaudet's)
- Membranous layer of perineal subcutaneous tissue (Colles' fascia)

- Round ligament of uterus
- Vagina
- Bulbospongiosus m.
- Bulb of vestibule

Male: frontal section

- Fundus of bladder
- Ductus deferens
- Interureteric crest
- Right ureteric orifice
- Trigone of bladder
- Neck of bladder
- Vesical venous plexus
- Tendinous arch of levator ani
- Uvula of bladder
- Obturator internus
- Levator ani
- Capsule of prostate
- Prostate
- Prostatic utricle
- Seminal colliculus
- Bulbourethral gland (Cowper's)
- External urethral sphincter
- Perineal membrane
- Spongy urethra
- Corpus spongiosum
- Perineal fascia (Gallaudet's)
- Bulbospongiosus m.

- Pelvic part of peritoneum
- Internal urethral sphincter
- Prostatic urethra
- Tendinous arch of pelvic fascia
- Anterior recess of ischioanal fossa
- Ischiopubic ramus
- Crus of penis
- Ischiocavernosus m.
- Membranous layer of perineal subcutaneous tissue (Colles' fascia)

Plate 371

Urinary Bladder

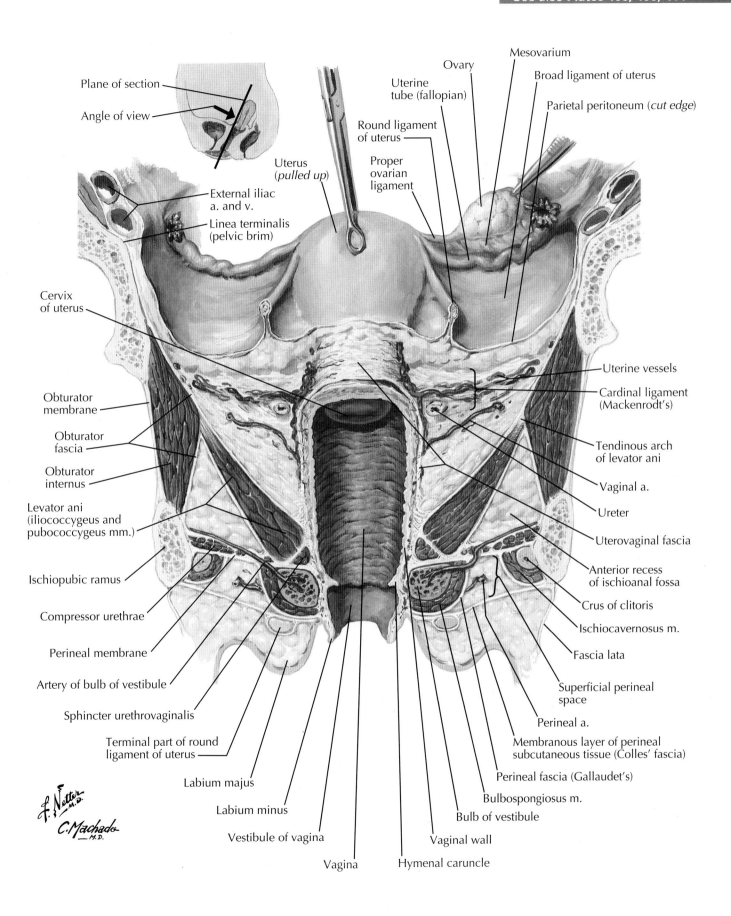

Plane of section

Angle of view

Ovary

Mesovarium

Broad ligament of uterus

Uterine tube (fallopian)

Parietal peritoneum (*cut edge*)

Round ligament of uterus

Uterus (*pulled up*)

Proper ovarian ligament

External iliac a. and v.

Linea terminalis (pelvic brim)

Cervix of uterus

Uterine vessels

Cardinal ligament (Mackenrodt's)

Obturator membrane

Tendinous arch of levator ani

Obturator fascia

Vaginal a.

Obturator internus

Ureter

Levator ani (iliococcygeus and pubococcygeus mm.)

Uterovaginal fascia

Anterior recess of ischioanal fossa

Ischiopubic ramus

Crus of clitoris

Compressor urethrae

Ischiocavernosus m.

Perineal membrane

Fascia lata

Artery of bulb of vestibule

Superficial perineal space

Sphincter urethrovaginalis

Perineal a.

Terminal part of round ligament of uterus

Membranous layer of perineal subcutaneous tissue (Colles' fascia)

Perineal fascia (Gallaudet's)

Labium majus

Bulbospongiosus m.

Labium minus

Bulb of vestibule

Vestibule of vagina

Vaginal wall

Vagina

Hymenal caruncle

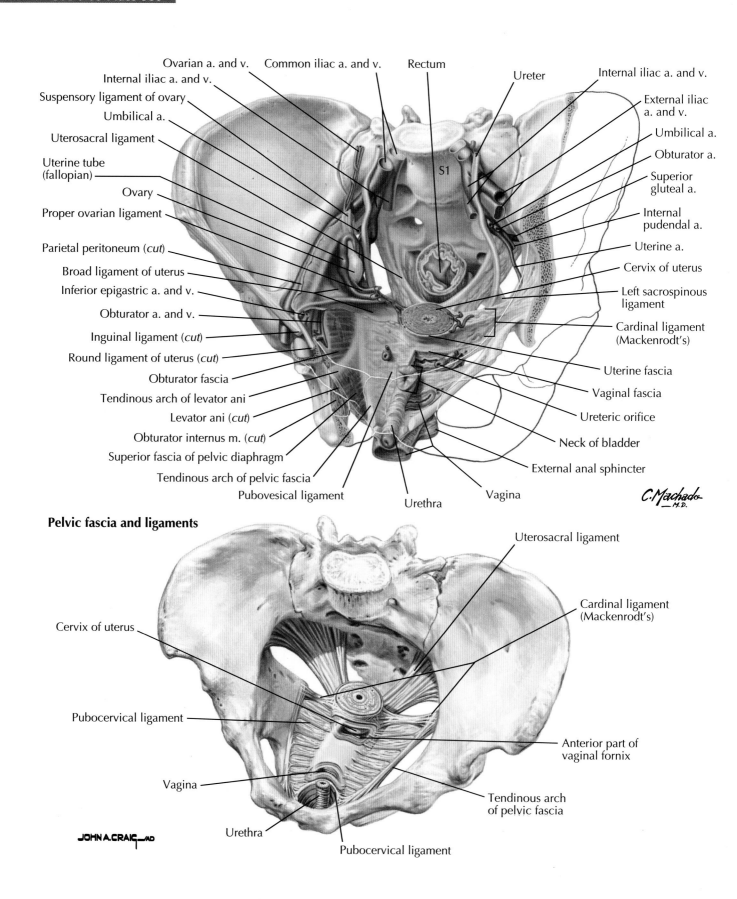

Ovarian a. and v.

Common iliac a. and v.

Rectum

Ureter

Internal iliac a. and v.

Internal iliac a. and v.

External iliac a. and v.

Suspensory ligament of ovary

Umbilical a.

Umbilical a.

Uterosacral ligament

Obturator a.

Uterine tube (fallopian)

Superior gluteal a.

Ovary

Internal pudendal a.

Proper ovarian ligament

Uterine a.

Parietal peritoneum (cut)

Cervix of uterus

Broad ligament of uterus

Left sacrospinous ligament

Inferior epigastric a. and v.

Cardinal ligament (Mackenrodt's)

Obturator a. and v.

Inguinal ligament (cut)

Uterine fascia

Round ligament of uterus (cut)

Vaginal fascia

Obturator fascia

Ureteric orifice

Tendinous arch of levator ani

Neck of bladder

Levator ani (cut)

Obturator internus m. (cut)

External anal sphincter

Superior fascia of pelvic diaphragm

Tendinous arch of pelvic fascia

Pubovesical ligament

Urethra

Vagina

S1

C. Machado M.D.

Pelvic fascia and ligaments

Uterosacral ligament

Cardinal ligament (Mackenrodt's)

Cervix of uterus

Pubocervical ligament

Anterior part of vaginal fornix

Vagina

Tendinous arch of pelvic fascia

Urethra

Pubocervical ligament

JOHN A. CRAIG—MD

Plate 373

Female Internal Genitalia

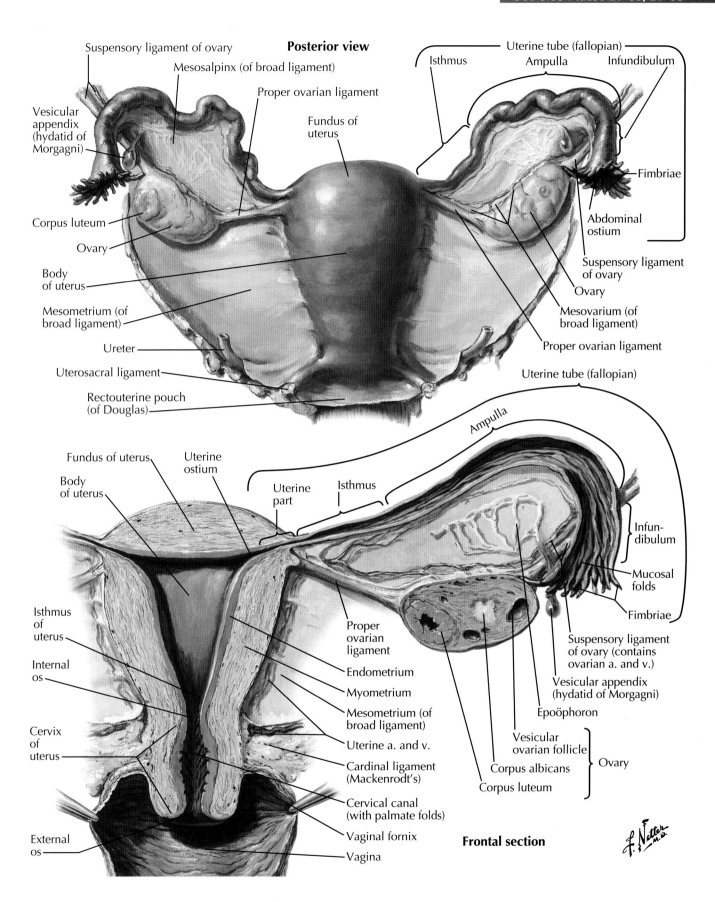

Posterior view

Suspensory ligament of ovary

Mesosalpinx (of broad ligament)

Proper ovarian ligament

Vesicular appendix (hydatid of Morgagni)

Fundus of uterus

Uterine tube (fallopian)

Isthmus

Ampulla

Infundibulum

Corpus luteum

Ovary

Fimbriae

Body of uterus

Abdominal ostium

Mesometrium (of broad ligament)

Suspensory ligament of ovary

Ovary

Ureter

Mesovarium (of broad ligament)

Uterosacral ligament

Proper ovarian ligament

Rectouterine pouch (of Douglas)

Uterine tube (fallopian)

Fundus of uterus

Uterine ostium

Ampulla

Body of uterus

Uterine part

Isthmus

Isthmus of uterus

Infun-dibulum

Internal os

Mucosal folds

Fimbriae

Proper ovarian ligament

Endometrium

Suspensory ligament of ovary (contains ovarian a. and v.)

Cervix of uterus

Myometrium

Mesometrium (of broad ligament)

Vesicular appendix (hydatid of Morgagni)

Uterine a. and v.

Epoöphoron

Cardinal ligament (Mackenrodt's)

Vesicular ovarian follicle

Cervical canal (with palmate folds)

Corpus albicans

Ovary

External os

Corpus luteum

Vaginal fornix

Frontal section

Vagina

Female Internal Genitalia

Plate 374

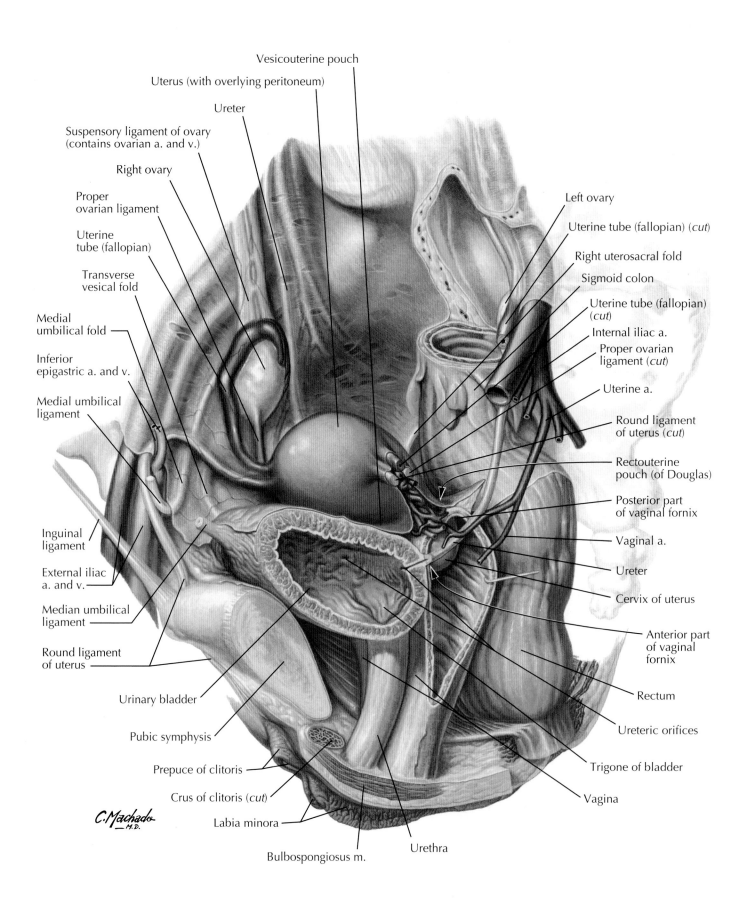

Vesicouterine pouch

Uterus (with overlying peritoneum)

Ureter

Suspensory ligament of ovary
(contains ovarian a. and v.)

Right ovary

Proper
ovarian ligament

Uterine
tube (fallopian)

Transverse
vesical fold

Medial
umbilical fold

Inferior
epigastric a. and v.

Medial umbilical
ligament

Inguinal
ligament

External iliac
a. and v.

Median umbilical
ligament

Round ligament
of uterus

Urinary bladder

Pubic symphysis

Prepuce of clitoris

Crus of clitoris (cut)

Labia minora

C. Machado
— M.D.

Bulbospongiosus m.

Left ovary

Uterine tube (fallopian) (cut)

Right uterosacral fold

Sigmoid colon

Uterine tube (fallopian)
(cut)

Internal iliac a.

Proper ovarian
ligament (cut)

Uterine a.

Round ligament
of uterus (cut)

Rectouterine
pouch (of Douglas)

Posterior part
of vaginal fornix

Vaginal a.

Ureter

Cervix of uterus

Anterior part
of vaginal
fornix

Rectum

Ureteric orifices

Trigone of bladder

Vagina

Urethra

Plate 375

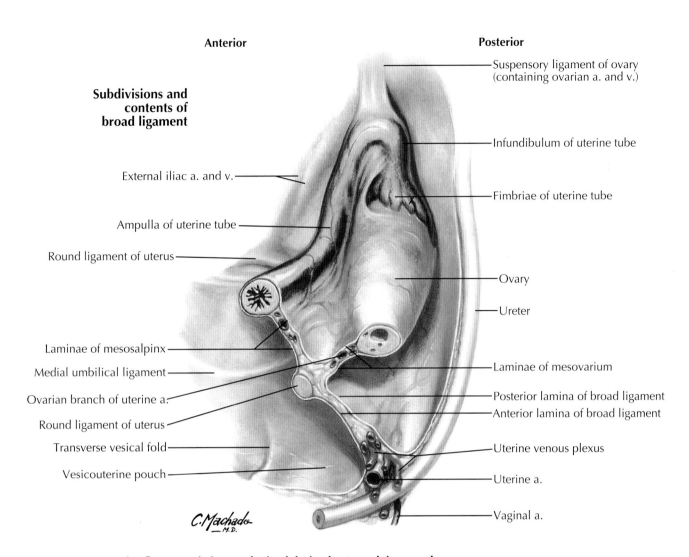

Anterior **Posterior**

Subdivisions and contents of broad ligament

Suspensory ligament of ovary (containing ovarian a. and v.)

Infundibulum of uterine tube

External iliac a. and v.

Fimbriae of uterine tube

Ampulla of uterine tube

Round ligament of uterus

Ovary

Ureter

Laminae of mesosalpinx

Medial umbilical ligament

Laminae of mesovarium

Ovarian branch of uterine a.

Posterior lamina of broad ligament

Round ligament of uterus

Anterior lamina of broad ligament

Transverse vesical fold

Uterine venous plexus

Vesicouterine pouch

Uterine a.

Vaginal a.

C. Machado — M.D.

Anteroposterior fluoroscopic image obtained during hysterosalpingography

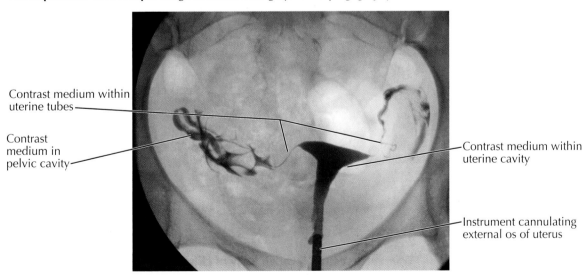

Contrast medium within uterine tubes

Contrast medium in pelvic cavity

Contrast medium within uterine cavity

Instrument cannulating external os of uterus

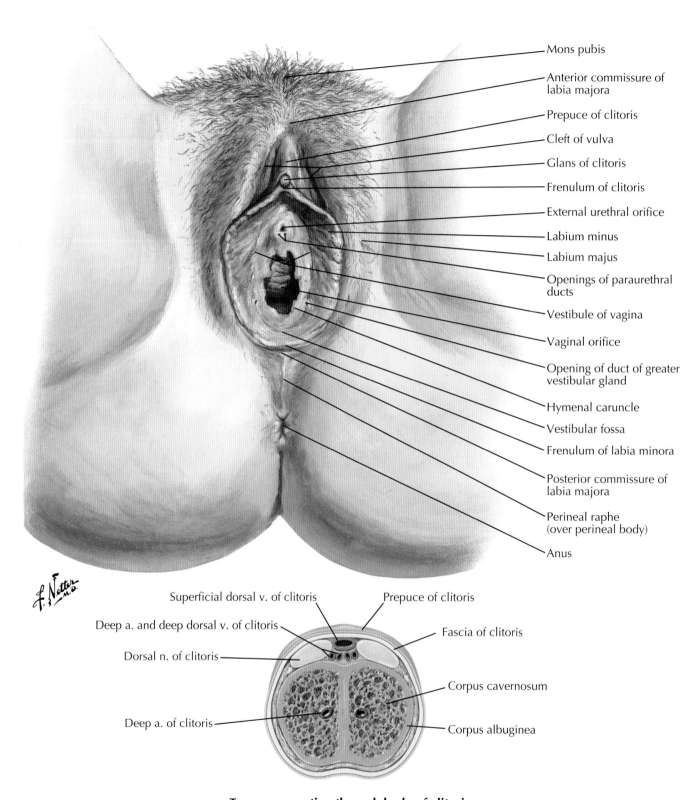

Mons pubis

Anterior commissure of labia majora

Prepuce of clitoris

Cleft of vulva

Glans of clitoris

Frenulum of clitoris

External urethral orifice

Labium minus

Labium majus

Openings of paraurethral ducts

Vestibule of vagina

Vaginal orifice

Opening of duct of greater vestibular gland

Hymenal caruncle

Vestibular fossa

Frenulum of labia minora

Posterior commissure of labia majora

Perineal raphe (over perineal body)

Anus

Superficial dorsal v. of clitoris

Prepuce of clitoris

Deep a. and deep dorsal v. of clitoris

Fascia of clitoris

Dorsal n. of clitoris

Corpus cavernosum

Deep a. of clitoris

Corpus albuginea

Transverse section through body of clitoris

Plate 377

Female Perineum and External Genitalia

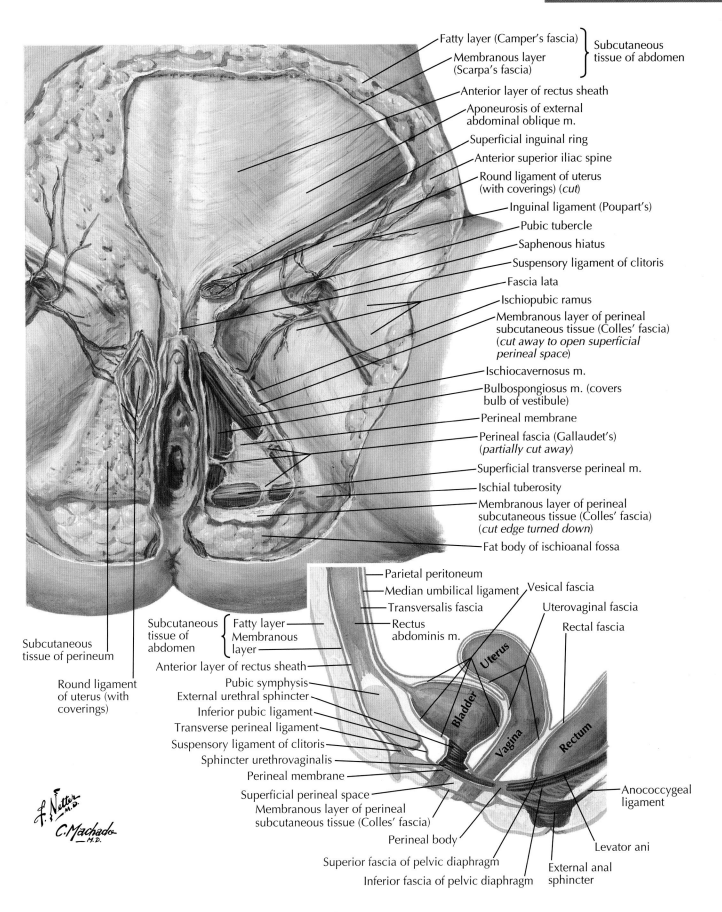

Fatty layer (Camper's fascia) ⎫ Subcutaneous
Membranous layer (Scarpa's fascia) ⎭ tissue of abdomen

Anterior layer of rectus sheath

Aponeurosis of external abdominal oblique m.

Superficial inguinal ring

Anterior superior iliac spine

Round ligament of uterus (with coverings) (cut)

Inguinal ligament (Poupart's)

Pubic tubercle

Saphenous hiatus

Suspensory ligament of clitoris

Fascia lata

Ischiopubic ramus

Membranous layer of perineal subcutaneous tissue (Colles' fascia) (cut away to open superficial perineal space)

Ischiocavernosus m.

Bulbospongiosus m. (covers bulb of vestibule)

Perineal membrane

Perineal fascia (Gallaudet's) (partially cut away)

Superficial transverse perineal m.

Ischial tuberosity

Membranous layer of perineal subcutaneous tissue (Colles' fascia) (cut edge turned down)

Fat body of ischioanal fossa

Subcutaneous tissue of perineum

Round ligament of uterus (with coverings)

Subcutaneous tissue of abdomen ⎰ Fatty layer / Membranous layer

Anterior layer of rectus sheath

Pubic symphysis

External urethral sphincter

Inferior pubic ligament

Transverse perineal ligament

Suspensory ligament of clitoris

Sphincter urethrovaginalis

Perineal membrane

Superficial perineal space

Membranous layer of perineal subcutaneous tissue (Colles' fascia)

Perineal body

Superior fascia of pelvic diaphragm

Inferior fascia of pelvic diaphragm

Parietal peritoneum

Median umbilical ligament

Transversalis fascia

Rectus abdominis m.

Vesical fascia

Uterovaginal fascia

Rectal fascia

Uterus

Bladder

Vagina

Rectum

Anococcygeal ligament

Levator ani

External anal sphincter

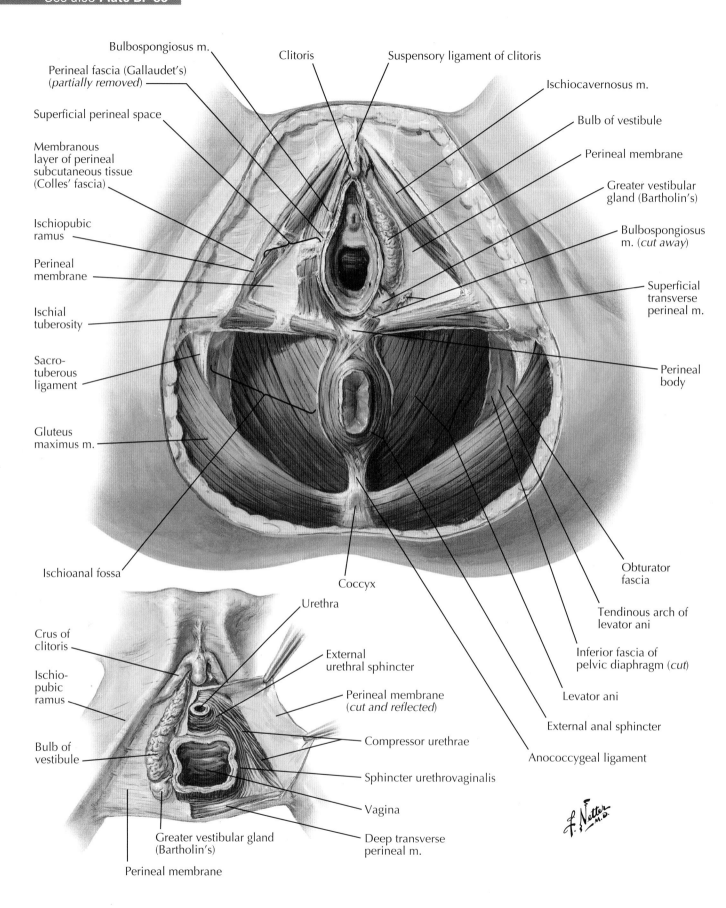

Bulbospongiosus m.

Perineal fascia (Gallaudet's) (*partially removed*)

Superficial perineal space

Membranous layer of perineal subcutaneous tissue (Colles' fascia)

Ischiopubic ramus

Perineal membrane

Ischial tuberosity

Sacro-tuberous ligament

Gluteus maximus m.

Ischioanal fossa

Clitoris

Suspensory ligament of clitoris

Ischiocavernosus m.

Bulb of vestibule

Perineal membrane

Greater vestibular gland (Bartholin's)

Bulbospongiosus m. (*cut away*)

Superficial transverse perineal m.

Perineal body

Obturator fascia

Tendinous arch of levator ani

Inferior fascia of pelvic diaphragm (*cut*)

Levator ani

External anal sphincter

Anococcygeal ligament

Coccyx

Crus of clitoris

Ischio-pubic ramus

Bulb of vestibule

Urethra

External urethral sphincter

Perineal membrane (*cut and reflected*)

Compressor urethrae

Sphincter urethrovaginalis

Vagina

Deep transverse perineal m.

Greater vestibular gland (Bartholin's)

Perineal membrane

Plate 379

Female Perineum and External Genitalia

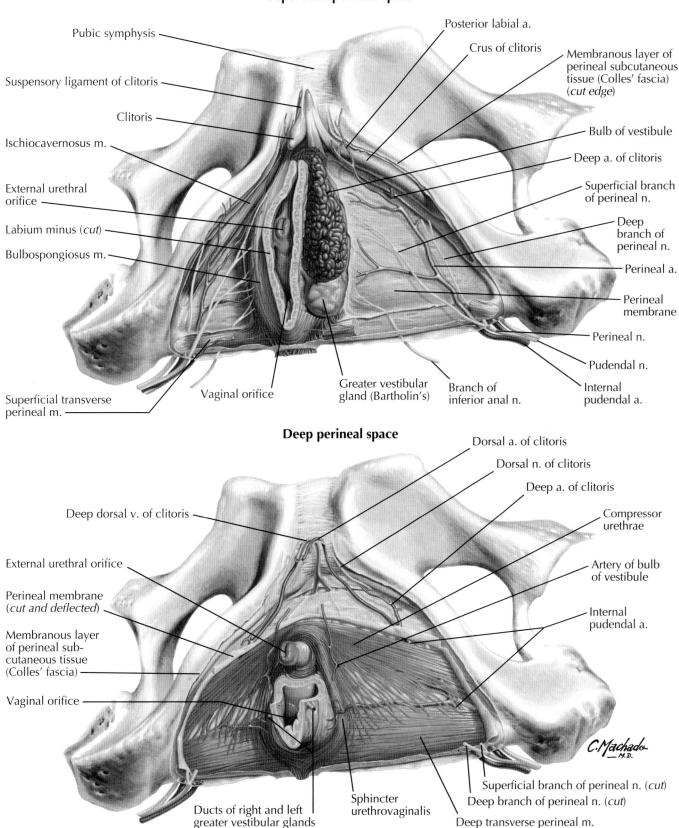

Superficial perineal space

Pubic symphysis

Posterior labial a.

Crus of clitoris

Membranous layer of perineal subcutaneous tissue (Colles' fascia) (*cut edge*)

Suspensory ligament of clitoris

Clitoris

Bulb of vestibule

Deep a. of clitoris

Ischiocavernosus m.

Superficial branch of perineal n.

External urethral orifice

Deep branch of perineal n.

Labium minus (*cut*)

Perineal a.

Bulbospongiosus m.

Perineal membrane

Perineal n.

Pudendal n.

Superficial transverse perineal m.

Vaginal orifice

Greater vestibular gland (Bartholin's)

Branch of inferior anal n.

Internal pudendal a.

Deep perineal space

Dorsal a. of clitoris

Dorsal n. of clitoris

Deep a. of clitoris

Deep dorsal v. of clitoris

Compressor urethrae

External urethral orifice

Artery of bulb of vestibule

Perineal membrane (*cut and deflected*)

Internal pudendal a.

Membranous layer of perineal subcutaneous tissue (Colles' fascia)

Vaginal orifice

Ducts of right and left greater vestibular glands

Sphincter urethrovaginalis

Superficial branch of perineal n. (*cut*)

Deep branch of perineal n. (*cut*)

Deep transverse perineal m.

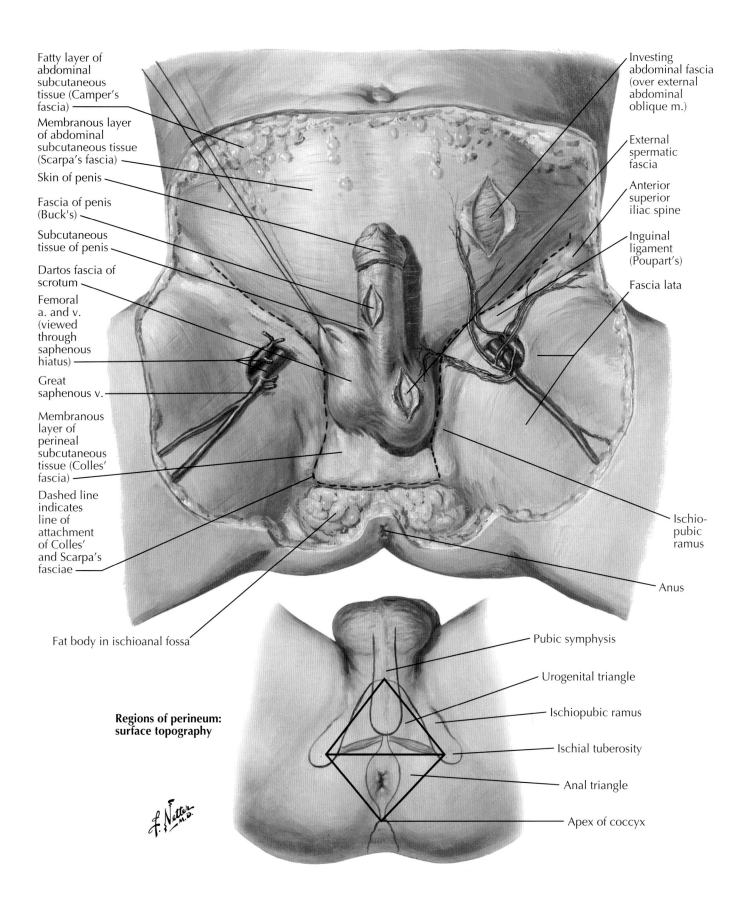

Fatty layer of abdominal subcutaneous tissue (Camper's fascia)

Membranous layer of abdominal subcutaneous tissue (Scarpa's fascia)

Skin of penis

Fascia of penis (Buck's)

Subcutaneous tissue of penis

Dartos fascia of scrotum

Femoral a. and v. (viewed through saphenous hiatus)

Great saphenous v.

Membranous layer of perineal subcutaneous tissue (Colles' fascia)

Dashed line indicates line of attachment of Colles' and Scarpa's fasciae

Investing abdominal fascia (over external abdominal oblique m.)

External spermatic fascia

Anterior superior iliac spine

Inguinal ligament (Poupart's)

Fascia lata

Ischio-pubic ramus

Anus

Fat body in ischioanal fossa

Regions of perineum: surface topography

Pubic symphysis

Urogenital triangle

Ischiopubic ramus

Ischial tuberosity

Anal triangle

Apex of coccyx

Plate 381

Male Perineum and External Genitalia

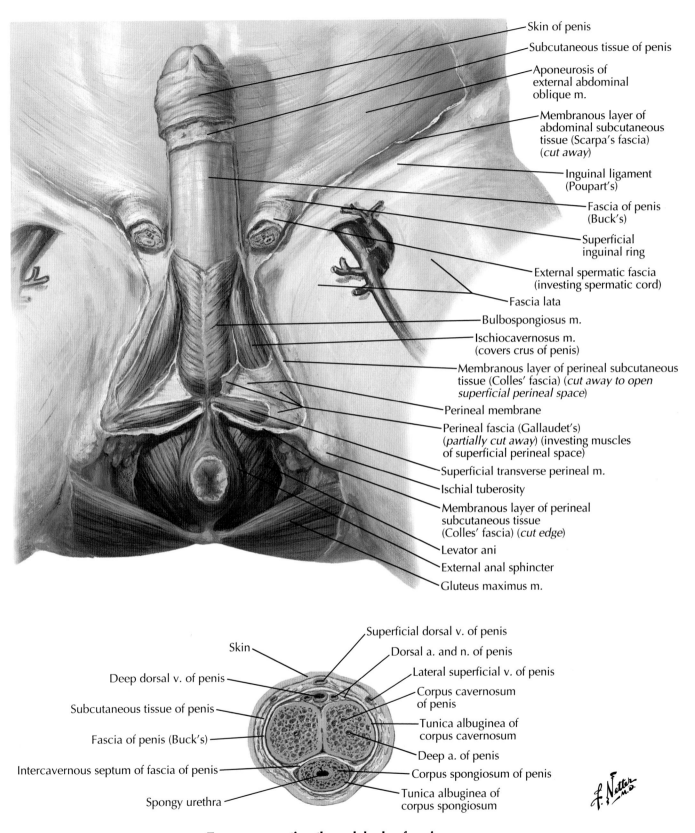

Skin of penis

Subcutaneous tissue of penis

Aponeurosis of external abdominal oblique m.

Membranous layer of abdominal subcutaneous tissue (Scarpa's fascia) (*cut away*)

Inguinal ligament (Poupart's)

Fascia of penis (Buck's)

Superficial inguinal ring

External spermatic fascia (investing spermatic cord)

Fascia lata

Bulbospongiosus m.

Ischiocavernosus m. (covers crus of penis)

Membranous layer of perineal subcutaneous tissue (Colles' fascia) (*cut away to open superficial perineal space*)

Perineal membrane

Perineal fascia (Gallaudet's) (*partially cut away*) (investing muscles of superficial perineal space)

Superficial transverse perineal m.

Ischial tuberosity

Membranous layer of perineal subcutaneous tissue (Colles' fascia) (*cut edge*)

Levator ani

External anal sphincter

Gluteus maximus m.

Superficial dorsal v. of penis

Dorsal a. and n. of penis

Lateral superficial v. of penis

Corpus cavernosum of penis

Tunica albuginea of corpus cavernosum

Deep a. of penis

Corpus spongiosum of penis

Tunica albuginea of corpus spongiosum

Skin

Deep dorsal v. of penis

Subcutaneous tissue of penis

Fascia of penis (Buck's)

Intercavernous septum of fascia of penis

Spongy urethra

Transverse section through body of penis

Male Perineum and External Genitalia

Plate 382

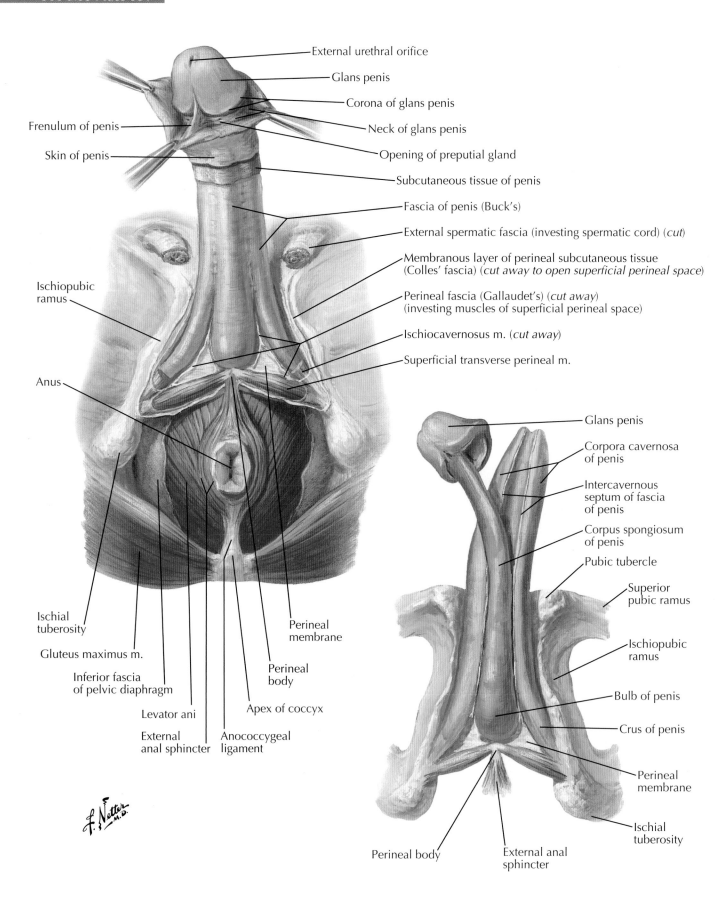

External urethral orifice

Glans penis

Corona of glans penis

Neck of glans penis

Frenulum of penis

Opening of preputial gland

Skin of penis

Subcutaneous tissue of penis

Fascia of penis (Buck's)

External spermatic fascia (investing spermatic cord) (*cut*)

Membranous layer of perineal subcutaneous tissue
(Colles' fascia) (*cut away to open superficial perineal space*)

Ischiopubic
ramus

Perineal fascia (Gallaudet's) (*cut away*)
(investing muscles of superficial perineal space)

Ischiocavernosus m. (*cut away*)

Superficial transverse perineal m.

Anus

Glans penis

Corpora cavernosa
of penis

Intercavernous
septum of fascia
of penis

Corpus spongiosum
of penis

Pubic tubercle

Superior
pubic ramus

Ischial
tuberosity

Ischiopubic
ramus

Gluteus maximus m.

Perineal
membrane

Bulb of penis

Inferior fascia
of pelvic diaphragm

Crus of penis

Perineal
body

Levator ani

Apex of coccyx

Perineal
membrane

External
anal sphincter

Anococcygeal
ligament

Ischial
tuberosity

Perineal body

External anal
sphincter

Plate 383

Male Perineum and External Genitalia

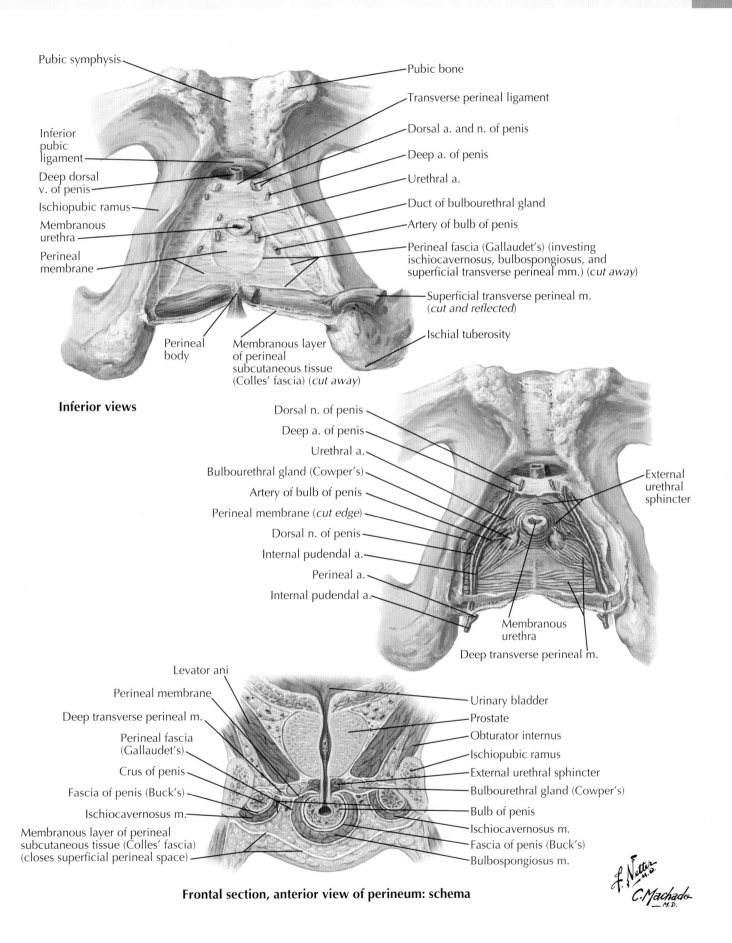

Pubic symphysis

Inferior pubic ligament

Deep dorsal v. of penis

Ischiopubic ramus

Membranous urethra

Perineal membrane

Pubic bone

Transverse perineal ligament

Dorsal a. and n. of penis

Deep a. of penis

Urethral a.

Duct of bulbourethral gland

Artery of bulb of penis

Perineal fascia (Gallaudet's) (investing ischiocavernosus, bulbospongiosus, and superficial transverse perineal mm.) (cut away)

Superficial transverse perineal m. (cut and reflected)

Ischial tuberosity

Perineal body

Membranous layer of perineal subcutaneous tissue (Colles' fascia) (cut away)

Inferior views

Dorsal n. of penis

Deep a. of penis

Urethral a.

Bulbourethral gland (Cowper's)

Artery of bulb of penis

Perineal membrane (cut edge)

Dorsal n. of penis

Internal pudendal a.

Perineal a.

Internal pudendal a.

External urethral sphincter

Membranous urethra

Deep transverse perineal m.

Levator ani

Perineal membrane

Deep transverse perineal m.

Perineal fascia (Gallaudet's)

Crus of penis

Fascia of penis (Buck's)

Ischiocavernosus m.

Membranous layer of perineal subcutaneous tissue (Colles' fascia) (closes superficial perineal space)

Urinary bladder

Prostate

Obturator internus

Ischiopubic ramus

External urethral sphincter

Bulbourethral gland (Cowper's)

Bulb of penis

Ischiocavernosus m.

Fascia of penis (Buck's)

Bulbospongiosus m.

Frontal section, anterior view of perineum: schema

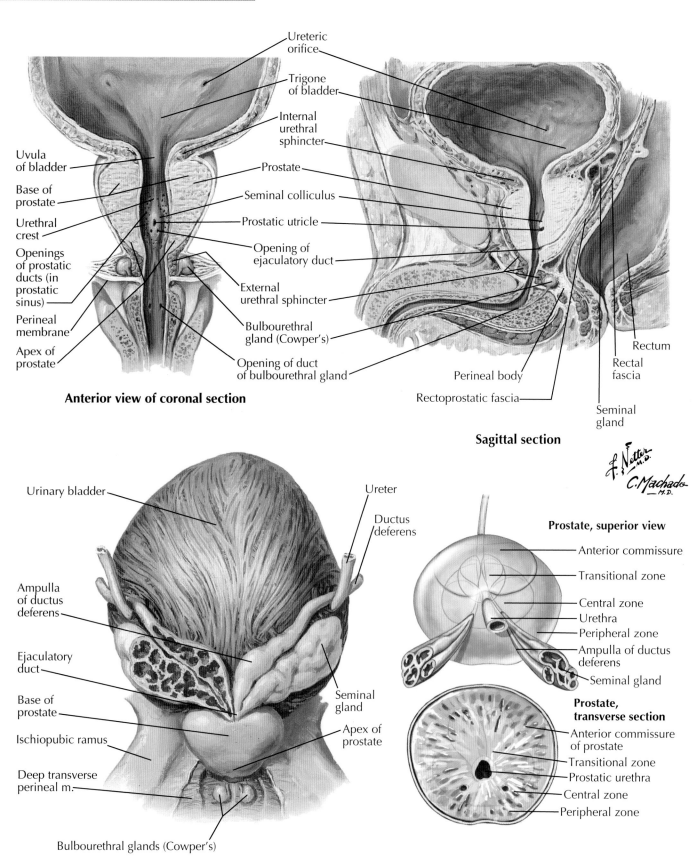

Ureteric orifice

Trigone of bladder

Internal urethral sphincter

Uvula of bladder

Base of prostate

Urethral crest

Openings of prostatic ducts (in prostatic sinus)

Perineal membrane

Apex of prostate

Prostate

Seminal colliculus

Prostatic utricle

Opening of ejaculatory duct

External urethral sphincter

Bulbourethral gland (Cowper's)

Opening of duct of bulbourethral gland

Anterior view of coronal section

Rectum

Rectal fascia

Seminal gland

Perineal body

Rectoprostatic fascia

Sagittal section

Urinary bladder

Ureter

Ductus deferens

Ampulla of ductus deferens

Ejaculatory duct

Base of prostate

Ischiopubic ramus

Deep transverse perineal m.

Bulbourethral glands (Cowper's)

Seminal gland

Apex of prostate

Posterior view

Prostate, superior view

Anterior commissure

Transitional zone

Central zone

Urethra

Peripheral zone

Ampulla of ductus deferens

Seminal gland

Prostate, transverse section

Anterior commissure of prostate

Transitional zone

Prostatic urethra

Central zone

Peripheral zone

Plate 385

Male Perineum and External Genitalia

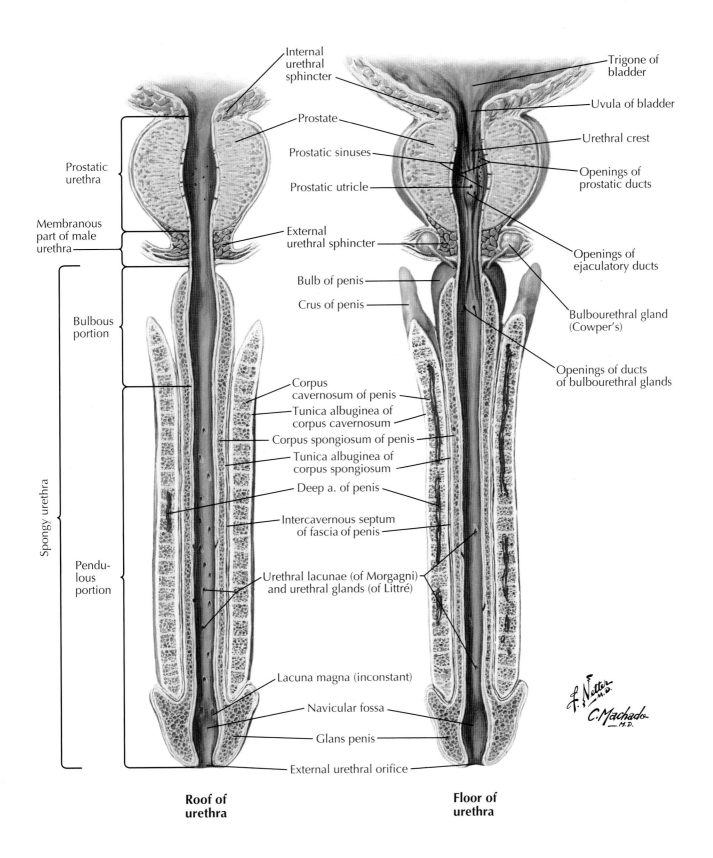

Internal urethral sphincter

Prostate

Prostatic sinuses

Prostatic utricle

External urethral sphincter

Bulb of penis

Crus of penis

Corpus cavernosum of penis

Tunica albuginea of corpus cavernosum

Corpus spongiosum of penis

Tunica albuginea of corpus spongiosum

Deep a. of penis

Intercavernous septum of fascia of penis

Urethral lacunae (of Morgagni) and urethral glands (of Littré)

Lacuna magna (inconstant)

Navicular fossa

Glans penis

External urethral orifice

Prostatic urethra

Membranous part of male urethra

Bulbous portion

Spongy urethra

Pendulous portion

Trigone of bladder

Uvula of bladder

Urethral crest

Openings of prostatic ducts

Openings of ejaculatory ducts

Bulbourethral gland (Cowper's)

Openings of ducts of bulbourethral glands

Roof of urethra

Floor of urethra

Male Perineum and External Genitalia

Plate 386

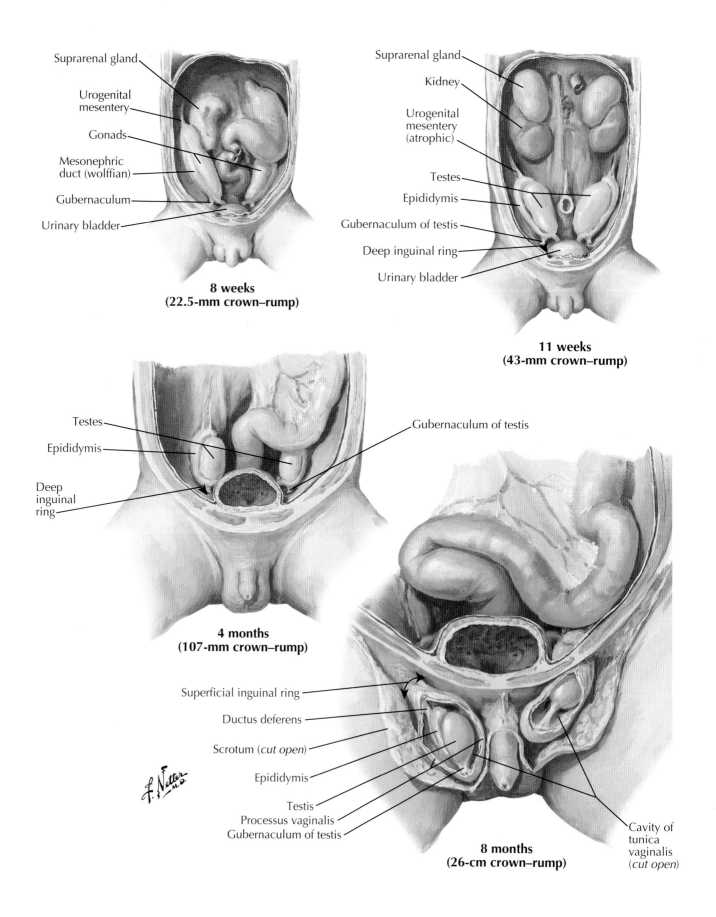

Suprarenal gland

Urogenital mesentery

Gonads

Mesonephric duct (wolffian)

Gubernaculum

Urinary bladder

8 weeks
(22.5-mm crown–rump)

Suprarenal gland

Kidney

Urogenital mesentery (atrophic)

Testes

Epididymis

Gubernaculum of testis

Deep inguinal ring

Urinary bladder

11 weeks
(43-mm crown–rump)

Testes

Epididymis

Deep inguinal ring

Gubernaculum of testis

4 months
(107-mm crown–rump)

Superficial inguinal ring

Ductus deferens

Scrotum (*cut open*)

Epididymis

Testis

Processus vaginalis

Gubernaculum of testis

Cavity of tunica vaginalis (*cut open*)

8 months
(26-cm crown–rump)

Plate 387

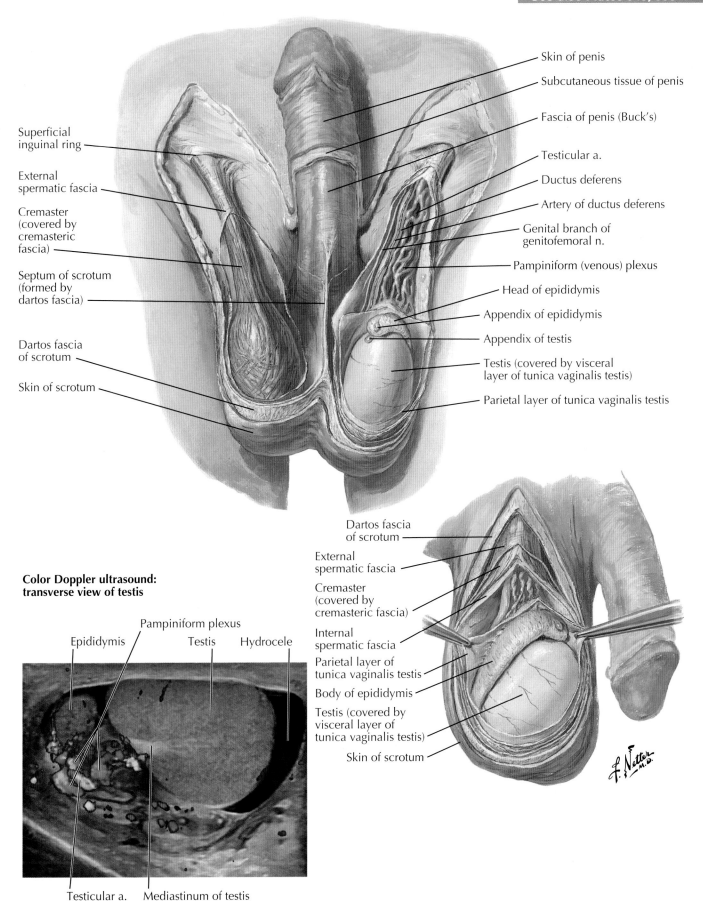

Skin of penis

Subcutaneous tissue of penis

Fascia of penis (Buck's)

Testicular a.

Ductus deferens

Artery of ductus deferens

Genital branch of
genitofemoral n.

Pampiniform (venous) plexus

Head of epididymis

Appendix of epididymis

Appendix of testis

Testis (covered by visceral
layer of tunica vaginalis testis)

Parietal layer of tunica vaginalis testis

Superficial
inguinal ring

External
spermatic fascia

Cremaster
(covered by
cremasteric
fascia)

Septum of scrotum
(formed by
dartos fascia)

Dartos fascia
of scrotum

Skin of scrotum

Dartos fascia
of scrotum

External
spermatic fascia

Cremaster
(covered by
cremasteric fascia)

Internal
spermatic fascia

Parietal layer of
tunica vaginalis testis

Body of epididymis

Testis (covered by
visceral layer of
tunica vaginalis testis)

Skin of scrotum

**Color Doppler ultrasound:
transverse view of testis**

Pampiniform plexus

Epididymis   Testis   Hydrocele

Testicular a.   Mediastinum of testis

Male Perineum and External Genitalia   Plate 388

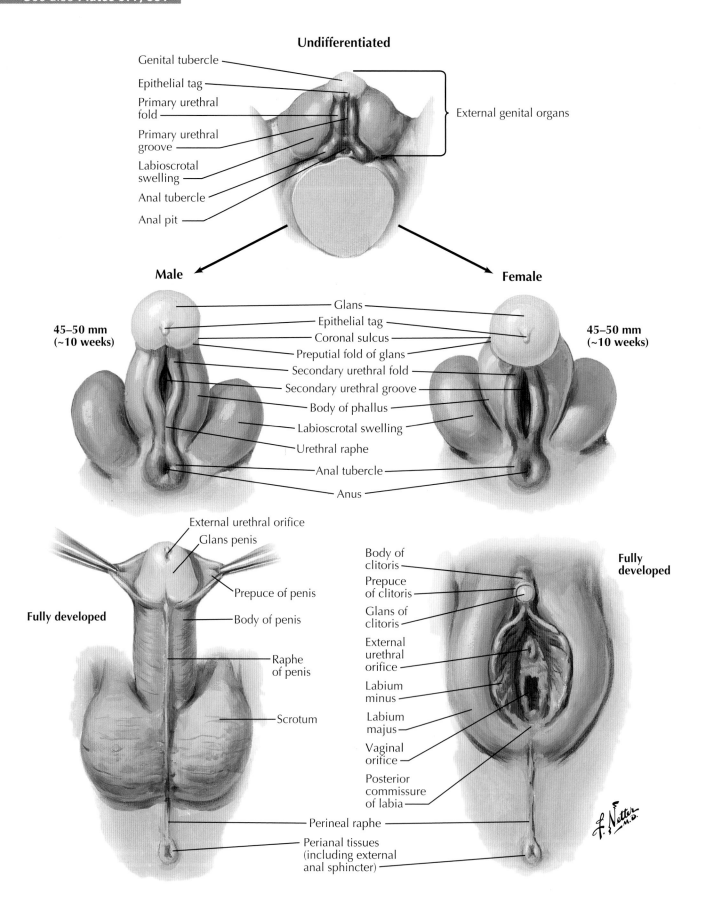

Undifferentiated

Genital tubercle

Epithelial tag

Primary urethral fold

Primary urethral groove

Labioscrotal swelling

Anal tubercle

Anal pit

External genital organs

Male

Female

45–50 mm (~10 weeks)

45–50 mm (~10 weeks)

Glans

Epithelial tag

Coronal sulcus

Preputial fold of glans

Secondary urethral fold

Secondary urethral groove

Body of phallus

Labioscrotal swelling

Urethral raphe

Anal tubercle

Anus

External urethral orifice

Glans penis

Prepuce of penis

Body of penis

Fully developed

Raphe of penis

Scrotum

Body of clitoris

Prepuce of clitoris

Glans of clitoris

External urethral orifice

Labium minus

Labium majus

Vaginal orifice

Posterior commissure of labia

Fully developed

Perineal raphe

Perianal tissues (including external anal sphincter)

Plate 389

Homologies of Male and Female Genitalia

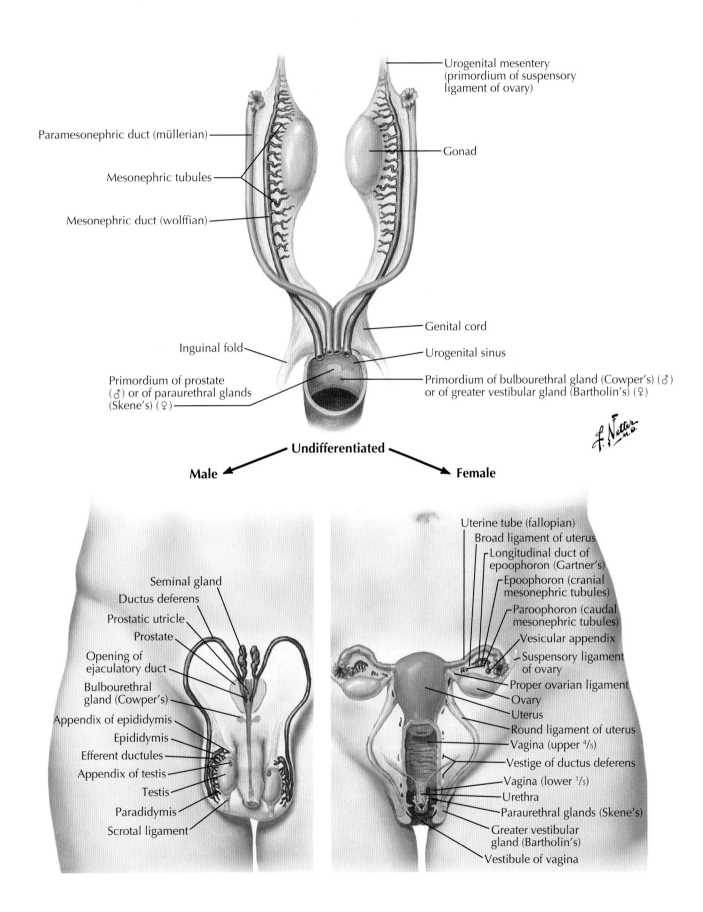

Urogenital mesentery (primordium of suspensory ligament of ovary)

Paramesonephric duct (müllerian)

Gonad

Mesonephric tubules

Mesonephric duct (wolffian)

Genital cord

Inguinal fold

Urogenital sinus

Primordium of prostate (♂) or of paraurethral glands (Skene's) (♀)

Primordium of bulbourethral gland (Cowper's) (♂) or of greater vestibular gland (Bartholin's) (♀)

Undifferentiated

Male

Female

Seminal gland

Ductus deferens

Prostatic utricle

Prostate

Opening of ejaculatory duct

Bulbourethral gland (Cowper's)

Appendix of epididymis

Epididymis

Efferent ductules

Appendix of testis

Testis

Paradidymis

Scrotal ligament

Uterine tube (fallopian)

Broad ligament of uterus

Longitudinal duct of epoophoron (Gartner's)

Epoophoron (cranial mesonephric tubules)

Paroophoron (caudal mesonephric tubules)

Vesicular appendix

Suspensory ligament of ovary

Proper ovarian ligament

Ovary

Uterus

Round ligament of uterus

Vagina (upper ⁴/₅)

Vestige of ductus deferens

Vagina (lower ¹/₅)

Urethra

Paraurethral glands (Skene's)

Greater vestibular gland (Bartholin's)

Vestibule of vagina

Spermatogenic epithelium: spermatogenesis

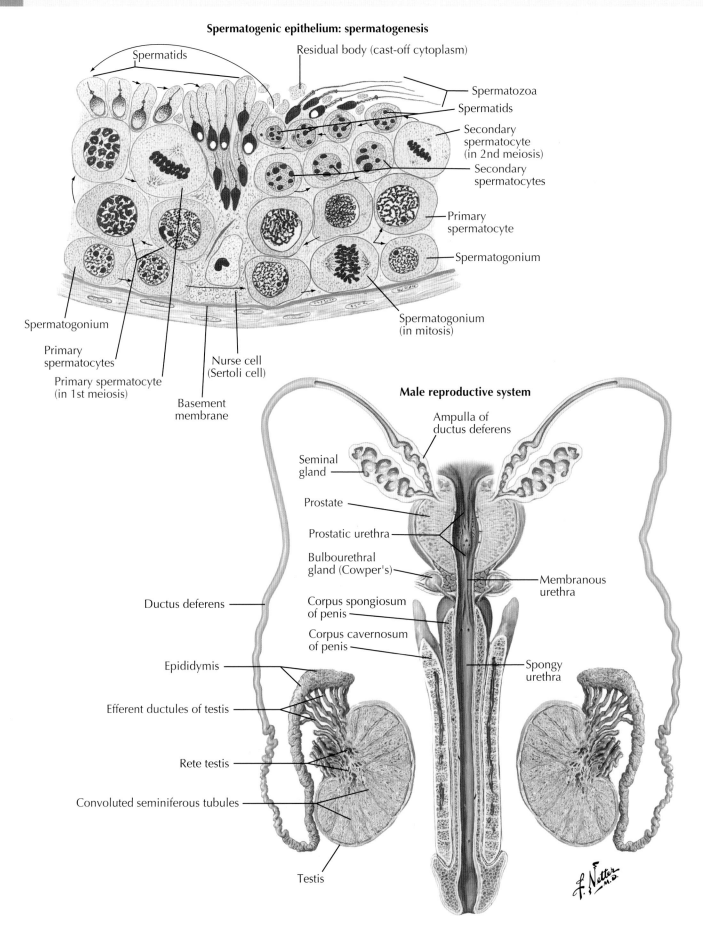

Spermatids

Residual body (cast-off cytoplasm)

Spermatozoa

Spermatids

Secondary spermatocyte (in 2nd meiosis)

Secondary spermatocytes

Primary spermatocyte

Spermatogonium

Spermatogonium (in mitosis)

Spermatogonium

Primary spermatocytes

Primary spermatocyte (in 1st meiosis)

Nurse cell (Sertoli cell)

Basement membrane

Male reproductive system

Ampulla of ductus deferens

Seminal gland

Prostate

Prostatic urethra

Bulbourethral gland (Cowper's)

Membranous urethra

Corpus spongiosum of penis

Corpus cavernosum of penis

Spongy urethra

Ductus deferens

Epididymis

Efferent ductules of testis

Rete testis

Convoluted seminiferous tubules

Testis

Plate 391   **Male Internal Genitalia**

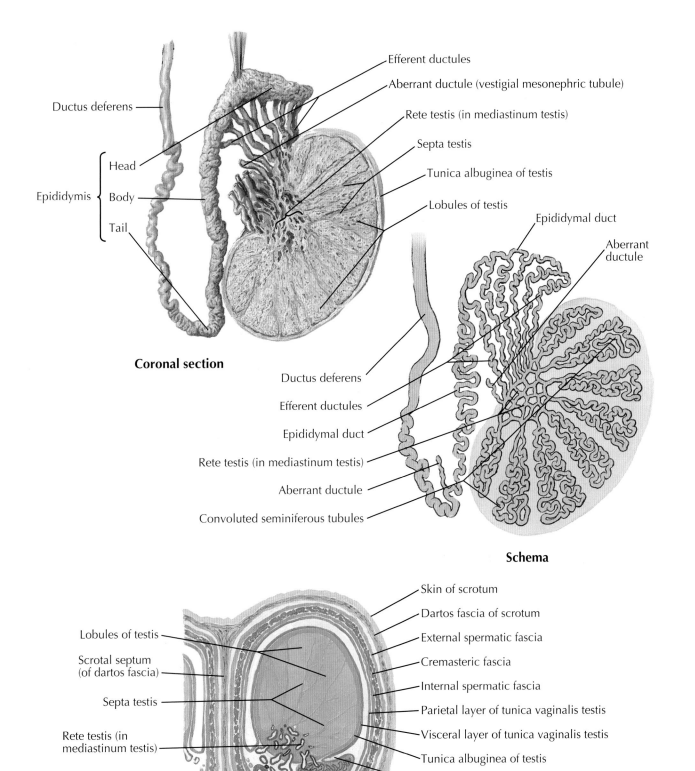

Ductus deferens

Epididymis
- Head
- Body
- Tail

Coronal section

Efferent ductules

Aberrant ductule (vestigial mesonephric tubule)

Rete testis (in mediastinum testis)

Septa testis

Tunica albuginea of testis

Lobules of testis

Epididymal duct

Aberrant ductule

Ductus deferens

Efferent ductules

Epididymal duct

Rete testis (in mediastinum testis)

Aberrant ductule

Convoluted seminiferous tubules

Schema

Lobules of testis

Scrotal septum (of dartos fascia)

Septa testis

Rete testis (in mediastinum testis)

Ductus deferens

Skin of scrotum

Dartos fascia of scrotum

External spermatic fascia

Cremasteric fascia

Internal spermatic fascia

Parietal layer of tunica vaginalis testis

Visceral layer of tunica vaginalis testis

Tunica albuginea of testis

Sinus of epididymis

Epididymis

Horizontal section through scrotum and testis

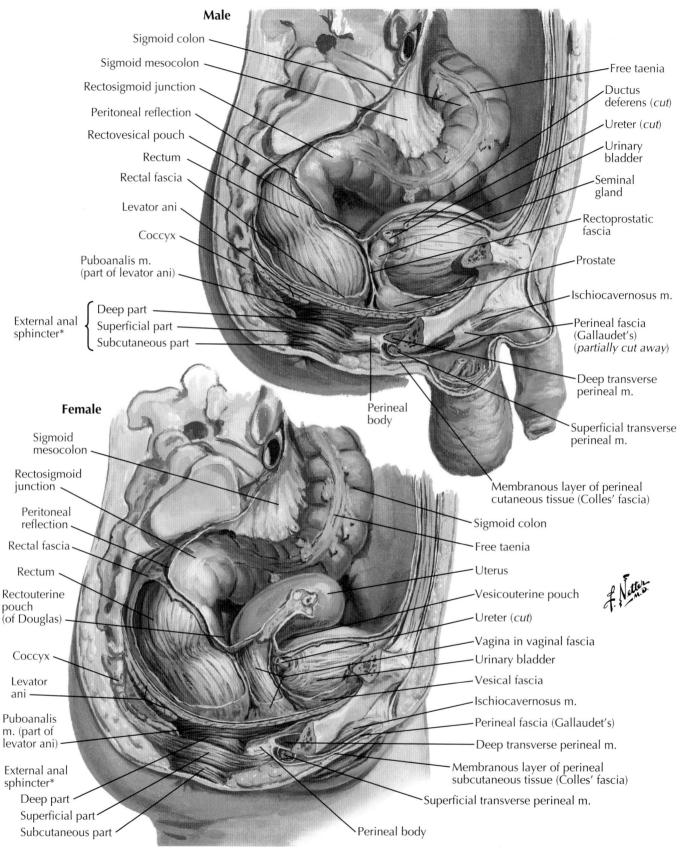

Male

Sigmoid colon

Sigmoid mesocolon

Rectosigmoid junction

Peritoneal reflection

Rectovesical pouch

Rectum

Rectal fascia

Levator ani

Coccyx

Puboanalis m. (part of levator ani)

External anal sphincter*
- Deep part
- Superficial part
- Subcutaneous part

Free taenia

Ductus deferens (*cut*)

Ureter (*cut*)

Urinary bladder

Seminal gland

Rectoprostatic fascia

Prostate

Ischiocavernosus m.

Perineal fascia (Gallaudet's) (*partially cut away*)

Deep transverse perineal m.

Superficial transverse perineal m.

Perineal body

Membranous layer of perineal cutaneous tissue (Colles' fascia)

Female

Sigmoid mesocolon

Rectosigmoid junction

Peritoneal reflection

Rectal fascia

Rectum

Rectouterine pouch (of Douglas)

Coccyx

Levator ani

Puboanalis m. (part of levator ani)

External anal sphincter*
- Deep part
- Superficial part
- Subcutaneous part

Sigmoid colon

Free taenia

Uterus

Vesicouterine pouch

Ureter (*cut*)

Vagina in vaginal fascia

Urinary bladder

Vesical fascia

Ischiocavernosus m.

Perineal fascia (Gallaudet's)

Deep transverse perineal m.

Membranous layer of perineal subcutaneous tissue (Colles' fascia)

Superficial transverse perineal m.

Perineal body

*Parts variable and often indistinct

Plate 393

Rectum and Anal Canal

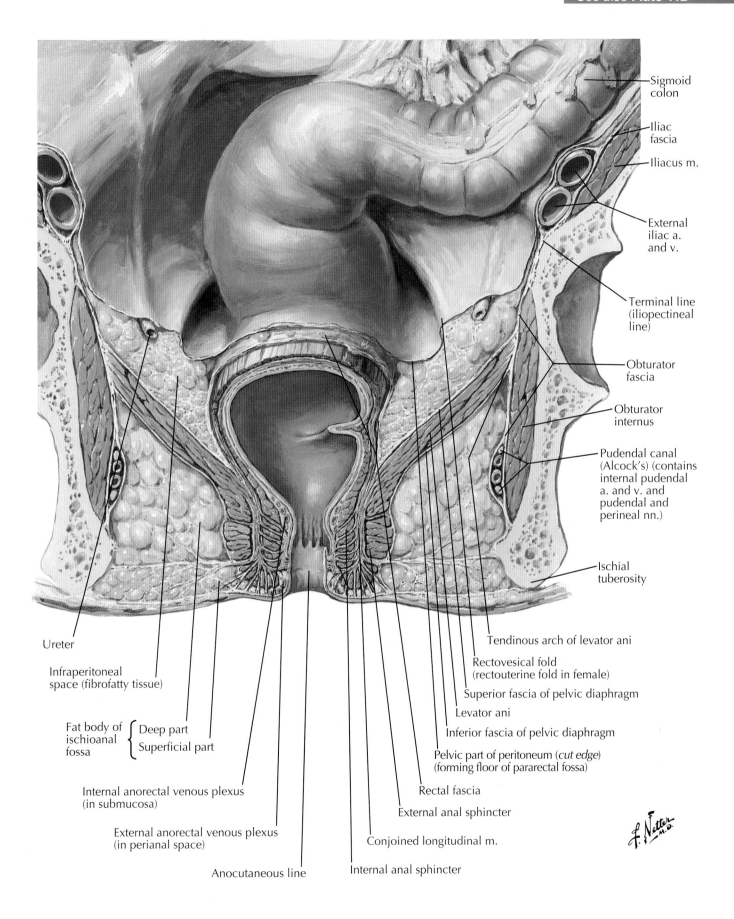

Sigmoid colon

Iliac fascia

Iliacus m.

External iliac a. and v.

Terminal line (iliopectineal line)

Obturator fascia

Obturator internus

Pudendal canal (Alcock's) (contains internal pudendal a. and v. and pudendal and perineal nn.)

Ischial tuberosity

Tendinous arch of levator ani

Rectovesical fold (rectouterine fold in female)

Superior fascia of pelvic diaphragm

Levator ani

Inferior fascia of pelvic diaphragm

Pelvic part of peritoneum (*cut edge*) (forming floor of pararectal fossa)

Rectal fascia

External anal sphincter

Conjoined longitudinal m.

Internal anal sphincter

Anocutaneous line

External anorectal venous plexus (in perianal space)

Internal anorectal venous plexus (in submucosa)

Fat body of ischioanal fossa { Deep part / Superficial part }

Infraperitoneal space (fibrofatty tissue)

Ureter

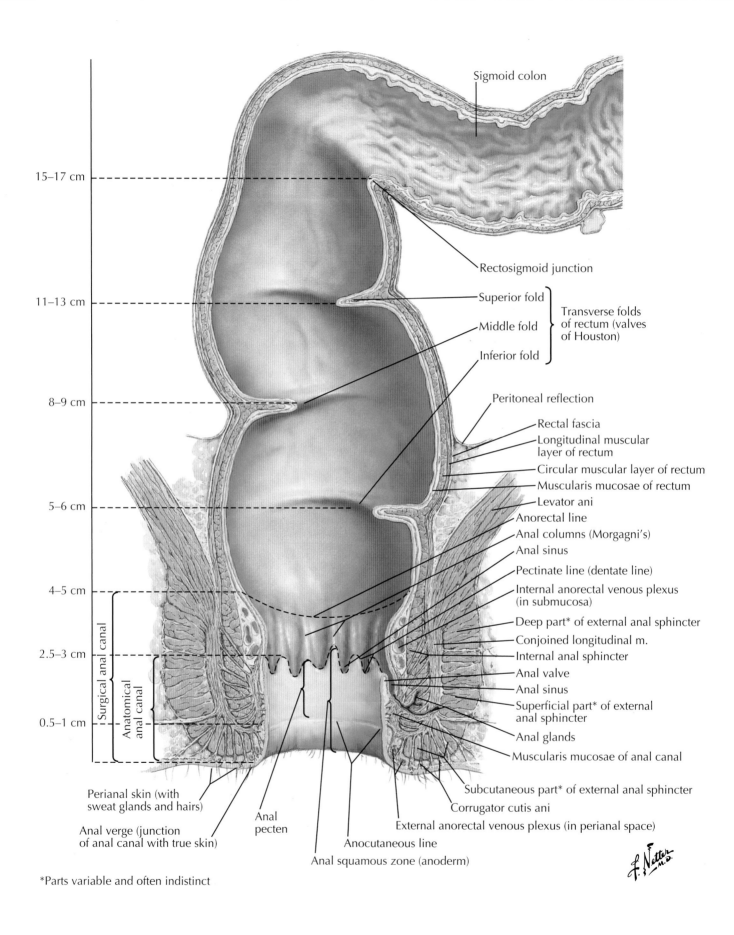

Sigmoid colon

15–17 cm

Rectosigmoid junction

11–13 cm

Superior fold

Middle fold } Transverse folds of rectum (valves of Houston)

Inferior fold

Peritoneal reflection

8–9 cm

Rectal fascia

Longitudinal muscular layer of rectum

Circular muscular layer of rectum

Muscularis mucosae of rectum

5–6 cm

Levator ani

Anorectal line

Anal columns (Morgagni's)

Anal sinus

Pectinate line (dentate line)

Internal anorectal venous plexus (in submucosa)

4–5 cm

Deep part* of external anal sphincter

Conjoined longitudinal m.

Internal anal sphincter

2.5–3 cm

Anal valve

Anal sinus

Superficial part* of external anal sphincter

0.5–1 cm

Anal glands

Muscularis mucosae of anal canal

Surgical anal canal

Anatomical anal canal

Subcutaneous part* of external anal sphincter

Perianal skin (with sweat glands and hairs)

Anal pecten

Corrugator cutis ani

External anorectal venous plexus (in perianal space)

Anal verge (junction of anal canal with true skin)

Anocutaneous line

Anal squamous zone (anoderm)

*Parts variable and often indistinct

Plate 395

Rectum and Anal Canal

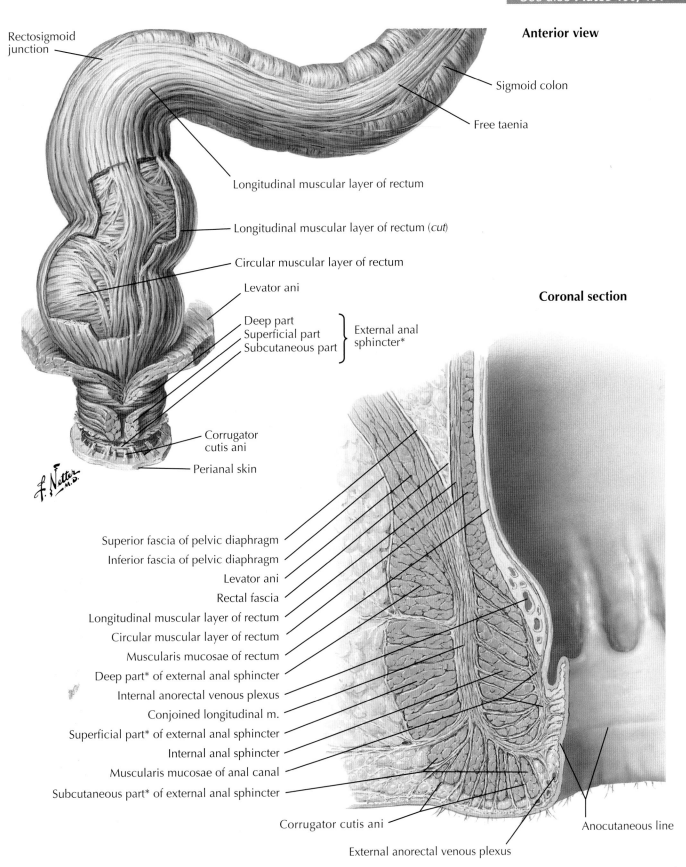

Anterior view

Rectosigmoid junction

Sigmoid colon

Free taenia

Longitudinal muscular layer of rectum

Longitudinal muscular layer of rectum (*cut*)

Circular muscular layer of rectum

Levator ani

Deep part
Superficial part } External anal sphincter*
Subcutaneous part

Corrugator cutis ani

Perianal skin

Coronal section

Superior fascia of pelvic diaphragm
Inferior fascia of pelvic diaphragm
Levator ani
Rectal fascia
Longitudinal muscular layer of rectum
Circular muscular layer of rectum
Muscularis mucosae of rectum
Deep part* of external anal sphincter
Internal anorectal venous plexus
Conjoined longitudinal m.
Superficial part* of external anal sphincter
Internal anal sphincter
Muscularis mucosae of anal canal
Subcutaneous part* of external anal sphincter

Corrugator cutis ani

External anorectal venous plexus

Anocutaneous line

*Parts variable and often indistinct

Rectum and Anal Canal

Plate 396

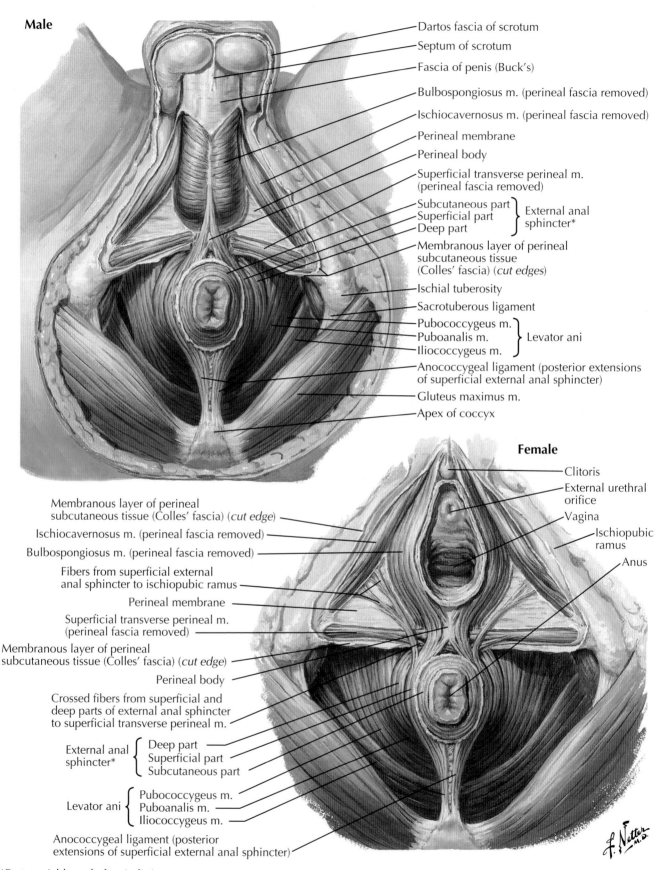

Male

Dartos fascia of scrotum

Septum of scrotum

Fascia of penis (Buck's)

Bulbospongiosus m. (perineal fascia removed)

Ischiocavernosus m. (perineal fascia removed)

Perineal membrane

Perineal body

Superficial transverse perineal m. (perineal fascia removed)

Subcutaneous part
Superficial part } External anal
Deep part } sphincter*

Membranous layer of perineal subcutaneous tissue (Colles' fascia) (cut edges)

Ischial tuberosity

Sacrotuberous ligament

Pubococcygeus m.
Puboanalis m. } Levator ani
Iliococcygeus m.

Anococcygeal ligament (posterior extensions of superficial external anal sphincter)

Gluteus maximus m.

Apex of coccyx

Female

Clitoris

External urethral orifice

Vagina

Ischiopubic ramus

Anus

Membranous layer of perineal subcutaneous tissue (Colles' fascia) (cut edge)

Ischiocavernosus m. (perineal fascia removed)

Bulbospongiosus m. (perineal fascia removed)

Fibers from superficial external anal sphincter to ischiopubic ramus

Perineal membrane

Superficial transverse perineal m. (perineal fascia removed)

Membranous layer of perineal subcutaneous tissue (Colles' fascia) (cut edge)

Perineal body

Crossed fibers from superficial and deep parts of external anal sphincter to superficial transverse perineal m.

External anal sphincter* {
Deep part
Superficial part
Subcutaneous part

Levator ani {
Pubococcygeus m.
Puboanalis m.
Iliococcygeus m.

Anococcygeal ligament (posterior extensions of superficial external anal sphincter)

*Parts variable and often indistinct

Plate 397

Rectum and Anal Canal

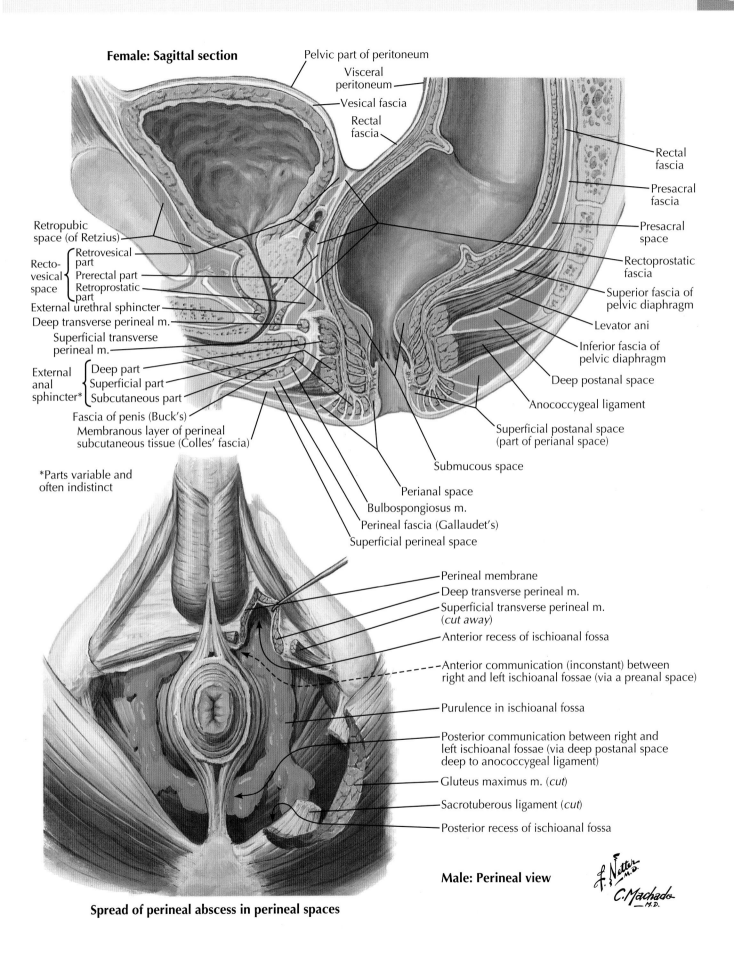

Female: Sagittal section

Pelvic part of peritoneum

Visceral peritoneum

Vesical fascia

Rectal fascia

Rectal fascia

Presacral fascia

Presacral space

Rectoprostatic fascia

Superior fascia of pelvic diaphragm

Levator ani

Inferior fascia of pelvic diaphragm

Deep postanal space

Anococcygeal ligament

Superficial postanal space (part of perianal space)

Retropubic space (of Retzius)

Recto-vesical space { Retrovesical part / Prerectal part / Retroprostatic part

External urethral sphincter

Deep transverse perineal m.

Superficial transverse perineal m.

External anal sphincter* { Deep part / Superficial part / Subcutaneous part

Fascia of penis (Buck's)

Membranous layer of perineal subcutaneous tissue (Colles' fascia)

*Parts variable and often indistinct

Submucous space

Perianal space

Bulbospongiosus m.

Perineal fascia (Gallaudet's)

Superficial perineal space

Perineal membrane

Deep transverse perineal m.

Superficial transverse perineal m. (*cut away*)

Anterior recess of ischioanal fossa

Anterior communication (inconstant) between right and left ischioanal fossae (via a preanal space)

Purulence in ischioanal fossa

Posterior communication between right and left ischioanal fossae (via deep postanal space deep to anococcygeal ligament)

Gluteus maximus m. (*cut*)

Sacrotuberous ligament (*cut*)

Posterior recess of ischioanal fossa

Male: Perineal view

Spread of perineal abscess in perineal spaces

MRI of female pelvis (without intravenous contrast medium)

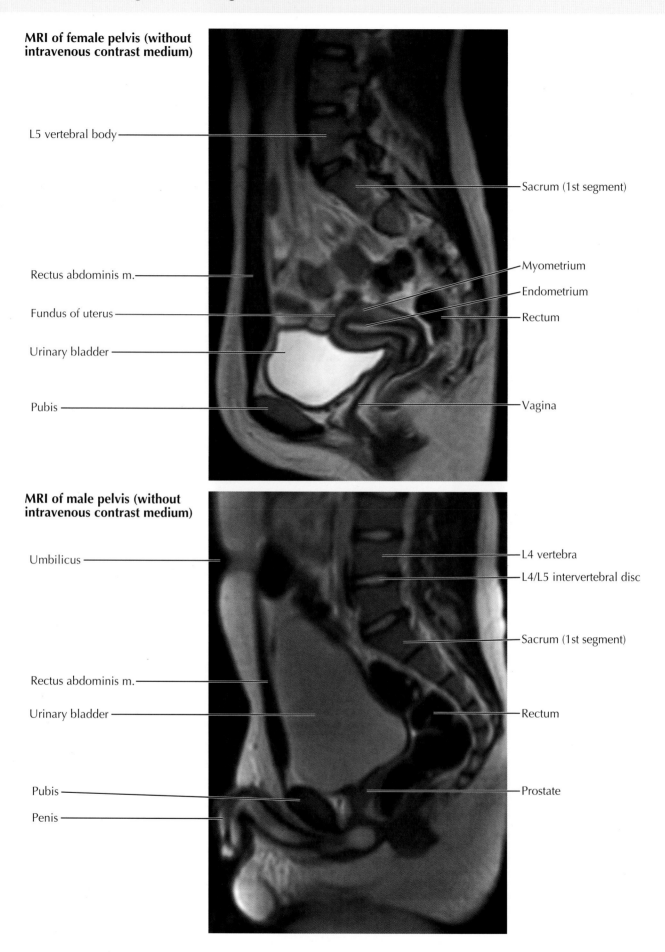

L5 vertebral body

Sacrum (1st segment)

Rectus abdominis m.

Myometrium

Endometrium

Fundus of uterus

Rectum

Urinary bladder

Pubis

Vagina

MRI of male pelvis (without intravenous contrast medium)

Umbilicus

L4 vertebra

L4/L5 intervertebral disc

Sacrum (1st segment)

Rectus abdominis m.

Urinary bladder

Rectum

Pubis

Penis

Prostate

Plate 399

Rectum and Anal Canal

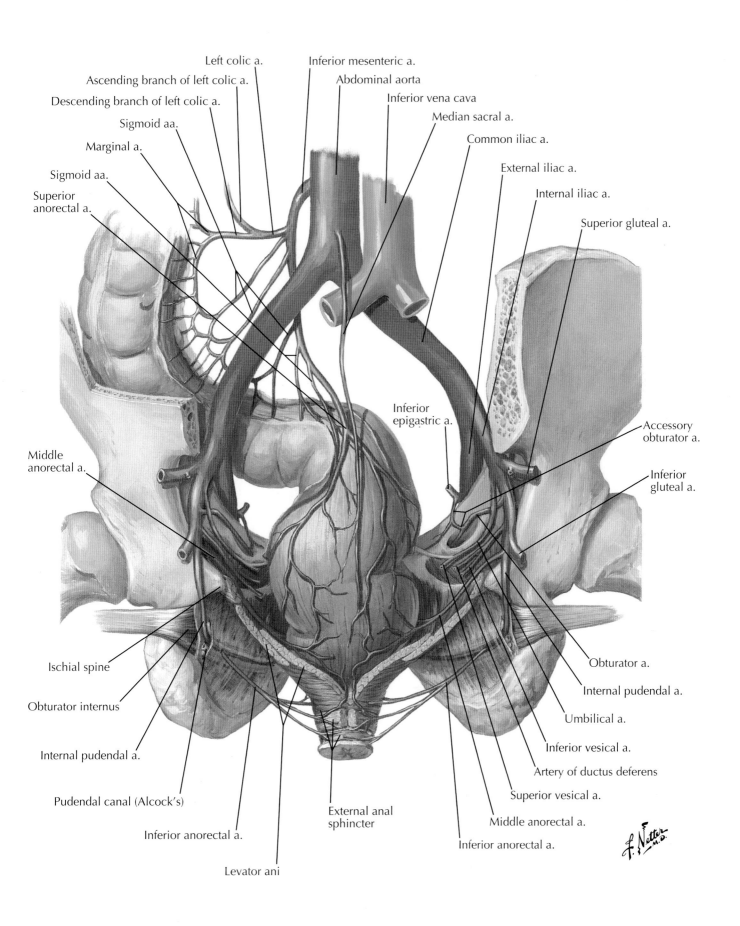

Left colic a.

Ascending branch of left colic a.

Descending branch of left colic a.

Sigmoid aa.

Marginal a.

Sigmoid aa.

Superior anorectal a.

Inferior mesenteric a.

Abdominal aorta

Inferior vena cava

Median sacral a.

Common iliac a.

External iliac a.

Internal iliac a.

Superior gluteal a.

Inferior epigastric a.

Accessory obturator a.

Middle anorectal a.

Inferior gluteal a.

Ischial spine

Obturator internus

Internal pudendal a.

Pudendal canal (Alcock's)

Inferior anorectal a.

Levator ani

External anal sphincter

Inferior anorectal a.

Middle anorectal a.

Superior vesical a.

Artery of ductus deferens

Inferior vesical a.

Umbilical a.

Internal pudendal a.

Obturator a.

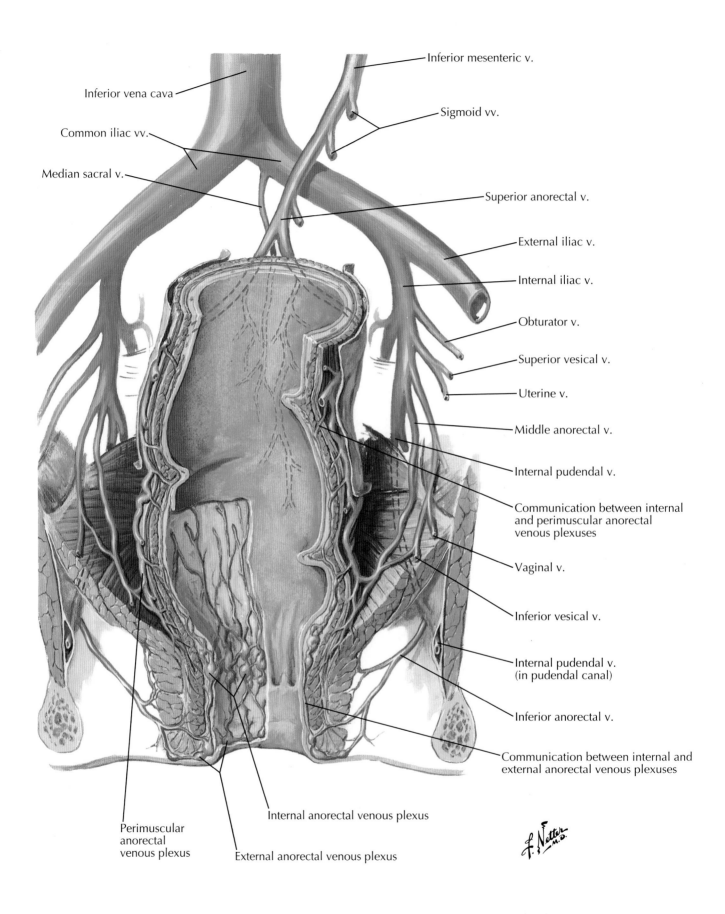

Inferior mesenteric v.

Inferior vena cava

Sigmoid vv.

Common iliac vv.

Median sacral v.

Superior anorectal v.

External iliac v.

Internal iliac v.

Obturator v.

Superior vesical v.

Uterine v.

Middle anorectal v.

Internal pudendal v.

Communication between internal and perimuscular anorectal venous plexuses

Vaginal v.

Inferior vesical v.

Internal pudendal v. (in pudendal canal)

Inferior anorectal v.

Communication between internal and external anorectal venous plexuses

Internal anorectal venous plexus

Perimuscular anorectal venous plexus

External anorectal venous plexus

Plate 401

Vasculature

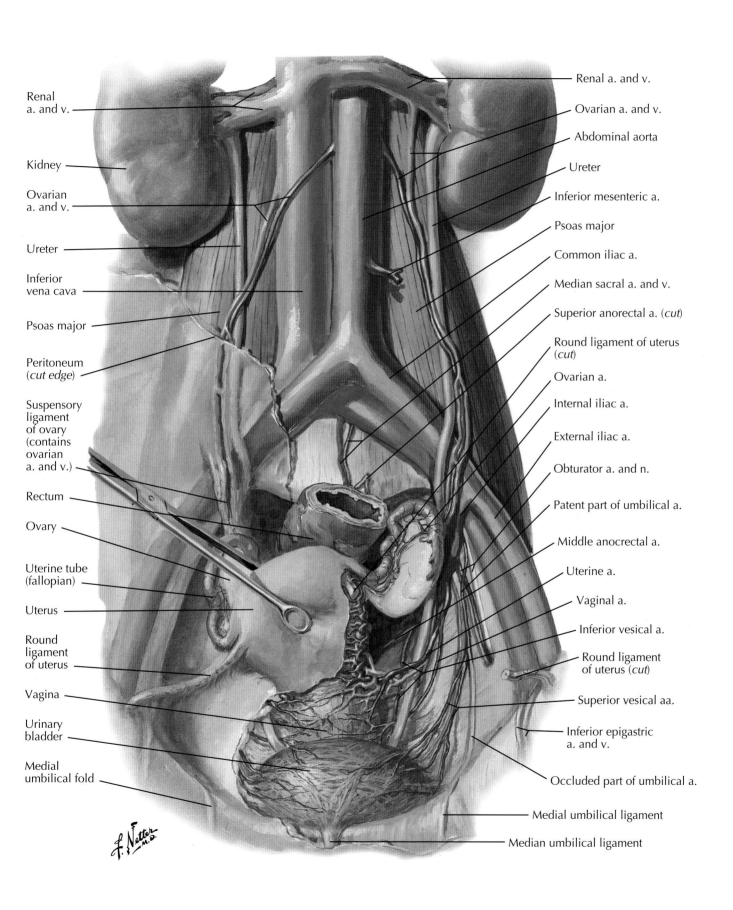

Renal a. and v.

Kidney

Ovarian a. and v.

Ureter

Inferior vena cava

Psoas major

Peritoneum (*cut edge*)

Suspensory ligament of ovary (contains ovarian a. and v.)

Rectum

Ovary

Uterine tube (fallopian)

Uterus

Round ligament of uterus

Vagina

Urinary bladder

Medial umbilical fold

Renal a. and v.

Ovarian a. and v.

Abdominal aorta

Ureter

Inferior mesenteric a.

Psoas major

Common iliac a.

Median sacral a. and v.

Superior anorectal a. (*cut*)

Round ligament of uterus (*cut*)

Ovarian a.

Internal iliac a.

External iliac a.

Obturator a. and n.

Patent part of umbilical a.

Middle anorectal a.

Uterine a.

Vaginal a.

Inferior vesical a.

Round ligament of uterus (*cut*)

Superior vesical aa.

Inferior epigastric a. and v.

Occluded part of umbilical a.

Medial umbilical ligament

Median umbilical ligament

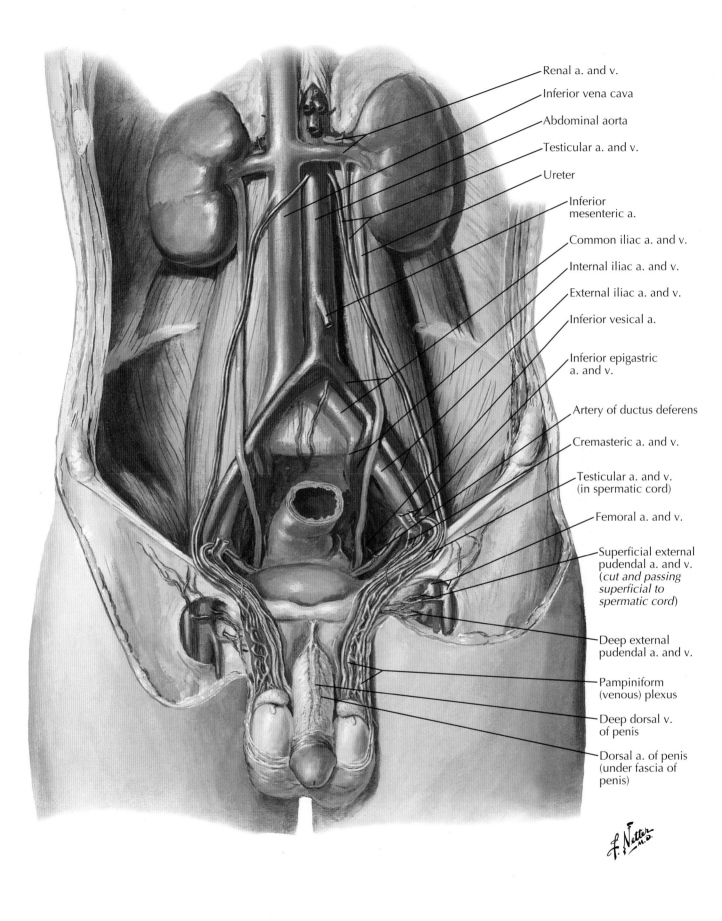

Renal a. and v.

Inferior vena cava

Abdominal aorta

Testicular a. and v.

Ureter

Inferior mesenteric a.

Common iliac a. and v.

Internal iliac a. and v.

External iliac a. and v.

Inferior vesical a.

Inferior epigastric a. and v.

Artery of ductus deferens

Cremasteric a. and v.

Testicular a. and v. (in spermatic cord)

Femoral a. and v.

Superficial external pudendal a. and v. (*cut and passing superficial to spermatic cord*)

Deep external pudendal a. and v.

Pampiniform (venous) plexus

Deep dorsal v. of penis

Dorsal a. of penis (under fascia of penis)

Plate 403

Vasculature

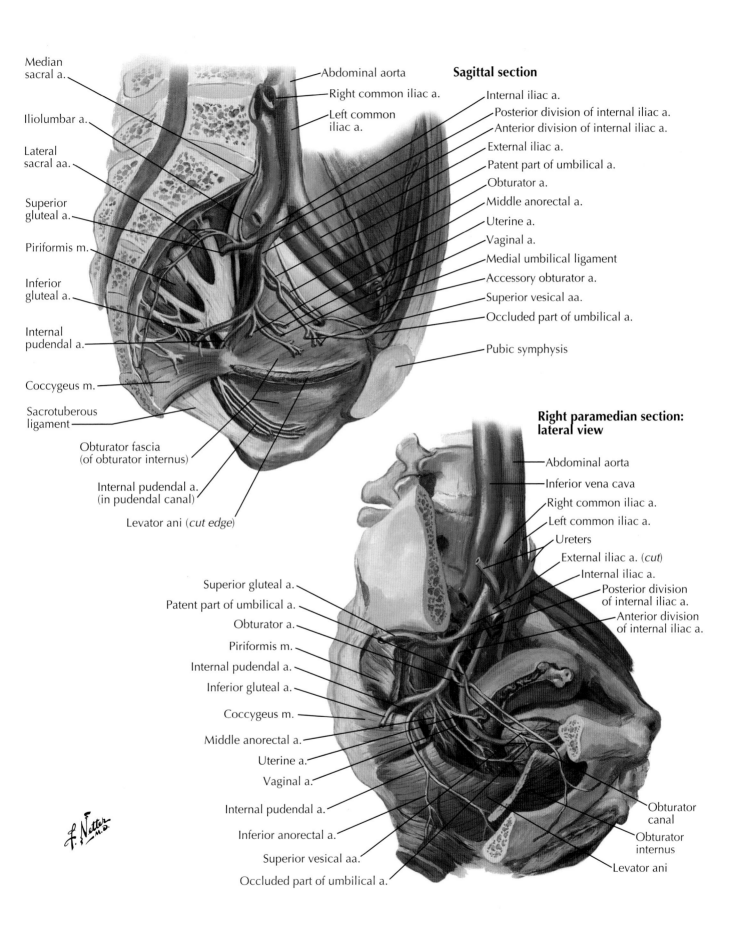

Median sacral a.

Iliolumbar a.

Lateral sacral aa.

Superior gluteal a.

Piriformis m.

Inferior gluteal a.

Internal pudendal a.

Coccygeus m.

Sacrotuberous ligament

Obturator fascia (of obturator internus)

Internal pudendal a. (in pudendal canal)

Levator ani (cut edge)

Abdominal aorta

Right common iliac a.

Left common iliac a.

Sagittal section

Internal iliac a.

Posterior division of internal iliac a.

Anterior division of internal iliac a.

External iliac a.

Patent part of umbilical a.

Obturator a.

Middle anorectal a.

Uterine a.

Vaginal a.

Medial umbilical ligament

Accessory obturator a.

Superior vesical aa.

Occluded part of umbilical a.

Pubic symphysis

Right paramedian section: lateral view

Abdominal aorta

Inferior vena cava

Right common iliac a.

Left common iliac a.

Ureters

External iliac a. (cut)

Internal iliac a.

Posterior division of internal iliac a.

Anterior division of internal iliac a.

Superior gluteal a.

Patent part of umbilical a.

Obturator a.

Piriformis m.

Internal pudendal a.

Inferior gluteal a.

Coccygeus m.

Middle anorectal a.

Uterine a.

Vaginal a.

Internal pudendal a.

Inferior anorectal a.

Superior vesical aa.

Occluded part of umbilical a.

Obturator canal

Obturator internus

Levator ani

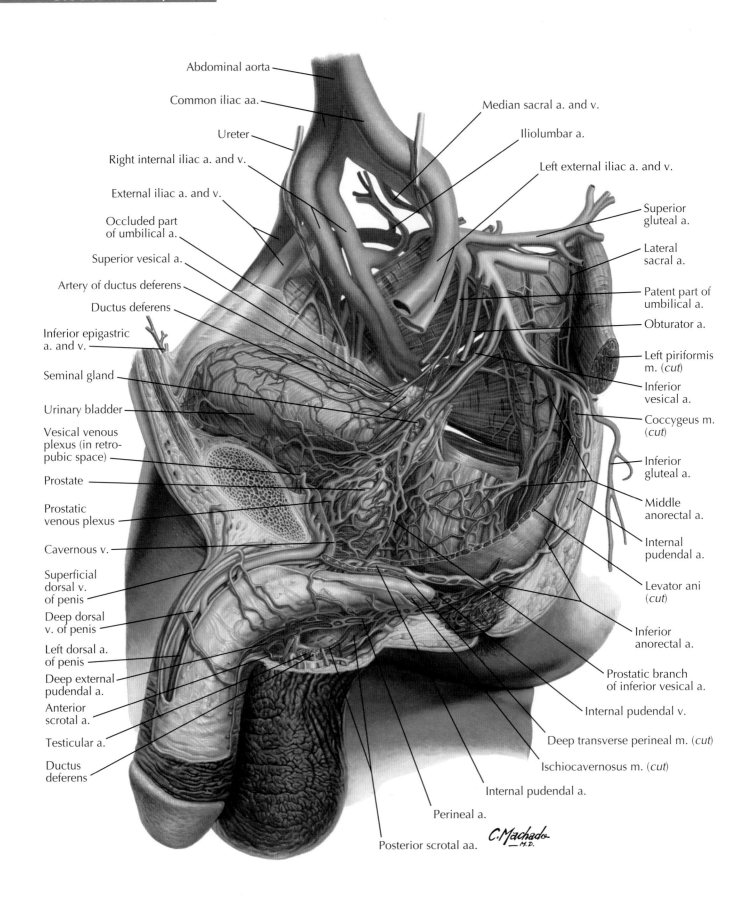

Abdominal aorta

Common iliac aa.

Ureter

Right internal iliac a. and v.

External iliac a. and v.

Occluded part of umbilical a.

Superior vesical a.

Artery of ductus deferens

Ductus deferens

Inferior epigastric a. and v.

Seminal gland

Urinary bladder

Vesical venous plexus (in retro-pubic space)

Prostate

Prostatic venous plexus

Cavernous v.

Superficial dorsal v. of penis

Deep dorsal v. of penis

Left dorsal a. of penis

Deep external pudendal a.

Anterior scrotal a.

Testicular a.

Ductus deferens

Median sacral a. and v.

Iliolumbar a.

Left external iliac a. and v.

Superior gluteal a.

Lateral sacral a.

Patent part of umbilical a.

Obturator a.

Left piriformis m. (*cut*)

Inferior vesical a.

Coccygeus m. (*cut*)

Inferior gluteal a.

Middle anorectal a.

Internal pudendal a.

Levator ani (*cut*)

Inferior anorectal a.

Prostatic branch of inferior vesical a.

Internal pudendal v.

Deep transverse perineal m. (*cut*)

Ischiocavernosus m. (*cut*)

Internal pudendal a.

Perineal a.

Posterior scrotal aa.

C. Machado
_M.D._

Plate 405

Vasculature

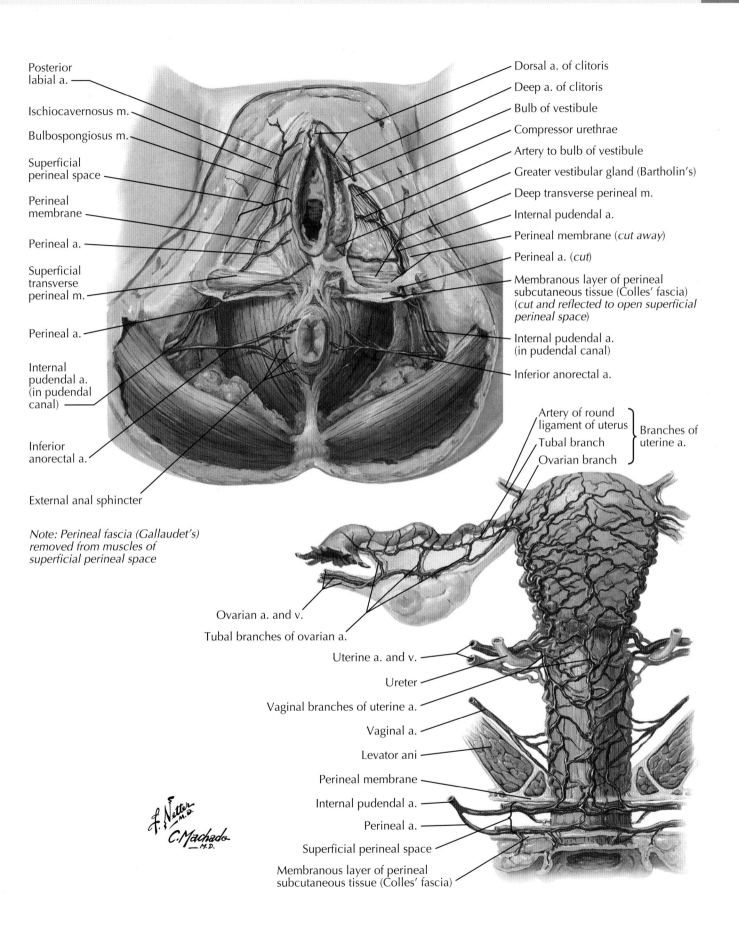

Posterior labial a.

Ischiocavernosus m.

Bulbospongiosus m.

Superficial perineal space

Perineal membrane

Perineal a.

Superficial transverse perineal m.

Perineal a.

Internal pudendal a. (in pudendal canal)

Inferior anorectal a.

External anal sphincter

Note: Perineal fascia (Gallaudet's) removed from muscles of superficial perineal space

Dorsal a. of clitoris

Deep a. of clitoris

Bulb of vestibule

Compressor urethrae

Artery to bulb of vestibule

Greater vestibular gland (Bartholin's)

Deep transverse perineal m.

Internal pudendal a.

Perineal membrane (*cut away*)

Perineal a. (*cut*)

Membranous layer of perineal subcutaneous tissue (Colles' fascia) (*cut and reflected to open superficial perineal space*)

Internal pudendal a. (in pudendal canal)

Inferior anorectal a.

Artery of round ligament of uterus ⎫
Tubal branch ⎬ Branches of uterine a.
Ovarian branch ⎭

Ovarian a. and v.

Tubal branches of ovarian a.

Uterine a. and v.

Ureter

Vaginal branches of uterine a.

Vaginal a.

Levator ani

Perineal membrane

Internal pudendal a.

Perineal a.

Superficial perineal space

Membranous layer of perineal subcutaneous tissue (Colles' fascia)

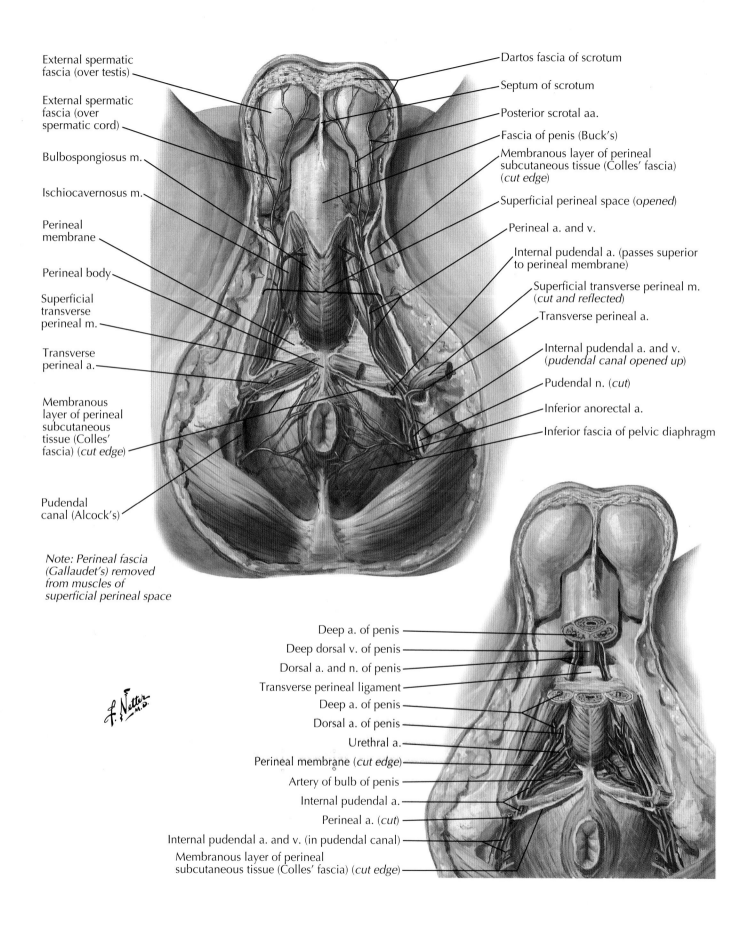

External spermatic fascia (over testis)

External spermatic fascia (over spermatic cord)

Bulbospongiosus m.

Ischiocavernosus m.

Perineal membrane

Perineal body

Superficial transverse perineal m.

Transverse perineal a.

Membranous layer of perineal subcutaneous tissue (Colles' fascia) (cut edge)

Pudendal canal (Alcock's)

Note: Perineal fascia (Gallaudet's) removed from muscles of superficial perineal space

Dartos fascia of scrotum

Septum of scrotum

Posterior scrotal aa.

Fascia of penis (Buck's)

Membranous layer of perineal subcutaneous tissue (Colles' fascia) (cut edge)

Superficial perineal space (opened)

Perineal a. and v.

Internal pudendal a. (passes superior to perineal membrane)

Superficial transverse perineal m. (cut and reflected)

Transverse perineal a.

Internal pudendal a. and v. (pudendal canal opened up)

Pudendal n. (cut)

Inferior anorectal a.

Inferior fascia of pelvic diaphragm

Deep a. of penis

Deep dorsal v. of penis

Dorsal a. and n. of penis

Transverse perineal ligament

Deep a. of penis

Dorsal a. of penis

Urethral a.

Perineal membrane (cut edge)

Artery of bulb of penis

Internal pudendal a.

Perineal a. (cut)

Internal pudendal a. and v. (in pudendal canal)

Membranous layer of perineal subcutaneous tissue (Colles' fascia) (cut edge)

Plate 407

Vasculature

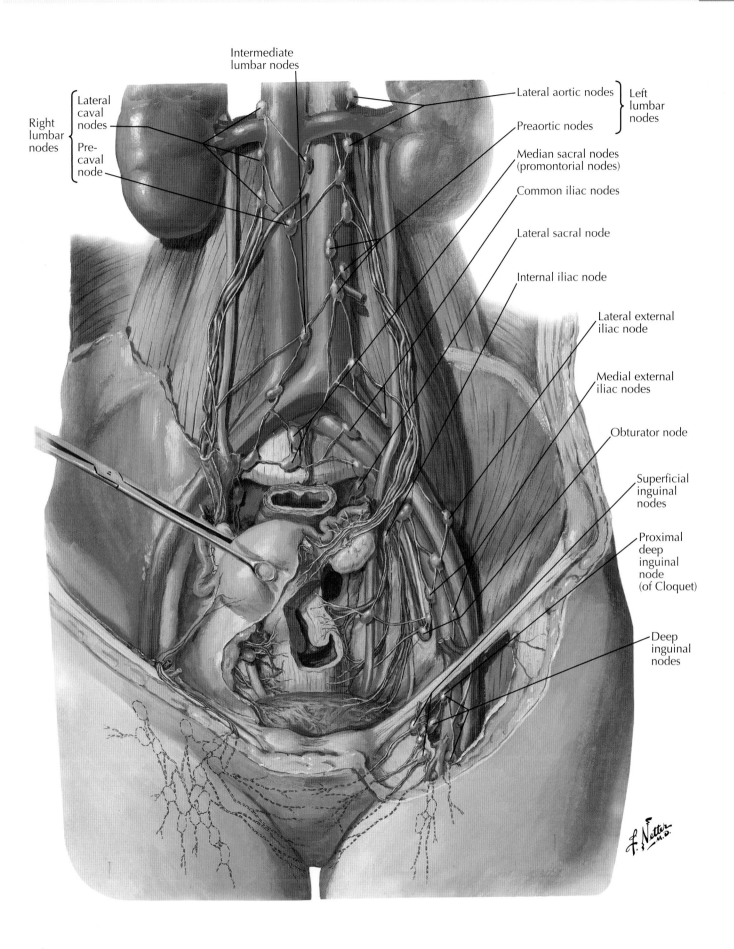

Intermediate lumbar nodes

Lateral caval nodes

Right lumbar nodes

Pre-caval node

Lateral aortic nodes — Left lumbar nodes

Preaortic nodes

Median sacral nodes (promontorial nodes)

Common iliac nodes

Lateral sacral node

Internal iliac node

Lateral external iliac node

Medial external iliac nodes

Obturator node

Superficial inguinal nodes

Proximal deep inguinal node (of Cloquet)

Deep inguinal nodes

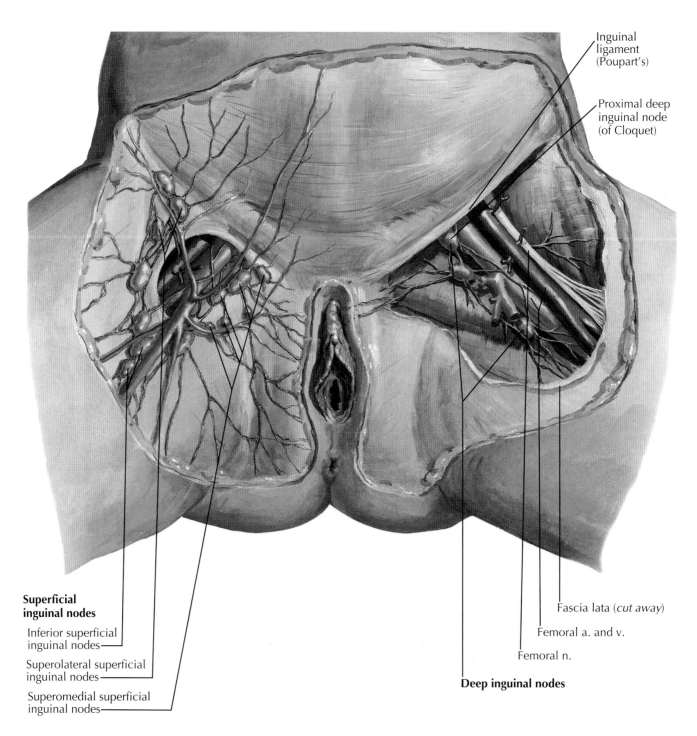

Inguinal
ligament
(Poupart's)

Proximal deep
inguinal node
(of Cloquet)

**Superficial
inguinal nodes**

Inferior superficial
inguinal nodes

Superolateral superficial
inguinal nodes

Superomedial superficial
inguinal nodes

Fascia lata (*cut away*)

Femoral a. and v.

Femoral n.

Deep inguinal nodes

Plate 409

Vasculature

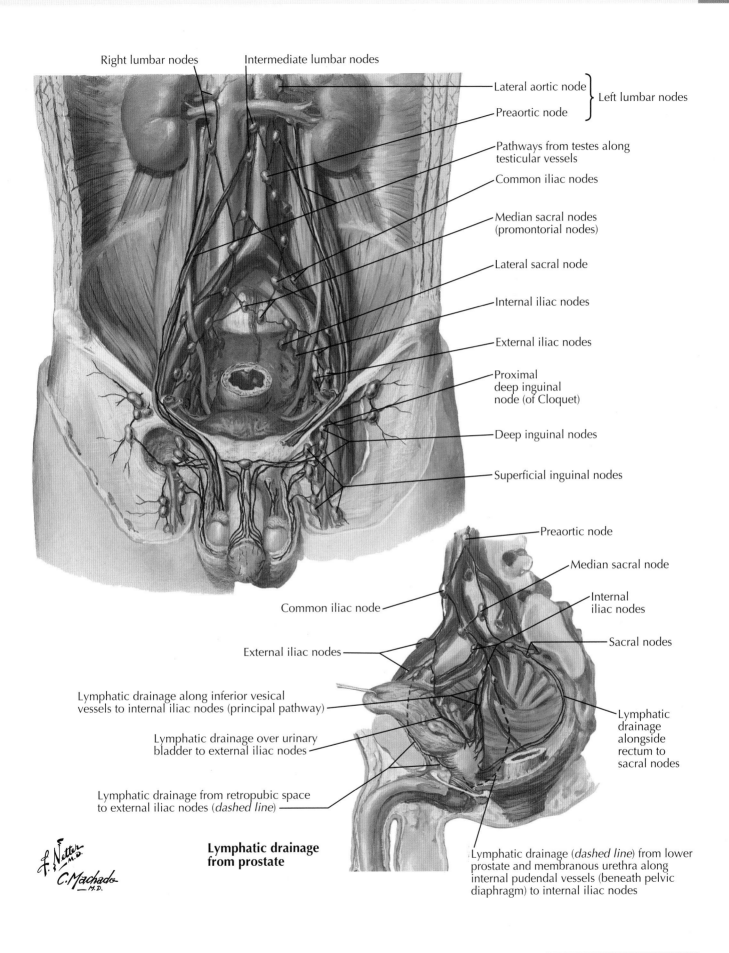

Right lumbar nodes

Intermediate lumbar nodes

Lateral aortic node
Preaortic node
} Left lumbar nodes

Pathways from testes along testicular vessels

Common iliac nodes

Median sacral nodes (promontorial nodes)

Lateral sacral node

Internal iliac nodes

External iliac nodes

Proximal deep inguinal node (of Cloquet)

Deep inguinal nodes

Superficial inguinal nodes

Preaortic node

Median sacral node

Internal iliac nodes

Common iliac node

Sacral nodes

External iliac nodes

Lymphatic drainage along inferior vesical vessels to internal iliac nodes (principal pathway)

Lymphatic drainage over urinary bladder to external iliac nodes

Lymphatic drainage alongside rectum to sacral nodes

Lymphatic drainage from retropubic space to external iliac nodes (*dashed line*)

Lymphatic drainage from prostate

Lymphatic drainage (*dashed line*) from lower prostate and membranous urethra along internal pudendal vessels (beneath pelvic diaphragm) to internal iliac nodes

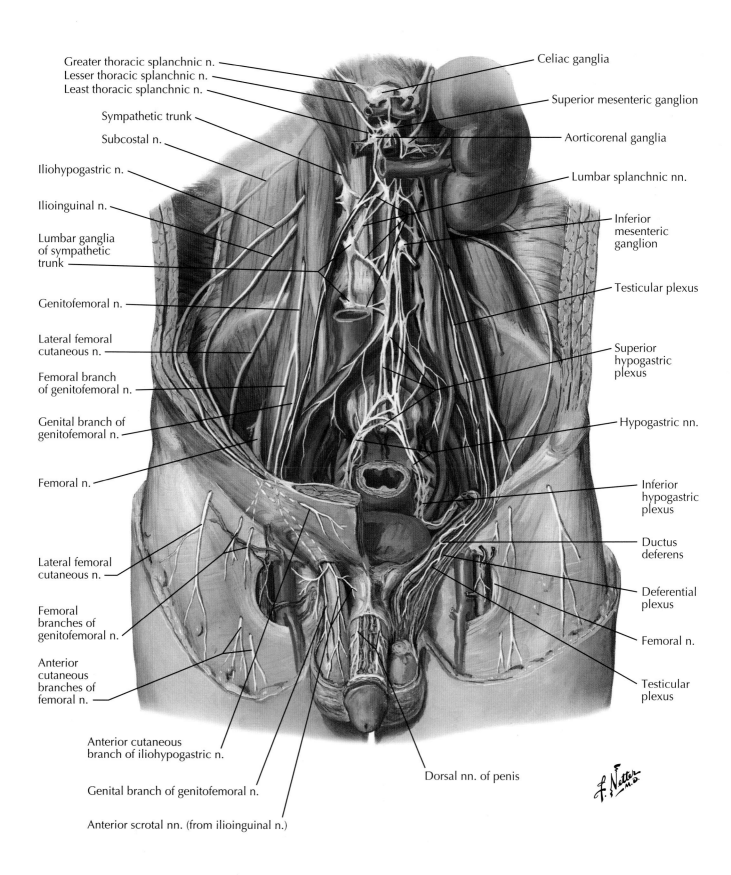

Greater thoracic splanchnic n.
Lesser thoracic splanchnic n.
Least thoracic splanchnic n.

Sympathetic trunk

Subcostal n.

Iliohypogastric n.

Ilioinguinal n.

Lumbar ganglia
of sympathetic
trunk

Genitofemoral n.

Lateral femoral
cutaneous n.

Femoral branch
of genitofemoral n.

Genital branch of
genitofemoral n.

Femoral n.

Lateral femoral
cutaneous n.

Femoral
branches of
genitofemoral n.

Anterior
cutaneous
branches of
femoral n.

Anterior cutaneous
branch of iliohypogastric n.

Genital branch of genitofemoral n.

Anterior scrotal nn. (from ilioinguinal n.)

Celiac ganglia

Superior mesenteric ganglion

Aorticorenal ganglia

Lumbar splanchnic nn.

Inferior
mesenteric
ganglion

Testicular plexus

Superior
hypogastric
plexus

Hypogastric nn.

Inferior
hypogastric
plexus

Ductus
deferens

Deferential
plexus

Femoral n.

Testicular
plexus

Dorsal nn. of penis

Plate 411

Nerves of Perineum and Pelvic Viscera

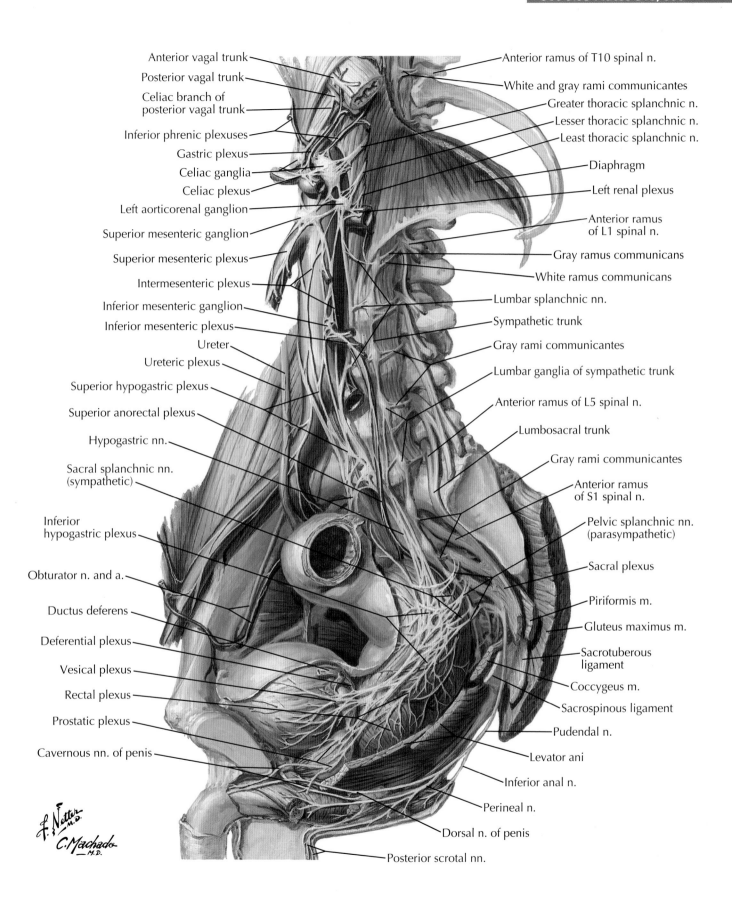

Anterior vagal trunk

Posterior vagal trunk

Celiac branch of posterior vagal trunk

Inferior phrenic plexuses

Gastric plexus

Celiac ganglia

Celiac plexus

Left aorticorenal ganglion

Superior mesenteric ganglion

Superior mesenteric plexus

Intermesenteric plexus

Inferior mesenteric ganglion

Inferior mesenteric plexus

Ureter

Ureteric plexus

Superior hypogastric plexus

Superior anorectal plexus

Hypogastric nn.

Sacral splanchnic nn. (sympathetic)

Inferior hypogastric plexus

Obturator n. and a.

Ductus deferens

Deferential plexus

Vesical plexus

Rectal plexus

Prostatic plexus

Cavernous nn. of penis

Anterior ramus of T10 spinal n.

White and gray rami communicantes

Greater thoracic splanchnic n.

Lesser thoracic splanchnic n.

Least thoracic splanchnic n.

Diaphragm

Left renal plexus

Anterior ramus of L1 spinal n.

Gray ramus communicans

White ramus communicans

Lumbar splanchnic nn.

Sympathetic trunk

Gray rami communicantes

Lumbar ganglia of sympathetic trunk

Anterior ramus of L5 spinal n.

Lumbosacral trunk

Gray rami communicantes

Anterior ramus of S1 spinal n.

Pelvic splanchnic nn. (parasympathetic)

Sacral plexus

Piriformis m.

Gluteus maximus m.

Sacrotuberous ligament

Coccygeus m.

Sacrospinous ligament

Pudendal n.

Levator ani

Inferior anal n.

Perineal n.

Dorsal n. of penis

Posterior scrotal nn.

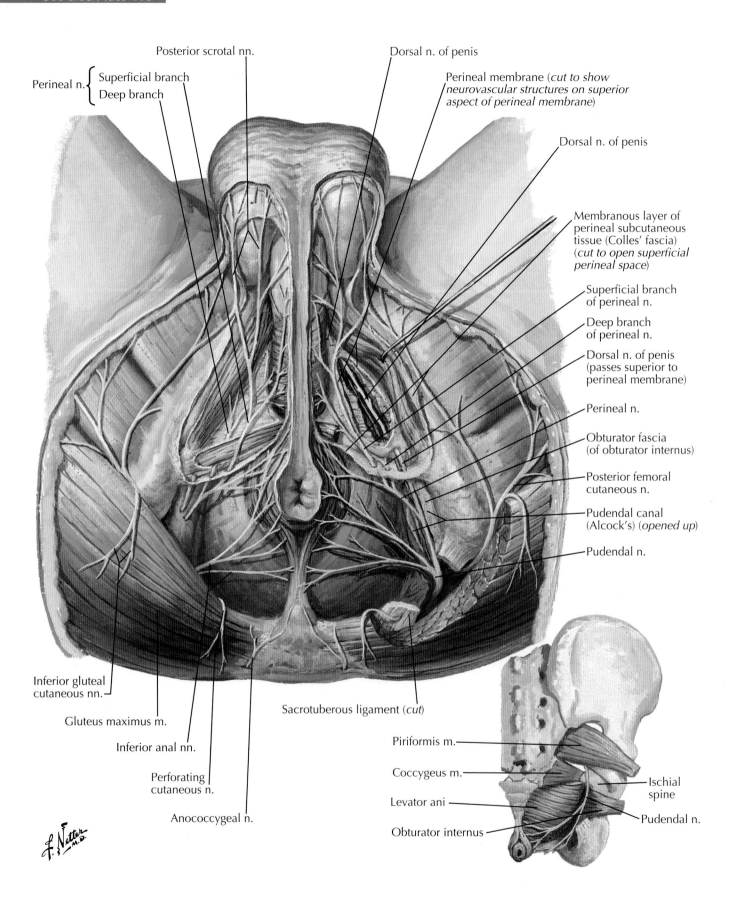

Posterior scrotal nn.

Dorsal n. of penis

Perineal n. { Superficial branch
Deep branch }

Perineal membrane (*cut to show neurovascular structures on superior aspect of perineal membrane*)

Dorsal n. of penis

Membranous layer of perineal subcutaneous tissue (Colles' fascia) (*cut to open superficial perineal space*)

Superficial branch of perineal n.

Deep branch of perineal n.

Dorsal n. of penis (passes superior to perineal membrane)

Perineal n.

Obturator fascia (of obturator internus)

Posterior femoral cutaneous n.

Pudendal canal (Alcock's) (*opened up*)

Pudendal n.

Inferior gluteal cutaneous nn.

Gluteus maximus m.

Inferior anal nn.

Perforating cutaneous n.

Anococcygeal n.

Sacrotuberous ligament (*cut*)

Piriformis m.

Coccygeus m.

Levator ani

Obturator internus

Ischial spine

Pudendal n.

Plate 413

Nerves of Perineum and Pelvic Viscera

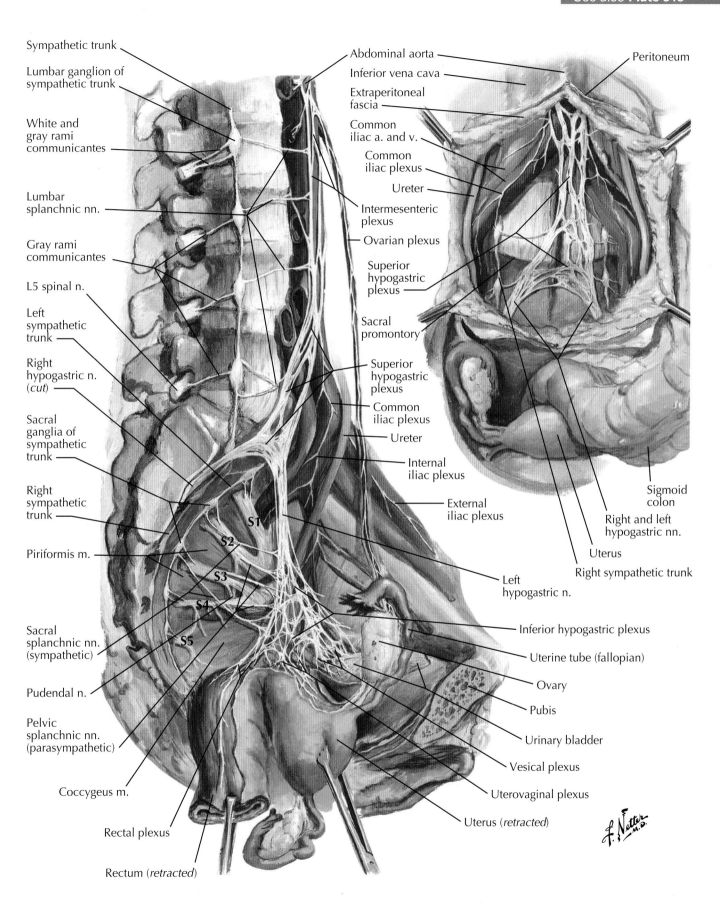

Sympathetic trunk

Lumbar ganglion of sympathetic trunk

White and gray rami communicantes

Lumbar splanchnic nn.

Gray rami communicantes

L5 spinal n.

Left sympathetic trunk

Right hypogastric n. (*cut*)

Sacral ganglia of sympathetic trunk

Right sympathetic trunk

Piriformis m.

Sacral splanchnic nn. (sympathetic)

Pudendal n.

Pelvic splanchnic nn. (parasympathetic)

Coccygeus m.

Rectal plexus

Rectum (*retracted*)

Abdominal aorta

Inferior vena cava

Extraperitoneal fascia

Common iliac a. and v.

Common iliac plexus

Ureter

Intermesenteric plexus

Ovarian plexus

Superior hypogastric plexus

Sacral promontory

Superior hypogastric plexus

Common iliac plexus

Ureter

Internal iliac plexus

External iliac plexus

Left hypogastric n.

Peritoneum

Sigmoid colon

Right and left hypogastric nn.

Uterus

Right sympathetic trunk

Inferior hypogastric plexus

Uterine tube (fallopian)

Ovary

Pubis

Urinary bladder

Vesical plexus

Uterovaginal plexus

Uterus (*retracted*)

S1
S2
S3
S4
S5

F. Netter
M.D.

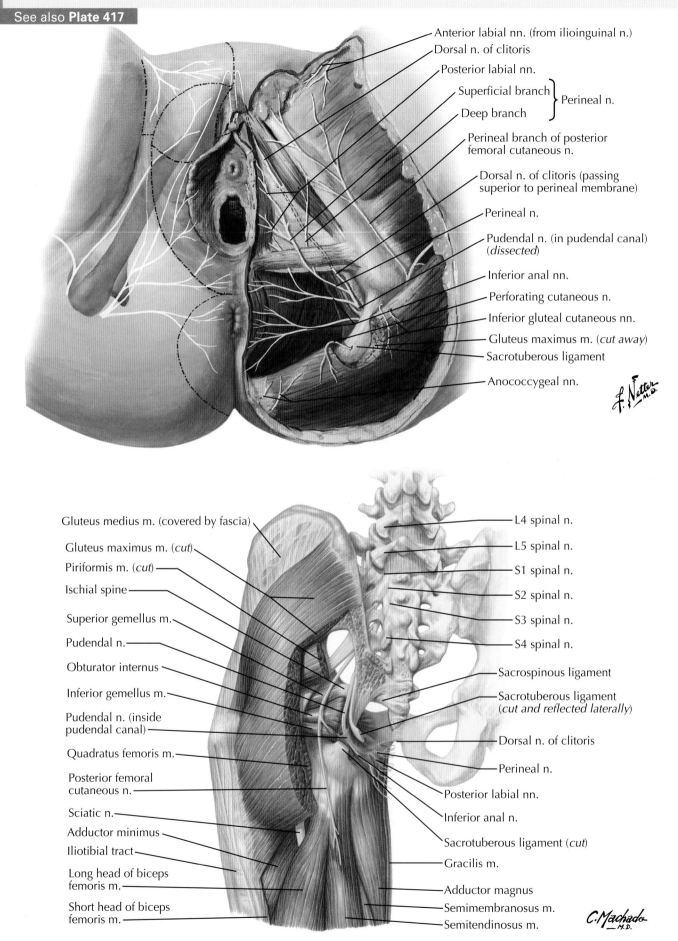

Anterior labial nn. (from ilioinguinal n.)

Dorsal n. of clitoris

Posterior labial nn.

Superficial branch ⎫
⎬ Perineal n.
Deep branch ⎭

Perineal branch of posterior femoral cutaneous n.

Dorsal n. of clitoris (passing superior to perineal membrane)

Perineal n.

Pudendal n. (in pudendal canal) (*dissected*)

Inferior anal nn.

Perforating cutaneous n.

Inferior gluteal cutaneous nn.

Gluteus maximus m. (*cut away*)

Sacrotuberous ligament

Anococcygeal nn.

Gluteus medius m. (covered by fascia)

Gluteus maximus m. (*cut*)

Piriformis m. (*cut*)

Ischial spine

Superior gemellus m.

Pudendal n.

Obturator internus

Inferior gemellus m.

Pudendal n. (inside pudendal canal)

Quadratus femoris m.

Posterior femoral cutaneous n.

Sciatic n.

Adductor minimus

Iliotibial tract

Long head of biceps femoris m.

Short head of biceps femoris m.

L4 spinal n.

L5 spinal n.

S1 spinal n.

S2 spinal n.

S3 spinal n.

S4 spinal n.

Sacrospinous ligament

Sacrotuberous ligament (*cut and reflected laterally*)

Dorsal n. of clitoris

Perineal n.

Posterior labial nn.

Inferior anal n.

Sacrotuberous ligament (*cut*)

Gracilis m.

Adductor magnus

Semimembranosus m.

Semitendinosus m.

Plate 415

Nerves of Perineum and Pelvic Viscera

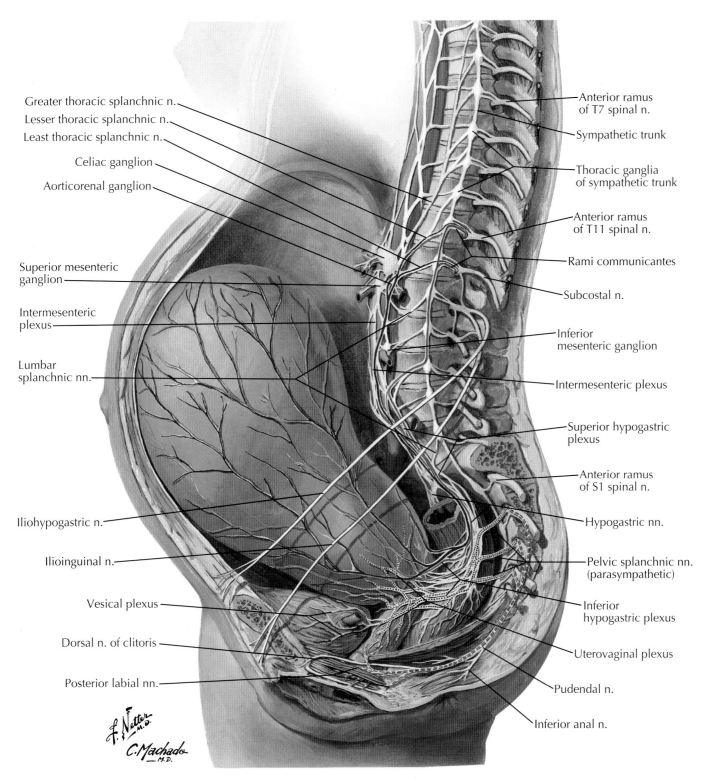

Greater thoracic splanchnic n.

Lesser thoracic splanchnic n.

Least thoracic splanchnic n.

Celiac ganglion

Aorticorenal ganglion

Superior mesenteric ganglion

Intermesenteric plexus

Lumbar splanchnic nn.

Iliohypogastric n.

Ilioinguinal n.

Vesical plexus

Dorsal n. of clitoris

Posterior labial nn.

Anterior ramus of T7 spinal n.

Sympathetic trunk

Thoracic ganglia of sympathetic trunk

Anterior ramus of T11 spinal n.

Rami communicantes

Subcostal n.

Inferior mesenteric ganglion

Intermesenteric plexus

Superior hypogastric plexus

Anterior ramus of S1 spinal n.

Hypogastric nn.

Pelvic splanchnic nn. (parasympathetic)

Inferior hypogastric plexus

Uterovaginal plexus

Pudendal n.

Inferior anal n.

———— Sensory fibers from body and fundus of uterus accompany sympathetic fibers to lower thoracic part of spinal cord via hypogastric plexuses

———— Sympathetic fibers to body and fundus of uterus

·········· Sensory fibers from cervix and upper vagina accompany parasympathetic fibers to sacral part of spinal cord via pelvic splanchnic nerves

·········· Parasympathetic fibers to lower uterine segment, cervix, and upper vagina

– – – – Sensory fibers from lower vagina and perineum accompany somatic motor fibers to sacral part of spinal cord via pudendal nerve

– – – – – Somatic motor fibers to lower vagina and perineum via pudendal nerve

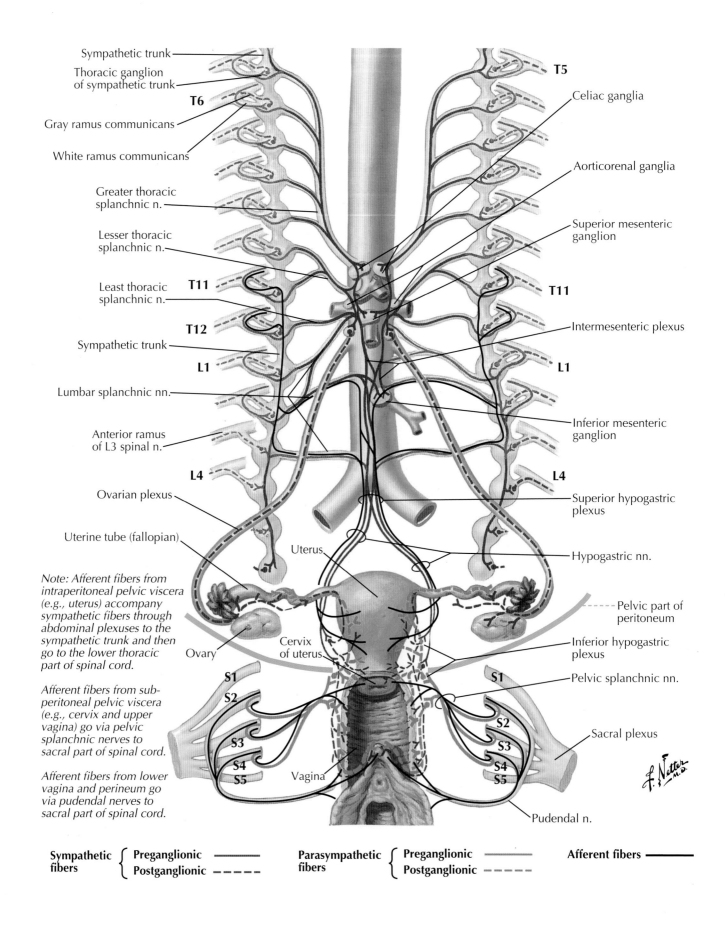

Sympathetic trunk

Thoracic ganglion of sympathetic trunk

T6

Gray ramus communicans

White ramus communicans

Greater thoracic splanchnic n.

Lesser thoracic splanchnic n.

Least thoracic splanchnic n.

T11

T12

Sympathetic trunk

L1

Lumbar splanchnic nn.

Anterior ramus of L3 spinal n.

L4

Ovarian plexus

Uterine tube (fallopian)

Note: Afferent fibers from intraperitoneal pelvic viscera (e.g., uterus) accompany sympathetic fibers through abdominal plexuses to the sympathetic trunk and then go to the lower thoracic part of spinal cord.

Afferent fibers from sub-peritoneal pelvic viscera (e.g., cervix and upper vagina) go via pelvic splanchnic nerves to sacral part of spinal cord.

Afferent fibers from lower vagina and perineum go via pudendal nerves to sacral part of spinal cord.

T5

Celiac ganglia

Aorticorenal ganglia

Superior mesenteric ganglion

T11

Intermesenteric plexus

L1

Inferior mesenteric ganglion

L4

Superior hypogastric plexus

Uterus

Hypogastric nn.

Pelvic part of peritoneum

Inferior hypogastric plexus

Ovary

Cervix of uterus

S1

S2

S3

S4

S5

Vagina

Pelvic splanchnic nn.

Sacral plexus

S1

S2

S3

S4

S5

Pudendal n.

Sympathetic fibers { Preganglionic ——— Postganglionic – – – –

Parasympathetic fibers { Preganglionic ——— Postganglionic – – – –

Afferent fibers ———

Plate 417   **Nerves of Perineum and Pelvic Viscera**

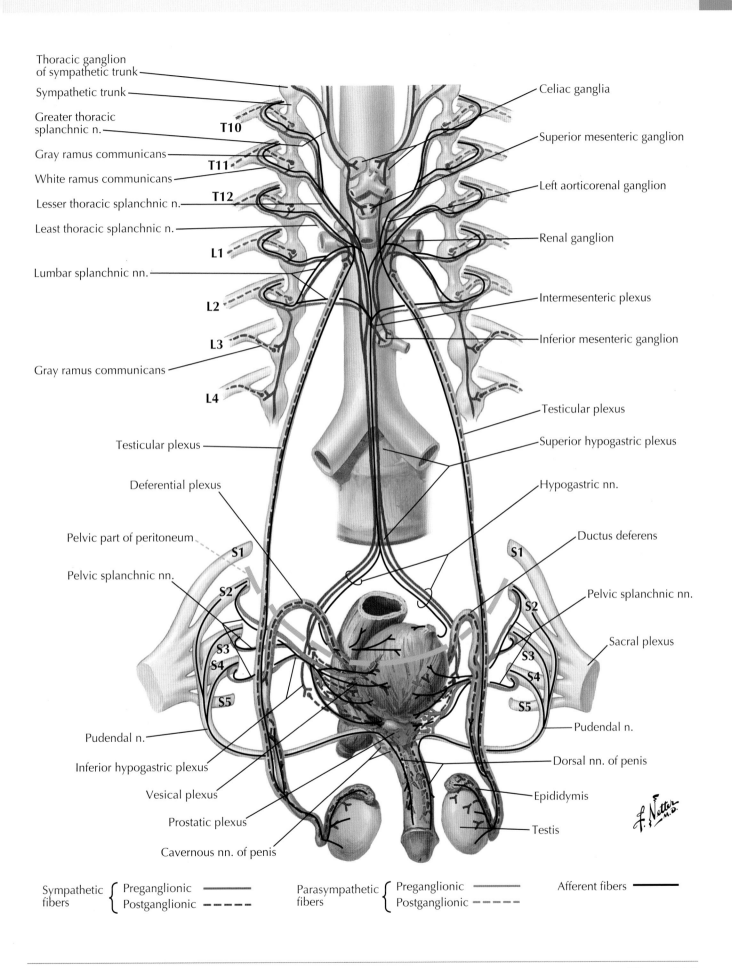

Thoracic ganglion of sympathetic trunk

Sympathetic trunk

Greater thoracic splanchnic n.

Gray ramus communicans

White ramus communicans

Lesser thoracic splanchnic n.

Least thoracic splanchnic n.

Lumbar splanchnic nn.

Gray ramus communicans

Testicular plexus

Deferential plexus

Pelvic part of peritoneum

Pelvic splanchnic nn.

Pudendal n.

Inferior hypogastric plexus

Vesical plexus

Prostatic plexus

Cavernous nn. of penis

Celiac ganglia

Superior mesenteric ganglion

Left aorticorenal ganglion

Renal ganglion

Intermesenteric plexus

Inferior mesenteric ganglion

Testicular plexus

Superior hypogastric plexus

Hypogastric nn.

Ductus deferens

Pelvic splanchnic nn.

Sacral plexus

Pudendal n.

Dorsal nn. of penis

Epididymis

Testis

T10, T11, T12, L1, L2, L3, L4

S1, S2, S3, S4, S5

| Sympathetic fibers | Preganglionic —— | Parasympathetic fibers | Preganglionic —— | Afferent fibers —— |
| | Postganglionic - - - | | Postganglionic - - - | |

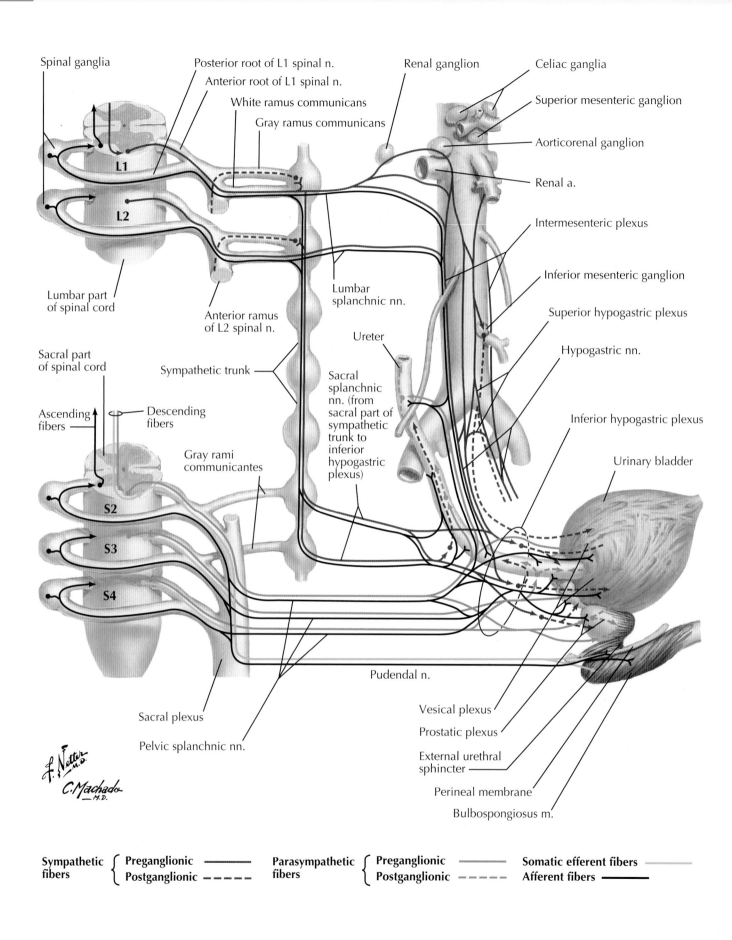

Spinal ganglia

Posterior root of L1 spinal n.

Anterior root of L1 spinal n.

White ramus communicans

Gray ramus communicans

Renal ganglion

Celiac ganglia

Superior mesenteric ganglion

Aorticorenal ganglion

Renal a.

L1

L2

Lumbar part of spinal cord

Anterior ramus of L2 spinal n.

Lumbar splanchnic nn.

Intermesenteric plexus

Inferior mesenteric ganglion

Superior hypogastric plexus

Hypogastric nn.

Ureter

Sacral part of spinal cord

Ascending fibers

Descending fibers

Sympathetic trunk

Sacral splanchnic nn. (from sacral part of sympathetic trunk to inferior hypogastric plexus)

Inferior hypogastric plexus

Urinary bladder

Gray rami communicantes

S2

S3

S4

Pudendal n.

Sacral plexus

Pelvic splanchnic nn.

Vesical plexus

Prostatic plexus

External urethral sphincter

Perineal membrane

Bulbospongiosus m.

| Sympathetic fibers | Preganglionic ——— | Parasympathetic fibers | Preganglionic ——— | Somatic efferent fibers ——— |
| | Postganglionic - - - - | | Postganglionic - - - - | Afferent fibers ——— |

Plate 419

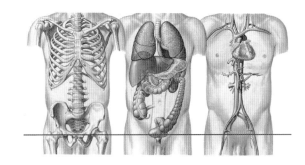

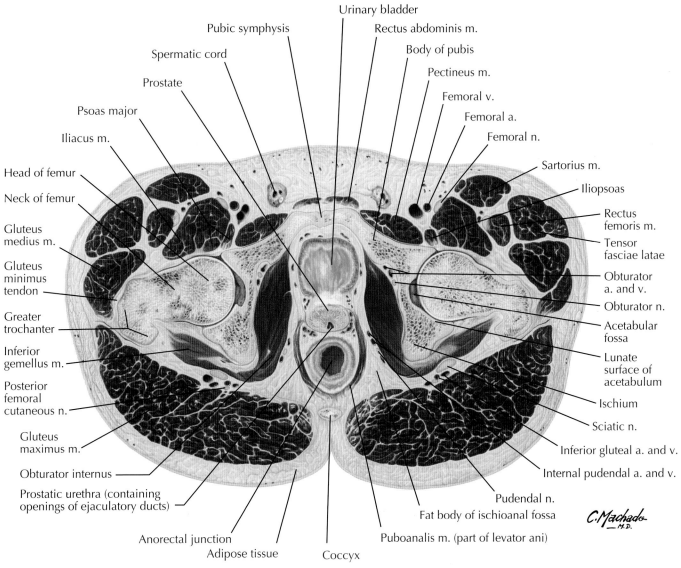

Urinary bladder

Pubic symphysis

Rectus abdominis m.

Spermatic cord

Body of pubis

Prostate

Pectineus m.

Psoas major

Femoral v.

Iliacus m.

Femoral a.

Femoral n.

Head of femur

Sartorius m.

Neck of femur

Iliopsoas

Gluteus medius m.

Rectus femoris m.

Gluteus minimus tendon

Tensor fasciae latae

Greater trochanter

Obturator a. and v.

Inferior gemellus m.

Obturator n.

Posterior femoral cutaneous n.

Acetabular fossa

Gluteus maximus m.

Lunate surface of acetabulum

Obturator internus

Ischium

Prostatic urethra (containing openings of ejaculatory ducts)

Sciatic n.

Anorectal junction

Inferior gluteal a. and v.

Adipose tissue

Internal pudendal a. and v.

Coccyx

Pudendal n.

Fat body of ischioanal fossa

Puboanalis m. (part of levator ani)

C. Machado
_M.D.

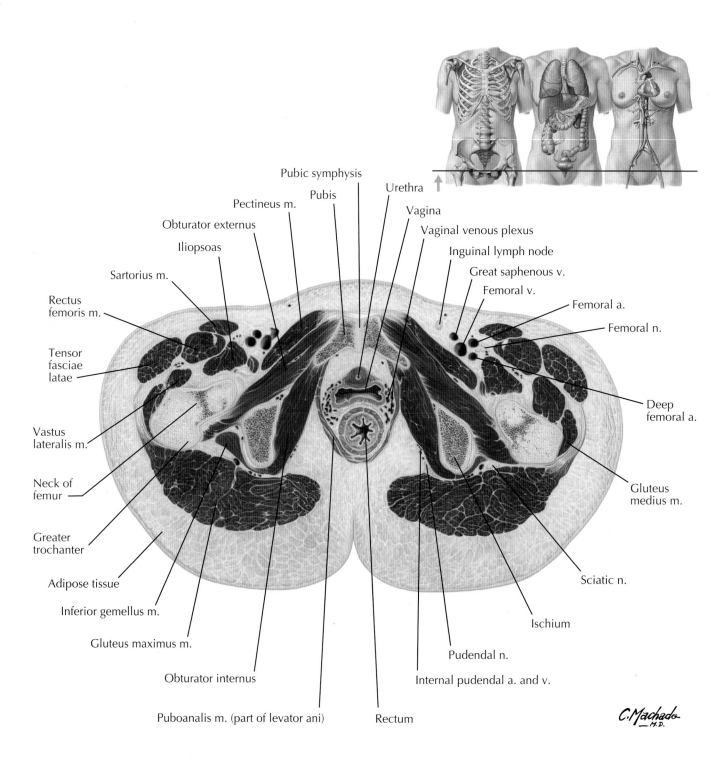

Pubic symphysis

Pubis

Urethra

Pectineus m.

Vagina

Obturator externus

Vaginal venous plexus

Iliopsoas

Inguinal lymph node

Great saphenous v.

Sartorius m.

Femoral v.

Rectus femoris m.

Femoral a.

Femoral n.

Tensor fasciae latae

Deep femoral a.

Vastus lateralis m.

Neck of femur

Gluteus medius m.

Greater trochanter

Sciatic n.

Adipose tissue

Inferior gemellus m.

Ischium

Gluteus maximus m.

Pudendal n.

Obturator internus

Internal pudendal a. and v.

Puboanalis m. (part of levator ani)

Rectum

Plate 421

Cross-Sectional Anatomy

ANATOMICAL STRUCTURE	CLINICAL COMMENT	PLATE NUMBERS
Nervous System and Sense Organs		
Pudendal nerve	Pudendal block is performed to anesthetize the perineum for childbirth or minor surgical procedures in the perineum	415
Inferior anal (rectal) nerve	Anesthetized in ischioanal fossa for surgical excision of external hemorrhoids	413
Prostatic plexus and cavernous nerves	Disruption of these nerves during procedures (e.g., prostate surgery) can produce inability to achieve erection	412
Skeletal System		
Pubic symphysis	Palpable landmark used to obtain pelvic measurements (e.g., diagonal conjugate) that may be used to assess adequacy of pelvis for childbirth; during prenatal examinations, used for estimating fetal growth (symphysis–fundal height measurement); injury may result in widening on x-ray	355
Ischial spine	Palpable landmark used to estimate interspinous diameter for childbirth and to locate pudendal nerve for pudendal nerve block	355
Ischial tuberosity	Palpable landmark used to estimate width of pelvic outlet for childbirth; proximal attachment site of hamstring muscles	355
Superior pubic ramus	Often fractured by lateral compression injury of pelvis in anteroposterior plane by crush injury or falls in elderly with osteoporosis	353
Sacroiliac joint	Stiffening, sclerosis, and fusion occur in autoimmune disease known as ankylosing spondylitis; difficult diagnosis to make and known to refer pain to adjacent joints	355
Muscular System		
Pelvic diaphragm (levator ani and coccygeus muscle)	Provides support to urethrovesical angle, helping to maintain urinary continence; weakness or injury during childbirth can lead to stress urinary incontinence in women	359, 370
Pelvic extraperitoneal (endopelvic) fascia	Weakness or tearing of endopelvic fascial ligaments (e.g., pubovesical or cardinal ligaments) may result in prolapse of pelvic organs	366, 373
Perineal body	Tearing of perineal body may occur during childbirth; prophylactic incision into or lateral to perineal body, known as episiotomy, may be performed to facilitate vaginal delivery in some circumstances	379
Cardiovascular System		
Pampiniform plexus	Dilation of these veins can cause testicular varicocele, affecting testicular temperature regulation and potentially contributing to infertility; most commonly occurs on left side due to longer course of left gonadal vein and angle of drainage into renal vein	403
Uterine artery	Ligated or cauterized during hysterectomy; selective embolization of branches may be performed to treat uterine fibroids	402, 406
Deep and dorsal arteries of penis, and cavernous tissue	Blockage or loss of vascular smooth muscle function can lead to erectile dysfunction, treated with vasodilators	407
Internal iliac veins	Provide communication between prostatic venous plexus and veins draining vertebral column, which is route of spread for prostate cancer	405
Anorectal (rectal) veins	Portal hypertension may cause dilated anorectal veins if portosystemic anastomoses develop between superior anorectal veins (portal drainage) and middle and/or inferior anorectal veins (systemic drainage)	318, 401
Internal and external anorectal (rectal) venous plexuses	Enlargement may result in painful condition known as hemorrhoids	395, 401

Structures with High Clinical Significance

Table 6.1

ANATOMICAL STRUCTURE	CLINICAL COMMENT	PLATE NUMBERS
Lymph Vessels and Lymphoid Organs		
Pelvic and lumbar nodes	Spread of ovarian cancer cells via venous drainage to inferior vena cava and lungs or via lymphatics	408
Lumbar and tracheobronchial nodes	Prostate cancer cells may spread via lymphatics to retroperitoneum and mediastinum	260, 410
Lumbar (e.g., lateral aortic, preaortic, lateral caval) nodes	Receive lymphatic drainage from ovary, uterine tube, and fundus of uterus in women and from testis in men; cancers in these organs may therefore spread to retroperitoneum	408, 410
Pelvic nodes	Sampling or dissection is performed to assess spread of gynecological malignancies	408
Digestive System		
Rectum and anal canal	Examined by digital rectal examination to detect internal hemorrhoids, fecal impaction, and rectal cancer	301, 393, 395
Peritoneum	Common site for metastatic spread of ovarian cancer via fluid in peritoneal cavity	363, 364
Urinary System		
Urinary bladder	Degree of filling is readily assessed with ultrasound; in patients with poor urine output, this can establish diagnosis of bladder outlet obstruction	368, 369
Ureteric orifice	Abnormal reflux of urine from urinary bladder to ureters (vesicoureteral reflux) may occur in children, contributing to recurrent urinary infections and progressive renal fibrosis	371
Ureter	Enlargement indicates obstruction of ureter or urinary bladder; impaction of renal stone in ureter causes severe pain and, in some cases, hematuria; ureter may be injured during hysterectomy because of its close relationship to uterine artery	335, 336, 402
Reproductive System		
Rectouterine pouch (of Douglas)	Region examined with ultrasound to detect presence of abdominopelvic fluid; common site of ectopic pregnancy; may be accessed via posterior vaginal fornix; normally contains small, physiologic amount of peritoneal fluid	364, 365
Uterus	Site of fetal gestation; palpated during prenatal examinations to assess fetal growth; can also contain large, sometimes painful growths known as leiomyomas (fibroids)	364, 374
Uterine tubes (fallopian)	Common site of ectopic pregnancy; inflammation (salpingitis) may occur in pelvic inflammatory disease (PID), the result of sexually transmitted infection, possibly leading to fibrosis and infertility; surgical occlusion (tubal ligation) is performed when women desire permanent contraception	363, 364, 375
Cervix of uterus	Epithelium of transformation zone of cervix is prone to dysplasia and malignancy; cells are sampled from this region during Pap smear examination and tested for infection with human papillomavirus, the leading risk factor for cervical malignancy	372, 374
Vagina	Posterior part of vaginal fornix allows access to rectouterine pouch (of Douglas)	364
Ovary	Examined with ultrasound to identify cysts, or for oocyte collection; torsion is painful condition that occurs when ovary twists on axis of suspensory ligament of ovary, occluding ovarian vessels and causing engorgement and ischemia	363, 374, 375
Testis	Torsion is painful condition that occurs when testis twists on axis of testicular vasculature, causing engorgement and ischemia	388

Table 6.2

Structures with High Clinical Significance

ANATOMICAL STRUCTURE	CLINICAL COMMENT	PLATE NUMBERS
Prostate	Prone to benign hypertrophy with aging, which results in urinary outflow obstruction; prostate cancer is second most common cancer in men	368, 385
Ductus deferens	Ligation (vasectomy) performed when men desire permanent contraception	368, 388

*Selections were based largely on clinical data and commonly discussed clinical correlations in macroscopic ("gross") anatomy courses.

MUSCLE	MUSCLE GROUP	ORIGIN ATTACHMENT	INSERTION ATTACHMENT	INNERVATION	BLOOD SUPPLY	MAIN ACTIONS
Bulbospongiosus m.	Perineal	*Male:* perineal body	*Male:* perineal membrane, corpus cavernosum, bulb of penis	Perineal nerve	Perineal artery	*Male:* compresses bulb of penis, forces blood into body of penis during erection, propels urine and semen through urethra
		Female: perineal body	*Female:* dorsum of clitoris, perineal membrane, bulb of vestibule, pubic arch			*Female:* constricts vaginal orifice, assists in expressing secretions of greater vestibular gland, forces blood into body of clitoris
Coccygeus m.	Pelvic diaphragm	Ischial spine	Inferior sacrum, coccyx	Nerve to coccygeus muscle	Inferior gluteal artery	Supports pelvic viscera
Compressor urethrae (female only)	Perineal	Ischiopubic ramus	Merges with contralateral partner anterior to urethra	Perineal nerve	Perineal artery	Sphincter of urethra
Cremaster	Spermatic cord	Inferior edge of internal abdominal oblique muscle, middle of inguinal ligament	Pubic tubercle, pubic crest	Genital branch of genitofemoral nerve	Cremasteric artery	Retracts testis
Deep transverse perineal m.	Perineal	Internal surface of ramus of ischium, ischial tuberosity	Perineal body	Perineal nerve	Perineal artery	Stabilizes perineal body, supports prostate/vagina
External anal sphincter	Perineal	Anococcygeal ligament	Perineal body	Perineal and inferior anal nerves	Inferior anorectal and perineal arteries	Closes anal orifice
External urethral sphincter	Perineal	Ischiopubic ramus	*Male:* median raphe in front and behind urethra	Perineal nerve	Perineal artery	Compresses urethra at end of micturition; in female also compresses distal vagina
			Female: encloses urethra, attaches to sides of vagina			
Ischiocavernosus m.	Perineal	Inferior internal surface of ramus of ischium, ischial tuberosity	*Male:* anterior end of crus of penis	Perineal nerve	Perineal artery	Forces blood into body of penis and clitoris during erection
			Female: anterior end of crus of clitoris			
Levator ani (iliococcygeus, pubococcygeus, and puboanalis) mm.	Pelvic diaphragm	Body of pubis, tendinous arch of levator ani (on obturator fascia), ischial spine	Perineal body, coccyx, raphe of pelvic diaphragm, walls of prostate or vagina, anorectal junction	Nerve to levator ani, perineal nerve	Inferior gluteal artery, internal pudendal artery and its branches (inferior anorectal and perineal arteries)	Supports pelvic viscera, elevates pelvic floor
Sphincter urethrovaginalis (female only)	Perineal	Perineal body	Passes anteriorly around urethra and merges with its contralateral partner	Perineal nerve	Perineal artery	Sphincter of urethra and vagina
Superficial transverse perineal m.	Perineal	Ramus of ischium, ischial tuberosity	Perineal body	Perineal nerve	Perineal artery	Stabilizes perineal body

Variations in spinal nerve contributions to the innervation of muscles, their arterial supply, their attachments, and their actions are common themes in human anatomy. Therefore, expect differences between texts and realize that anatomical variation is normal.

Table 6.4　　　　**Muscles**

UPPER LIMB 7

ELECTRONIC BONUS PLATES

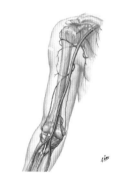

BP 96 Arteries of Arm and Proximal Forearm

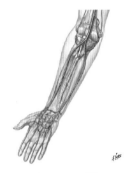

BP 97 Arteries of Forearm and Hand

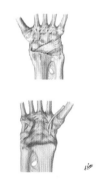

BP 98 Ligaments of Wrist: Posterior and Anterior Views

BP 99 Flexor and Extensor Zones of Hand

BP 100 Cross Sections Through Metacarpal and Distal Carpal Bones

BP 101 Cross Section of Hand: Axial View

BP 102 Cross Section of Hand: Axial View (Continued)

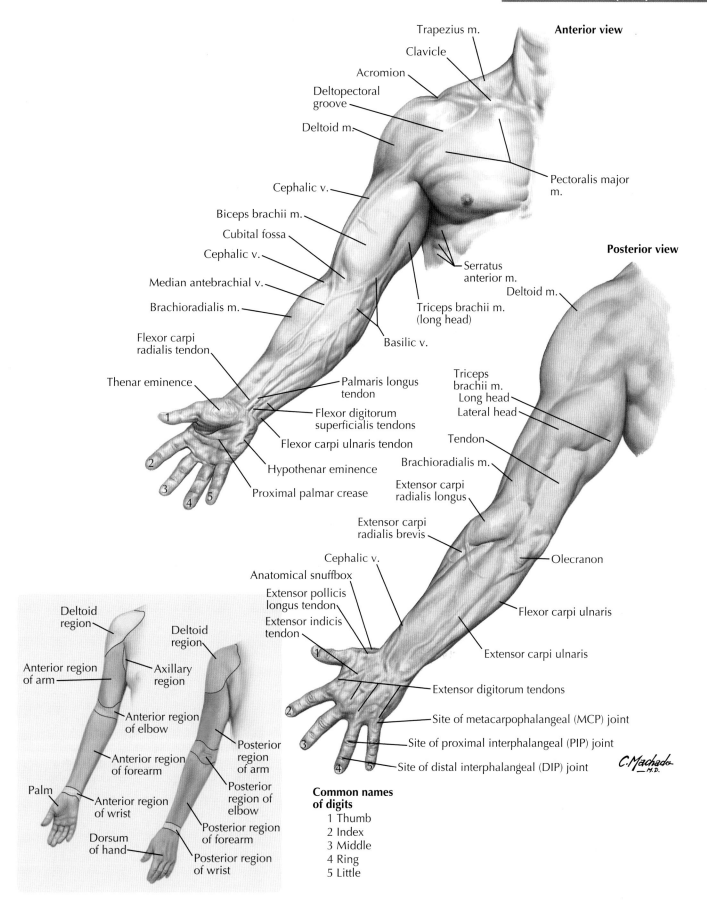

Anterior view

Trapezius m.

Clavicle

Acromion

Deltopectoral groove

Deltoid m.

Cephalic v.

Biceps brachii m.

Cubital fossa

Cephalic v.

Median antebrachial v.

Brachioradialis m.

Flexor carpi radialis tendon

Thenar eminence

Palmaris longus tendon

Flexor digitorum superficialis tendons

Flexor carpi ulnaris tendon

Hypothenar eminence

Proximal palmar crease

Pectoralis major m.

Serratus anterior m.

Triceps brachii m. (long head)

Basilic v.

Posterior view

Deltoid m.

Triceps brachii m.
Long head
Lateral head

Tendon

Brachioradialis m.

Extensor carpi radialis longus

Extensor carpi radialis brevis

Cephalic v.

Anatomical snuffbox

Extensor pollicis longus tendon

Extensor indicis tendon

Olecranon

Flexor carpi ulnaris

Extensor carpi ulnaris

Extensor digitorum tendons

Site of metacarpophalangeal (MCP) joint

Site of proximal interphalangeal (PIP) joint

Site of distal interphalangeal (DIP) joint

Deltoid region

Deltoid region

Anterior region of arm

Axillary region

Anterior region of elbow

Anterior region of forearm

Posterior region of arm

Palm

Anterior region of wrist

Posterior region of elbow

Posterior region of forearm

Dorsum of hand

Posterior region of wrist

Common names of digits
1 Thumb
2 Index
3 Middle
4 Ring
5 Little

C.Machado
_M.D.

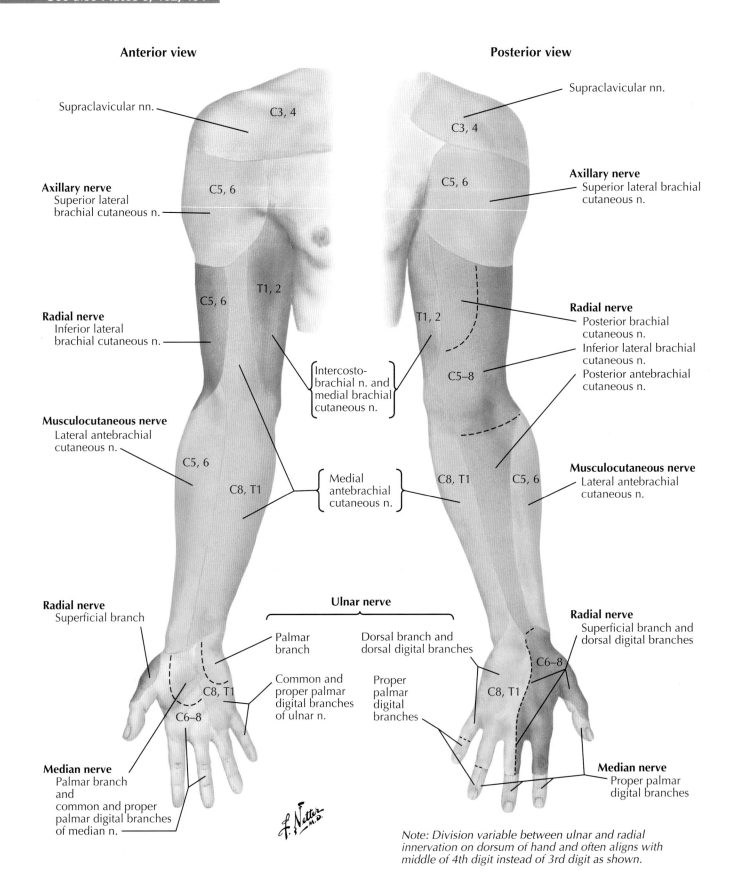

Anterior view

Posterior view

Supraclavicular nn.

C3, 4

Supraclavicular nn.

C3, 4

C5, 6

Axillary nerve
Superior lateral
brachial cutaneous n.

C5, 6

Axillary nerve
Superior lateral brachial
cutaneous n.

C5, 6

T1, 2

Radial nerve
Inferior lateral
brachial cutaneous n.

T1, 2

Radial nerve
Posterior brachial
cutaneous n.
Inferior lateral brachial
cutaneous n.
Posterior antebrachial
cutaneous n.

C5–8

Intercosto-
brachial n. and
medial brachial
cutaneous n.

Musculocutaneous nerve
Lateral antebrachial
cutaneous n.

C5, 6

C8, T1

Medial
antebrachial
cutaneous n.

C8, T1

C5, 6

Musculocutaneous nerve
Lateral antebrachial
cutaneous n.

Ulnar nerve

Radial nerve
Superficial branch

Palmar
branch

Dorsal branch and
dorsal digital branches

Radial nerve
Superficial branch and
dorsal digital branches

C6–8

C8, T1

Common and
proper palmar
digital branches
of ulnar n.

Proper
palmar
digital
branches

C8, T1

C6–8

Median nerve
Palmar branch
and
common and proper
palmar digital branches
of median n.

Median nerve
Proper palmar
digital branches

*Note: Division variable between ulnar and radial
innervation on dorsum of hand and often aligns with
middle of 4th digit instead of 3rd digit as shown.*

Plate 423

Surface Anatomy

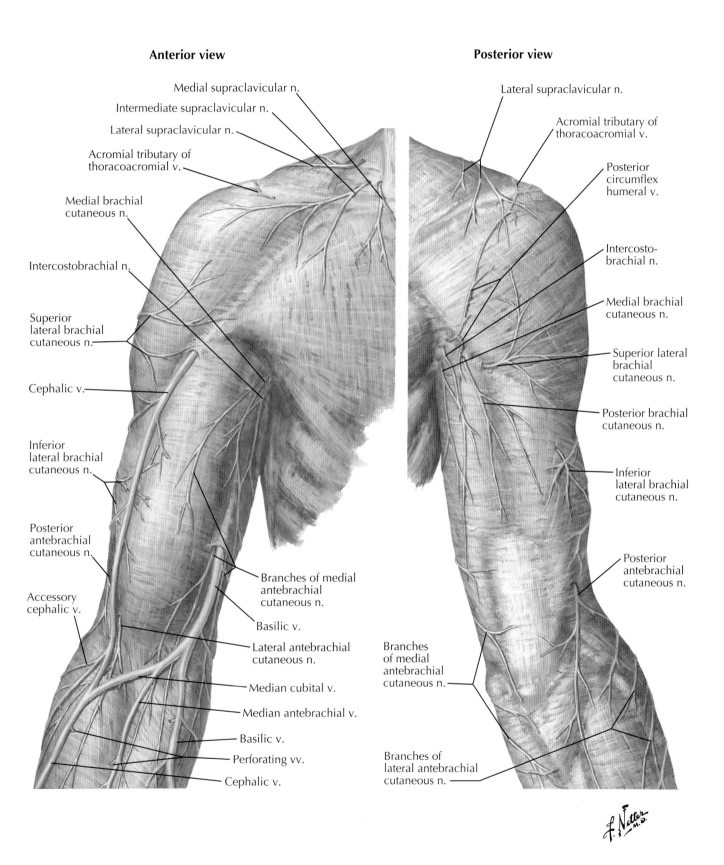

Anterior view

Medial supraclavicular n.

Intermediate supraclavicular n.

Lateral supraclavicular n.

Acromial tributary of thoracoacromial v.

Medial brachial cutaneous n.

Intercostobrachial n.

Superior lateral brachial cutaneous n.

Cephalic v.

Inferior lateral brachial cutaneous n.

Posterior antebrachial cutaneous n.

Accessory cephalic v.

Branches of medial antebrachial cutaneous n.

Basilic v.

Lateral antebrachial cutaneous n.

Median cubital v.

Median antebrachial v.

Basilic v.

Perforating vv.

Cephalic v.

Posterior view

Lateral supraclavicular n.

Acromial tributary of thoracoacromial v.

Posterior circumflex humeral v.

Intercosto-brachial n.

Medial brachial cutaneous n.

Superior lateral brachial cutaneous n.

Posterior brachial cutaneous n.

Inferior lateral brachial cutaneous n.

Posterior antebrachial cutaneous n.

Branches of medial antebrachial cutaneous n.

Branches of lateral antebrachial cutaneous n.

Surface Anatomy

Plate 424

Anterior view

Posterior view

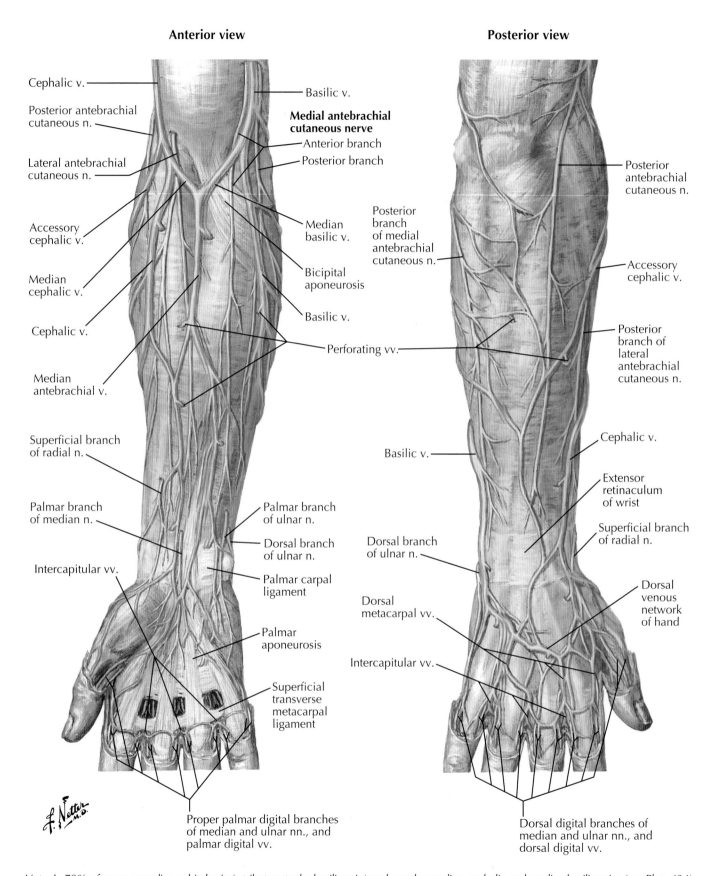

Cephalic v.

Posterior antebrachial cutaneous n.

Lateral antebrachial cutaneous n.

Accessory cephalic v.

Median cephalic v.

Cephalic v.

Median antebrachial v.

Superficial branch of radial n.

Palmar branch of median n.

Intercapitular vv.

Basilic v.

Medial antebrachial cutaneous nerve

Anterior branch

Posterior branch

Median basilic v.

Bicipital aponeurosis

Basilic v.

Perforating vv.

Palmar branch of ulnar n.

Dorsal branch of ulnar n.

Palmar carpal ligament

Palmar aponeurosis

Superficial transverse metacarpal ligament

Posterior branch of medial antebrachial cutaneous n.

Posterior antebrachial cutaneous n.

Accessory cephalic v.

Posterior branch of lateral antebrachial cutaneous n.

Basilic v.

Cephalic v.

Extensor retinaculum of wrist

Superficial branch of radial n.

Dorsal branch of ulnar n.

Dorsal metacarpal vv.

Intercapitular vv.

Dorsal venous network of hand

Proper palmar digital branches of median and ulnar nn., and palmar digital vv.

Dorsal digital branches of median and ulnar nn., and dorsal digital vv.

Note: In 70% of cases, a median cubital vein (a tributary to the basilic vein) replaces the median cephalic and median basilic veins (see Plate 424).

Plate 425

Surface Anatomy

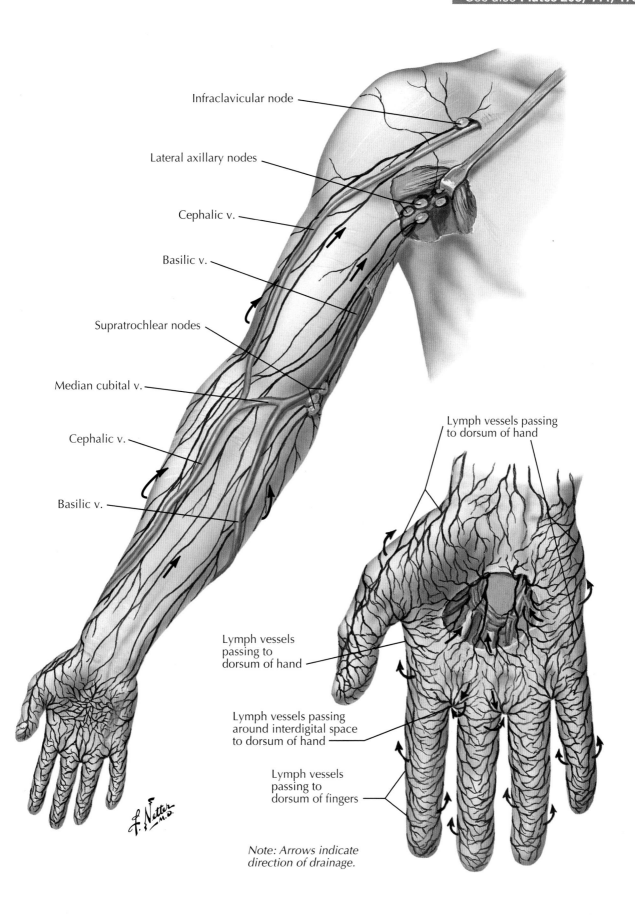

Infraclavicular node

Lateral axillary nodes

Cephalic v.

Basilic v.

Supratrochlear nodes

Median cubital v.

Cephalic v.

Basilic v.

Lymph vessels passing to dorsum of hand

Lymph vessels passing to dorsum of hand

Lymph vessels passing around interdigital space to dorsum of hand

Lymph vessels passing to dorsum of fingers

Note: Arrows indicate direction of drainage.

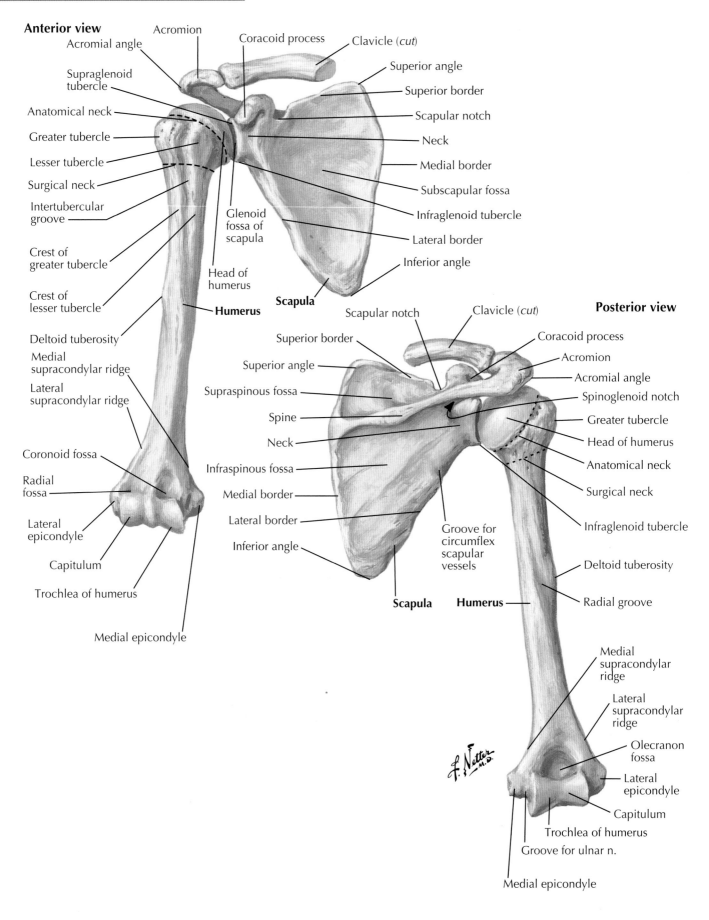

Anterior view

Acromion
Acromial angle
Supraglenoid tubercle
Anatomical neck
Greater tubercle
Lesser tubercle
Surgical neck
Intertubercular groove
Crest of greater tubercle
Crest of lesser tubercle
Deltoid tuberosity
Medial supracondylar ridge
Lateral supracondylar ridge
Coronoid fossa
Radial fossa
Lateral epicondyle
Capitulum
Trochlea of humerus
Medial epicondyle

Coracoid process
Clavicle (*cut*)
Superior angle
Superior border
Scapular notch
Neck
Medial border
Subscapular fossa
Infraglenoid tubercle
Lateral border
Inferior angle
Glenoid fossa of scapula
Head of humerus
Humerus
Scapula

Posterior view

Scapular notch
Clavicle (*cut*)
Superior border
Coracoid process
Superior angle
Acromion
Supraspinous fossa
Acromial angle
Spine
Spinoglenoid notch
Neck
Greater tubercle
Infraspinous fossa
Head of humerus
Medial border
Anatomical neck
Lateral border
Surgical neck
Inferior angle
Infraglenoid tubercle
Groove for circumflex scapular vessels
Deltoid tuberosity
Radial groove
Scapula
Humerus
Medial supracondylar ridge
Lateral supracondylar ridge
Olecranon fossa
Lateral epicondyle
Capitulum
Trochlea of humerus
Groove for ulnar n.
Medial epicondyle

Plate 427

Shoulder and Axilla

See also **Plates 203, 427, 429**

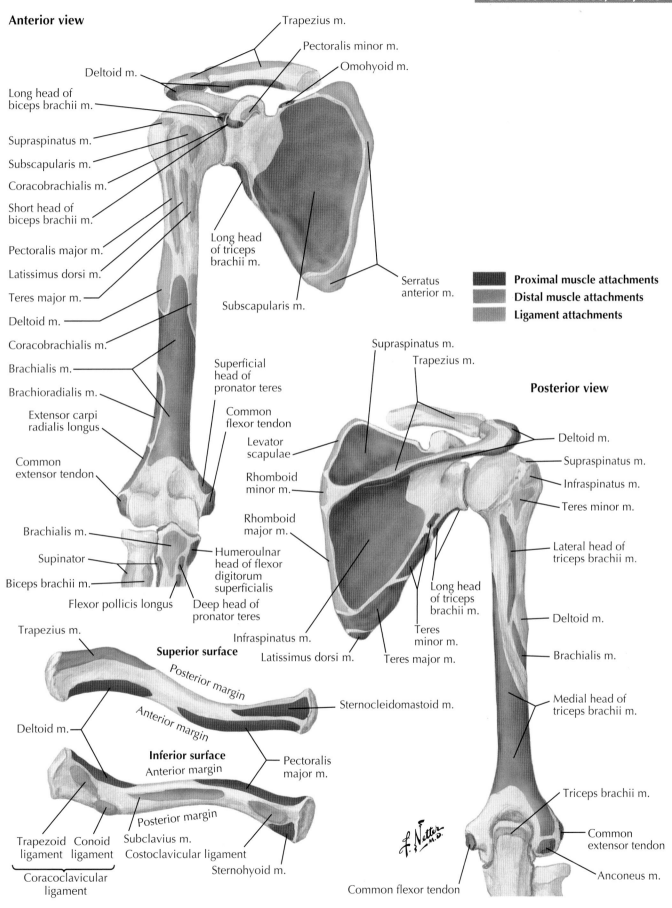

Anterior view

Trapezius m.

Pectoralis minor m.

Omohyoid m.

Deltoid m.

Long head of biceps brachii m.

Supraspinatus m.

Subscapularis m.

Coracobrachialis m.

Short head of biceps brachii m.

Pectoralis major m.

Latissimus dorsi m.

Teres major m.

Deltoid m.

Coracobrachialis m.

Brachialis m.

Brachioradialis m.

Extensor carpi radialis longus

Common extensor tendon

Brachialis m.

Supinator

Biceps brachii m.

Flexor pollicis longus

Long head of triceps brachii m.

Subscapularis m.

Serratus anterior m.

Superficial head of pronator teres

Common flexor tendon

Levator scapulae

Rhomboid minor m.

Rhomboid major m.

Humeroulnar head of flexor digitorum superficialis

Deep head of pronator teres

Infraspinatus m.

Latissimus dorsi m.

Supraspinatus m.

Trapezius m.

Posterior view

Deltoid m.

Supraspinatus m.

Infraspinatus m.

Teres minor m.

Lateral head of triceps brachii m.

Deltoid m.

Brachialis m.

Medial head of triceps brachii m.

Triceps brachii m.

Common extensor tendon

Anconeus m.

Common flexor tendon

Long head of triceps brachii m.

Teres minor m.

Teres major m.

Proximal muscle attachments

Distal muscle attachments

Ligament attachments

Trapezius m.

Deltoid m.

Superior surface

Posterior margin

Anterior margin

Inferior surface

Anterior margin

Posterior margin

Sternocleidomastoid m.

Pectoralis major m.

Subclavius m.

Costoclavicular ligament

Sternohyoid m.

Trapezoid ligament

Conoid ligament

Coracoclavicular ligament

F. Netter m.d.

Right clavicle

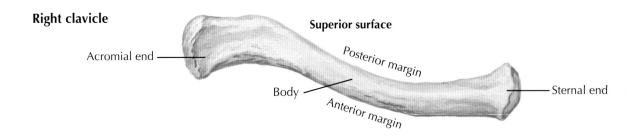

Superior surface

Acromial end

Posterior margin

Body

Sternal end

Anterior margin

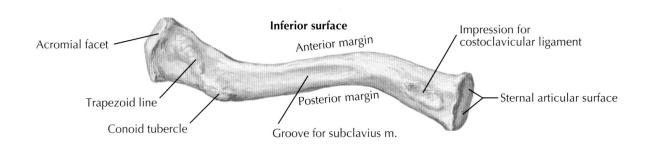

Inferior surface

Acromial facet

Anterior margin

Impression for costoclavicular ligament

Trapezoid line

Conoid tubercle

Posterior margin

Sternal articular surface

Groove for subclavius m.

Sternoclavicular joint

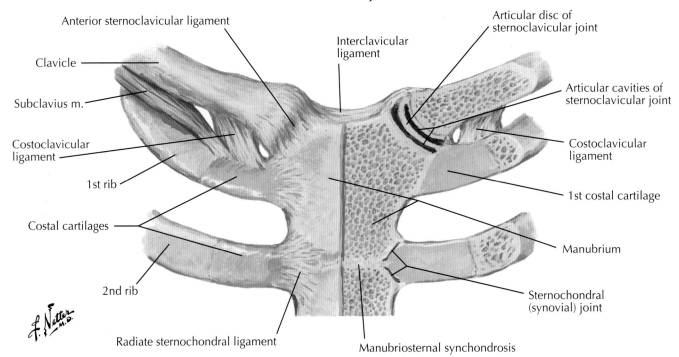

Anterior sternoclavicular ligament

Interclavicular ligament

Articular disc of sternoclavicular joint

Clavicle

Articular cavities of sternoclavicular joint

Subclavius m.

Costoclavicular ligament

Costoclavicular ligament

1st rib

1st costal cartilage

Costal cartilages

Manubrium

2nd rib

Sternochondral (synovial) joint

Radiate sternochondral ligament

Manubriosternal synchondrosis

Plate 429

Shoulder and Axilla

Anteroposterior radiograph of right shoulder

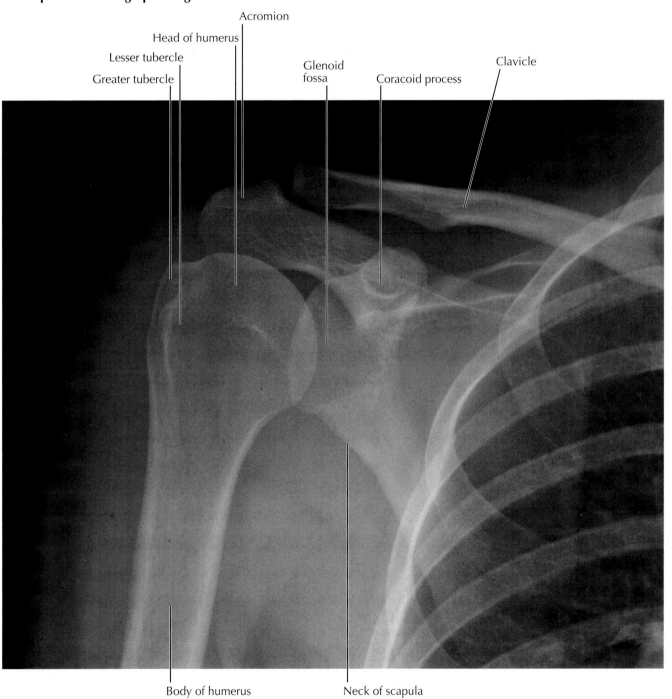

Acromion

Head of humerus

Lesser tubercle

Glenoid fossa

Clavicle

Greater tubercle

Coracoid process

Body of humerus

Neck of scapula

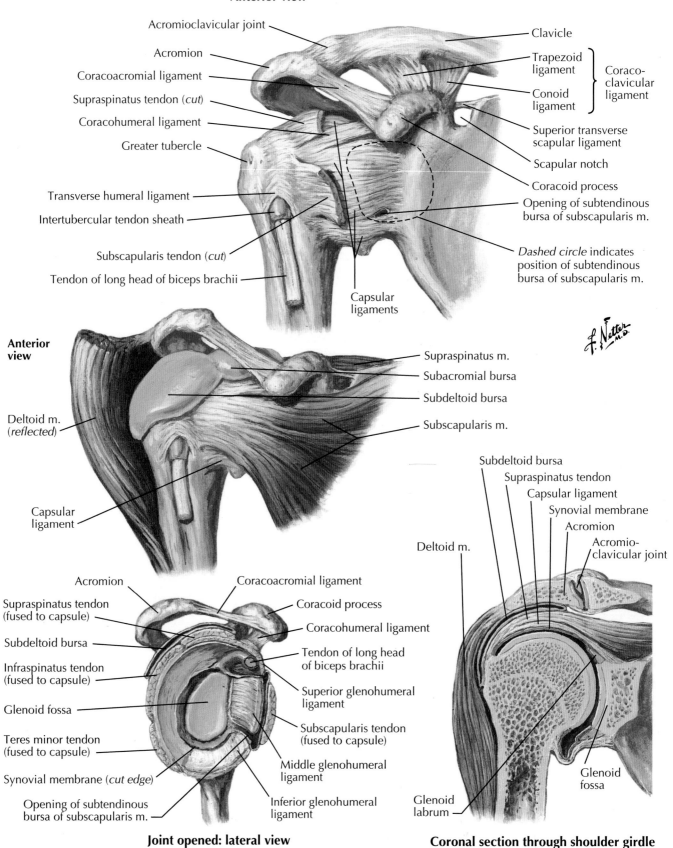

Anterior view

Acromioclavicular joint

Acromion

Coracoacromial ligament

Supraspinatus tendon (*cut*)

Coracohumeral ligament

Greater tubercle

Transverse humeral ligament

Intertubercular tendon sheath

Subscapularis tendon (*cut*)

Tendon of long head of biceps brachii

Clavicle

Trapezoid ligament

Conoid ligament

} Coraco-clavicular ligament

Superior transverse scapular ligament

Scapular notch

Coracoid process

Opening of subtendinous bursa of subscapularis m.

Dashed circle indicates position of subtendinous bursa of subscapularis m.

Capsular ligaments

Anterior view

Deltoid m. (*reflected*)

Capsular ligament

Supraspinatus m.

Subacromial bursa

Subdeltoid bursa

Subscapularis m.

Subdeltoid bursa

Supraspinatus tendon

Capsular ligament

Synovial membrane

Acromion

Acromio-clavicular joint

Deltoid m.

Glenoid fossa

Glenoid labrum

Acromion

Supraspinatus tendon (fused to capsule)

Subdeltoid bursa

Infraspinatus tendon (fused to capsule)

Glenoid fossa

Teres minor tendon (fused to capsule)

Synovial membrane (*cut edge*)

Opening of subtendinous bursa of subscapularis m.

Coracoacromial ligament

Coracoid process

Coracohumeral ligament

Tendon of long head of biceps brachii

Superior glenohumeral ligament

Subscapularis tendon (fused to capsule)

Middle glenohumeral ligament

Inferior glenohumeral ligament

Joint opened: lateral view

Coronal section through shoulder girdle

Plate 431

Shoulder and Axilla

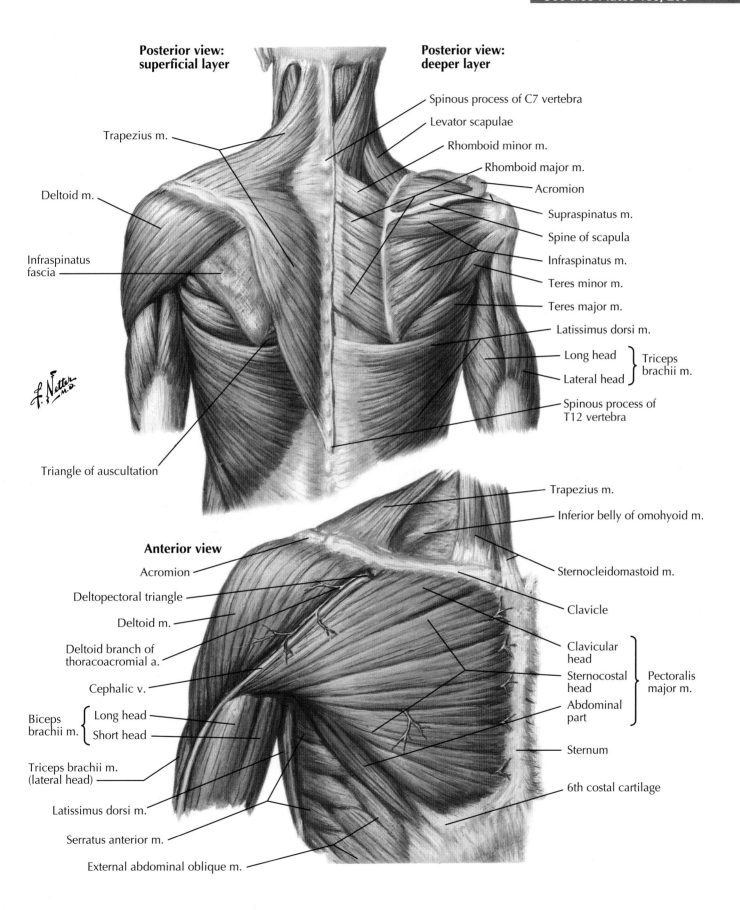

**Posterior view:
superficial layer**

**Posterior view:
deeper layer**

Trapezius m.

Deltoid m.

Infraspinatus
fascia

Triangle of auscultation

Spinous process of C7 vertebra

Levator scapulae

Rhomboid minor m.

Rhomboid major m.

Acromion

Supraspinatus m.

Spine of scapula

Infraspinatus m.

Teres minor m.

Teres major m.

Latissimus dorsi m.

Long head } Triceps
Lateral head } brachii m.

Spinous process of
T12 vertebra

Trapezius m.

Inferior belly of omohyoid m.

Sternocleidomastoid m.

Clavicle

Clavicular
head

Sternocostal
head   } Pectoralis
major m.
Abdominal
part

Sternum

6th costal cartilage

Anterior view

Acromion

Deltopectoral triangle

Deltoid m.

Deltoid branch of
thoracoacromial a.

Cephalic v.

Biceps { Long head
brachii m. { Short head

Triceps brachii m.
(lateral head)

Latissimus dorsi m.

Serratus anterior m.

External abdominal oblique m.

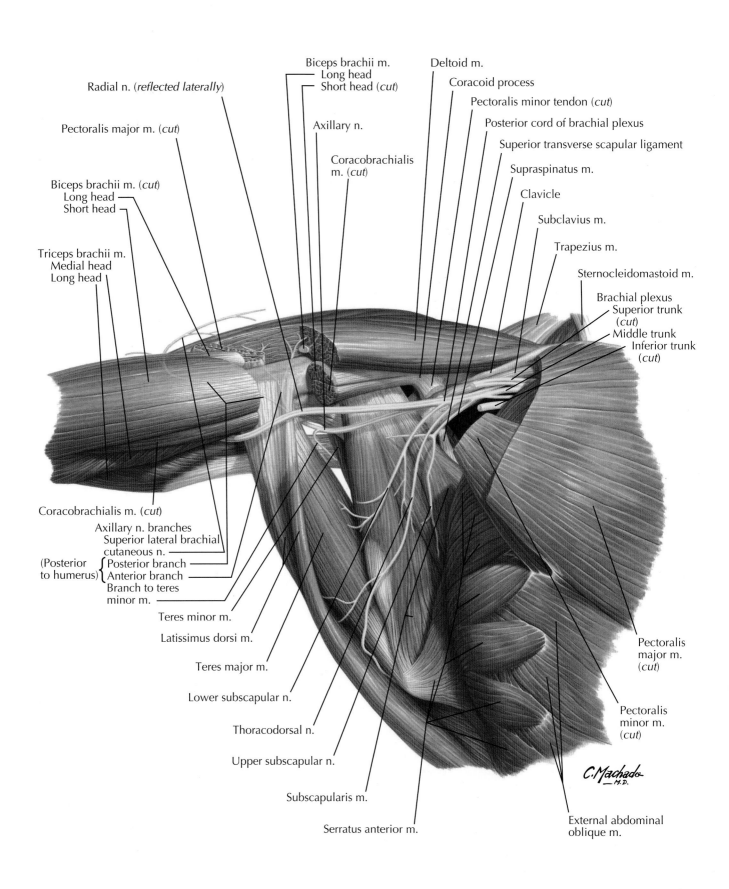

Biceps brachii m.
Long head
Short head (*cut*)

Radial n. (*reflected laterally*)

Axillary n.

Pectoralis major m. (*cut*)

Coracobrachialis
m. (*cut*)

Deltoid m.

Coracoid process

Pectoralis minor tendon (*cut*)

Posterior cord of brachial plexus

Superior transverse scapular ligament

Supraspinatus m.

Biceps brachii m. (*cut*)
Long head
Short head

Clavicle

Subclavius m.

Trapezius m.

Triceps brachii m.
Medial head
Long head

Sternocleidomastoid m.

Brachial plexus
Superior trunk
(*cut*)
Middle trunk
Inferior trunk
(*cut*)

Coracobrachialis m. (*cut*)

Axillary n. branches
Superior lateral brachial
cutaneous n.
(Posterior { Posterior branch
to humerus) { Anterior branch
Branch to teres
minor m.

Teres minor m.

Latissimus dorsi m.

Teres major m.

Lower subscapular n.

Thoracodorsal n.

Upper subscapular n.

Subscapularis m.

Serratus anterior m.

Pectoralis
major m.
(*cut*)

Pectoralis
minor m.
(*cut*)

C. Machado
M.D.

External abdominal
oblique m.

Plate 433

Shoulder and Axilla

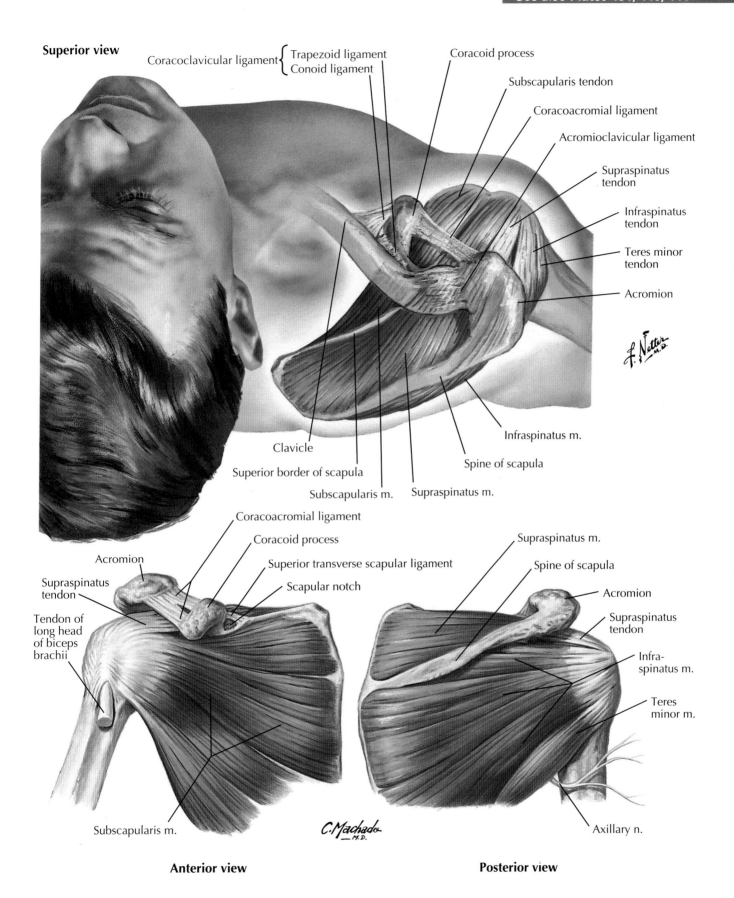

Superior view

Coracoclavicular ligament { Trapezoid ligament / Conoid ligament

Coracoid process

Subscapularis tendon

Coracoacromial ligament

Acromioclavicular ligament

Supraspinatus tendon

Infraspinatus tendon

Teres minor tendon

Acromion

Clavicle

Superior border of scapula

Subscapularis m.

Supraspinatus m.

Spine of scapula

Infraspinatus m.

Coracoacromial ligament

Coracoid process

Superior transverse scapular ligament

Scapular notch

Acromion

Supraspinatus tendon

Tendon of long head of biceps brachii

Subscapularis m.

Supraspinatus m.

Spine of scapula

Acromion

Supraspinatus tendon

Infra-spinatus m.

Teres minor m.

Axillary n.

Anterior view

Posterior view

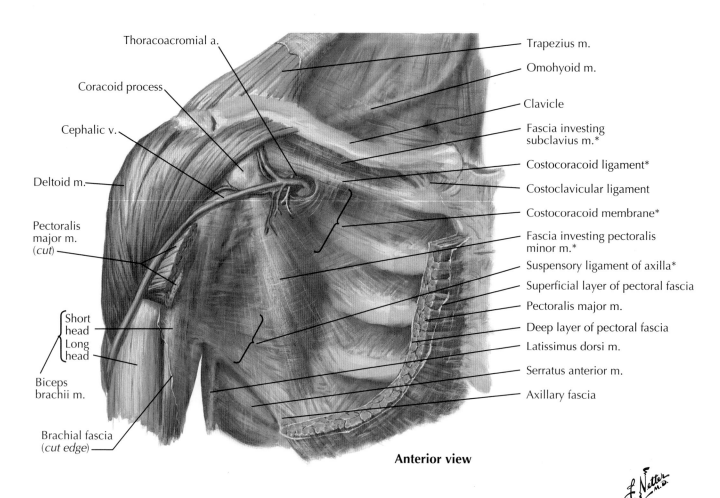

Thoracoacromial a.

Coracoid process

Cephalic v.

Deltoid m.

Pectoralis major m. (*cut*)

Short head
Long head

Biceps brachii m.

Brachial fascia (*cut edge*)

Trapezius m.

Omohyoid m.

Clavicle

Fascia investing subclavius m.*

Costocoracoid ligament*

Costoclavicular ligament

Costocoracoid membrane*

Fascia investing pectoralis minor m.*

Suspensory ligament of axilla*

Superficial layer of pectoral fascia

Pectoralis major m.

Deep layer of pectoral fascia

Latissimus dorsi m.

Serratus anterior m.

Axillary fascia

Anterior view

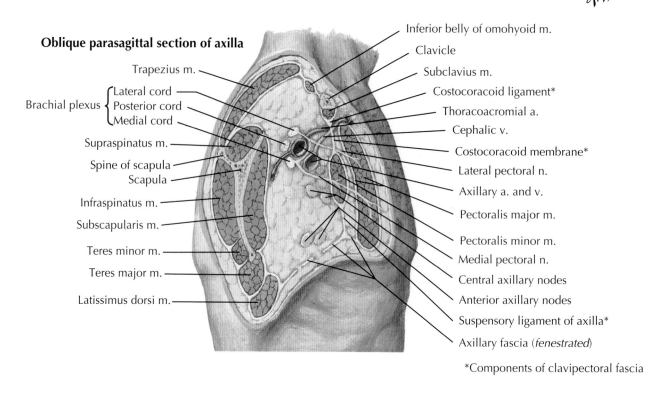

Oblique parasagittal section of axilla

Trapezius m.

Brachial plexus
Lateral cord
Posterior cord
Medial cord

Supraspinatus m.

Spine of scapula

Scapula

Infraspinatus m.

Subscapularis m.

Teres minor m.

Teres major m.

Latissimus dorsi m.

Inferior belly of omohyoid m.

Clavicle

Subclavius m.

Costocoracoid ligament*

Thoracoacromial a.

Cephalic v.

Costocoracoid membrane*

Lateral pectoral n.

Axillary a. and v.

Pectoralis major m.

Pectoralis minor m.

Medial pectoral n.

Central axillary nodes

Anterior axillary nodes

Suspensory ligament of axilla*

Axillary fascia (*fenestrated*)

*Components of clavipectoral fascia

Plate 435

Shoulder and Axilla

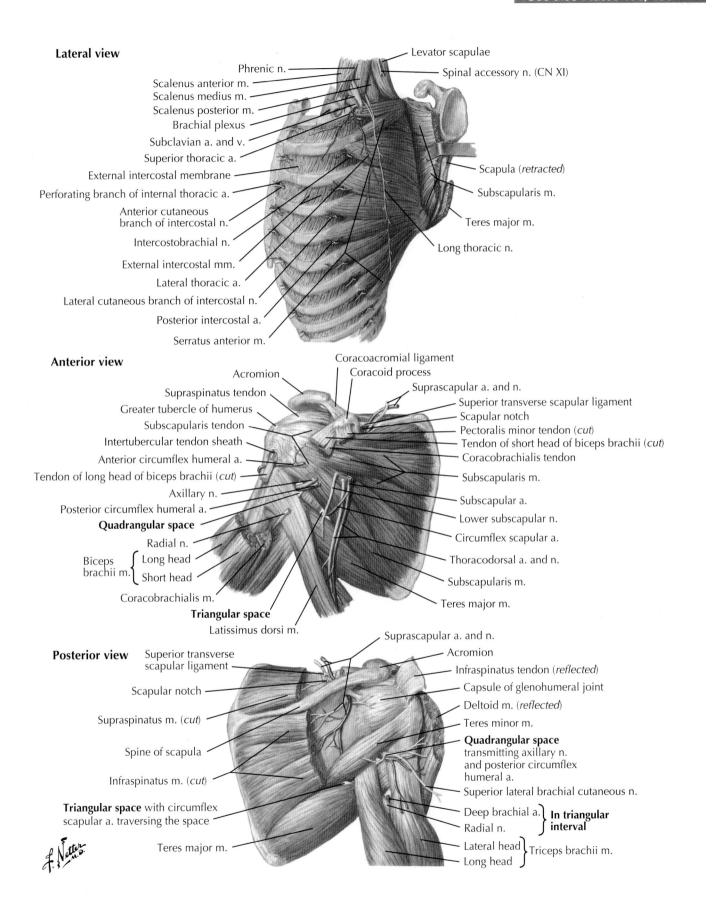

Lateral view

Levator scapulae
Phrenic n.
Spinal accessory n. (CN XI)
Scalenus anterior m.
Scalenus medius m.
Scalenus posterior m.
Brachial plexus
Subclavian a. and v.
Superior thoracic a.
Scapula (*retracted*)
External intercostal membrane
Perforating branch of internal thoracic a.
Subscapularis m.
Anterior cutaneous
branch of intercostal n.
Teres major m.
Intercostobrachial n.
Long thoracic n.
External intercostal mm.
Lateral thoracic a.
Lateral cutaneous branch of intercostal n.
Posterior intercostal a.
Serratus anterior m.

Anterior view

Coracoacromial ligament
Coracoid process
Acromion
Suprascapular a. and n.
Supraspinatus tendon
Superior transverse scapular ligament
Greater tubercle of humerus
Scapular notch
Subscapularis tendon
Pectoralis minor tendon (*cut*)
Intertubercular tendon sheath
Tendon of short head of biceps brachii (*cut*)
Anterior circumflex humeral a.
Coracobrachialis tendon
Tendon of long head of biceps brachii (*cut*)
Subscapularis m.
Axillary n.
Subscapular a.
Posterior circumflex humeral a.
Lower subscapular n.
Quadrangular space
Circumflex scapular a.
Radial n.
Thoracodorsal a. and n.
Biceps { Long head
brachii m. { Short head
Subscapularis m.
Coracobrachialis m.
Teres major m.
Triangular space
Latissimus dorsi m.

Posterior view
Suprascapular a. and n.
Superior transverse
scapular ligament
Acromion
Infraspinatus tendon (*reflected*)
Scapular notch
Capsule of glenohumeral joint
Supraspinatus m. (*cut*)
Deltoid m. (*reflected*)
Teres minor m.
Spine of scapula
Quadrangular space
transmitting axillary n.
and posterior circumflex
humeral a.
Infraspinatus m. (*cut*)
Superior lateral brachial cutaneous n.
Triangular space with circumflex
scapular a. traversing the space
Deep brachial a. } **In triangular**
Radial n. } **interval**
Lateral head } Triceps brachii m.
Teres major m.
Long head }

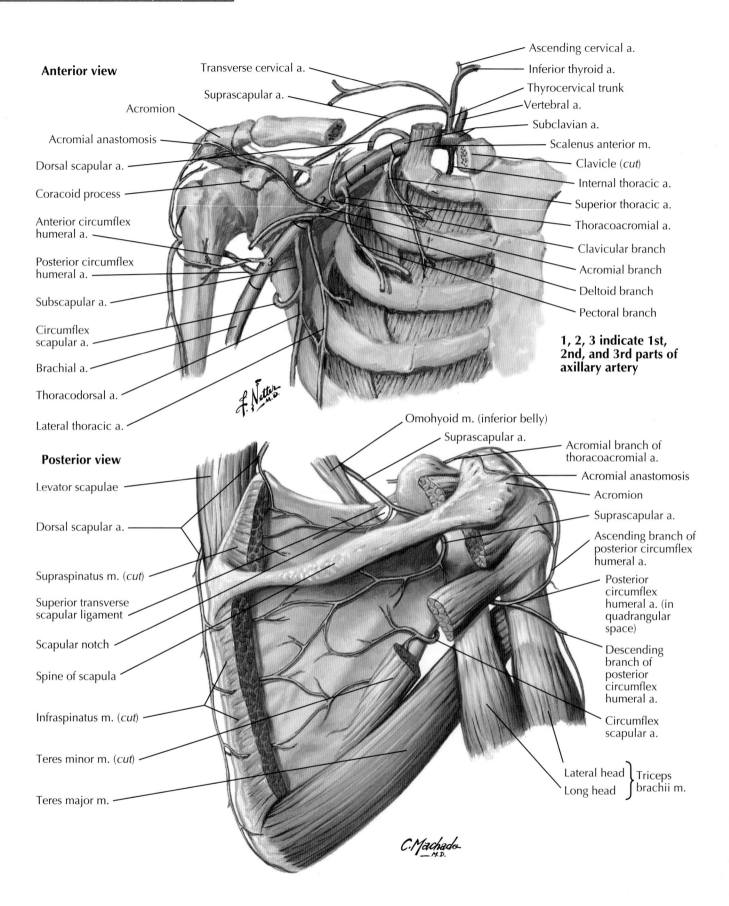

Anterior view

Transverse cervical a.

Suprascapular a.

Acromion

Acromial anastomosis

Dorsal scapular a.

Coracoid process

Anterior circumflex humeral a.

Posterior circumflex humeral a.

Subscapular a.

Circumflex scapular a.

Brachial a.

Thoracodorsal a.

Lateral thoracic a.

Ascending cervical a.

Inferior thyroid a.

Thyrocervical trunk

Vertebral a.

Subclavian a.

Scalenus anterior m.

Clavicle (*cut*)

Internal thoracic a.

Superior thoracic a.

Thoracoacromial a.

Clavicular branch

Acromial branch

Deltoid branch

Pectoral branch

1, 2, 3 indicate 1st, 2nd, and 3rd parts of axillary artery

Posterior view

Levator scapulae

Dorsal scapular a.

Supraspinatus m. (*cut*)

Superior transverse scapular ligament

Scapular notch

Spine of scapula

Infraspinatus m. (*cut*)

Teres minor m. (*cut*)

Teres major m.

Omohyoid m. (inferior belly)

Suprascapular a.

Acromial branch of thoracoacromial a.

Acromial anastomosis

Acromion

Suprascapular a.

Ascending branch of posterior circumflex humeral a.

Posterior circumflex humeral a. (in quadrangular space)

Descending branch of posterior circumflex humeral a.

Circumflex scapular a.

Lateral head } Triceps
Long head } brachii m.

Plate 437

Shoulder and Axilla

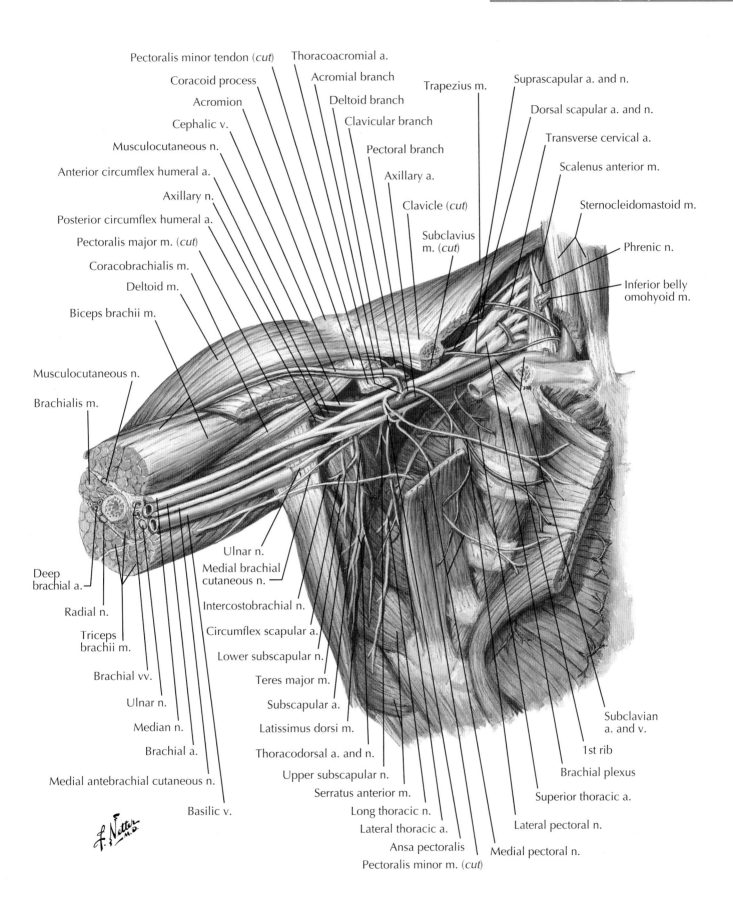

Pectoralis minor tendon (*cut*)
Coracoid process
Acromion
Cephalic v.
Musculocutaneous n.
Anterior circumflex humeral a.
Axillary n.
Posterior circumflex humeral a.
Pectoralis major m. (*cut*)
Coracobrachialis m.
Deltoid m.
Biceps brachii m.
Musculocutaneous n.
Brachialis m.
Deep brachial a.
Radial n.
Triceps brachii m.
Brachial vv.
Ulnar n.
Median n.
Brachial a.
Medial antebrachial cutaneous n.
Basilic v.

Thoracoacromial a.
Acromial branch
Deltoid branch
Clavicular branch
Pectoral branch
Axillary a.
Clavicle (*cut*)
Subclavius m. (*cut*)

Trapezius m.
Suprascapular a. and n.
Dorsal scapular a. and n.
Transverse cervical a.
Scalenus anterior m.
Sternocleidomastoid m.
Phrenic n.
Inferior belly omohyoid m.

Ulnar n.
Medial brachial cutaneous n.
Intercostobrachial n.
Circumflex scapular a.
Lower subscapular n.
Teres major m.
Subscapular a.
Latissimus dorsi m.
Thoracodorsal a. and n.
Upper subscapular n.
Serratus anterior m.
Long thoracic n.
Lateral thoracic a.
Ansa pectoralis
Pectoralis minor m. (*cut*)

Lateral pectoral n.
Medial pectoral n.
Superior thoracic a.
Brachial plexus
1st rib
Subclavian a. and v.

Shoulder and Axilla

Plate 438

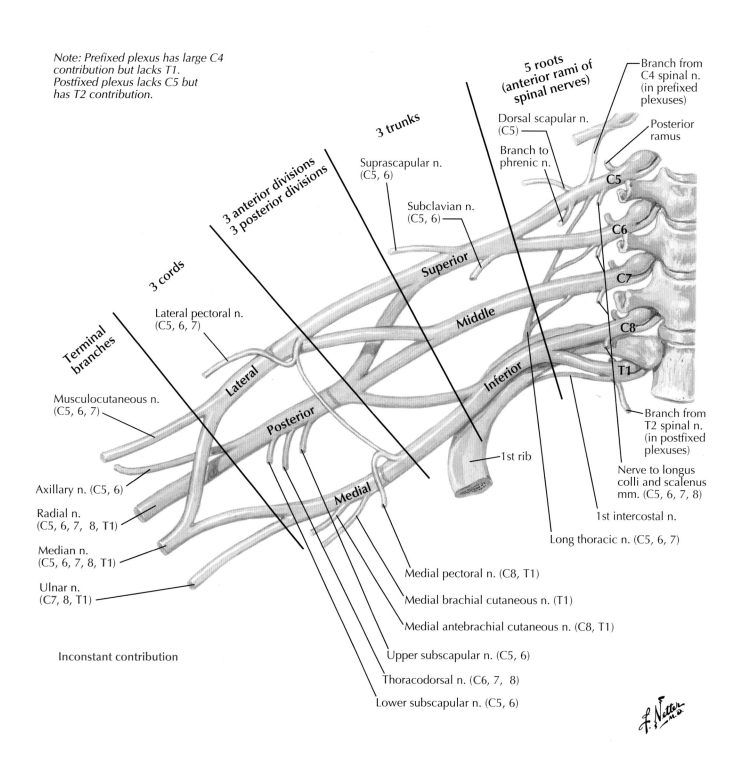

Note: Prefixed plexus has large C4 contribution but lacks T1. Postfixed plexus lacks C5 but has T2 contribution.

5 roots (anterior rami of spinal nerves)

Branch from C4 spinal n. (in prefixed plexuses)

Posterior ramus

Dorsal scapular n. (C5)

Branch to phrenic n.

3 trunks

Suprascapular n. (C5, 6)

C5

Subclavian n. (C5, 6)

C6

3 anterior divisions
3 posterior divisions

Superior

C7

Middle

Lateral pectoral n. (C5, 6, 7)

C8

3 cords

Lateral

Inferior

T1

Terminal branches

Musculocutaneous n. (C5, 6, 7)

Posterior

Branch from T2 spinal n. (in postfixed plexuses)

Axillary n. (C5, 6)

Medial

1st rib

Nerve to longus colli and scalenus mm. (C5, 6, 7, 8)

Radial n. (C5, 6, 7, 8, T1)

1st intercostal n.

Median n. (C5, 6, 7, 8, T1)

Long thoracic n. (C5, 6, 7)

Ulnar n. (C7, 8, T1)

Medial pectoral n. (C8, T1)

Medial brachial cutaneous n. (T1)

Medial antebrachial cutaneous n. (C8, T1)

Inconstant contribution

Upper subscapular n. (C5, 6)

Thoracodorsal n. (C6, 7, 8)

Lower subscapular n. (C5, 6)

Plate 439

Shoulder and Axilla

See also **Plates 434, 485**

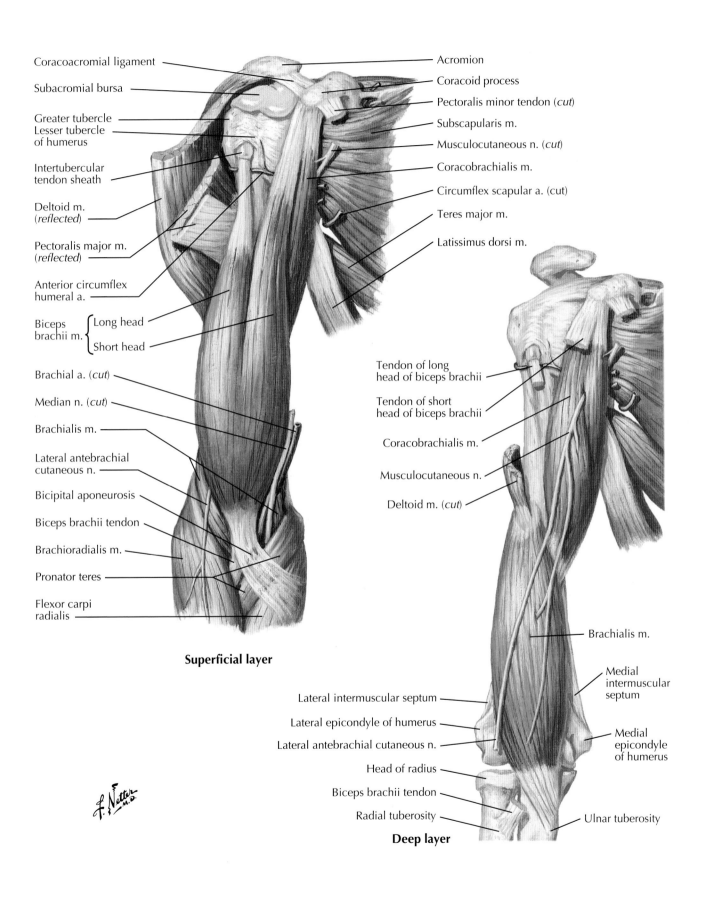

Coracoacromial ligament

Subacromial bursa

Greater tubercle
Lesser tubercle
of humerus

Intertubercular
tendon sheath

Deltoid m.
(*reflected*)

Pectoralis major m.
(*reflected*)

Anterior circumflex
humeral a.

Biceps
brachii m. { Long head

Short head

Brachial a. (*cut*)

Median n. (*cut*)

Brachialis m.

Lateral antebrachial
cutaneous n.

Bicipital aponeurosis

Biceps brachii tendon

Brachioradialis m.

Pronator teres

Flexor carpi
radialis

Acromion

Coracoid process

Pectoralis minor tendon (*cut*)

Subscapularis m.

Musculocutaneous n. (*cut*)

Coracobrachialis m.

Circumflex scapular a. (*cut*)

Teres major m.

Latissimus dorsi m.

Tendon of long
head of biceps brachii

Tendon of short
head of biceps brachii

Coracobrachialis m.

Musculocutaneous n.

Deltoid m. (*cut*)

Superficial layer

Brachialis m.

Medial
intermuscular
septum

Lateral intermuscular septum

Lateral epicondyle of humerus

Lateral antebrachial cutaneous n.

Head of radius

Biceps brachii tendon

Radial tuberosity

Medial
epicondyle
of humerus

Ulnar tuberosity

Deep layer

Arm

Plate 440

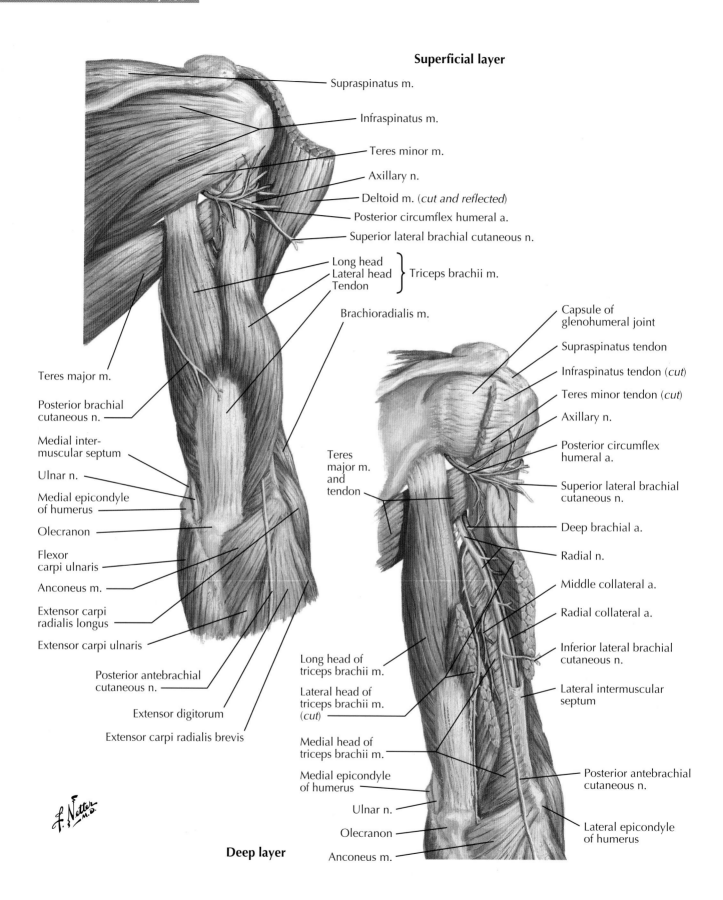

Superficial layer

Supraspinatus m.

Infraspinatus m.

Teres minor m.

Axillary n.

Deltoid m. (*cut and reflected*)

Posterior circumflex humeral a.

Superior lateral brachial cutaneous n.

Long head
Lateral head } Triceps brachii m.
Tendon

Brachioradialis m.

Capsule of glenohumeral joint

Supraspinatus tendon

Infraspinatus tendon (*cut*)

Teres minor tendon (*cut*)

Axillary n.

Posterior circumflex humeral a.

Superior lateral brachial cutaneous n.

Deep brachial a.

Radial n.

Middle collateral a.

Radial collateral a.

Inferior lateral brachial cutaneous n.

Lateral intermuscular septum

Teres major m.

Posterior brachial cutaneous n.

Medial intermuscular septum

Ulnar n.

Medial epicondyle of humerus

Olecranon

Flexor carpi ulnaris

Anconeus m.

Extensor carpi radialis longus

Extensor carpi ulnaris

Posterior antebrachial cutaneous n.

Extensor digitorum

Extensor carpi radialis brevis

Teres major m. and tendon

Long head of triceps brachii m.

Lateral head of triceps brachii m. (*cut*)

Medial head of triceps brachii m.

Medial epicondyle of humerus

Ulnar n.

Olecranon

Anconeus m.

Posterior antebrachial cutaneous n.

Lateral epicondyle of humerus

Deep layer

Plate 441

Arm

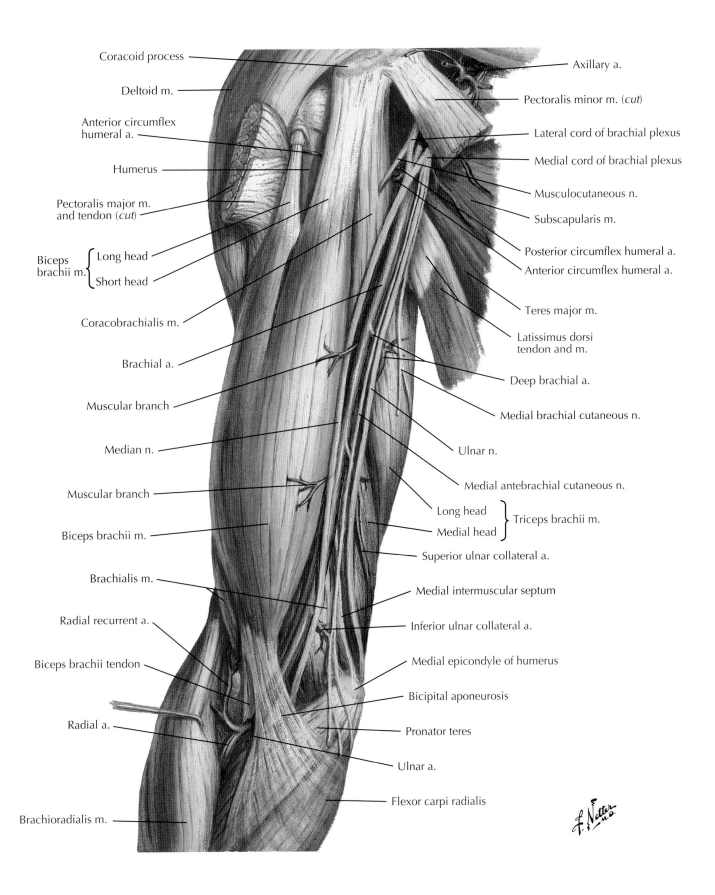

Coracoid process

Deltoid m.

Anterior circumflex
humeral a.

Humerus

Pectoralis major m.
and tendon (*cut*)

Biceps
brachii m. {
Long head

Short head

Coracobrachialis m.

Brachial a.

Muscular branch

Median n.

Muscular branch

Biceps brachii m.

Brachialis m.

Radial recurrent a.

Biceps brachii tendon

Radial a.

Brachioradialis m.

Axillary a.

Pectoralis minor m. (*cut*)

Lateral cord of brachial plexus

Medial cord of brachial plexus

Musculocutaneous n.

Subscapularis m.

Posterior circumflex humeral a.

Anterior circumflex humeral a.

Teres major m.

Latissimus dorsi
tendon and m.

Deep brachial a.

Medial brachial cutaneous n.

Ulnar n.

Medial antebrachial cutaneous n.

Long head
Medial head
} Triceps brachii m.

Superior ulnar collateral a.

Medial intermuscular septum

Inferior ulnar collateral a.

Medial epicondyle of humerus

Bicipital aponeurosis

Pronator teres

Ulnar a.

Flexor carpi radialis

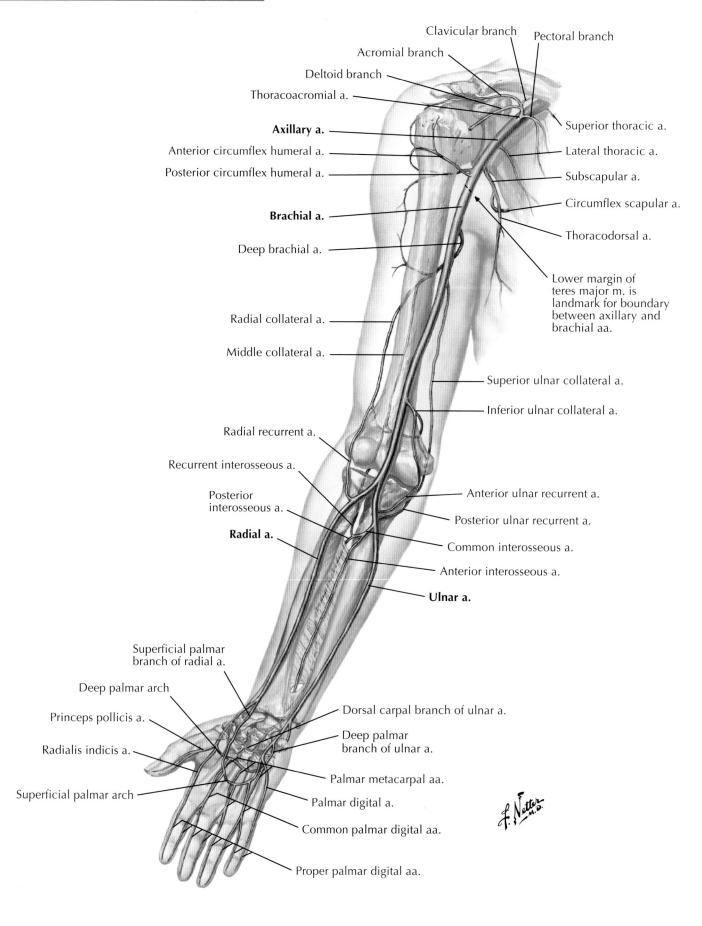

Clavicular branch

Pectoral branch

Acromial branch

Deltoid branch

Thoracoacromial a.

Axillary a.

Anterior circumflex humeral a.

Posterior circumflex humeral a.

Brachial a.

Deep brachial a.

Superior thoracic a.

Lateral thoracic a.

Subscapular a.

Circumflex scapular a.

Thoracodorsal a.

Lower margin of teres major m. is landmark for boundary between axillary and brachial aa.

Radial collateral a.

Middle collateral a.

Superior ulnar collateral a.

Inferior ulnar collateral a.

Radial recurrent a.

Recurrent interosseous a.

Posterior interosseous a.

Radial a.

Anterior ulnar recurrent a.

Posterior ulnar recurrent a.

Common interosseous a.

Anterior interosseous a.

Ulnar a.

Superficial palmar branch of radial a.

Deep palmar arch

Princeps pollicis a.

Radialis indicis a.

Superficial palmar arch

Dorsal carpal branch of ulnar a.

Deep palmar branch of ulnar a.

Palmar metacarpal aa.

Palmar digital a.

Common palmar digital aa.

Proper palmar digital aa.

Plate 443

Arm

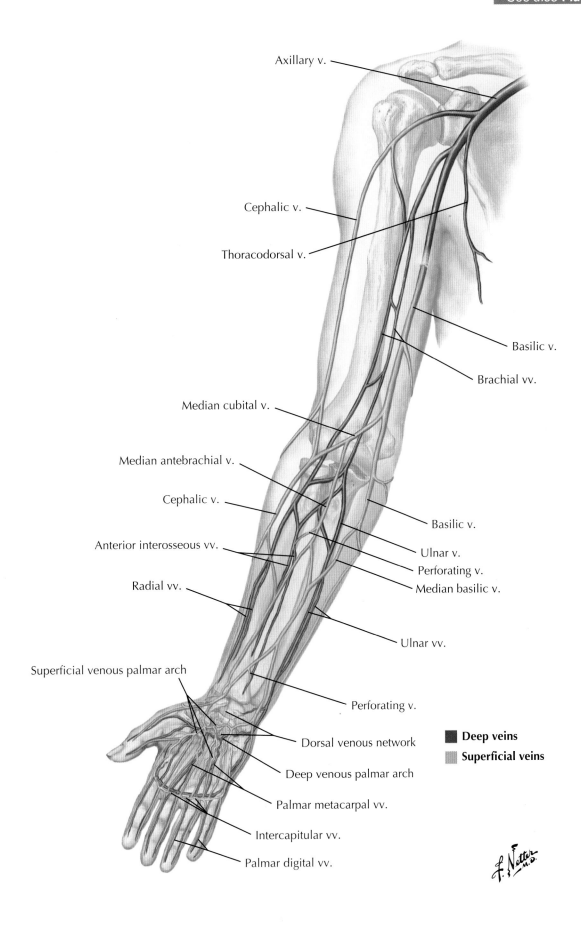

Axillary v.

Cephalic v.

Thoracodorsal v.

Basilic v.

Brachial vv.

Median cubital v.

Median antebrachial v.

Cephalic v.

Basilic v.

Anterior interosseous vv.

Ulnar v.

Perforating v.

Median basilic v.

Radial vv.

Ulnar vv.

Superficial venous palmar arch

Perforating v.

Dorsal venous network

Deep venous palmar arch

Palmar metacarpal vv.

Intercapitular vv.

Palmar digital vv.

Deep veins
Superficial veins

F. Netter, M.D.

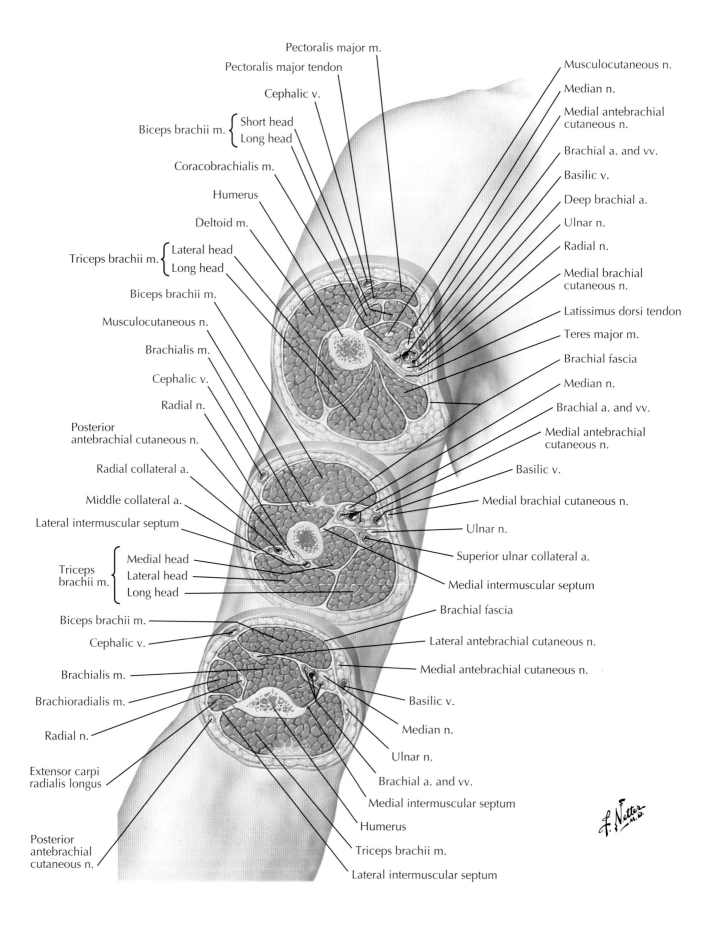

Pectoralis major m.

Pectoralis major tendon

Cephalic v.

Biceps brachii m. { Short head / Long head

Coracobrachialis m.

Humerus

Deltoid m.

Triceps brachii m. { Lateral head / Long head

Biceps brachii m.

Musculocutaneous n.

Brachialis m.

Cephalic v.

Radial n.

Posterior antebrachial cutaneous n.

Radial collateral a.

Middle collateral a.

Lateral intermuscular septum

Triceps brachii m. { Medial head / Lateral head / Long head

Biceps brachii m.

Cephalic v.

Brachialis m.

Brachioradialis m.

Radial n.

Extensor carpi radialis longus

Posterior antebrachial cutaneous n.

Musculocutaneous n.

Median n.

Medial antebrachial cutaneous n.

Brachial a. and vv.

Basilic v.

Deep brachial a.

Ulnar n.

Radial n.

Medial brachial cutaneous n.

Latissimus dorsi tendon

Teres major m.

Brachial fascia

Median n.

Brachial a. and vv.

Medial antebrachial cutaneous n.

Basilic v.

Medial brachial cutaneous n.

Ulnar n.

Superior ulnar collateral a.

Medial intermuscular septum

Brachial fascia

Lateral antebrachial cutaneous n.

Medial antebrachial cutaneous n.

Basilic v.

Median n.

Ulnar n.

Brachial a. and vv.

Medial intermuscular septum

Humerus

Triceps brachii m.

Lateral intermuscular septum

Plate 445

Arm

Right elbow

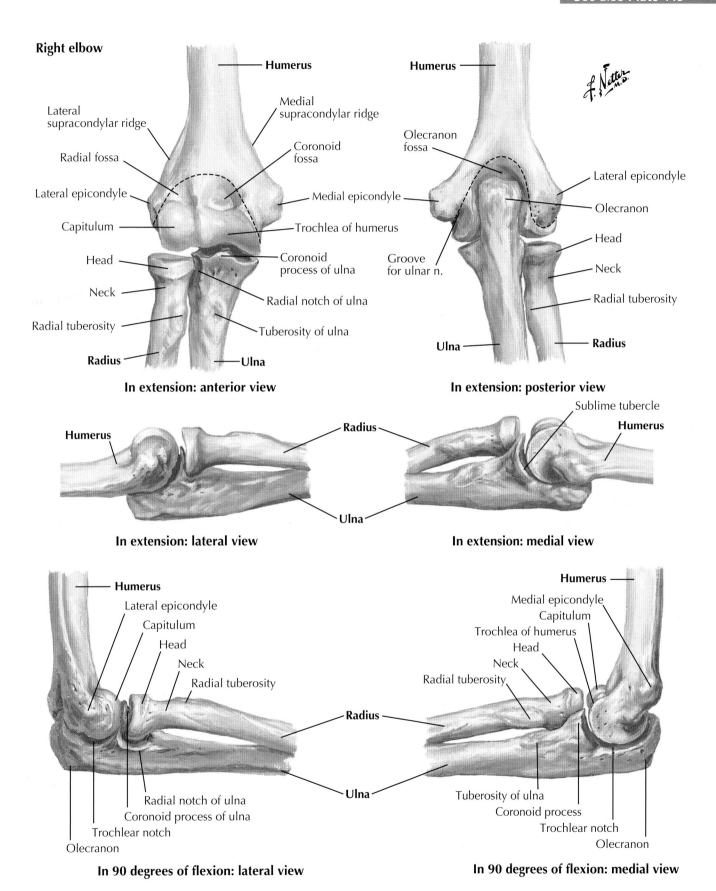

Humerus

Medial
supracondylar ridge

Lateral
supracondylar ridge

Coronoid
fossa

Radial fossa

Lateral epicondyle

Medial epicondyle

Capitulum

Trochlea of humerus

Head

Coronoid
process of ulna

Neck

Radial notch of ulna

Radial tuberosity

Tuberosity of ulna

Radius

Ulna

In extension: anterior view

Humerus

Olecranon
fossa

Lateral epicondyle

Olecranon

Head

Groove
for ulnar n.

Neck

Radial tuberosity

Ulna

Radius

In extension: posterior view

Humerus

Radius

Ulna

In extension: lateral view

Sublime tubercle

Humerus

Radius

Ulna

In extension: medial view

Humerus

Lateral epicondyle

Capitulum

Head

Neck

Radial tuberosity

Radius

Ulna

Radial notch of ulna

Coronoid process of ulna

Trochlear notch

Olecranon

In 90 degrees of flexion: lateral view

Humerus

Medial epicondyle

Capitulum

Trochlea of humerus

Head

Neck

Radial tuberosity

Radius

Ulna

Tuberosity of ulna

Coronoid process

Trochlear notch

Olecranon

In 90 degrees of flexion: medial view

Elbow and Forearm

Plate 446

Anteroposterior view

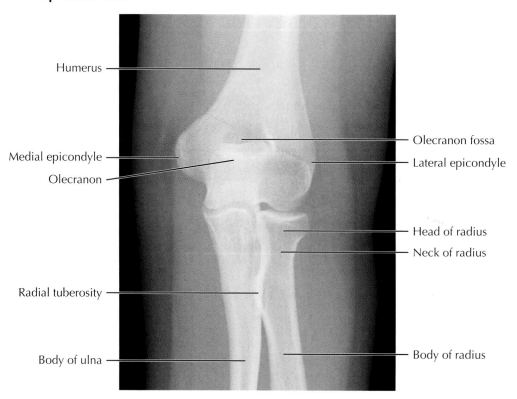

Humerus

Olecranon fossa

Medial epicondyle

Lateral epicondyle

Olecranon

Head of radius

Neck of radius

Radial tuberosity

Body of ulna

Body of radius

Lateral view

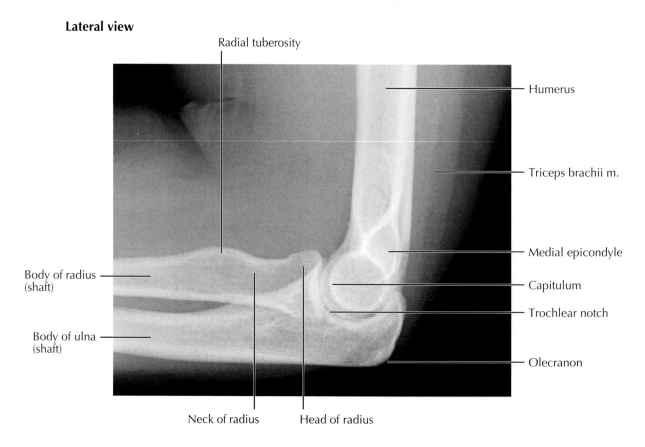

Radial tuberosity

Humerus

Triceps brachii m.

Body of radius (shaft)

Medial epicondyle

Capitulum

Trochlear notch

Body of ulna (shaft)

Olecranon

Neck of radius

Head of radius

Plate 447

Elbow and Forearm

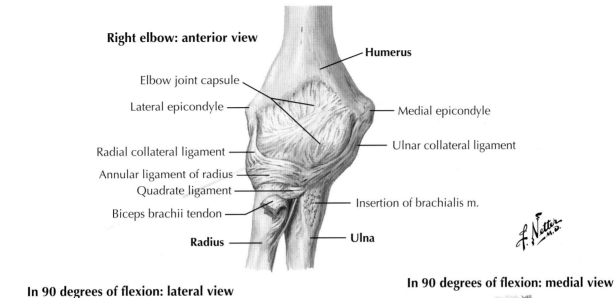

Right elbow: anterior view

Humerus

Elbow joint capsule

Lateral epicondyle

Medial epicondyle

Radial collateral ligament

Ulnar collateral ligament

Annular ligament of radius

Quadrate ligament

Insertion of brachialis m.

Biceps brachii tendon

Radius

Ulna

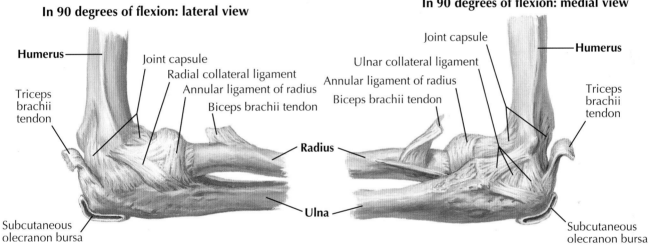

In 90 degrees of flexion: lateral view

Humerus

Joint capsule

Radial collateral ligament

Annular ligament of radius

Biceps brachii tendon

Triceps brachii tendon

Radius

Ulna

Subcutaneous olecranon bursa

In 90 degrees of flexion: medial view

Joint capsule

Humerus

Ulnar collateral ligament

Annular ligament of radius

Biceps brachii tendon

Triceps brachii tendon

Radius

Subcutaneous olecranon bursa

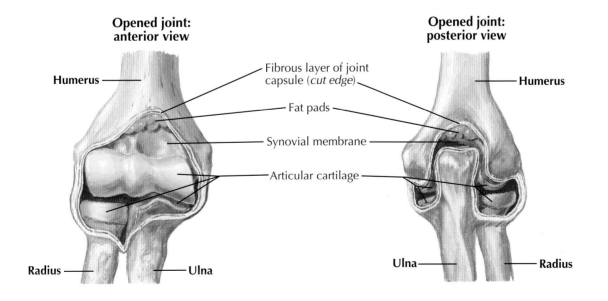

Opened joint: anterior view

Humerus

Fibrous layer of joint capsule (*cut edge*)

Fat pads

Synovial membrane

Articular cartilage

Radius

Ulna

Opened joint: posterior view

Humerus

Ulna

Radius

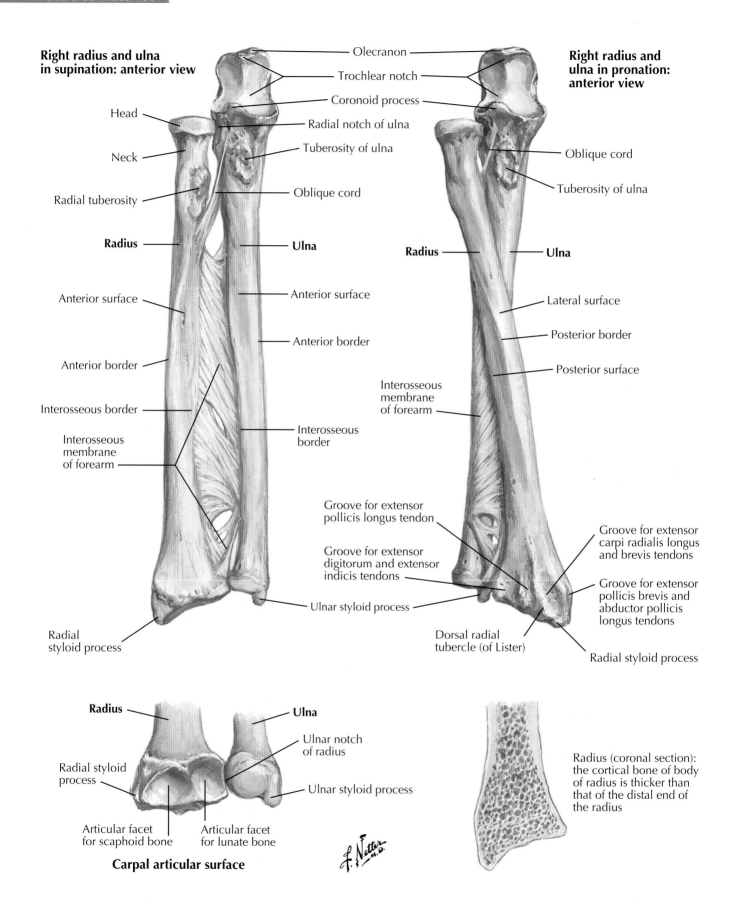

Right radius and ulna in supination: anterior view

Olecranon

Trochlear notch

Coronoid process

Head

Neck

Radial notch of ulna

Tuberosity of ulna

Radial tuberosity

Oblique cord

Radius

Ulna

Anterior surface

Anterior surface

Anterior border

Anterior border

Interosseous border

Interosseous membrane of forearm

Interosseous border

Groove for extensor pollicis longus tendon

Groove for extensor digitorum and extensor indicis tendons

Ulnar styloid process

Radial styloid process

Right radius and ulna in pronation: anterior view

Oblique cord

Tuberosity of ulna

Radius

Ulna

Lateral surface

Posterior border

Posterior surface

Interosseous membrane of forearm

Groove for extensor carpi radialis longus and brevis tendons

Groove for extensor pollicis brevis and abductor pollicis longus tendons

Dorsal radial tubercle (of Lister)

Radial styloid process

Radius

Ulna

Ulnar notch of radius

Radial styloid process

Ulnar styloid process

Articular facet for scaphoid bone

Articular facet for lunate bone

Carpal articular surface

Radius (coronal section): the cortical bone of body of radius is thicker than that of the distal end of the radius

Plate 449

Elbow and Forearm

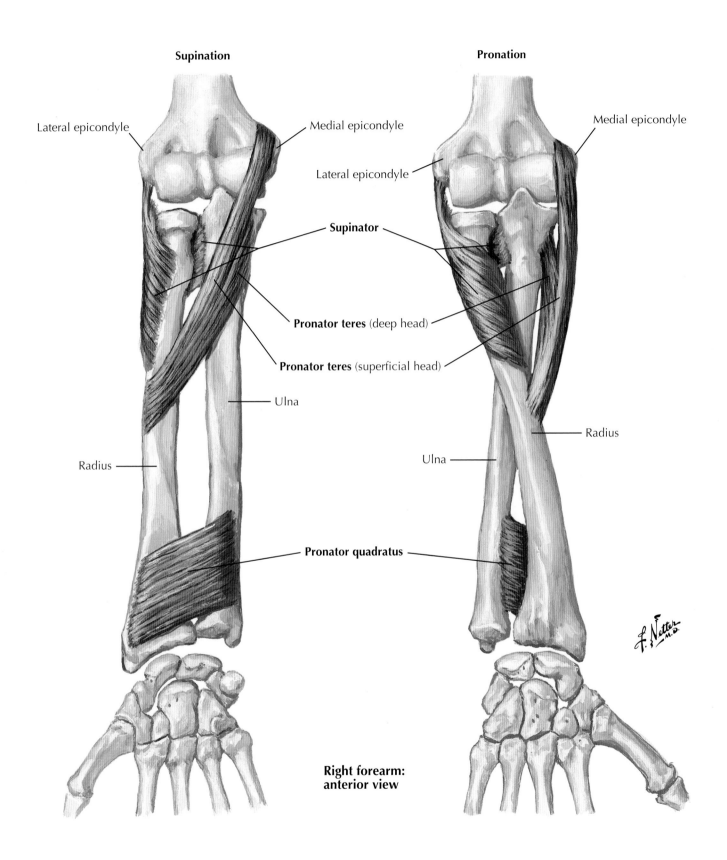

Supination

Pronation

Lateral epicondyle

Medial epicondyle

Medial epicondyle

Lateral epicondyle

Supinator

Pronator teres (deep head)

Pronator teres (superficial head)

Ulna

Radius

Radius

Ulna

Pronator quadratus

**Right forearm:
anterior view**

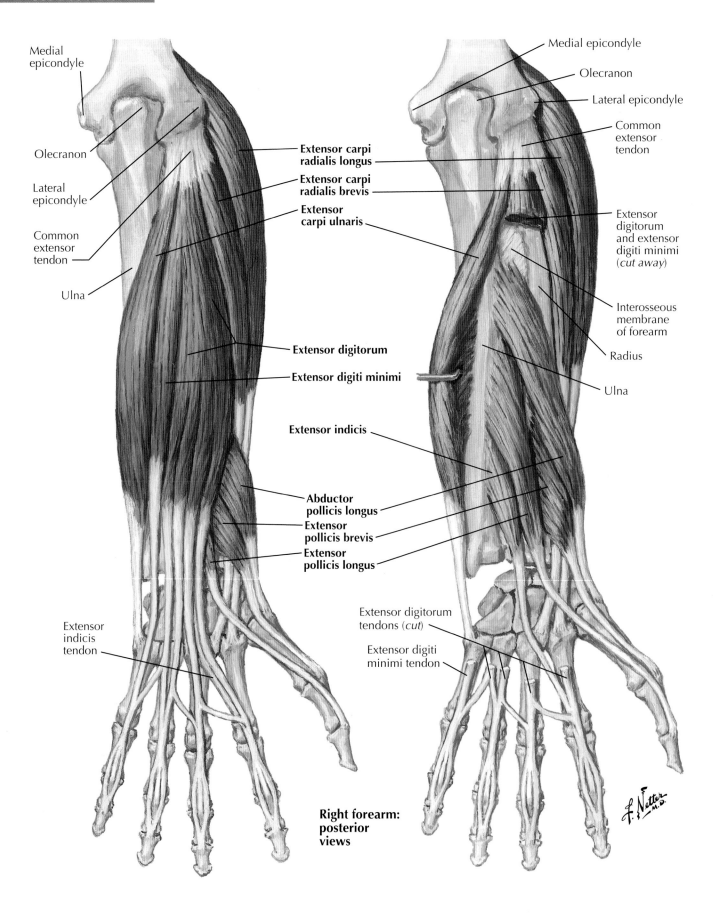

Medial epicondyle

Olecranon

Lateral epicondyle

Common extensor tendon

Ulna

Extensor carpi radialis longus

Extensor carpi radialis brevis

Extensor carpi ulnaris

Extensor digitorum

Extensor digiti minimi

Extensor indicis

Abductor pollicis longus

Extensor pollicis brevis

Extensor pollicis longus

Extensor indicis tendon

Medial epicondyle

Olecranon

Lateral epicondyle

Common extensor tendon

Extensor digitorum and extensor digiti minimi (*cut away*)

Interosseous membrane of forearm

Radius

Ulna

Extensor digitorum tendons (*cut*)

Extensor digiti minimi tendon

Right forearm: posterior views

Plate 451

Elbow and Forearm

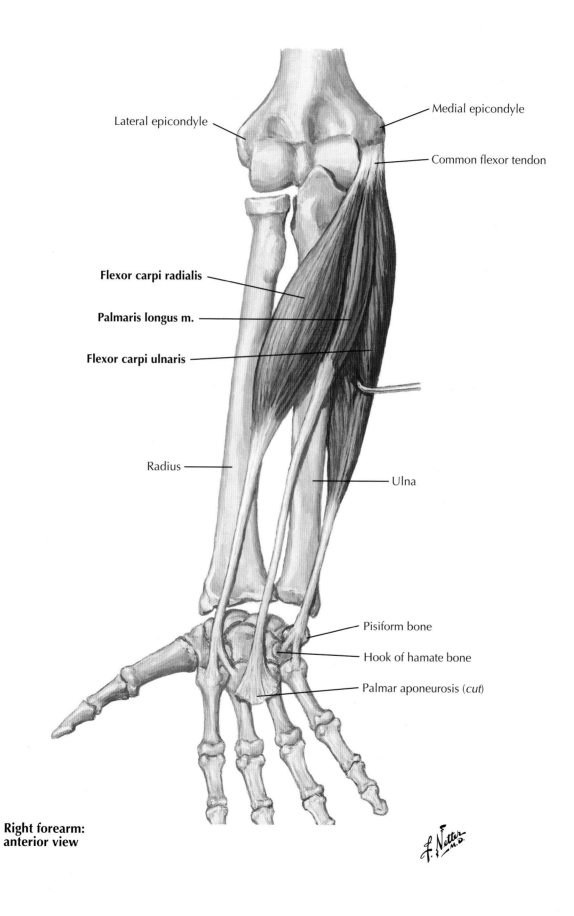

Lateral epicondyle

Medial epicondyle

Common flexor tendon

Flexor carpi radialis

Palmaris longus m.

Flexor carpi ulnaris

Radius

Ulna

Pisiform bone

Hook of hamate bone

Palmar aponeurosis (*cut*)

**Right forearm:
anterior view**

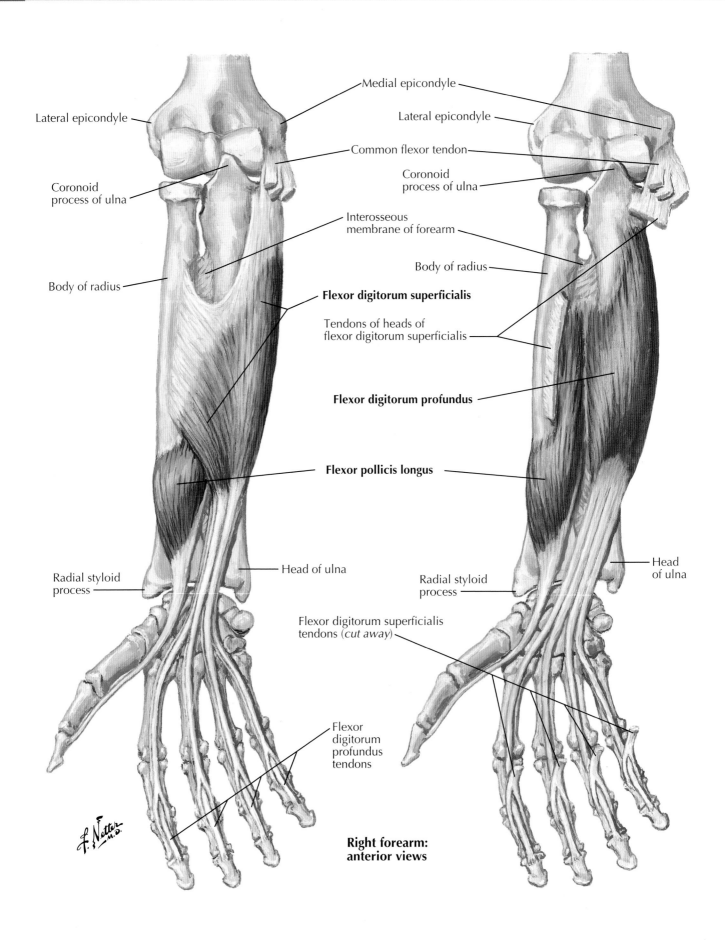

Lateral epicondyle

Medial epicondyle

Lateral epicondyle

Common flexor tendon

Coronoid process of ulna

Coronoid process of ulna

Interosseous membrane of forearm

Body of radius

Body of radius

Flexor digitorum superficialis

Tendons of heads of flexor digitorum superficialis

Flexor digitorum profundus

Flexor pollicis longus

Radial styloid process

Head of ulna

Radial styloid process

Head of ulna

Flexor digitorum superficialis tendons (*cut away*)

Flexor digitorum profundus tendons

Right forearm: anterior views

Plate 453

Elbow and Forearm

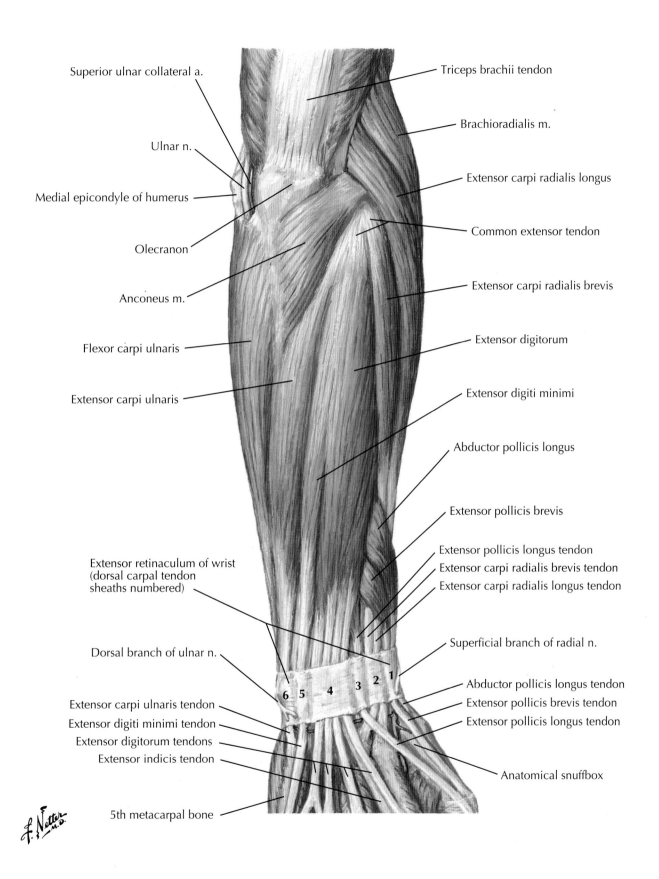

Superior ulnar collateral a.

Ulnar n.

Medial epicondyle of humerus

Olecranon

Anconeus m.

Flexor carpi ulnaris

Extensor carpi ulnaris

Extensor retinaculum of wrist
(dorsal carpal tendon
sheaths numbered)

Dorsal branch of ulnar n.

Extensor carpi ulnaris tendon

Extensor digiti minimi tendon

Extensor digitorum tendons

Extensor indicis tendon

5th metacarpal bone

Triceps brachii tendon

Brachioradialis m.

Extensor carpi radialis longus

Common extensor tendon

Extensor carpi radialis brevis

Extensor digitorum

Extensor digiti minimi

Abductor pollicis longus

Extensor pollicis brevis

Extensor pollicis longus tendon
Extensor carpi radialis brevis tendon
Extensor carpi radialis longus tendon

Superficial branch of radial n.

Abductor pollicis longus tendon
Extensor pollicis brevis tendon
Extensor pollicis longus tendon

Anatomical snuffbox

6 5 4 3 2 1

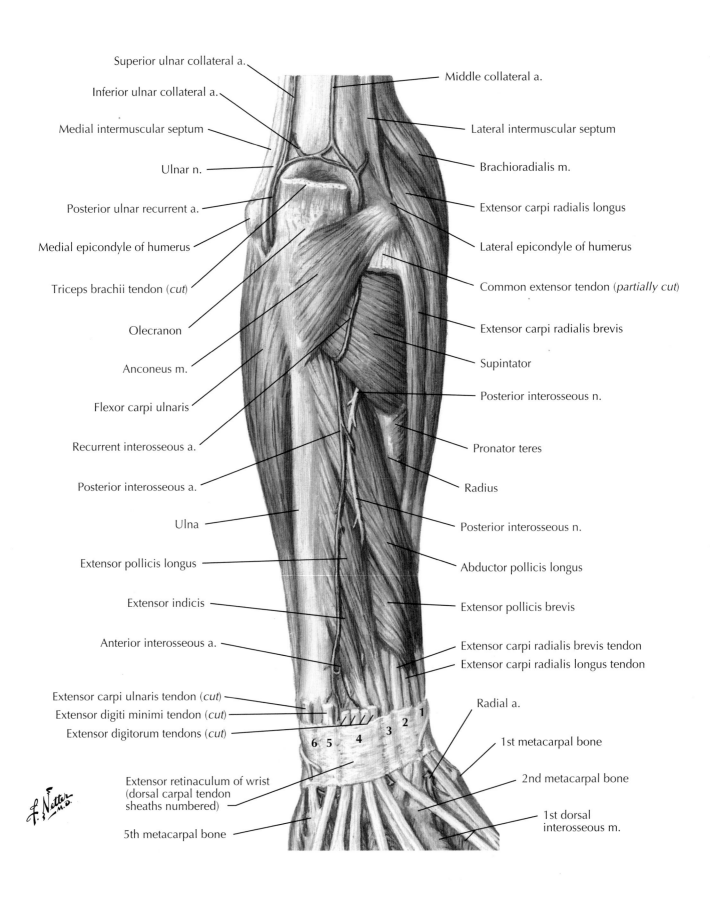

Superior ulnar collateral a.

Inferior ulnar collateral a.

Medial intermuscular septum

Ulnar n.

Posterior ulnar recurrent a.

Medial epicondyle of humerus

Triceps brachii tendon (*cut*)

Olecranon

Anconeus m.

Flexor carpi ulnaris

Recurrent interosseous a.

Posterior interosseous a.

Ulna

Extensor pollicis longus

Extensor indicis

Anterior interosseous a.

Extensor carpi ulnaris tendon (*cut*)
Extensor digiti minimi tendon (*cut*)
Extensor digitorum tendons (*cut*)

Extensor retinaculum of wrist
(dorsal carpal tendon
sheaths numbered)

5th metacarpal bone

Middle collateral a.

Lateral intermuscular septum

Brachioradialis m.

Extensor carpi radialis longus

Lateral epicondyle of humerus

Common extensor tendon (*partially cut*)

Extensor carpi radialis brevis

Supinator

Posterior interosseous n.

Pronator teres

Radius

Posterior interosseous n.

Abductor pollicis longus

Extensor pollicis brevis

Extensor carpi radialis brevis tendon
Extensor carpi radialis longus tendon

Radial a.

1st metacarpal bone

2nd metacarpal bone

1st dorsal
interosseous m.

1 2 3 4 5 6

Plate 455

Elbow and Forearm

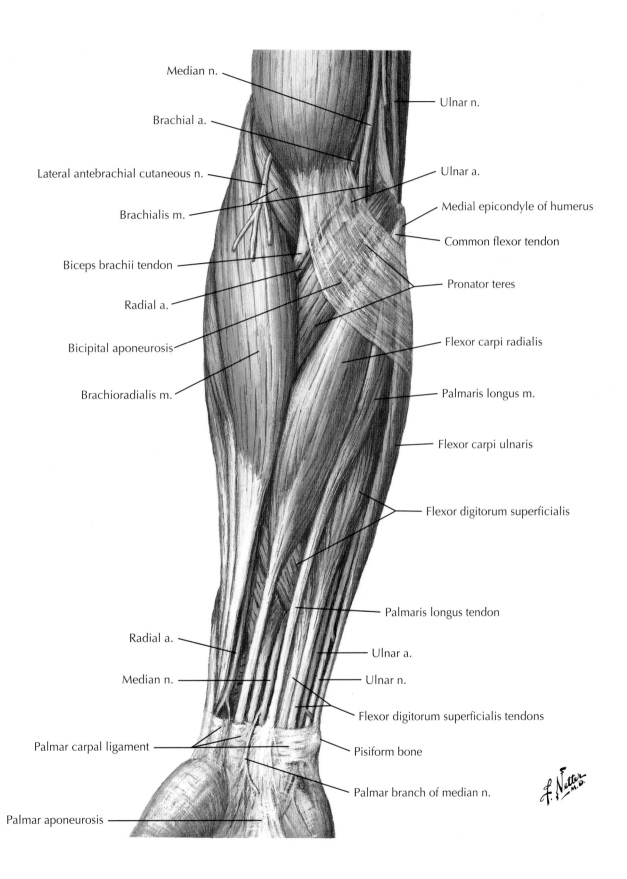

Median n.

Brachial a.

Lateral antebrachial cutaneous n.

Brachialis m.

Biceps brachii tendon

Radial a.

Bicipital aponeurosis

Brachioradialis m.

Radial a.

Median n.

Palmar carpal ligament

Palmar aponeurosis

Ulnar n.

Ulnar a.

Medial epicondyle of humerus

Common flexor tendon

Pronator teres

Flexor carpi radialis

Palmaris longus m.

Flexor carpi ulnaris

Flexor digitorum superficialis

Palmaris longus tendon

Ulnar a.

Ulnar n.

Flexor digitorum superficialis tendons

Pisiform bone

Palmar branch of median n.

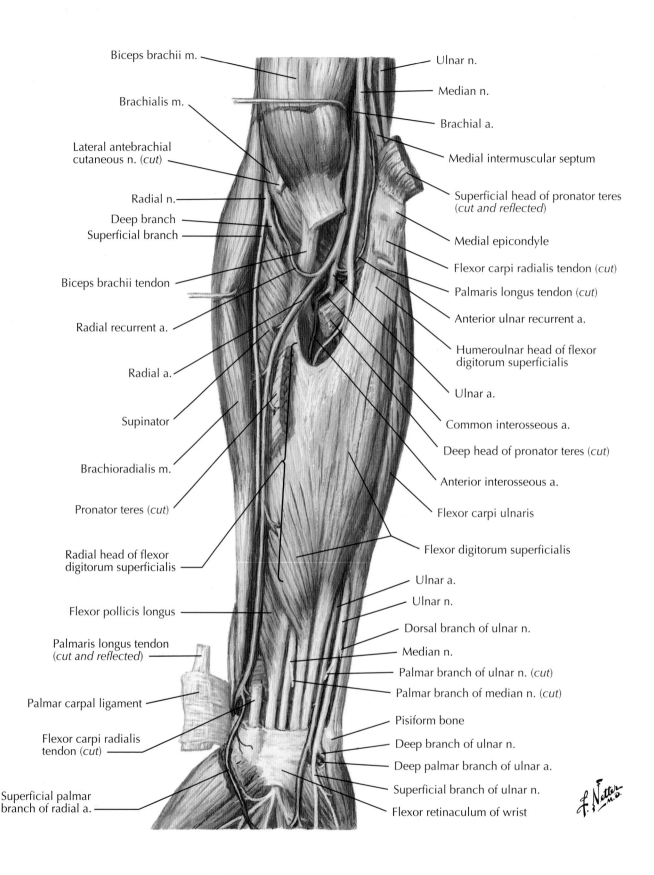

Biceps brachii m.

Brachialis m.

Lateral antebrachial
cutaneous n. (*cut*)

Radial n.

Deep branch

Superficial branch

Biceps brachii tendon

Radial recurrent a.

Radial a.

Supinator

Brachioradialis m.

Pronator teres (*cut*)

Radial head of flexor
digitorum superficialis

Flexor pollicis longus

Palmaris longus tendon
(*cut and reflected*)

Palmar carpal ligament

Flexor carpi radialis
tendon (*cut*)

Superficial palmar
branch of radial a.

Ulnar n.

Median n.

Brachial a.

Medial intermuscular septum

Superficial head of pronator teres
(*cut and reflected*)

Medial epicondyle

Flexor carpi radialis tendon (*cut*)

Palmaris longus tendon (*cut*)

Anterior ulnar recurrent a.

Humeroulnar head of flexor
digitorum superficialis

Ulnar a.

Common interosseous a.

Deep head of pronator teres (*cut*)

Anterior interosseous a.

Flexor carpi ulnaris

Flexor digitorum superficialis

Ulnar a.

Ulnar n.

Dorsal branch of ulnar n.

Median n.

Palmar branch of ulnar n. (*cut*)

Palmar branch of median n. (*cut*)

Pisiform bone

Deep branch of ulnar n.

Deep palmar branch of ulnar a.

Superficial branch of ulnar n.

Flexor retinaculum of wrist

Plate 457

Elbow and Forearm

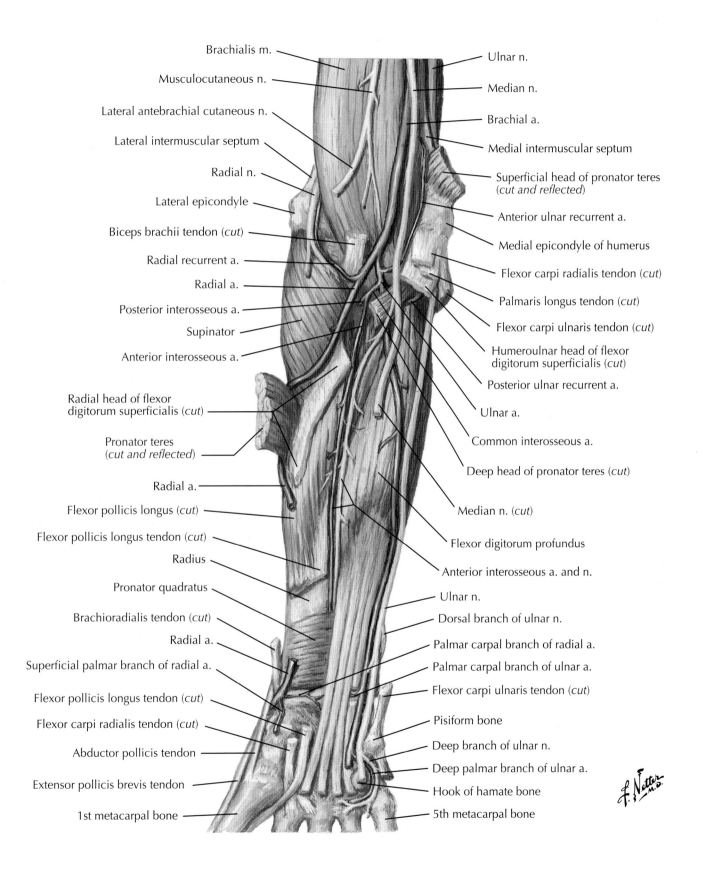

Brachialis m.

Musculocutaneous n.

Lateral antebrachial cutaneous n.

Lateral intermuscular septum

Radial n.

Lateral epicondyle

Biceps brachii tendon (*cut*)

Radial recurrent a.

Radial a.

Posterior interosseous a.

Supinator

Anterior interosseous a.

Radial head of flexor digitorum superficialis (*cut*)

Pronator teres (*cut and reflected*)

Radial a.

Flexor pollicis longus (*cut*)

Flexor pollicis longus tendon (*cut*)

Radius

Pronator quadratus

Brachioradialis tendon (*cut*)

Radial a.

Superficial palmar branch of radial a.

Flexor pollicis longus tendon (*cut*)

Flexor carpi radialis tendon (*cut*)

Abductor pollicis tendon

Extensor pollicis brevis tendon

1st metacarpal bone

Ulnar n.

Median n.

Brachial a.

Medial intermuscular septum

Superficial head of pronator teres (*cut and reflected*)

Anterior ulnar recurrent a.

Medial epicondyle of humerus

Flexor carpi radialis tendon (*cut*)

Palmaris longus tendon (*cut*)

Flexor carpi ulnaris tendon (*cut*)

Humeroulnar head of flexor digitorum superficialis (*cut*)

Posterior ulnar recurrent a.

Ulnar a.

Common interosseous a.

Deep head of pronator teres (*cut*)

Median n. (*cut*)

Flexor digitorum profundus

Anterior interosseous a. and n.

Ulnar n.

Dorsal branch of ulnar n.

Palmar carpal branch of radial a.

Palmar carpal branch of ulnar a.

Flexor carpi ulnaris tendon (*cut*)

Pisiform bone

Deep branch of ulnar n.

Deep palmar branch of ulnar a.

Hook of hamate bone

5th metacarpal bone

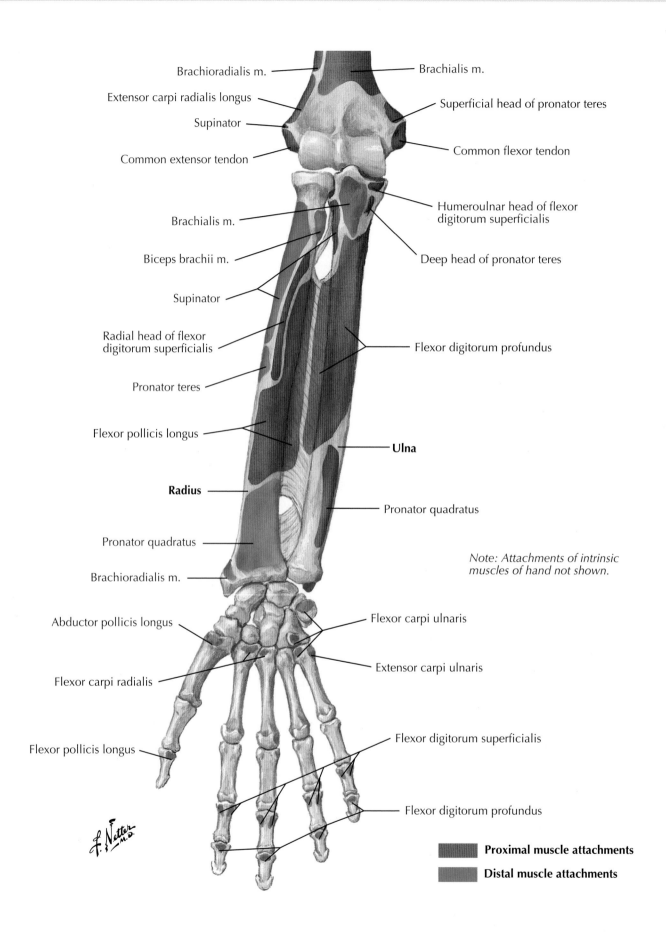

Brachioradialis m.

Brachialis m.

Extensor carpi radialis longus

Superficial head of pronator teres

Supinator

Common extensor tendon

Common flexor tendon

Humeroulnar head of flexor
digitorum superficialis

Brachialis m.

Biceps brachii m.

Deep head of pronator teres

Supinator

Radial head of flexor
digitorum superficialis

Flexor digitorum profundus

Pronator teres

Flexor pollicis longus

Ulna

Radius

Pronator quadratus

Pronator quadratus

*Note: Attachments of intrinsic
muscles of hand not shown.*

Brachioradialis m.

Abductor pollicis longus

Flexor carpi ulnaris

Extensor carpi ulnaris

Flexor carpi radialis

Flexor pollicis longus

Flexor digitorum superficialis

Flexor digitorum profundus

Proximal muscle attachments

Distal muscle attachments

Plate 459

Elbow and Forearm

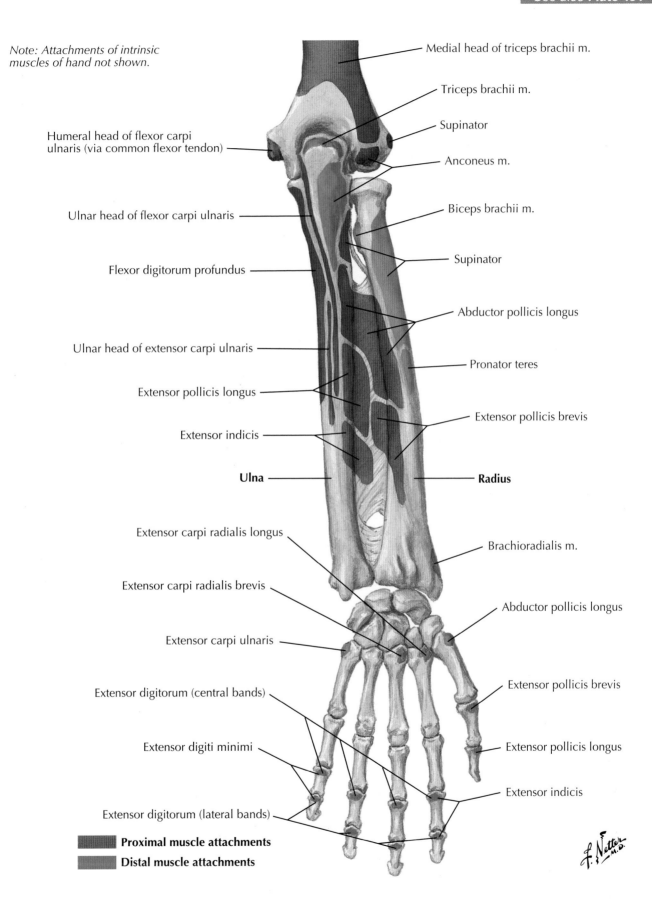

Note: Attachments of intrinsic
muscles of hand not shown.

Medial head of triceps brachii m.

Triceps brachii m.

Supinator

Humeral head of flexor carpi
ulnaris (via common flexor tendon)

Anconeus m.

Ulnar head of flexor carpi ulnaris

Biceps brachii m.

Flexor digitorum profundus

Supinator

Abductor pollicis longus

Ulnar head of extensor carpi ulnaris

Pronator teres

Extensor pollicis longus

Extensor pollicis brevis

Extensor indicis

Ulna

Radius

Extensor carpi radialis longus

Brachioradialis m.

Extensor carpi radialis brevis

Abductor pollicis longus

Extensor carpi ulnaris

Extensor pollicis brevis

Extensor digitorum (central bands)

Extensor pollicis longus

Extensor digiti minimi

Extensor indicis

Extensor digitorum (lateral bands)

Proximal muscle attachments

Distal muscle attachments

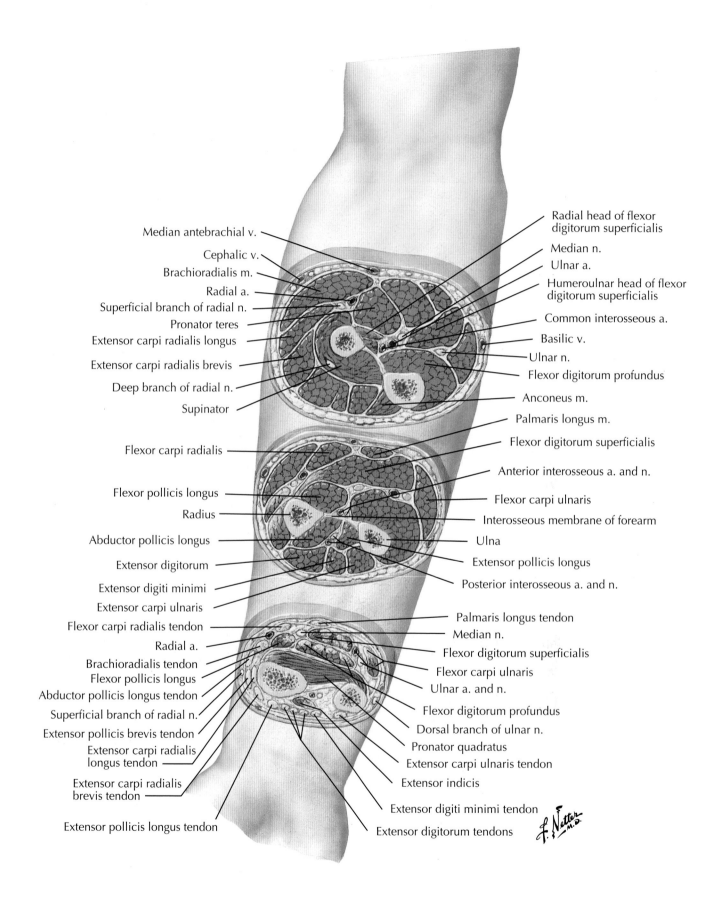

Median antebrachial v.
Cephalic v.
Brachioradialis m.
Radial a.
Superficial branch of radial n.
Pronator teres
Extensor carpi radialis longus
Extensor carpi radialis brevis
Deep branch of radial n.
Supinator

Radial head of flexor digitorum superficialis
Median n.
Ulnar a.
Humeroulnar head of flexor digitorum superficialis
Common interosseous a.
Basilic v.
Ulnar n.
Flexor digitorum profundus
Anconeus m.
Palmaris longus m.

Flexor carpi radialis

Flexor pollicis longus
Radius
Abductor pollicis longus
Extensor digitorum
Extensor digiti minimi
Extensor carpi ulnaris
Flexor carpi radialis tendon
Radial a.
Brachioradialis tendon
Flexor pollicis longus
Abductor pollicis longus tendon
Superficial branch of radial n.
Extensor pollicis brevis tendon
Extensor carpi radialis longus tendon
Extensor carpi radialis brevis tendon

Extensor pollicis longus tendon

Flexor digitorum superficialis
Anterior interosseous a. and n.
Flexor carpi ulnaris
Interosseous membrane of forearm
Ulna
Extensor pollicis longus
Posterior interosseous a. and n.
Palmaris longus tendon
Median n.
Flexor digitorum superficialis
Flexor carpi ulnaris
Ulnar a. and n.
Flexor digitorum profundus
Dorsal branch of ulnar n.
Pronator quadratus
Extensor carpi ulnaris tendon
Extensor indicis
Extensor digiti minimi tendon
Extensor digitorum tendons

Plate 461

Elbow and Forearm

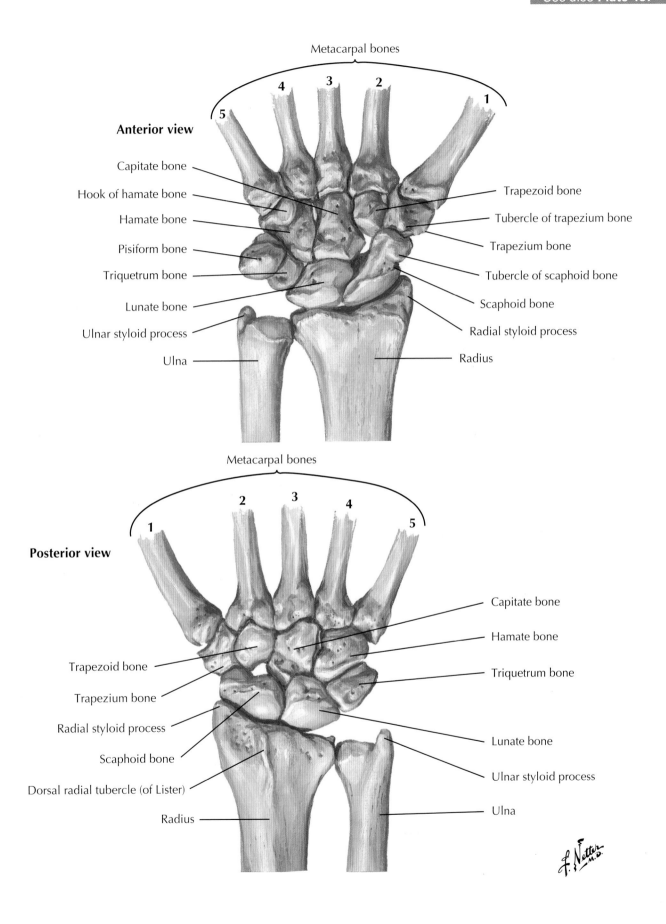

Metacarpal bones

5 4 3 2 1

Anterior view

Capitate bone

Hook of hamate bone

Hamate bone

Pisiform bone

Triquetrum bone

Lunate bone

Ulnar styloid process

Ulna

Trapezoid bone

Tubercle of trapezium bone

Trapezium bone

Tubercle of scaphoid bone

Scaphoid bone

Radial styloid process

Radius

Metacarpal bones

1 2 3 4 5

Posterior view

Trapezoid bone

Trapezium bone

Radial styloid process

Scaphoid bone

Dorsal radial tubercle (of Lister)

Radius

Capitate bone

Hamate bone

Triquetrum bone

Lunate bone

Ulnar styloid process

Ulna

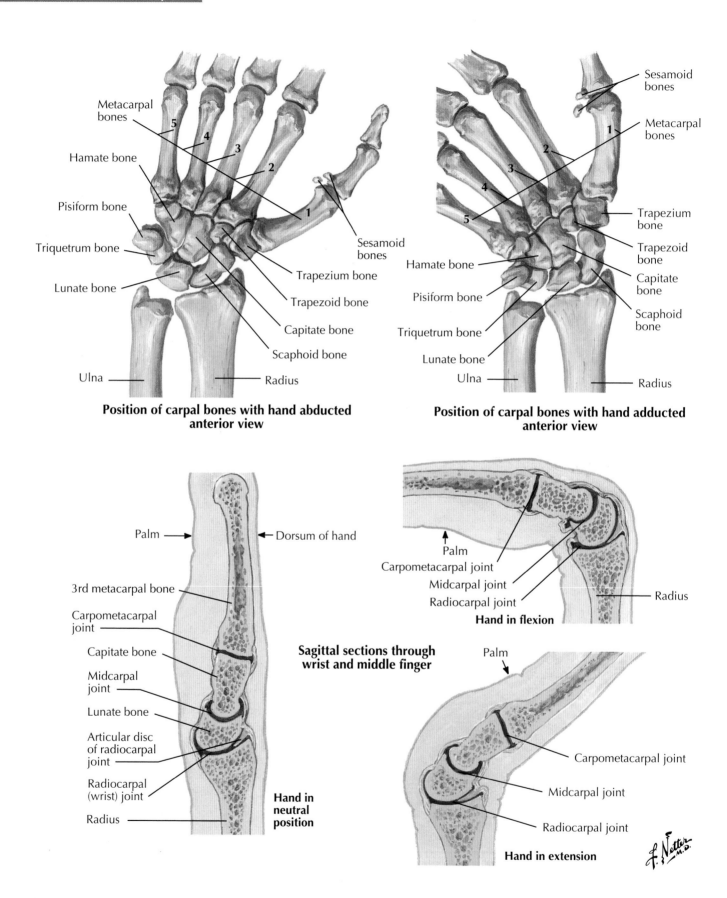

Metacarpal bones

Hamate bone

Pisiform bone

Triquetrum bone

Lunate bone

Ulna — Radius

Sesamoid bones

Trapezium bone

Trapezoid bone

Capitate bone

Scaphoid bone

Position of carpal bones with hand abducted anterior view

Sesamoid bones

Metacarpal bones

Hamate bone

Pisiform bone

Triquetrum bone

Lunate bone

Ulna — Radius

Trapezium bone

Trapezoid bone

Capitate bone

Scaphoid bone

Position of carpal bones with hand adducted anterior view

Palm → ← Dorsum of hand

3rd metacarpal bone

Carpometacarpal joint

Capitate bone

Midcarpal joint

Lunate bone

Articular disc of radiocarpal joint

Radiocarpal (wrist) joint

Radius

Hand in neutral position

Palm

Carpometacarpal joint

Midcarpal joint

Radiocarpal joint

Radius

Hand in flexion

Sagittal sections through wrist and middle finger

Palm

Carpometacarpal joint

Midcarpal joint

Radiocarpal joint

Hand in extension

Plate 463

Wrist and Hand

Deep palm

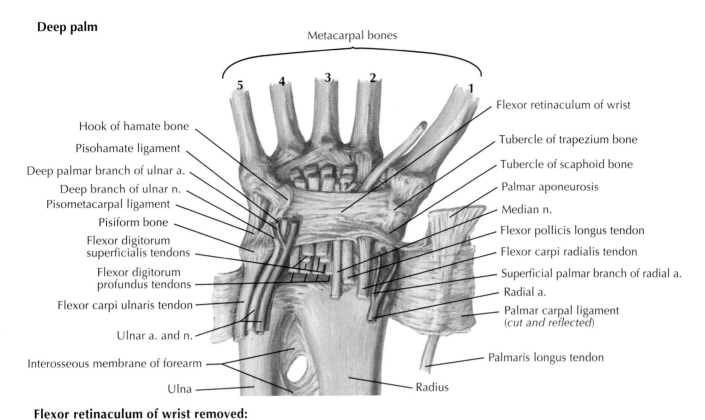

Metacarpal bones

5   4   3   2   1

Hook of hamate bone
Pisohamate ligament
Deep palmar branch of ulnar a.
Deep branch of ulnar n.
Pisometacarpal ligament
Pisiform bone
Flexor digitorum superficialis tendons
Flexor digitorum profundus tendons
Flexor carpi ulnaris tendon
Ulnar a. and n.
Interosseous membrane of forearm
Ulna

Flexor retinaculum of wrist
Tubercle of trapezium bone
Tubercle of scaphoid bone
Palmar aponeurosis
Median n.
Flexor pollicis longus tendon
Flexor carpi radialis tendon
Superficial palmar branch of radial a.
Radial a.
Palmar carpal ligament (*cut and reflected*)
Palmaris longus tendon
Radius

Flexor retinaculum of wrist removed: anterior view

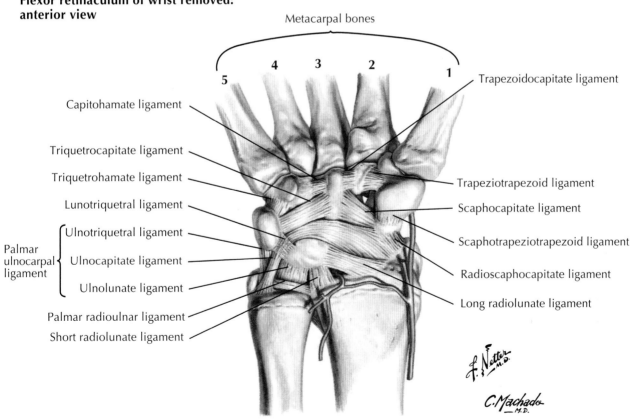

Metacarpal bones

5   4   3   2   1

Capitohamate ligament
Triquetrocapitate ligament
Triquetrohamate ligament
Lunotriquetral ligament
Ulnotriquetral ligament
Palmar ulnocarpal ligament { Ulnocapitate ligament
Ulnolunate ligament
Palmar radioulnar ligament
Short radiolunate ligament

Trapezoidocapitate ligament
Trapeziotrapezoid ligament
Scaphocapitate ligament
Scaphotrapeziotrapezoid ligament
Radioscaphocapitate ligament
Long radiolunate ligament

Wrist and Hand

Plate 464

Posterior view

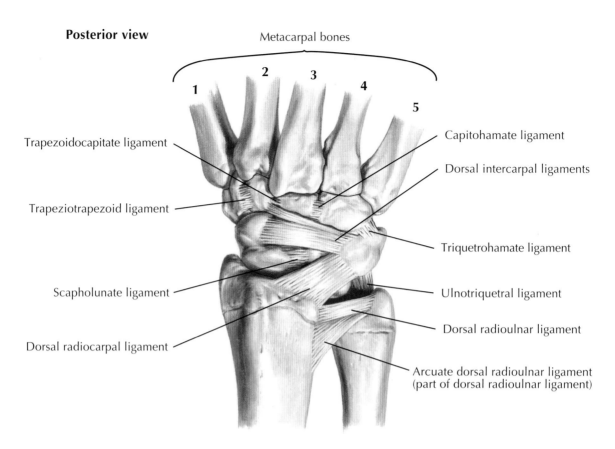

Metacarpal bones

1   2   3   4   5

Trapezoidocapitate ligament

Trapeziotrapezoid ligament

Scapholunate ligament

Dorsal radiocarpal ligament

Capitohamate ligament

Dorsal intercarpal ligaments

Triquetrohamate ligament

Ulnotriquetral ligament

Dorsal radioulnar ligament

Arcuate dorsal radioulnar ligament
(part of dorsal radioulnar ligament)

Coronal section: posterior view

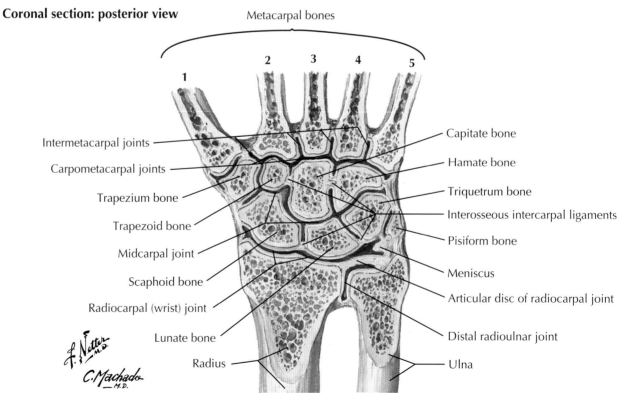

Metacarpal bones

1   2   3   4   5

Intermetacarpal joints

Carpometacarpal joints

Trapezium bone

Trapezoid bone

Midcarpal joint

Scaphoid bone

Radiocarpal (wrist) joint

Lunate bone

Radius

Capitate bone

Hamate bone

Triquetrum bone

Interosseous intercarpal ligaments

Pisiform bone

Meniscus

Articular disc of radiocarpal joint

Distal radioulnar joint

Ulna

Plate 465

Wrist and Hand

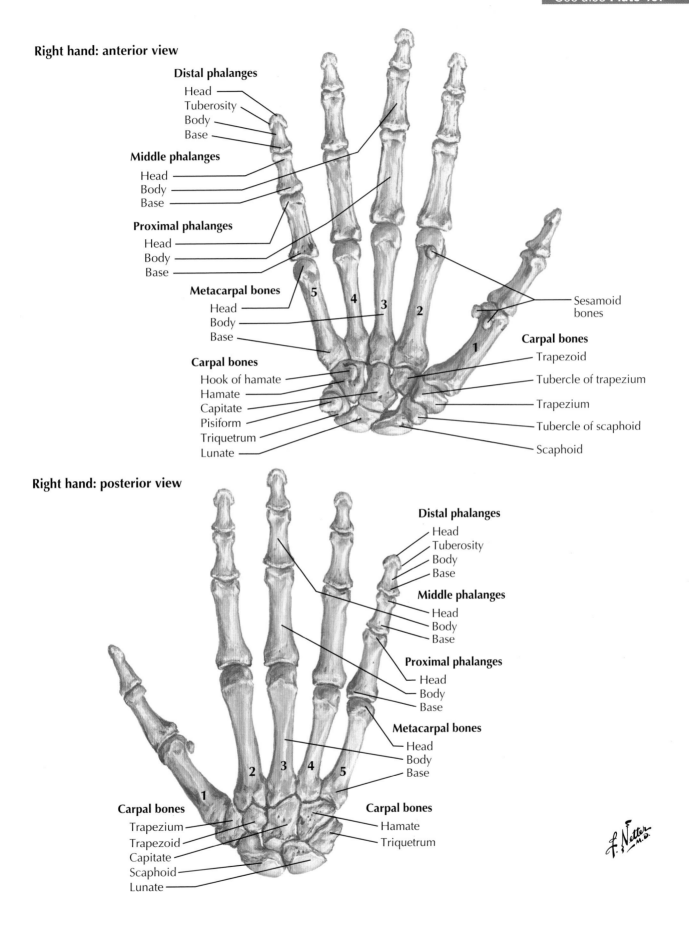

Right hand: anterior view

Distal phalanges
Head
Tuberosity
Body
Base

Middle phalanges
Head
Body
Base

Proximal phalanges
Head
Body
Base

Metacarpal bones
Head
Body
Base

Carpal bones
Hook of hamate
Hamate
Capitate
Pisiform
Triquetrum
Lunate

5   4   3   2

1

Sesamoid bones

Carpal bones
Trapezoid
Tubercle of trapezium
Trapezium
Tubercle of scaphoid
Scaphoid

Right hand: posterior view

Distal phalanges
Head
Tuberosity
Body
Base

Middle phalanges
Head
Body
Base

Proximal phalanges
Head
Body
Base

Metacarpal bones
Head
Body
Base

2   3   4   5

1

Carpal bones
Trapezium
Trapezoid
Capitate
Scaphoid
Lunate

Carpal bones
Hamate
Triquetrum

Anteroposterior view

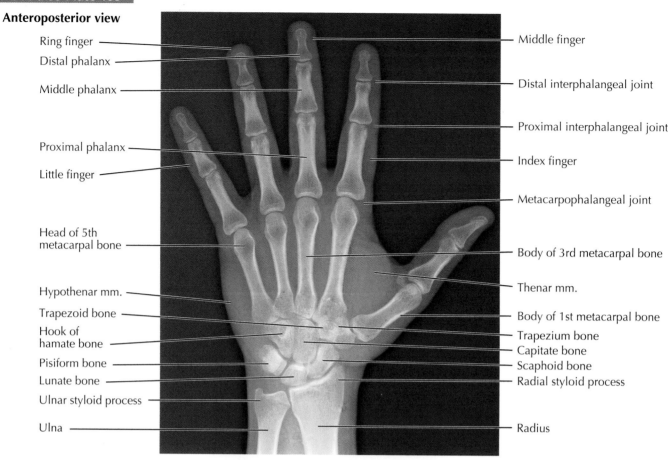

Ring finger

Distal phalanx

Middle phalanx

Proximal phalanx

Little finger

Head of 5th metacarpal bone

Hypothenar mm.

Trapezoid bone

Hook of hamate bone

Pisiform bone

Lunate bone

Ulnar styloid process

Ulna

Middle finger

Distal interphalangeal joint

Proximal interphalangeal joint

Index finger

Metacarpophalangeal joint

Body of 3rd metacarpal bone

Thenar mm.

Body of 1st metacarpal bone

Trapezium bone

Capitate bone

Scaphoid bone

Radial styloid process

Radius

Lateral view

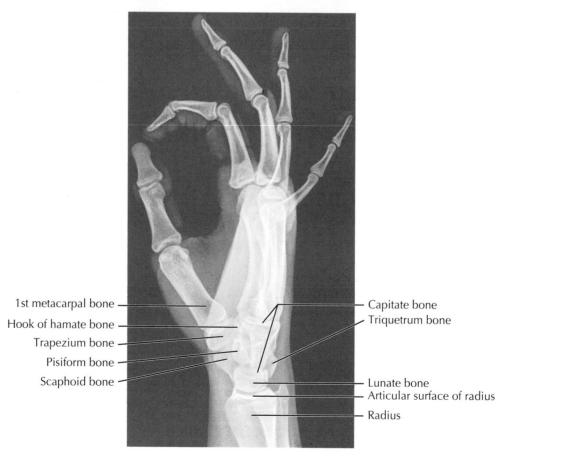

1st metacarpal bone

Hook of hamate bone

Trapezium bone

Pisiform bone

Scaphoid bone

Capitate bone

Triquetrum bone

Lunate bone

Articular surface of radius

Radius

Plate 467

Wrist and Hand

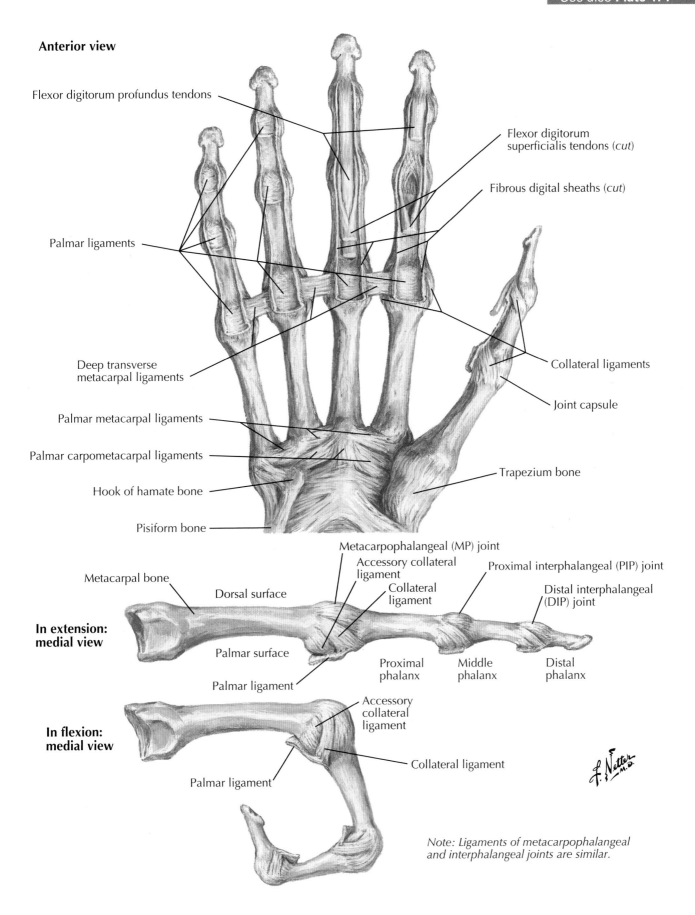

Anterior view

Flexor digitorum profundus tendons

Flexor digitorum superficialis tendons (*cut*)

Fibrous digital sheaths (*cut*)

Palmar ligaments

Deep transverse metacarpal ligaments

Collateral ligaments

Joint capsule

Palmar metacarpal ligaments

Palmar carpometacarpal ligaments

Hook of hamate bone

Trapezium bone

Pisiform bone

Metacarpophalangeal (MP) joint

Accessory collateral ligament

Proximal interphalangeal (PIP) joint

Metacarpal bone

Dorsal surface

Collateral ligament

Distal interphalangeal (DIP) joint

In extension: medial view

Palmar surface

Proximal phalanx

Middle phalanx

Distal phalanx

Palmar ligament

Accessory collateral ligament

In flexion: medial view

Collateral ligament

Palmar ligament

Note: Ligaments of metacarpophalangeal and interphalangeal joints are similar.

F. Netter
M.D.

Anterior views

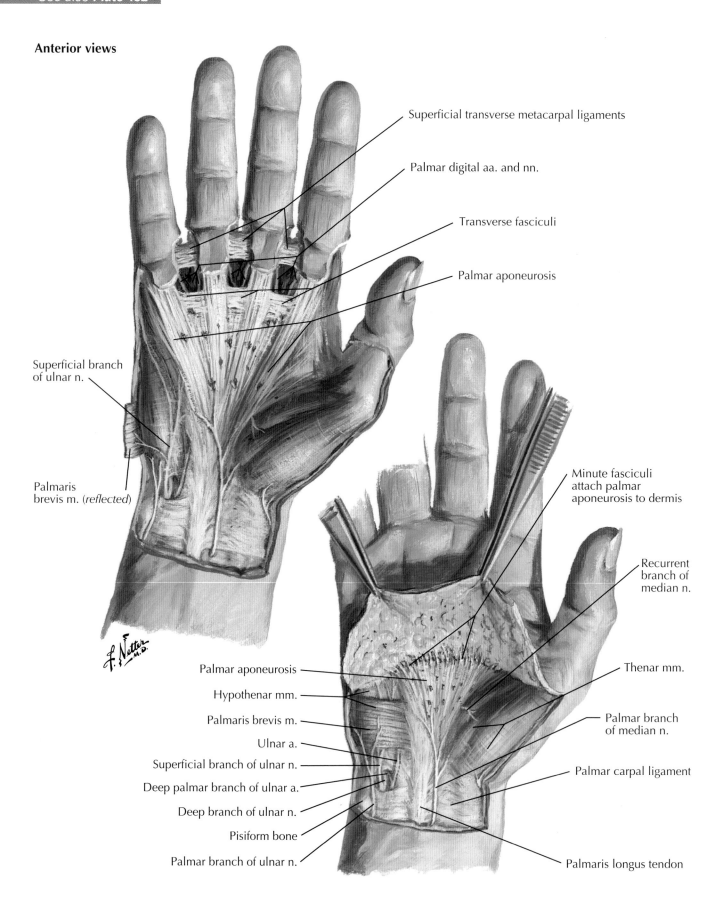

Superficial transverse metacarpal ligaments

Palmar digital aa. and nn.

Transverse fasciculi

Palmar aponeurosis

Superficial branch of ulnar n.

Palmaris brevis m. (*reflected*)

Minute fasciculi attach palmar aponeurosis to dermis

Recurrent branch of median n.

Palmar aponeurosis

Hypothenar mm.

Palmaris brevis m.

Ulnar a.

Superficial branch of ulnar n.

Deep palmar branch of ulnar a.

Deep branch of ulnar n.

Pisiform bone

Palmar branch of ulnar n.

Thenar mm.

Palmar branch of median n.

Palmar carpal ligament

Palmaris longus tendon

Plate 469

Wrist and Hand

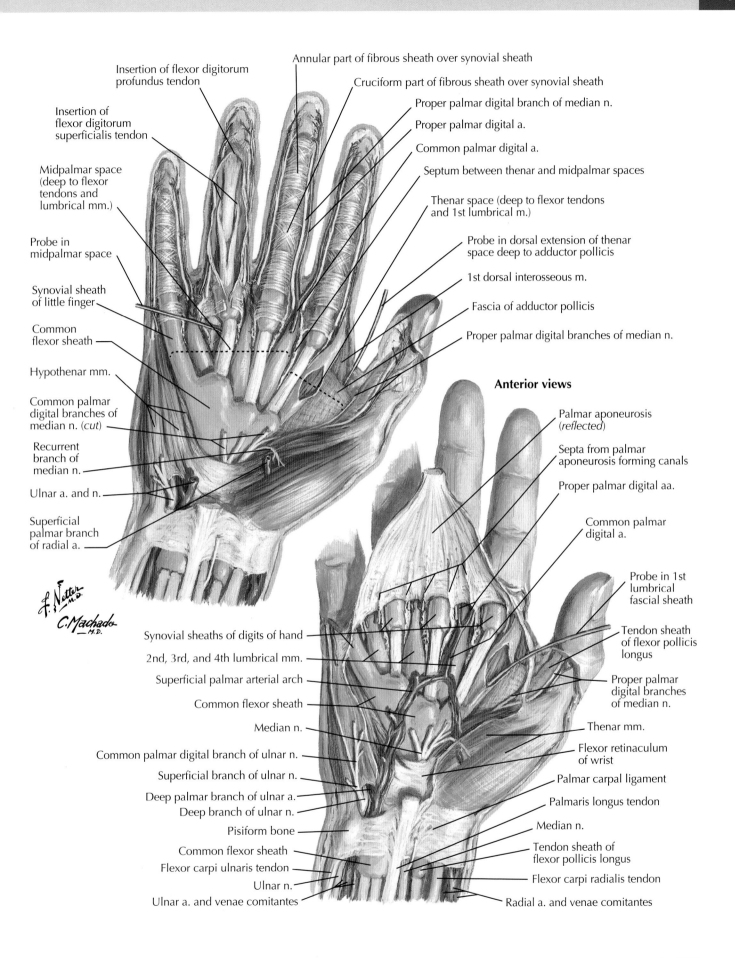

Insertion of flexor digitorum profundus tendon

Insertion of flexor digitorum superficialis tendon

Midpalmar space (deep to flexor tendons and lumbrical mm.)

Probe in midpalmar space

Synovial sheath of little finger

Common flexor sheath

Hypothenar mm.

Common palmar digital branches of median n. (*cut*)

Recurrent branch of median n.

Ulnar a. and n.

Superficial palmar branch of radial a.

Annular part of fibrous sheath over synovial sheath

Cruciform part of fibrous sheath over synovial sheath

Proper palmar digital branch of median n.

Proper palmar digital a.

Common palmar digital a.

Septum between thenar and midpalmar spaces

Thenar space (deep to flexor tendons and 1st lumbrical m.)

Probe in dorsal extension of thenar space deep to adductor pollicis

1st dorsal interosseous m.

Fascia of adductor pollicis

Proper palmar digital branches of median n.

Anterior views

Palmar aponeurosis (*reflected*)

Septa from palmar aponeurosis forming canals

Proper palmar digital aa.

Common palmar digital a.

Probe in 1st lumbrical fascial sheath

Tendon sheath of flexor pollicis longus

Proper palmar digital branches of median n.

Thenar mm.

Flexor retinaculum of wrist

Palmar carpal ligament

Palmaris longus tendon

Median n.

Tendon sheath of flexor pollicis longus

Flexor carpi radialis tendon

Radial a. and venae comitantes

Synovial sheaths of digits of hand

2nd, 3rd, and 4th lumbrical mm.

Superficial palmar arterial arch

Common flexor sheath

Median n.

Common palmar digital branch of ulnar n.

Superficial branch of ulnar n.

Deep palmar branch of ulnar a.

Deep branch of ulnar n.

Pisiform bone

Common flexor sheath

Flexor carpi ulnaris tendon

Ulnar n.

Ulnar a. and venae comitantes

Common variation

Usual arrangement

Synovial sheaths
of fingers

Lumbrical mm.

Midpalmar space

Thenar space

Common flexor sheath

Tendon sheath of
flexor pollicis longus

Intermediate bursa
(communication
between common
flexor sheath and
tendon sheath of
flexor pollicis longus)

Lumbrical muscles: schema

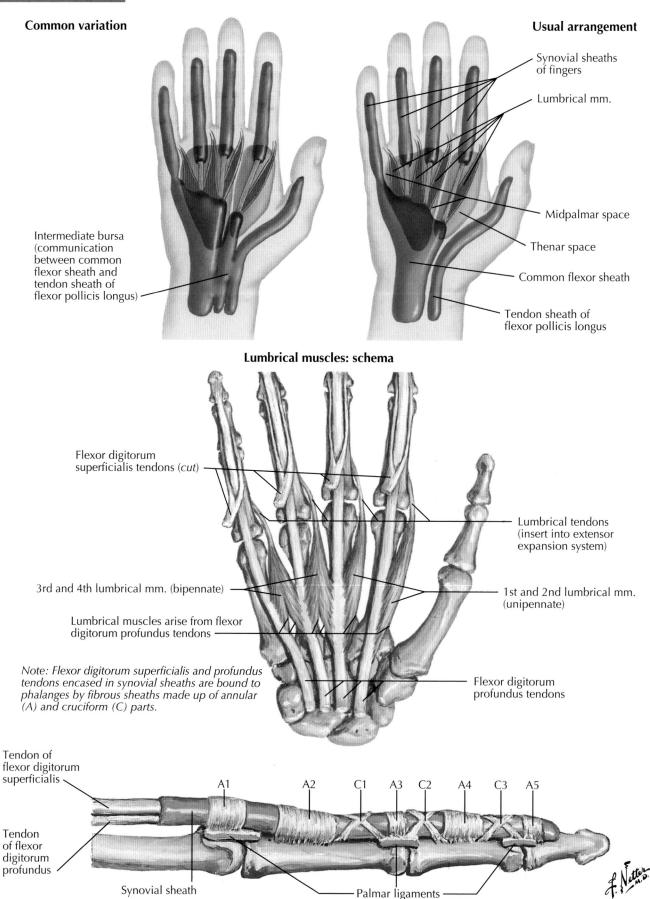

Flexor digitorum
superficialis tendons (*cut*)

Lumbrical tendons
(insert into extensor
expansion system)

3rd and 4th lumbrical mm. (bipennate)

1st and 2nd lumbrical mm.
(unipennate)

Lumbrical muscles arise from flexor
digitorum profundus tendons

Flexor digitorum
profundus tendons

*Note: Flexor digitorum superficialis and profundus
tendons encased in synovial sheaths are bound to
phalanges by fibrous sheaths made up of annular
(A) and cruciform (C) parts.*

Tendon of
flexor digitorum
superficialis

Tendon
of flexor
digitorum
profundus

A1   A2   C1   A3   C2   A4   C3   A5

Synovial sheath

Palmar ligaments

Plate 471

Wrist and Hand

Anterior view

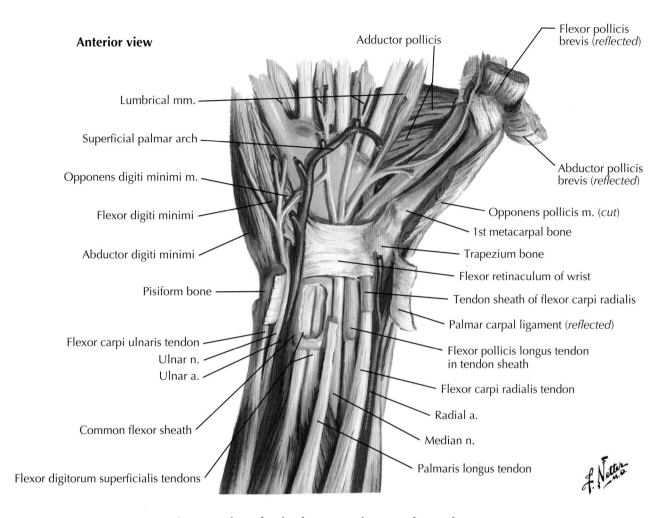

Adductor pollicis

Flexor pollicis brevis (*reflected*)

Lumbrical mm.

Superficial palmar arch

Opponens digiti minimi m.

Flexor digiti minimi

Abductor digiti minimi

Pisiform bone

Flexor carpi ulnaris tendon

Ulnar n.

Ulnar a.

Common flexor sheath

Flexor digitorum superficialis tendons

Abductor pollicis brevis (*reflected*)

Opponens pollicis m. (*cut*)

1st metacarpal bone

Trapezium bone

Flexor retinaculum of wrist

Tendon sheath of flexor carpi radialis

Palmar carpal ligament (*reflected*)

Flexor pollicis longus tendon in tendon sheath

Flexor carpi radialis tendon

Radial a.

Median n.

Palmaris longus tendon

Cross section of wrist demonstrating carpal tunnel

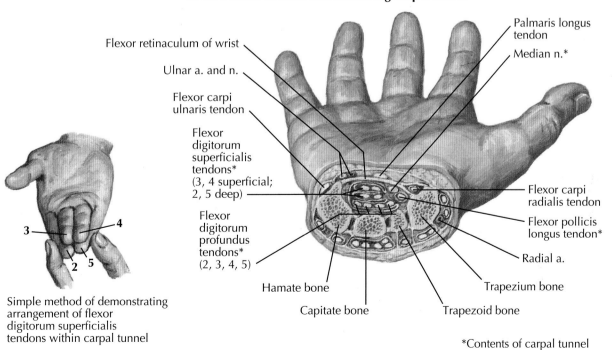

Flexor retinaculum of wrist

Ulnar a. and n.

Flexor carpi ulnaris tendon

Flexor digitorum superficialis tendons* (3, 4 superficial; 2, 5 deep)

Flexor digitorum profundus tendons* (2, 3, 4, 5)

Palmaris longus tendon

Median n.*

Flexor carpi radialis tendon

Flexor pollicis longus tendon*

Radial a.

Trapezium bone

Trapezoid bone

Capitate bone

Hamate bone

Simple method of demonstrating arrangement of flexor digitorum superficialis tendons within carpal tunnel

3   4

2   5

*Contents of carpal tunnel

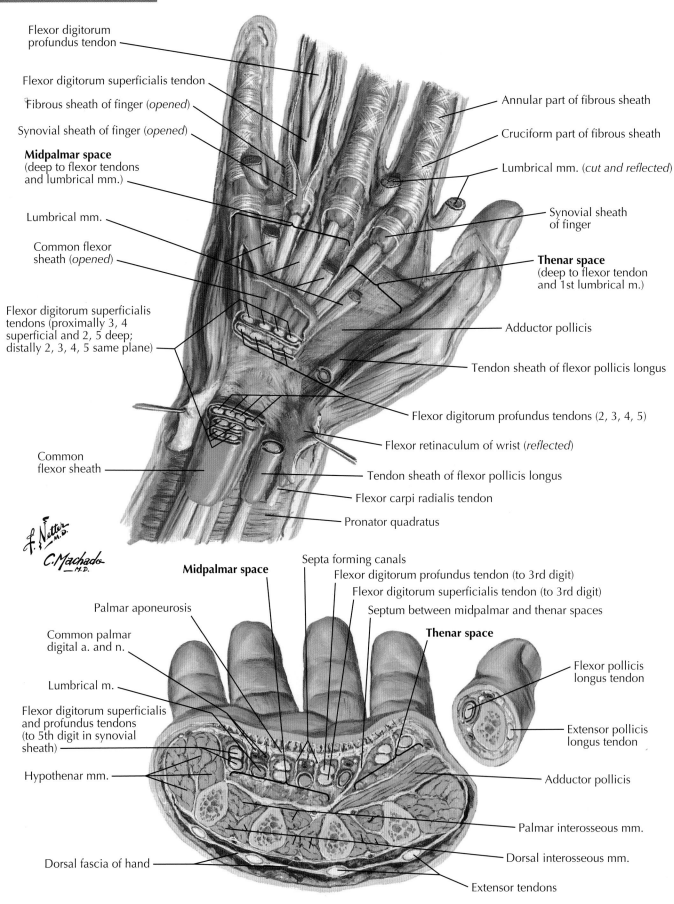

Flexor digitorum profundus tendon

Flexor digitorum superficialis tendon

Fibrous sheath of finger (*opened*)

Synovial sheath of finger (*opened*)

Midpalmar space
(deep to flexor tendons and lumbrical mm.)

Lumbrical mm.

Common flexor sheath (*opened*)

Flexor digitorum superficialis tendons (proximally 3, 4 superficial and 2, 5 deep; distally 2, 3, 4, 5 same plane)

Common flexor sheath

Annular part of fibrous sheath

Cruciform part of fibrous sheath

Lumbrical mm. (*cut and reflected*)

Synovial sheath of finger

Thenar space
(deep to flexor tendon and 1st lumbrical m.)

Adductor pollicis

Tendon sheath of flexor pollicis longus

Flexor digitorum profundus tendons (2, 3, 4, 5)

Flexor retinaculum of wrist (*reflected*)

Tendon sheath of flexor pollicis longus

Flexor carpi radialis tendon

Pronator quadratus

Septa forming canals

Midpalmar space

Flexor digitorum profundus tendon (to 3rd digit)

Flexor digitorum superficialis tendon (to 3rd digit)

Septum between midpalmar and thenar spaces

Thenar space

Palmar aponeurosis

Common palmar digital a. and n.

Lumbrical m.

Flexor digitorum superficialis and profundus tendons (to 5th digit in synovial sheath)

Hypothenar mm.

Dorsal fascia of hand

Flexor pollicis longus tendon

Extensor pollicis longus tendon

Adductor pollicis

Palmar interosseous mm.

Dorsal interosseous mm.

Extensor tendons

Plate 473

Wrist and Hand

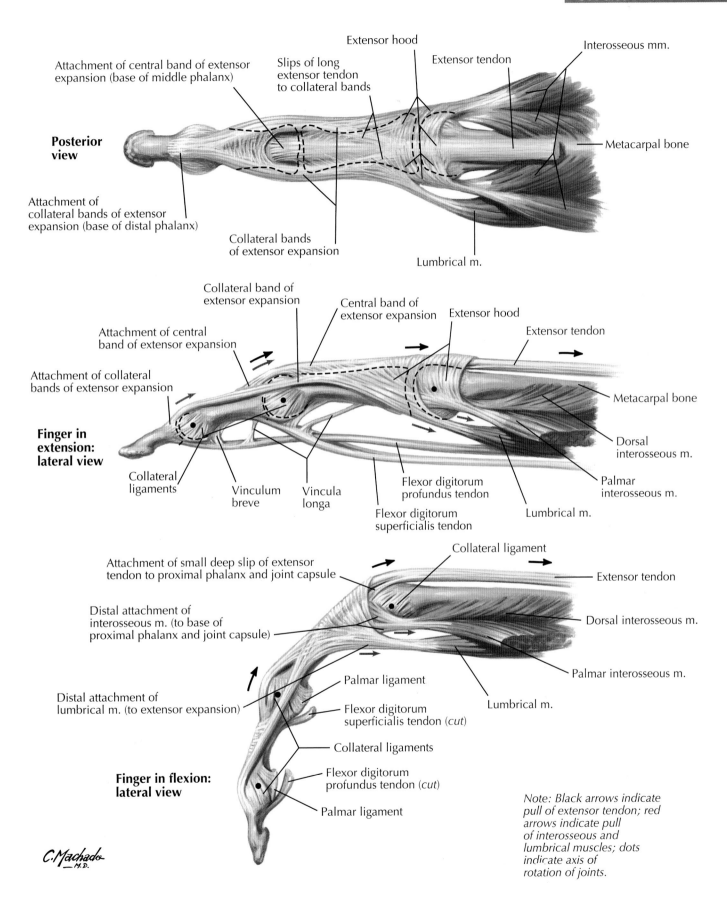

Extensor hood

Extensor tendon

Interosseous mm.

Attachment of central band of extensor expansion (base of middle phalanx)

Slips of long extensor tendon to collateral bands

Posterior view

Metacarpal bone

Attachment of collateral bands of extensor expansion (base of distal phalanx)

Collateral bands of extensor expansion

Lumbrical m.

Collateral band of extensor expansion

Central band of extensor expansion

Extensor hood

Extensor tendon

Attachment of central band of extensor expansion

Attachment of collateral bands of extensor expansion

Metacarpal bone

Finger in extension: lateral view

Collateral ligaments

Vinculum breve

Vincula longa

Flexor digitorum profundus tendon

Flexor digitorum superficialis tendon

Dorsal interosseous m.

Palmar interosseous m.

Lumbrical m.

Attachment of small deep slip of extensor tendon to proximal phalanx and joint capsule

Collateral ligament

Extensor tendon

Distal attachment of interosseous m. (to base of proximal phalanx and joint capsule)

Dorsal interosseous m.

Palmar interosseous m.

Distal attachment of lumbrical m. (to extensor expansion)

Palmar ligament

Flexor digitorum superficialis tendon (cut)

Lumbrical m.

Collateral ligaments

Finger in flexion: lateral view

Flexor digitorum profundus tendon (cut)

Palmar ligament

Note: Black arrows indicate pull of extensor tendon; red arrows indicate pull of interosseous and lumbrical muscles; dots indicate axis of rotation of joints.

C. Machado ⎯M.D.

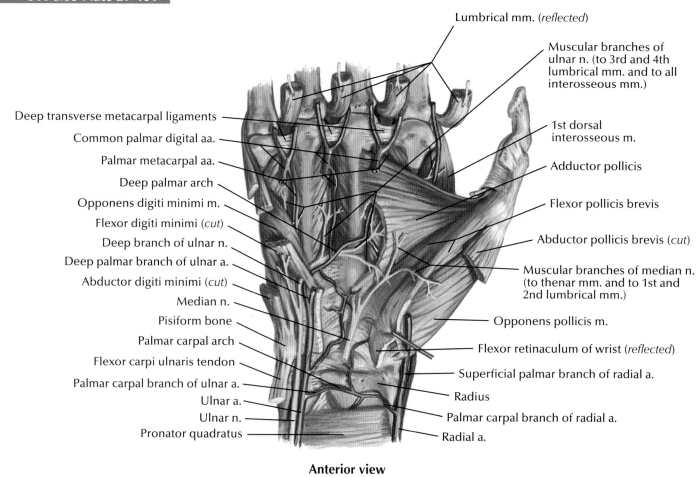

Lumbrical mm. (*reflected*)

Muscular branches of ulnar n. (to 3rd and 4th lumbrical mm. and to all interosseous mm.)

Deep transverse metacarpal ligaments

Common palmar digital aa.

Palmar metacarpal aa.

Deep palmar arch

Opponens digiti minimi m.

Flexor digiti minimi (*cut*)

Deep branch of ulnar n.

Deep palmar branch of ulnar a.

Abductor digiti minimi (*cut*)

Median n.

Pisiform bone

Palmar carpal arch

Flexor carpi ulnaris tendon

Palmar carpal branch of ulnar a.

Ulnar a.

Ulnar n.

Pronator quadratus

1st dorsal interosseous m.

Adductor pollicis

Flexor pollicis brevis

Abductor pollicis brevis (*cut*)

Muscular branches of median n. (to thenar mm. and to 1st and 2nd lumbrical mm.)

Opponens pollicis m.

Flexor retinaculum of wrist (*reflected*)

Superficial palmar branch of radial a.

Radius

Palmar carpal branch of radial a.

Radial a.

Anterior view

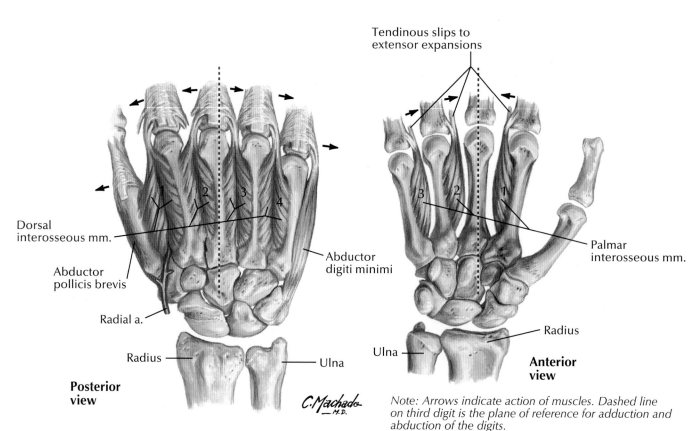

Tendinous slips to extensor expansions

Dorsal interosseous mm.

Abductor pollicis brevis

Radial a.

Radius

Ulna

Abductor digiti minimi

Palmar interosseous mm.

Radius

Ulna

Anterior view

Posterior view

C. Machado M.D.

Note: *Arrows indicate action of muscles. Dashed line on third digit is the plane of reference for adduction and abduction of the digits.*

Plate 475

Wrist and Hand

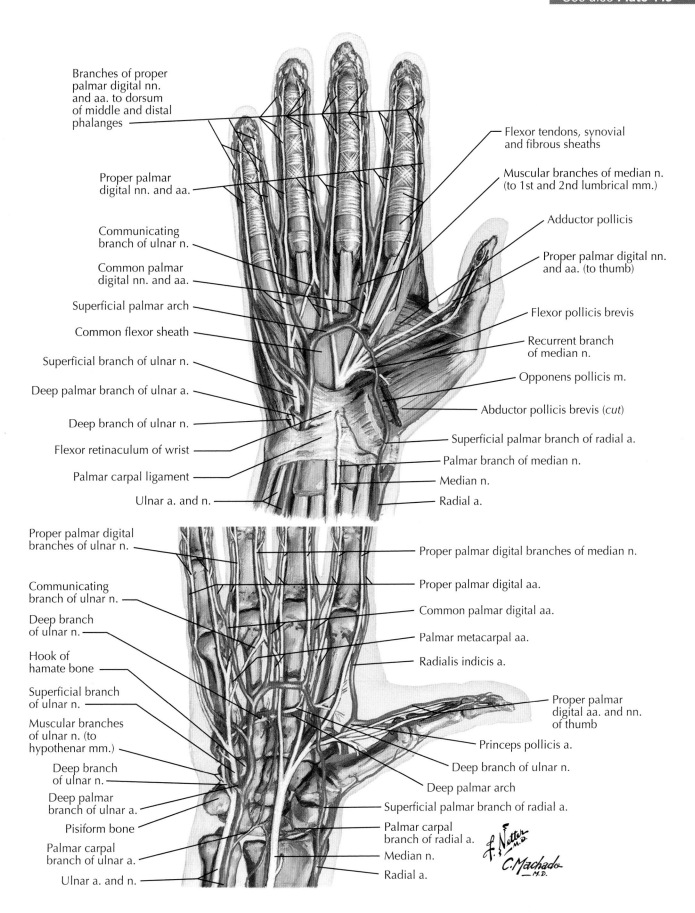

Branches of proper palmar digital nn. and aa. to dorsum of middle and distal phalanges

Proper palmar digital nn. and aa.

Communicating branch of ulnar n.

Common palmar digital nn. and aa.

Superficial palmar arch

Common flexor sheath

Superficial branch of ulnar n.

Deep palmar branch of ulnar a.

Deep branch of ulnar n.

Flexor retinaculum of wrist

Palmar carpal ligament

Ulnar a. and n.

Flexor tendons, synovial and fibrous sheaths

Muscular branches of median n. (to 1st and 2nd lumbrical mm.)

Adductor pollicis

Proper palmar digital nn. and aa. (to thumb)

Flexor pollicis brevis

Recurrent branch of median n.

Opponens pollicis m.

Abductor pollicis brevis (cut)

Superficial palmar branch of radial a.

Palmar branch of median n.

Median n.

Radial a.

Proper palmar digital branches of ulnar n.

Communicating branch of ulnar n.

Deep branch of ulnar n.

Hook of hamate bone

Superficial branch of ulnar n.

Muscular branches of ulnar n. (to hypothenar mm.)

Deep branch of ulnar n.

Deep palmar branch of ulnar a.

Pisiform bone

Palmar carpal branch of ulnar a.

Ulnar a. and n.

Proper palmar digital branches of median n.

Proper palmar digital aa.

Common palmar digital aa.

Palmar metacarpal aa.

Radialis indicis a.

Proper palmar digital aa. and nn. of thumb

Princeps pollicis a.

Deep branch of ulnar n.

Deep palmar arch

Superficial palmar branch of radial a.

Palmar carpal branch of radial a.

Median n.

Radial a.

f. Netter M.D.
C. Machado M.D.

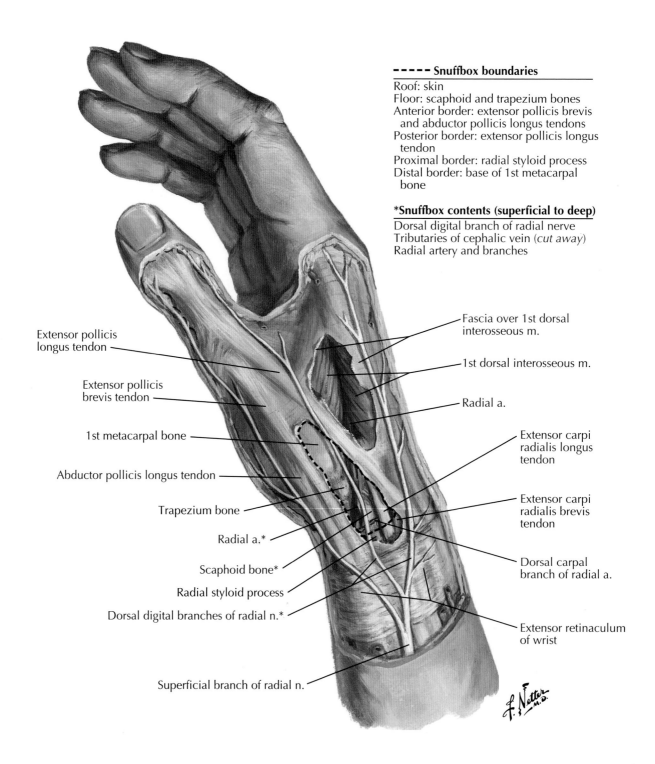

- - - - - Snuffbox boundaries

Roof: skin
Floor: scaphoid and trapezium bones
Anterior border: extensor pollicis brevis
  and abductor pollicis longus tendons
Posterior border: extensor pollicis longus
  tendon
Proximal border: radial styloid process
Distal border: base of 1st metacarpal
  bone

***Snuffbox contents (superficial to deep)**

Dorsal digital branch of radial nerve
Tributaries of cephalic vein (cut away)
Radial artery and branches

Fascia over 1st dorsal
interosseous m.

1st dorsal interosseous m.

Radial a.

Extensor carpi
radialis longus
tendon

Extensor carpi
radialis brevis
tendon

Dorsal carpal
branch of radial a.

Extensor retinaculum
of wrist

Extensor pollicis
longus tendon

Extensor pollicis
brevis tendon

1st metacarpal bone

Abductor pollicis longus tendon

Trapezium bone

Radial a.*

Scaphoid bone*

Radial styloid process

Dorsal digital branches of radial n.*

Superficial branch of radial n.

Plate 477

Wrist and Hand

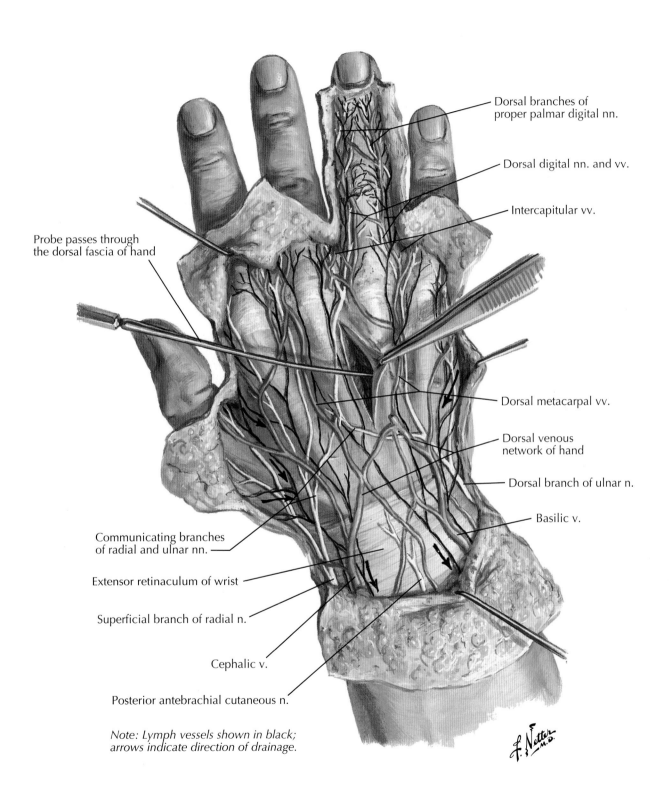

Dorsal branches of
proper palmar digital nn.

Dorsal digital nn. and vv.

Intercapitular vv.

Probe passes through
the dorsal fascia of hand

Dorsal metacarpal vv.

Dorsal venous
network of hand

Dorsal branch of ulnar n.

Basilic v.

Communicating branches
of radial and ulnar nn.

Extensor retinaculum of wrist

Superficial branch of radial n.

Cephalic v.

Posterior antebrachial cutaneous n.

*Note: Lymph vessels shown in black;
arrows indicate direction of drainage.*

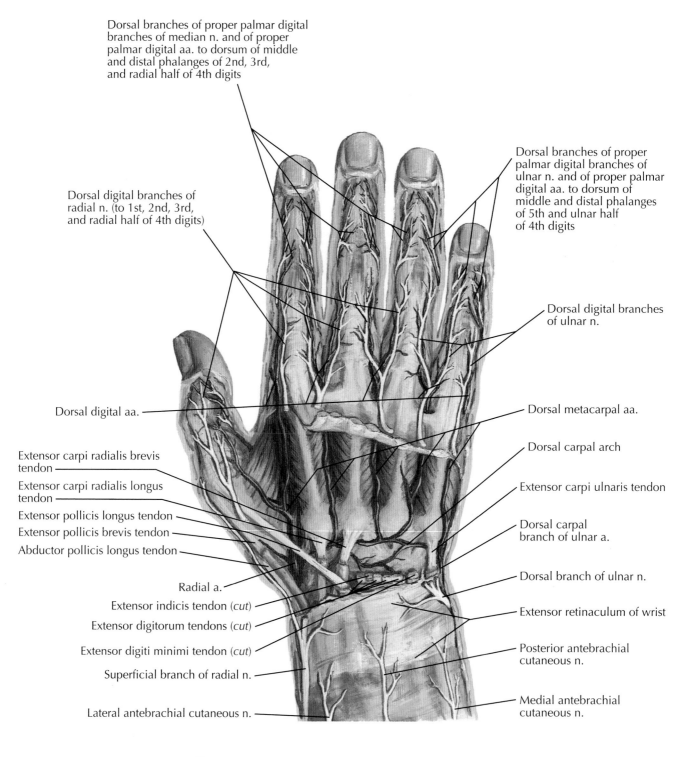

Dorsal branches of proper palmar digital branches of median n. and of proper palmar digital aa. to dorsum of middle and distal phalanges of 2nd, 3rd, and radial half of 4th digits

Dorsal branches of proper palmar digital branches of ulnar n. and of proper palmar digital aa. to dorsum of middle and distal phalanges of 5th and ulnar half of 4th digits

Dorsal digital branches of radial n. (to 1st, 2nd, 3rd, and radial half of 4th digits)

Dorsal digital branches of ulnar n.

Dorsal digital aa.

Dorsal metacarpal aa.

Extensor carpi radialis brevis tendon

Dorsal carpal arch

Extensor carpi radialis longus tendon

Extensor carpi ulnaris tendon

Extensor pollicis longus tendon

Extensor pollicis brevis tendon

Dorsal carpal branch of ulnar a.

Abductor pollicis longus tendon

Radial a.

Dorsal branch of ulnar n.

Extensor indicis tendon (*cut*)

Extensor digitorum tendons (*cut*)

Extensor retinaculum of wrist

Extensor digiti minimi tendon (*cut*)

Posterior antebrachial cutaneous n.

Superficial branch of radial n.

Medial antebrachial cutaneous n.

Lateral antebrachial cutaneous n.

Plate 479

Wrist and Hand

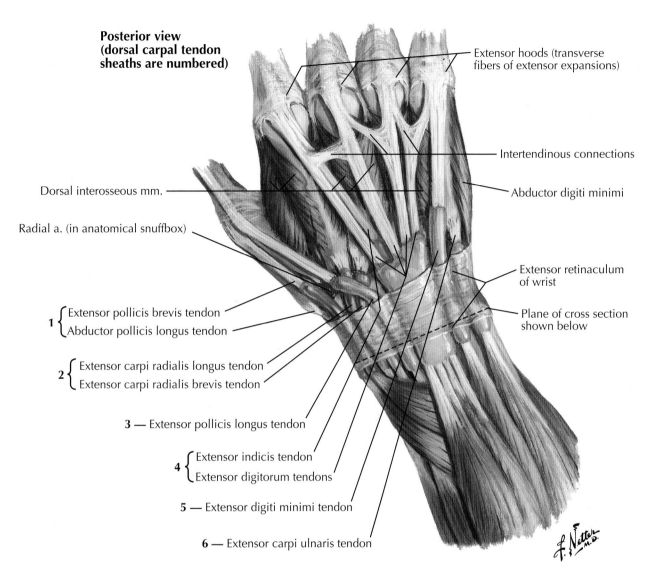

**Posterior view
(dorsal carpal tendon
sheaths are numbered)**

Extensor hoods (transverse
fibers of extensor expansions)

Intertendinous connections

Abductor digiti minimi

Dorsal interosseous mm.

Radial a. (in anatomical snuffbox)

Extensor retinaculum
of wrist

Plane of cross section
shown below

1 { Extensor pollicis brevis tendon
  Abductor pollicis longus tendon

2 { Extensor carpi radialis longus tendon
  Extensor carpi radialis brevis tendon

3 — Extensor pollicis longus tendon

4 { Extensor indicis tendon
  Extensor digitorum tendons

5 — Extensor digiti minimi tendon

6 — Extensor carpi ulnaris tendon

Cross section of distal forearm

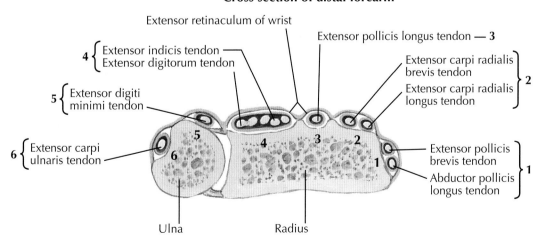

Extensor retinaculum of wrist

Extensor pollicis longus tendon — 3

4 { Extensor indicis tendon
  Extensor digitorum tendon

Extensor carpi radialis
brevis tendon

5 { Extensor digiti
  minimi tendon

Extensor carpi radialis
longus tendon

6 { Extensor carpi
  ulnaris tendon

Extensor pollicis
brevis tendon

Abductor pollicis
longus tendon

Ulna

Radius

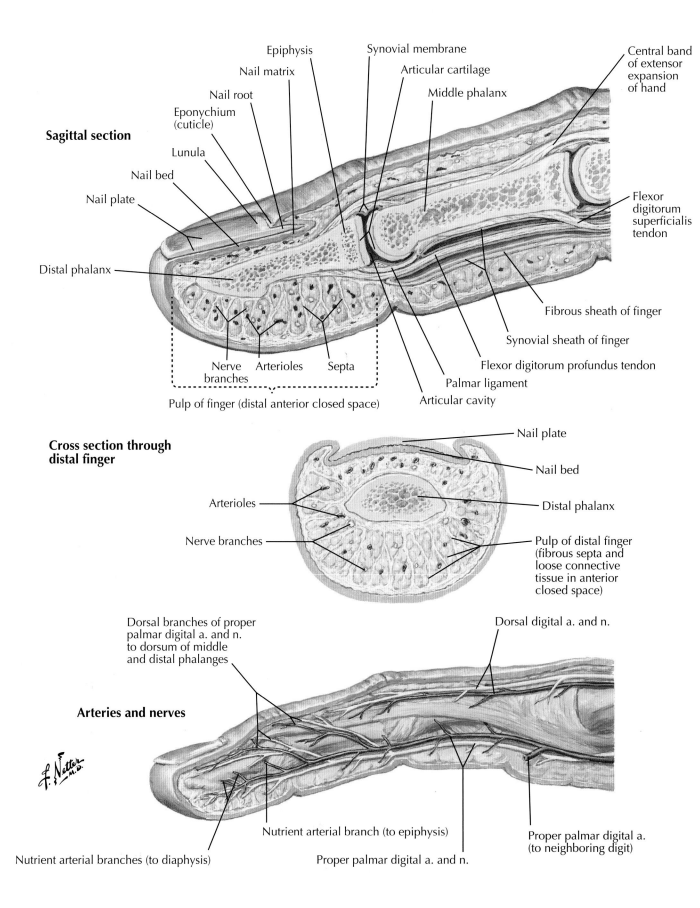

Sagittal section

Epiphysis

Synovial membrane

Nail matrix

Articular cartilage

Nail root

Middle phalanx

Eponychium (cuticle)

Central band of extensor expansion of hand

Lunula

Nail bed

Nail plate

Distal phalanx

Flexor digitorum superficialis tendon

Fibrous sheath of finger

Synovial sheath of finger

Nerve branches

Arterioles

Septa

Flexor digitorum profundus tendon

Palmar ligament

Articular cavity

Pulp of finger (distal anterior closed space)

Cross section through distal finger

Nail plate

Nail bed

Arterioles

Distal phalanx

Nerve branches

Pulp of distal finger (fibrous septa and loose connective tissue in anterior closed space)

Dorsal branches of proper palmar digital a. and n. to dorsum of middle and distal phalanges

Dorsal digital a. and n.

Arteries and nerves

Nutrient arterial branch (to epiphysis)

Nutrient arterial branches (to diaphysis)

Proper palmar digital a. and n.

Proper palmar digital a. (to neighboring digit)

Plate 481

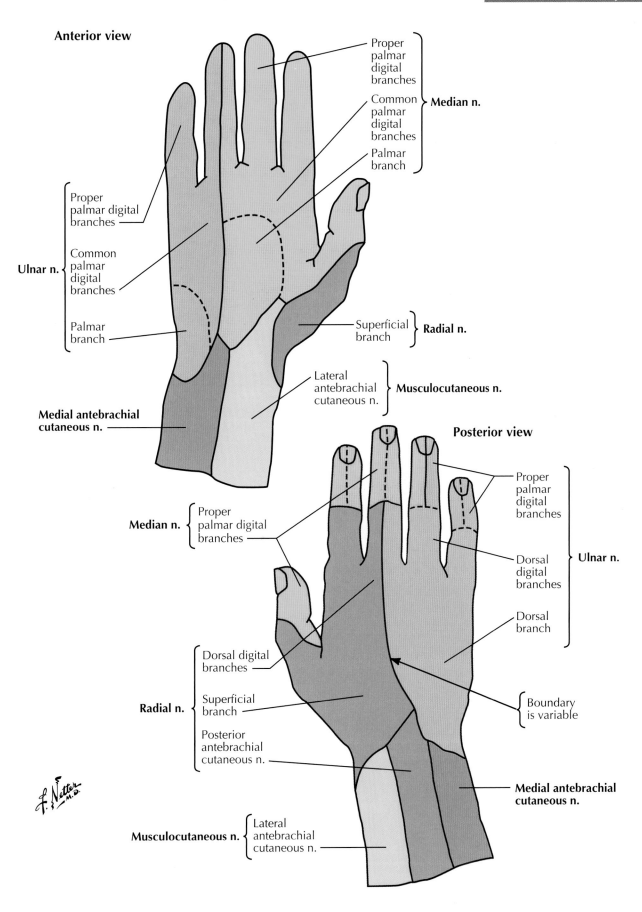

Anterior view

Proper palmar digital branches
Common palmar digital branches
Palmar branch
⎫ **Median n.**

Proper palmar digital branches
Common palmar digital branches
Palmar branch
Ulnar n. ⎰

Superficial branch ⎱ **Radial n.**

Lateral antebrachial cutaneous n. ⎱ **Musculocutaneous n.**

Medial antebrachial cutaneous n.

Posterior view

Median n. ⎰ Proper palmar digital branches

Proper palmar digital branches
Dorsal digital branches
Dorsal branch
⎫ **Ulnar n.**

Radial n. ⎰ Dorsal digital branches
Superficial branch
Posterior antebrachial cutaneous n.

Boundary is variable

Medial antebrachial cutaneous n.

Musculocutaneous n. ⎰ Lateral antebrachial cutaneous n.

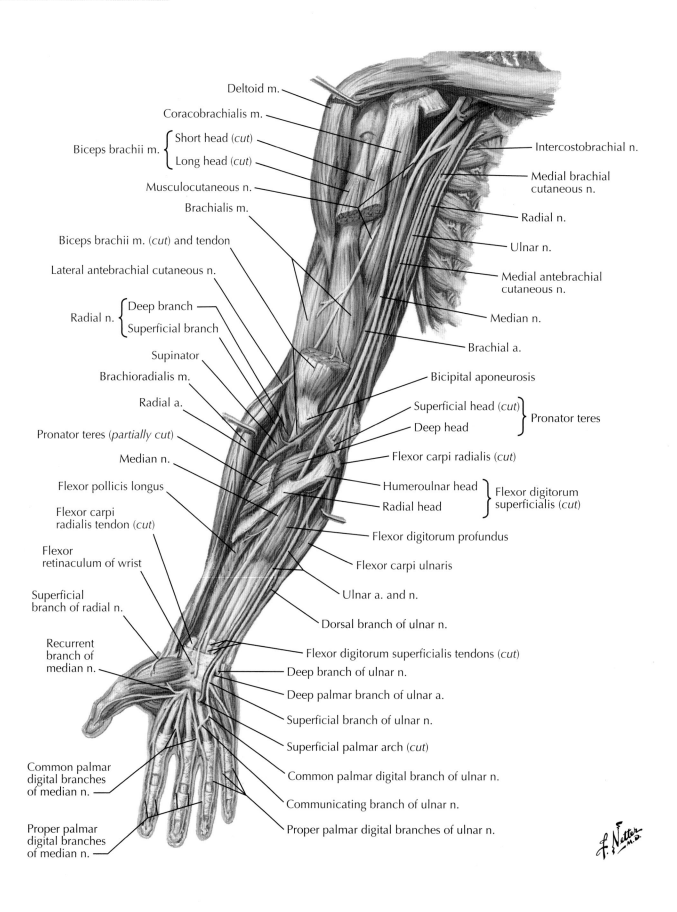

Deltoid m.

Coracobrachialis m.

Biceps brachii m. { Short head (*cut*)
                    Long head (*cut*)

Musculocutaneous n.

Brachialis m.

Biceps brachii m. (*cut*) and tendon

Lateral antebrachial cutaneous n.

Radial n. { Deep branch
            Superficial branch

Supinator

Brachioradialis m.

Radial a.

Pronator teres (*partially cut*)

Median n.

Flexor pollicis longus

Flexor carpi radialis tendon (*cut*)

Flexor retinaculum of wrist

Superficial branch of radial n.

Recurrent branch of median n.

Common palmar digital branches of median n.

Proper palmar digital branches of median n.

Intercostobrachial n.

Medial brachial cutaneous n.

Radial n.

Ulnar n.

Medial antebrachial cutaneous n.

Median n.

Brachial a.

Bicipital aponeurosis

Superficial head (*cut*)
Deep head } Pronator teres

Flexor carpi radialis (*cut*)

Humeroulnar head
Radial head } Flexor digitorum superficialis (*cut*)

Flexor digitorum profundus

Flexor carpi ulnaris

Ulnar a. and n.

Dorsal branch of ulnar n.

Flexor digitorum superficialis tendons (*cut*)

Deep branch of ulnar n.

Deep palmar branch of ulnar a.

Superficial branch of ulnar n.

Superficial palmar arch (*cut*)

Common palmar digital branch of ulnar n.

Communicating branch of ulnar n.

Proper palmar digital branches of ulnar n.

Plate 483

Nerves and Vasculature

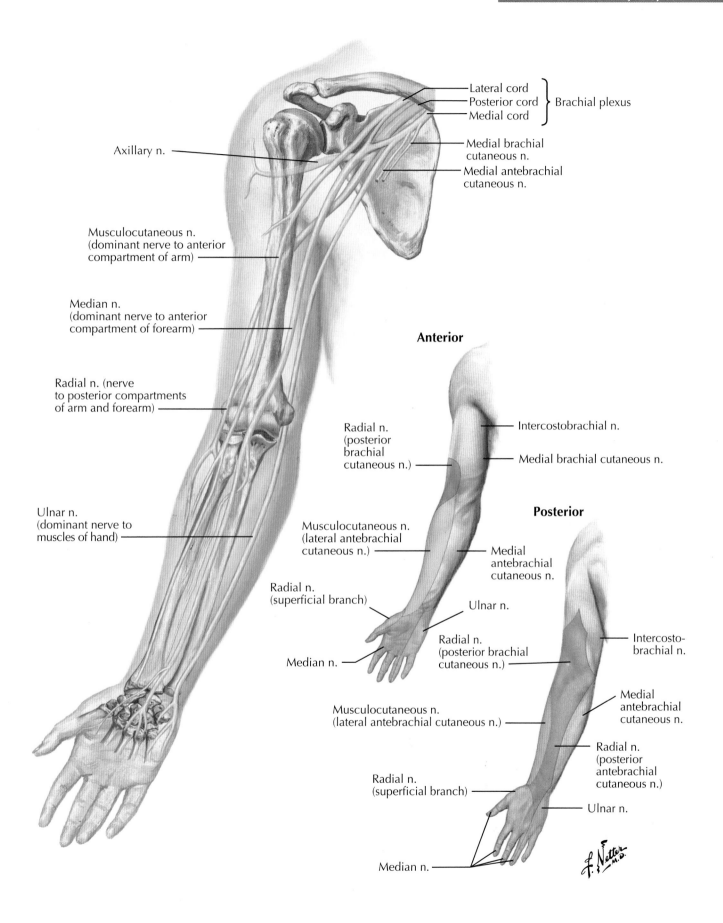

Lateral cord
Posterior cord
Medial cord
} Brachial plexus

Axillary n.

Medial brachial cutaneous n.

Medial antebrachial cutaneous n.

Musculocutaneous n. (dominant nerve to anterior compartment of arm)

Median n. (dominant nerve to anterior compartment of forearm)

Anterior

Radial n. (nerve to posterior compartments of arm and forearm)

Radial n. (posterior brachial cutaneous n.)

Intercostobrachial n.

Medial brachial cutaneous n.

Posterior

Ulnar n. (dominant nerve to muscles of hand)

Musculocutaneous n. (lateral antebrachial cutaneous n.)

Medial antebrachial cutaneous n.

Radial n. (superficial branch)

Ulnar n.

Radial n. (posterior brachial cutaneous n.)

Intercosto-brachial n.

Median n.

Medial antebrachial cutaneous n.

Musculocutaneous n. (lateral antebrachial cutaneous n.)

Radial n. (posterior antebrachial cutaneous n.)

Radial n. (superficial branch)

Ulnar n.

Median n.

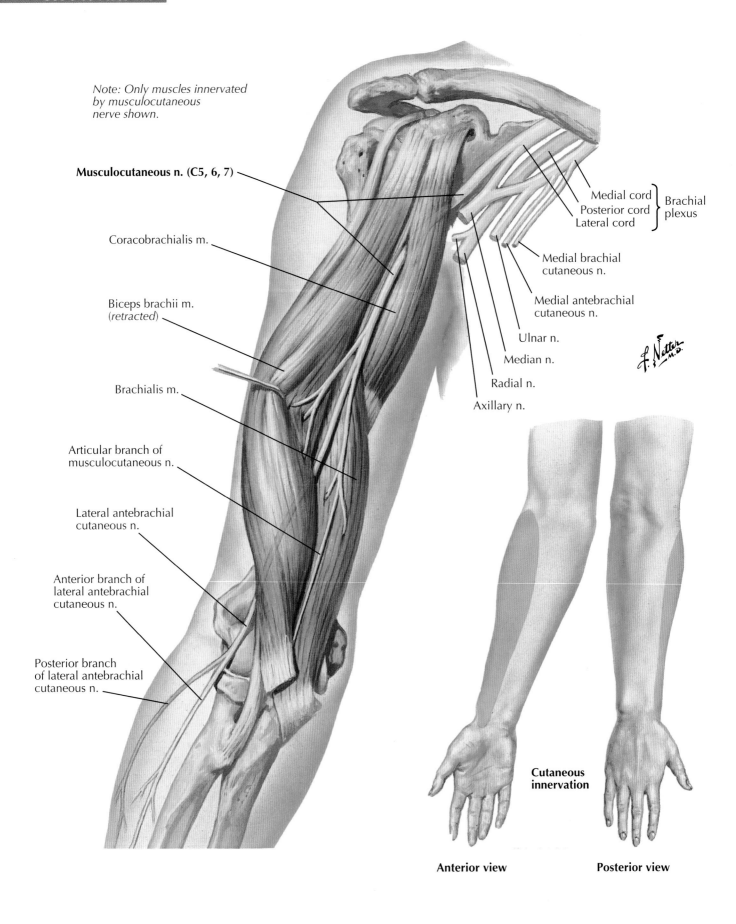

Note: Only muscles innervated by musculocutaneous nerve shown.

Musculocutaneous n. (C5, 6, 7)

Coracobrachialis m.

Biceps brachii m. (*retracted*)

Brachialis m.

Articular branch of musculocutaneous n.

Lateral antebrachial cutaneous n.

Anterior branch of lateral antebrachial cutaneous n.

Posterior branch of lateral antebrachial cutaneous n.

Medial cord
Posterior cord } Brachial plexus
Lateral cord

Medial brachial cutaneous n.

Medial antebrachial cutaneous n.

Ulnar n.

Median n.

Radial n.

Axillary n.

Cutaneous innervation

Anterior view

Posterior view

Plate 485

Nerves and Vasculature

See also **Plates 423, 456, 457**

Note: Only muscles innervated by median nerve shown.

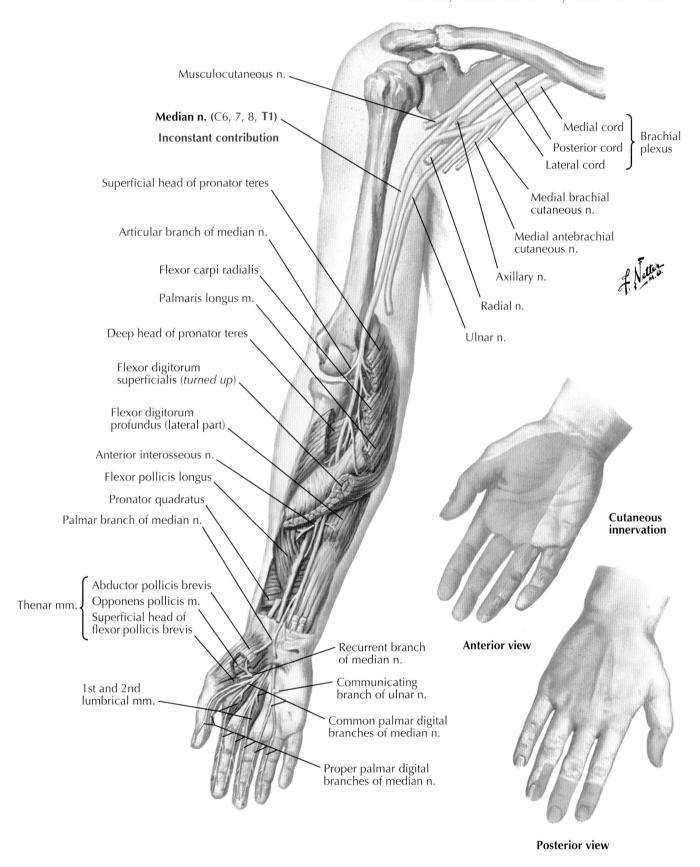

Musculocutaneous n.

Median n. (C6, 7, 8, **T1**)
Inconstant contribution

Superficial head of pronator teres

Articular branch of median n.

Flexor carpi radialis

Palmaris longus m.

Deep head of pronator teres

Flexor digitorum superficialis *(turned up)*

Flexor digitorum profundus (lateral part)

Anterior interosseous n.

Flexor pollicis longus

Pronator quadratus

Palmar branch of median n.

Thenar mm. {
Abductor pollicis brevis
Opponens pollicis m.
Superficial head of flexor pollicis brevis

1st and 2nd lumbrical mm.

Medial cord
Posterior cord } Brachial plexus
Lateral cord

Medial brachial cutaneous n.

Medial antebrachial cutaneous n.

Axillary n.

Radial n.

Ulnar n.

f. Netter M.D.

Cutaneous innervation

Recurrent branch of median n.

Communicating branch of ulnar n.

Common palmar digital branches of median n.

Proper palmar digital branches of median n.

Anterior view

Posterior view

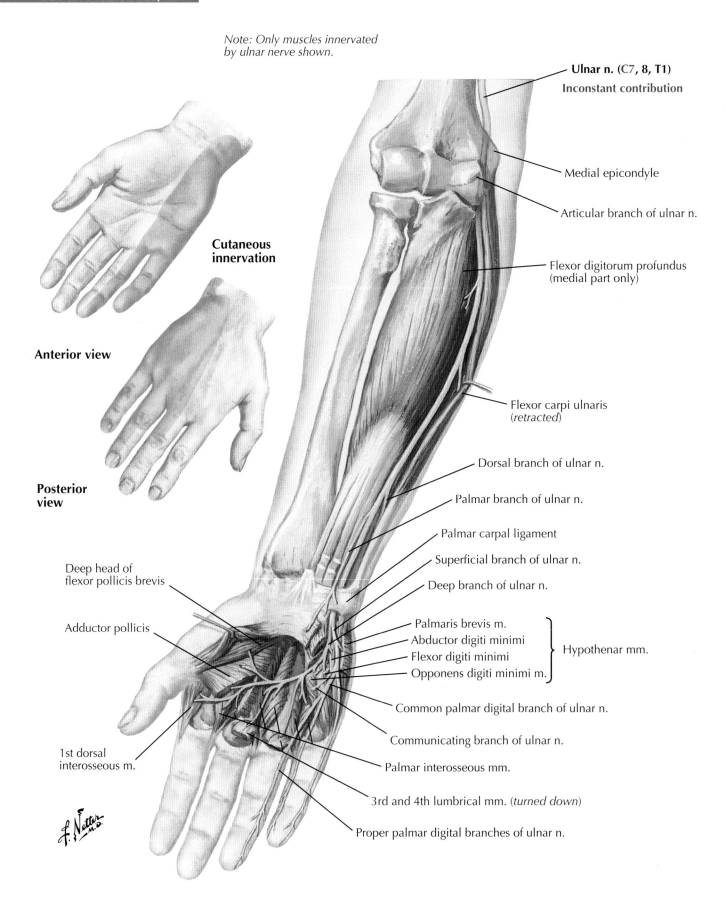

Note: Only muscles innervated by ulnar nerve shown.

Cutaneous innervation

Anterior view

Posterior view

Ulnar n. (C7, 8, T1)
Inconstant contribution

Medial epicondyle

Articular branch of ulnar n.

Flexor digitorum profundus (medial part only)

Flexor carpi ulnaris (*retracted*)

Dorsal branch of ulnar n.

Palmar branch of ulnar n.

Palmar carpal ligament

Superficial branch of ulnar n.

Deep branch of ulnar n.

Deep head of flexor pollicis brevis

Adductor pollicis

Palmaris brevis m.
Abductor digiti minimi
Flexor digiti minimi
Opponens digiti minimi m.

Hypothenar mm.

Common palmar digital branch of ulnar n.

Communicating branch of ulnar n.

Palmar interosseous mm.

1st dorsal interosseous m.

3rd and 4th lumbrical mm. (*turned down*)

Proper palmar digital branches of ulnar n.

Plate 487

Nerves and Vasculature

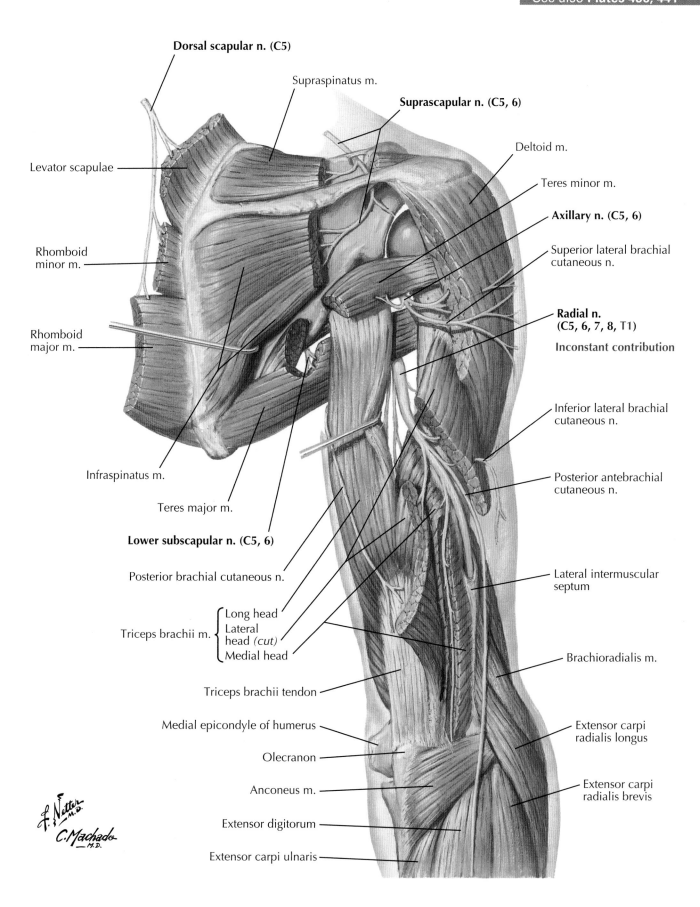

Dorsal scapular n. (C5)

Supraspinatus m.

Suprascapular n. (C5, 6)

Deltoid m.

Levator scapulae

Teres minor m.

Axillary n. (C5, 6)

Superior lateral brachial cutaneous n.

Rhomboid minor m.

Radial n. (C5, 6, 7, 8, T1)

Inconstant contribution

Rhomboid major m.

Inferior lateral brachial cutaneous n.

Infraspinatus m.

Posterior antebrachial cutaneous n.

Teres major m.

Lower subscapular n. (C5, 6)

Posterior brachial cutaneous n.

Lateral intermuscular septum

Triceps brachii m. { Long head Lateral head (cut) Medial head

Triceps brachii tendon

Brachioradialis m.

Medial epicondyle of humerus

Extensor carpi radialis longus

Olecranon

Anconeus m.

Extensor carpi radialis brevis

Extensor digitorum

Extensor carpi ulnaris

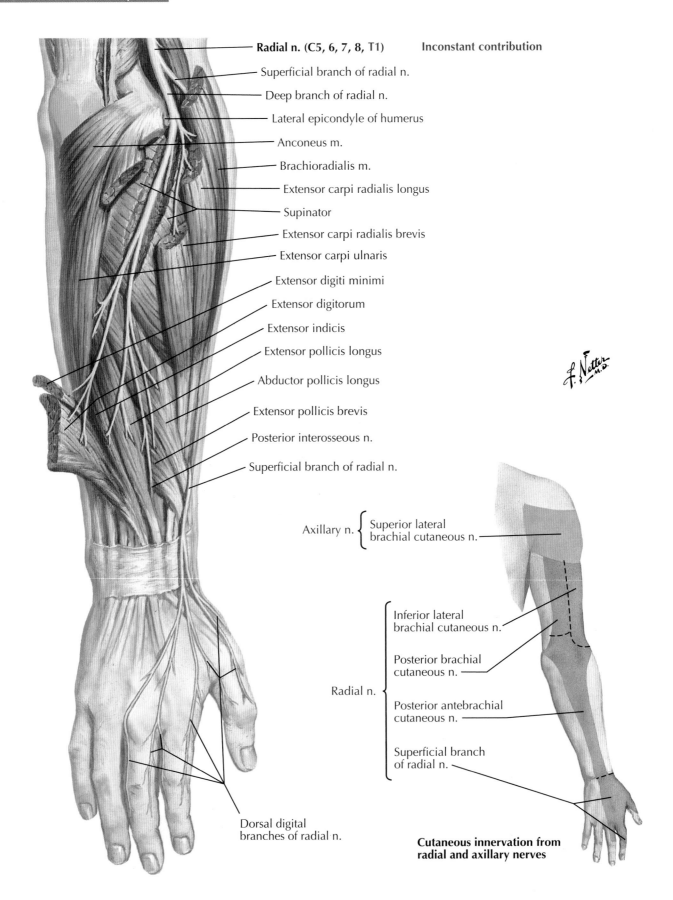

Radial n. (C5, 6, 7, 8, T1) Inconstant contribution

Superficial branch of radial n.

Deep branch of radial n.

Lateral epicondyle of humerus

Anconeus m.

Brachioradialis m.

Extensor carpi radialis longus

Supinator

Extensor carpi radialis brevis

Extensor carpi ulnaris

Extensor digiti minimi

Extensor digitorum

Extensor indicis

Extensor pollicis longus

Abductor pollicis longus

Extensor pollicis brevis

Posterior interosseous n.

Superficial branch of radial n.

Axillary n. { Superior lateral brachial cutaneous n.

Inferior lateral brachial cutaneous n.

Posterior brachial cutaneous n.

Radial n. { Posterior antebrachial cutaneous n.

Superficial branch of radial n.

Dorsal digital branches of radial n.

Cutaneous innervation from radial and axillary nerves

Plate 489

Nerves and Vasculature

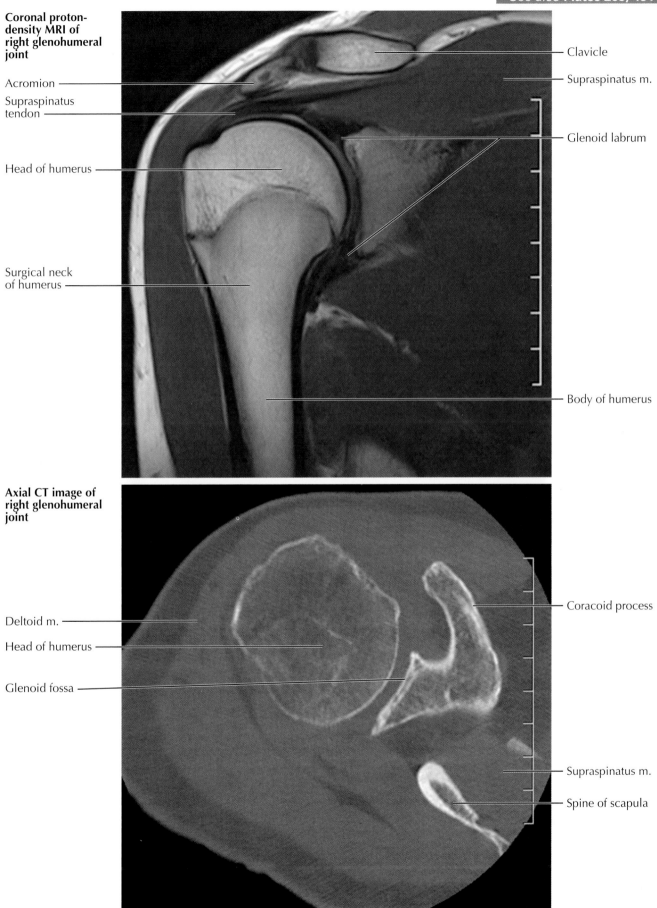

Coronal proton-density MRI of right glenohumeral joint

Acromion

Supraspinatus tendon

Head of humerus

Surgical neck of humerus

Clavicle

Supraspinatus m.

Glenoid labrum

Body of humerus

Axial CT image of right glenohumeral joint

Deltoid m.

Head of humerus

Glenoid fossa

Coracoid process

Supraspinatus m.

Spine of scapula

Structures with High* Clinical Significance

ANATOMICAL STRUCTURE	CLINICAL COMMENT	PLATE NUMBERS
Nervous System and Sense Organs		
Long thoracic nerve	Injury may produce "winged scapula" caused by denervation of serratus anterior muscle; can be injured with repetitive overhead motion	436, 438
Axillary nerve	Position of nerve close to surgical neck of humerus makes it vulnerable to injury with fractures or dislocations of humerus; poorly fitting crutches can also compress axillary nerve	441
Median nerve	Compressed in carpal tunnel syndrome, producing pain and paresthesia in lateral three and one-half digits; major risk factors include obesity, pregnancy, diabetes, and hypothyroidism	470, 486
Recurrent branch of median nerve	May be injured in superficial lacerations of palm over thenar eminence	469
Ulnar nerve	Vulnerable to compression or injury where it passes posterior to medial epicondyle of humerus, and at wrist in ulnar tunnel (Guyon's canal)	483, 487
Radial nerve	Vulnerable to compression or injury where it lies against humerus in radial groove (e.g., with humerus fracture); common symptom is wrist drop due to weakness of wrist extensors; poorly fitting crutches can also compress radial nerve	488, 489
Skeletal System		
Clavicle	Most clavicular fractures are caused from fall on outstretched arm or direct trauma delivered to lateral side of shoulder; middle third of clavicle is most commonly fractured; supraclavicular nerve block relieves acute pain associated with fracture	427, 429
Humerus	Proximal humerus, especially surgical neck, is fractured due to low-energy falls in elderly persons and high-energy trauma in young persons; axillary nerve and circumflex humeral arteries can be injured; hematoma from anterior/posterior circumflex humeral artery damage as result of dislocation may complicate reductions; midbody fractures are also relatively common and may affect radial nerve and/or deep brachial artery; distal humerus fractures may affect ulnar nerve medially and radial nerve laterally	427, 428, 430
Ulna	Subcutaneous location of olecranon makes it vulnerable to fracture by direct trauma, especially when elbow is flexed; ulnar styloid process may also be fractured with distal radial fractures	446, 449
Radius	Fractures of distal radius are most common fracture of upper limb (Colles' fracture), typically caused by fall on outstretched hand (FOOSH)	449
Scaphoid bone	Most commonly fractured carpal bone, typically from fall on outstretched hand (FOOSH)	459, 460, 462
Muscular System		
Palmar aponeurosis	Progressive fibrosis may result in nodules and eventually a palpable cord that limits finger extension (Dupuytren's contracture)	469
Rotator cuff muscles	Injuries to this group of muscles can result from acute injury or chronic overuse and are a common cause of shoulder pain and disability	431, 434, 441
Supraspinatus tendon	Most commonly torn rotator cuff tendon	434–436, 441
Biceps brachii tendon	Can rupture from sudden load on muscle when contracting; used in flexor compartment reflex assessing C5 and C6 spinal nerves	440, 442
Long head of biceps brachii muscle	Tendon of long head of biceps brachii muscle can cause shoulder pain from tendinosis of intraarticular portion and can rupture in elderly persons from falls on outstretched arm; when long head has been ruptured, it usually tears from supraglenoid tubercle and retracts down into arm; muscle commonly bulges (Popeye deformity) at midbody of humerus; spontaneous rupture may occur in amyloidosis, infiltrative disease that also causes cardiomyopathy	440

Table 7.1　　　　　　　　　　　　　　　　　　　　　　**Structures With High Clinical Significance**

ANATOMICAL STRUCTURE	CLINICAL COMMENT	PLATE NUMBERS
Muscles of posterior compartment of forearm	Repetitive use of muscles arising from common extensor origin can damage tendons and produce pain over lateral epicondyle region (epicondylitis); activities such as swinging tennis racquet and poor technique with hammer can result in "tennis elbow"; muscle most likely involved is extensor carpi radialis brevis	451
Muscles of anterior compartment of forearm	Repetitive use of muscles arising from common flexor origin can damage tendons and produce pain over medial epicondyle region (golfer's elbow)	452, 453
Cardiovascular System		
Median cubital vein	Accessed in cubital fossa for venipuncture	424
Suprascapular, dorsal scapular, and circumflex scapular arteries	Provide collateral circulation around scapula, allowing blood to reach distal part of upper limb if axillary artery is blocked or compressed	437
Brachial artery	During deflation of sphygmomanometer on upper arm, brachial artery is auscultated for Korotkoff sounds to measure systolic and diastolic blood pressure; identified medial to biceps brachii tendon and deep to bicipital aponeurosis in cubital fossa	442, 443
Radial artery	Palpated at lateral aspect of wrist to assess radial pulse; common site of vascular access for percutaneous cardiac procedures, such as angioplasty, and for sampling of arterial blood	14, 443
Ulnar artery	Provides important collateral circulation to hand via palmar arch during catheterization of radial artery; patency is assessed prior to procedure using Allen test, in which both radial and ulnar arteries are compressed, then pressure over ulnar artery is released; return of color to wrist within a few seconds indicates patent ulnar artery	443, 476

*Selections were based largely on clinical data and commonly discussed clinical correlations in macroscopic ("gross") anatomy courses.

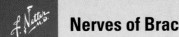

Nerves of Brachial Plexus

The roots of the brachial plexus are typically the anterior rami of the C5–T1 spinal nerves. Variation in the spinal nerve contributions to the plexus, and the nerves that arise from this plexus, is common, due to prefixed (high) and postfixed (low) plexuses.

NERVE	ORIGIN	COURSE	BRANCHES	MOTOR	CUTANEOUS
Dorsal scapular nerve	Anterior ramus of C5 spinal nerve	Pierces scalenus medius muscle to run posteriorly and inferiorly on levator scapulae along vertebral border of scapula		Rhomboid major and minor muscles, levator scapulae	
Long thoracic nerve	Anterior rami of C5–C7 spinal nerves	C5–C6 join within scalenus medius muscle, and at 1st rib are joined by C7; runs inferiorly and posterior to brachial plexus and axillary vessels; follows midaxillary line on surface of serratus anterior muscle		Serratus anterior muscle	
Suprascapular nerve	Superior trunk (C5–C6)	Traverses posterior cervical triangle, coursing posterior to inferior belly of omohyoid muscle and border of trapezius muscle to pass through scapular notch deep to superior transverse scapular ligament; continues laterally and then through spinoglenoid notch into infraspinous fossa		Supraspinatus and infraspinatus muscles	
Subclavian nerve	Superior trunk (C5–C6)	Runs inferiorly at distal aspect of anterior rami		Subclavius muscle	
Lateral pectoral nerve	Lateral cord (C5–C7)	Emerges lateral and superficial to axillary artery and vein, coursing just medial to pectoralis minor muscle		Pectoralis major and minor muscles	
Musculocutaneous nerve	Lateral cord (C5–C7)	Emerges at inferior border of pectoralis minor muscle, pierces coracobrachialis muscle to run between brachialis and biceps brachii muscles; just proximal to elbow, pierces deep fascia to continue as lateral antebrachial cutaneous nerve	Muscular branches, lateral antebrachial cutaneous nerve	Anterior compartment of arm	See lateral antebrachial cutaneous nerve
Lateral antebrachial cutaneous nerve	Musculocutaneous nerve	Runs posterior to cephalic vein and divides at elbow joint into anterior and posterior branches that travel along lateral surface of forearm	Anterior and posterior branches		Lateral forearm
Subscapular nerves	Posterior cord (C5–C6)	Upper and lower subscapular nerves emerge to traverse anterior surface of sub-scapularis muscle; lower subscapular nerve ends in teres major muscle		Teres major and subscapularis muscles	
Thoracodorsal nerve	Posterior cord (C6–C8)	Emerges between upper and lower subscapular nerves, courses with thoracodorsal artery along posterior axillary wall, diving deep to latissimus dorsi muscle		Latissimus dorsi muscle	
Radial nerve	Posterior cord (C5–T1)	Runs anterior to latissimus dorsi muscle to inferior border of teres major muscle, where it accompanies deep brachial artery along radial groove of humerus to course between medial and lateral heads of triceps brachii muscle	Posterior and inferior lateral brachial cutaneous nerves, posterior antebrachial cutaneous nerve, muscular, deep, and superficial branches and posterior interosseous nerve	Triceps brachii, anconeus, and brachioradialis muscles, extensores carpi radiales longus and brevis, supinator (also see posterior interosseous nerve)	Lateral part of dorsum of hand (also see cutaneous branches)
Posterior brachial cutaneous nerve	Radial nerve	Emerges from radial nerve in medial axilla			Posterior part of medial arm

Table 7.3

Nerves of Brachial Plexus

NERVE	ORIGIN	COURSE	BRANCHES	MOTOR	CUTANEOUS
Inferior lateral brachial cutaneous nerve	Radial nerve	Perforates lateral head of triceps brachii muscle below deltoid tuberosity, coursing anteriorly with cephalic vein			Distal part of lateral arm
Posterior antebrachial cutaneous nerve	Radial nerve	Emerges from plane between lateral and medial heads of triceps brachii muscle to become cutaneous			Posterior part of lateral forearm
Posterior interosseous nerve	Deep branch of radial nerve	Continuation of deep radial nerve courses under cover of supinator distally along posterior surface of interosseous membrane of forearm		Posterior compartment of forearm (some exceptions)	
Axillary nerve	Posterior cord (C5–C6)	Passes anterior to subscapularis muscle to exit axilla with posterior circumflex humeral artery through quadrangular space	Muscular branches, superior lateral brachial cutaneous nerve	Deltoid and teres minor muscles	See superior lateral brachial cutaneous nerve
Superior lateral brachial cutaneous nerve	Axillary nerve	Pierces deep fascia at posteroinferior edge of deltoid muscle to become cutaneous			Proximal part of lateral arm
Medial pectoral nerve	Medial cord (C8–T1)	Emerges and runs between axillary artery and vein to pierce pectoralis minor muscle en route to pectoralis major muscle		Pectoralis minor and major muscles	
Medial brachial cutaneous nerve	Medial cord (T1)	Emerges and traverses axilla anterior to latissimus dorsi muscle, running posteromedial with axillary vein, piercing deep fascia to descend with basilic vein	Anterior and posterior branches		Anterior part of medial arm
Medial antebrachial cutaneous nerve	Medial cord (C8–T1)	Emerges medial to axillary artery, traverses axilla to pierce deep fascia supplying anterior arm, and continues on ulnar side of forearm with basilic vein	Anterior and posterior branches		Anterior arm, medial part of forearm
Ulnar nerve	Medial cord (C7–T1)	Emerges medial to axillary artery, continuing medial to brachial artery along medial head of triceps brachii muscle in groove for ulnar nerve between olecranon and medial epicondyle; enters forearm between heads of flexor carpi ulnaris; runs distally between flexor carpi ulnaris and flexor digitorum profundus, giving off dorsal branch before entering hand	Muscular, dorsal, palmar, superficial, and deep branches	Flexor carpi ulnaris, flexor digitorum profundus (medial half), adductor pollicis, hypothenar muscles, dorsal and palmar interosseous muscles, lumbrical muscles (medial two)	Medial part of palm and dorsum of hand, 5th finger and part of 4th
Median nerve	Medial and lateral cords (C6–T1)	Emerges and runs distally with brachial artery to enter forearm between heads of pronator teres; courses distally on deep surface of flexor digitorum superficialis to become superficial at flexor retinaculum of wrist; traverses carpal tunnel deep to flexor retinaculum of wrist	Anterior interosseous nerve, muscular, palmar, recurrent, and common palmar digital branches	Anterior compartment of forearm (*some exceptions*), lumbricals (lateral two), and thenar muscles (also see anterior interosseous nerve)	Lateral part of palm, thumb, 2nd and 3rd fingers, and part of 4th finger
Anterior interosseous nerve	Median nerve	At elbow runs distally with anterior interosseous artery along anterior surface of interosseous membrane of forearm		Flexor pollicis longus, pronator quadratus, flexor digitorum profundus (lateral half)	

MUSCLE	MUSCLE GROUP	PROXIMAL ATTACHMENT	DISTAL ATTACHMENT	INNERVATION	BLOOD SUPPLY	MAIN ACTIONS
Abductor digiti minimi of hand	Hand	Pisiform bone, tendon of flexor carpi ulnaris	Medial surface of base of proximal phalanx of little finger (5th digit)	Ulnar nerve (deep branch)	Deep palmar branch of ulnar artery	Abducts little finger
Abductor pollicis brevis	Hand	Flexor retinaculum of wrist, tubercles of scaphoid and trapezium bones	Base of proximal phalanx of thumb	Median nerve (recurrent branch)	Superficial palmar branch of radial artery	Abducts thumb
Abductor pollicis longus	Posterior forearm	Posterior surfaces of ulna, radius, and interosseous membrane of forearm	Base of 1st metacarpal	Posterior interosseous nerve	Posterior interosseous artery	Abducts and extends thumb
Adductor pollicis	Hand	*Oblique head:* bases of 2nd and 3rd metacarpals, capitate and adjacent carpal bones *Transverse head:* anterior surface of 3rd metacarpal	Base of proximal phalanx of thumb	Ulnar nerve (deep branch)	Deep palmar arch	Adducts thumb
Anconeus m.	Posterior forearm	Posterior surface of lateral epicondyle of humerus	Lateral surface of olecranon, posterior surface of proximal ulna	Radial nerve	Deep brachial artery	Assists triceps brachii m. in extending elbow
Biceps brachii m.	Anterior arm	*Long head:* supraglenoid tubercle of scapula *Short head:* coracoid process of scapula	Radial tuberosity, fascia of forearm (via bicipital aponeurosis)	Musculocutaneous nerve	Brachial artery	Flexes and supinates forearm
Brachialis m.	Anterior arm	Distal half of anterior surface of humerus	Coronoid process and tuberosity of ulna	Musculocutaneous and radial nerves	Radial recurrent and brachial arteries	Flexes forearm
Brachioradialis m.	Posterior forearm	Proximal two thirds of lateral supracondylar ridge of humerus	Lateral surface of distal end of radius	Radial nerve	Radial recurrent artery	Weak flexion of forearm when forearm is in midpronation
Coracobrachialis m.	Anterior arm	Coracoid process of scapula	Middle one third of medial surface of humerus	Musculocutaneous nerve	Brachial artery	Flexes and adducts arm
Deltoid m.	Shoulder	*Clavicular part:* lateral one third of clavicle *Acromial part:* acromion *Spinous part:* spine of scapula	Deltoid tuberosity of humerus	Axillary nerve	Posterior circumflex humeral artery Deltoid branch of thoraco-acromial artery	*Clavicular part:* flexes and medially rotates arm *Acromial part:* abducts arm beyond initial 15 degrees done by supraspinatus m. *Spinous part:* extends and laterally rotates arm
Dorsal interosseous mm. of hand	Hand	Facing surfaces of adjacent metacarpal bones	Bases of proximal phalanges, and extensor expansions of digits 2–4	Ulnar nerve (deep branch)	Deep palmar arch	Abduct digits; flex digits at metacarpophalangeal joint and extend interphalangeal joints
Extensor carpi radialis brevis	Posterior forearm	Lateral epicondyle of humerus	Bases of 3rd and 2nd metacarpal bones	Radial nerve (deep branch)	Radial and radial recurrent arteries	Extends and abducts hand
Extensor carpi radialis longus	Posterior forearm	Distal one third of lateral supracondylar ridge of humerus	Base of 2nd metacarpal	Radial nerve	Radial and radial recurrent arteries	Extends and abducts hand

Table 7.5

Muscles

MUSCLE	MUSCLE GROUP	PROXIMAL ATTACHMENT	DISTAL ATTACHMENT	INNERVATION	BLOOD SUPPLY	MAIN ACTIONS
Extensor carpi ulnaris	Posterior forearm	Lateral epicondyle of humerus, posterior border of ulna	Base of 5th metacarpal bone	Posterior interosseous nerve	Posterior interosseous artery	Extends and adducts hand
Extensor digiti minimi	Posterior forearm	Lateral epicondyle of humerus	Extensor expansion of little finger	Posterior interosseous nerve	Posterior interosseous artery	Extends 5th digit
Extensor digitorum	Posterior forearm	Lateral epicondyle of humerus	Extensor expansions of digits 2–5	Posterior interosseous nerve	Posterior interosseous artery	Extends medial four metacarpophalangeal joints, assists in wrist extension
Extensor indicis	Posterior forearm	Posterior surfaces of ulna and interosseous membrane of forearm	Extensor expansion of 2nd digit	Posterior interosseous nerve	Posterior interosseous artery	Extends 2nd digit and helps extend hand
Extensor pollicis brevis	Posterior forearm	Posterior surfaces of radius and interosseous membrane of forearm	Dorsal surface of base of proximal phalanx of thumb	Posterior interosseous nerve	Posterior interosseous artery	Extends proximal phalanx of thumb
Extensor pollicis longus	Posterior forearm	Posterior surfaces of middle one third of ulna and interosseous membrane of forearm	Dorsal surface of base of distal phalanx of thumb	Posterior interosseous nerve	Posterior interosseous artery	Extends distal phalanx of thumb
Flexor carpi radialis	Anterior forearm	Medial epicondyle of humerus	Base of 2nd metacarpal bone	Median nerve	Radial artery	Flexes and abducts hand
Flexor carpi ulnaris	Anterior forearm	*Superficial head:* medial epicondyle of humerus *Deep head:* olecranon and posterior border of ulna	Pisiform bone, hook of hamate bone, base of 5th metacarpal bone	Ulnar nerve	Posterior ulnar recurrent artery	Flexes and adducts hand
Flexor digiti minimi of hand	Hand	Flexor retinaculum of wrist, hook of hamate bone	Medial surface of base of proximal phalanx of little finger	Ulnar nerve (deep branch)	Deep palmar branch of ulnar artery	Flexes proximal phalanx of little finger
Flexor digitorum profundus	Anterior forearm	Medial and anterior surfaces of proximal three fourths of ulna, interosseous membrane of forearm	Palmar surfaces of bases of distal phalanges of digits 2–5	Medial part: ulnar nerve Lateral part: median nerve	Anterior interosseous and ulnar arteries	Flexes distal phalanges of medial four digits, assists with flexion of hand
Flexor digitorum superficialis	Anterior forearm	*Humeroulnar head:* medial epicondyle of humerus, coronoid process of ulna, ulnar collateral ligament *Radial head:* anterior surface of proximal radius	Bodies of middle phalanges of medial four digits	Median nerve	Ulnar and radial arteries	Flexes middle and proximal phalanges of medial four digits, flexes hand
Flexor pollicis brevis	Hand	*Superficial head:* flexor retinaculum, tubercle of trapezium bone *Deep head:* trapezoid and capitate bones	Lateral surface of base of proximal phalanx of thumb	Superficial head: median nerve (recurrent branch) Deep head: ulnar nerve (deep branch)	Superficial palmar branch of radial artery	Flexes proximal phalanx of thumb
Flexor pollicis longus	Anterior forearm	Anterior surfaces of radius and interosseous membrane	Palmar base of distal phalanx of thumb	Anterior interosseous nerve	Anterior interosseous artery	Flexes thumb

Continued

MUSCLE	MUSCLE GROUP	PROXIMAL ATTACHMENT	DISTAL ATTACHMENT	INNERVATION	BLOOD SUPPLY	MAIN ACTIONS
Infraspinatus m.	Shoulder	Infraspinous fossa of scapula, infraspinatus fascia	Greater tubercle of humerus	Suprascapular nerve	Suprascapular artery	Lateral rotation of arm
Lumbrical mm. of hand	Hand	Tendons of flexor digitorum profundus	Lateral sides of extensor expansions of digits 2–5	*Lateral two:* median nerve (digital branches) / *Medial two:* ulnar nerve (deep branch)	Superficial and deep palmar arches	Extend digits, flex metacarpophalangeal joints
Opponens digiti minimi m. of hand	Hand	Flexor retinaculum of wrist, hook of hamate bone	Palmar surface of 5th metacarpal bone	Ulnar nerve (deep branch)	Deep palmar branch of ulnar artery	Draws 5th metacarpal bone anteriorly and rotates it to face thumb
Opponens pollicis m.	Hand	Flexor retinaculum of wrist, tubercle of trapezium bone	Lateral surface of 1st metacarpal bone	Median nerve (recurrent branch)	Superficial palmar branch of radial artery	Draws 1st metacarpal forward and rotates it medially
Palmar interosseous mm.	Hand	Palmar surfaces of metacarpal bones 2, 4, and 5	Bases of proximal phalanges and extensor expansions of digits 2, 4, and 5	Ulnar nerve (deep branch)	Deep palmar arch	Adduct digits; flex digits and extend interphalangeal joints
Palmaris brevis m.	Hand	Palmar aponeurosis, flexor retinaculum	Skin of medial border of palm	Ulnar nerve (superficial branch)	Superficial palmar arch	Deepens hollow of hand, assists grip
Palmaris longus m.	Anterior forearm	Medial epicondyle of humerus	Distal half of flexor retinaculum of wrist, palmar aponeurosis	Median nerve	Posterior ulnar recurrent artery	Flexes hand and tenses palmar aponeurosis
Pronator quadratus	Anterior forearm	Distal one fourth of anterior surface of ulna	Distal one fourth of anterior surface of radius	Anterior interosseous nerve	Anterior interosseous artery	Pronates forearm
Pronator teres	Anterior forearm	*Humeral head:* medial epicondyle of humerus / *Ulnar head:* coronoid process of ulna	Middle part of lateral surface of radius	Median nerve	Anterior ulnar recurrent artery	Pronates forearm and flexes elbow
Subscapularis m.	Shoulder	Subscapular fossa	Lesser tubercle of humerus	Upper and lower subscapular nerves	Subscapular and lateral thoracic arteries	Medially rotates and adducts arm; helps hold humeral head in glenoid fossa
Supinator	Posterior forearm	Lateral epicondyle of humerus, radial collateral and annular ligaments, supinator fossa, and crest of ulna	Lateral, posterior, and anterior surfaces of proximal one third of radius	Radial nerve	Radial recurrent and posterior interosseous arteries	Supinates forearm
Supraspinatus m.	Shoulder	Supraspinous fossa of scapula, supraspinatus fascia	Greater tubercle of humerus	Suprascapular nerve	Suprascapular artery	Initiates arm abduction
Teres major m.	Shoulder	Posterior surface of inferior angle of scapula	Medial lip of intertubercular sulcus of humerus	Lower subscapular nerve	Circumflex scapular artery	Adducts and medially rotates arm
Teres minor m.	Shoulder	Superior two-thirds of posterior surface of lateral border of scapula	Greater tubercle of humerus	Axillary nerve	Circumflex scapular artery	Laterally rotates arm

MUSCLE	MUSCLE GROUP	PROXIMAL ATTACHMENT	DISTAL ATTACHMENT	INNERVATION	BLOOD SUPPLY	MAIN ACTIONS
Triceps brachii m.	Posterior arm	*Long head:* infraglenoid tubercle of scapula	Posterior surface of olecranon	Radial nerve	Deep brachial artery	Extends forearm; long head stabilizes head of abducted humerus and extends and adducts arm
		Lateral head: proximal half of posterior humerus				
		Medial head: distal two thirds of medial and posterior humerus				

Variations in spinal nerve contributions to the innervation of muscles, their arterial supply, their attachments, and their actions are common themes in human anatomy. Therefore, expect differences between texts and realize that anatomical variation is normal.

LOWER LIMB 8

ELECTRONIC BONUS PLATES

BP 103 Arteries of Knee and Foot

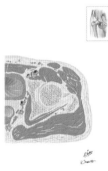

BP 104 Cross-Sectional Anatomy of Hip: Axial View

BP 105 Arteries of Thigh and Knee

BP 106 Leg: Serial Cross Sections

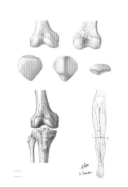

BP 107 Osteology of Knee

BP 108 Knee Radiograph: Lateral View

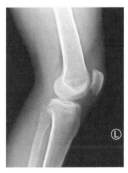

BP 109 Foot: Nerves and Arteries

BP 110 Cross-Sectional Anatomy of Ankle and Foot

ELECTRONIC BONUS PLATES—*cont'd*

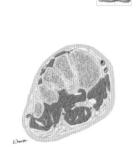

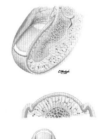

BP 111 Cross-Sectional Anatomy of Ankle and Foot (continued)

BP 112 Anatomy of Toenail

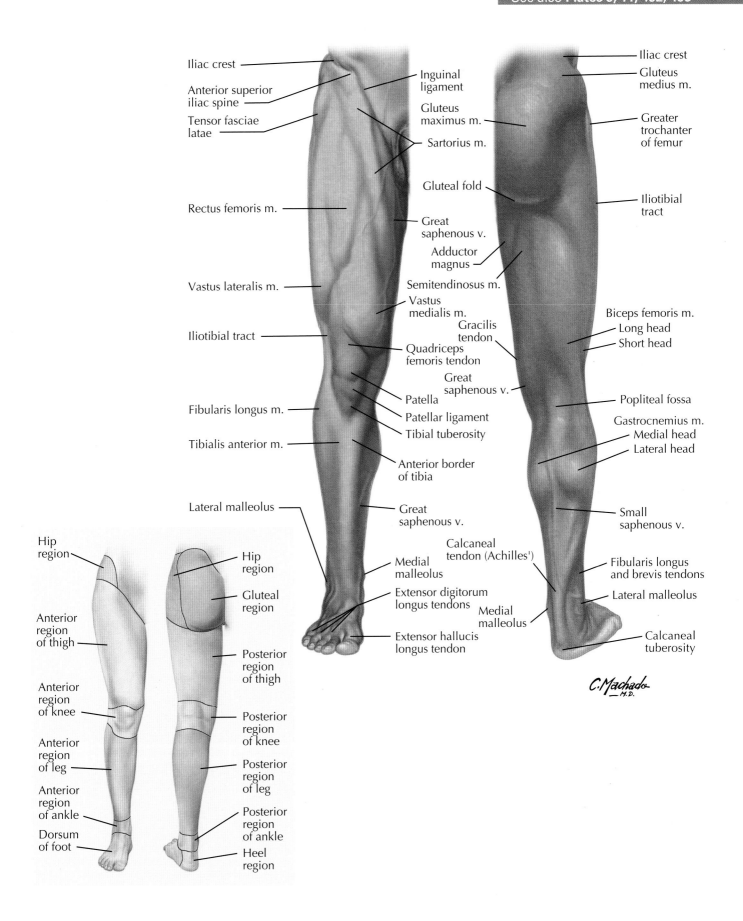

Iliac crest

Anterior superior iliac spine

Tensor fasciae latae

Rectus femoris m.

Vastus lateralis m.

Iliotibial tract

Fibularis longus m.

Tibialis anterior m.

Lateral malleolus

Inguinal ligament

Gluteus maximus m.

Sartorius m.

Gluteal fold

Great saphenous v.

Adductor magnus

Semitendinosus m.

Vastus medialis m.

Gracilis tendon

Quadriceps femoris tendon

Great saphenous v.

Patella

Patellar ligament

Tibial tuberosity

Anterior border of tibia

Great saphenous v.

Medial malleolus

Extensor digitorum longus tendons

Extensor hallucis longus tendon

Iliac crest

Gluteus medius m.

Greater trochanter of femur

Iliotibial tract

Biceps femoris m.
Long head
Short head

Popliteal fossa

Gastrocnemius m.
Medial head
Lateral head

Small saphenous v.

Calcaneal tendon (Achilles')

Medial malleolus

Fibularis longus and brevis tendons

Lateral malleolus

Calcaneal tuberosity

C. Machado — M.D.

Hip region

Anterior region of thigh

Anterior region of knee

Anterior region of leg

Anterior region of ankle

Dorsum of foot

Hip region

Gluteal region

Posterior region of thigh

Posterior region of knee

Posterior region of leg

Posterior region of ankle

Heel region

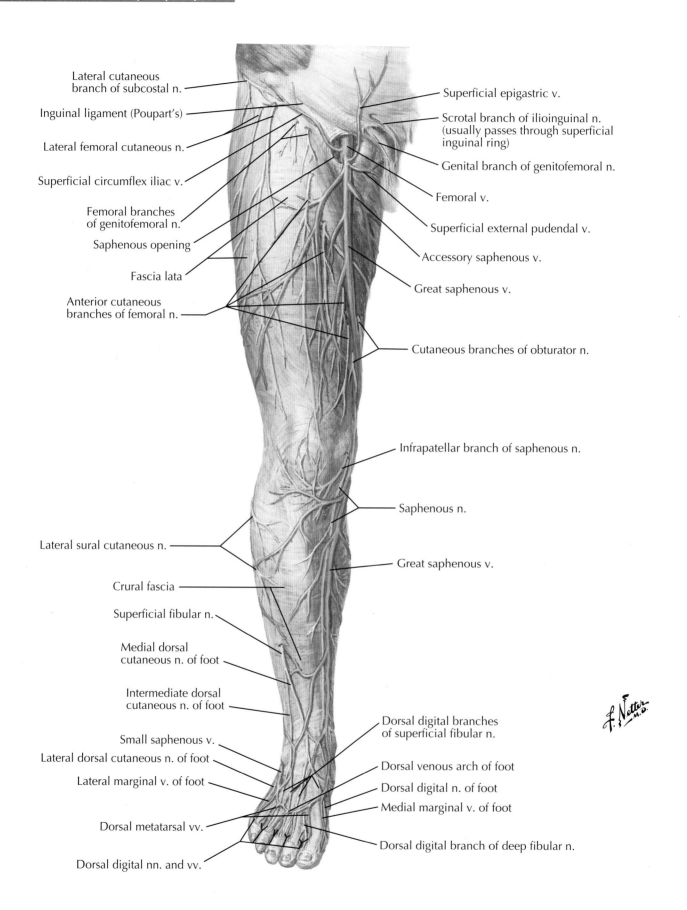

Lateral cutaneous branch of subcostal n.

Inguinal ligament (Poupart's)

Lateral femoral cutaneous n.

Superficial circumflex iliac v.

Femoral branches of genitofemoral n.

Saphenous opening

Fascia lata

Anterior cutaneous branches of femoral n.

Superficial epigastric v.

Scrotal branch of ilioinguinal n. (usually passes through superficial inguinal ring)

Genital branch of genitofemoral n.

Femoral v.

Superficial external pudendal v.

Accessory saphenous v.

Great saphenous v.

Cutaneous branches of obturator n.

Infrapatellar branch of saphenous n.

Saphenous n.

Lateral sural cutaneous n.

Great saphenous v.

Crural fascia

Superficial fibular n.

Medial dorsal cutaneous n. of foot

Intermediate dorsal cutaneous n. of foot

Dorsal digital branches of superficial fibular n.

Small saphenous v.

Lateral dorsal cutaneous n. of foot

Lateral marginal v. of foot

Dorsal metatarsal vv.

Dorsal digital nn. and vv.

Dorsal venous arch of foot

Dorsal digital n. of foot

Medial marginal v. of foot

Dorsal digital branch of deep fibular n.

Plate 492

Surface Anatomy

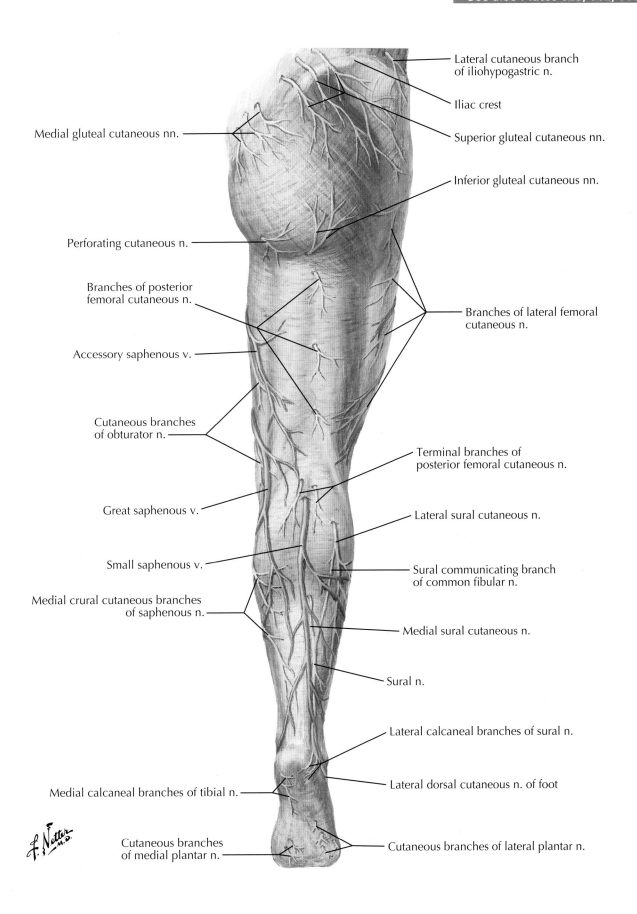

Lateral cutaneous branch
of iliohypogastric n.

Iliac crest

Superior gluteal cutaneous nn.

Medial gluteal cutaneous nn.

Inferior gluteal cutaneous nn.

Perforating cutaneous n.

Branches of posterior
femoral cutaneous n.

Branches of lateral femoral
cutaneous n.

Accessory saphenous v.

Cutaneous branches
of obturator n.

Terminal branches of
posterior femoral cutaneous n.

Great saphenous v.

Lateral sural cutaneous n.

Small saphenous v.

Sural communicating branch
of common fibular n.

Medial crural cutaneous branches
of saphenous n.

Medial sural cutaneous n.

Sural n.

Lateral calcaneal branches of sural n.

Lateral dorsal cutaneous n. of foot

Medial calcaneal branches of tibial n.

Cutaneous branches
of medial plantar n.

Cutaneous branches of lateral plantar n.

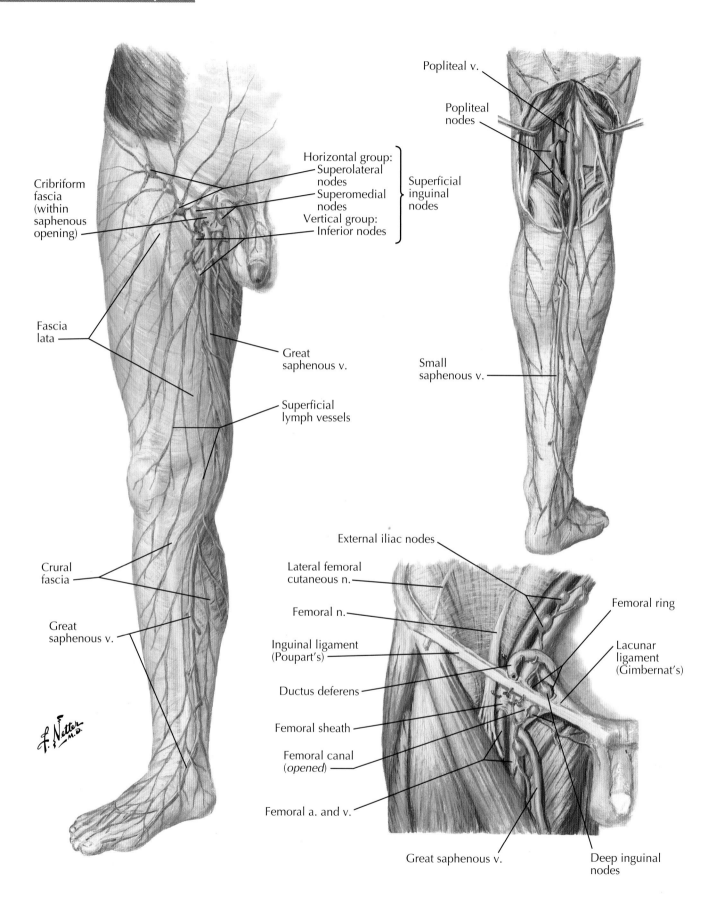

Cribriform
fascia
(within
saphenous
opening)

Horizontal group:
Superolateral
nodes
Superomedial
nodes
Vertical group:
Inferior nodes

Superficial
inguinal
nodes

Popliteal v.

Popliteal
nodes

Fascia
lata

Great
saphenous v.

Superficial
lymph vessels

Small
saphenous v.

Crural
fascia

Great
saphenous v.

External iliac nodes

Lateral femoral
cutaneous n.

Femoral ring

Femoral n.

Inguinal ligament
(Poupart's)

Lacunar
ligament
(Gimbernat's)

Ductus deferens

Femoral sheath

Femoral canal
(*opened*)

Femoral a. and v.

Great saphenous v.

Deep inguinal
nodes

Plate 494

Surface Anatomy

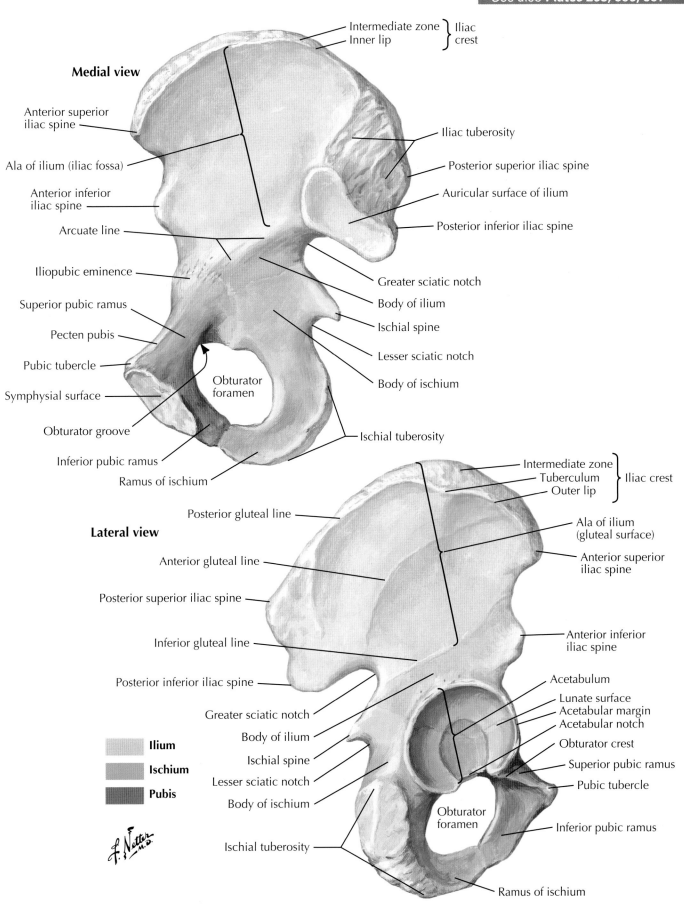

Medial view

Intermediate zone ⎱ Iliac
Inner lip ⎰ crest

Anterior superior iliac spine

Iliac tuberosity

Ala of ilium (iliac fossa)

Posterior superior iliac spine

Anterior inferior iliac spine

Auricular surface of ilium

Arcuate line

Posterior inferior iliac spine

Iliopubic eminence

Greater sciatic notch

Superior pubic ramus

Body of ilium

Pecten pubis

Ischial spine

Pubic tubercle

Lesser sciatic notch

Symphysial surface

Body of ischium

Obturator foramen

Obturator groove

Inferior pubic ramus

Ischial tuberosity

Ramus of ischium

Posterior gluteal line

Intermediate zone ⎱
Tuberculum ⎰ Iliac crest
Outer lip

Lateral view

Ala of ilium (gluteal surface)

Anterior gluteal line

Anterior superior iliac spine

Posterior superior iliac spine

Inferior gluteal line

Anterior inferior iliac spine

Posterior inferior iliac spine

Acetabulum

Greater sciatic notch

Lunate surface

Body of ilium

Acetabular margin

Ischial spine

Acetabular notch

Lesser sciatic notch

Obturator crest

Body of ischium

Superior pubic ramus

Pubic tubercle

Ilium

Ischium

Pubis

Obturator foramen

Inferior pubic ramus

Ischial tuberosity

Ramus of ischium

Hip, Buttock, and Thigh

Plate 495

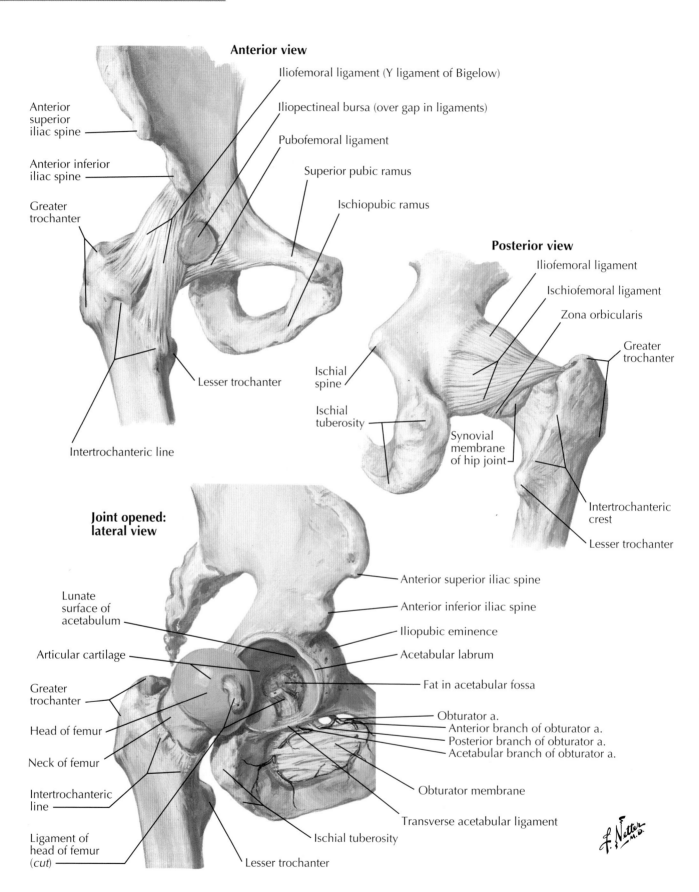

Anterior view

Anterior superior iliac spine

Anterior inferior iliac spine

Greater trochanter

Iliofemoral ligament (Y ligament of Bigelow)

Iliopectineal bursa (over gap in ligaments)

Pubofemoral ligament

Superior pubic ramus

Ischiopubic ramus

Lesser trochanter

Intertrochanteric line

Posterior view

Iliofemoral ligament

Ischiofemoral ligament

Zona orbicularis

Greater trochanter

Ischial spine

Ischial tuberosity

Synovial membrane of hip joint

Intertrochanteric crest

Lesser trochanter

Joint opened: lateral view

Lunate surface of acetabulum

Articular cartilage

Greater trochanter

Head of femur

Neck of femur

Intertrochanteric line

Ligament of head of femur (*cut*)

Anterior superior iliac spine

Anterior inferior iliac spine

Iliopubic eminence

Acetabular labrum

Fat in acetabular fossa

Obturator a.

Anterior branch of obturator a.

Posterior branch of obturator a.

Acetabular branch of obturator a.

Obturator membrane

Transverse acetabular ligament

Ischial tuberosity

Lesser trochanter

Plate 496

Hip, Buttock, and Thigh

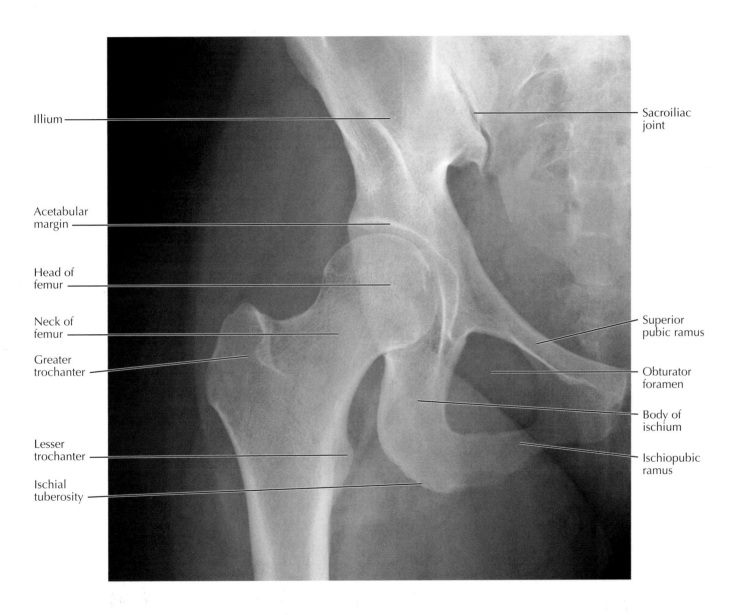

Illium

Acetabular margin

Head of femur

Neck of femur

Greater trochanter

Lesser trochanter

Ischial tuberosity

Sacroiliac joint

Superior pubic ramus

Obturator foramen

Body of ischium

Ischiopubic ramus

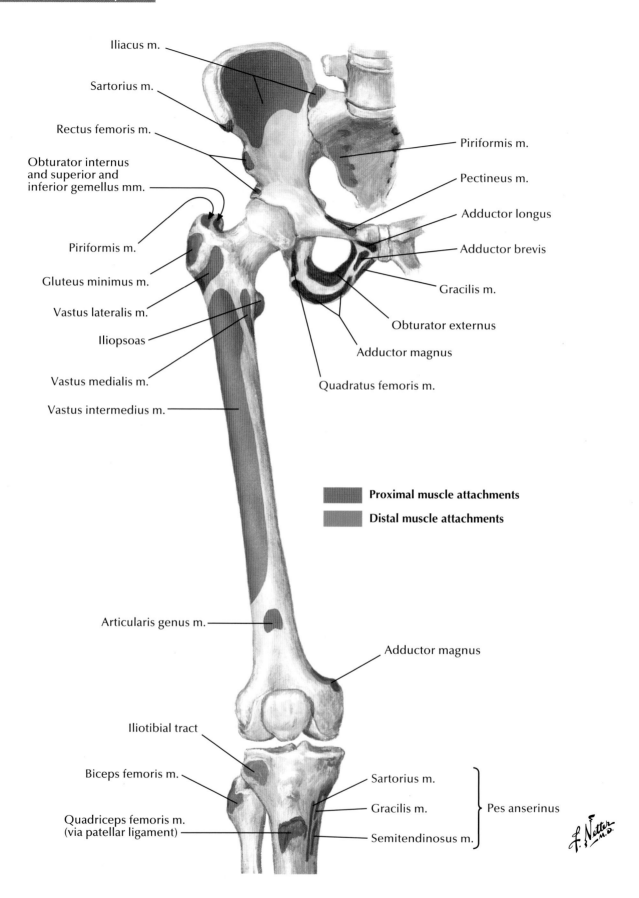

Iliacus m.

Sartorius m.

Rectus femoris m.

Obturator internus
and superior and
inferior gemellus mm.

Piriformis m.

Gluteus minimus m.

Vastus lateralis m.

Iliopsoas

Vastus medialis m.

Vastus intermedius m.

Piriformis m.

Pectineus m.

Adductor longus

Adductor brevis

Gracilis m.

Obturator externus

Adductor magnus

Quadratus femoris m.

Proximal muscle attachments

Distal muscle attachments

Articularis genus m.

Adductor magnus

Iliotibial tract

Biceps femoris m.

Sartorius m.

Gracilis m.

Pes anserinus

Quadriceps femoris m.
(via patellar ligament)

Semitendinosus m.

Plate 498

Hip, Buttock, and Thigh

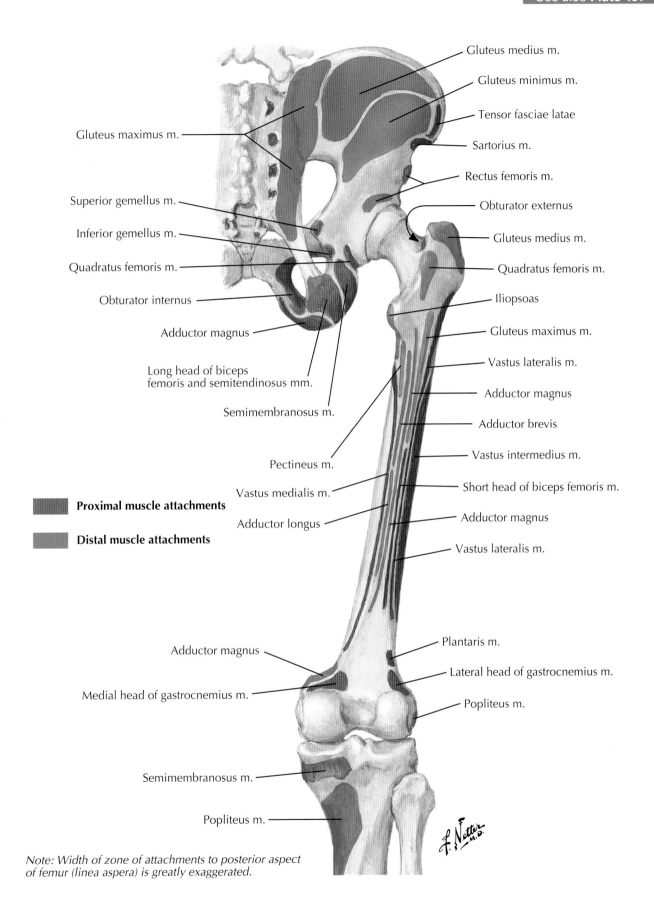

Gluteus medius m.

Gluteus minimus m.

Tensor fasciae latae

Sartorius m.

Rectus femoris m.

Obturator externus

Gluteus medius m.

Quadratus femoris m.

Iliopsoas

Gluteus maximus m.

Vastus lateralis m.

Adductor magnus

Adductor brevis

Vastus intermedius m.

Short head of biceps femoris m.

Adductor magnus

Vastus lateralis m.

Gluteus maximus m.

Superior gemellus m.

Inferior gemellus m.

Quadratus femoris m.

Obturator internus

Adductor magnus

Long head of biceps femoris and semitendinosus mm.

Semimembranosus m.

Pectineus m.

Vastus medialis m.

Adductor longus

Proximal muscle attachments

Distal muscle attachments

Plantaris m.

Lateral head of gastrocnemius m.

Popliteus m.

Adductor magnus

Medial head of gastrocnemius m.

Semimembranosus m.

Popliteus m.

Note: Width of zone of attachments to posterior aspect of femur (linea aspera) is greatly exaggerated.

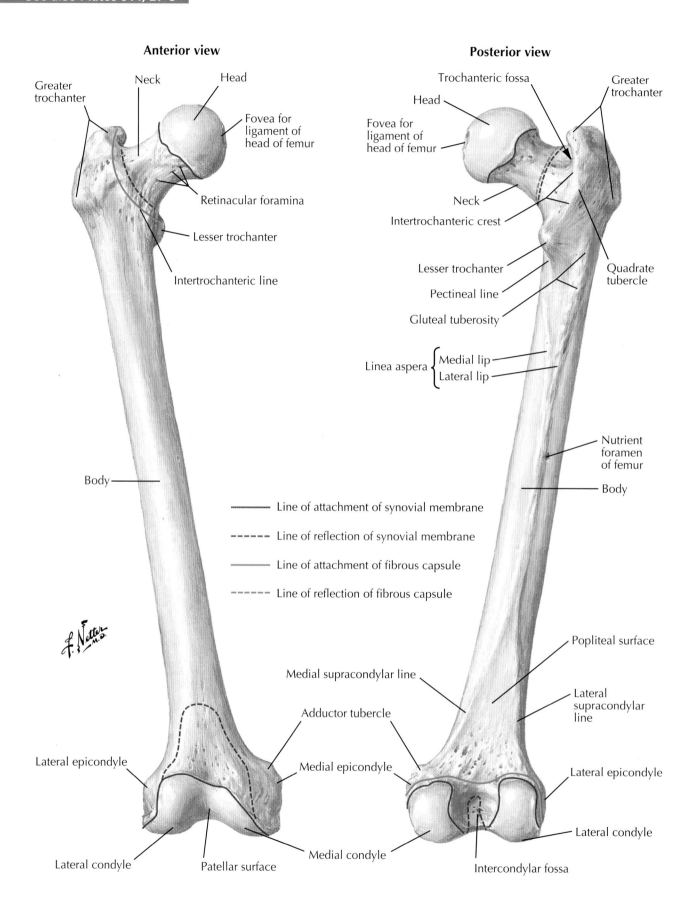

Anterior view

Greater trochanter
Neck
Head
Fovea for ligament of head of femur
Retinacular foramina
Lesser trochanter
Intertrochanteric line
Body
Lateral epicondyle
Lateral condyle
Patellar surface

Posterior view

Trochanteric fossa
Head
Fovea for ligament of head of femur
Greater trochanter
Neck
Intertrochanteric crest
Lesser trochanter
Pectineal line
Gluteal tuberosity
Quadrate tubercle
Linea aspera { Medial lip / Lateral lip
Nutrient foramen of femur
Body
Popliteal surface
Lateral supracondylar line
Lateral epicondyle
Lateral condyle
Intercondylar fossa

Medial supracondylar line
Adductor tubercle
Medial epicondyle
Medial condyle

——— Line of attachment of synovial membrane

- - - - - Line of reflection of synovial membrane

——— Line of attachment of fibrous capsule

- - - - - Line of reflection of fibrous capsule

Plate 500

Hip, Buttock, and Thigh

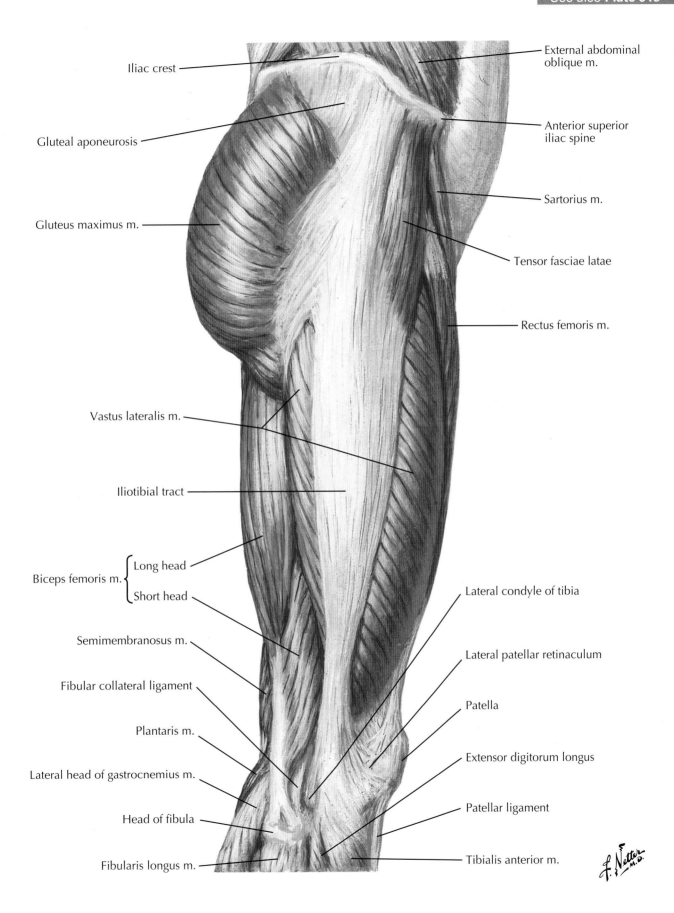

Iliac crest

External abdominal oblique m.

Gluteal aponeurosis

Anterior superior iliac spine

Sartorius m.

Gluteus maximus m.

Tensor fasciae latae

Rectus femoris m.

Vastus lateralis m.

Iliotibial tract

Biceps femoris m. { Long head / Short head

Lateral condyle of tibia

Semimembranosus m.

Lateral patellar retinaculum

Fibular collateral ligament

Patella

Plantaris m.

Extensor digitorum longus

Lateral head of gastrocnemius m.

Patellar ligament

Head of fibula

Fibularis longus m.

Tibialis anterior m.

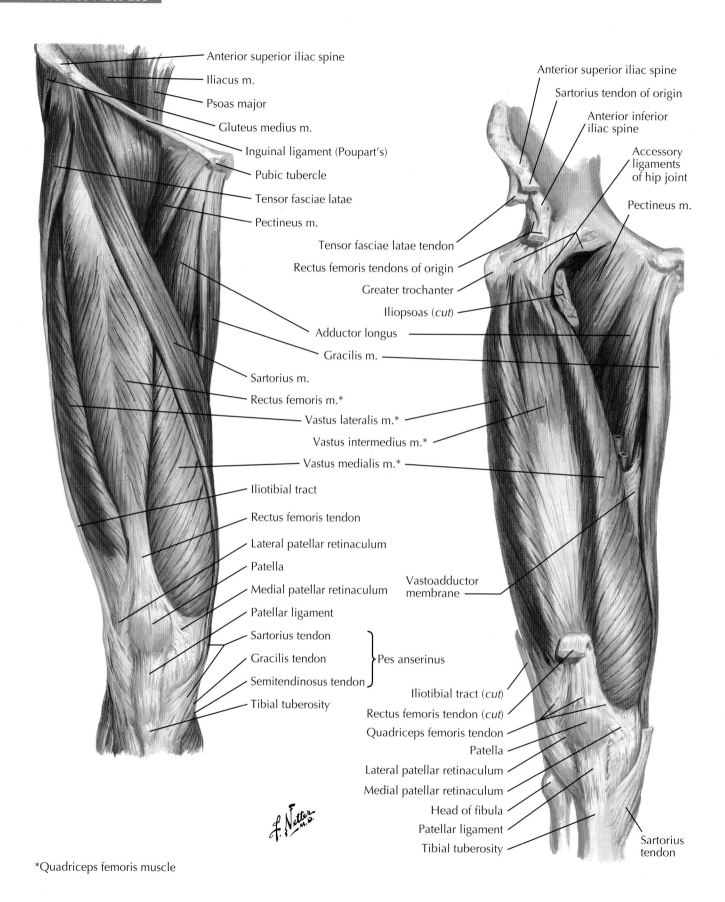

Anterior superior iliac spine

Iliacus m.

Psoas major

Gluteus medius m.

Inguinal ligament (Poupart's)

Pubic tubercle

Tensor fasciae latae

Pectineus m.

Sartorius m.

Rectus femoris m.*

Vastus lateralis m.*

Vastus intermedius m.*

Vastus medialis m.*

Iliotibial tract

Rectus femoris tendon

Lateral patellar retinaculum

Patella

Medial patellar retinaculum

Patellar ligament

Sartorius tendon

Gracilis tendon

Semitendinosus tendon

Pes anserinus

Tibial tuberosity

Anterior superior iliac spine

Sartorius tendon of origin

Anterior inferior iliac spine

Accessory ligaments of hip joint

Pectineus m.

Tensor fasciae latae tendon

Rectus femoris tendons of origin

Greater trochanter

Iliopsoas (*cut*)

Adductor longus

Gracilis m.

Vastoadductor membrane

Iliotibial tract (*cut*)

Rectus femoris tendon (*cut*)

Quadriceps femoris tendon

Patella

Lateral patellar retinaculum

Medial patellar retinaculum

Head of fibula

Patellar ligament

Tibial tuberosity

Sartorius tendon

*Quadriceps femoris muscle

Plate 502

Hip, Buttock, and Thigh

Deep dissection

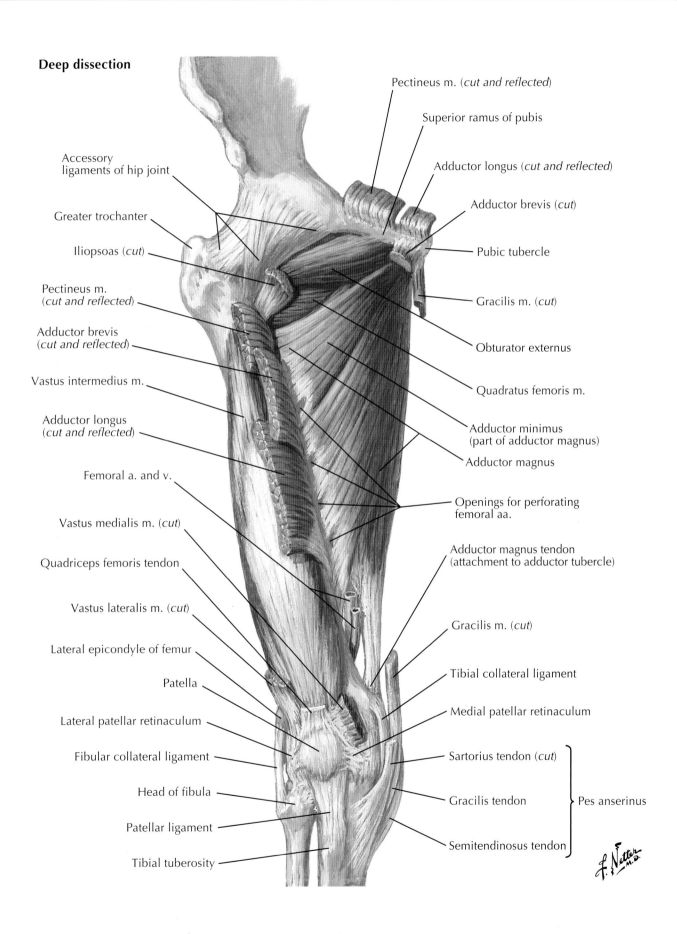

Pectineus m. (*cut and reflected*)

Superior ramus of pubis

Adductor longus (*cut and reflected*)

Adductor brevis (*cut*)

Pubic tubercle

Gracilis m. (*cut*)

Obturator externus

Quadratus femoris m.

Adductor minimus (part of adductor magnus)

Adductor magnus

Openings for perforating femoral aa.

Adductor magnus tendon (attachment to adductor tubercle)

Gracilis m. (*cut*)

Tibial collateral ligament

Medial patellar retinaculum

Sartorius tendon (*cut*)

Gracilis tendon

Semitendinosus tendon

Pes anserinus

Accessory ligaments of hip joint

Greater trochanter

Iliopsoas (*cut*)

Pectineus m. (*cut and reflected*)

Adductor brevis (*cut and reflected*)

Vastus intermedius m.

Adductor longus (*cut and reflected*)

Femoral a. and v.

Vastus medialis m. (*cut*)

Quadriceps femoris tendon

Vastus lateralis m. (*cut*)

Lateral epicondyle of femur

Patella

Lateral patellar retinaculum

Fibular collateral ligament

Head of fibula

Patellar ligament

Tibial tuberosity

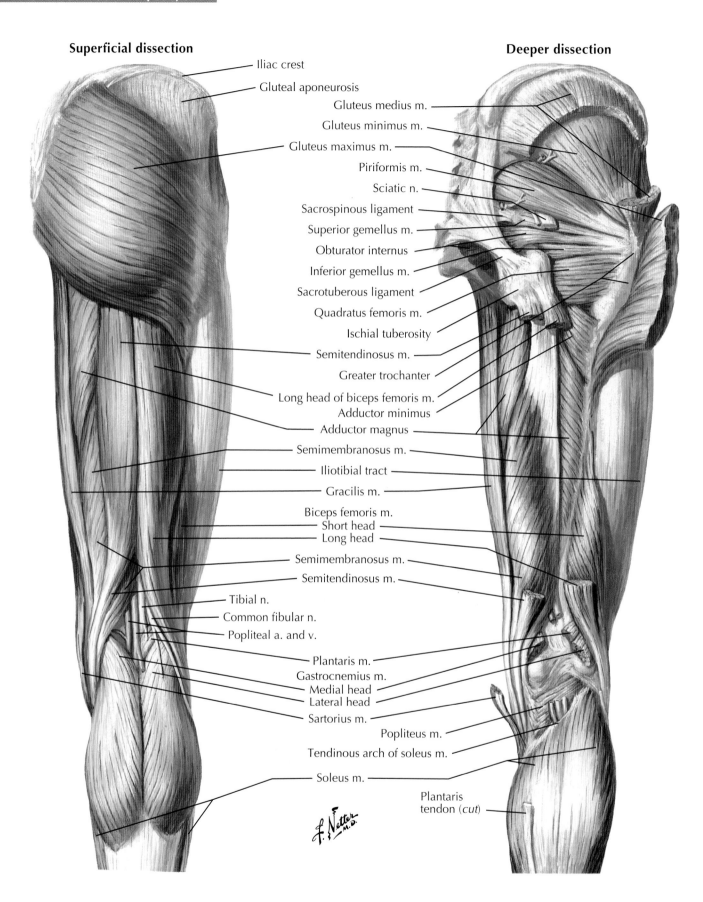

Superficial dissection

Deeper dissection

Iliac crest

Gluteal aponeurosis

Gluteus medius m.

Gluteus minimus m.

Gluteus maximus m.

Piriformis m.

Sciatic n.

Sacrospinous ligament

Superior gemellus m.

Obturator internus

Inferior gemellus m.

Sacrotuberous ligament

Quadratus femoris m.

Ischial tuberosity

Semitendinosus m.

Greater trochanter

Long head of biceps femoris m.

Adductor minimus

Adductor magnus

Semimembranosus m.

Iliotibial tract

Gracilis m.

Biceps femoris m.
Short head
Long head

Semimembranosus m.

Semitendinosus m.

Tibial n.

Common fibular n.

Popliteal a. and v.

Plantaris m.

Gastrocnemius m.
Medial head
Lateral head

Sartorius m.

Popliteus m.

Tendinous arch of soleus m.

Soleus m.

Plantaris
tendon (*cut*)

Plate 504

Hip, Buttock, and Thigh

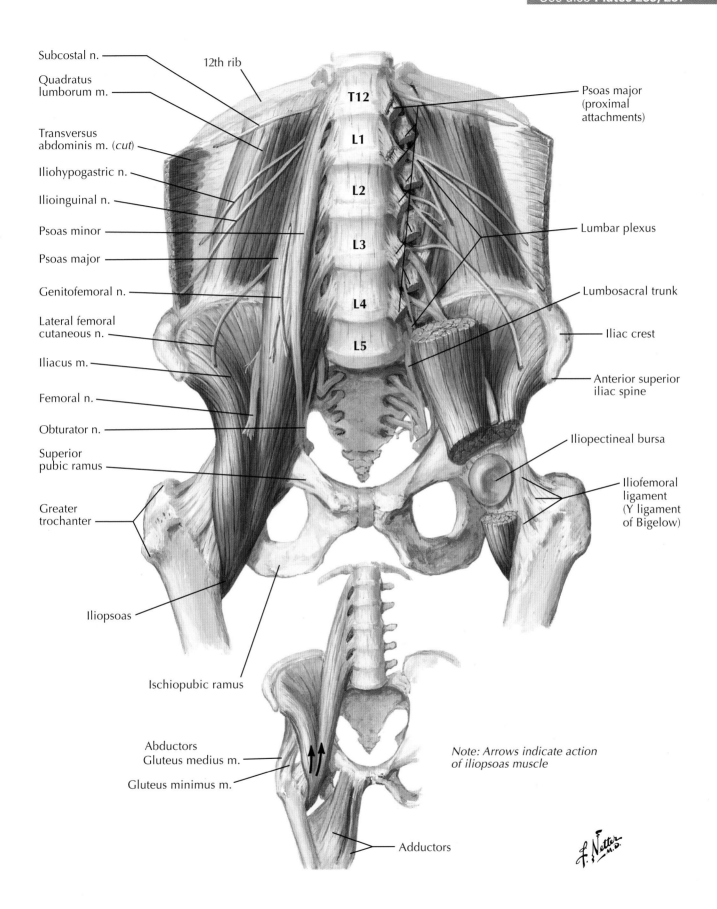

Subcostal n.

12th rib

Quadratus lumborum m.

Transversus abdominis m. (*cut*)

Iliohypogastric n.

Ilioinguinal n.

Psoas minor

Psoas major

Genitofemoral n.

Lateral femoral cutaneous n.

Iliacus m.

Femoral n.

Obturator n.

Superior pubic ramus

Greater trochanter

Iliopsoas

Ischiopubic ramus

Abductors Gluteus medius m.

Gluteus minimus m.

Adductors

T12

L1

L2

L3

L4

L5

Psoas major (proximal attachments)

Lumbar plexus

Lumbosacral trunk

Iliac crest

Anterior superior iliac spine

Iliopectineal bursa

Iliofemoral ligament (Y ligament of Bigelow)

Note: Arrows indicate action of iliopsoas muscle

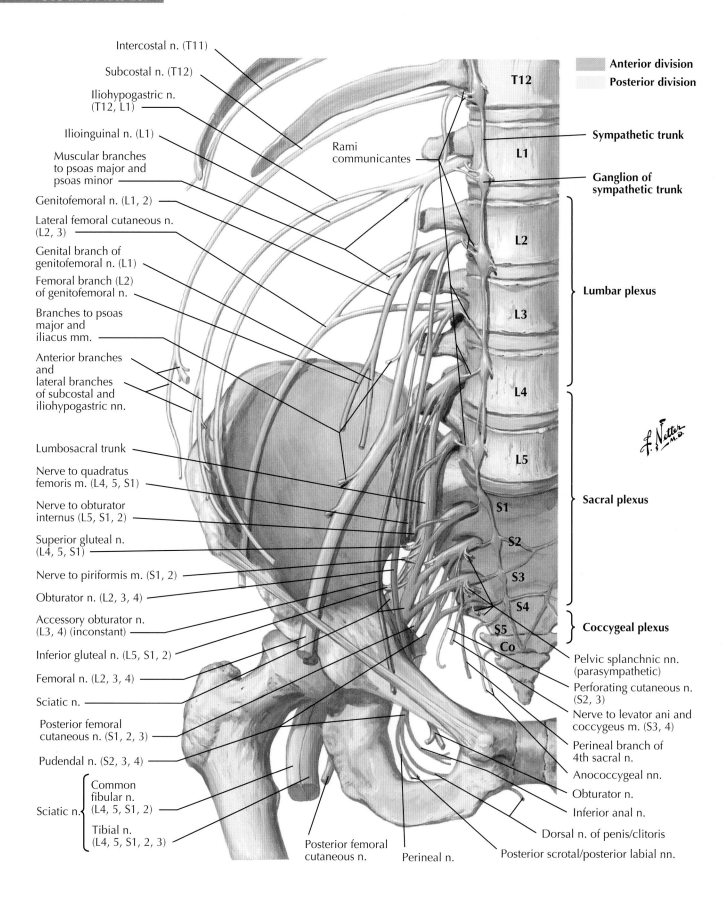

Intercostal n. (T11)

Subcostal n. (T12)

Iliohypogastric n. (T12, L1)

Ilioinguinal n. (L1)

Muscular branches to psoas major and psoas minor

Genitofemoral n. (L1, 2)

Lateral femoral cutaneous n. (L2, 3)

Genital branch of genitofemoral n. (L1)

Femoral branch (L2) of genitofemoral n.

Branches to psoas major and iliacus mm.

Anterior branches and lateral branches of subcostal and iliohypogastric nn.

Lumbosacral trunk

Nerve to quadratus femoris m. (L4, 5, S1)

Nerve to obturator internus (L5, S1, 2)

Superior gluteal n. (L4, 5, S1)

Nerve to piriformis m. (S1, 2)

Obturator n. (L2, 3, 4)

Accessory obturator n. (L3, 4) (inconstant)

Inferior gluteal n. (L5, S1, 2)

Femoral n. (L2, 3, 4)

Sciatic n.

Posterior femoral cutaneous n. (S1, 2, 3)

Pudendal n. (S2, 3, 4)

Sciatic n. {
Common fibular n. (L4, 5, S1, 2)

Tibial n. (L4, 5, S1, 2, 3)
}

Rami communicantes

Anterior division
Posterior division

T12

Sympathetic trunk

L1

Ganglion of sympathetic trunk

L2

Lumbar plexus

L3

L4

L5

S1

Sacral plexus

S2

S3

S4

Coccygeal plexus

S5
Co

Pelvic splanchnic nn. (parasympathetic)

Perforating cutaneous n. (S2, 3)

Nerve to levator ani and coccygeus m. (S3, 4)

Perineal branch of 4th sacral n.

Anococcygeal nn.

Obturator n.

Inferior anal n.

Dorsal n. of penis/clitoris

Posterior scrotal/posterior labial nn.

Posterior femoral cutaneous n.

Perineal n.

Plate 506

Hip, Buttock, and Thigh

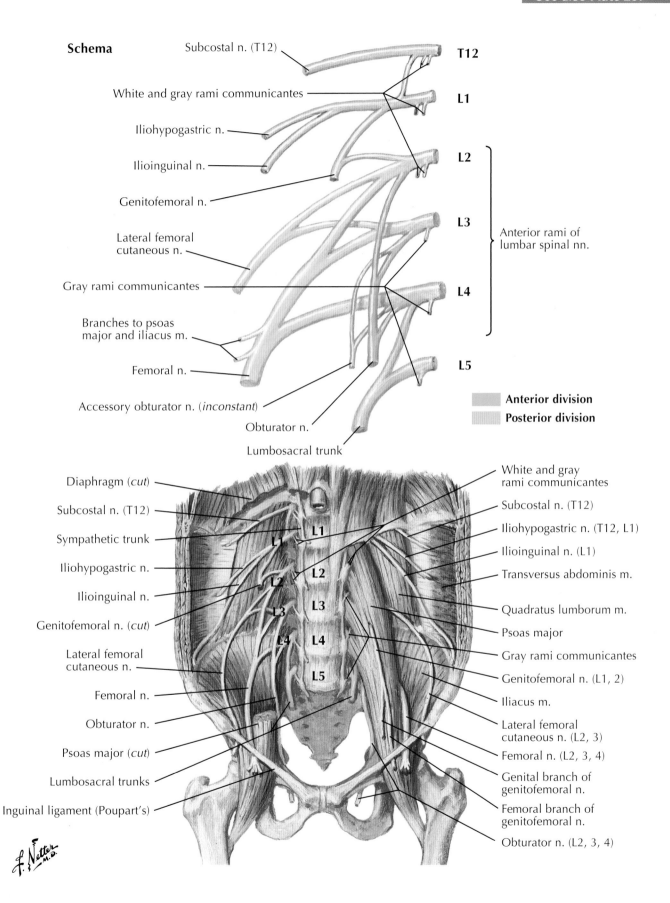

Schema

Subcostal n. (T12)

White and gray rami communicantes

Iliohypogastric n.

Ilioinguinal n.

Genitofemoral n.

Lateral femoral cutaneous n.

Gray rami communicantes

Branches to psoas major and iliacus m.

Femoral n.

Accessory obturator n. (*inconstant*)

Obturator n.

Lumbosacral trunk

T12

L1

L2

L3

L4

L5

Anterior rami of lumbar spinal nn.

Anterior division
Posterior division

Diaphragm (*cut*)

Subcostal n. (T12)

Sympathetic trunk

Iliohypogastric n.

Ilioinguinal n.

Genitofemoral n. (*cut*)

Lateral femoral cutaneous n.

Femoral n.

Obturator n.

Psoas major (*cut*)

Lumbosacral trunks

Inguinal ligament (Poupart's)

L1

L2

L3

L4

L5

White and gray rami communicantes

Subcostal n. (T12)

Iliohypogastric n. (T12, L1)

Ilioinguinal n. (L1)

Transversus abdominis m.

Quadratus lumborum m.

Psoas major

Gray rami communicantes

Genitofemoral n. (L1, 2)

Iliacus m.

Lateral femoral cutaneous n. (L2, 3)

Femoral n. (L2, 3, 4)

Genital branch of genitofemoral n.

Femoral branch of genitofemoral n.

Obturator n. (L2, 3, 4)

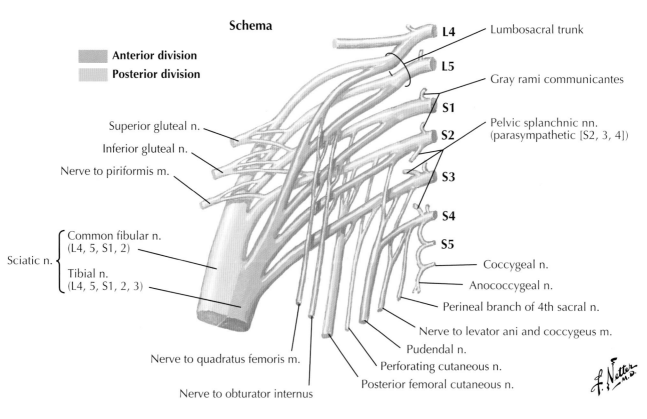

Schema

Anterior division
Posterior division

Superior gluteal n.

Inferior gluteal n.

Nerve to piriformis m.

Sciatic n. { Common fibular n. (L4, 5, S1, 2)

Tibial n. (L4, 5, S1, 2, 3)

L4 — Lumbosacral trunk
L5
Gray rami communicantes
S1
S2 — Pelvic splanchnic nn. (parasympathetic [S2, 3, 4])
S3
S4
S5
Coccygeal n.
Anococcygeal n.
Perineal branch of 4th sacral n.
Nerve to levator ani and coccygeus m.
Pudendal n.

Nerve to quadratus femoris m.

Nerve to obturator internus

Perforating cutaneous n.
Posterior femoral cutaneous n.

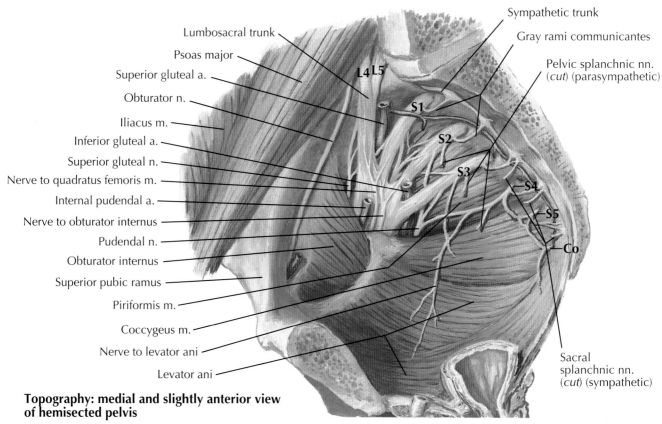

Lumbosacral trunk
Psoas major
Superior gluteal a.
Obturator n.
Iliacus m.
Inferior gluteal a.
Superior gluteal n.
Nerve to quadratus femoris m.
Internal pudendal a.
Nerve to obturator internus
Pudendal n.
Obturator internus
Superior pubic ramus
Piriformis m.
Coccygeus m.
Nerve to levator ani
Levator ani

Sympathetic trunk
Gray rami communicantes
Pelvic splanchnic nn. (*cut*) (parasympathetic)

L4 L5
S1
S2
S3
S4
S5
Co

Sacral splanchnic nn. (*cut*) (sympathetic)

Topography: medial and slightly anterior view of hemisected pelvis

Plate 508

Hip, Buttock, and Thigh

Superficial dissections

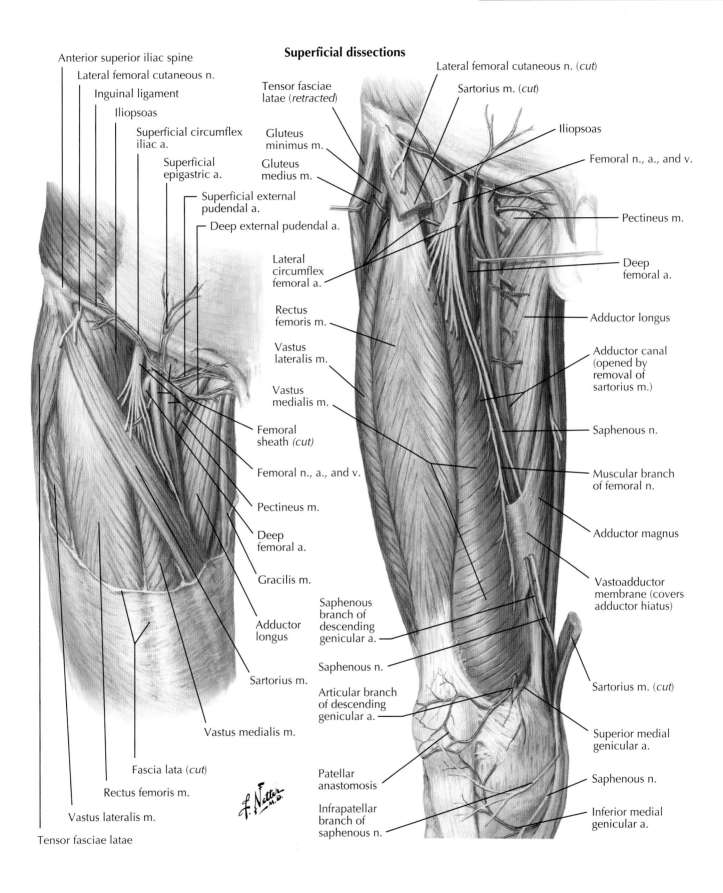

Anterior superior iliac spine

Lateral femoral cutaneous n.

Inguinal ligament

Iliopsoas

Superficial circumflex iliac a.

Superficial epigastric a.

Superficial external pudendal a.

Deep external pudendal a.

Tensor fasciae latae (*retracted*)

Gluteus minimus m.

Gluteus medius m.

Lateral circumflex femoral a.

Rectus femoris m.

Vastus lateralis m.

Vastus medialis m.

Femoral sheath (*cut*)

Femoral n., a., and v.

Pectineus m.

Deep femoral a.

Gracilis m.

Adductor longus

Sartorius m.

Vastus medialis m.

Fascia lata (*cut*)

Rectus femoris m.

Vastus lateralis m.

Tensor fasciae latae

Lateral femoral cutaneous n. (*cut*)

Sartorius m. (*cut*)

Iliopsoas

Femoral n., a., and v.

Pectineus m.

Deep femoral a.

Adductor longus

Adductor canal (opened by removal of sartorius m.)

Saphenous n.

Muscular branch of femoral n.

Adductor magnus

Vastoadductor membrane (covers adductor hiatus)

Sartorius m. (*cut*)

Superior medial genicular a.

Saphenous n.

Inferior medial genicular a.

Saphenous branch of descending genicular a.

Saphenous n.

Articular branch of descending genicular a.

Patellar anastomosis

Infrapatellar branch of saphenous n.

F. Netter M.D.

Deep dissection

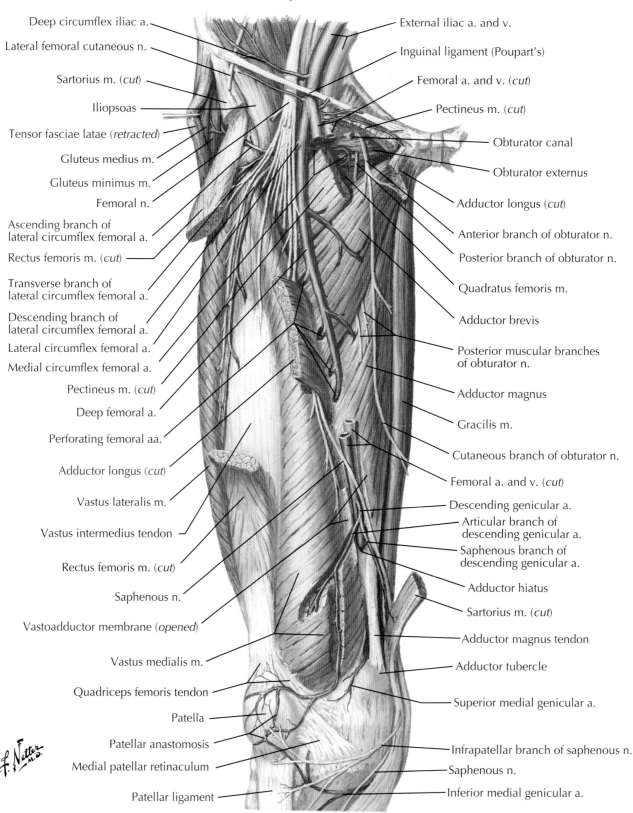

Deep circumflex iliac a.

Lateral femoral cutaneous n.

Sartorius m. (*cut*)

Iliopsoas

Tensor fasciae latae (*retracted*)

Gluteus medius m.

Gluteus minimus m.

Femoral n.

Ascending branch of
lateral circumflex femoral a.

Rectus femoris m. (*cut*)

Transverse branch of
lateral circumflex femoral a.

Descending branch of
lateral circumflex femoral a.

Lateral circumflex femoral a.

Medial circumflex femoral a.

Pectineus m. (*cut*)

Deep femoral a.

Perforating femoral aa.

Adductor longus (*cut*)

Vastus lateralis m.

Vastus intermedius tendon

Rectus femoris m. (*cut*)

Saphenous n.

Vastoadductor membrane (*opened*)

Vastus medialis m.

Quadriceps femoris tendon

Patella

Patellar anastomosis

Medial patellar retinaculum

Patellar ligament

External iliac a. and v.

Inguinal ligament (Poupart's)

Femoral a. and v. (*cut*)

Pectineus m. (*cut*)

Obturator canal

Obturator externus

Adductor longus (*cut*)

Anterior branch of obturator n.

Posterior branch of obturator n.

Quadratus femoris m.

Adductor brevis

Posterior muscular branches
of obturator n.

Adductor magnus

Gracilis m.

Cutaneous branch of obturator n.

Femoral a. and v. (*cut*)

Descending genicular a.

Articular branch of
descending genicular a.

Saphenous branch of
descending genicular a.

Adductor hiatus

Sartorius m. (*cut*)

Adductor magnus tendon

Adductor tubercle

Superior medial genicular a.

Infrapatellar branch of saphenous n.

Saphenous n.

Inferior medial genicular a.

Plate 510

Hip, Buttock, and Thigh

Deep dissection

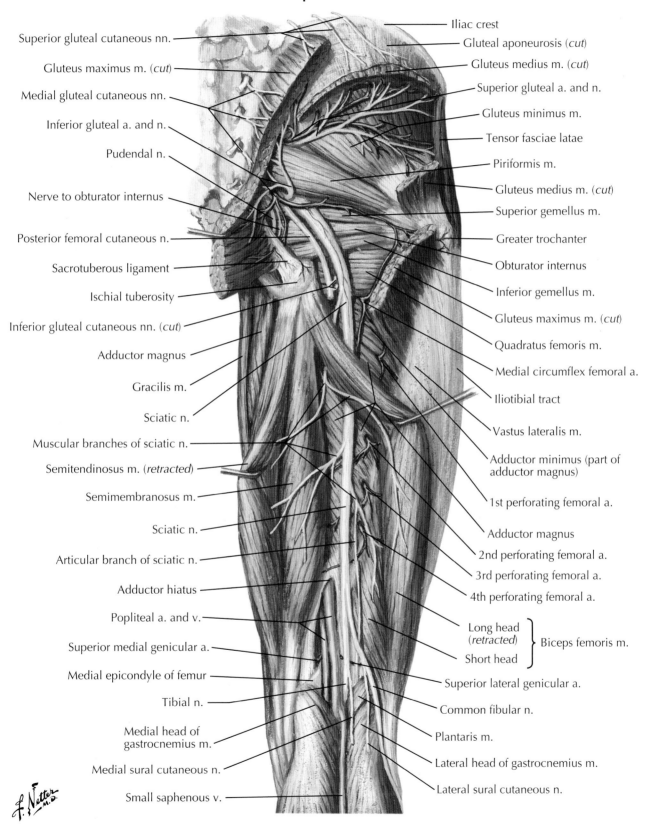

Superior gluteal cutaneous nn.

Gluteus maximus m. (*cut*)

Medial gluteal cutaneous nn.

Inferior gluteal a. and n.

Pudendal n.

Nerve to obturator internus

Posterior femoral cutaneous n.

Sacrotuberous ligament

Ischial tuberosity

Inferior gluteal cutaneous nn. (*cut*)

Adductor magnus

Gracilis m.

Sciatic n.

Muscular branches of sciatic n.

Semitendinosus m. (*retracted*)

Semimembranosus m.

Sciatic n.

Articular branch of sciatic n.

Adductor hiatus

Popliteal a. and v.

Superior medial genicular a.

Medial epicondyle of femur

Tibial n.

Medial head of gastrocnemius m.

Medial sural cutaneous n.

Small saphenous v.

Iliac crest

Gluteal aponeurosis (*cut*)

Gluteus medius m. (*cut*)

Superior gluteal a. and n.

Gluteus minimus m.

Tensor fasciae latae

Piriformis m.

Gluteus medius m. (*cut*)

Superior gemellus m.

Greater trochanter

Obturator internus

Inferior gemellus m.

Gluteus maximus m. (*cut*)

Quadratus femoris m.

Medial circumflex femoral a.

Iliotibial tract

Vastus lateralis m.

Adductor minimus (part of adductor magnus)

1st perforating femoral a.

Adductor magnus

2nd perforating femoral a.

3rd perforating femoral a.

4th perforating femoral a.

Long head (*retracted*)
Short head
} Biceps femoris m.

Superior lateral genicular a.

Common fibular n.

Plantaris m.

Lateral head of gastrocnemius m.

Lateral sural cutaneous n.

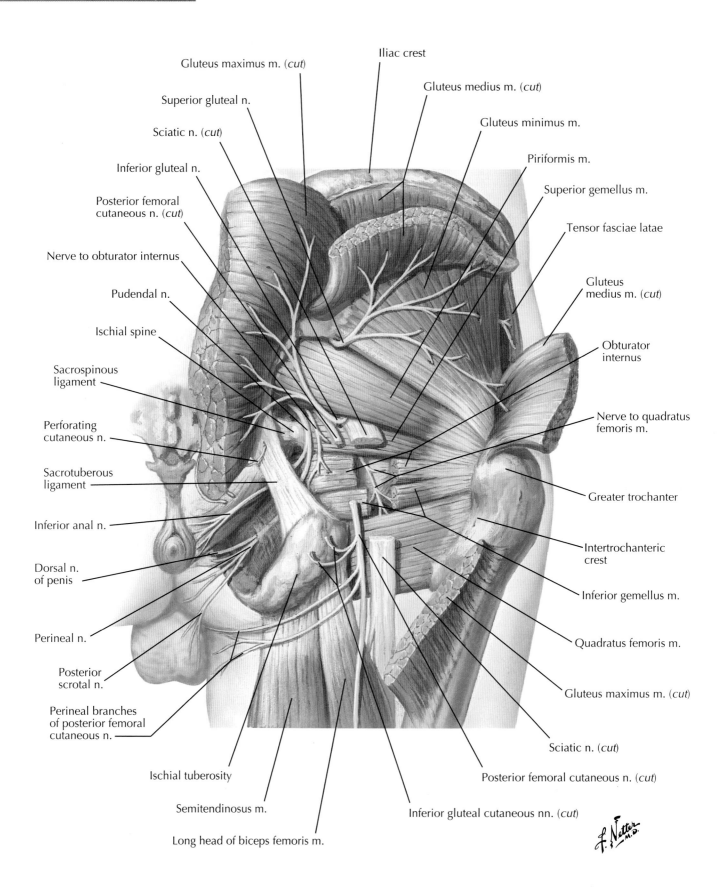

Gluteus maximus m. (*cut*)

Iliac crest

Gluteus medius m. (*cut*)

Superior gluteal n.

Gluteus minimus m.

Sciatic n. (*cut*)

Piriformis m.

Inferior gluteal n.

Superior gemellus m.

Posterior femoral cutaneous n. (*cut*)

Tensor fasciae latae

Nerve to obturator internus

Gluteus medius m. (*cut*)

Pudendal n.

Obturator internus

Ischial spine

Sacrospinous ligament

Nerve to quadratus femoris m.

Perforating cutaneous n.

Sacrotuberous ligament

Greater trochanter

Inferior anal n.

Intertrochanteric crest

Dorsal n. of penis

Inferior gemellus m.

Perineal n.

Quadratus femoris m.

Posterior scrotal n.

Gluteus maximus m. (*cut*)

Perineal branches of posterior femoral cutaneous n.

Sciatic n. (*cut*)

Posterior femoral cutaneous n. (*cut*)

Ischial tuberosity

Inferior gluteal cutaneous nn. (*cut*)

Semitendinosus m.

Long head of biceps femoris m.

Plate 512

Hip, Buttock, and Thigh

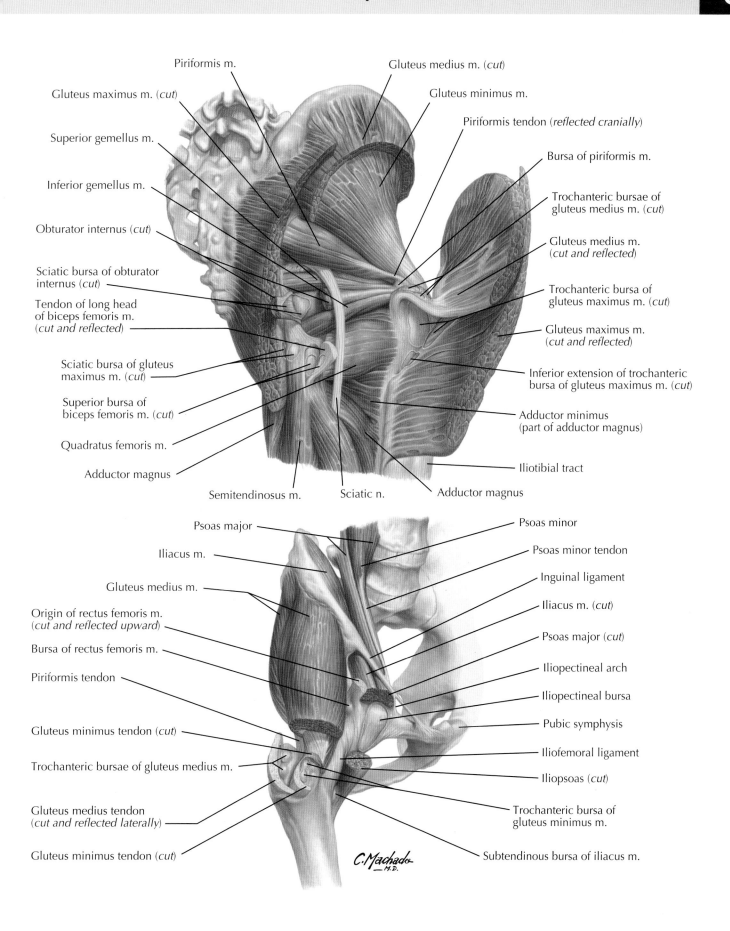

Piriformis m.

Gluteus maximus m. (*cut*)

Superior gemellus m.

Inferior gemellus m.

Obturator internus (*cut*)

Sciatic bursa of obturator internus (*cut*)

Tendon of long head of biceps femoris m. (*cut and reflected*)

Sciatic bursa of gluteus maximus m. (*cut*)

Superior bursa of biceps femoris m. (*cut*)

Quadratus femoris m.

Adductor magnus

Semitendinosus m.

Gluteus medius m. (*cut*)

Gluteus minimus m.

Piriformis tendon (*reflected cranially*)

Bursa of piriformis m.

Trochanteric bursae of gluteus medius m. (*cut*)

Gluteus medius m. (*cut and reflected*)

Trochanteric bursa of gluteus maximus m. (*cut*)

Gluteus maximus m. (*cut and reflected*)

Inferior extension of trochanteric bursa of gluteus maximus m. (*cut*)

Adductor minimus (part of adductor magnus)

Iliotibial tract

Sciatic n.

Adductor magnus

Psoas major

Iliacus m.

Gluteus medius m.

Origin of rectus femoris m. (*cut and reflected upward*)

Bursa of rectus femoris m.

Piriformis tendon

Gluteus minimus tendon (*cut*)

Trochanteric bursae of gluteus medius m.

Gluteus medius tendon (*cut and reflected laterally*)

Gluteus minimus tendon (*cut*)

Psoas minor

Psoas minor tendon

Inguinal ligament

Iliacus m. (*cut*)

Psoas major (*cut*)

Iliopectineal arch

Iliopectineal bursa

Pubic symphysis

Iliofemoral ligament

Iliopsoas (*cut*)

Trochanteric bursa of gluteus minimus m.

Subtendinous bursa of iliacus m.

C. Machado _M.D.

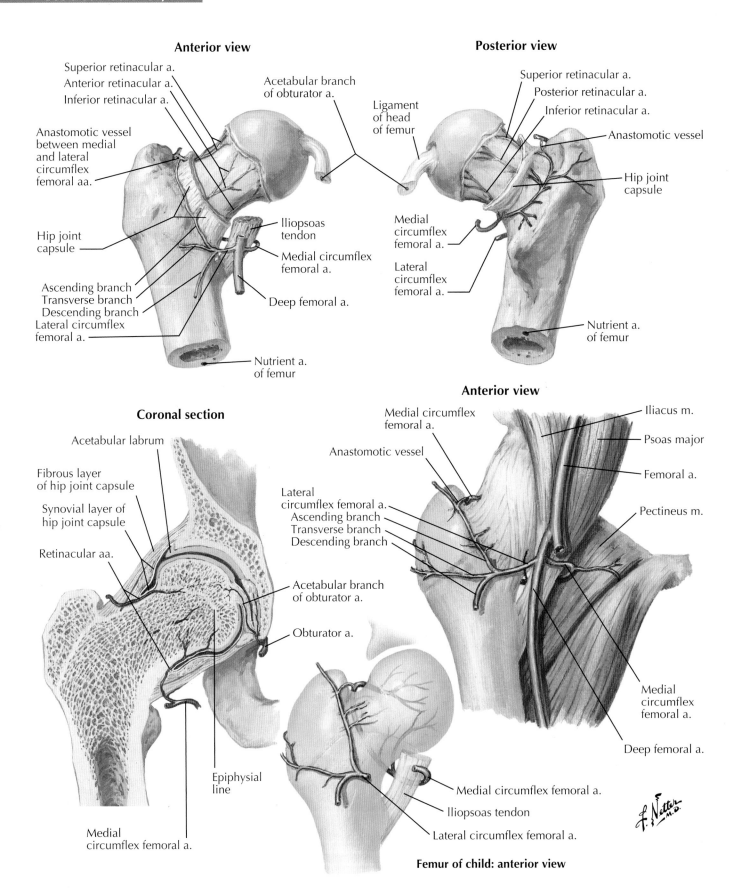

Anterior view

Superior retinacular a.
Anterior retinacular a.
Inferior retinacular a.

Acetabular branch of obturator a.

Ligament of head of femur

Anastomotic vessel between medial and lateral circumflex femoral aa.

Hip joint capsule

Iliopsoas tendon

Medial circumflex femoral a.

Ascending branch
Transverse branch
Descending branch
Lateral circumflex femoral a.

Deep femoral a.

Nutrient a. of femur

Posterior view

Superior retinacular a.
Posterior retinacular a.
Inferior retinacular a.

Anastomotic vessel

Hip joint capsule

Medial circumflex femoral a.

Lateral circumflex femoral a.

Nutrient a. of femur

Coronal section

Acetabular labrum

Fibrous layer of hip joint capsule

Synovial layer of hip joint capsule

Retinacular aa.

Lateral circumflex femoral a.
Ascending branch
Transverse branch
Descending branch

Acetabular branch of obturator a.

Obturator a.

Epiphysial line

Medial circumflex femoral a.

Anterior view

Medial circumflex femoral a.

Anastomotic vessel

Iliacus m.

Psoas major

Femoral a.

Pectineus m.

Medial circumflex femoral a.

Deep femoral a.

Medial circumflex femoral a.

Iliopsoas tendon

Lateral circumflex femoral a.

Femur of child: anterior view

Plate 514

Hip, Buttock, and Thigh

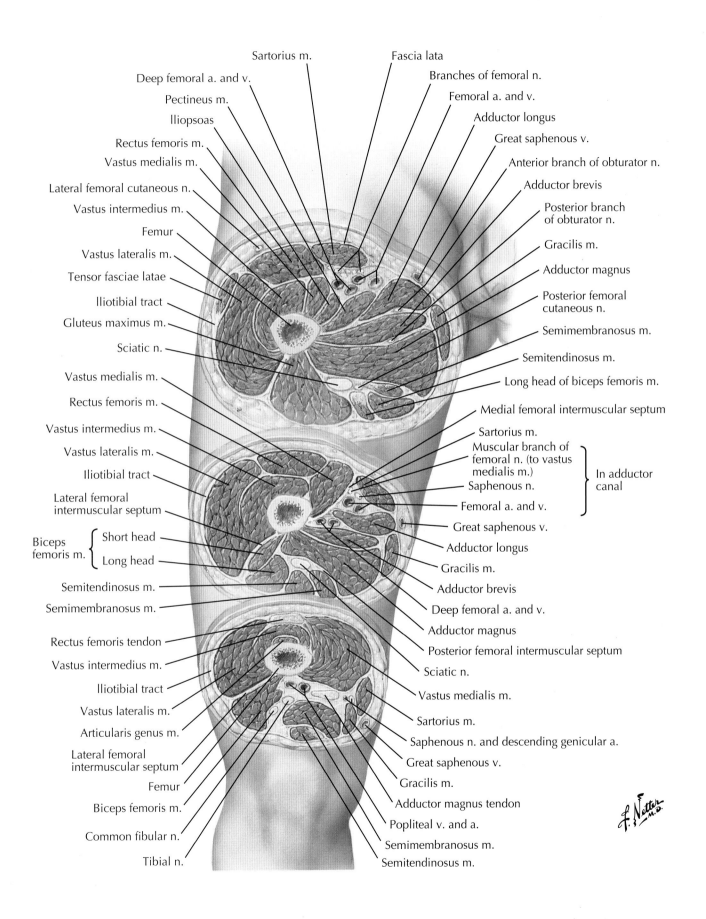

Sartorius m.

Deep femoral a. and v.

Pectineus m.

Iliopsoas

Rectus femoris m.

Vastus medialis m.

Lateral femoral cutaneous n.

Vastus intermedius m.

Femur

Vastus lateralis m.

Tensor fasciae latae

Iliotibial tract

Gluteus maximus m.

Sciatic n.

Vastus medialis m.

Rectus femoris m.

Vastus intermedius m.

Vastus lateralis m.

Iliotibial tract

Lateral femoral
intermuscular septum

Biceps
femoris m. { Short head

Long head

Semitendinosus m.

Semimembranosus m.

Rectus femoris tendon

Vastus intermedius m.

Iliotibial tract

Vastus lateralis m.

Articularis genus m.

Lateral femoral
intermuscular septum

Femur

Biceps femoris m.

Common fibular n.

Tibial n.

Fascia lata

Branches of femoral n.

Femoral a. and v.

Adductor longus

Great saphenous v.

Anterior branch of obturator n.

Adductor brevis

Posterior branch
of obturator n.

Gracilis m.

Adductor magnus

Posterior femoral
cutaneous n.

Semimembranosus m.

Semitendinosus m.

Long head of biceps femoris m.

Medial femoral intermuscular septum

Sartorius m.

Muscular branch of
femoral n. (to vastus
medialis m.)

Saphenous n.

Femoral a. and v.

In adductor
canal

Great saphenous v.

Adductor longus

Gracilis m.

Adductor brevis

Deep femoral a. and v.

Adductor magnus

Posterior femoral intermuscular septum

Sciatic n.

Vastus medialis m.

Sartorius m.

Saphenous n. and descending genicular a.

Great saphenous v.

Gracilis m.

Adductor magnus tendon

Popliteal v. and a.

Semimembranosus m.

Semitendinosus m.

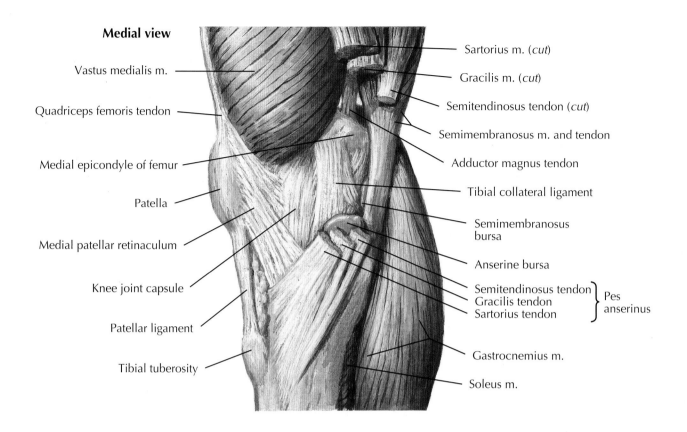

Medial view

Vastus medialis m.

Quadriceps femoris tendon

Medial epicondyle of femur

Patella

Medial patellar retinaculum

Knee joint capsule

Patellar ligament

Tibial tuberosity

Sartorius m. (*cut*)

Gracilis m. (*cut*)

Semitendinosus tendon (*cut*)

Semimembranosus m. and tendon

Adductor magnus tendon

Tibial collateral ligament

Semimembranosus bursa

Anserine bursa

Semitendinosus tendon
Gracilis tendon } Pes anserinus
Sartorius tendon

Gastrocnemius m.

Soleus m.

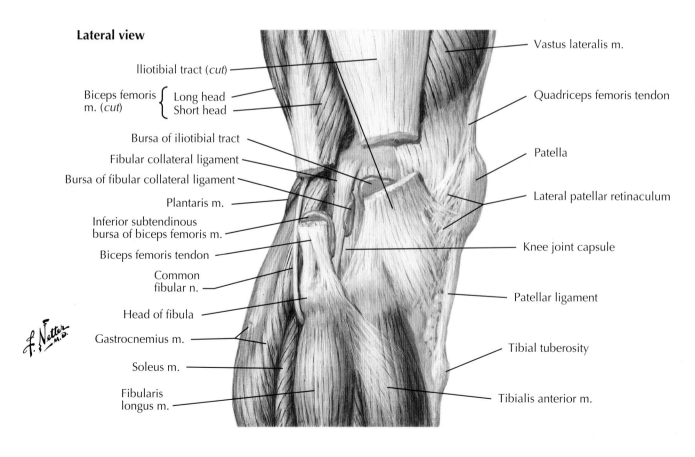

Lateral view

Iliotibial tract (*cut*)

Biceps femoris m. (*cut*) { Long head
Short head

Bursa of iliotibial tract

Fibular collateral ligament

Bursa of fibular collateral ligament

Plantaris m.

Inferior subtendinous bursa of biceps femoris m.

Biceps femoris tendon

Common fibular n.

Head of fibula

Gastrocnemius m.

Soleus m.

Fibularis longus m.

Vastus lateralis m.

Quadriceps femoris tendon

Patella

Lateral patellar retinaculum

Knee joint capsule

Patellar ligament

Tibial tuberosity

Tibialis anterior m.

Plate 516

Knee

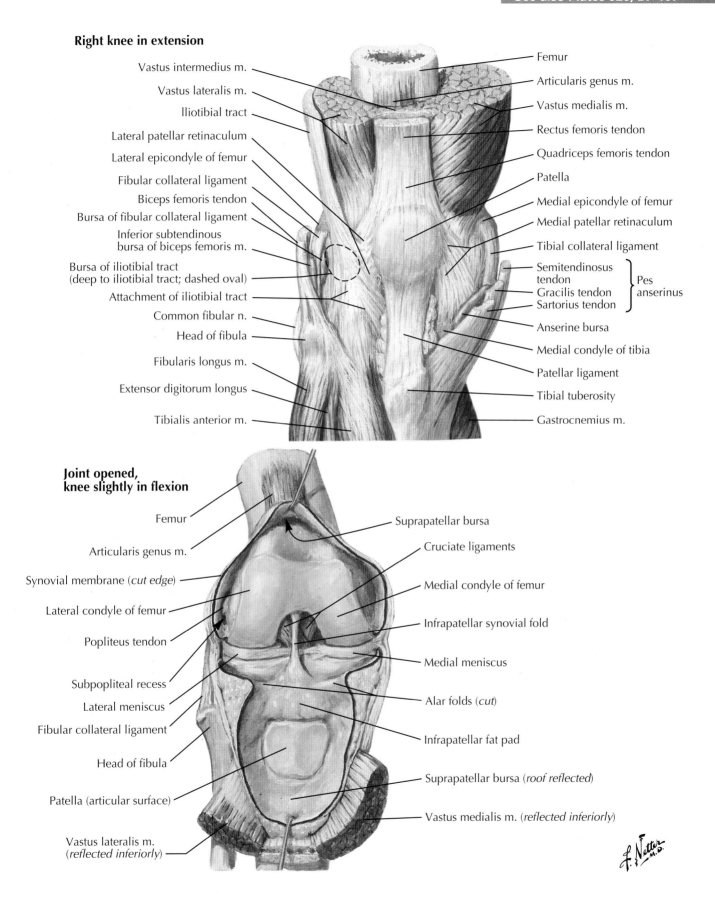

Right knee in extension

Vastus intermedius m.
Vastus lateralis m.
Iliotibial tract
Lateral patellar retinaculum
Lateral epicondyle of femur
Fibular collateral ligament
Biceps femoris tendon
Bursa of fibular collateral ligament
Inferior subtendinous
bursa of biceps femoris m.
Bursa of iliotibial tract
(deep to iliotibial tract; dashed oval)
Attachment of iliotibial tract
Common fibular n.
Head of fibula
Fibularis longus m.
Extensor digitorum longus
Tibialis anterior m.

Femur
Articularis genus m.
Vastus medialis m.
Rectus femoris tendon
Quadriceps femoris tendon
Patella
Medial epicondyle of femur
Medial patellar retinaculum
Tibial collateral ligament
Semitendinosus
tendon
Gracilis tendon   } Pes
Sartorius tendon   anserinus
Anserine bursa
Medial condyle of tibia
Patellar ligament
Tibial tuberosity
Gastrocnemius m.

**Joint opened,
knee slightly in flexion**

Femur
Articularis genus m.
Synovial membrane (*cut edge*)
Lateral condyle of femur
Popliteus tendon
Subpopliteal recess
Lateral meniscus
Fibular collateral ligament
Head of fibula
Patella (articular surface)
Vastus lateralis m.
(*reflected inferiorly*)

Suprapatellar bursa
Cruciate ligaments
Medial condyle of femur
Infrapatellar synovial fold
Medial meniscus
Alar folds (*cut*)
Infrapatellar fat pad
Suprapatellar bursa (*roof reflected*)
Vastus medialis m. (*reflected inferiorly*)

Inferior view

Iliotibial tract

Bursa of iliotibial tract

Subpopliteal recess

Popliteus tendon

Fibular collateral ligament

Bursa of fibular collateral ligament

Lateral condyle of femur

Anterior cruciate ligament

Arcuate popliteal ligament

Patellar ligament

Medial patellar retinaculum

Suprapatellar bursa

Synovial membrane of knee joint capsule (*cut edge*)

Infrapatellar synovial fold

Posterior cruciate ligament

Tibial collateral ligament (superficial and deep parts)

Medial condyle of femur

Oblique popliteal ligament

Semimembranosus tendon

Posterior aspect

Superior view

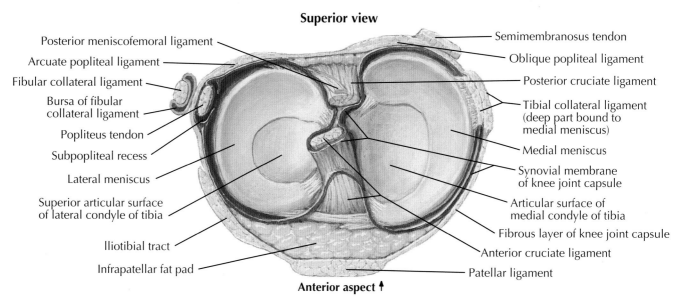

Posterior meniscofemoral ligament

Arcuate popliteal ligament

Fibular collateral ligament

Bursa of fibular collateral ligament

Popliteus tendon

Subpopliteal recess

Lateral meniscus

Superior articular surface of lateral condyle of tibia

Iliotibial tract

Infrapatellar fat pad

Semimembranosus tendon

Oblique popliteal ligament

Posterior cruciate ligament

Tibial collateral ligament (deep part bound to medial meniscus)

Medial meniscus

Synovial membrane of knee joint capsule

Articular surface of medial condyle of tibia

Fibrous layer of knee joint capsule

Anterior cruciate ligament

Patellar ligament

Anterior aspect

Superior view: ligaments and cartilage removed

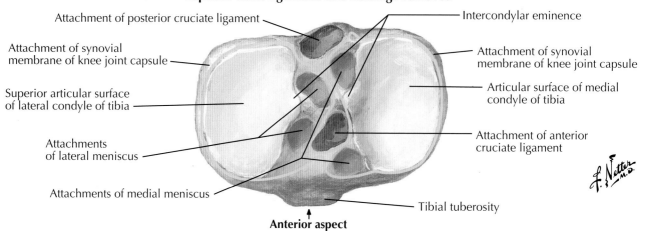

Attachment of posterior cruciate ligament

Attachment of synovial membrane of knee joint capsule

Superior articular surface of lateral condyle of tibia

Attachments of lateral meniscus

Attachments of medial meniscus

Intercondylar eminence

Attachment of synovial membrane of knee joint capsule

Articular surface of medial condyle of tibia

Attachment of anterior cruciate ligament

Tibial tuberosity

Anterior aspect

Plate 518

Knee

Right knee in flexion: anterior view

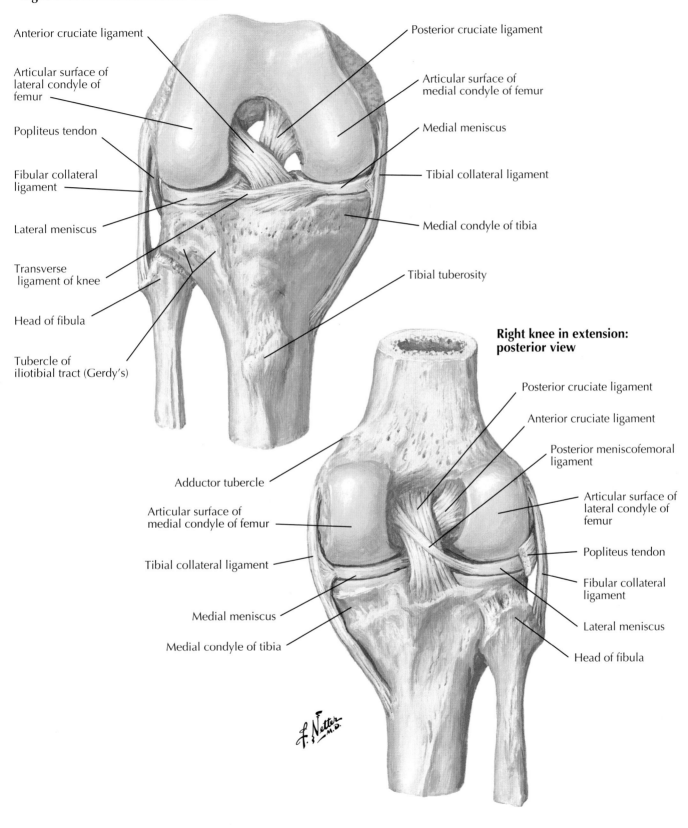

Anterior cruciate ligament

Articular surface of lateral condyle of femur

Popliteus tendon

Fibular collateral ligament

Lateral meniscus

Transverse ligament of knee

Head of fibula

Tubercle of iliotibial tract (Gerdy's)

Posterior cruciate ligament

Articular surface of medial condyle of femur

Medial meniscus

Tibial collateral ligament

Medial condyle of tibia

Tibial tuberosity

Right knee in extension: posterior view

Adductor tubercle

Articular surface of medial condyle of femur

Tibial collateral ligament

Medial meniscus

Medial condyle of tibia

Posterior cruciate ligament

Anterior cruciate ligament

Posterior meniscofemoral ligament

Articular surface of lateral condyle of femur

Popliteus tendon

Fibular collateral ligament

Lateral meniscus

Head of fibula

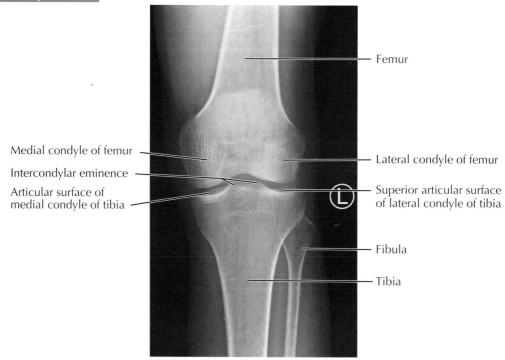

Femur

Medial condyle of femur

Lateral condyle of femur

Intercondylar eminence

Superior articular surface
of lateral condyle of tibia

Articular surface of
medial condyle of tibia

Fibula

Tibia

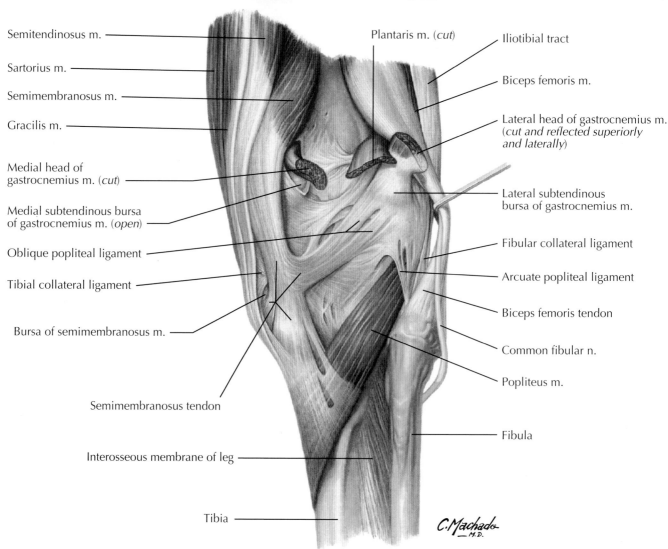

Semitendinosus m.

Plantaris m. (*cut*)

Iliotibial tract

Sartorius m.

Biceps femoris m.

Semimembranosus m.

Lateral head of gastrocnemius m.
(*cut and reflected superiorly
and laterally*)

Gracilis m.

Medial head of
gastrocnemius m. (*cut*)

Lateral subtendinous
bursa of gastrocnemius m.

Medial subtendinous bursa
of gastrocnemius m. (*open*)

Oblique popliteal ligament

Fibular collateral ligament

Tibial collateral ligament

Arcuate popliteal ligament

Biceps femoris tendon

Bursa of semimembranosus m.

Common fibular n.

Popliteus m.

Semimembranosus tendon

Fibula

Interosseous membrane of leg

Tibia

C. Machado
M.D.

Plate 520

Knee

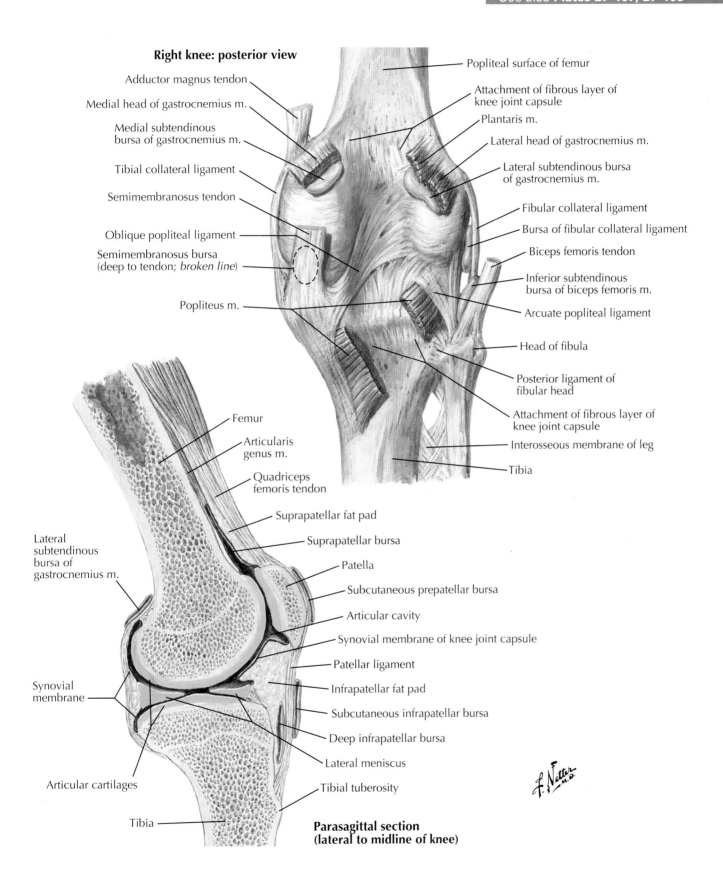

Right knee: posterior view

Adductor magnus tendon

Medial head of gastrocnemius m.

Medial subtendinous bursa of gastrocnemius m.

Tibial collateral ligament

Semimembranosus tendon

Oblique popliteal ligament

Semimembranosus bursa (deep to tendon; *broken line*)

Popliteus m.

Popliteal surface of femur

Attachment of fibrous layer of knee joint capsule

Plantaris m.

Lateral head of gastrocnemius m.

Lateral subtendinous bursa of gastrocnemius m.

Fibular collateral ligament

Bursa of fibular collateral ligament

Biceps femoris tendon

Inferior subtendinous bursa of biceps femoris m.

Arcuate popliteal ligament

Head of fibula

Posterior ligament of fibular head

Attachment of fibrous layer of knee joint capsule

Interosseous membrane of leg

Tibia

Femur

Articularis genus m.

Quadriceps femoris tendon

Suprapatellar fat pad

Suprapatellar bursa

Patella

Subcutaneous prepatellar bursa

Articular cavity

Synovial membrane of knee joint capsule

Patellar ligament

Infrapatellar fat pad

Subcutaneous infrapatellar bursa

Deep infrapatellar bursa

Lateral meniscus

Tibial tuberosity

Lateral subtendinous bursa of gastrocnemius m.

Synovial membrane

Articular cartilages

Tibia

Parasagittal section (lateral to midline of knee)

Knee

Plate 521

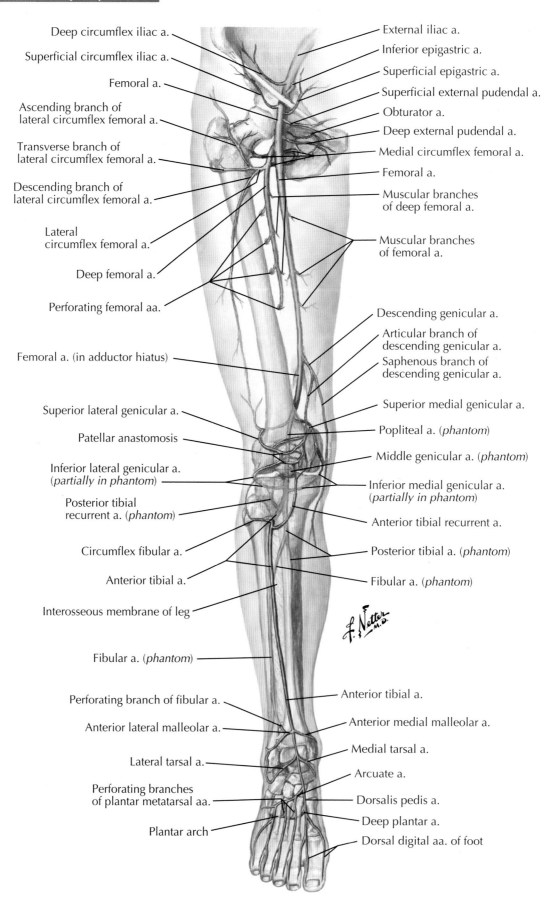

Deep circumflex iliac a.

Superficial circumflex iliac a.

Femoral a.

Ascending branch of
lateral circumflex femoral a.

Transverse branch of
lateral circumflex femoral a.

Descending branch of
lateral circumflex femoral a.

Lateral
circumflex femoral a.

Deep femoral a.

Perforating femoral aa.

Femoral a. (in adductor hiatus)

Superior lateral genicular a.

Patellar anastomosis

Inferior lateral genicular a.
(*partially in phantom*)

Posterior tibial
recurrent a. (*phantom*)

Circumflex fibular a.

Anterior tibial a.

Interosseous membrane of leg

Fibular a. (*phantom*)

Perforating branch of fibular a.

Anterior lateral malleolar a.

Lateral tarsal a.

Perforating branches
of plantar metatarsal aa.

Plantar arch

External iliac a.

Inferior epigastric a.

Superficial epigastric a.

Superficial external pudendal a.

Obturator a.

Deep external pudendal a.

Medial circumflex femoral a.

Femoral a.

Muscular branches
of deep femoral a.

Muscular branches
of femoral a.

Descending genicular a.

Articular branch of
descending genicular a.

Saphenous branch of
descending genicular a.

Superior medial genicular a.

Popliteal a. (*phantom*)

Middle genicular a. (*phantom*)

Inferior medial genicular a.
(*partially in phantom*)

Anterior tibial recurrent a.

Posterior tibial a. (*phantom*)

Fibular a. (*phantom*)

Anterior tibial a.

Anterior medial malleolar a.

Medial tarsal a.

Arcuate a.

Dorsalis pedis a.

Deep plantar a.

Dorsal digital aa. of foot

Plate 522

Knee

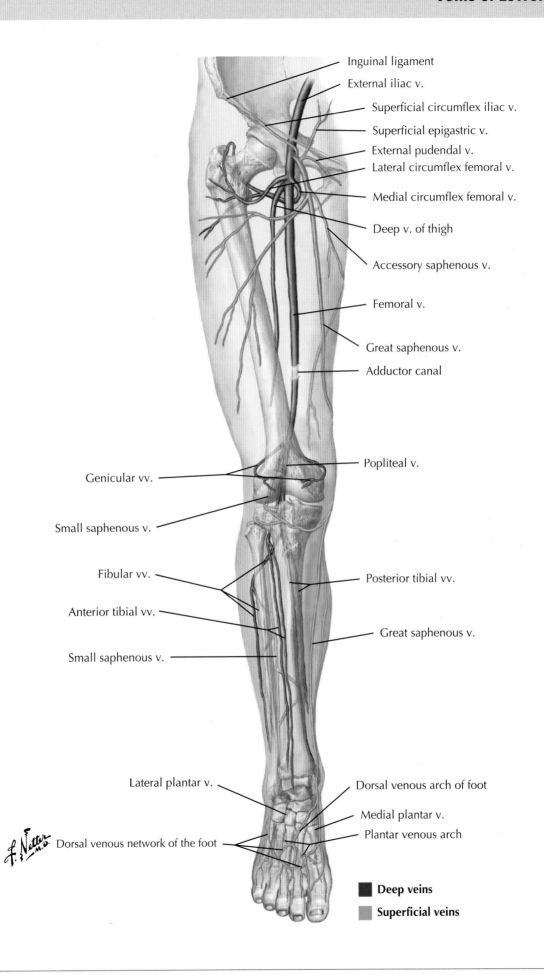

Inguinal ligament

External iliac v.

Superficial circumflex iliac v.

Superficial epigastric v.

External pudendal v.

Lateral circumflex femoral v.

Medial circumflex femoral v.

Deep v. of thigh

Accessory saphenous v.

Femoral v.

Great saphenous v.

Adductor canal

Popliteal v.

Genicular vv.

Small saphenous v.

Fibular vv.

Anterior tibial vv.

Posterior tibial vv.

Great saphenous v.

Small saphenous v.

Lateral plantar v.

Dorsal venous arch of foot

Medial plantar v.

Plantar venous arch

Dorsal venous network of the foot

■ Deep veins

■ Superficial veins

Knee

Plate 523

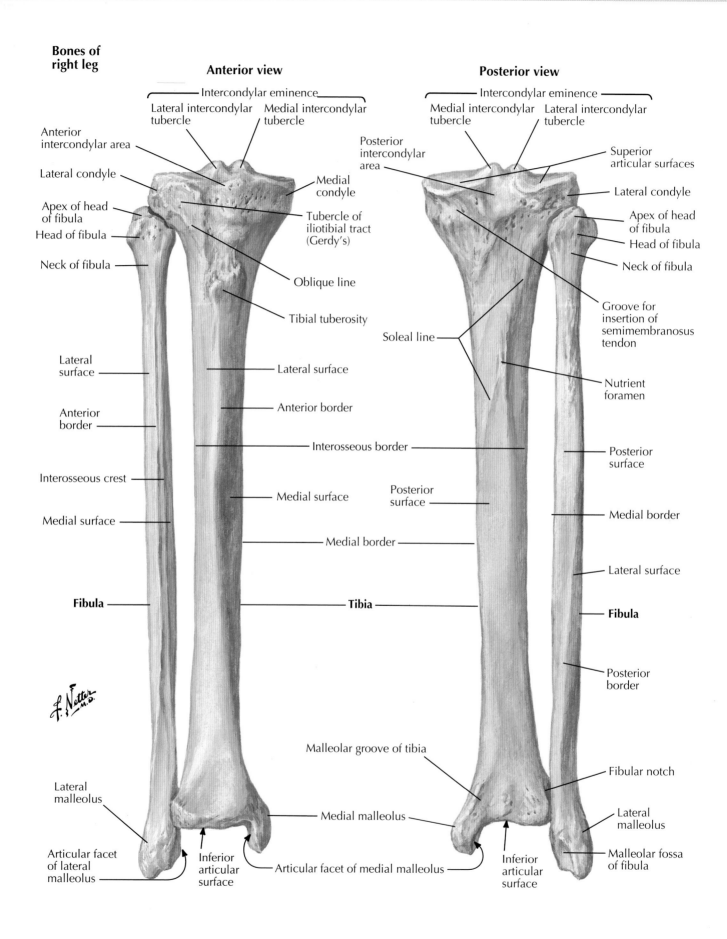

Bones of right leg

Anterior view

Intercondylar eminence

Lateral intercondylar tubercle   Medial intercondylar tubercle

Anterior intercondylar area

Lateral condyle

Apex of head of fibula

Head of fibula

Neck of fibula

Medial condyle

Tubercle of iliotibial tract (Gerdy's)

Oblique line

Tibial tuberosity

Lateral surface

Lateral surface

Anterior border

Anterior border

Interosseous border

Interosseous crest

Medial surface

Medial surface

Medial border

Fibula

Tibia

Lateral malleolus

Articular facet of lateral malleolus

Inferior articular surface

Malleolar groove of tibia

Medial malleolus

Articular facet of medial malleolus

Posterior view

Intercondylar eminence

Medial intercondylar tubercle   Lateral intercondylar tubercle

Posterior intercondylar area

Superior articular surfaces

Lateral condyle

Apex of head of fibula

Head of fibula

Neck of fibula

Groove for insertion of semimembranosus tendon

Soleal line

Nutrient foramen

Posterior surface

Posterior surface

Medial border

Lateral surface

Fibula

Posterior border

Fibular notch

Lateral malleolus

Inferior articular surface

Malleolar fossa of fibula

Plate 524

Leg

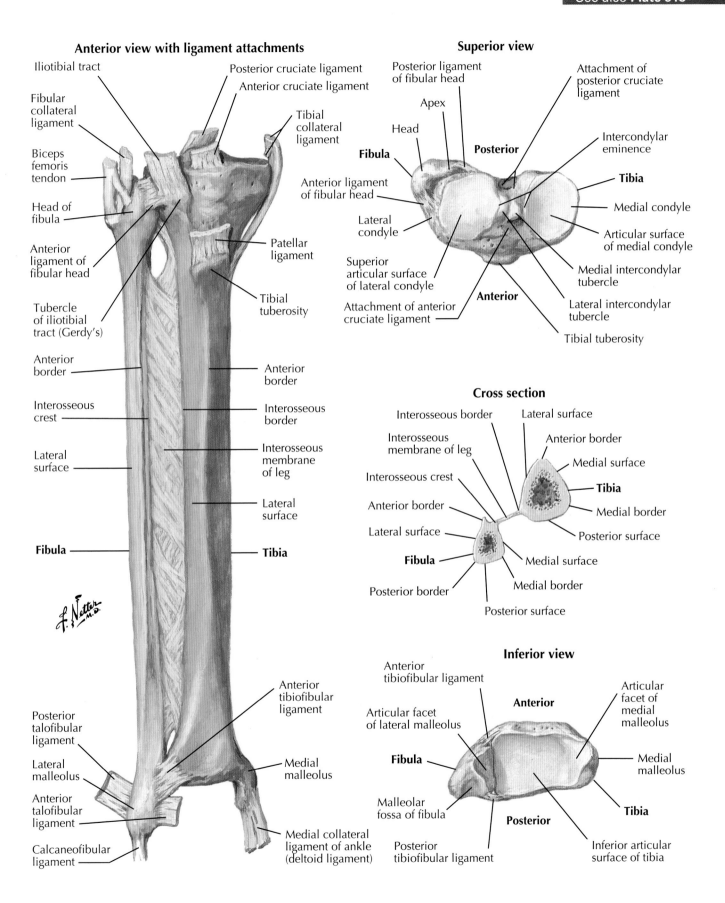

Anterior view with ligament attachments

Iliotibial tract

Fibular collateral ligament

Biceps femoris tendon

Head of fibula

Anterior ligament of fibular head

Tubercle of iliotibial tract (Gerdy's)

Anterior border

Interosseous crest

Lateral surface

Fibula

Posterior cruciate ligament

Anterior cruciate ligament

Tibial collateral ligament

Patellar ligament

Tibial tuberosity

Anterior border

Interosseous border

Interosseous membrane of leg

Lateral surface

Tibia

Posterior talofibular ligament

Lateral malleolus

Anterior talofibular ligament

Calcaneofibular ligament

Anterior tibiofibular ligament

Medial malleolus

Medial collateral ligament of ankle (deltoid ligament)

Superior view

Posterior ligament of fibular head

Apex

Head

Fibula

Anterior ligament of fibular head

Lateral condyle

Superior articular surface of lateral condyle

Attachment of anterior cruciate ligament

Attachment of posterior cruciate ligament

Intercondylar eminence

Posterior

Tibia

Medial condyle

Articular surface of medial condyle

Medial intercondylar tubercle

Lateral intercondylar tubercle

Tibial tuberosity

Anterior

Cross section

Interosseous border

Interosseous membrane of leg

Interosseous crest

Anterior border

Lateral surface

Fibula

Posterior border

Lateral surface

Anterior border

Medial surface

Tibia

Medial border

Posterior surface

Medial surface

Medial border

Posterior surface

Inferior view

Anterior tibiofibular ligament

Articular facet of lateral malleolus

Fibula

Malleolar fossa of fibula

Posterior tibiofibular ligament

Anterior

Posterior

Articular facet of medial malleolus

Medial malleolus

Tibia

Inferior articular surface of tibia

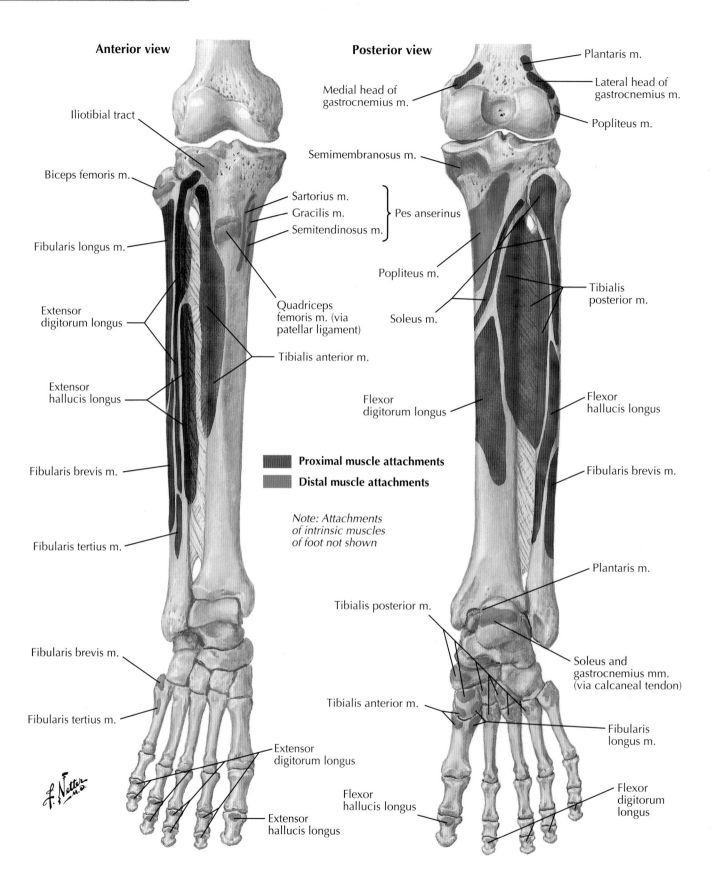

Anterior view

Iliotibial tract

Biceps femoris m.

Fibularis longus m.

Extensor digitorum longus

Extensor hallucis longus

Fibularis brevis m.

Fibularis tertius m.

Fibularis brevis m.

Fibularis tertius m.

Extensor digitorum longus

Extensor hallucis longus

Posterior view

Plantaris m.

Lateral head of gastrocnemius m.

Popliteus m.

Medial head of gastrocnemius m.

Semimembranosus m.

Sartorius m.
Gracilis m.
Semitendinosus m.

Pes anserinus

Quadriceps femoris m. (via patellar ligament)

Tibialis anterior m.

Popliteus m.

Soleus m.

Tibialis posterior m.

Flexor digitorum longus

Flexor hallucis longus

Fibularis brevis m.

Plantaris m.

Tibialis posterior m.

Soleus and gastrocnemius mm. (via calcaneal tendon)

Tibialis anterior m.

Fibularis longus m.

Flexor hallucis longus

Flexor digitorum longus

Proximal muscle attachments

Distal muscle attachments

Note: Attachments of intrinsic muscles of foot not shown

Plate 526

Leg

Right leg

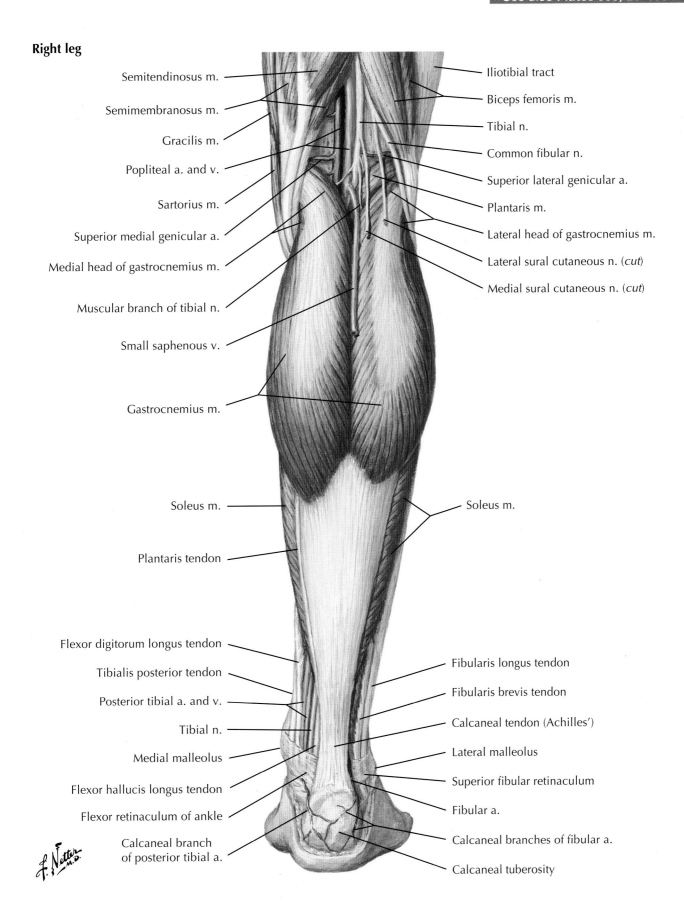

Semitendinosus m.

Semimembranosus m.

Gracilis m.

Popliteal a. and v.

Sartorius m.

Superior medial genicular a.

Medial head of gastrocnemius m.

Muscular branch of tibial n.

Small saphenous v.

Gastrocnemius m.

Soleus m.

Plantaris tendon

Flexor digitorum longus tendon

Tibialis posterior tendon

Posterior tibial a. and v.

Tibial n.

Medial malleolus

Flexor hallucis longus tendon

Flexor retinaculum of ankle

Calcaneal branch
of posterior tibial a.

Iliotibial tract

Biceps femoris m.

Tibial n.

Common fibular n.

Superior lateral genicular a.

Plantaris m.

Lateral head of gastrocnemius m.

Lateral sural cutaneous n. (*cut*)

Medial sural cutaneous n. (*cut*)

Soleus m.

Fibularis longus tendon

Fibularis brevis tendon

Calcaneal tendon (Achilles')

Lateral malleolus

Superior fibular retinaculum

Fibular a.

Calcaneal branches of fibular a.

Calcaneal tuberosity

Right leg

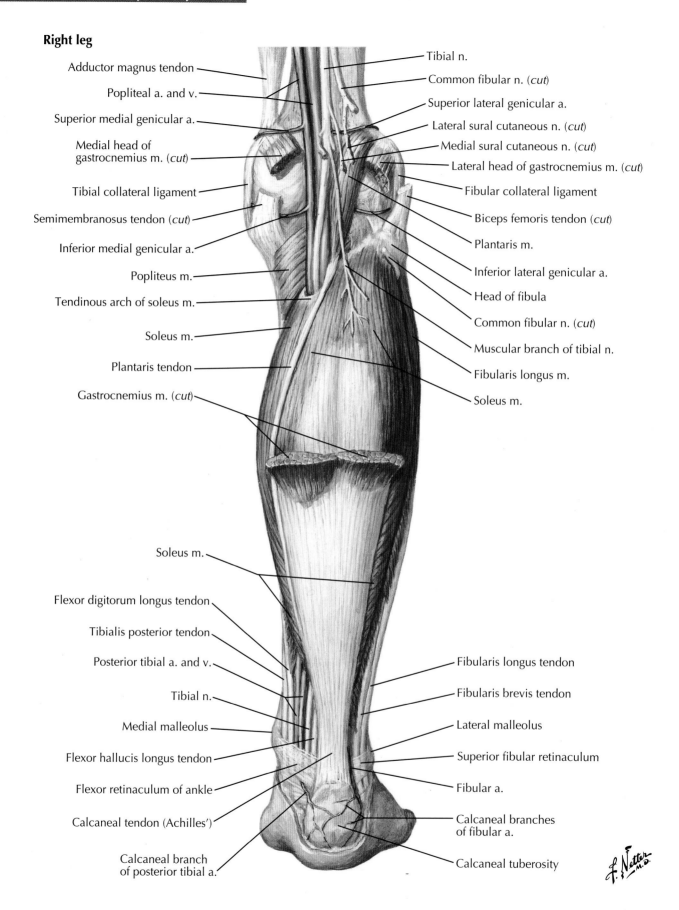

Adductor magnus tendon

Popliteal a. and v.

Superior medial genicular a.

Medial head of gastrocnemius m. (*cut*)

Tibial collateral ligament

Semimembranosus tendon (*cut*)

Inferior medial genicular a.

Popliteus m.

Tendinous arch of soleus m.

Soleus m.

Plantaris tendon

Gastrocnemius m. (*cut*)

Soleus m.

Flexor digitorum longus tendon

Tibialis posterior tendon

Posterior tibial a. and v.

Tibial n.

Medial malleolus

Flexor hallucis longus tendon

Flexor retinaculum of ankle

Calcaneal tendon (Achilles')

Calcaneal branch of posterior tibial a.

Tibial n.

Common fibular n. (*cut*)

Superior lateral genicular a.

Lateral sural cutaneous n. (*cut*)

Medial sural cutaneous n. (*cut*)

Lateral head of gastrocnemius m. (*cut*)

Fibular collateral ligament

Biceps femoris tendon (*cut*)

Plantaris m.

Inferior lateral genicular a.

Head of fibula

Common fibular n. (*cut*)

Muscular branch of tibial n.

Fibularis longus m.

Soleus m.

Fibularis longus tendon

Fibularis brevis tendon

Lateral malleolus

Superior fibular retinaculum

Fibular a.

Calcaneal branches of fibular a.

Calcaneal tuberosity

Plate 528

Leg

Right leg

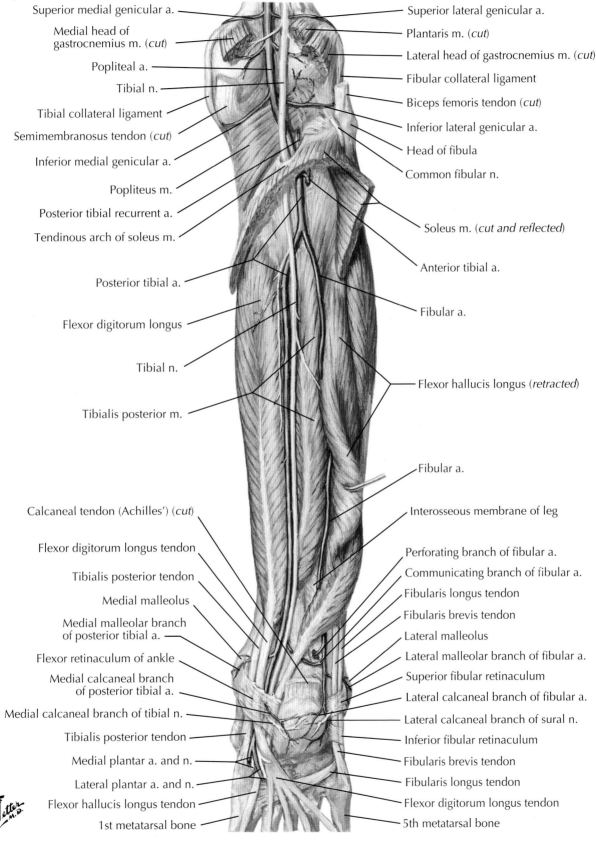

Superior medial genicular a.

Medial head of gastrocnemius m. (*cut*)

Popliteal a.

Tibial n.

Tibial collateral ligament

Semimembranosus tendon (*cut*)

Inferior medial genicular a.

Popliteus m.

Posterior tibial recurrent a.

Tendinous arch of soleus m.

Posterior tibial a.

Flexor digitorum longus

Tibial n.

Tibialis posterior m.

Calcaneal tendon (Achilles') (*cut*)

Flexor digitorum longus tendon

Tibialis posterior tendon

Medial malleolus

Medial malleolar branch of posterior tibial a.

Flexor retinaculum of ankle

Medial calcaneal branch of posterior tibial a.

Medial calcaneal branch of tibial n.

Tibialis posterior tendon

Medial plantar a. and n.

Lateral plantar a. and n.

Flexor hallucis longus tendon

1st metatarsal bone

Superior lateral genicular a.

Plantaris m. (*cut*)

Lateral head of gastrocnemius m. (*cut*)

Fibular collateral ligament

Biceps femoris tendon (*cut*)

Inferior lateral genicular a.

Head of fibula

Common fibular n.

Soleus m. (*cut and reflected*)

Anterior tibial a.

Fibular a.

Flexor hallucis longus (*retracted*)

Fibular a.

Interosseous membrane of leg

Perforating branch of fibular a.

Communicating branch of fibular a.

Fibularis longus tendon

Fibularis brevis tendon

Lateral malleolus

Lateral malleolar branch of fibular a.

Superior fibular retinaculum

Lateral calcaneal branch of fibular a.

Lateral calcaneal branch of sural n.

Inferior fibular retinaculum

Fibularis brevis tendon

Fibularis longus tendon

Flexor digitorum longus tendon

5th metatarsal bone

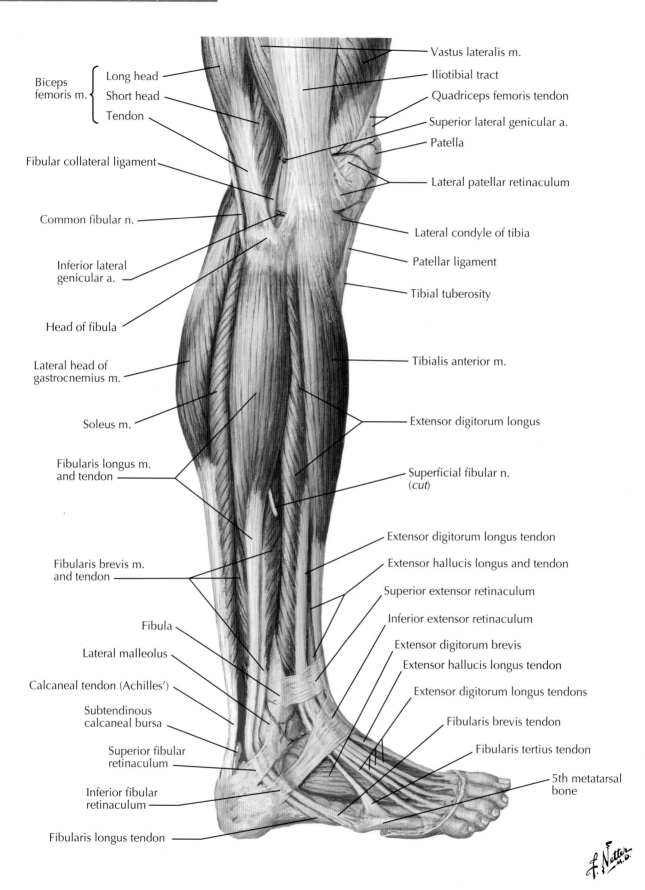

Biceps femoris m. { Long head
Short head
Tendon }

Vastus lateralis m.

Iliotibial tract

Quadriceps femoris tendon

Superior lateral genicular a.

Patella

Fibular collateral ligament

Lateral patellar retinaculum

Common fibular n.

Lateral condyle of tibia

Patellar ligament

Inferior lateral genicular a.

Tibial tuberosity

Head of fibula

Lateral head of gastrocnemius m.

Tibialis anterior m.

Soleus m.

Extensor digitorum longus

Fibularis longus m. and tendon

Superficial fibular n. (*cut*)

Extensor digitorum longus tendon

Fibularis brevis m. and tendon

Extensor hallucis longus and tendon

Superior extensor retinaculum

Inferior extensor retinaculum

Fibula

Extensor digitorum brevis

Lateral malleolus

Extensor hallucis longus tendon

Calcaneal tendon (Achilles')

Extensor digitorum longus tendons

Subtendinous calcaneal bursa

Fibularis brevis tendon

Superior fibular retinaculum

Fibularis tertius tendon

5th metatarsal bone

Inferior fibular retinaculum

Fibularis longus tendon

Plate 530

Leg

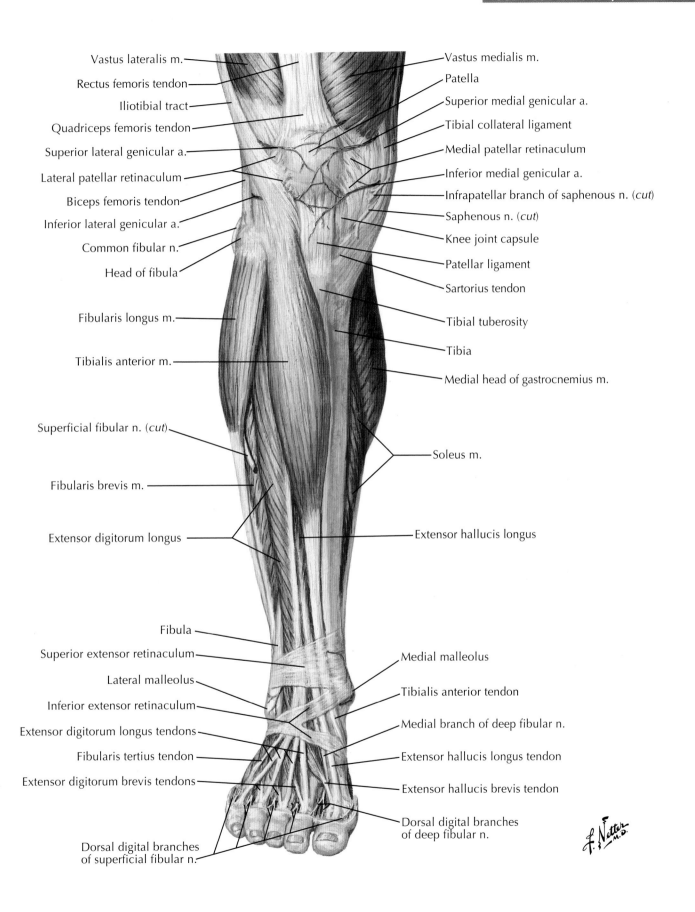

Vastus lateralis m.

Rectus femoris tendon

Iliotibial tract

Quadriceps femoris tendon

Superior lateral genicular a.

Lateral patellar retinaculum

Biceps femoris tendon

Inferior lateral genicular a.

Common fibular n.

Head of fibula

Fibularis longus m.

Tibialis anterior m.

Superficial fibular n. (*cut*)

Fibularis brevis m.

Extensor digitorum longus

Fibula

Superior extensor retinaculum

Lateral malleolus

Inferior extensor retinaculum

Extensor digitorum longus tendons

Fibularis tertius tendon

Extensor digitorum brevis tendons

Dorsal digital branches
of superficial fibular n.

Vastus medialis m.

Patella

Superior medial genicular a.

Tibial collateral ligament

Medial patellar retinaculum

Inferior medial genicular a.

Infrapatellar branch of saphenous n. (*cut*)

Saphenous n. (*cut*)

Knee joint capsule

Patellar ligament

Sartorius tendon

Tibial tuberosity

Tibia

Medial head of gastrocnemius m.

Soleus m.

Extensor hallucis longus

Medial malleolus

Tibialis anterior tendon

Medial branch of deep fibular n.

Extensor hallucis longus tendon

Extensor hallucis brevis tendon

Dorsal digital branches
of deep fibular n.

f. Netter
m.d.

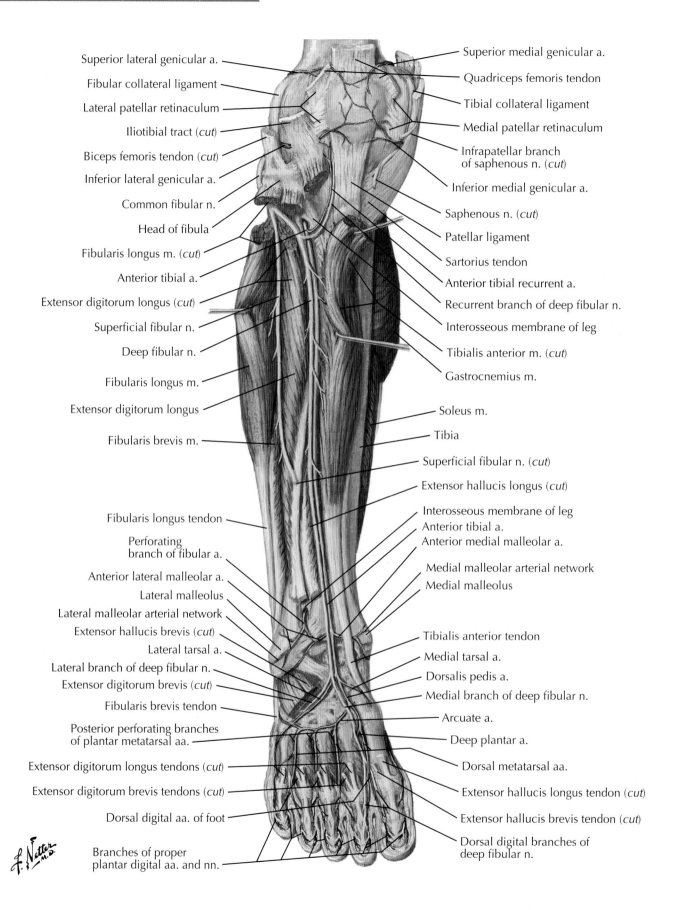

Superior lateral genicular a.

Fibular collateral ligament

Lateral patellar retinaculum

Iliotibial tract (*cut*)

Biceps femoris tendon (*cut*)

Inferior lateral genicular a.

Common fibular n.

Head of fibula

Fibularis longus m. (*cut*)

Anterior tibial a.

Extensor digitorum longus (*cut*)

Superficial fibular n.

Deep fibular n.

Fibularis longus m.

Extensor digitorum longus

Fibularis brevis m.

Fibularis longus tendon

Perforating
branch of fibular a.

Anterior lateral malleolar a.

Lateral malleolus

Lateral malleolar arterial network

Extensor hallucis brevis (*cut*)

Lateral tarsal a.

Lateral branch of deep fibular n.

Extensor digitorum brevis (*cut*)

Fibularis brevis tendon

Posterior perforating branches
of plantar metatarsal aa.

Extensor digitorum longus tendons (*cut*)

Extensor digitorum brevis tendons (*cut*)

Dorsal digital aa. of foot

Branches of proper
plantar digital aa. and nn.

Superior medial genicular a.

Quadriceps femoris tendon

Tibial collateral ligament

Medial patellar retinaculum

Infrapatellar branch
of saphenous n. (*cut*)

Inferior medial genicular a.

Saphenous n. (*cut*)

Patellar ligament

Sartorius tendon

Anterior tibial recurrent a.

Recurrent branch of deep fibular n.

Interosseous membrane of leg

Tibialis anterior m. (*cut*)

Gastrocnemius m.

Soleus m.

Tibia

Superficial fibular n. (*cut*)

Extensor hallucis longus (*cut*)

Interosseous membrane of leg
Anterior tibial a.
Anterior medial malleolar a.

Medial malleolar arterial network

Medial malleolus

Tibialis anterior tendon

Medial tarsal a.

Dorsalis pedis a.

Medial branch of deep fibular n.

Arcuate a.

Deep plantar a.

Dorsal metatarsal aa.

Extensor hallucis longus tendon (*cut*)

Extensor hallucis brevis tendon (*cut*)

Dorsal digital branches of
deep fibular n.

Plate 532

Leg

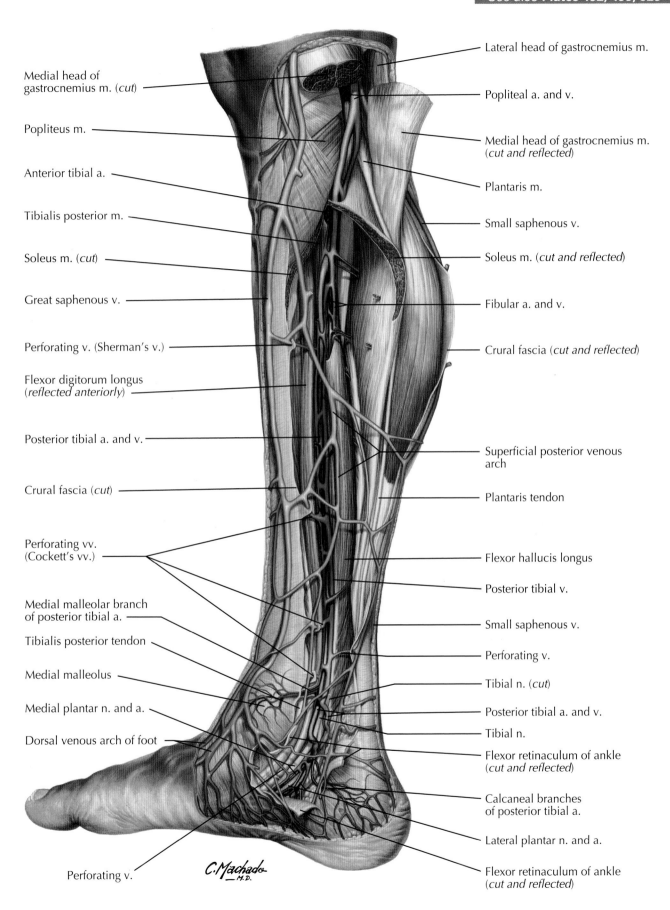

Medial head of
gastrocnemius m. (*cut*)

Popliteus m.

Anterior tibial a.

Tibialis posterior m.

Soleus m. (*cut*)

Great saphenous v.

Perforating v. (Sherman's v.)

Flexor digitorum longus
(*reflected anteriorly*)

Posterior tibial a. and v.

Crural fascia (*cut*)

Perforating vv.
(Cockett's vv.)

Medial malleolar branch
of posterior tibial a.

Tibialis posterior tendon

Medial malleolus

Medial plantar n. and a.

Dorsal venous arch of foot

Perforating v.

Lateral head of gastrocnemius m.

Popliteal a. and v.

Medial head of gastrocnemius m.
(*cut and reflected*)

Plantaris m.

Small saphenous v.

Soleus m. (*cut and reflected*)

Fibular a. and v.

Crural fascia (*cut and reflected*)

Superficial posterior venous
arch

Plantaris tendon

Flexor hallucis longus

Posterior tibial v.

Small saphenous v.

Perforating v.

Tibial n. (*cut*)

Posterior tibial a. and v.

Tibial n.

Flexor retinaculum of ankle
(*cut and reflected*)

Calcaneal branches
of posterior tibial a.

Lateral plantar n. and a.

Flexor retinaculum of ankle
(*cut and reflected*)

C. Machado
M.D.

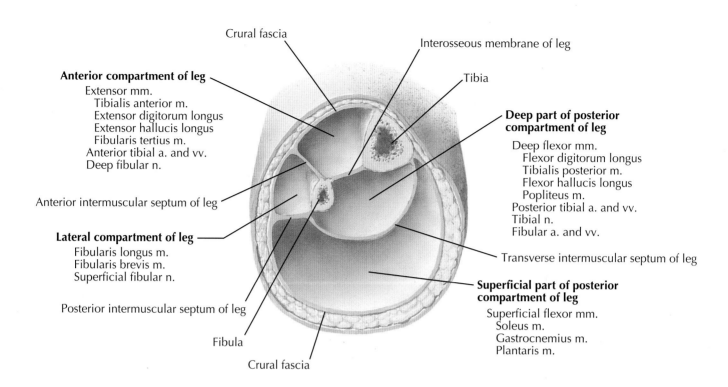

Crural fascia

Interosseous membrane of leg

Tibia

Anterior compartment of leg
Extensor mm.
Tibialis anterior m.
Extensor digitorum longus
Extensor hallucis longus
Fibularis tertius m.
Anterior tibial a. and vv.
Deep fibular n.

Deep part of posterior compartment of leg
Deep flexor mm.
Flexor digitorum longus
Tibialis posterior m.
Flexor hallucis longus
Popliteus m.
Posterior tibial a. and vv.
Tibial n.
Fibular a. and vv.

Anterior intermuscular septum of leg

Transverse intermuscular septum of leg

Lateral compartment of leg
Fibularis longus m.
Fibularis brevis m.
Superficial fibular n.

Superficial part of posterior compartment of leg
Superficial flexor mm.
Soleus m.
Gastrocnemius m.
Plantaris m.

Posterior intermuscular septum of leg

Fibula

Crural fascia

Cross section just above middle of leg

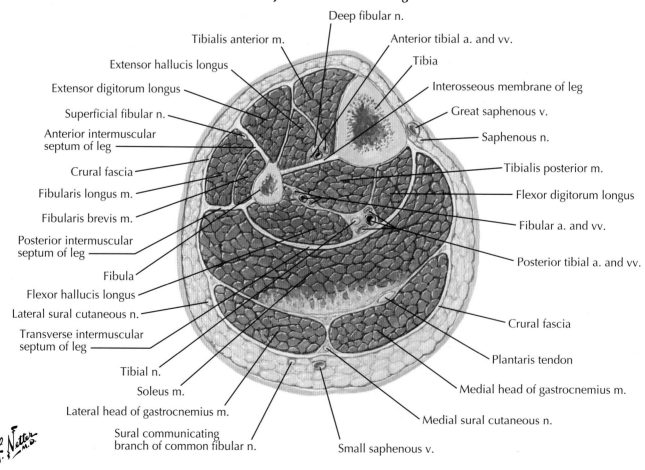

Deep fibular n.

Tibialis anterior m.

Anterior tibial a. and vv.

Extensor hallucis longus

Tibia

Extensor digitorum longus

Interosseous membrane of leg

Superficial fibular n.

Great saphenous v.

Anterior intermuscular septum of leg

Saphenous n.

Crural fascia

Tibialis posterior m.

Fibularis longus m.

Flexor digitorum longus

Fibularis brevis m.

Fibular a. and vv.

Posterior intermuscular septum of leg

Posterior tibial a. and vv.

Fibula

Flexor hallucis longus

Crural fascia

Lateral sural cutaneous n.

Transverse intermuscular septum of leg

Plantaris tendon

Tibial n.

Medial head of gastrocnemius m.

Soleus m.

Medial sural cutaneous n.

Lateral head of gastrocnemius m.

Sural communicating branch of common fibular n.

Small saphenous v.

Plate 534

Leg

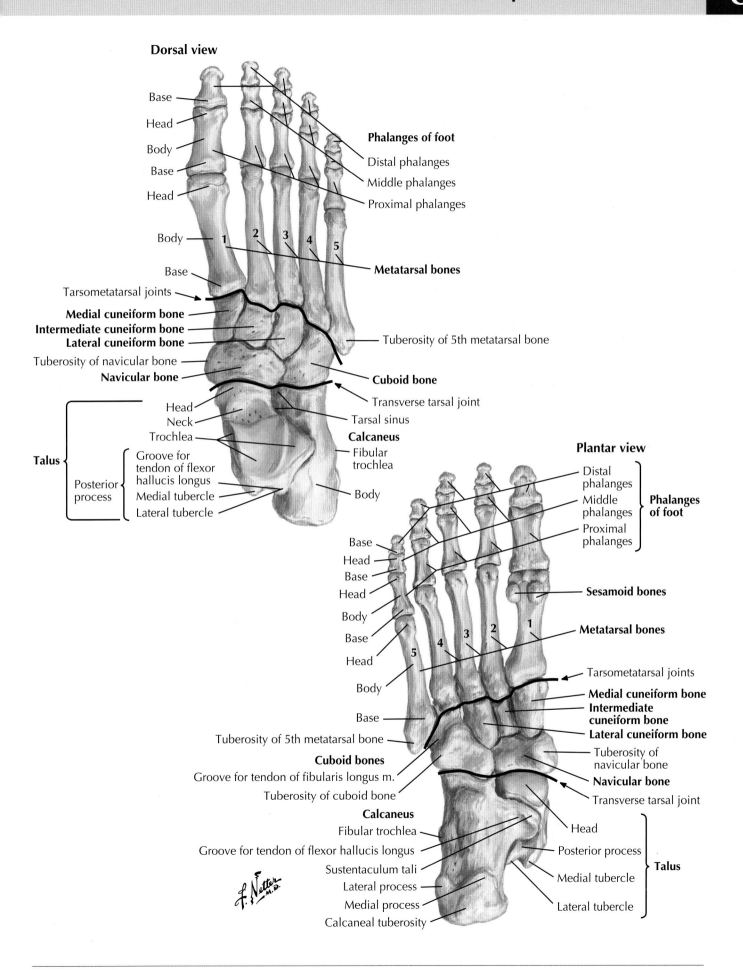

Dorsal view

Base
Head
Body
Base
Head

Phalanges of foot

Distal phalanges
Middle phalanges
Proximal phalanges

Body   1   2   3   4   5

Base

Metatarsal bones

Tarsometatarsal joints

Medial cuneiform bone
Intermediate cuneiform bone
Lateral cuneiform bone

Tuberosity of 5th metatarsal bone

Tuberosity of navicular bone
Navicular bone

Cuboid bone

Transverse tarsal joint

Head
Neck
Trochlea

Tarsal sinus

Calcaneus

Fibular
trochlea

Talus

Posterior
process

Groove for
tendon of flexor
hallucis longus
Medial tubercle
Lateral tubercle

Body

Plantar view

Distal
phalanges
Middle
phalanges
Proximal
phalanges

**Phalanges
of foot**

Base
Head
Base
Head

Sesamoid bones

Body
Base
Head

5   4   3   2   1

Metatarsal bones

Body

Tarsometatarsal joints

Medial cuneiform bone
**Intermediate
cuneiform bone**
Lateral cuneiform bone

Base

Tuberosity of 5th metatarsal bone

Cuboid bones

Groove for tendon of fibularis longus m.

Tuberosity of cuboid bone

Tuberosity of
navicular bone

Navicular bone

Transverse tarsal joint

Calcaneus

Fibular trochlea

Groove for tendon of flexor hallucis longus

Sustentaculum tali

Lateral process

Medial process

Calcaneal tuberosity

Head

Posterior process

Medial tubercle

Lateral tubercle

Talus

f. Netter.
M.D.

Lateral view

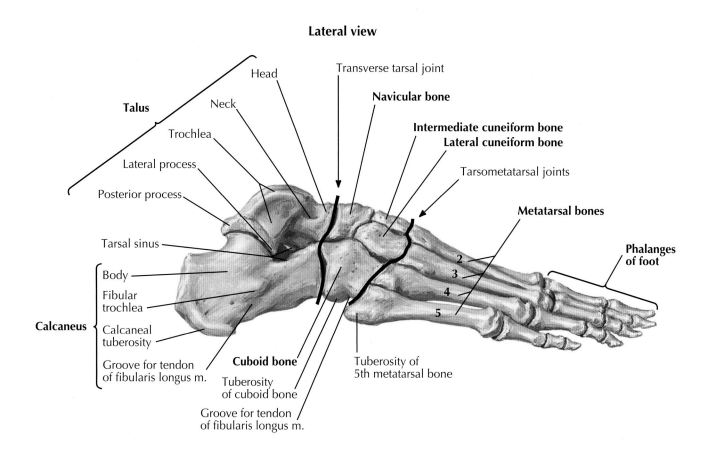

Head

Transverse tarsal joint

Talus

Neck

Navicular bone

Trochlea

Intermediate cuneiform bone
Lateral cuneiform bone

Lateral process

Tarsometatarsal joints

Posterior process

Metatarsal bones

Tarsal sinus

**Phalanges
of foot**

Body

2

Fibular
trochlea

3

Calcaneus

4

Calcaneal
tuberosity

5

Groove for tendon
of fibularis longus m.

Cuboid bone

Tuberosity of
5th metatarsal bone

Tuberosity
of cuboid bone

Groove for tendon
of fibularis longus m.

Medial view

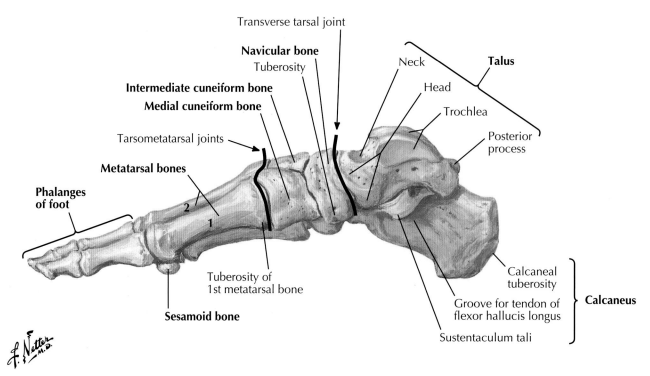

Transverse tarsal joint

Navicular bone

Neck

Talus

Tuberosity

Head

Intermediate cuneiform bone

Trochlea

Medial cuneiform bone

Posterior
process

Tarsometatarsal joints

Metatarsal bones

2

**Phalanges
of foot**

1

Tuberosity of
1st metatarsal bone

Calcaneal
tuberosity

Sesamoid bone

Groove for tendon of
flexor hallucis longus

Calcaneus

Sustentaculum tali

Plate 536

Ankle and Foot

Right foot

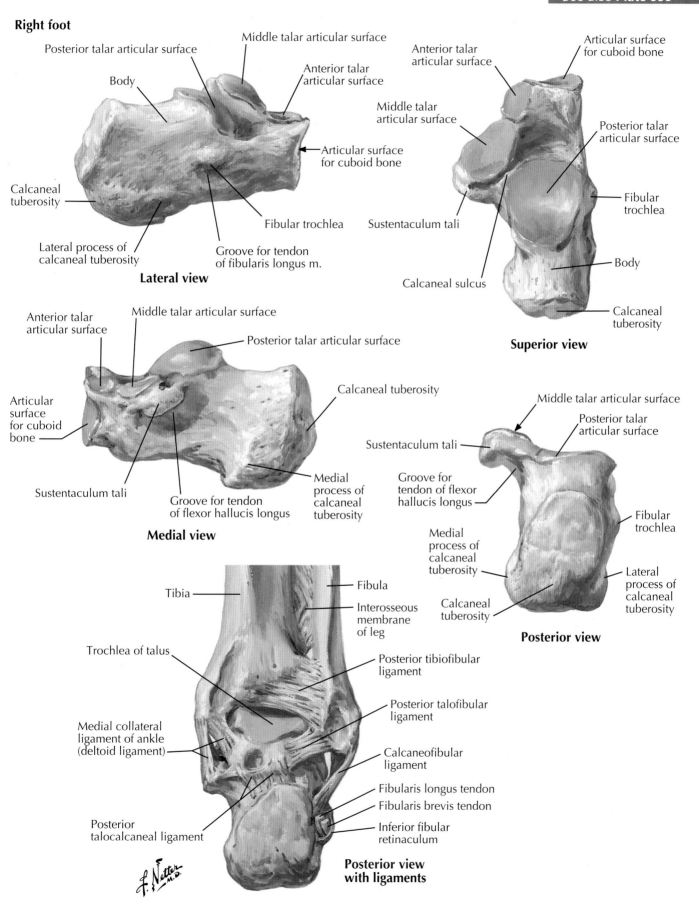

Posterior talar articular surface

Body

Middle talar articular surface

Anterior talar articular surface

Articular surface for cuboid bone

Calcaneal tuberosity

Fibular trochlea

Lateral process of calcaneal tuberosity

Groove for tendon of fibularis longus m.

Lateral view

Anterior talar articular surface

Articular surface for cuboid bone

Middle talar articular surface

Middle talar articular surface

Posterior talar articular surface

Calcaneal tuberosity

Sustentaculum tali

Groove for tendon of flexor hallucis longus

Medial process of calcaneal tuberosity

Medial view

Anterior talar articular surface

Articular surface for cuboid bone

Middle talar articular surface

Posterior talar articular surface

Fibular trochlea

Sustentaculum tali

Body

Calcaneal sulcus

Calcaneal tuberosity

Superior view

Middle talar articular surface

Posterior talar articular surface

Sustentaculum tali

Groove for tendon of flexor hallucis longus

Medial process of calcaneal tuberosity

Calcaneal tuberosity

Fibular trochlea

Lateral process of calcaneal tuberosity

Posterior view

Tibia

Fibula

Interosseous membrane of leg

Trochlea of talus

Posterior tibiofibular ligament

Posterior talofibular ligament

Medial collateral ligament of ankle (deltoid ligament)

Calcaneofibular ligament

Fibularis longus tendon

Fibularis brevis tendon

Inferior fibular retinaculum

Posterior talocalcaneal ligament

Posterior view with ligaments

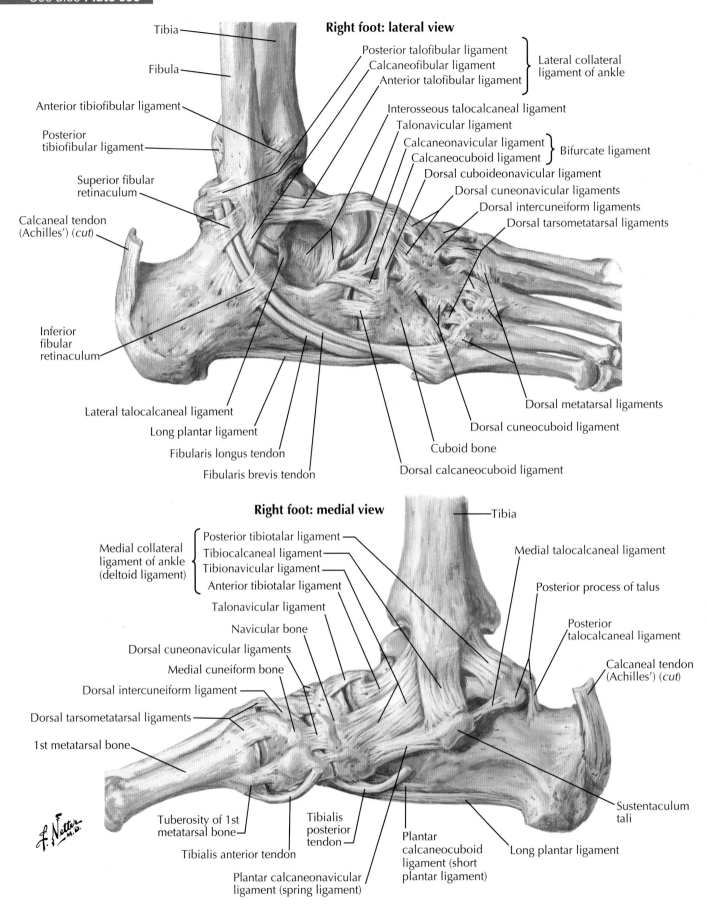

Right foot: lateral view

Tibia

Fibula

Anterior tibiofibular ligament

Posterior tibiofibular ligament

Superior fibular retinaculum

Calcaneal tendon (Achilles') (cut)

Inferior fibular retinaculum

Posterior talofibular ligament
Calcaneofibular ligament
Anterior talofibular ligament

Lateral collateral ligament of ankle

Interosseous talocalcaneal ligament

Talonavicular ligament

Calcaneonavicular ligament
Calcaneocuboid ligament

Bifurcate ligament

Dorsal cuboideonavicular ligament

Dorsal cuneonavicular ligaments

Dorsal intercuneiform ligaments

Dorsal tarsometatarsal ligaments

Lateral talocalcaneal ligament

Long plantar ligament

Fibularis longus tendon

Fibularis brevis tendon

Dorsal metatarsal ligaments

Dorsal cuneocuboid ligament

Cuboid bone

Dorsal calcaneocuboid ligament

Right foot: medial view

Tibia

Medial talocalcaneal ligament

Posterior process of talus

Posterior talocalcaneal ligament

Calcaneal tendon (Achilles') (cut)

Medial collateral ligament of ankle (deltoid ligament)

Posterior tibiotalar ligament
Tibiocalcaneal ligament
Tibionavicular ligament
Anterior tibiotalar ligament

Talonavicular ligament

Navicular bone

Dorsal cuneonavicular ligaments

Medial cuneiform bone

Dorsal intercuneiform ligament

Dorsal tarsometatarsal ligaments

1st metatarsal bone

Tuberosity of 1st metatarsal bone

Tibialis anterior tendon

Plantar calcaneonavicular ligament (spring ligament)

Tibialis posterior tendon

Plantar calcaneocuboid ligament (short plantar ligament)

Long plantar ligament

Sustentaculum tali

Plate 538

Ankle and Foot

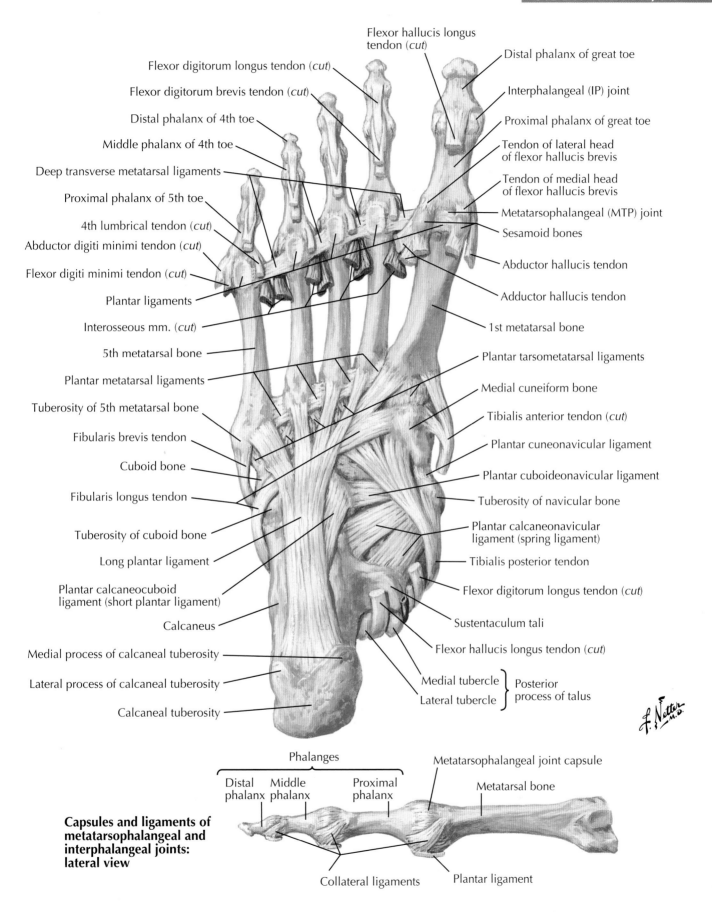

Flexor hallucis longus
tendon (*cut*)

Flexor digitorum longus tendon (*cut*)

Flexor digitorum brevis tendon (*cut*)

Distal phalanx of 4th toe

Middle phalanx of 4th toe

Deep transverse metatarsal ligaments

Proximal phalanx of 5th toe

4th lumbrical tendon (*cut*)

Abductor digiti minimi tendon (*cut*)

Flexor digiti minimi tendon (*cut*)

Plantar ligaments

Interosseous mm. (*cut*)

5th metatarsal bone

Plantar metatarsal ligaments

Tuberosity of 5th metatarsal bone

Fibularis brevis tendon

Cuboid bone

Fibularis longus tendon

Tuberosity of cuboid bone

Long plantar ligament

Plantar calcaneocuboid
ligament (short plantar ligament)

Calcaneus

Medial process of calcaneal tuberosity

Lateral process of calcaneal tuberosity

Calcaneal tuberosity

Distal phalanx of great toe

Interphalangeal (IP) joint

Proximal phalanx of great toe

Tendon of lateral head
of flexor hallucis brevis

Tendon of medial head
of flexor hallucis brevis

Metatarsophalangeal (MTP) joint

Sesamoid bones

Abductor hallucis tendon

Adductor hallucis tendon

1st metatarsal bone

Plantar tarsometatarsal ligaments

Medial cuneiform bone

Tibialis anterior tendon (*cut*)

Plantar cuneonavicular ligament

Plantar cuboideonavicular ligament

Tuberosity of navicular bone

Plantar calcaneonavicular
ligament (spring ligament)

Tibialis posterior tendon

Flexor digitorum longus tendon (*cut*)

Sustentaculum tali

Flexor hallucis longus tendon (*cut*)

Medial tubercle ⎫ Posterior
Lateral tubercle ⎭ process of talus

Phalanges

Distal   Middle   Proximal
phalanx phalanx   phalanx

Metatarsophalangeal joint capsule

Metatarsal bone

**Capsules and ligaments of
metatarsophalangeal and
interphalangeal joints:
lateral view**

Collateral ligaments   Plantar ligament

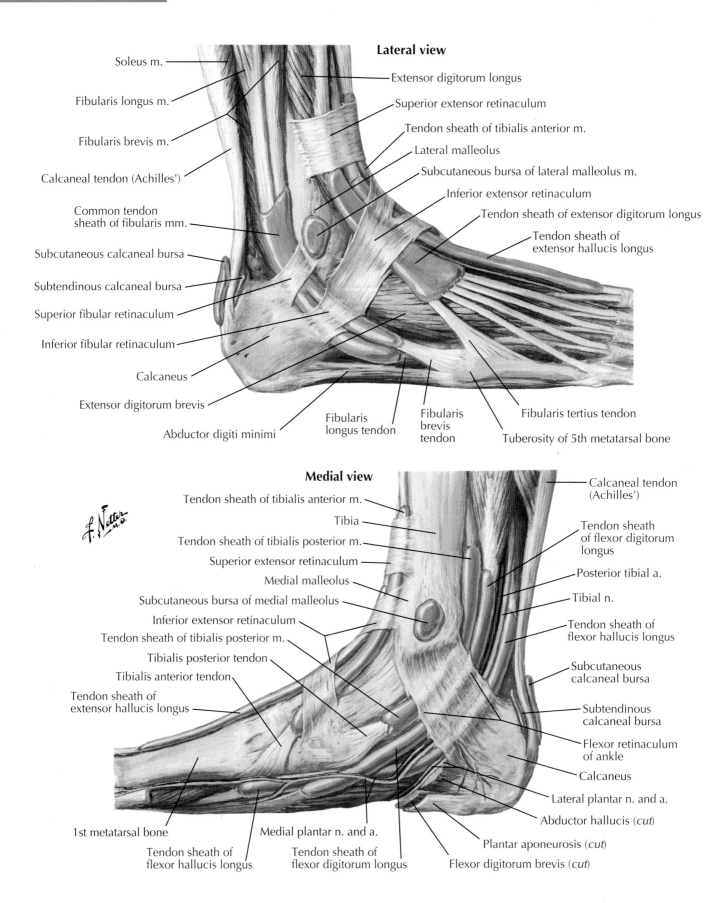

Lateral view

Soleus m.

Fibularis longus m.

Fibularis brevis m.

Calcaneal tendon (Achilles')

Common tendon sheath of fibularis mm.

Subcutaneous calcaneal bursa

Subtendinous calcaneal bursa

Superior fibular retinaculum

Inferior fibular retinaculum

Calcaneus

Extensor digitorum brevis

Abductor digiti minimi

Fibularis longus tendon

Fibularis brevis tendon

Extensor digitorum longus

Superior extensor retinaculum

Tendon sheath of tibialis anterior m.

Lateral malleolus

Subcutaneous bursa of lateral malleolus m.

Inferior extensor retinaculum

Tendon sheath of extensor digitorum longus

Tendon sheath of extensor hallucis longus

Fibularis tertius tendon

Tuberosity of 5th metatarsal bone

Medial view

Tendon sheath of tibialis anterior m.

Tibia

Tendon sheath of tibialis posterior m.

Superior extensor retinaculum

Medial malleolus

Subcutaneous bursa of medial malleolus

Inferior extensor retinaculum

Tendon sheath of tibialis posterior m.

Tibialis posterior tendon

Tibialis anterior tendon

Tendon sheath of extensor hallucis longus

1st metatarsal bone

Tendon sheath of flexor hallucis longus

Medial plantar n. and a.

Tendon sheath of flexor digitorum longus

Calcaneal tendon (Achilles')

Tendon sheath of flexor digitorum longus

Posterior tibial a.

Tibial n.

Tendon sheath of flexor hallucis longus

Subcutaneous calcaneal bursa

Subtendinous calcaneal bursa

Flexor retinaculum of ankle

Calcaneus

Lateral plantar n. and a.

Abductor hallucis (cut)

Plantar aponeurosis (cut)

Flexor digitorum brevis (cut)

Plate 540

Ankle and Foot

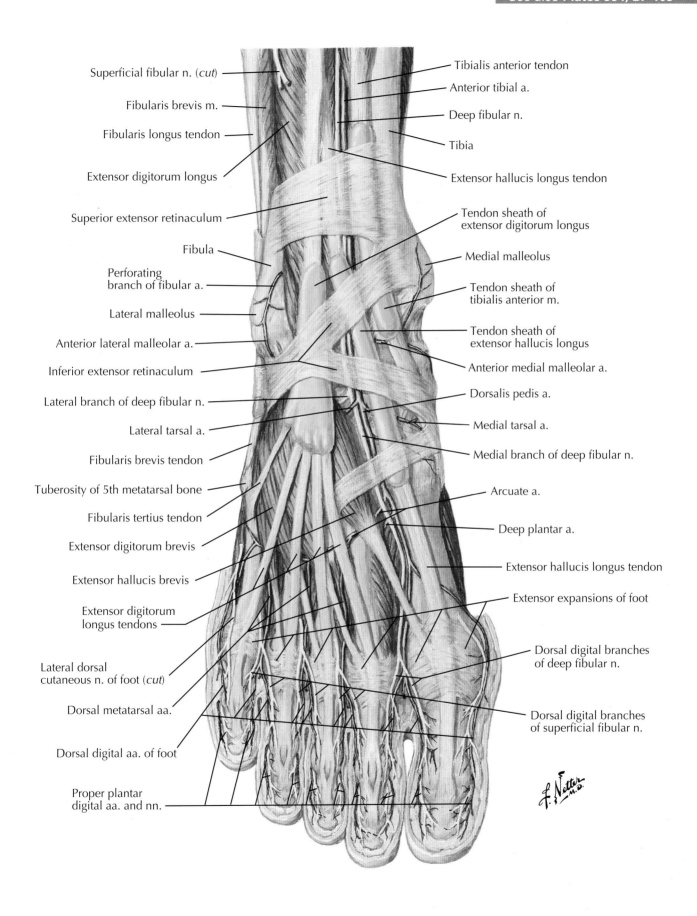

Superficial fibular n. (*cut*)

Fibularis brevis m.

Fibularis longus tendon

Extensor digitorum longus

Superior extensor retinaculum

Fibula

Perforating branch of fibular a.

Lateral malleolus

Anterior lateral malleolar a.

Inferior extensor retinaculum

Lateral branch of deep fibular n.

Lateral tarsal a.

Fibularis brevis tendon

Tuberosity of 5th metatarsal bone

Fibularis tertius tendon

Extensor digitorum brevis

Extensor hallucis brevis

Extensor digitorum longus tendons

Lateral dorsal cutaneous n. of foot (*cut*)

Dorsal metatarsal aa.

Dorsal digital aa. of foot

Proper plantar digital aa. and nn.

Tibialis anterior tendon

Anterior tibial a.

Deep fibular n.

Tibia

Extensor hallucis longus tendon

Tendon sheath of extensor digitorum longus

Medial malleolus

Tendon sheath of tibialis anterior m.

Tendon sheath of extensor hallucis longus

Anterior medial malleolar a.

Dorsalis pedis a.

Medial tarsal a.

Medial branch of deep fibular n.

Arcuate a.

Deep plantar a.

Extensor hallucis longus tendon

Extensor expansions of foot

Dorsal digital branches of deep fibular n.

Dorsal digital branches of superficial fibular n.

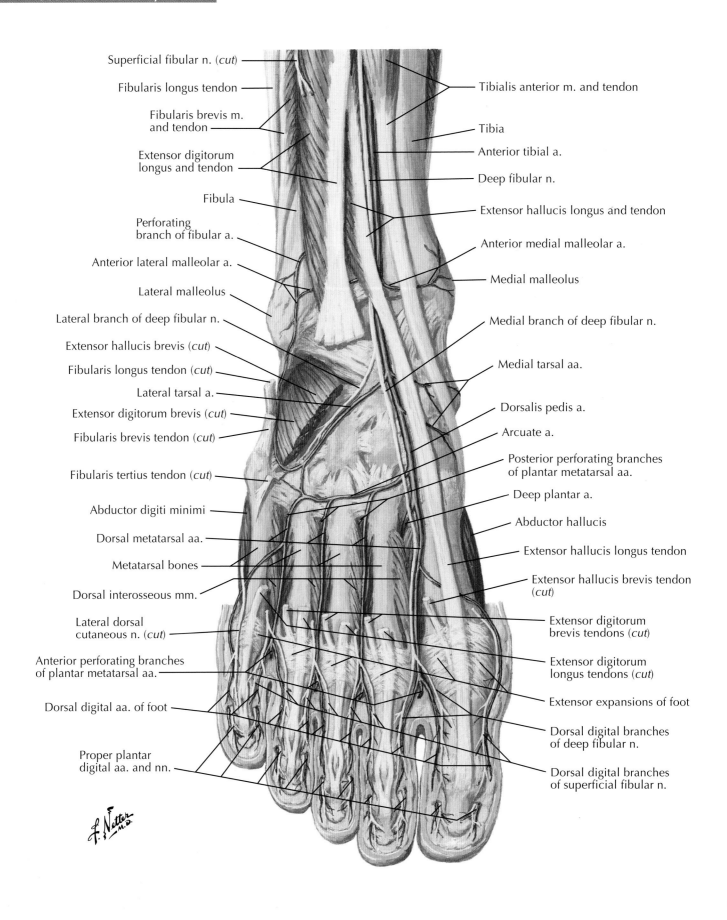

Superficial fibular n. (*cut*)

Fibularis longus tendon

Fibularis brevis m. and tendon

Extensor digitorum longus and tendon

Fibula

Perforating branch of fibular a.

Anterior lateral malleolar a.

Lateral malleolus

Lateral branch of deep fibular n.

Extensor hallucis brevis (*cut*)

Fibularis longus tendon (*cut*)

Lateral tarsal a.

Extensor digitorum brevis (*cut*)

Fibularis brevis tendon (*cut*)

Fibularis tertius tendon (*cut*)

Abductor digiti minimi

Dorsal metatarsal aa.

Metatarsal bones

Dorsal interosseous mm.

Lateral dorsal cutaneous n. (*cut*)

Anterior perforating branches of plantar metatarsal aa.

Dorsal digital aa. of foot

Proper plantar digital aa. and nn.

Tibialis anterior m. and tendon

Tibia

Anterior tibial a.

Deep fibular n.

Extensor hallucis longus and tendon

Anterior medial malleolar a.

Medial malleolus

Medial branch of deep fibular n.

Medial tarsal aa.

Dorsalis pedis a.

Arcuate a.

Posterior perforating branches of plantar metatarsal aa.

Deep plantar a.

Abductor hallucis

Extensor hallucis longus tendon

Extensor hallucis brevis tendon (*cut*)

Extensor digitorum brevis tendons (*cut*)

Extensor digitorum longus tendons (*cut*)

Extensor expansions of foot

Dorsal digital branches of deep fibular n.

Dorsal digital branches of superficial fibular n.

Plate 542

Ankle and Foot

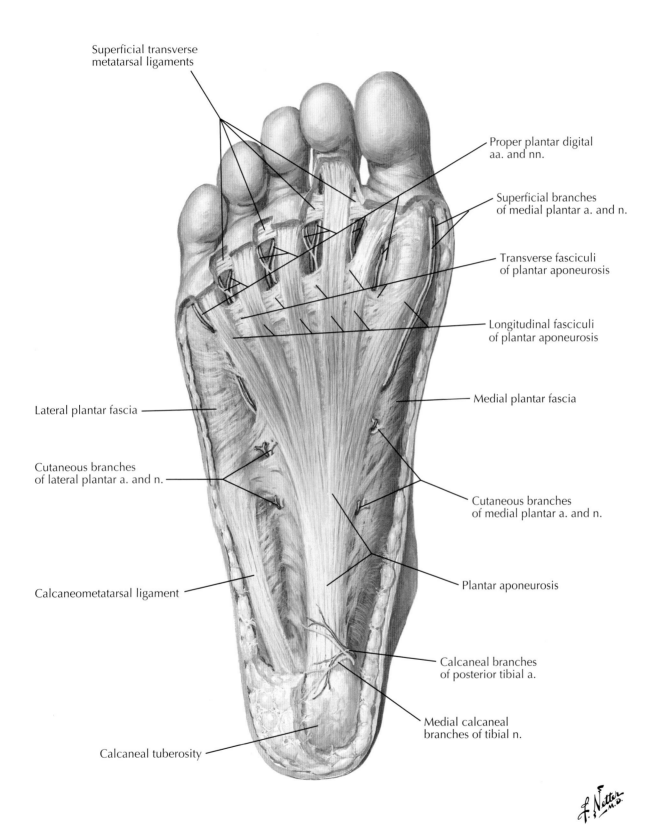

Superficial transverse metatarsal ligaments

Proper plantar digital aa. and nn.

Superficial branches of medial plantar a. and n.

Transverse fasciculi of plantar aponeurosis

Longitudinal fasciculi of plantar aponeurosis

Medial plantar fascia

Lateral plantar fascia

Cutaneous branches of lateral plantar a. and n.

Cutaneous branches of medial plantar a. and n.

Plantar aponeurosis

Calcaneometatarsal ligament

Calcaneal branches of posterior tibial a.

Medial calcaneal branches of tibial n.

Calcaneal tuberosity

First layer of muscles in bold

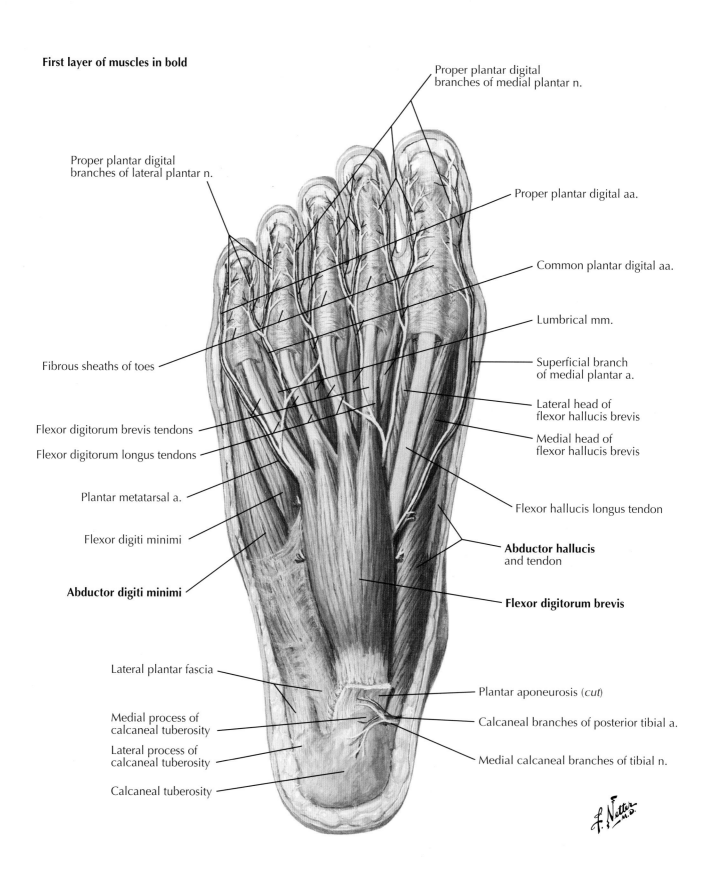

Proper plantar digital
branches of medial plantar n.

Proper plantar digital
branches of lateral plantar n.

Proper plantar digital aa.

Common plantar digital aa.

Lumbrical mm.

Superficial branch
of medial plantar a.

Fibrous sheaths of toes

Lateral head of
flexor hallucis brevis

Flexor digitorum brevis tendons

Medial head of
flexor hallucis brevis

Flexor digitorum longus tendons

Flexor hallucis longus tendon

Plantar metatarsal a.

Abductor hallucis
and tendon

Flexor digiti minimi

Flexor digitorum brevis

Abductor digiti minimi

Lateral plantar fascia

Plantar aponeurosis (*cut*)

Medial process of
calcaneal tuberosity

Calcaneal branches of posterior tibial a.

Lateral process of
calcaneal tuberosity

Medial calcaneal branches of tibial n.

Calcaneal tuberosity

Plate 544

Ankle and Foot

Second layer of tendons and muscles in bold

Proper plantar digital branches of medial plantar n.

Proper plantar digital branches of lateral plantar n.

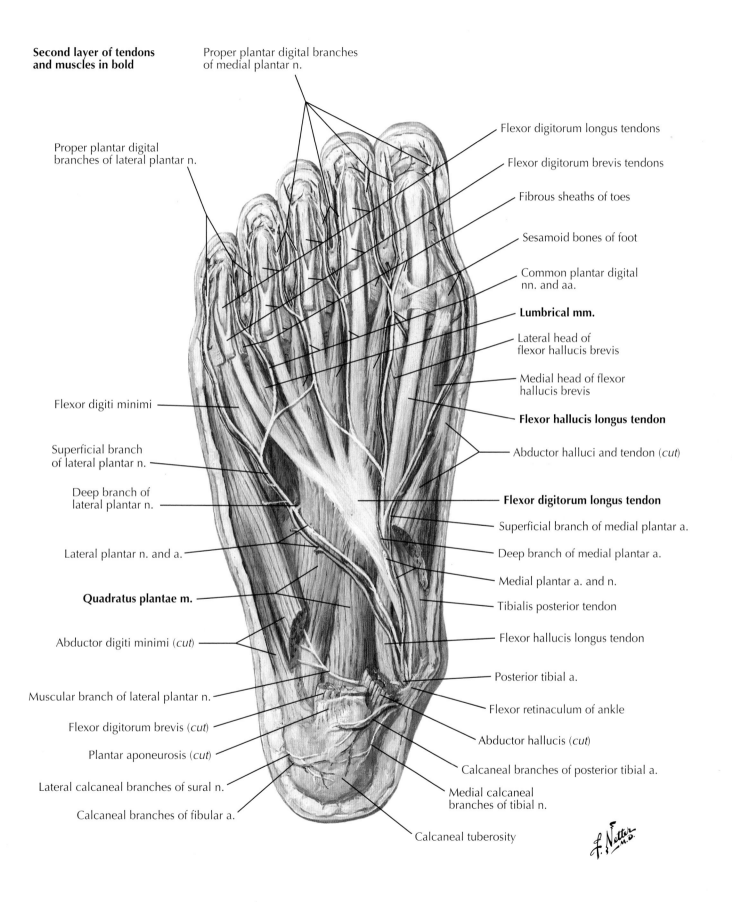

Flexor digitorum longus tendons

Flexor digitorum brevis tendons

Fibrous sheaths of toes

Sesamoid bones of foot

Common plantar digital nn. and aa.

Lumbrical mm.

Lateral head of flexor hallucis brevis

Medial head of flexor hallucis brevis

Flexor hallucis longus tendon

Abductor halluci and tendon (*cut*)

Flexor digiti minimi

Superficial branch of lateral plantar n.

Deep branch of lateral plantar n.

Lateral plantar n. and a.

Quadratus plantae m.

Abductor digiti minimi (*cut*)

Muscular branch of lateral plantar n.

Flexor digitorum brevis (*cut*)

Plantar aponeurosis (*cut*)

Lateral calcaneal branches of sural n.

Calcaneal branches of fibular a.

Flexor digitorum longus tendon

Superficial branch of medial plantar a.

Deep branch of medial plantar a.

Medial plantar a. and n.

Tibialis posterior tendon

Flexor hallucis longus tendon

Posterior tibial a.

Flexor retinaculum of ankle

Abductor hallucis (*cut*)

Calcaneal branches of posterior tibial a.

Medial calcaneal branches of tibial n.

Calcaneal tuberosity

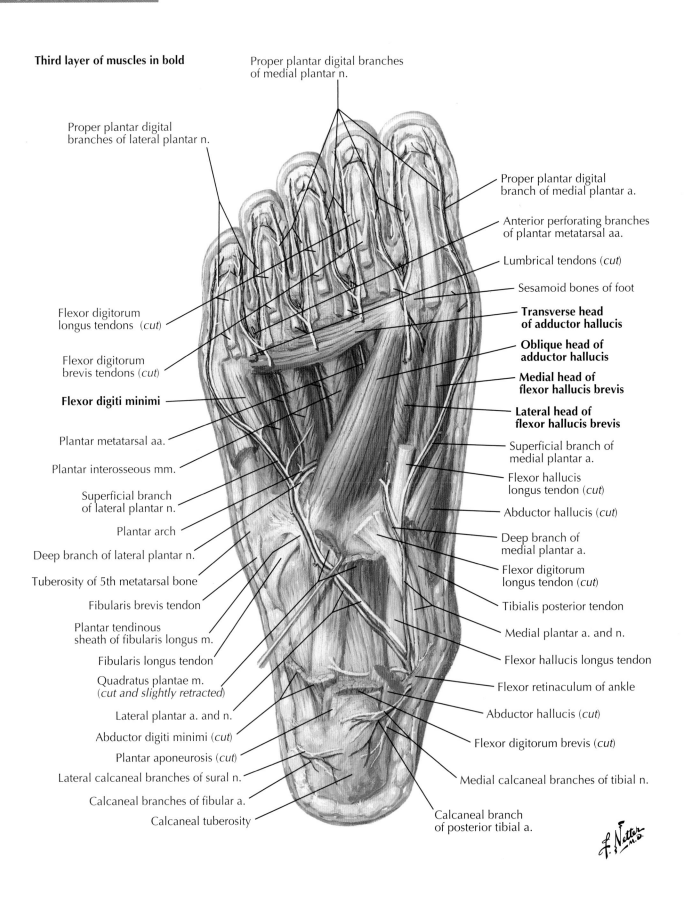

Third layer of muscles in bold

Proper plantar digital branches
of medial plantar n.

Proper plantar digital
branches of lateral plantar n.

Proper plantar digital
branch of medial plantar a.

Anterior perforating branches
of plantar metatarsal aa.

Lumbrical tendons (*cut*)

Sesamoid bones of foot

**Transverse head
of adductor hallucis**

**Oblique head of
adductor hallucis**

**Medial head of
flexor hallucis brevis**

**Lateral head of
flexor hallucis brevis**

Superficial branch of
medial plantar a.

Flexor hallucis
longus tendon (*cut*)

Abductor hallucis (*cut*)

Deep branch of
medial plantar a.

Flexor digitorum
longus tendon (*cut*)

Tibialis posterior tendon

Medial plantar a. and n.

Flexor hallucis longus tendon

Flexor retinaculum of ankle

Abductor hallucis (*cut*)

Flexor digitorum brevis (*cut*)

Medial calcaneal branches of tibial n.

Calcaneal branch
of posterior tibial a.

Flexor digitorum
longus tendons (*cut*)

Flexor digitorum
brevis tendons (*cut*)

Flexor digiti minimi

Plantar metatarsal aa.

Plantar interosseous mm.

Superficial branch
of lateral plantar n.

Plantar arch

Deep branch of lateral plantar n.

Tuberosity of 5th metatarsal bone

Fibularis brevis tendon

Plantar tendinous
sheath of fibularis longus m.

Fibularis longus tendon

Quadratus plantae m.
(*cut and slightly retracted*)

Lateral plantar a. and n.

Abductor digiti minimi (*cut*)

Plantar aponeurosis (*cut*)

Lateral calcaneal branches of sural n.

Calcaneal branches of fibular a.

Calcaneal tuberosity

Plate 546

Ankle and Foot

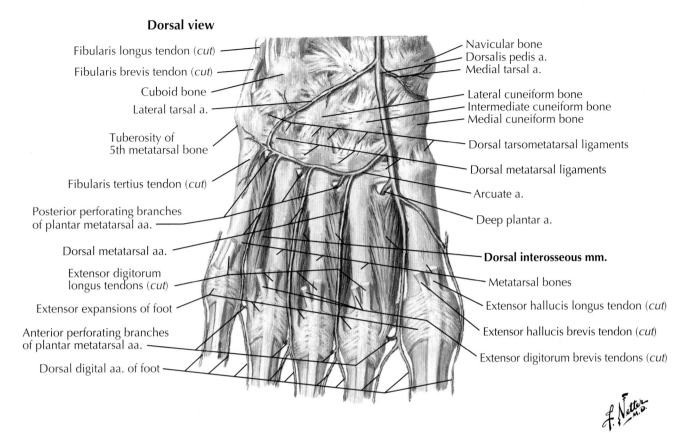

Dorsal view

Fibularis longus tendon (*cut*)

Fibularis brevis tendon (*cut*)

Cuboid bone

Lateral tarsal a.

Tuberosity of
5th metatarsal bone

Fibularis tertius tendon (*cut*)

Posterior perforating branches
of plantar metatarsal aa.

Dorsal metatarsal aa.

Extensor digitorum
longus tendons (*cut*)

Extensor expansions of foot

Anterior perforating branches
of plantar metatarsal aa.

Dorsal digital aa. of foot

Navicular bone

Dorsalis pedis a.

Medial tarsal a.

Lateral cuneiform bone

Intermediate cuneiform bone

Medial cuneiform bone

Dorsal tarsometatarsal ligaments

Dorsal metatarsal ligaments

Arcuate a.

Deep plantar a.

Dorsal interosseous mm.

Metatarsal bones

Extensor hallucis longus tendon (*cut*)

Extensor hallucis brevis tendon (*cut*)

Extensor digitorum brevis tendons (*cut*)

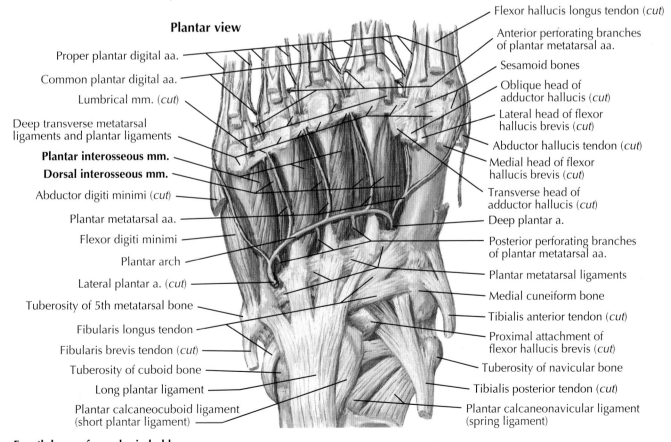

Plantar view

Proper plantar digital aa.

Common plantar digital aa.

Lumbrical mm. (*cut*)

Deep transverse metatarsal
ligaments and plantar ligaments

Plantar interosseous mm.

Dorsal interosseous mm.

Abductor digiti minimi (*cut*)

Plantar metatarsal aa.

Flexor digiti minimi

Plantar arch

Lateral plantar a. (*cut*)

Tuberosity of 5th metatarsal bone

Fibularis longus tendon

Fibularis brevis tendon (*cut*)

Tuberosity of cuboid bone

Long plantar ligament

Plantar calcaneocuboid ligament
(short plantar ligament)

Flexor hallucis longus tendon (*cut*)

Anterior perforating branches
of plantar metatarsal aa.

Sesamoid bones

Oblique head of
adductor hallucis (*cut*)

Lateral head of flexor
hallucis brevis (*cut*)

Abductor hallucis tendon (*cut*)

Medial head of flexor
hallucis brevis (*cut*)

Transverse head of
adductor hallucis (*cut*)

Deep plantar a.

Posterior perforating branches
of plantar metatarsal aa.

Plantar metatarsal ligaments

Medial cuneiform bone

Tibialis anterior tendon (*cut*)

Proximal attachment of
flexor hallucis brevis (*cut*)

Tuberosity of navicular bone

Tibialis posterior tendon (*cut*)

Plantar calcaneonavicular ligament
(spring ligament)

Fourth layer of muscles in bold

Dorsal view

Distal phalanx of great toe

Proximal phalanx of great toe

Distal phalanx ⎱
Middle phalanx ⎬ Little toe
Proximal phalanx ⎰

1st metatarsal bone

5th metatarsal bone

Dorsal interosseous mm.

Medial cuneiform bone

Intermediate cuneiform bone

Lateral cuneiform bone

Tuberosity of 5th metatarsal bone

Navicular bone

Cuboid bone

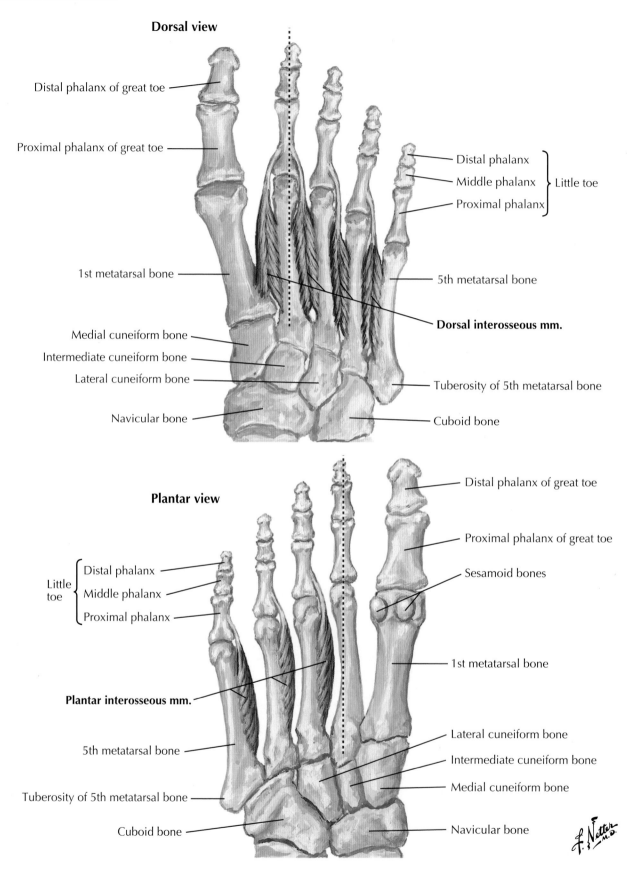

Plantar view

Distal phalanx of great toe

Proximal phalanx of great toe

Sesamoid bones

Distal phalanx
Little toe Middle phalanx
Proximal phalanx

Plantar interosseous mm.

1st metatarsal bone

5th metatarsal bone

Lateral cuneiform bone

Intermediate cuneiform bone

Tuberosity of 5th metatarsal bone

Medial cuneiform bone

Cuboid bone

Navicular bone

Note: dashed line is the line of reference for abduction and adduction of the toes

Plate 548

Ankle and Foot

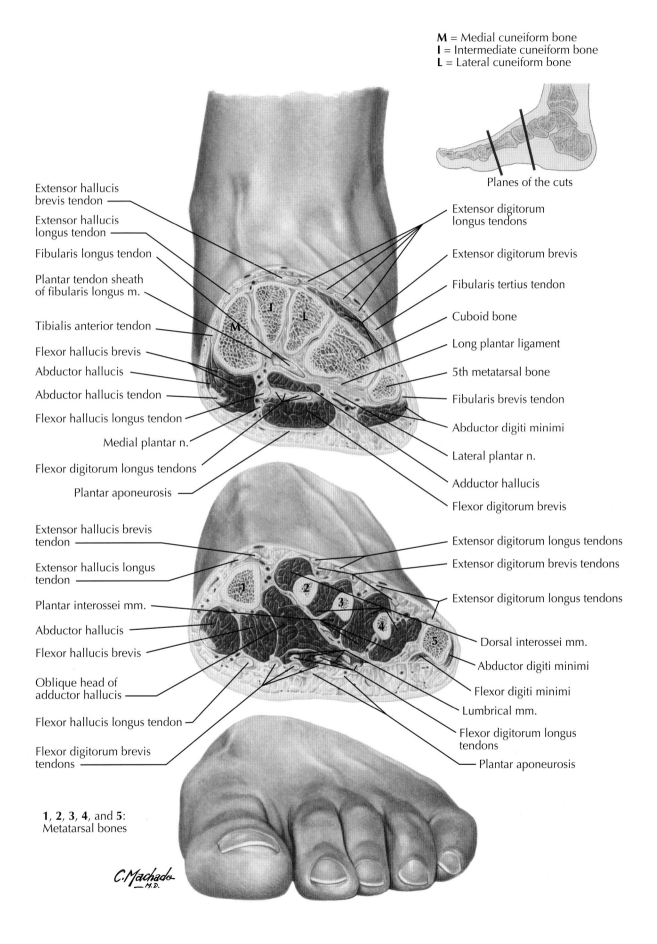

M = Medial cuneiform bone
I = Intermediate cuneiform bone
L = Lateral cuneiform bone

Planes of the cuts

Extensor hallucis brevis tendon

Extensor hallucis longus tendon

Fibularis longus tendon

Plantar tendon sheath of fibularis longus m.

Tibialis anterior tendon

Flexor hallucis brevis

Abductor hallucis

Abductor hallucis tendon

Flexor hallucis longus tendon

Medial plantar n.

Flexor digitorum longus tendons

Plantar aponeurosis

Extensor digitorum longus tendons

Extensor digitorum brevis

Fibularis tertius tendon

Cuboid bone

Long plantar ligament

5th metatarsal bone

Fibularis brevis tendon

Abductor digiti minimi

Lateral plantar n.

Adductor hallucis

Flexor digitorum brevis

Extensor hallucis brevis tendon

Extensor hallucis longus tendon

Plantar interossei mm.

Abductor hallucis

Flexor hallucis brevis

Oblique head of adductor hallucis

Flexor hallucis longus tendon

Flexor digitorum brevis tendons

Extensor digitorum longus tendons

Extensor digitorum brevis tendons

Extensor digitorum longus tendons

Dorsal interossei mm.

Abductor digiti minimi

Flexor digiti minimi

Lumbrical mm.

Flexor digitorum longus tendons

Plantar aponeurosis

1, 2, 3, 4, and **5:** Metatarsal bones

C. Machado M.D.

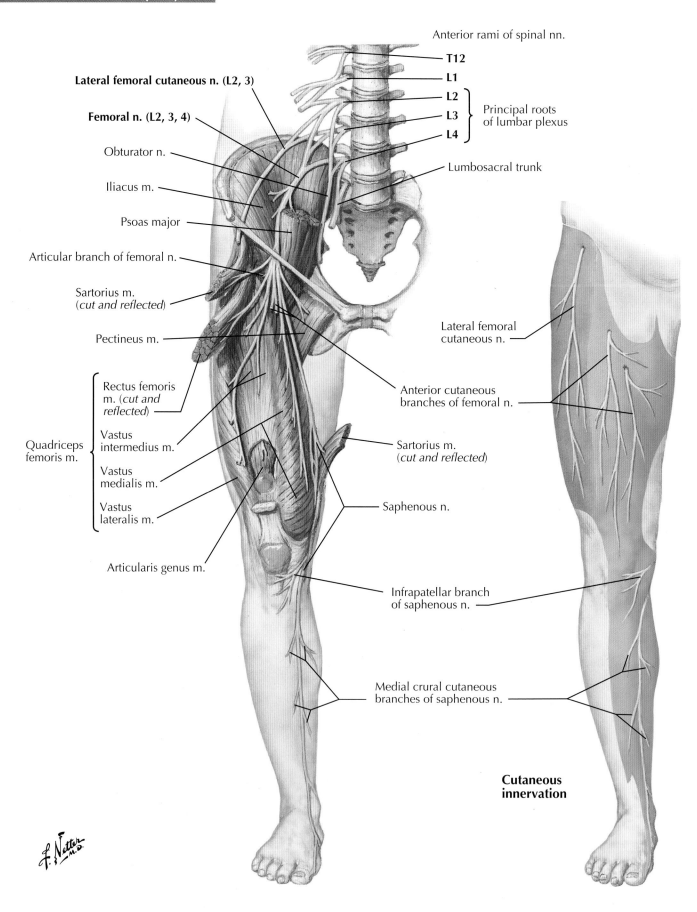

Anterior rami of spinal nn.

T12

L1

Lateral femoral cutaneous n. (L2, 3)

L2

L3 } Principal roots of lumbar plexus

L4

Femoral n. (L2, 3, 4)

Obturator n.

Lumbosacral trunk

Iliacus m.

Psoas major

Articular branch of femoral n.

Sartorius m. (*cut and reflected*)

Lateral femoral cutaneous n.

Pectineus m.

Rectus femoris m. (*cut and reflected*)

Anterior cutaneous branches of femoral n.

Vastus intermedius m.

Sartorius m. (*cut and reflected*)

Quadriceps femoris m. {

Vastus medialis m.

Saphenous n.

Vastus lateralis m.

Articularis genus m.

Infrapatellar branch of saphenous n.

Medial crural cutaneous branches of saphenous n.

Cutaneous innervation

Plate 550

Nerves

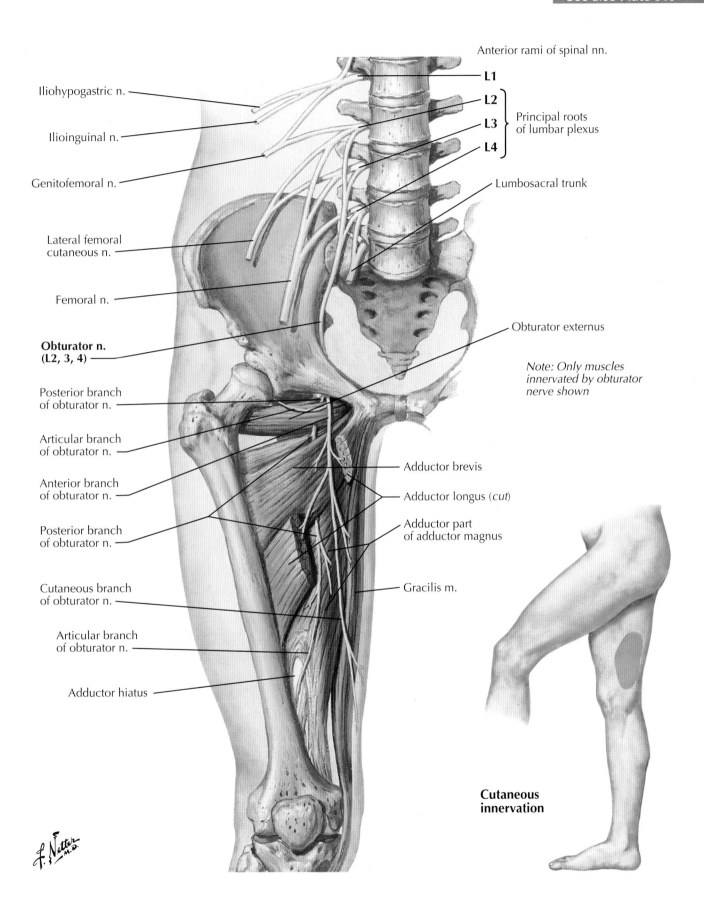

Anterior rami of spinal nn.

L1

L2
L3 } Principal roots of lumbar plexus
L4

Iliohypogastric n.

Ilioinguinal n.

Genitofemoral n.

Lumbosacral trunk

Lateral femoral cutaneous n.

Femoral n.

Obturator externus

Obturator n. (L2, 3, 4)

Note: Only muscles innervated by obturator nerve shown

Posterior branch of obturator n.

Articular branch of obturator n.

Anterior branch of obturator n.

Adductor brevis

Adductor longus (*cut*)

Posterior branch of obturator n.

Adductor part of adductor magnus

Cutaneous branch of obturator n.

Gracilis m.

Articular branch of obturator n.

Adductor hiatus

Cutaneous innervation

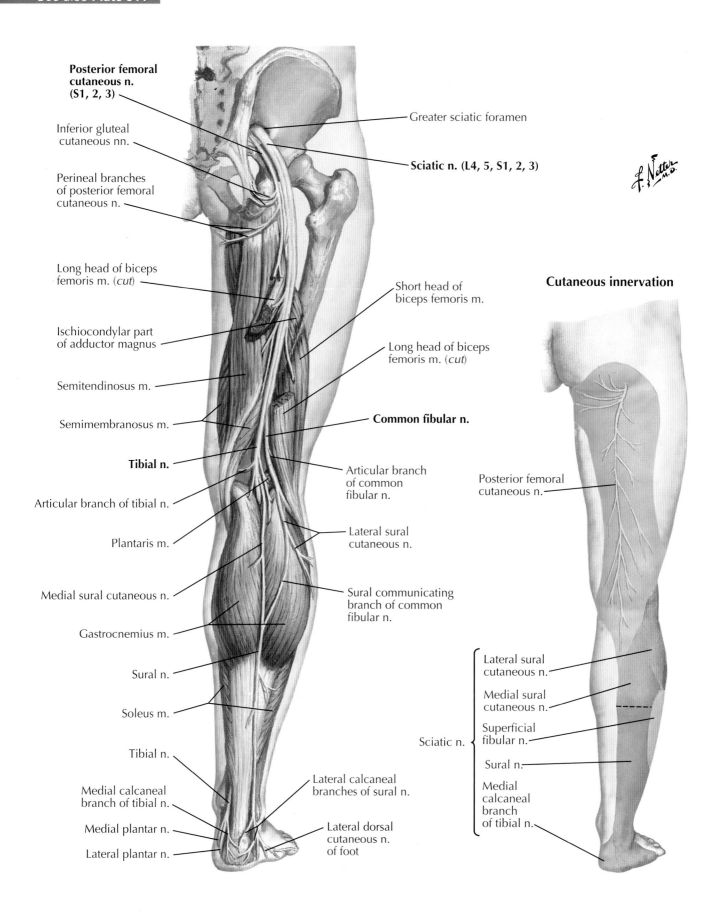

Posterior femoral
cutaneous n.
(S1, 2, 3)

Inferior gluteal
cutaneous nn.

Perineal branches
of posterior femoral
cutaneous n.

Long head of biceps
femoris m. (cut)

Ischiocondylar part
of adductor magnus

Semitendinosus m.

Semimembranosus m.

Tibial n.

Articular branch of tibial n.

Plantaris m.

Medial sural cutaneous n.

Gastrocnemius m.

Sural n.

Soleus m.

Tibial n.

Medial calcaneal
branch of tibial n.

Medial plantar n.

Lateral plantar n.

Greater sciatic foramen

Sciatic n. (L4, 5, S1, 2, 3)

Short head of
biceps femoris m.

Long head of biceps
femoris m. (cut)

Common fibular n.

Articular branch
of common
fibular n.

Lateral sural
cutaneous n.

Sural communicating
branch of common
fibular n.

Lateral calcaneal
branches of sural n.

Lateral dorsal
cutaneous n.
of foot

Cutaneous innervation

Posterior femoral
cutaneous n.

Lateral sural
cutaneous n.

Medial sural
cutaneous n.

Superficial
fibular n.

Sural n.

Medial
calcaneal
branch
of tibial n.

Sciatic n.

Plate 552

Nerves

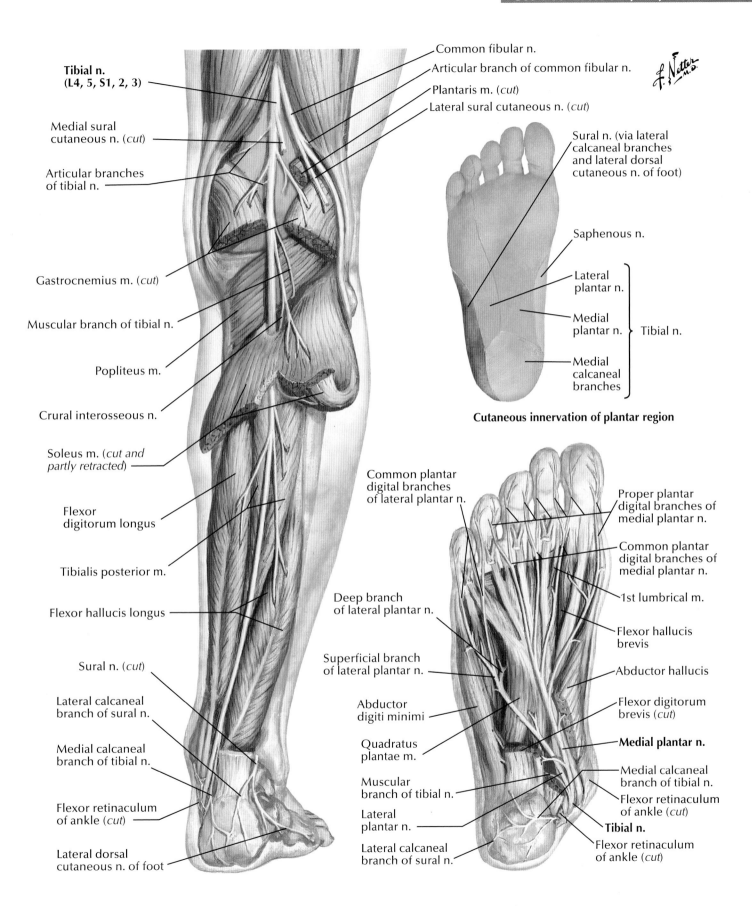

Tibial n.
(L4, 5, S1, 2, 3)

Medial sural
cutaneous n. (*cut*)

Articular branches
of tibial n.

Gastrocnemius m. (*cut*)

Muscular branch of tibial n.

Popliteus m.

Crural interosseous n.

Soleus m. (*cut and
partly retracted*)

Flexor
digitorum longus

Tibialis posterior m.

Flexor hallucis longus

Sural n. (*cut*)

Lateral calcaneal
branch of sural n.

Medial calcaneal
branch of tibial n.

Flexor retinaculum
of ankle (*cut*)

Lateral dorsal
cutaneous n. of foot

Common fibular n.

Articular branch of common fibular n.

Plantaris m. (*cut*)

Lateral sural cutaneous n. (*cut*)

Sural n. (via lateral
calcaneal branches
and lateral dorsal
cutaneous n. of foot)

Saphenous n.

Lateral
plantar n.

Medial
plantar n. ⎫ Tibial n.

Medial
calcaneal
branches

Cutaneous innervation of plantar region

Common plantar
digital branches
of lateral plantar n.

Deep branch
of lateral plantar n.

Superficial branch
of lateral plantar n.

Abductor
digiti minimi

Quadratus
plantae m.

Muscular
branch of tibial n.

Lateral
plantar n.

Lateral calcaneal
branch of sural n.

Proper plantar
digital branches of
medial plantar n.

Common plantar
digital branches of
medial plantar n.

1st lumbrical m.

Flexor hallucis
brevis

Abductor hallucis

Flexor digitorum
brevis (*cut*)

Medial plantar n.

Medial calcaneal
branch of tibial n.

Flexor retinaculum
of ankle (*cut*)

Tibial n.

Flexor retinaculum
of ankle (*cut*)

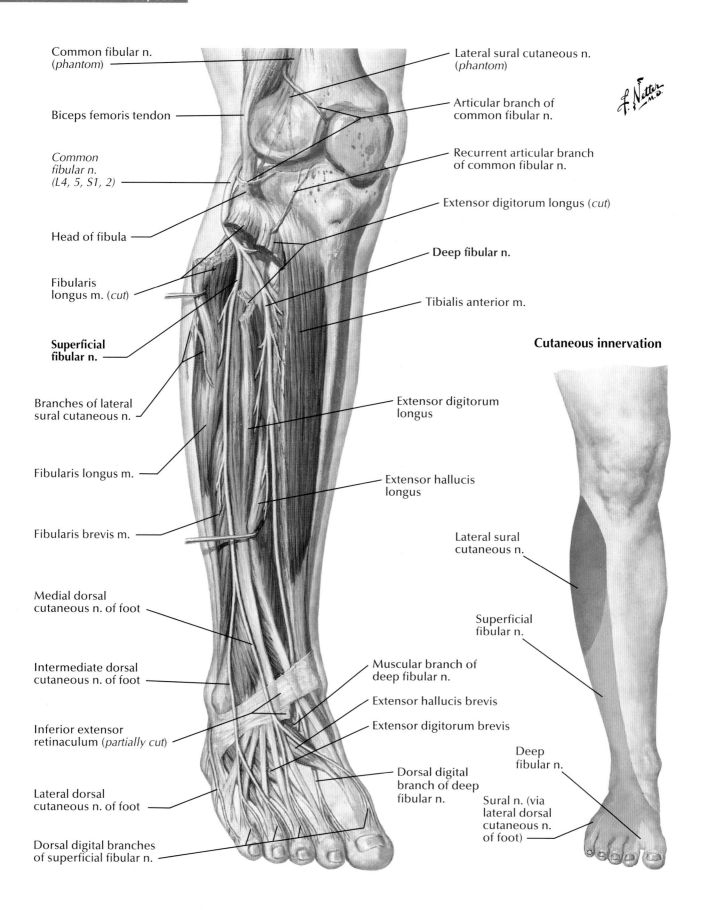

Common fibular n.
(*phantom*)

Biceps femoris tendon

*Common
fibular n.*
(*L4, 5, S1, 2*)

Head of fibula

Fibularis
longus m. (*cut*)

**Superficial
fibular n.**

Branches of lateral
sural cutaneous n.

Fibularis longus m.

Fibularis brevis m.

Medial dorsal
cutaneous n. of foot

Intermediate dorsal
cutaneous n. of foot

Inferior extensor
retinaculum (*partially cut*)

Lateral dorsal
cutaneous n. of foot

Dorsal digital branches
of superficial fibular n.

Lateral sural cutaneous n.
(*phantom*)

Articular branch of
common fibular n.

Recurrent articular branch
of common fibular n.

Extensor digitorum longus (*cut*)

Deep fibular n.

Tibialis anterior m.

Extensor digitorum
longus

Extensor hallucis
longus

Muscular branch of
deep fibular n.

Extensor hallucis brevis

Extensor digitorum brevis

Dorsal digital
branch of deep
fibular n.

Cutaneous innervation

Lateral sural
cutaneous n.

Superficial
fibular n.

Deep
fibular n.

Sural n. (via
lateral dorsal
cutaneous n.
of foot)

Plate 554

Nerves

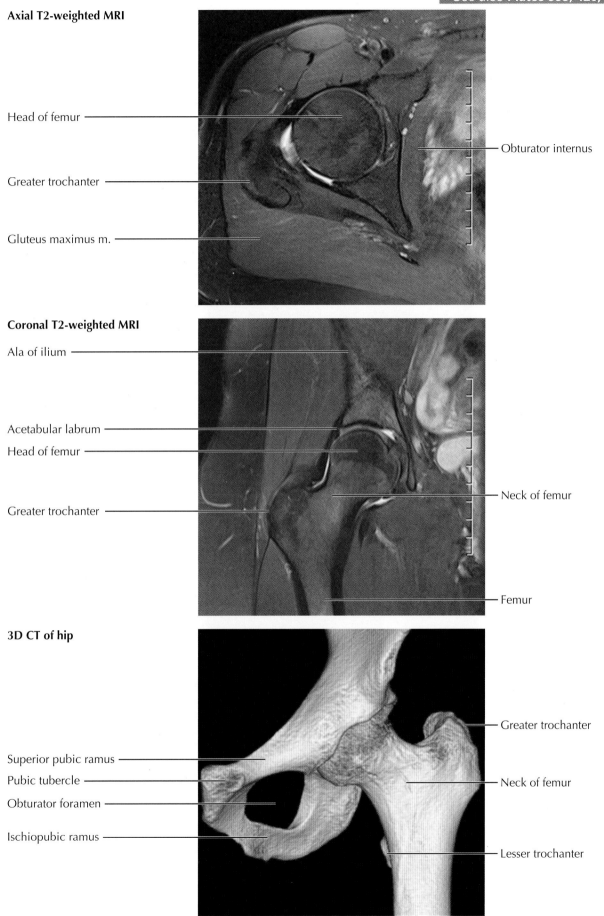

Axial T2-weighted MRI

Head of femur

Greater trochanter

Gluteus maximus m.

Obturator internus

Coronal T2-weighted MRI

Ala of ilium

Acetabular labrum

Head of femur

Greater trochanter

Neck of femur

Femur

3D CT of hip

Superior pubic ramus

Pubic tubercle

Obturator foramen

Ischiopubic ramus

Greater trochanter

Neck of femur

Lesser trochanter

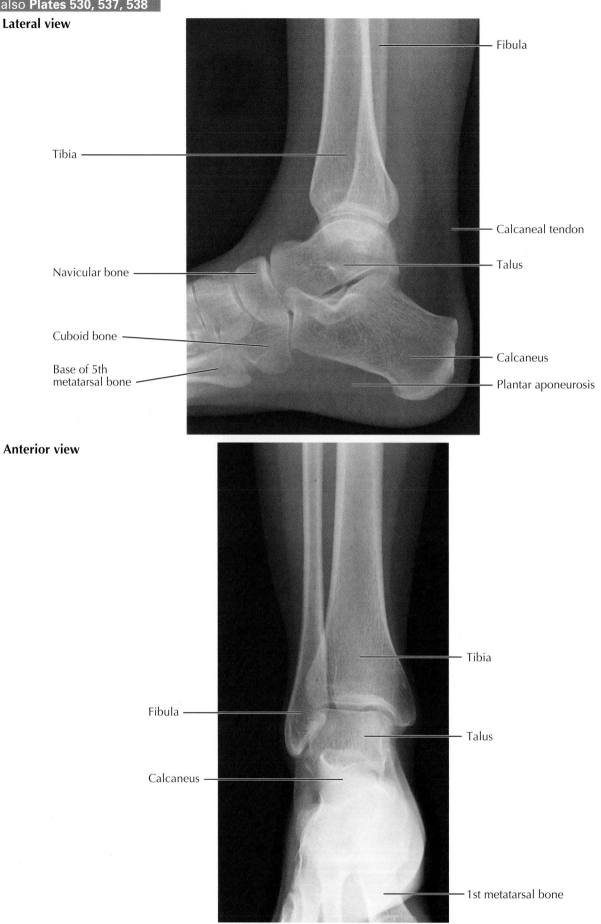

Lateral view

Fibula

Tibia

Calcaneal tendon

Talus

Navicular bone

Cuboid bone

Calcaneus

Base of 5th metatarsal bone

Plantar aponeurosis

Anterior view

Tibia

Fibula

Talus

Calcaneus

1st metatarsal bone

Plate 556

Regional Imaging

ANATOMICAL STRUCTURE	CLINICAL COMMENT	PLATE NUMBERS
Nervous System and Sense Organs		
Sural nerve	Nerve is commonly biopsied for peripheral neuropathies and commonly used as donor graft in neurotization procedures	493, 552
Common fibular nerve	Injury to this nerve from blunt trauma or compression by leg cast weakens dorsiflexion and results in foot drop	550, 552, 554
Obturator nerve	Nerve is blocked or transected for adductor muscle spasticity in cerebral palsy; may be injured during pelvic fractures or pelvic surgical procedures such as lymphadenectomy	551
Femoral nerve	Can be compressed from femoral artery hematoma and can be anesthetized for procedures of lower limb just below inguinal ligament	507, 509, 550
Saphenous nerve	Can be anesthetized in adductor canal to provide pain relief after knee replacement surgery	492, 509, 550
Lateral femoral cutaneous nerve	Compression at inguinal ligament leads to meralgia paresthetica, a pain and paresthesia syndrome of anterolateral thigh; risk factors include obesity, pregnancy, and tight-fitting waistbands	509, 550
Skeletal System		
Neck of femur	Common fracture in elderly persons from falls; can lead to avascular necrosis of head of femur because of disrupted blood supply	497, 500, 514
Body of femur	Midbody is common fracture site in high-energy trauma (motor vehicle collisions)	500
Hip joint	Potential for avascular necrosis of head of femur in hip dislocations or fractures	496, 514
Anterior cruciate ligament	Most commonly injured knee ligament, typically from sudden pivot of knee causing excessive valgus stress coupled with medial rotation of tibia	517-519
Tibial (medial) collateral and anterior cruciate ligaments and medial meniscus	"Unhappy triad of the knee"; damage to these structures can result from blow to lateral aspect of joint in extension	517-519
Tibia and fibula	High-energy fractures of body (boot-top skiing fracture) from falling forward at high speed	524
First metatarsophalangeal joint	Joint misalignment leads to hallux valgus (bunion); wearing narrow shoes can contribute, though there is also a strong genetic component	535
Calcaneus	Most common tarsal bone fracture, usually caused by landing forcefully on heel after falling from height	537
Ankle joint	Most sprains are inversion injuries that occur when foot is plantar flexed, placing stress on lateral collateral ligaments of ankle; fractures often occur to lateral malleolus of fibula and inferior articular surface of tibia	538
Muscular System		
Muscles of medial compartment of thigh	Excessive stretching or tearing of adductor and gracilis muscles is common in sports that require repeated sprints or quick changes in direction (e.g., soccer, hockey)	502, 503
Patellar ligament	Striking patellar ligament with reflex hammer elicits patellar (knee jerk) reflex to test L3 to L4 spinal cord levels (innervation of quadriceps femoris muscle by femoral nerve)	502, 517
Iliotibial tract	Can cause lateral knee pain in runners when tight iliotibial tract rubs repetitively across lateral epicondyle of femur (iliotibial band syndrome)	501, 516

Structures with High* Clinical Significance

ANATOMICAL STRUCTURE	CLINICAL COMMENT	PLATE NUMBERS
Muscular System—Continued		
Semitendinosus and semimembranosus muscles and long head of biceps femoris muscle	Excessive stretching or tearing of hamstring muscles occurs most often during high-speed running or activities with high kicks	504
Piriformis muscle	Piriformis muscle strain or structural variations (e.g., split piriformis muscle) may produce compression of sciatic nerve	511
Gluteus medius and minimus muscles	Paralysis results in contralateral pelvic dip due to weakened hip abduction when standing on affected limb (Trendelenburg sign or gait)	512
Calcaneal (Achilles') tendon	Inflammation results from repetitive stress on tendon, often from running on uneven surfaces; extreme stress may cause tendon to rupture; striking calcaneal tendon with reflex hammer elicits ankle (ankle jerk) reflex to test S1–S2 spinal cord levels (innervation of superficial calf muscles by tibial nerve)	527, 528, 538
Compartments of leg	Acute compartment syndrome may occur after trauma, such as long bone fracture, that increases compartment pressure and thereby compromises vascular flow; symptoms include pain, paresthesias, pallor, pulselessness, and paralysis (five P's); fasciotomy often required to relieve pressure	534
Plantar aponeurosis	Inflammation results from increased tension, weight, or overuse, causing heel and foot pain (plantar fasciitis)	543
Cardiovascular System		
Femoral vein	Common site of vascular access for central venous catheters; however, risk of catheter infection is greater than with jugular or subclavian vein access	509
Deep veins of lower limb	Venous thrombosis of deep leg veins is due to venous stasis, vessel injury, and/or coagulation disorders (Virchow's triad); can lead to thrombus formation and thromboemboli, such as to the lungs	533
Great saphenous vein	Often used as graft in coronary artery bypass surgery	492
Superficial veins of lower limb	Varicose veins are dilated, tortuous superficial veins, often associated with reflux of superficial and/or deep veins; progressive venous disease may cause edema, pain, and ulceration	492, 533
Femoral artery	Common site of vascular access for percutaneous cardiac and vascular procedures; target is segment between origins of inferior epigastric and deep femoral arteries, generally identified using fluoroscopy by its location alongside the femoral head; "high stick" (too cephalad) may result in retroperitoneal hemorrhage	522
Femoral, popliteal, tibial, and fibular arteries	Peripheral arterial disease due to atherosclerosis may occur in major arteries of lower limb, resulting in reduced blood flow; patients experience claudication (cramping pain in thigh or calf) upon exertion	522
Arteries of lower limb	Pulse points: femoral artery in femoral triangle, popliteal artery in deep popliteal region of knee, anterior tibial artery between extensor hallucis longus and extensor digitorum longus at ankle joint, dorsalis pedis artery on dorsum of foot, and posterior tibial artery in tarsal tunnel posterior to medial malleolus	14, 509, 527, 540, 541
Lymph Vessels and Lymphoid Organs		
Superficial inguinal nodes	Superficial inguinal nodes drain lower limb, gluteal region, lower abdominal region, and perineum; are palpable when enlarged	494
Lymph vessels of lower limb	Lymphedema (stasis of lymph flow in lymph vessels obstructed by inflammation, fibrosis, tumor, or abnormally small diameter)	494

*Selections were based largely on clinical data and commonly discussed clinical correlations in macroscopic ("gross") anatomy courses.

Table 8.2   **Structures with High* Clinical Significance**

The lumbosacral plexus includes the lumbar, sacral, and coccygeal plexuses, whose roots are the anterior rami of spinal nerves (lumbar plexus: typically L2–L4 with a small contribution from L1; sacral plexus: L4–S4; coccygeal plexus: S4–Co). Variation in the spinal nerve contributions to the plexuses, and the nerves that arise from these plexuses, is common and can be due to a prefixed (high) or postfixed (low) plexus.

NERVE	ORIGIN	COURSE	BRANCHES	MOTOR	CUTANEOUS
Lateral femoral cutaneous nerve	Posterior divisions of lumbar plexus (L2–L3)	Courses posterior to or through inguinal ligament medial to anterior superior iliac spine, then travels superficially in lateral thigh			Lateral thigh
Femoral nerve	Posterior divisions of lumbar plexus (L2–L4)	Passes posterior to inguinal ligament to lie on iliacus muscle in femoral triangle	Muscular and anterior cutaneous branches, saphenous nerve	Anterior compartment of thigh, iliacus and pectineus muscles	Anterior thigh (see also saphenous nerve)
Saphenous nerve	Femoral nerve	Leaves femoral nerve in adductor canal, pierces fascia lata to travel superficially with great saphenous vein	Infrapatellar and medial crural cutaneous branches		Medial knee, leg, ankle, and foot
Genitofemoral nerve	Anterior divisions of lumbar plexus (L1–L2)	Runs on anterior surface of psoas major; genital branch traverses inguinal canal; femoral branch courses into femoral triangle by passing deep to inguinal ligament	Genital and femoral branches	Cremaster	Lateral part of femoral triangle, anterior scrotum/vulva
Obturator nerve	Anterior divisions of lumbar plexus (L2–L4)	At L5 vertebra passes deep to common iliac vessels, enters medial thigh via obturator foramen	Anterior and posterior branches	Medial compartment of thigh	Medial thigh
Superior gluteal nerve	Posterior divisions of sacral plexus (L4–S1)	Exits pelvis via greater sciatic foramen superior to piriformis muscle		Gluteus medius and minimus muscles, tensor fasciae latae	
Inferior gluteal nerve	Posterior divisions of sacral plexus (L5–S2)	Exits pelvis via greater sciatic foramen inferior to piriformis muscle		Gluteus maximus muscle	
Nerve to piriformis muscle	Posterior divisions of sacral plexus (S1–S2)	Innervates piriformis muscle near where it exits pelvis via greater sciatic foramen		Piriformis muscle	
Perforating cutaneous nerve	Posterior divisions of sacral plexus (S2–S3)	Exits pelvis and pierces sacrotuberous ligament			Inferomedial gluteal region
Nerve to obturator internus	Anterior divisions of sacral plexus (L5–S2)	Exits pelvis via greater sciatic foramen, inferior to piriformis muscle; re-enters pelvis via lesser sciatic foramen to innervate obturator internus		Obturator internus and superior gemellus muscle	
Nerve to quadratus femoris muscle	Anterior divisions of sacral plexus (L4–S1)	Exits pelvis via greater sciatic foramen inferior to piriformis muscle		Quadratus femoris and inferior gemellus muscles	
Nerve to levator ani	Anterior divisions of sacral plexus (S3–S4)	From its source, travels on superior surface of levator ani		Iliococcygeus and pubococcygeus muscles	
Nerve to coccygeus muscle	Anterior divisions of sacral plexus (S3–S4)	From its source, travels on superior surface of coccygeus muscle		Coccygeus (ischiococcygeus) muscle	
Pudendal nerve	Anterior divisions of sacral plexus (S2–S4)	Exits pelvis via greater sciatic foramen inferior to piriformis muscle; enters perineum via lesser sciatic foramen, coursing through ischioanal fossa and pudendal (Alcock's) canal	Inferior anal and perineal nerves, dorsal nerve of clitoris/penis	Perineal muscles, external anal sphincter, pubococcygeus muscle	Posterior scrotum/vulva, clitoris/penis
Posterior femoral cutaneous nerve	Anterior and posterior divisions of sacral plexus (S1–S3)	Exits pelvis via greater sciatic foramen inferior to piriformis muscle; runs inferiorly to gluteus maximus muscle and continues distally to popliteal fossa	Inferior gluteal cutaneous nerves, perineal branches		Inferior gluteal region, posterior thigh, popliteal region

Continued

Nerves of Lumbosacral Plexus

Table 8.3

Nerves of Lumbosacral Plexus

NERVE	ORIGIN	COURSE	BRANCHES	MOTOR	CUTANEOUS
Sciatic nerve	Anterior and posterior divisions of sacral plexus (L4–S3)	Exits pelvis via greater sciatic foramen inferior to piriformis muscle, passing superficial to lateral rotators and deep to gluteus medius muscle to access thigh between greater trochanter of femur and ischial tuberosity	Muscular branches, tibial and common fibular nerves	Posterior compartment of thigh (also see branches)	See branches
Tibial nerve	Sciatic nerve	Courses through popliteal fossa and sural region deep to soleus muscle, traversing tarsal tunnel posterior to medial malleolus	Muscular and medial calcaneal branches, crural interosseous nerve, medial sural cutaneous nerve, medial and lateral plantar nerves	Posterior compartment of leg (also see branches)	Heel (also see branches)
Common fibular nerve	Sciatic nerve	Courses medial to biceps femoris muscle and lateral to lateral head of gastrocnemius muscle and neck of fibula	Superficial and deep fibular nerves, lateral sural cutaneous nerve, sural communicating branch	*See branches*	*See branches*
Deep fibular nerve	Common fibular nerve	Courses deep to fibularis longus muscle and extensor digitorum longus on surface of interosseous membrane of leg deep to extensor retinaculum	Muscular and dorsal digital branches	Anterior compartment of leg	Dorsal aspect of adjacent parts of toes 1 and 2
Superficial fibular nerve	Common fibular nerve	Courses between fibularis longus and brevis muscles in lateral compartment of leg, piercing crural fascia distally	Muscular branches, medial and intermediate dorsal cutaneous nerves of foot	Lateral compartment of leg	Inferior part of anterior leg, dorsum of foot and toes
Lateral sural cutaneous nerve	Common fibular nerve	Branches from common fibular nerve just proximal to plantaris muscle			Lateral leg
Medial sural cutaneous nerve	Tibial nerve	Branches from tibial nerve just proximal to plantaris muscle, runs between both heads of gastrocnemius muscle, pierces crural fascia	Sural nerve		Proximal posterolateral leg (also see sural nerve)
Sural nerve	Union of medial sural cutaneous nerve and sural communicating branch of common fibular nerve	Descends sural region along lateral calcaneal tendon and into foot between lateral malleolus and calcaneus	Lateral dorsal cutaneous nerve of foot		Distal posterolateral leg, lateral foot
Lateral plantar nerve	Tibial nerve	Passes deep to abductor hallucis brevis and travels between flexor digitorum brevis and quadratus plantae muscle	Superficial and deep branches	Abductor digiti minimi, adductor hallucis, flexor digiti minimi, quadratus plantae muscle, dorsal and plantar interossei muscles, lumbrical muscles (lateral three)	Lateral sole, plantar aspect of 5th toe and part of 4th toe
Medial plantar nerve	Tibial nerve	Passes deep to abductor hallucis brevis and runs anteriorly along medial border of flexor digitorum brevis	Muscular and common plantar digital branches	Abductor hallucis, flexor digitorum brevis, flexor hallucis brevis, lumbrical muscles (most medial one)	Medial sole, plantar aspect of toes 1–3, and part of 4th toe

Table 8.4   **Nerves of Lumbosacral Plexus**

MUSCLE	MUSCLE GROUP	PROXIMAL ATTACHMENT	DISTAL ATTACHMENT	INNERVATION	BLOOD SUPPLY	MAIN ACTIONS
Abductor digiti minimi of foot	Foot	Medial and lateral processes of tuberosity of calcaneus, plantar aponeurosis, intermuscular septum	Lateral surface of base of proximal phalanx of 5th digit	Lateral plantar nerve	Lateral plantar artery, plantar metatarsal and plantar digital arteries to 5th digit	Abducts and flexes 5th digit
Abductor hallucis	Foot	Medial process of tuberosity of calcaneus, flexor retinaculum, plantar aponeurosis	Medial surface of base of proximal phalanx of 1st digit	Medial plantar nerve	Medial plantar and 1st plantar metatarsal arteries	Abducts and flexes 1st digit
Adductor brevis	Medial thigh	Body and inferior ramus of pubis	Pectineal line and proximal part of linea aspera of femur	Obturator nerve	Deep femoral, medial circumflex femoral, and obturator arteries	Adducts thigh at hip, weak hip flexor
Adductor hallucis	Foot	*Oblique head:* bases of 2nd through 4th metatarsals	Lateral surface of base of proximal phalanx of 1st digit	Lateral plantar nerve	Medial and lateral plantar arteries, plantar arch, plantar metatarsal arteries	Adducts 1st digit, maintains transverse arch of foot
		Transverse head: ligaments of metatarsophalangeal joints of digits 3–5				
Adductor longus	Medial thigh	Body of pubis inferior to pubic crest	Middle one-third of linea aspera of femur	Obturator nerve	Deep femoral and medial circumflex femoral arteries	Adducts thigh at hip
Adductor magnus	Medial thigh	*Adductor part:* ischiopubic ramus	*Adductor part:* gluteal tuberosity, linea aspera, medial supracondylar line	*Adductor part:* obturator nerve	Femoral, deep femoral, and obturator arteries	*Adductor part:* adducts and flexes thigh
		Hamstring part: ischial tuberosity	*Hamstring part:* adductor tubercle of femur	*Hamstring part:* sciatic nerve (tibial division)		*Hamstring part:* extends thigh
Articularis genus m.	Anterior thigh	Anterior surface of distal femur	Suprapatellar bursa	Femoral nerve	Femoral artery	Pulls suprapatellar bursa superiorly with extension of knee
Biceps femoris m.	Posterior thigh	*Long head:* ischial tuberosity	Lateral surface of head of fibula	*Long head:* sciatic nerve (tibial division)	Perforating femoral arteries, inferior gluteal and medial circumflex femoral arteries	Flexes and laterally rotates leg, extends thigh
		Short head: linea aspera and lateral supracondylar line of femur		*Short head:* sciatic nerve (common fibular division)		
Dorsal interossei mm. of foot	Foot	Adjacent surfaces of 1st through 5th metatarsals	*Medial one:* medial surface of proximal phalanx of 2nd digit	Lateral plantar nerve	Arcuate artery, dorsal and plantar metatarsal arteries	Abduct 2nd through 4th digits of foot, flex metatarsophalangeal joints, and extend phalanges
			Lateral three: lateral surfaces of proximal phalanges of digits 2–4			
Extensor digitorum brevis	Foot	Superolateral surface of calcaneus, lateral talocalcaneal ligament, deep surface of inferior extensor retinaculum	Lateral sides of tendons of extensor digitorum longus to digits 2–4	Deep fibular nerve	Dorsalis pedis, lateral tarsal, arcuate, and fibular arteries	Extends 2nd through 4th digits at metatarsophalangeal and interphalangeal joints
Extensor digitorum longus	Anterior leg	Lateral condyle of tibia, proximal three-fourths of anterior surfaces of interosseous membrane and fibula	Middle and distal phalanges of digits 2–5	Deep fibular nerve	Anterior tibial artery	Extends lateral four digits and dorsiflexes foot
Extensor hallucis brevis	Foot	Superolateral surface of calcaneus	Dorsal surface of proximal phalanx of great toe	Deep fibular nerve	Dorsalis pedis, lateral tarsal, arcuate, and fibular arteries	Extends great toe at metatarsophalangeal and interphalangeal joints

Continued

MUSCLE	MUSCLE GROUP	PROXIMAL ATTACHMENT	DISTAL ATTACHMENT	INNERVATION	BLOOD SUPPLY	MAIN ACTIONS
Extensor hallucis longus	Anterior leg	Middle part of anterior surfaces of fibula and interosseous membrane of leg	Dorsal surface of base of distal phalanx of great toe	Deep fibular nerve	Anterior tibial artery	Extends great toe, dorsiflexes foot
Fibularis brevis m.	Lateral leg	Distal two-thirds of lateral surface of fibula	Dorsal surface of tuberosity on lateral side of 5th metatarsal bone	Superficial fibular nerve	Anterior tibial and fibular arteries	Everts and plantar flexes foot
Fibularis longus m.	Lateral leg	Head of fibula, proximal two-thirds of lateral surface of fibula	Plantar surfaces of base of 1st metatarsal and of medial cuneiform bones	Superficial fibular nerve	Anterior tibial and fibular arteries	Everts and plantar flexes foot
Fibularis tertius m.	Anterior leg	Distal one-third of anterior surfaces of fibula and interosseous membrane of leg	Dorsal surface of base of 5th metatarsal bone	Deep fibular nerve	Anterior tibial artery	Dorsiflexes and everts foot
Flexor digiti minimi of foot	Foot	Base of 5th metatarsal bone	Lateral surface of base of proximal phalanx of 5th digit	Lateral plantar nerve	Lateral plantar artery, plantar digital artery to 5th digit, arcuate artery	Flexes proximal phalanx of 5th digit
Flexor digitorum brevis	Foot	Medial process of tuberosity of calcaneus, plantar aponeurosis, intermuscular septum	Both sides of middle phalangeal bones of digits 2–5	Medial plantar nerve	Medial and lateral plantar arteries, plantar arch, plantar metatarsal and plantar digital arteries	Flexes lateral four digits
Flexor digitorum longus	Posterior leg	Medial part of posterior tibia inferior to soleal line	Plantar surfaces of bases of distal phalanges of digits 2–5	Tibial nerve	Posterior tibial artery	Flexes lateral four digits and plantar flexes foot; supports longitudinal arches of foot
Flexor hallucis brevis	Foot	Plantar surfaces of cuboid bone and lateral cuneiform bone	Both sides of base of proximal phalanx of 1st digit	Medial plantar nerve	Medial plantar and 1st plantar metatarsal arteries	Flexes proximal phalanx of 1st digit
Flexor hallucis longus	Posterior leg	Distal two-thirds of posterior surfaces of fibula and interosseous membrane of leg	Base of distal phalanx of great toe	Tibial nerve	Fibular artery	Flexes all joints of great toe, plantar flexes foot
Gastrocnemius m.	Posterior leg	*Lateral head:* lateral surface of lateral condyle of femur *Medial head:* popliteal surface above medial condyle of femur	Posterior surface of calcaneus (via calcaneal tendon)	Tibial nerve	Popliteal and posterior tibial arteries	Plantar flexes foot, assists in flexion of knee
Gluteus maximus m.	Superficial gluteal	Ilium posterior to posterior gluteal line, dorsal surfaces of sacrum and coccyx, sacrotuberous ligament	Lateral condyle of tibia (via iliotibial tract), gluteal tuberosity of femur	Inferior gluteal nerve	Inferior and superior gluteal arteries	Extends flexed thigh, assists in lateral rotation, and abducts thigh
Gluteus medius m.	Superficial gluteal	Gluteal surface of ilium between anterior and posterior gluteal lines	Lateral surface of greater trochanter of femur	Superior gluteal nerve	Superior gluteal artery	Abducts and medially rotates thigh
Gluteus minimus m.	Superficial gluteal	Gluteal surface of ilium between anterior and inferior gluteal lines	Anterior surface of greater trochanter of femur	Superior gluteal nerve	Superior gluteal artery	Abducts and medially rotates thigh
Gracilis m.	Medial thigh	Body and inferior ramus of pubis	Superior part of medial surface of tibia	Obturator nerve	Deep femoral and medial circumflex femoral arteries	Adducts thigh, flexes and medially rotates leg

MUSCLE	MUSCLE GROUP	PROXIMAL ATTACHMENT	DISTAL ATTACHMENT	INNERVATION	BLOOD SUPPLY	MAIN ACTIONS
Iliacus m.	Iliopsoas	Superior two-thirds of iliac fossa, iliac crest, ala of sacrum, anterior sacroiliac ligament	Lesser trochanter and body of femur	Femoral nerve	Iliac branch of iliolumbar artery	Flexes thigh
Inferior gemellus m.	Deep gluteal	Ischial tuberosity	Greater trochanter of femur	Nerve to quadratus femoris muscle	Medial circumflex femoral artery	Laterally rotates extended thigh and abducts flexed thigh
Lumbrical mm. of foot	Foot	Tendons of flexor digitorum longus	Medial side of dorsal digital expansions of digits 2–5	*Medial one:* medial plantar nerve *Lateral three:* lateral plantar nerve	Lateral plantar artery, plantar metatarsal arteries	Flex proximal phalanges at metatarsopha-langeal joint, extend phalanges at proximal and distal joints
Obturator externus	Medial thigh	Margins of obturator foramen, external surface of obturator membrane	Trochanteric fossa of femur	Obturator nerve	Medial circumflex femoral and obturator arteries	Laterally rotates thigh
Obturator internus	Deep gluteal	Pelvic surface of obturator membrane, margins of obturator foramen	Greater trochanter of femur	Nerve to obturator internus muscle	Internal pudendal and obturator arteries	Laterally rotates extended thigh, abducts flexed thigh
Pectineus m.	Medial thigh	Superior ramus of pubis	Pectineal line of femur	Femoral nerve (sometimes also obturator nerve)	Medial circumflex femoral and obturator arteries	Adducts and flexes thigh
Piriformis m.	Deep gluteal	Anterior surface of sacral segments 2–4, sacrotuberous ligament (inconstant)	Superior border of greater trochanter of femur	Nerve to piriformis muscle	Superior and inferior gluteal arteries, internal pudendal artery	Laterally rotates extended thigh, abducts flexed thigh
Plantar interossei mm.	Foot	Bases and medial sides of 3rd through 5th metatarsals	Medial surfaces of bases of proximal phalangeal bones of digits 3–5	Lateral plantar nerve	Lateral plantar artery, plantar arch, plantar metatarsal and plantar digital arteries	Adduct digits (3–5), flex metatarsopha-langeal joint, and extend phalanges
Plantaris m.	Posterior leg	Inferior end of lateral supracondylar line of femur, oblique popliteal ligament	Posterior surface of calcaneus (via calcaneal tendon)	Tibial nerve	Popliteal artery	Weakly assists gastrocnemius
Popliteus m.	Posterior leg	Lateral surface of lateral condyle of femur, lateral meniscus	Posterior tibia superior to soleal line	Tibial nerve	Inferior medial and lateral genicular arteries	Flexes knee
Psoas major	Iliopsoas	Transverse processes of lumbar vertebrae, sides of bodies of T12–L5 vertebrae, intervening intervertebral discs	Lesser trochanter of femur	Anterior rami of L1–L3 spinal nerves	Lumbar branch of iliolumbar artery	Acting superiorly with iliacus, flexes hip; acting inferiorly, flexes vertebral column laterally; used to balance trunk in sitting position; acting inferiorly with iliacus, flexes trunk
Psoas minor	Iliopsoas	Lateral surfaces of bodies of T12 and L1 vertebrae, T12–L1 intervertebral disc	Pectineal line, iliopubic eminence	Anterior ramus of L1 spinal nerve	Lumbar branch of iliolumbar artery	Flexes pelvis on vertebral column
Quadratus femoris m.	Deep gluteal	Lateral margin of ischial tuberosity	Quadrate tubercle	Nerve to quadratus femoris muscle	Medial circumflex femoral artery	Laterally rotates thigh
Quadratus plantae m.	Foot	Medial and lateral margins of plantar surface of calcaneus	Posterolateral edge of flexor digitorum longus tendon	Lateral plantar nerve	Medial and lateral plantar arteries, plantar arch	Corrects for oblique pull of flexor digitorum longus tendon, thus assisting in flexion of digits of foot

Continued

MUSCLE	MUSCLE GROUP	PROXIMAL ATTACHMENT	DISTAL ATTACHMENT	INNERVATION	BLOOD SUPPLY	MAIN ACTIONS
Rectus femoris m.	Anterior thigh (quadriceps femoris)	Anterior inferior iliac spine, ilium superior to acetabulum	Tibial tuberosity (via patellar ligament)	Femoral nerve	Deep femoral and lateral circumflex femoral arteries	Extends leg and flexes thigh
Sartorius m.	Anterior thigh	Anterior superior iliac spine, ilium inferior to that spine	Superior part of medial surface of tibia	Femoral nerve	Femoral artery	Abducts, laterally rotates, and flexes thigh; flexes knee
Semimembranosus m.	Posterior thigh	Ischial tuberosity	Posterior part of medial condyle of tibia	Sciatic nerve (tibial division)	Perforating femoral arteries, medial circumflex femoral artery	Flexes leg, extends thigh
Semitendinosus m.	Posterior thigh	Ischial tuberosity	Superior part of medial surface of tibia	Sciatic nerve (tibial division)	Perforating femoral arteries, medial circumflex femoral artery	Flexes leg, extends thigh
Soleus m.	Posterior leg	Posterior surface of head of fibula, proximal one-fourth of posterior surface of fibula, soleal line of tibia	Posterior surface of calcaneus (via calcaneal tendon)	Tibial nerve	Popliteal, posterior tibial, and fibular arteries	Plantar flexes foot
Superior gemellus m.	Deep gluteal	External surface of ischial spine	Medial surface of greater trochanter of femur	Nerve to obturator internus muscle	Inferior gluteal and internal pudendal arteries	Laterally rotates extended thigh and abducts flexed thigh
Tensor fasciae latae	Superficial gluteal	Anterior superior iliac spine, anterior part of iliac crest	Lateral condyle of tibia (via iliotibial tract)	Superior gluteal nerve	Ascending branch of lateral circumflex femoral artery	Abducts, medially rotates, and flexes thigh; helps to keep knee extended
Tibialis anterior m.	Anterior leg	Lateral condyle of tibia, proximal half of lateral tibia, interosseous membrane of leg	Medial cuneiform bone and base of 1st metatarsal bone	Deep fibular nerve	Anterior tibial artery	Dorsiflexes foot and inverts foot
Tibialis posterior m.	Posterior leg	Posterior tibia below soleal line, interosseous membrane, proximal half of posterior fibula	Tuberosity of navicular bone, all cuneiform bones, cuboid bone, bases of 2nd through 4th metatarsal bones	Tibial nerve	Fibular artery	Plantar flexes foot and inverts foot
Vastus intermedius m.	Anterior thigh (quadriceps femoris)	Anterior and lateral surfaces of body of femur	Tibial tuberosity (via patellar ligament)	Femoral nerve	Lateral circumflex femoral and deep femoral arteries	Extends leg
Vastus lateralis m.	Anterior thigh (quadriceps femoris)	Greater trochanter, gluteal tuberosity, lateral lip of linea aspera	Tibial tuberosity (via patellar ligament)	Femoral nerve	Lateral circumflex femoral and deep femoral arteries	Extends leg
Vastus medialis m.	Anterior thigh (quadriceps femoris)	Intertrochanteric line, greater trochanter, gluteal tuberosity, lateral lip of linea aspera	Tibial tuberosity (via patellar ligament)	Femoral nerve	Femoral and deep femoral arteries	Extends leg

Variations in spinal nerve contributions to the innervation of muscles, their arterial supply, their attachments, and their actions are common themes in human anatomy. Therefore, expect differences between texts and realize that anatomical variation is normal.

Table 8.8   **Muscles**

REFERENCES

Plates 5, 188
Lee MWL, McPhee RW, Stringer MD. An evidence-based approach to human dermatomes. *Clin Anat.* 2008;21(5): 363–373.

Plates 29, 61–63, 67–69
Lang J. *Clinical Anatomy of the Nose, Nasal Cavity, and Paranasal Sinuses.* Thieme; 1989.

Plates 40, 85
Benninger B, Andrews K, Carter W. Clinical measurements of hard palate and implications for subepithelial connective tissue grafts with suggestions for palatal nomenclature. *J Oral Maxillofac Surg.* 2012;70(1):149–153.

Plates 43–45
Baccetti T, Franchi L, McNamara J Jr. The cervical vertebral maturation (CVM) method for the assessment of optimal treatment timing in dentofacial orthopedics. *Semin Orthod.* 2005;11(3):119–129.

San Roman P, Palma JC, Oteo MD, Nevado E. Skeletal maturation determined by cervical vertebrae development. *Eur J Orthod.* 2002;24(3):303–311.

Plate 47
Tubbs RS, Kelly DR, Humphrey ER, et al. The tectorial membrane: anatomical, biomechanical, and histological analysis. *Clin Anat.* 2007;20(4):382–386.

Plates 48, 49, 51–54
Noden DM, Francis-West P. The differentiation and morphogenesis of craniofacial muscles. *Dev Dyn.* 2006;235(5):1194–1218.

Plate 52
Feigl G. Fascia and spaces on the neck: myths and reality. *Medicina Fluminensis.* 2015;51(4):430–439.

Jain M, Dhall U. Morphometry of the thyroid and cricoid cartilages in adults. *J Anat Soc India.* 2008;57(2):119–123.

Plates 56, 58, 152–158
Chang KV, Lin CP, Hung CY, et al. Sonographic nerve tracking in the cervical region: a pictorial essay and video demonstration. *Am J Phys Med Rehabil.* 2016;95(11):862–870.

Tubbs RS, Salter EG, Oakes WJ. Anatomic landmarks for nerves of the neck: a vade mecum for neurosurgeons. *Neurosurgery.* 2005;56(2 Suppl):256–260.

Plate 69
de Miranda CMNR, Maranhão CPM, Padilha IG, et al. Anatomical variations of paranasal sinuses at multislice computed tomography: what to look for. *Radiol Bras.* 2011;44(4):256–262.

Souza SA, Idagawa M, Wolosker AMB, et al. Computed tomography assessment of the ethmoid roof: a relevant region at risk in endoscopic sinus surgery. *Radiol Bras.* 2008;41(3):143–147.

Plate 72
Benninger B, Lee BI. Clinical importance of morphology and nomenclature of distal attachment of temporalis tendon. *J Oral Maxillofac Surg.* 2012;70(3):557–561.

Plate 76
Alomar X, Medrano J, Cabratosa J, et al. Anatomy of the temporomandibular joint. *Semin Ultrasound CT MR.* 2007;28(3):170–183.

Campos PSF, Reis FP, Aragão JA. Morphofunctional features of the temporomandibular joint. *Int J Morphol.* 2011;29(4):1394–1397.

Cristo JA, Bennett S, Wilkinson TM, Townsend GC. Discal attachments of the human temporomandibular joint. *Aust Dent J.* 2005;50(3):152–160.

Cuccia AM, Caradonna C, Caradonna D, et al. The arterial blood supply of the temporomandibular joint: an anatomical study and clinical implications. *Imaging Sci Dent.* 2013;43(1):37–44.

Langdon JD, Berkovitz BKV, Moxham BJ, eds. *Surgical Anatomy of the Infratemporal Fossa.* Martin Dunitz; 2003.

Schmolke C. The relationship between the temporomandibular joint capsule, articular disc and jaw muscles. *J Anat.* 1994;184(Pt 2):335–345.

Siéssere S, Vitti M, de Sousa LG, et al. Bilaminar zone: anatomical aspects, irrigation, and innervation. *Braz J Morphol Sci.* 2004;21(4):217–220.

Plates 77, 87
Benninger B, Kloenne J, Horn JL. Clinical anatomy of the lingual nerve and identification with ultrasonography. *Br J Oral Maxillofac Surg.* 2013;51(6):541–544.

Plate 78
Joo W, Yoshioka F, Funaki T, et al. Microsurgical anatomy of the trigeminal nerve. *Clin Anat.* 2014;27(1):61–88.

Joo W, Funaki T, Yoshioka F, Rhoton AL Jr. Microsurgical anatomy of the infratemporal fossa. *Clin Anat.* 2013;26(4):455–469.

Plate 84
Fawcett E. The structure of the inferior maxilla, with special reference to the position of the inferior dental canal. *J Anat Physiol.* 1895;29(Pt 3):355–366.

He P, Truong MK, Adeeb N, et al. Clinical anatomy and surgical significance of the lingual foramina and their canals. *Clin Anat.* 2017;30(2):194–204.

Iwanaga J. The clinical view for dissection of the lingual nerve with application to minimizing iatrogenic injury. *Clin Anat.* 2017;30(4):467–469.

Otake I, Kageyama I, Mataga I. Clinical anatomy of the maxillary artery. *Okajimas Folia Anat Jpn.* 2011;87(4):155–164.

Siéssere S, Vitti M, de Souza LG, et al. Anatomic variation of cranial parasympathetic ganglia. *Braz Oral Res*. 2008;22(2):101–105.

Plates 96, 122, 123
Kierner AC, Mayer R, v Kirschhofer K. Do the tensor tympani and tensor veli palatini muscles of man form a functional unit? A histochemical investigation of their putative connections. *Hear Res*. 2002;165(1–2):48–52.

Plates 101, 102
Benninger B, Barrett R. A head and neck lymph node classification using an anatomical grid system while maintaining clinical relevance. *J Oral Maxillofac Surg*. 2011;69(10):2670–2673.

Plates 107–109
Ludlow CL. Central nervous system control of the laryngeal muscles in humans. *Respir Physiol Neurobiol*. 2005;147(2–3):205–222.

Plate 114
Cornelius CP, Mayer P, Ehrenfeld M, Metzger MC. The orbits—anatomical features in view of innovative surgical methods. *Facial Plast Surg*. 2014;30(5):487–508.

Sherman DD, Burkat CN, Lemke BN. Orbital anatomy and its clinical applications. In: Tasman W, Jaeger EA, eds. *Duane's Ophthalmology*. Lippincott Williams & Wilkins; 2006.

Plates 128–141
Rhoton AL Jr. *Congress of Neurological Surgeons. Cranial Anatomy and Surgical Approaches*. Lippincott Williams & Wilkins; 2003.

Plates 131, 167
Tubbs RS, Hansasuta A, Loukas M, et al. Branches of the petrous and cavernous segments of the internal carotid artery. *Clin Anat*. 2007;20(6):596–601.

Plates 143–145, 151
Schrott-Fischer A, Kammen-Jolly K, Scholtz AW, et al. Patterns of GABA-like immunoreactivity in efferent fibers of the human cochlea. *Hear Res*. 2002;174(1–2):75–85.

Plates 186, 199
Tubbs RS, Loukas M, Slappey JB, et al. Clinical anatomy of the C1 dorsal root, ganglion, and ramus: a review and anatomical study. *Clin Anat*. 2007;20(6):624–627.

Plate 192
Bosmia AN, Hogan E, Loukas M, et al. Blood supply to the human spinal cord: part I. Anatomy and hemodynamics. *Clin Anat*. 2015;28(1):52–64.

Plate 193
Stringer MD, Restieaux M, Fisher AL, Crosado B. The vertebral venous plexuses: the internal veins are muscular and external veins have valves. *Clin Anat*. 2012;25(5):609–618.

Plates 198, 199
Tubbs RS, Mortazavi MM, Loukas M, et al. Anatomical study of the third occipital nerve and its potential role in occipital headache/neck pain following midline dissections of the craniocervical junction. *J Neurosurg Spine*. 2011;15(1):71–75.

Vanderhoek MD, Hoang HT, Goff B. Ultrasound-guided greater occipital nerve blocks and pulsed radiofrequency ablation for diagnosis and treatment of occipital neuralgia. *Anesth Pain Med*. 2013;3(2):256–259.

Plates 205–207
Hassiotou F, Geddes D. Anatomy of the human mammary gland: current status of knowledge. *Clin Anat*. 2013;26(1):29–48.

Plates 223, 224
Hyde DM, Hamid Q, Irvin CG. Anatomy, pathology, and physiology of the tracheobronchial tree: emphasis on the distal airways. *J Allergy Clin Immunol*. 2009;124(6 Suppl):S72–S77.

Plate 226
Ikeda S, Ono Y, Miyazawa S, et al. Flexible bronchofiberscope. *Otolaryngology (Tokyo)*. 1970;42(10):855–861.

Plate 239
Angelini P, Velasco JA, Flamm S. Coronary anomalies: incidence, pathophysiology, and clinical relevance. *Circulation*. 2002;105(20):2449–2454.

Plates 239, 240
Chiu IS, Anderson RH. Can we better understand the known variations in coronary arterial anatomy? *Ann Thorac Surg*. 2012;94(5):1751–1760.

Plate 248
James TN. The internodal pathways of the human heart. *Prog Cardiovasc Dis*. 2001;43(6):495–535.

Plates 248–250
Hildreth V, Anderson RH, Henderson DJ. Autonomic innervation of the developing heart: origins and function. *Clin Anat*. 2009;22(1):36–46.

Plate 261
Yang HJ, Gil YC, Lee WJ, et al. Anatomy of thoracic splanchnic nerves for surgical resection. *Clin Anat*. 2008;21(2):171–177.

Plate 304
MacSween RNM, Burt AD, Portmann BC, et al., eds. *Pathology of the Liver*. Churchill Livingstone; 2002.

Robinson PJA, Ward J. *MRI of the Liver: A Practical Guide*. CRC Press; 2006.

Plates 308, 310
Odze RD, Goldblum JR, Crawford JM. *Surgical Pathology of the GI Tract, Liver, Biliary Tract, and Pancreas*. Saunders-Elsevier; 2004.

Plate 327
Thomas MD. *The Ciba Collection of Medical Illustrations. Part 2: Digestive System: Lower Digestive Tract*. CIBA. 1970;Vol. 3:78.

Plates 343, 368, 385, 398, 416
Stormont TJ, Cahill DR, King BF, Myers RP. Fascias of the male external genitalia and perineum. *Clin Anat.* 1994;7(3):115–124.

Plates 358, 364, 369, 372, 379
Oelrich TM. The striated urogenital sphincter muscle in the female. *Anat Rec.* 1983;205:223–232.

Plochocki JH, Rodriguez-Sosa JR, Adrian B, et al. A functional and clinical reinterpretation of human perineal neuromuscular anatomy: application to sexual function and continence. *Clin Anat.* 2016;29(8):1053–1058.

Plates 362, 368
Myers RP, Goellner JR, Cahill DR. Prostate shape, external striated urethral sphincter and radical prostatectomy: the apical dissection. *J Urol.* 1987;138(3):543–550.

Plates 362, 368, 384, 385
Oelrich TM. The urethral sphincter muscle in the male. *Am J Anat.* 1980;158(2):229–246.

Plate 375
Feil P, Sora MC. A 3D reconstruction model of the female pelvic floor by using plastinated cross sections. *Austin J Anat.* 2014;1(5):1022.

Shin DS, Jang HG, Hwang SB, et al. Two-dimensional sectioned images and three-dimensional surface models for learning the anatomy of the female pelvis. *Anat Sci Educ.* 2013;6(5):316–323.

Plate 405
Nathoo N, Caris EC, Wiener JA, Mendel E. History of the vertebral venous plexus and the significant contributions of Breschet and Batson. *Neurosurgery.* 2011;69(5):1007–1014.

Pai MM, Krishnamurthy A, Prabhu LV, et al. Variability in the origin of the obturator artery. *Clinics (Sao Paulo).* 2009;64(9):897–901.

Park YH, Jeong CW. Lee SE. A comprehensive review of neuroanatomy of the prostate. *Prostate Int.* 2013;1(4):139–145.

Raychaudhuri B, Cahill D. Pelvic fasciae in urology. *Ann R Coll Surg Engl.* 2008;90(8):633–637.

Stoney RA. The anatomy of the visceral pelvic fascia. *J Anat Physiol.* 1904;38(Pt 4):438–447.

Walz J, Burnett AL, Costello AJ, et al. A critical analysis of the current knowledge of surgical anatomy related to optimization of cancer control and preservation of continence and erection in candidates for radical prostatectomy. *Eur Urol.* 2010;57(2):179–192.

Plate 513
Beck M, Sledge JB, Gautier E, et al. The anatomy and function of the gluteus minimus muscle. *J Bone Joint Surg Br.* 2000;82(3):358–363.

Woodley SJ, Mercer SR, Nicholson HD. Morphology of the bursae associated with the greater trochanter of the femur. *J Bone Joint Surg Am.* 2008;90(2):284–294.

Plate 533
Aragão JA, Reis FP, de Figueiredo LFP, et al. The anatomy of the gastrocnemius veins and trunks in adult human cadavers. *J Vasc Br.* 2004;3(4):297–303.

Plate 537
Lee MWL, McPhee RW, Stringer MD. An evidence-based approach to human dermatomes. *Clin Anat.* 2008;21(5):363–373.

INDEX

Alveolar cell
type I, BP45
type II, BP45
Alveolar ducts, BP44
opening of, BP44
Alveolar foramina, 27
Alveolar macrophage, BP45
Alveolar nerve
anterior, middle, and posterior superior, 98
dental and gingival branches of superior, 81
inferior, 42, 73–75, 77–78, 84, 87, 98, 149, 158, 161
posterior superior, 81
right, 75
superior
anterior, 84
infraorbital, 78
middle, 84
posterior, 75, 78
Alveolar pores (of Kohn), BP44
Alveolar process, of maxillary bone, 25, 27, 29, 38, 62, 69, BP23
Alveolar sac, BP44
Alveolar vein
inferior, 100
posterior superior, 100
Alveolar wall
capillary bed within, 227
capillary plexuses within, 227
Alveoli, BP44, BP45
Amacrine cells, 147
Ammon's horn, 139
Ampulla, 374
of ductus deferens, 385, 391
of ear, 125
anterior, 124
lateral, 124
posterior, 124
of rectum, 369
of uterine tube, 376
Amygdaloid body, 137, 139, 146, 162
Anal canal, 395, T6.1–T6.3
arteries of, 400
muscularis mucosae of, 395, 396
veins of, 401
Anal columns, Morgagni's, 395
Anal crypt, 395
Anal glands, 395
Anal nerves, inferior, 380, 324, 412, 413, 415, 416, 506, 512
Anal pit, 389
Anal sinus, 395
Anal sphincter muscle
external, 301, 324, 358, 364, 369, 373, 378–379, 382, 383, 393–394, 396–397, 400, 406, T6.4
attachment of, 360
deep, 343, 368, 393, 396–398
subcutaneous, 343, 368, 393, 396–398
superficial, 343, 368, 393, 396–398
internal, 394, 396
Anal triangle, 352, 381
Anal tubercle, 389
Anal valve, 395
Anal verge, 395
Anastomosis
between medial and lateral circumflex femoral arteries, 514
paravertebral, 192
patellar, 522
prevertebral, 192
Anastomotic vein, 100
inferior (of Labbé), 129, 172
superior (of Trolard), 129
Anastomotic vessel, 514
to anterior spinal artery, 191
to posterior spinal artery, 191

Anatomical snuffbox, 422, 454
Anconeus muscle, 428, 441, 454, 455, 460, 461, 488, 489, T7.5–T7.8
Angle of mandible, 22, 26, 36, 39, 91, 93
Angular artery, 24, 59, 60, 74, 99, 115, 165
Angular gyrus, 132
artery to, 168, 169
Angular vein, 24, 100, 115
Angulus oris, modiolus of, T2.1–T2.4
Ankle, 2, 3, BP15
anterior region of, 491
cross-sectional anatomy of, BP110
joint of, T8.1–T8.2
ligaments of, 538
posterior region of, 491
radiographs of, 556
tendon sheaths of, 540
Annular hymen, BP91
Anococcygeal body, 360, 378–379, 397–398
Anococcygeal ligament, 360, 378, 383, 398
Anococcygeal nerves, 413, 415, 506, 508
Anoderm, 395
Anorectal artery
inferior, 314, 317, 404
middle, 284, 314, 317, 335, 402, 404, BP93
superior, 284, 293, 314, 317, 335
Anorectal flexure, 420
Anorectal hiatus, 361
Anorectal junction, 362, BP94
Anorectal line, 395
Anorectal musculature, 396
Anorectal plexus
middle, 323
superior, 323–324
Anorectal veins, T6.1–T6.3
inferior, 318
middle, 285, 318
superior, 318, T5.1–T5.3
Anorectal (rectal) venous plexuses, 285
external, T6.1–T6.3
internal, T6.1–T6.3
Ansa cervicalis, 56–58, 75, 155, 156
inferior root of, 56–58, 75, 98, 103, 155–156, T2.9
infrahyoid branches of, 98
superior root of, 56–58, 75, 98, 103, 155, 156
Ansa of Galen, 108
Ansa pectoralis, 438
Ansa subclavia, 229, 249, 250, 261, 157
Anserine bursa, 516, 517
Antebrachial cutaneous nerve
lateral, 423–425, 440, 445, 456, 458, 461, 479, 482–485, T7.3–T7.4
medial, 423–424, 438–439, 442, 445, 461, 479, 482–486, T7.3–T7.4
posterior, 423–425, 441, 445, 461, 478, 479, 482, 484, 488, 489, T7.3–T7.4
Antebrachial fascia, 461
Antebrachial vein, median, 422, 424, 425, 444, 461
Anterior chamber
endothelium of, 117
of eye, 110, 116, 118, 119
of eyeball, 117
Anterior ethmoidal artery, external nasal branch of, 60
Anterior ethmoidal nerve, external nasal branch of, 60
Anterior plane, 1
Anterior ramus, 190, 253
Anterior root, 190, 212, 326, BP38
rootlets of, 189
Anterior root ganglion, 419
Anterior tibial artery, 14, 532, 533, BP110
Antidromic conduction, BP8

Antihelix, 22, 122
crura of, 122
Antitragus, 22, 122
Anular ligaments, 225
Anulus fibrosus, 181, 185
Anus, 360, 364, 377, 381, 383, 389, 397
Aorta, 13, 50, 165, 190, 222, 234, 240, 241, 247, 262, 283, 328, 331, 344, 346, 405, BP38, BP64, BP82
abdominal, 200, 284, 289, 292–294, 296, 308, 310, 324, 330, 332, 336, 343, 349, 350, 351, 400, 402–404, 414, T5.1–T5.3, BP79
ascending, 165, 236, 238, 241, 244, 245, 248, 262, 265, BP51
descending, 14, 17, 165
esophageal branch of, 222, 258
inferior left bronchial branch of, 222, 258
posterior intercostal branch of, 258
right bronchial branch of, 222, 258
superior left bronchial branch of, 222, 258
thoracic, 192
ureteric branch from, 336
Aortic arch, 14, 17, 51, 103–104, 165, 233, 234, 253, BP24, BP51
Aortic arch lymph node of ligamentum arteriosum, 228
Aortic heart valve, 233, 236, 238, 242–245, 248, T4.1–T4.3, BP51
left coronary leaflet of, 242–245, 248
noncoronary leaflet of, 242–245, 248
right coronary leaflet of, 242–245, 248
Aortic hiatus, 216, 253
Aortic nodes, lateral, 408, 410
Aortic plexus, abdominal, 417
Aortic sinuses (of Valsalva), 244
Aorticorenal ganglion, 6, 287, 321, 322, 324, 325, 339–341, 411, 416, 417, 419, BP6
left, 319, 321, 323, 412, 418
right, 319, 321, 323
Apical foramina, 41
Apical ligament, of dens, 47, 95, 96, BP18
Aponeurosis, of external abdominal oblique muscle, 378, 382
Appendicular artery, 298, 313, 314, BP61
Appendicular nodes, BP77
Appendicular skeleton, 8
Appendicular vein, 317, 319
Appendix, 300, BP57
of epididymis, 388
of testis, 388
Aqueduct of midbrain (of Sylvius), 135
Arachnoid, 127, 129
Arachnoid-dura interface, 129
Arachnoid granulations, 127, 128, 129, 136
Arachnoid mater, 189, 190, 136
Arch of aorta, 217, 220, 231, 232, 235, 236, 242, 248, 252, 254, 255, 258, 264, BP50
lung groove for, 221
Arcuate artery, 334, 522, 532, 541, 542, 547, BP105, BP109, BP73
Arcuate eminence, 32, 121
Arcuate ligament
inferior, 362
lateral, 216, 283
medial, 216, 283
median, 216, 283
Arcuate line, 268, 274, 353, 357, 358, 361, 495
of ilium, 359
of rectus sheath, 272, 276, 280
Arcuate nucleus, 174
Arcuate popliteal ligament, 518
Arcuate pubic ligament, 378, 384
Areola, 205
Areolar glands (of Montgomery), 205
Areolar tissue, loose, 129

Netter Atlas of Human Anatomy: Classic Regional Approach

Areolar venous plexus, 277
Arm, 2, 3
  anterior, 10
  anterior region of, 422
  arteries of, BP96
  muscles of
    anterior views of, 440
    posterior views of, 441
  posterior, 10
  posterior region of, 422
  radial nerve in, 488
  serial cross sections of, 445
Arrector pili muscle, 12, BP1, BP8
Arteriae rectae, 313, 314
Arterial arches, superficial palmar,
    470
Arterial rete, marginal, 123
Arterial wall, BP13
  avascular zone, BP13
  vascular zone, BP13
Arteries, 12, 13
  of anal canal, 400
  of brain
    frontal section of, 168
    frontal view of, 168
    inferior view of, 166
    lateral view of, 169
    medial view of, 169
  of breast, 206
  of ductus deferens, 400, 403,
    BP93
  of duodenum, 310, 312
  of esophagus, 258, BP49
  of eyelids, 115
  of large intestine, 314
  of liver, 308, 310
  major, 14
  of malleolar stria, 123
  of mammary gland, 206
  of orbit, 115
  of pancreas, 310, 312
  of pelvic organs, female, 402
  of pelvis
    female, 403
    male, 405, BP93
  of perineum, 406
    male, 407
  of rectum, 400
  of round ligament, 402
  of small intestine, 313
  of spinal cord
    intrinsic distribution, 192
    schema of, 191
  of spleen, 308, 310
  of stomach, 308
  of testis, 403
Arterioles, 13, 481
  intralobular, 304
  macular
    inferior, 119
    superior, 119
  nasal retinal
    inferior, 119
    superior, 119
  periportal, 304
  portal, 304
  temporal retinal
    inferior, 119
    superior, 119
Artery of Adamkiewicz, 191
Articular branch, 550
Articular cartilage, 9, 496, 448, 481, 521,
    BP102
Articular cavity, 204, 481, 521, BP18
  of sternoclavicular joint, 429
Articular disc, 42, 78, 204, BP98, BP25
  of sternoclavicular joint, 429

Articular facet
  cervical, 43
    inferior, 44
    superior, 43
  of dens, posterior, 47
  inferior, 185
  lumbar, inferior, 181
  thoracic, superior, 180
  vertebral body for, 204
    inferior, 204
    superior, 204
Articular nerve, recurrent, 554
Articular process
  cervical
    inferior, 43–45
    superior, 44, 45, BP17
  lumbar
    inferior, 181, 183, 185
    superior, 181, 183, 185
  sacral, superior, 183
  thoracic
    inferior, 180
    superior, 180
Articular surface, superior, 525
Articular tubercle, 27, 31, 42, 73
Articularis genu muscle, 498, 515, 517, 521,
    550, T8.5–T8.8
Aryepiglottic fold, 92, 93, 107
Arytenoid cartilage, 96, 106
  muscular process of, 106, 107
  vocal process of, 106, 107
Arytenoid muscle
  oblique, 92, 94, 107, 108, T2.10–T2.14
    action of, 109
    aryepiglottic part of, 107, 108
  transverse, 92, 94, 95, 107, 108, T2.10–T2.14
    action of, 109
Ascending cervical artery, 191
Ascending colon, 301, 331, 351
  as site of referred visceral pain, BP5
Ascending fibers, 419, 340
Asterion, 27, T2.1–T2.4
Atlantoaxial joint, BP34, BP35
  lateral, 46
    capsule of, 46, 47
  median, 47
    articular cavity of, BP18
Atlantoaxial ligament, 47
Atlantoaxial membrane, posterior, BP18
Atlantooccipital joint, capsule of, 46, 47
Atlantooccipital junction, BP18
Atlantooccipital ligament, anterior, 47
Atlantooccipital membrane
  anterior, 46, 95, 96, BP18
  posterior, 46, BP18
Atlas (C1), 9, 36, 45–47, 186, BP17
  anterior arch of, BP18, BP34, BP35,
    BP39
  anterior tubercle of, 45
  anterior view of, 179
  arch of
    anterior, 28, 43, 45, 95, BP16, BP18
      imaging of, 177
    posterior, 43
  dens of, articular facet for, 43
  groove for vertebral artery of, 43
  inferior articular surface of, BP35
  inferior longitudinal band of cruciate
    ligament of, BP18
  inferior view of, 43
  lateral mass of, 26, 43
  left lateral view of, 179
  posterior arch of, 199, BP34, BP16, BP18
  posterior tubercle of, 197
  posterior view of, 179
  superior articular facet for, 43
  superior articular surface of, BP35

Atlas (C1) (Continued)
  superior longitudinal band of cruciate
    ligament of, BP18
  superior view of, 43
  transverse foramen of, 43
  transverse ligament of, 47, BP18
    tubercle for, 43
  transverse process of, 43, 46, 55, 197
  tubercle of
    anterior, 43, 47
    posterior, 43, 196
  vertebral foramen of, 43
Atonic stomach, BP55
Atrioventricular (AV) bundle (of His), 248
Atrioventricular (AV) node, 248, T4.1–T4.3
Atrioventricular septum, 241–244, 248
Atrioventricular valve
  left, 236–238, 242–245, 248, 266
    anterior cusp of, 242–245
    commissural leaflets of, 243–244
    left fibrous ring of, 243
    posterior cusp of, 242–245
  right, 236–238, 241–246, 266
    inferior leaflet (posterior cusp) of, 241,
      243–245
    right fibrous ring of, 248
    septal leaflet (cusp) of, 241, 243–245
    superior leaflet of, 241, 243–245
Atrium, 245
  left, 234–237, 241–245, 262, 266, BP50
    oblique vein of (of Marshall), 235, 236,
      239, 242
  right, 232–238, 241, 244–246, 251, 262, 266,
    BP51
Auditory canal, internal, imaging of, BP32
Auditory ossicles, 8, T2.1–T2.4
Auditory tube, 121–123, 125, 153, BP28, BP29,
    BP30
  cartilage of, 92, BP22
  cartilaginous part of, 73, 85, 90, 96, BP30
    lateral lamina of, BP30
    medial lamina of, BP30
  groove for, 31
  opening of, 61
  pharyngeal opening of, 17, 90, 93, 95
Auerbach's plexus, BP48. See Myenteric
    (Auerbach's) plexus
Auricle (atrial appendage), 28, 121, 126,
    BP29
  concha of, 122
  left, 232, 235, 236, 239, 240, 242, 245, 248,
    BP50, BP51
  lobule of, 122
  right, 232, 235, 241, 245
Auricular artery
  anterior, 24
  deep, 74, 75, 76, 99, 123
  posterior, 58, 74, 78, 99, 123, 128, 163, 165,
    199
Auricular cartilage, BP22
Auricular muscle, branches to, 150
Auricular nerve
  great, 56, 57, 156, 198
  posterior, 71, 75, 77, 150, BP20
    occipital branch of, 150
Auricular surface, 183
  of ilium, 495
  of sacrum, 183
Auricular tubercle (of Darwin), 122
Auricular vein, posterior, 24, 56, 100
Auricularis muscle
  anterior, 48, T2.10–T2.14, BP19
  posterior, 48, T2.10–T2.14
  superior, 48, T2.10–T2.14
Auriculotemporal nerve, 23, 42, 70, 73–75,
    77–78, 98–99, 149, 152, 161
  roots of, 78

Cauda equina, 186, 187, 190, 200, T3.1
  radiology of, 182
  within subarachnoid space, BP38
Caudal, as term of relationship, 1
Caudate nucleus, 137
  anterior vein of, 172, 173
  body of, 133, 135, 137
  head of, 137–139, 170, 172
    imaging of, 177, BP32
  posterior terminal vein of, 173
  posterior vein of, 172
  tail of, 135, 137, 139
  transverse veins of, 172, 173
Caval foramen, 216, 283, BP78
Cave of Retzius, 343
Cavernous nerves, BP7, T6.1–T6.3
  of penis, 412, 418
Cavernous plexus, 148
Cavernous sinus, 65, 115, 131, 163, 167,
  T2.1–T2.4
Cavernous tissue, T6.1–T6.3
Cavernous vein, 405
Cavernous venous plexus, of urethra, BP86
Cavum septum pellucidum, 127
Cecal artery
  anterior, 298, 313, 314, BP61
  posterior, 298, 313, 314, BP61
Cecal folds, 363, 367, 298
Cecal plexuses, 323
Cecal vein
  anterior, 317, 318
  posterior, 317, 318
Cecum, 153, 288, 290, 298, 299, 300, 301, 363,
  367, BP57, BP61, BP80, BP81
  peritoneal attachment of, BP61
  as site of referred visceral pain, BP5
  vascular fold of, 298
Celiac branch, 412
Celiac ganglion, 6, 7, 261, 319–329, 339–341,
  411, 412, 416–419, T5.1–T5.3, BP6, BP8,
  BP79
Celiac nodes, 286, 260, BP74, BP75, BP76,
  BP78
Celiac plexus, 261, 320, 321, 322, 323, 339, 340,
  341, 412, 417, BP6, BP7
Celiac trunk, 14, 216, 247, 253, 254, 258, 284,
  289, 293, 296, 306, 308–313, 324, 325, 329,
  330, 332, 412, BP51, BP64, BP67, BP69,
  BP83
  arteriogram of, 311
  variations in, BP55, BP67
Celiacomesenteric trunk, BP67
Cement, 41
Central arteries
  anterolateral, 164, 166, 167, 168, 170
  anteromedial, 167
  posteromedial, 167
Central axillary nodes, 207
Central canal, 144
  of spinal cord, 136, 141
Central incisor, 40
Central nervous system, 4
Central sulcus (of Rolando), 132, 133
Central superior mesenteric nodes, BP77
Central tendon, of diaphragm, 215, 253
Central vein, 303, 304
Cephalic vein, 15, 202, 209–210, 238, 277, 422,
  424–426, 432, 435, 438, 444, 445, 461, 478
  accessory, 424, 425
  median, 425
Cerebellar artery
  anterior inferior, 170, 176
  inferior
    anterior, 163, 165–168, 191
    posterior, 163, 165, 166, 168, 191
  posterior, 191
    lateral posterior choroidal branch of, 170

Cerebellar artery *(Continued)*
    medial posterior choroidal branch of, 170
  posterior inferior, 170
    cerebellar tonsillar branch of, 170
    choroidal branch to, 170
    inferior vermian branch to, 170
  superior, 191, 163, 165, 166, 167, 168, 170,
    176
    imaging of, BP32
    lateral branch of, 170
    superior vermian branch of, 170
Cerebellar hemispheric veins, inferior, 171
Cerebellar notch
  anterior, 142
  posterior, 142
Cerebellar nuclei, 142
  dentate of, 142
  emboliform of, 142
  fastigial, 142
  globose of, 142
Cerebellar peduncle, 141
  inferior, 140, 141, 142, 151
  middle, 140, 141, 142, 171
    imaging of, 177, BP32
  superior, 140, 141, 142
    decussation of, 142
Cerebellar vein
  precentral, 171
  superior, 171
Cerebellomedullary cistern, posterior, 136
Cerebellum, 126, 133, 138, 141, 142, 173, BP35
  anterior lobe of, 142
  biventer lobule of, 142
  flocculonodular lobe of, 142
  flocculus of, 140, 142
  fourth ventricle of, 142
  horizontal fissure of, 142
  imaging of, 177, BP32
  lingula of, 142
  medullary velum of
    inferior, 142
    superior, 142
  posterior lobe of, 142
  posterolateral fissure of, 142
  postlunate fissure of, 142
  primary fissure of, 142
  quadrangular lobule of, 142
  retrotonsillar fissure of, 142
  secondary fissure of, 142
  semilunar lobule of
    inferior, 142
    superior, 142
  simple lobule of, 142
  tonsil of, 142
  vermis of, 141
    central lobule of, 141
    culmen of, 141
    declive of, 141
    folium of, 141
    lingula of, 141
    nodulus of, 141
    pyramid of, 141
    tuber of, 141
    uvula of, 141
Cerebral aqueduct, 69, 133, 134, 136, 141, 144,
  173, 174
  imaging of, 177
Cerebral arterial circle (of Willis), 166, 167,
  T2.1–T2.4
Cerebral arteries, 129
  anterior, 163, 164, 165, 166, 167, 168, 170,
    172
    cingular branches of, 169
    imaging of, BP31, BP32
    left, 169
    medial frontal branches of, 168, 169
    postcommunicating part of, 176

Cerebral arteries *(Continued)*
    precommunicating part of, 176
    terminal branches of, 169
  anterolateral, 170
  middle, 163–168, 170, 176
    branches of, 164
    imaging of, BP31, BP32
    left, 169
    occipitotemporal branches of, 169
    temporal branches of, 168
  posterior, 163–168, 170, 176
    calcarine branch of, 169
    imaging of, BP31, BP32
    parietooccipital branch of, 169
    right, 169
    temporal branches of, 170
    terminal branches of, 169
Cerebral cortex, postcentral gyrus, BP14
Cerebral crus, 134, 140, 142, 166
Cerebral fissure, longitudinal, 134, 170,
  172
  imaging of, 177
Cerebral hemisphere, 129
Cerebral peduncle, 69, 133, 141
Cerebral sulcus, lateral, 168
Cerebral veins
  anterior, 171, 172, 173
  great (of Galen), 176
  inferior, 129, 131
  internal, 127, 133, 135, 138, 171, 172, 173,
    176
    imaging of, BP31
    left, 173
    right, 173
  middle
    deep, 127, 171, 172, 173
    superficial, 127, 129, 131, 172
  superficial, 129
  superior, 129, 176, T2.1–T2.4
    opening of, 128
Cerebrospinal fluid
  circulation of, 136
  radiology of, 182
  within subarachnoid space, BP35
Cerebrum
  frontal lobe of, 132
  frontal pole of, 132, 134
  inferior (inferolateral) margin of, 134
  occipital lobe of, 132
  occipital pole of, 132, 134
  parietal lobe of, 132
  temporal lobe of, 132
  temporal pole of, 132, 134
Cervical artery
  ascending, 57, 98, 99, 103–105, 165, 191,
    437, BP24
  deep, 165, 191, BP24
  transverse, 56, 57, 103, 105, 437, 438, BP24
    deep branch of, BP24
    superficial branch of, 198, BP24
Cervical canal, 374
Cervical cardiac branch, BP7
Cervical cardiac nerves, 229, 261, BP52
  inferior, 249, 250, 157
  middle, 249, 250, 157
  superior, 249, 250, 91, 157
Cervical fasciae, 52
  deep
    prevertebral layer of, 50
    superficial layer of, 432, 48, 49, 50, 95
  deep investing, 51, 96
  superficial investing, 51, 52
Cervical ganglion
  middle, 57, 91, 98, 155, 157, BP6
  superior, 6, 80, 82, 91, 152, 155, 157–161,
    249, 250, 261, BP6
Cervical lordosis, 179

Clivus, 32, 47, BP35
  imaging of, 177
Clunial nerves, inferior, 413, 415
CMC joint radial styloid process, BP99
CN VI. See Abducens nerve
CN VII. See Facial nerve
CN IX. See Glossopharyngeal nerve
CN XI. See Accessory nerve
CN XII. See Hypoglossal nerve
Coccygeal horn, 183
Coccygeal nerve, 186, 187, 508
Coccygeal plexuses, 506, 508
Coccygeus muscle, 283, 358–362, 404, 405,
    412–414, T6.1–T6.4
  nerve to, 506, 508, T8.3–T8.4
Coccyx, 8, 186, 187, 268, 183, 353, 356–359,
    379, 393, BP104, BP94
  anterior view of, 179
  apex of, 352, 355, 360, 362
  lateral view of, 179
  posterior view of, 179
  radiograph of, 354
  tip of, 360, 362, 381, 383, 397, 420
  transverse process of, 183
Cochlea, 82, 124, 126, BP29
  course of sound in, 121
  duct of, BP29
  helicotrema of, 121, 124, 125, BP29
  modiolus of, 125
  scala tympani of, 121, BP29
  scala vestibuli of, 121, BP29
  section through turn of, 125
Cochlear aqueduct, 125
Cochlear cupula, 124
Cochlear duct, 124, 125, 126
Cochlear (spiral) ganglion, 151
Cochlear nerve, 121, 124–126, 151, T2.5–T2.8,
    BP29
Cochlear nucleus, 151
  anterior, 143, 144, 151
  posterior, 143, 144, 151
Cochlear recess, 124
Cochlear (round) window, 35, 121, 124, BP29
  fossa of, 122, BP28
Colic artery
  ascending branch, 400
  descending branch, 400
  left, 284, 400, 314, 335, BP68
  middle, 306, 309, 310, 313, 314, BP68, BP69
  right, 313, 314, 335, BP68, BP69
  variations in, BP68, BP69
Colic flexure
  left, splenic, 288–291, 294, 296, 300, 301,
      303, 306, 342
  right, hepatic, 288, 290, 291, 294, 296, 301,
      303, 306, 337
Colic impression, 302, 307
Colic nodes, right, BP77
Colic vein
  left, 317, 318
  middle, 316–318
  right, 316–318, BP66
Collagen
  in arterial wall, BP13
  fibers, in connective tissues and cartilage,
      BP11
  lamellae of anulus fibrosus, 185
Collateral artery
  inferior ulnar, 443, BP96
  middle, 441, 443, 445, 455, BP96
  radial, 441, 443, 445, BP96
  superior ulnar, 443, BP96
Collateral eminence, 138
Collateral ligament, 468, 474, BP102
  accessory, 468
  of ankle, medial, 525
  fibular, 516–521, 525, 529, 530, 532

Collateral ligament (Continued)
  of foot, 539
  radial, 448, BP98
  tibial, 516–521, 525, 529, 531, T8.1–T8.2
  ulnar, 448, BP98
Collateral sulcus, 133, 134
Collateral trigone, 138
Collecting duct, BP72
Colles' fascia, 84, 368, 371, 372, 378–384, 393,
    397, 398, 406, 407, 413, BP86
Colliculus
  facial, 141
  inferior, 133, 138, 140, 141
    brachium of, 138
    left, 171
  superior, 133, 134, 138, 140, 141, 143, 159,
      170, T2.1–T2.4
    brachium of, 138
    left, 171
Colon, T5.1–T5.3
  area for, 330
  ascending, 363, 367, 288, 290, 296, 153
    bed, 293
    as site of referred visceral pain, BP5
  circular muscle, 299
  descending, 6, 7, 296, 301, 330, 346, 349,
      350, 363, 367, 412, BP80
  sigmoid, 6, 7, 393–396, 412, 414
    as site of referred visceral pain, BP5
  transverse, 288, 289, 290, 292, 296, 301, 305–
      307, 314, 316, 349, 350
Columns of fornix, 170
Commissural fibers, 146
Commissure, BP40
  anterior, 133, 141, 146, 173, 174
  posterior, 133, 138, 141
  of semilunar valve cusps, 244
Common bony limb, 125
Common carotid artery, 51, 55, 57
Common iliac arteries, 14, BP38
Common iliac nodes, 286, 338
Common membranous limb, 124, 125
Common tendinous ring (of Zinn), 112, 113,
    148, BP27
Communicating artery
  anterior, 163–169, 176
    imaging of, BP31, BP32
  posterior, 131, 163–170, 176
    imaging of, BP31, BP32
Communicating vein, 65, 194, BP39
Compact bone, BP9
Compressor urethrae muscle, 358, 370, 372,
    379, 380, 406, T6.4
Concentric lamellae, of osteon, BP9
Concha, inferior, 38
Condylar canal, 31, 34
Condylar fossa, 31
Condylar process
  head of, BP25
  of mandible, 39
    head of, 27
Condyles
  of femur
    lateral, 500, 517–520
    medial, 500, 517–520
  of fibula
    lateral, 525
    medial, 525
  of knee
    lateral, BP107
    medial, BP107
  mandibular, 28
  occipital, 29, 31, 32, 38
  of tibia
    lateral, 501, 520, 524
    medial, 517, 519, 520
Condyloid joint, 9

Cone of light, 122
Cones, 147
Confluence of sinuses, 126, 130, 131, 171,
    172
Conjoined longitudinal muscle, 362, 396
Conjunctiva, 120
Conjunctival artery, 119
  posterior, 120
Conjunctival fornix
  inferior, 110
  superior, 110
Conjunctival vein, 119
  posterior, 120
Connecting tubules, BP72
Connective tissues, BP11
  dense, BP11
  of skull, 129
Conoid ligament, 429, 431, 434
Conoid tubercle, 429
Conus arteriosus, 232, 233, 241, 242, 243, 245
Conus elasticus, 106, 107, 108
Conus medullaris, 186, 187, 350, T3.1
  radiology of, 182
Cooper's ligament. See Pectineal ligament
    (Cooper's)
Coracoacromial ligament, 431, 434, 436, 440
Coracobrachialis muscle, 264, 265, 428, 433,
    436, 438, 440, 442, 445, 483, 485, T7.5–T7.8
Coracobrachialis tendon, 433, 436
Coracoclavicular ligament, 431, 434
Coracohumeral ligament, 431
Coracoid process, 209, 210, 427, 430–438, 440,
    442, 490
  of scapula, 203
Cornea, 110, 116, 117, 119, 120, BP27
Corneoscleral junction, 110
Corniculate cartilage, 96, 106
Corniculate tubercle, 92, 93, 107
Corona radiata, BP90
Coronal plane, 1
Coronal sulcus, 389
Coronal suture, 25–30, 35, BP16
Coronary arteries, 239, T4.1–T4.3
  imaging of, 240
  left, 239, 240, BP46
    anterolateral views with arteriograms,
        BP46
    opening of, 244
  opening of, 245
  right, 232, 233, 235, 239, 240, 243, BP46
    anterolateral views with arteriograms,
        BP46
    atrial branch of, 239
    AV nodal branch of, 240, 243, BP46
    conal branch of, BP46
    opening of, 244
    right inferolateral branches of, BP46
    right marginal branch of, 239, 240, BP46
    SA nodal branch of, 240, 248, BP46
    sinuatrial nodal branch of, 239
  variations of, BP47
Coronary ligament, of liver, 293, 302, 343
Coronary sinus, 235–237, 239, 242, 266, BP50
  opening of, 238, 241, 244
  valve (thebesian) of, 241
Coronary sulcus, 232, 235
Coronoid fossa, 427, 446
Coronoid process, 449, 453, 78
  of mandible, 27, 28, 39
Corpora cavernosa, of penis, 383
Corpus albicans, 365, 374
Corpus albuginea, 377
Corpus callosum, 133, 135, 138, 168, 170
  body of, imaging of, 177, BP32
  dorsal vein of, 171, 173
  genu of, 133, 134, 137, 139, 173
    imaging of, 177

Deltoid tuberosity, 427
Deltopectoral groove, 422
Deltopectoral triangle, 432
Dendrite, BP3
Denonvilliers' fascia, 385, 393, 398
Dens, 47, BP39
  abnormalities of, BP18
  apical ligament of, 47, 95, 96, BP18
  of atlas, 43
  of axis, 26, 28, 43, 45
  posterior articular facet of, 47
Dens axis, 9, BP34, BP35
Dental plexus, inferior, 149
Dental pulp, 41
Dentate gyrus, 133, 135, 138, 139, 146
Dentate line, 395
Dentate nucleus, 141
Denticulate ligament, 189, 190
Dentin, 41
Depressor anguli oris muscle, 48, 72, 150,
  T2.10–T2.14, BP19, BP20
Depressor labii inferioris muscle, 48, 72, 150,
  T2.10–T2.14, BP19, BP20
Depressor septi nasi muscle, 48, 59, 60, 150,
  T2.10–T2.14, BP19
Dermatomes, 188
  levels of, 188
  of lower limb, 5
  of upper limb, 5
Dermis, 12
Descemet's membrane, 117
Descending aorta, 14, 17. See also Thoracic
  aorta.
Descending colon, 6, 7, 290, 412, 337
Descending fibers, 419, 340
Desmosome, BP13
Detrusor muscle, of bladder wall, BP86
Diaphragm, 17, 211, 219, 220, 283, 285, 291,
  229, 292, 231, 294, 234, 238, 274, 276, 251–
  255, 257, 258, 259, 302, 307, 328, 330, 331,
  337, 344, 347, 412, 507, T4.1–T4.4, BP2,
  BP36, BP43, BP50, BP51, BP58
  abdominal surface of, 216
  area for, 331
  central tendon of, 216, 283
  costal part of, 211, 216
  crura of, 200
    left, 216, 253, 254
    right, 216, 253, 254, BP50
  crus of, BP38
  dome of
    left, 217, 218
    right, 217, 218
  esophageal hiatus of, 287
  left crus of, 283, 289, 337
  lumbar part of, 216
  right crus of, 283, 289, 337
  as site of referred visceral pain, BP5
  sternal part of, 211, 216
  thoracic surface of, 215
Diaphragmatic constriction, 255
Diaphragmatic fascia, 257, 337
Diaphragmatic grooves, BP59
Diaphragmatic nodes
  inferior, 286
    lymph vessels to, 207
  superior, 207
Diaphragmatic recess, right, 234
Digastric fossa, 39
Digastric muscle, T2.10–T2.14
  anterior belly of, 53, 54, 58, 69, 70, 77, 87,
    97, BP23
  intermediate tendon of, 88
  phantom, 58
  posterior belly of, 49, 53, 54, 58, 70, 74, 77,
    86, 88, 92, 97, 102, 150
  trochlea for intermediate tendon of, 53, 87

Digastric tendon, intermediate, 86
  fibrous loop for, 49, 88
Digestive system, 18, T5.1–T5.3
Digestive tract, 21
Digital arteries
  dorsal, 479, 481, 522, 532, 541, 542, 547,
    BP105, BP109
  palmar, 469, BP97
    common, 470, 473, 476, BP97
    proper, 470, 476, 479, 481, BP97
  plantar, branches of, proper, 532, 541,
    542
Digital fibrous sheaths, 468
Digital nerve
  dorsal, 492, 425, 478
    of foot, 554
  palmar, 425, 469, 470
    common, 476, 486, 487
    proper, 470, 476, 478, 479, 481, 486,
      487
  proper plantar, branches of, 532, 541,
    542
Digital vein
  dorsal, 492, 425
  palmar, 15, 425
    dorsal, 478
Digits
  extensors of, 451
  flexors of, 453
Dilator pupillae muscle, 117, 148, 159,
  T2.1–T2.4, T2.10–T2.14
Diploë, 30
Diploic vein, 127, 129
  anterior temporal, 127
  frontal, 127
  occipital, 127
  posterior temporal, 127
Direct vein, lateral, 172, 173
Distal, as term of relationship, 1
Distal convoluted tubule, BP71, BP72
Distal interphalangeal (DIP) joint, 422, 467,
  468, BP98, BP99
Dorsal arteries
  of clitoris, 406
  of penis, 382, 384, 403
    deep, BP93
Dorsal nasal artery, 60
Dorsal nerves
  of clitoris, 415
  of penis, 382, 384, 407, 411–413, 415, 418
Dorsal veins
  deep
    of clitoris, 358–360, 364, 366, 369, 380
    of penis, 361, 362, 384, 403, 405, 407,
      BP93
  superficial, of penis, 382, 405, BP93
Dorsal venous arch, 15
Dorsal venous network, 425, 444
Dorsalis pedis artery, 14, 532, 541, 542, 547,
  BP105, BP111
  pulse point, 14
Dorsiflexion, of lower limbs movements, 11
Dorsum, 463
  of foot, 541, 542
Duct of gland, 89
Duct systems, of liver, 303
Ductus arteriosus, 247
Ductus (vas) deferens, 20, 274, 280, 282, 284,
  335, 371, 385, 387, 388, 390, 392, 393, 405,
  411, 412, 418, 494, T6.1–T6.3, BP104, BP93,
  BP54
  ampulla of, 385
  arteries of, 388, 400, 403, 405, BP93
  covered by peritoneum, 281
  in peritoneal fold, 367
  vestige of, 390
Ductus reuniens, 125

Ductus venosus, 247
Duodenal bulb, 295, 297
Duodenal flexure
  inferior, 297
  superior, 297
Duodenal fossa
  inferior, 289
  superior, 289
Duodenal impression, 302
Duodenal papilla
  major, 297, 305, T5.1–T5.3, BP62
  minor, 297, 305, BP62
Duodenal wall, layers of, BP56
Duodenojejunal flexure, 289, 296, 297, 306,
  337, 342
Duodenojejunal fold, 289
Duodenojejunal junction, 200
Duodenomesocolic fold, 289, 296
Duodenum, 153, 289, 292–294, 296, 302,
  303, 306, 330, 335, 337, 342, 345, 346,
  T5.1–T5.3, BP57
  arteries of, 310
  ascending part of, 289, 296, 297
  autonomic innervation of, 320, 321, 325
  circular muscular layer of, 295, 305
  descending part of, 200, 289, 291, 296, 297,
    305
  horizontal part of, 289, 296, 297
  inferior part of, 296
  longitudinal muscle of, 305, BP56
  longitudinal muscular layer of, 295
  as site of referred visceral pain, BP5
  in situ, 296
  superior part of, 295, 296, 305, 348
  suspensory muscle of, 289
  veins of, 315
Dura mater, 63, 122, 125, 127, 128, 136, 146,
  190, 192, 194, BP35, BP39
  meningeal layer of, 127, 129
  periosteal layer of, 127, 129
  radiology of, 182
  spinal, 200
Dura-skull interface, 127, 129
Dural sac, termination of, 182, 186, 187
Dural venous sinuses, T2.1–T2.4
  sagittal section of, 130
  superior view of, 131

E

Ear, 3
  ampulla of, 124, 125
  bony and membranous labyrinths of, 124,
    125
  course of sound in cochlea and, 121
  external, 122
  pediatric, anatomy of, BP29
  vestibule of, 124, 125
Effectors, 4
Efferent ductules, 392
Efferent glomerular arteriole, BP71,
  BP73
Ejaculatory ducts, 385, 386, BP92
  opening of, 390
  in prostatic urethra, 420
Elastic fibers, 12, BP11
Elastic laminae, BP13
Elastic membrane, of arterial wall
  external, BP13
  internal, BP13
Elbow, 2, 3, BP15
  anterior region of, 422
  bones of, 446
  ligaments of, 448
  posterior region of, 422
  radiographs of, 447
Elliptical recess, 124

Emissary vein, 34, 127, 129, T2.1–T2.4
  condylar, 194
  mastoid, 24, 194, 127
    in mastoid foramen, 33
  occipital, 127
  parietal, 24, 127
Enamel, 41
Endocrine glands, BP95
Endocrine system, 21, T5.1–T5.3
Endolymphatic duct, 34, 124
Endolymphatic sac, 125
Endometrium, 374
  of uterus, 399
Endoneurium, BP4
Endopelvic fascia, 366
  deep perineal pouch with, 84
  paravesical, 371
Endosteum, BP9
Endothelial cell, BP11, BP45
Endothelial cell junctions, BP45
Endothelium, BP71
  of arterial wall, BP13
Enteric plexuses, 326
Eosinophils, BP11, BP12
Epicondyles
  of femur
    lateral, 500, 503, 517
    medial, 500, 511, 516, 517
  of knee, medial, BP107
    lateral, 427
    medial, 427
Epicranial aponeurosis, 24, 48, 129, 199, BP19
Epicranius muscle, frontal belly (frontalis) of, BP19, BP20
Epidermis, 12, BP1
Epididymal duct, 392
Epididymis, 20, 387, 388, 391, 392, 418
  appendix of, 388, 390
  body, 392
  head, 392
  sinus of, 388
  tail, 392
Epidural fat, BP39
  radiology of, 182
Epidural space, fat in, 190
Epigastric artery
  inferior, 272, 276, 281, 284, 363, 366, 367, 373, 375, 336, 400, 402, 522, BP103, BP54
    cremasteric branches of, 274
    pubic branches of, 272, 274, 280, 367, 284
    pubic tributary of, 280
  superficial, 270, 272, 276, 284, 522, BP103
  superior, 209, 272, 210, 211, 212, 276, 231, BP83
Epigastric node, inferior, 286
Epigastric region, 267, 269
Epigastric veins
  inferior, 277, 281, 285, 363, 367, 373, 375, 402, 403, 405, BP93
  superficial, 277, 285, 492, 523
  superior, 210, 211, 213, 277
Epigastric vessels
  inferior, 272, 274, 280, 282
  superficial, 270, 272
  superior, 272
Epiglottic cartilage, 36, 106
Epiglottis, 36, 17, 255, 89, 90, 92–95, 105–108, 162, T2.1–T2.4
Epimysium, BP10
Epineurium
  inner, BP4
  outer, BP4
Epiphysial line, 514
Epiphysial union, BP95
Epiphysis, 481
  nutrient branch to, 481
Episcleral artery, 119

Episcleral space, 116, BP27
Episcleral vein, 119, 120
  segment of, 120
Epithelial tag, 389
Epitympanic recess, 121, 122, BP28, BP29
Eponychium, 481, BP112
Epoophoron
  cranial mesonephric tubules, 390
  Gartner's duct, longitudinal, 390
Erector spinae muscle, 178, 195, 197, 200–201, 212, 263, 279, 331, 351, T3.2–T3.4, BP80
Esophageal hiatus, 216, 253, T5.1–T5.3
Esophageal impression, 302
Esophageal mucosa, 94, 257
Esophageal muscle, 95
  circular, 257, 96
  longitudinal, 257, 96, 104
Esophageal plexus, 153, 220, 229, 237, 251, 252, 254, 261, 325, BP6, BP7
Esophageal submucosa, 94
Esophageal veins, 318, 259, T5.1–T5.3, BP66
Esophagogastric Junction, see
  Gastroesophageal junction, 257, 347
  cross-sectional anatomy, 347
Esophagus, 17, 18, 50–52, 91, 93, 95, 97, 105, 107, 153, 215, 220, 222, 236, 237, 238, 251, 252, 253, 255, 259, 261–266, 283, 285, 286, 289, 293, 323, 330, 332, 343, BP36, BP50
  abdominal part of, 254, 255, 258, 294
  arteries of, 258
    variations in, BP49
  autonomic innervation of, 325
  cervical part of, 254, 255, 258
  circular muscular layer of, 92, 94, 104
  constrictions and relations of, 255
  impression of, 221, 236
  longitudinal muscle of, 295
  longitudinal muscular layer of, 92, 94, 225
  lung groove for
    left, 221
    right, 221
  lymph vessels and nodes of, 260
  musculature of, 256
  nerves of, 261
    variations in, intrinsic and, BP48
  recurrent branch of, 216
  in situ, 254
  thoracic part of, 254, 255, 258
  veins of, 259
  ventral surface of, bare area on, 256
Estrogen, BP88
Ethmoid artery
  anterior, T2.1–T2.4
  anterior lateral nasal branch of, BP21
  anterior septal branch of, BP21
  external nasal branch of, BP21
Ethmoid bone, 25, 27, 29, 32, 37, 62, 63
  anterior ethmoidal foramen of, 35
  cribriform plate of, 29, 32, 61–63, 146
  crista galli of, 29, 32, 63
  nasal concha of
    highest, 62
    middle, 25, 29, 62
    superior, 29, 62
  of newborn, 35
  orbital plate of, 25, 27, 35
  perpendicular plate of, 25, 29, 63
  uncinate process of, 62, 67, 68
Ethmoid sinus, imaging of, 177
  Ethmoidal air cells, 69, BP23, BP27
    anterior, 114
    middle, opening of, 61, 62
    posterior, openings of, 62
  Ethmoidal artery
    anterior, 34, 65, 115
      anterior meningeal branch of, 128
      lateral nasal branch of, 65
    posterior, 34, 65, 115, 128

Ethmoid sinus, imaging of (Continued)
  Ethmoidal bulla, 61, 62
  Ethmoidal cells, 26, 37, 67
  Ethmoidal foramen
    anterior, 34
    posterior, 34
  Ethmoidal nerve
    anterior, 34, 66, 79, 113, 114, 148, 149
      external nasal branch of, 23, 64, 79
      lateral internal nasal branch of, 64
      medial internal nasal branch of, 64, 114
    posterior, 34, 79, 113, 114, 148, 149
  Ethmoidal vein
    anterior, 34, 65
    posterior, 34, 65
Eversion, of lower limbs movements, 11
Extension
  of lower limbs movements, 11
  of upper limbs movements, 11
Extensor carpi radialis brevis muscle, 422, 441, 451, 454, 455, 459, 460, 461, 488, 489, T7.5–T7.8
Extensor carpi radialis brevis tendon, 454, 455, 461, 477, 479, 480
Extensor carpi radialis longus muscle, 422, 428, 441, 445, 451, 454, 455, 459, 460, 461, 488, 489, T7.5–T7.8
Extensor carpi radialis longus tendon, 454, 455, 461, 477, 479, 480
Extensor carpi ulnaris muscle, 422, 441, 451, 454, 459, 460, 461, 488, 489, T7.5–T7.8
Extensor carpi ulnaris tendon, 454, 455, 461, 479, 480
Extensor digiti minimi muscle, 451, 454, 459–461, 489, T7.5–T7.8
Extensor digiti minimi tendon, 451, 454, 455, 461, 479, 480
Extensor digitorum brevis muscle, 530, 532, 540, 541, 542, 549, 554, T8.5–T8.8, BP111
Extensor digitorum brevis tendon, 531, 532, 542, 547, 549, BP111
Extensor digitorum longus muscle, 501, 517, 526, 531, 532, 534, 540–542, 554, T8.5–T8.8, BP106, BP110
Extensor digitorum longus tendon, 491, 530, 531, 532, 541, 542, 547, 549, BP110, BP111
Extensor digitorum muscle, 441, 451, 454, 459, 460, 461, 488, 489, T7.5–T7.8
Extensor digitorum tendon, 422, 451, 454, 455, 461, 479, 480, BP101, BP102
Extensor expansion, 474, 480, BP101, BP102
Extensor hallucis brevis muscle, 531, 532, 541, 542, 554, T8.5–T8.8, BP111
Extensor hallucis brevis tendon, 532, 542, 547, 549, BP111
Extensor hallucis longus muscle, 526, 530, 531, 532, 534, 542, 554, T8.5–T8.8, BP106, BP110
Extensor hallucis longus tendon, 491, 530, 531, 532, 541, 542, 547, 549, BP110, BP111
  tendinous sheath of, 540, 541
Extensor indicis muscle, 451, 455, 460, 461, 489, T7.5–T7.8
Extensor indicis tendon, 422, 451, 454, 461, 479, 480, BP101, BP102
Extensor pollicis brevis muscle, 451, 454, 455, 460, 489, T7.5–T7.8
Extensor pollicis brevis tendon, 454, 458, 461, 477, 479, 480
Extensor pollicis longus muscle, 451, 455, 460, 461, 489, T7.5–T7.8
Extensor pollicis longus tendon, 422, 454, 461, 473, 477, 479, 480
Extensor retinaculum, 425, 454, 455, 479, 480
  inferior, 531, 540, 541, 554

Hepatogastric ligament, 292, 294, 303
Hepatogastric trunk, BP67
Hepatopancreatic ampulla (of Vater), 305
   sphincter of, 305, 328
Hepatorenal ligament, 302
Hepatorenal recess (Morison's pouch), 296, BP79
Hepatosplenomesenteric trunk, BP67
Hering-Breuer reflex, 230
Hernia sac, BP54
   neck of, BP54
Hesselbach's triangle. *See* Inguinal
   (Hesselbach's) triangle
Hilar lymph nodes, 221
Hilum, 221, 307, 333
Hinge joint, 9
Hip, 2, 3, BP15
   bone, ligaments of, 357
   cross-sectional anatomy of, BP104
   MRI of, 555
   muscle of, 501, 504
     bony attachments of, 498, 499
   3D CT of, 555
Hip bone, 8, 495
Hip bursa, 513
Hip joint, 496, T8.1–T8.2
   accessory ligaments of, 502
   anteroposterior radiograph of, 497
   iliofemoral ligament of, 505
   ligaments of, 503
Hip joint capsule, 514
Hippocampal commissure, 139
Hippocampal sulcus, 139
Hippocampus, 135, 138, 139
   alveus of, 139
   fimbria of, 133, 135, 137–139, 146
   pes of, 138, 139
Homologous chromosomes, BP87
Hook of hamate, 452, 458, 462, 464, 466, 467,
   468, BP98
Horizontal cells, 147
Horizontal fissure, of lungs, right, 221
Horizontal plate, of palatine bone, 31, 38, 61,
   62, 63, 85, BP30
Hormones, BP95
Horn
   anterior, BP40
   lateral, BP40
   posterior, 189, BP40
Humeral artery
   anterior circumflex, 436–438, 440, 442, 443,
     BP96
   posterior circumflex, 436–438, 441, 442, 443,
     BP96
Humeral ligament, transverse, 431
Humeral vein, posterior circumflex, 424
Humeroulnar head, 483
Humerus, 8, 9, 427, 430, 442, 445–447, 448,
   T7.1–T7.2
   body of, 263, 264, 265, 430, 490
   epicondyles of
     lateral, 440, 441, 446–448, 450–453, 455,
      458, 489
     medial, 440, 441, 442, 446–448, 450–458,
      487, 488
   greater tubercle of, 430, 431, 436, 440
   head of, 427, 430, 234, 490
   lesser tubercle of, 440
   muscle attachment sites of, 428
   surgical neck of, 427, 263, 490
   trochlea of, 427
Huxley layer, BP1
Hyaloid canal, 116
Hydatid of Morgagni, 374
Hydrocele, 388
Hymen, variations of, BP91
Hymenal caruncle, 372, 377
Hyoepiglottic ligament, 95, 106

Hyoglossus muscle, 52, 53, 54, 58, 86–88, 90,
   96, 97, 155, T2.10–T2.14
Hyoid bone, 8, 36, 49, 53, 54, 58, 70, 86–88,
   95–97, 103, 105, 106, 108, T2.1–T2.4, BP34
   body of, 36, 87
   greater horn of, 36, 87, 92, 93, 97
     tip of, 104
   lesser horn of, 36, 87
Hyperplastic lateral lobe, BP93
Hyperplastic middle lobe, BP93
Hypertonic stomach, BP55
Hypochondriac region, 267
   left, 269
   right, 269
Hypogastric nerve, 319, 323, 324, 339, 340,
   411, 412, 416–419, BP6
   left, 414
   right, 414
Hypogastric plexus, 412
   inferior, 6, 323, 339, 340, 411, 412, 414, 416–
     419, BP6, BP7
   right inferior, 323
   superior, 6, 319, 323, 324, 339, 340, 363, 412,
     414, 416–419, BP6
   uterovaginal, 417
Hypogastric region, 267, 269
Hypogastric (neurovascular) sheath, 366
Hypoglossal canal, 29, 31, 33, 34, BP18
Hypoglossal nerve (CN XII), 56–58, 75, 77, 82,
   86, 91, 98–100, 102, 131, 140, 144, 155,
   156, T2.5–T2.8, BP22
   communicating branch of, T2.9
   distribution of, 145
   in hypoglossal canal, 33, 34
   meningeal branch of, 155
   nucleus of, 143, 144, 155
   schema of, 155
   thyrohyoid branch of, 98
   vena comitans of, 88, 100
Hypoglossal trigone, 141
Hyponychium, BP112
Hypophyseal artery
   inferior, 164, 167
   superior, 164, 167, 175
Hypophyseal veins, efferent, 167
Hypophysial fossa, 68, 95
Hypophysial portal system
   primary plexus of, 175
   secondary plexus of, 175
Hypophysial vein, 175
Hypophysis, 131, 133, 134, 174
   pars distalis of, 175
   portal veins of, 175
   in sella turcica, 61
   vasculature of, 175
Hypoplastic dens, BP18
Hypothalamic area, lateral, 162
Hypothalamic artery, 167
Hypothalamic nucleus
   paraventricular, 174
   supraoptic, 174
Hypothalamic sulcus, 133, 141, 174
Hypothalamohypophyseal tract, 174
Hypothalamus, 21, 135, 174, 175, 324
   principal nuclei of, 174
     arcuate (infundibular), 174
     dorsomedial, 174
     mammillary body, 174
     paraventricular, 174
     posterior, 174
     supraoptic, 174
     ventromedial, 174
   vasculature of, 175
Hypothenar eminence, 422
Hypothenar muscles, 467, 469, 470, 473, 487,
   BP100, BP101
Hypotonic stomach, BP55

I

Ileal arterial arcades, 297, 313
Ileal arteries, 313, 314
Ileal orifice, 299, 301
Ileal papilla, 299
Ileal veins, 316, 317, 318
   internal, 317
Ileal venous arcades, 316
Ileocecal fold (bloodless fold of Treves), 298
Ileocecal junction, 298, BP80
Ileocecal lips, 299
Ileocecal recess
   inferior, 298
   superior, 298
Ileocolic artery, 298, 313, 314, 335, BP61, BP68,
   BP69
   colic branch of, 298, 314
   ileal branch of, 298, 314
Ileocolic nodes, BP77
Ileocolic plexus, 322, 323
Ileocolic vein, 316–318
Ileum, 288, 297, 301, 316, 350, 351, BP80, BP81
   circular muscle layer of, 297
   terminal part of, 363, 290, 298, 299
Iliac artery
   common, 14, 284, 293, 330, 335, 336, 400,
     402, 403, 405, 414, BP80
     left, 404
     right, 404
   deep circumflex, 282, 510, 522, BP103
     ascending branch of, 276, 284
   external, 14, 284, 293, 317, 330, 335, 363,
     365, 367, 368, 372, 373, 375, 376, 400,
     402–404, 414, 510, 522, BP103, BP104,
     BP90, BP81
   internal, 14, 284, 314, 330, 335, 336, 373, 400,
     402–404, 414, T6.1–T6.3, BP81
     anterior division of, 404
     posterior division of, 404
   superficial circumflex, 270, 272, 276, 509,
     284, 522, BP103
Iliac crest, 178, 183, 195–198, 267, 268, 275,
   330, 331, 337, 352–353, 356, 357, 491, 493,
   495, 501, 504, 505, 511, 512, BP80
   iliac tubercle of, 357
   inner lip of, 268, 353, 357, 495
   intermediate zone of, 268, 353, 357, 495
   outer lip of, 268, 353, 357, 495
   tuberculum of, 268, 353, 357, 495
Iliac fascia, 282, 366
Iliac fossa, 356, 357, 363. *See also* Ilium;
   ala of.
Iliac lymph nodes, 16
   common, 286, 408, 410, BP77
   external, 286, 338, 410, 494, BP77
     inferior, 408
     medial, 408
     superior, 408
   internal, 286, 338, 408, 410
Iliac plexus
   common, 319, 414
   external, 319, 414
   internal, 319, 414
Iliac spine
   anterior
     inferior, 353, 268, 283, 356, 357, 361, 495–
      496, 502
     superior, 267, 268, 270–271, 280, 281, 283,
      300, 352, 353, 356, 357, 378, 381, 491,
      495, 496, 501, 502, 505, 509, T5.1–T5.3,
      BP53
   posterior
     inferior, 495, 183, 357
     superior, 178, 183, 356–357, 495
Iliac tubercle, 356
Iliac tuberosity, 353, 268, 495

               Netter Atlas of Human Anatomy: Classic Regional Approach

Iliac vein
common, 15, 285, 401, 403, BP81
deep circumflex, 285
tributaries to, 277
external, 15, 285, 317, 363, 365, 367, 368, 372, 373, 375, 376, 401, 403, 405, 510, 523, BP104, BP90
left, 405
internal, 15, 285, 373, 401, 403, T6.1–T6.3
right, 405
superficial circumflex, 277, 285, 492, 523
Iliac vessels
common, 290, 414, BP93
deep circumflex, 274, 366, 282, BP93
external, 274, 280, 282, 298, 364, 366, 376, 394, BP93
covered by peritoneum, 281
internal, 366, BP93
superficial circumflex, 270, BP53
Iliacus fascia, 394
Iliacus muscle, 366, 505, 280, 283, 330, 394, 420, 498, 502, 507, 513, 514, 550, T8.5–T8.8, BP94
muscular branches to, 507
nerve to, 506
Iliococcygeus muscle, 358–362, 372, 397, T6.4
Iliocostalis lumborum muscle, 196
Iliocostalis muscle, 196, T3.2–T3.4
Iliocostalis thoracis muscle, 196
Iliofemoral ligament, 496, 513, BP104
of hip joint, 505
Iliohypogastric nerve, 186, 198, 287, 330, 331, 411, 416, 505–507, 551, T5.1–T5.3
anterior branch of, 506
anterior cutaneous branch of, 278, 287, 411
lateral cutaneous branch of, 275, 278, 493, 506
Ilioinguinal nerve, 186, 278, 281, 287, 330, 331, 411, 416, 505–507, 551, T5.1–T5.3
anterior scrotal branches of, 278, 287, 411
labial branch of, anterior, 415
scrotal branches of, 492
Iliolumbar artery, 284, 336, 405, BP93
Iliolumbar ligament, 183, 331, 356
Iliolumbar vein, 285
Iliopectineal arch, 513
Iliopectineal bursa, 496, 505, 513
Iliopectineal line, 394
Iliopsoas muscle, 274, 420, 421, 498, 502, 503, 505, 509, 510, 513, 515, BP104, BP94
Iliopsoas tendon, 514
Iliopubic eminence, 353, 268, 356, 495, 357, 496, 361
Iliopubic tract, 363, 366, 280, 367
Iliotibial tract, 415, 491, 498, 501, 502, 504, 511, 513, 515, 516, 517, 518, 520, 525, 526, 527, 530–532, T8.1–T8.2
bursa deep to, 517
bursa of, 516
to Gerdy's tubercle, 519
Ilium, 497
ala of, 268, 353, 357, 361, 495, 555
arcuate line of, 359
auricular surface for, 495
body of, 357, 495
radiology of, 182, 354
Incisive canal, 29, 61, 63, 64, 66, 81, 95, BP21
anastomosis in, 74
Incisive fossa, 31, 33, 38, 40, 85
Incisive papilla, 85
Incisor teeth, 255
Incisors, 41
Incus, 78, 121, 122, 125, 151, BP29
lenticular process of, 122, BP28
long limb of, 122, BP28
posterior ligament of, BP28

Incus (Continued)
short limb of, 122, BP28
superior ligament of, BP28
Indirect inguinal hernia, BP54
Inferior, as term of relationship, 1
Inferior anterior segmental artery, 334
Inferior articular facet, of L4, BP36
Inferior articular process, 180, 181
of L1, BP37
of L2, 182
of L3, BP37
Inferior colliculus, 138
brachium of, 138
Inferior fascia, of pelvic diaphragm, 378
Inferior hypogastric plexus, 6, BP6, BP7
Inferior labial artery, 60
Inferior mediastinum, BP2
Inferior mesenteric artery, 14, 284
Inferior mesenteric ganglion, 6, BP6, BP8
Inferior mesenteric plexus, BP6
Inferior nasal concha bone, 37
Inferior nodes, 409
Inferior orbital fissure, 37
Inferior pancreaticoduodenal plexus, 322
Inferior pharyngeal constrictor, cricopharyngeal part of, 52
Inferior pubic ligament, 268
Inferior pubic ramus, 268
Inferior segmental artery, 334
Inferior vena cava, 13, 15, 285, 291–294, 296, 302, 306, 312, 315, 330–332, 337, 342, 345–351, 363, 366, 367, 401, 402–404, 414, BP38, BP79, BP82, BP93
Inferior vertebral notch, of L2 vertebra, BP37
Inferolateral lobe, BP92
Inferoposterior lobe, BP92
Infraclavicular node, 426
Infraglenoid tubercle, 427
Infraglottic cavity, 108
Infrahyoid artery, 103
Infrahyoid fascia, 50
Infrahyoid muscle, 219
fascia of, 49, 50
Infrahyoid muscles, 53
fascia of, 51, 52
Infraorbital artery, 24, 59, 60, 74, 75, 81, 99, 110, 115
Infraorbital foramen, 25, 26, 27, 59, BP16
of newborn, 35
Infraorbital groove, 25
Infraorbital margin, 22
Infraorbital nerve, 23, 59, 60, 79, 80–82, 84, 98, 110, 114, 148, 149, BP27
superior alveolar branches of, 149
Infraorbital vein, 24, 100
Infrapatellar bursa
deep, 521
subcutaneous, 521
Infrapatellar fat pads, 517, 518, 521
Infrapatellar synovial fold, 517, 518
Infraspinatus fascia, 275, 432, 195, 198
Infraspinatus muscle, 178, 212, 201, 263–265, 432, 434–437, 441, 488, T7.5–T7.8
Infraspinatus tendon, 431, 434, 436, 441
Infraspinous fossa, 427
of scapula, 203
Infratemporal crest, 27
Infratemporal fossa, 27, 38, 75, 78
Infratrochlear nerve, 59, 60, 79, 113, 149
left, 114
right, 114
Infundibular process, 174
Infundibular recess, 133, 135
Infundibular stem, 174
Infundibulum, 374, 140, 174
of uterine tube, 376
Inguinal canal, 281, 282, BP104

Inguinal falx, 271, 272, 274, 280, 281, BP53
Inguinal fold, 390
Inguinal ligament (Poupart's), 267, 270, 271, 272, 280, 281, 282, 283, 352, 359, 373, 375, 378, 381, 382, 409, 491, 492, 494, 502, 507, 509, 510, 513, 523, T5.1–T5.3, BP53
reflected, 271, 272, 280
Inguinal lymph nodes
deep, 494
superficial, 494, T8.1–T8.2
inferior, 494
superolateral, 494
superomedial, 494
Inguinal nodes, 16, BP77
of Cloquet, proximal deep, 408, 409, 410
deep, 286, 408, 409, 410
superficial, 286, 408, 409, 410
Inguinal region, BP53, 267
dissections of, 280
left, 269
right, 269
Inguinal ring
deep, 272, 274, 281, 280, 363, 365, 366, 367, 387, T5.1–T5.3, BP53
lateral crus, 280
medial crus, 280
superficial, 270, 280, 281, 378, 382, 387, 388, T5.1–T5.3, BP53
Inguinal (Hesselbach's) triangle, 274, 280, T5.1–T5.3
Inhibin, BP88
Inion, 27
Initial segment, BP3
Insula (island of Reil), 132, 137, 168
central sulcus of, 132
circular sulcus of, 132
limen of, 132
long gyrus of, 132
short gyrus of, 132
Interalveolar septa, 39
Interalveolar septum, BP45
Interarytenoid notch, 93
Interatrial septum, 241, 266
Intercapitular veins, 425, 444, 478
Intercarpal ligament, dorsal, 465
Intercavernous septum, of deep fascia, 382, 383, 386
Intercavernous sinus
anterior, 130, 131
posterior, 130, 131
Interchondral joints, 221
Interclavicular ligament, 221, 429
Intercondylar area, anterior, 524
Intercondylar eminence, 518, 520, 524, 525
Intercondylar fossa, 500
Intercondylar tubercles, 525
Intercostal artery, 205, 212, BP83
anterior, 211, 212, 276
collateral branches of, 211
lower, anastomoses, 276
posterior, 191, 209, 210, 212, 251, 252, 436
dorsal branch of, 192, 212
lateral cutaneous branches of, 212
lateral mammary branches of, 206
right, 222
spinal branch of, 212
supreme, 163, 165, BP24
Intercostal membrane
external, 212, 279, 201, 436
internal, 212, 279, 201
Intercostal muscles, 205, 276, 219, 238, 263–265, BP45, BP50
external, 197, 201, 209, 210, 212, 271, 276, 279, 436, T4.4, BP43
internal intercostal membrane deep to, 212

Intercostal muscles (Continued)
  innermost, 211, 212, 276, 279, 201, 251, 252, 253, T4.4
  internal, 209, 210, 211, 212, 276, 279, T4.4, BP43
Intercostal nerves, 186, 205, 210–212, 214, 251, 252, 279, 506, T4.1–T4.3, T5.1–T5.3, BP52
  first, 250, 439
  third, 261
  sixth, 229
  eight, 229
  anterior cutaneous branches of, 209, 211, 212, 436
  anterior ramus of, 201
  collateral branch of, 211, 279
  communicating branch of, 279
  cutaneous branches
    anterior, 278, 279
    lateral, 278, 279
  lateral cutaneous branches of, 209, 210, 212, 436
Intercostal nodes, 260
Intercostal spaces, 233, T4.1–T4.3
Intercostal veins, 15, 205
  anterior, 211, 213, 277
  collateral branches of, 211
  posterior, 213, 251, 252
    right, 259
  superior
    left, 252, 259
    right, 213, 251, 259
Intercostobrachial nerve, 210, 278, 423, 424, 436, 438, 483
Intercrural fibers, 270, 280, 281, BP53
Intercuneiform ligaments, dorsal, 538
Interdigitating fibers, of perineum, 360
Interfascicular fasciculus, BP40
Interfoveolar ligament, 274
Interglobular spaces, 41
Intergluteal cleft, 178
Interlobar artery, 334
Interlobular arteries, 304, 334, BP73
Interlobular bile ducts, 304
Interlobular lymph vessels, 228
Interlobular septum, 227
Interlobular veins, 304
Intermaxillary suture, 31, 59
Intermediate bronchus, BP50
Intermediate nerve (of Wrisberg), 123, 150, 158, 162
  motor root of, 151
Intermediate sulcus, posterior, BP40
Intermediolateral nucleus, 326, 341
Intermesenteric (abdominal aortic) plexus, 287, 319, 322–324, 339–340, 412, 414, 416–419
Intermetacarpal joints, 465
Intermuscular septum
  anterior, 534
  lateral, 440, 441, 445, 455, 458, 488
  medial, 440, 441, 442, 445, 455, 457, 458
  posterior, 534
  transverse, 534
Intermuscular stroma, 327
Internal abdominal oblique muscles, 200
Internal capsule, 135, 137, 168, BP15
  anterior limb of, 137
    imaging of, 177
  cleft for, 137
  genu of, 137
    imaging of, 177
  posterior limb of, 137, BP14
    imaging of, 177
  retrolenticular part of, 137
Internal carotid artery, 65
Internal carotid nerve, 6, BP8
Internal carotid plexus, 6

Internal intercostal membranes, 279
Internal jugular vein, 51, 53, 55, 57
Internal occipital protuberance, 32
Internal os, 374
Internal vertebral venous plexus, anterior, BP39
Internus muscle, obturator, 373
Interosseous artery
  anterior, 443, 455, 457, 458, 461, BP96, BP97
  common, 443, 457, 458, 461, BP96, BP97
  posterior, 443, 455, 458, 461, BP96, BP97
  recurrent, 443, 455, BP96, BP97
Interosseous fascia
  dorsal, 473
  palmar, 473
Interosseous intercarpal ligaments, 465
Interosseous membrane, 449, 451, 453, 461, 464, 520, 521–522, 537, 532, 534, BP103, BP105, BP106, BP110, BP98
Interosseous muscle, 474, 539, 547, BP111
  dorsal, 473–475, 480, 542, 547, 548, 549, T8.5–T8.8, BP100, BP102, T7.5–T7.8
    first, 455, 470, 475, 477, 487, BP101
    second, BP101
    third, BP101
    fourth, BP101
  palmar, 473, 487, T7.5–T7.8, BP100
    first, BP101
    second, BP101
    third, BP101
  plantar, 549, T8.5–T8.8
Interosseous nerve
  anterior, 458, 461, 486, T7.3–T7.4
  crural, 553
  posterior, 455, 461, 489, T7.3–T7.4
Interosseous veins, anterior, 444
Interpectoral (Rotter's) lymph nodes, 207, 208
Interpeduncular cistern, 136
  imaging of, BP32
Interpeduncular fossa, 69
Interphalangeal (IP) joint, 539
Interphalangeal ligaments, 468
Interproximal spaces, 41
Intersigmoid recess, 290, 335
Intersphincteric groove, 394–396
Interspinales muscle, T3.2–T3.4
Interspinalis colli muscle, 197
Interspinalis lumborum muscle, 197
Interspinous ligament, 183, 185
Interspinous plane, 269
Intertendinous connections, 480
Interthalamic adhesion, 133, 137, 138, 141, 173, 174
Intertragic notch, 122
Intertransversarii muscle, T3.2–T3.4
Intertransverse ligament, 221
Intertrochanteric crest, 496, 512
Intertrochanteric line, 496
Intertubercular groove, 427
Intertubercular plane, 269
Intertubercular tendon sheath, 431, 436, 440
Interureteric crest, 371
Interventricular artery
  anterior, 232, 233, 239, 240, BP46, BP47
    anterior ventricular branches of, 239
    diagonal branch of, 239, BP46
    septal branches of, 239, 240, BP46
  inferior, 235, 239, 240, 243, BP46, BP47
    septal branches of, 239
Interventricular foramen (of Monro), 133, 135, 136, 141, 172, 173
  imaging of, 177
  left, 133
Interventricular septum, 245, 248, 266, T4.1–T4.3
  membranous part of, 241, 242, 243, 244, 245, 248
  muscular part of, 241, 242, 244, 245, 248

Interventricular sulcus
  anterior, 232, 266
  inferior, 235, 236
Intervertebral disc space, BP37
  C6-C7, BP34
  L4-L5, BP36
Intervertebral discs, 181, 185, T3.1, BP33
  C6-C7, BP35
  cervical, 45, 46
  L1-2, cross section at, 350
  lumbar, 183
    L2-L3, 182
  lumbosacral, 357
  T3-T4, 264
  T4-T5, 265
  T12-L1, cross section at, 349
Intervertebral foramen, 45, 181, 183, 185, 201, T3.1, BP34, BP36, BP37
  narrowing of, BP17
Intervertebral vein, 193, BP39
Intestinal arteries, T5.1–T5.3
Intestinal trunk, 286, BP76
Intestines, 6, 7, 247
  autonomic innervation of, 324
  enteric plexus of, 327
Intraarticular ligament, BP33
  of head of rib, 221
Intraarticular sternocostal ligament, 221
Intraculminate vein, 171
Intralaminar nuclei, 138
Intralobular arteriole, 304
Intralobular bile ductules, 304
Intraparietal sulcus, 132
Intrapulmonary airways, 226
  schema of, BP44
Intrapulmonary blood circulation, schema of, 227
Intrapulmonary nodes, 228
Intrarenal arteries, 334
Intrinsic back muscles, T3.1
Intrinsic muscle
  of larynx, 107
    action of, 109
  of tongue, 89
Inversion, of lower limbs movements, 11
Investing fascia, 372, 378, 379, 382, 383, 384, 393, 397, BP86
  deep, 371
Iridocorneal angle, 116, 117, 119
Iris, 110, 116, 118, 119
  arteries and veins of, 120
  folds of, 117
  major arterial circle of, 117, 119, 120
  minor arterial circle of, 117, 119, 120
Irritant receptors, 230
Ischial spine, 183, 268, 357, 353, 355, 356, 358–361, 400, 413, 415, 495, 496, 512, 583, T6.1–T6.3
  radiograph of, 354
Ischial tuberosity, 183, 352, 353, 268, 355, 356, 357, 360, 362, 378, 379, 381–384, 394, 397, 495–496, 504, 512, T6.1–T6.3
  radiograph of, 354
Ischioanal fossa, 379, 394
  anterior recess of, 274, 371, 372, 398
  fat body of, 378, 381, 394, 420, BP94
  fibrous septum of, transverse, 396, 397
  left
    anterior communication, 398
    posterior communication, 398
  levator ani muscle roofing, 382
  posterior recess of, 398
  purulence in, 398
  right
    anterior communication, 398
    posterior communication, 398
  transverse fibrous septum of, 394

Ischiocavernosus muscle, 364, 371, 372, 378, 379, 380, 382–384, 393, 397, 406, 407, T6.4
Ischiofemoral ligament, 496, BP104
Ischiopubic ramus, 352, 356, 357, 362, 368, 372, 378, 379, 381, 383–385, 397, 496, 497, 505, 555
  radiograph of, 354
Ischium, 421
  body of, 495, 497
  radiograph of, 354
  ramus of, 495
Isthmus, 374
  of uterus, 374

J

Jejunal arterial arcade, 297, 313
Jejunal arteries, 313, 314
Jejunal veins, 316, 317, 318, BP66
Jejunal venous arcades, 316
Jejunum, 288, 289, 290, 296, 297, 306, 316, 346, 349, 350, 351
  circular muscle layer of, 297
  longitudinal muscle layer of, 297
  submucosa of, 297
Joint capsule, 448, 468, 518, BP11, BP111
  attachment of, 521
  fibrous layer of, 9
  of foot, 539
  synovial membrane of, 9
Joints, connective tissues and cartilage of, BP11
Jugular foramen, 29, 34, 131, 152, 153, 154
Jugular fossa, 31, 33, BP28
Jugular nerve, 157
Jugular notch, 22, 202, 217, 49
  of sternum, 203
Jugular process of occipital bone, 55
Jugular trunk, 101
Jugular vein
  anterior, 100, 103
  external, 15, 22, 24, 56, 70, 100, 103, 219, 220, 231, 238, 259, T2.1–T2.4
  internal, 15, 24, 49, 50, 57, 58, 70, 75, 82, 88, 91, 100, 102, 103, 104, 105, 121, 123, 126, 155, 176, 219, 220, 228, 231, 238, 254, 259, T2.1–T2.4, BP22
    imaging of, BP31
    inferior bulb of, 194, 104
    in jugular fossa, 33
    left, 232
    superior bulb of, 194
  right, T2.1–T2.4
Jugulodigastric node, 101, 102
Juguloomohyoid node, 101, 102
Juxtaesophageal nodes, 260
Juxtaglomerular cells, BP71
Juxtaintestinal mesenteric nodes, BP76
Juxtamedullary glomerulus, BP73
Juxtamedullary renal corpuscle, BP72

K

Kidneys, 6, 7, 13, 19, 21, 200, 218, 247, 291, 303, 387, 402, T5.1–T5.3, BP51, BP38
  autonomic nerves of, 339
  capsular branch of, 334
  fibrous capsule of, 337, BP72, BP73
  gross structure of, 333
  inferior pole of, 333
  innervation of, autonomic, 340
  lateral border of, 333
  left, 292, 296, 306, 307, 330, 335, 337, 342, BP79, BP82
  lymph vessels and nodes of, 338
  major calyces of, 333
  medial border of, 333

Kidneys (Continued)
  medulla (pyramids) of, 333
  minor calyces of, 333
  parenchyma of, blood vessels in, BP73
  perirenal branch of, 334
  pole of, 296
  renal capsule of, 333
  renal column (of Bertin) of, 333
  renal sinus of, 333
  right, 292, 293, 294, 296, 302, 306, 330, 331, 335, 337, 342, 345, 346
  as site of referred visceral pain, BP5
  in situ
    anterior view of, 330
    posterior view of, 331
  superior pole of, 333
Kiesselbach's plexus, 65
Knee, 2, 3, 516, 517, 518, 519, 521, BP15
  anterior region of, 491
  arteries of, BP103, BP105
  condyle of
    lateral, BP107
    medial, BP107
  epicondyle of, medial, BP107
  joint capsule of, 516, 531
  osteology of, BP107
  posterior region of, 491
  radiograph of, BP108, 520
  transverse ligament of, 519

L

L5/S1 intervertebral disc, T3.1
Labia, posterior commissure of, 389
Labia majora
  commissure of, anterior, 377
  posterior commissure of, 377
Labia minora, 375
  frenulum of, 377
Labia minus, 364
Labial artery
  inferior, 59, 99
  posterior, 380, 406
  superior, 59, 99
    nasal septal branch of, BP21
Labial nerves, posterior, 415, 416
Labial vein, inferior, 100
Labioscrotal swelling, 389
Labium majus, 352, 364, 369, 372, 377, 389, BP86
Labium minus, 364, 369, 372, 377, 380, 389, BP86
Labyrinthine artery, 34, 123, 166, 167, 168, 170, left, 163
Lacrimal apparatus, 111
Lacrimal artery, 115, 165
  recurrent meningeal branch of, 128
Lacrimal bone, 25, 27, 62, 68, 69
  of newborn, 35
Lacrimal canaliculi, 111
Lacrimal caruncle, 110, 111
Lacrimal gland, 6, 7, BP7, 69, 113–115, 160, BP27
  excretory ducts of, 111
  orbital part of, 111
  palpebral part of, 111
Lacrimal nerve, 79, 81, 113, 114, 148, 149, 158, BP27
  palpebral branch o, 23
  palpebral branch of, 79
Lacrimal papilla
  inferior, 110, 111
  superior, 110, 111
Lacrimal punctum
  inferior, 110, 111
  superior, 110, 111
Lacrimal sac, 110, 111
  fossa for, 25, 27

Lactiferous ducts, 205
Lactiferous sinus, 205
Lacuna (of Trolard), lateral (venous), 127, 128
Lacuna magna, 386
Lacuna vasorum, 283
Lacunar ligament (Gimbernat's), 271–274, 280, 282, 283, 494
Lacus lacrimalis, 110
Lambda, 30
Lambdoid suture, 26, 28, 29, 30, 35, BP16
Lamellar bodies, BP45
Lamellar corpuscle (Pacini's), 12
Lamina, 181, 183, 185, T3.1, BP38
  anterior, of broad ligament, 376
  of C6, BP34
  of L1, 182
  of L4 vertebra, BP37
  lumbar, 183
  of mesosalpinx, 376
  of mesovarium, 376
  posterior, of broad ligament, 376
  thoracic, 180
Lamina affixa, 138
Lamina terminalis, 133, 134, 141, 174
Large arteries, 13
Large intestine, 18, 301
  arteries of, 314
  autonomic innervation of, 323
  lymph vessels and nodes of, BP77
  veins of, 317
Laryngeal artery
  inferior, 108
  superior, 58, 94, 99, 103–105, 108, 163
Laryngeal cartilages, T2.1–T2.4
Laryngeal inlet, 93, 94
Laryngeal nerve
  fold of superior, 93
  recurrent, 50, 57, 91, 94, 98, 108, 157, 254, 261, T2.1–T2.8, T4.1–T4.3
    left, 103, 104, 105, 153, 214, 220, 229, 230, 232, 233, 249, 252, 254, 261, 263, 264
    right, 103–105, 108, 153, 232, 249, 261
  superior, 91, 94, 153, 157–158, 162, 231, 261, T2.5–T2.8
    external branch of, 98, 103–105, 108, 153
    internal branch of, 94, 96, 98, 103–105, 108, 153, 162
Laryngeal prominence, 106
Laryngeal vein, superior, 100
Laryngopharyngeal nerve, 157
Laryngopharynx, 17, 93, 95
Larynx, 6, 7, 17, 162, 230, BP8, BP34, BP95, BP15
  cartilages of, 106
  coronal section of, 108
  intrinsic muscles of, 107
    action of, 109
  nerves of, 108
  quadrangular membrane, 106
Lateral, as term of relationship, 1
Lateral aperture, 136
  left, 133
Lateral corticospinal tract, BP40
Lateral direct vein, 173
Lateral epicondyle, 427
Lateral incisors, 40
Lateral intertransversarii lumborum muscle, 197
Lateral ligament, of ankle, 538
Lateral node, 410
Lateral recess, 141
  left, 133
Lateral rectus (semilunar) plane, BP58
Lateral rotation, of lower limbs movements, 11
Lateral sulcus
  anterior, BP40
  posterior, BP40

Malleolus
   lateral, 491, 525, 527–532, 540–542
   medial, 491, 525, 527–529, 531–533, 540–542, BP110
Malleus, 122, 125
   anterior ligament of, 122
   anterior process of, BP28
   handle of, 122, 123, BP28
   head of, 78, 121, 122, 151, BP28, BP29
   lateral process of, 122
   superior ligament of, 122, BP28
Mammary gland, 20, 205, T4.1–T4.3
   anterolateral dissection of, 205
   arteries of, 206
   lymph nodes of, 207
   lymph vessels of, 207
   sagittal section of, 205
Mammillary bodies, 69, 133, 134, 135, 139, 140, 174, 175
   imaging of, 177, BP32
Mammillary process, 181
Mammillothalamic fasciculus, 133
Mammillothalamic tract (of Vicq d'Azyr), 174
Mandible, 8, 25–27, 36, 39, 50–52, 95
   of aged person, 39
   alveolar part of, 39
   angle of, 22, 26, 36, 39, 87, 91
   base of, 39
   body of, 25–27, 36, 39, 54, 69, 75, BP23
   condylar process of, 36, 39, 75
     head of, 27
   condyle of, 26, 28, BP16
   coronoid process of, 27, 28, 36, 39, BP16, BP22
   head of, 39
   mandibular foramen of, 39
   mandibular notch of, 27
   mental foramen of, 25, 27, 39
   mental protuberance of, 22, 25, 39
   mental spines of, 39
   mental tubercle of, 25, 39
   mylohyoid groove of, 39
   mylohyoid line of, 39
   neck of, 39, BP22
   oblique line of, 27, 39, 97
   opening of, BP25
   pterygoid fovea of, 39
   ramus of, 25–27, 36, 39, 54, 86, BP25
   retromolar fossa of, 39
   sublingual fossa of, 39
   submandibular fossa of, 39
Mandibular foramen, 39
Mandibular fossa, 27, 31, 35, 42, BP30
Mandibular nerve, 23, 33–34, 42, 75, 77, 84, 87, 98, 113, 114, 131, 148–149, 152, 158, 160–162, T2.5–T2.8, BP26
   anterior division of, 77
   auricular branches of, 149
   meningeal branch of, 33, 34, 77–78, 113, 149
   motor root of, 77
   parotid branches of, 149
   posterior division of, 77
   sensory root of, 77
   superficial temporal branches of, 149
Mandibular nodes, 101
Mandibular notch, 27, 36, 39, BP25
Manubriosternal joint, 221
Manubriosternal synchondrosis, 429
Manubrium, 429, 263, 264
   of sternum, 49, 50, 52, 54, 95, 203, 211, 219, 221
Marginal artery, 400, 314, BP68-BP69
   anastomosis, T5.1–T5.3
Marginal colic plexus, 323
Marginal sulcus, 133

Masseter muscle, 49, 52, 54, 70, 72, 78, 86, T2.9–T2.14, BP22
   deep part of, 72
   insertion of, 72
   superficial part of, 72
Masseteric artery, 72–74, 76, 78, 99
Masseteric fascia, 48
Masseteric nerve, 72–75, 77–78, 98
   nerve to, 149
Mast cell, BP11
Mastication, muscles involved in, 72, 73, T2.1–T2.4
Mastoid air cells, 26, 28, BP34
Mastoid angle, 32
Mastoid antrum, 123, BP28
Mastoid canaliculus, 31, 33
Mastoid cells, 26, 82, 123, BP22, BP28
Mastoid emissary vein, 24, 127
   in mastoid foramen, 33
Mastoid fontanelle, 35
Mastoid foramen, 31, 33, 34, 38
Mastoid nodes, 101
Mastoid notch, 31
Mastoid process, 27, 31, 36, 38, 49, 54, 55, 87–88, 197, BP25, BP30
Maturation, BP87
Mature follicle, BP88
Maxilla, 68, BP16, BP22
   sinus of, 68
Maxillary artery, 42, 58, 65, 72–78, 81, 87, 98–99, 123, 128, 158, 160–161, 163, 165, BP21
   posterior lateral nasal branch of, 74
   proximal, 76
Maxillary bone, 25, 27, 29, 31, 37, 62, 63
   alveolar process of, 25, 27, 29, 38, 62, 69, BP23
   anterior nasal spine of, 25, 27, 29, 59, 62, 63
   frontal process of, 25, 27, 59, 62, 110
   incisive canal of, 29, 62, 63
   incisive fossa of, 31
   infraorbital foramen of, 25, 27
   infratemporal surface of, 27
   intermaxillary suture of, 31
   nasal crest of, 63
   nasal surface of, 29
   of newborn, 35
   orbital surface of, 25
   palatine process of, 29, 31, 38, 40, 61–63, 69, 85, BP30
   tuberosity of, 27, 38
   zygomatic process of, 25, 31
Maxillary nerve, 23, 34, 64, 75, 77–79, 81–82, 84, 98, 113–114, 131, 148–149, 158, 160–162, T2.5–T2.8, BP26
   anterior superior alveolar branch of, 66
   lateral superior posterior nasal branch of, 66, 81
   meningeal branch of, 113, 149
   zygomaticofacial branch of, 149
   zygomaticotemporal branch of, 149
Maxillary ostium, 82
Maxillary sinus, 26, 28, 69, 80, 82, 112, BP22, BP23
   alveolar recess of, BP23
   imaging of, BP32
   infraorbital recess of, BP23
   opening of, 61–62, 67, 69, BP23
   postsynaptic fibers to, 80
   recesses of, 69
   transverse section of, BP22
   zygomatic recess of, BP23
Maxillary tuberosity, 27
Maxillary vein, 100, 115
McBurney's point, 300
McGregor's line, BP18
Medial, as term of relationship, 1
Medial eminence, 141

Medial ligament, of ankle, 538
Medial pterygoid muscle, 52
Medial rotation, of lower limbs movements, 11
Median aperture, 133, 136, 141, 144
Median cricothyroid ligament, 53
Median fissure, anterior, BP40
Median lobe, BP92
Median nerve, 423, 438–440, 442, 445, 456–458, 461, 464, 470, 472, 475, 482–486, T7.1–T7.2, T7.3–T7.4, BP100
   articular branch of, 486
   branches of, 476
   common palmar digital branch of, 483
   communicating branch of, 476
   muscular branches of, 475
   palmar branch of, 423, 425, 456–457, 469, 476, 482, 486
   palmar digital branches of, 423
     common, 482
     proper, 482–483
   proper palmar digital branches of, 470
   recurrent branch of, 469–470, 476, 483, T7.1–T7.2
Median nuclei, 138
Median raphe, of levator ani muscle, 360
Median sulcus, posterior, 141, BP40
Mediastinal lymph nodes, 16, 263
   anterior, pathway to, 207
Mediastinal pleura, 215, 232
Mediastinum
   anterior, 215
     lung area for, 221
   cross section of, 237
   great vessels of, 220
   lateral view
     left, 252
     right, 251
   lymph vessels of, 228
   of testis, 388, 392
Medulla oblongata, 7, 34, 17, 80, 133, 141, 151, 161–162, 250, 324, 340, BP8, BP22, BP35
   in foramen magnum, 33
   imaging of, 177
   lower part of, BP14
   reticular formation of, 230
Medullary artery, segmental, 191
Medullary capillary plexus, BP73
Medullary cavity, BP9
Medullary lamina
   external, 138
   internal, 138
Medullary vein, anteromedian, 171
Medullary velum
   inferior, 133, 141
   superior, 133, 140–141
Meiosis, BP87
Meissner's corpuscle, 12
Meissner's plexus, BP48. See also Submucosal (Meissner's) plexus.
Melanocyte, 12
Membranous labyrinth, 124–125
Membranous layer (Scarpa's fascia), 368, BP84
Membranous septum
   atrioventricular part of, 237
   interventricular part of, 237
Membranous urethra, 386, BP85
Mendelian inheritance, BP87
Meningeal artery, 128
   accessory, 33–34, 74, 78, 128
   anterior, 115
   deep, 75
   dorsal, 164
   middle, 34, 42, 73–78, 98–99, 123, 127–129, 131, 158, 163–165, T2.1–T2.4
     accessory branch of, 123
     branches of, 129

Mylohyoid raphe, 87
Myofibril, BP10
Myofilaments, BP10
Myometrium, 374
  of uterus, 399
Myosatellite cell, BP10

N

Nail bed, 481, BP112
Nail fold
  lateral, BP112
  proximal, BP112
Nail groove, lateral, BP112
Nail matrix, 481
Nail plate, 481, BP112
Nail root, 481
Nares, BP15
  anterior, 22
Nasal artery
  dorsal, 99, 115
  external, 59, 74, 115, 165
  lateral, 59
Nasal bone, 22, 25–27, 29, 59, 62–63, 68–69,
    BP16
  of newborn, 35
Nasal cartilage, accessory, 59
Nasal cavity, 37, 17, 68, 111, BP23,
    BP27
  arteries of, BP21
  autonomic innervation of, 80
  bones of, 68
  floor of, 61
  lateral wall of, 61–65
  medial wall of, 63
  nerves of, 64, 66
  postsynaptic fibers to, 80
  vasculature of, 65
Nasal concha
  bony highest, 68
  bony middle, 37, 68
  bony superior, 37, 68
  inferior, 25–26, 29, 17, 61–62, 67–68, 93, 111,
    BP22-BP23
    ethmoidal process of, 62
  middle, 25, 29, 17, 61, 62, 67, 111, BP23
  superior, 17, 61, 62
  supreme, 17
Nasal concha turbinate
  inferior, 69
  middle, 69
Nasal meatus
  inferior, 61, 69, 111, BP23
  middle, 69, BP23
    atrium of, 61
  superior, 61
Nasal nerve
  external, 59
  posterior, 160
Nasal retinal arteriole
  inferior, 119
  superior, 119
Nasal retinal venule
  inferior, 119
  superior, 119
Nasal septal cartilage, 59, 63, BP22
  lateral process of, 59, 62–63
Nasal septum, 26, 38, 64, 66, 69, 93, 95, 175,
    T2.1–T2.4, BP23
  bony part of, 37, 68, BP21
Nasal skeleton, 37
Nasal slit, 34
Nasal spine
  anterior, 25, 27, 29, 59, 63, BP16
  posterior, 31, BP18
Nasal vein, external, 100
Nasal vestibule, 17, 61, 63, BP22

Nasalis muscle, 48, 150, T2.9–T2.14, BP20
  alar part of, 48, 59–60, BP19
  transverse part of, 48, 59–60, BP19
Nasion, 25, BP16
Nasociliary nerve, 79, 113–114, 148–149, 158–
    159, BP27
Nasofrontal vein, 24, 65, 100, 115
Nasolabial lymph nodes, 101
Nasolacrimal canal, opening of, 62
Nasolacrimal duct, 111
  opening of, 61, 111
Nasopalatine nerve, 64, 66, 80–81, 84
  communication between, 81
  groove for, 63
  in incisive fossa, 33
Nasopalatine nerves, 66
Nasopalatine vessels, groove for, 63
Nasopharyngeal adenoids, imaging of, 177
Nasopharynx, 17, 93, 95, 121, 131, BP29
  airway to, 61
Navicular bones, 535, 536, 538, 547, 556
  tuberosity of, 539, 547
Navicular fossa, 368, 386
Neck, 2, 3, 10, 22–177, 427, BP15
  anatomical, 427
  arteries of, 99
  autonomic nerves in, 157
  bony framework of, 36
  cutaneous nerves of, 23
  fascial layers of, 50–51
  lymph nodes of, 101
  muscles of
    anterior view of, 49
    lateral view of, 54
  nerves of, 56–57
  posterior triangle of, 209
  of scapula, 203
  superficial veins of, 50
  surface anatomy of, 22
  triangle of, posterior, 195
  vessels of, 56–57
Nephron, BP72
Nephron loop (Henle's), BP72
Nerve fiber bundles, BP4
Nerve fibers, 125
Nerves
  of abdominal wall
    anterior, 278
    posterior, 287
  of back, 198
    of neck, 199
  of buttocks, 512
  of cranial base, 82
  of esophagus, 261
  of external genitalia
    female, 415
    male, 411
  of larynx, 108
  of nasal cavity, 64
  of neck, 56, 57
    regions, 56, 57
  of orbit, 113
  of perineum
    female, 415
    male, 413
Nervous system, T3.1–T3.3, T4.1–T4.3,
    T8.1–T8.2
  autonomic division of, 4
  overview of, 4
  sense organs and, T5.1–T5.3
Neural foramen. See Intervertebral foramen
Neurocranium, 8
Neuroendocrine G cell, BP55
Neurofilaments, BP3
Neurons, BP3
Neurotubules, BP3
Neurovascular compartment, 445

Neurovascular sheath. See Hypogastric
  (neurovascular) sheath
Neutrophils, BP12
Newborn, skull of, 35
Nipple, 202, 205
Node of Cloquet/Rosenmüller, 286
Node of Ranvier, BP3-BP4
Nose, 2, 59
  ala of, 22
  transverse section of, BP22
Nostril, 22
Nuchal ligament, BP18, 46, 178
Nuchal line
  inferior, 31, 38
  superior, 31, 38, 195
Nuclei, BP10
Nuclei of pulvinar, 138
Nucleus ambiguus, 143–144, 152–154
Nucleus pulposus, 181, 185, 45
Nurse cell (Sertoli cell), 391
Nutrient artery, of femur, 514
Nutrient foramen, of femur, 500

O

Obex, 141
Oblique cord, 449
Oblique fissure, of lungs, 219
  left, 221
  right, 221
Oblique line, 39, 106
Oblique muscle
  external, 267, 270–276, 278–281, 283, 331,
    349, 351, T5.4, BP53, BP80-BP81
    abdominal, 17, 202, 205, 209, 212, 195–
      196, 198, 200, 378, 381–382, BP43
    aponeurosis of, 272–273, 278, 280, 282,
      331, 351, BP53
    aponeurotic part, 270
    costal attachments of, 275
    muscular part of, 270
  inferior, 112, 114, 148, T2.9–T2.14
    tendon of, 114
  internal, 271–276, 278, 280–281, 283, 331,
    351, T5.4, BP81
    abdominal, 17
    aponeurosis of, 272–273, 278, 351
    internal abdominal, 209, 195, 196, 200,
      BP43
    in lumbar triangle (of Petit), 195
    tendon of origin of, 200
  superior, 112–114, T2.9–T2.14, BP27
Oblique popliteal ligament, 518, 520
Obliquus inferior capitis muscle, 196–197, 199,
  T3.4
Obliquus superior capitis muscle, 196–197,
  199, T3.4
Obturator artery, 274, 366, 284, 335–336, 373,
  400, 402, 404–405, 412, 420, 496, 514, 522,
  BP90, BP93-BP94, BP103
  accessory, 400, 404
  acetabular branch of, 496, 514
  anterior branch, 496
  posterior branch, 496
Obturator canal, 357–359, 361, 366, 369, 404,
  510, BP90
Obturator crest, 495, 357
Obturator externus muscle, 498, 503, 510, 551,
  421, T8.5–T8.8
Obturator fascia, 358, 359, 361, 372, 373, 379,
  404, 413, BP90
  over obturator internus muscle, 369
Obturator foramen, 353, 268, 495, 497, 555
  radiograph of, 354
Obturator groove, 495
Obturator internus fascia, 366, 394, 404,
  BP94

Obturator internus muscle, 358–359, 498, 360, 274, 361–362, 369, 283, 371–372, 384, 394, 400, 413, 415, 420–421, 504, 508, 511–513, 555, T8.5–T8.8, BP104
  nerve to, 506, 508, 511–512, T8.3–T8.4
  sciatic bursa of, 513
Obturator internus tendon, 360, 362
Obturator membrane, 357, 496, 372
Obturator nerve, 274, 287, 402, 412, 420, 505–508, 515, 550–551, T8.1–T8.4, BP94
  accessory, 287, 506–507
  adductor hiatus, 551
  anterior branch of, 510, 551
  articular branch of, 551
  cutaneous branch of, 492–493, 510, 551
  posterior branch of, 510, 551
Obturator node, 408
Obturator vein, 285, 317, 373, 401–420, BP94
Obturator vessels, 274, 280
  accessory, 280, 282
  right, BP93
Occipital artery, 24, 199, 58, 75, 91, 99, 100, 128, 163, 165
  descending branch of, 199, 58
  groove for, 31
  mastoid branch of, 128, 163
  medial, 169
  meningeal branch of, 24
  sternocleidomastoid branch of, 58, 99
Occipital bone, 27–32, 46, 186–187, BP16
  basilar part of, 29, 31–32, 46, 61–63, 85, 90, 92–93
  clivus of, 32
  condylar canal and fossa of, 31
  condyle of, 32
  foramen magnum of, 29, 31
  groove
    for inferior petrosal sinus, 29, 32
    for occipital sinus, 32
  hypoglossal canal of, 29, 31
  jugular foramen of, 29
  lowest level of, BP18
  of newborn, 35
  nuchal line of
    inferior, 31
    superior, 31
  occipital crest of
    external, 31
    internal, 32
  occipital protuberance of
    external, 29, 31
    internal, 32
  pharyngeal tubercle of, 31, 95
  posterior meningeal vessels of, groove for, 32
  superior sagittal sinus of, groove for, 32
  transverse sinus of, groove for, 29, 32
Occipital condyle, 29, 31, 38, 55, 155, BP30
  lateral mass for, superior articular surface of, 42
  superior articular surface for, 42
Occipital crest
  external, 31, 38
  internal, 32
Occipital (posterior) horn, 136
Occipital muscle, T2.9–T2.14
Occipital nerve
  greater, 23, 198–199, 156
  lesser, T2.9, 23, 56–57, 156, 198–199
  third, 23, 198, 199
Occipital nodes, 101
Occipital pole, of cerebrum, 132, 134
Occipital protuberance
  external, 27–29, 31, 38, 178
  internal, 32
Occipital sinus, 130
  groove for, 32

Occipital sulcus, transverse, 132
Occipital vein, 24, 100
  internal, 173
Occipitofrontalis muscle
  frontal belly of, 48, 59, 114, 150
  occipital belly of, 48, 150
  occipital muscle of, 199
Occipitomastoid suture, 38
Occipitotemporal gyrus
  lateral, 133–134
  medial, 133–134
Occipitotemporal sulcus, 133–134
Occiput, 3
Oculomotor nerve (CN III), 7, 78, 112–114, 131, 140, 143–144, 158, 171, T2.1–T2.4, T2.5–T2.8, BP7
  accessory nucleus of, 143, 159
  distribution of, 145
  inferior branch of, 113, 148, BP27
  schema of, 148
  superior branch of, 113, 148, BP27
  in superior orbital fissure, 34
Oculomotor nucleus, 143–144, 148
Odontoblast layer, 41
Olecranon, 441, 451, 454
Olecranon fossa, 427, 446–447
Olfactory bulb, 64, 69, 134, 146, BP23
  contralateral, 146
Olfactory bulb cells, 146
Olfactory cells, 146
Olfactory mucosa, 146
  distribution of, 64
Olfactory nerve (CN I), T2.1–T2.8, 34, 64, 66
  distribution of, 145
  schema of, 146
Olfactory nerve fibers, 146
Olfactory nucleus, anterior, 146
Olfactory stria
  lateral, 146
  medial, 146
Olfactory sulcus, 134
Olfactory tract, 64, 133–134, 140, 146
Olfactory tract nucleus, lateral, 146
Olfactory trigone, 146
Olfactory tubercle, 146
Oligodendrocyte, cell body of, BP4
Olivary complex, inferior, 144
Olive, 140
Omental (epiploic) appendices, 290, 301
Omental arterial arc, BP83
Omental artery, BP83
Omental bursa, 292, 302, 343
  cross section of, 292
  stomach reflected, 291
  superior recess of, 293, 302, 343
Omental foramen, 291–292, 294, 303, 343
Omental taenia, 298, 301
Omentum
  greater, 288, 290, 292, 294, 301, 303, 350, 351
  lesser, 292, 294, 303, 343
    anterior layer of, 305, 320
    free margin of, 306
    posterior layer of, 320
    right free margin of, 296
Omohyoid muscle, 50–51, 87, 209, 428, 432, 435, 438, T2.9–T2.14
  inferior belly of, 22, 54, 56, 155, 156
    nerve to, 57
  phantom, 58
  superior belly of, 49, 53–54, 155–156
    nerve to, 57
Oocytes, BP90
  discharged, 365
  primary, BP90
Oogenesis, BP87
Oogonium, BP87

Operculum
  frontal, 132
  parietal, 132
  temporal, 132
Ophthalmic artery, 34, 65, 74, 82, 115, 159, 163–165, 167, 170, T2.1–T2.4
  continuation of, 115
  in optic canal, BP27
Ophthalmic nerve (CN V₁), 23, 77–79, 82, 84, 113–114, 131, 148, 158–159, 160–162, T2.5–T2.8
  frontal branch of, 34
  lacrimal branch of, 34
  nasociliary branch of, 34
  tentorial (recurrent meningeal) branch of, 113, 149
Ophthalmic vein
  inferior, 65, 115, 120
  superior, 34, 65, 100, 115, 120, 131, BP27
Opisthion, BP18
Opponens digiti minimi muscle, 472, 475, 487, T7.5–T7.8
Opponens pollicis muscle, 472, 475, 476, 486, T7.5–T7.8
Opposition, of upper limbs movements, 11
Optic canal, 25, 29, 34
Optic chiasm, 69, 131, 133–134, 140, 147, 164, 167–168, 172–175, BP23
  imaging of, 177
Optic disc, 119, BP27
Optic nerve (CN II), 34, 69, 82, 112, 113, 114, 116, 119, 131, 133–134, 159, 170–171, T2.1–T2.8, BP23
  distribution of, 145
  inner sheath of, 116
  internal sheath of, vessels of, 120
  left, 164
  meningeal sheath of, 113, 116, 120
  in optic canal, BP27
  outer sheath of, 116
  right, 164
  schema of, 147
  subarachnoid space of, 116
  vessels of internal sheath of, 120
Optic nerve tract, imaging of, BP32
Optic radiation, 147
Optic tract, 69, 134, 135, 139, 140, 147, 166, 171
Ora serrata, 116, 118–119
Oral cavity, 17, 18, 63, 69, 95, BP23
  afferent innervation of, 84, BP26
  inspection of, 83
  roof of, 85
Oral region, nerves of, 98
Orbicularis oculi muscle, 48, 150, T2.9–T2.14, BP20
  orbital part of, 48, BP19-BP20
  palpebral part of, 48, 110, BP19-BP20
Orbicularis oris muscle, 48, 59–60, 72, 86, 150, T2.9–T2.14, BP19-BP20
Orbiculus ciliaris, 118
Orbit, T2.1–T2.4, 37
  anterior view of, 114
  arteries of, 115
  fasciae of, BP27
  medial wall of, 69, BP23
  nerves of, 113
  superior view of, 114
  surface of, 25
  veins of, 115
Orbital cavity, 67–68
Orbital fat, BP23
Orbital fat body, BP27
Orbital fissure
  branch to superior, 164
  inferior, 25, 27, 38, BP16-BP27
  superior, 25, 26, 34, BP16, BP27

Orbital gyri, 134
Orbital muscles, BP23
Orbital plate, of newborn, 35
Orbital process, of palatine bone, 25
Orbital septum, 110, T2.1–T2.4
Orbital sulcus, 134
Oropharynx, 17, 50–51, 255, 90, 93, 95
Orthotonic stomach, BP55
Osseous cochlea, 125
Osseous spiral lamina, 124–125
Osteoblasts, BP9
Osteoclasts, BP9
Osteocytes, BP9
Osteoid, BP9
Osteonic canals (Haversian), BP9
Otic capsule, 125
Otic ganglion, 7, 42, 73, 77–78, 123, 149–150,
    152, 158, 160–162
    schema of, 161
Oval (vestibular) window, 35, 124
Ovarian artery, 363, 365, 373, 402, 414, 417,
    336. See also Testicular artery.
    uterine, 406
Ovarian cycle, BP88
Ovarian follicles
    primary, 365
    primordial, 365, BP90
    ruptured, 365
    secondary, 365
    tertiary, 365
Ovarian hormone, BP88
Ovarian ligament, 375
    proper, 363, 365, 373, 375, 390
Ovarian plexus, 414, 417
Ovarian vein, 363, 365, 285, 373, 402. See also
    Testicular vein.
Ovarian vessels, 366, 406, 335
    tubal branches of, 406
Ovaries, 363–365, 20–21, 372–374, 390, 402,
    414, 417, 335, BP90, BP95
    corpus albicans of, 374
    corpus luteum of, 374
    cortex of, BP90
    follicle (graafian), 374
        vesicular, 374
    left, 375
    ligaments of
        proper, 364, 372, 374
        suspensory, 364, 366, 374, 376, 402
    right, 375
    superficial epithelium of, 365, BP90
    suspensory ligament of, 363, 365, 375, 390
Ovum, BP87-BP88

P

Pacini's corpuscle. See Lamellar corpuscle
    (Pacini's)
Pain, referred visceral, sites of, BP5
Palate, 7
    postsynaptic fibers to, 80
    uvula of, 83, 90, 93
Palatine aponeurosis, 85, 96
Palatine artery
    ascending, 74, 90, 99
        tonsillar branch of, 90
    descending, 65, 81–82, 99
    greater, 65, 74, 81–82, 85, BP21
    lesser, 65, 74, 82, 85, 90
        tonsillar branch of, 90
Palatine bone, 29, 31, 62, 68
    horizontal plate of, 31, 38, 40, 61–63, 85, BP30
    nasal crest of, 63
    nasal spine of, posterior, 31, 62–63
    of newborn, 35
    orbital process of, 25, 62
    palatine foramen of

Palatine bone *(Continued)*
    greater, 31, 63
    lesser, 31, 63
    perpendicular plate of, 62–63
    pyramidal process of, 31, 35, 38
    sphenoidal process of, 62
Palatine fold, transverse, 85
Palatine foramen
    greater, 33, 62, 85, BP21
    lesser, 33, 62, 85, BP21
Palatine gland, 85, 90, 95
Palatine nerve, 64, 160
    communication between, 81
    descending, 160
    greater, 64, 66, 80, 82, 85, 149, 158, 160
        in greater palatine foramen, 33
        posterior inferior lateral nasal branch of,
            64, 158
    inferior posterior nasal branches of greater, 81
    lesser, 64, 66, 80–82, 85, 149, 158, 160
        in lesser palatine foramen, 33
Palatine nerves, 98
    greater, 98
Palatine process, 29, 38
Palatine raphe, 85
Palatine tonsil, 83, 85, 86, 89, 90, 93, 95,
    T2.1–T2.4
Palatine vein
    descending, 65
    external, 100
Palatine vessels
    greater, 33
    lesser, 33
Palatoglossal arch, 83, 89–90
Palatoglossus muscle, 85, 88–90, 153,
    T2.9–T2.14
Palatomaxillary suture, 31
Palatopharyngeal arch, 83, 89–90, 93
Palatopharyngeal ridge (Passavant's), 96
Palatopharyngeus muscle, 85–90, 92, 94, 96,
    153, T2.9–T2.14
Palm, 422, 463
Palmar aponeurosis, 425, 452, 456, 464, 469–
    470, 473, T7.1–T7.2
    forming canals, 470
Palmar arch
    carpal, 475
    deep, 14, 443, 475, BP97
    superficial, 14, 443, 472, 476, 483, BP97
    superficial venous, 15
Palmar digital artery, 443
    common, 443, 475
    proper, 443
Palmar digital veins, 15, 444
Palmar ligament, 468, 471, 474, 481, BP102
Palmar longus muscle, 458
Palmar metacarpal arteries, 443, 475
Palmar metacarpal veins, 444
Palmar radioulnar ligament, 464
Palmaris brevis muscle, 469, 487, T7.5–T7.8
Palmaris longus muscle, 452, 456, 459, 461,
    486, T7.5–T7.8
Palmaris longus tendon, 422, 456, 457, 461,
    464, 469, 470, 472
Palpebral arterial arches
    inferior, 115
    superior, 115
Palpebral artery
    lateral, 115
        inferior, 115
        superior, 115
    medial, 115
        inferior, 115
        superior, 115
Palpebral conjunctiva, 110, BP27
    inferior, 110
    superior, 110

Palpebral ligament
    lateral, BP27
    medial, 110, BP27
Pampiniform (venous) plexus, 277, 388, 403,
    BP93, T6.1–T6.3
Pancreas, 6–7, 18, 153, 200, 292–294, 342–343,
    348, BP38, BP50, T5.1–T5.3
    autonomic innervation of, 329
    body of, 291, 294, 306
    head of, 291, 294, 296–297, 305, 346
        arteries of, 312
        as site of referred visceral pain, BP5
    lymph vessels and nodes of, BP74
    in situ, 306
    uncinate process of, 306
Pancreatic arteries, 310
    dorsal, 308, 309, 310, 312, 313, BP69, BP83
    great, 309, 310, 312
    inferior, 309, 310, 312, 313, BP83
Pancreatic duct, 305–306, BP62
    accessory (of Santorini), 297, 306, BP62
    sphincter of, 305
    variations in, BP63
    of Wirsung, 297
Pancreatic islets (Langerhans), 21
Pancreatic notch, 306
Pancreatic pain, areas of referred pain in, 329
Pancreatic tail, 291, 307, 309, 310, 330, BP82
Pancreatic veins, 315, 317
Pancreaticoduodenal arteries
    anterior
        inferior, 309–310, 312, 320–321, 325
        superior, 291, 308–309, 310, 312–313, 320–
            321, 325, BP83
    inferior, 309–310, 312–314, BP83
    posterior
        inferior, 309–310, 312, 314, 321, 325
        superior, 308, 309, 310, 312, 321, 325,
            BP83
Pancreaticoduodenal nodes, BP75
Pancreaticoduodenal vein
    anterior
        inferior, 315, 317
        superior, 315, 318
    posterior
        inferior, 315, 317
        superior, 315, 318
Papilla, keratinized tip of, 89
Papillary duct, BP72
Papillary muscle, 237, 266
    anterior, 241, 244, 248
        right, 245, 246
    inferior, 241, 242, 244, 248
        left, 245
        right, 245, 246
    septal, 241, 244–246
    superior, 242, 244, 248
        left, 245
Parabrachial nucleus, 162
Paracentral artery, 168, 169
Paracentral lobule, 133
Paracentral sulcus, 133
Paracolic gutter
    left, 363, 367, 290
    right, 363, 367, 290, 298
Paracolic nodes, BP77
Paradidymis, 390
Paraduodenal fossa, 289
Parahippocampal gyrus, 133–135, 146
Paramammary nodes, 208
Paramesonephric duct (müllerian), 390
Paranasal sinuses, 37, T2.1–T2.4, BP23
    changes with age, 68
    coronal section of, 69
    paramedian views, 67
    transverse section of, 69
Pararectal fossa, 363, 367, 394

Perforating branches
  anterior, 547
  posterior, 547
Perforating cutaneous nerves, 413, 415
Perforating radiate artery, 334
Perforating veins, 424–425, 444, 533
  anterior, 213
Perianal skin, 396
  sweat glands in, 395
Perianal space, 394–395, 398
Perianal tissues, 389
Peribiliary lymph vessel, BP78
Pericallosal artery, 168–169
Pericardiacophrenic artery, 211, 276, 214–215,
  220, 231–232, 236–237, 251–252
Pericardiacophrenic vein, 211, 215, 220, 231–
  232, 237, 251–252
Pericardiacophrenic vessels, 236
Pericardial cavity, 236–237, BP2
Pericardial sac, 214–215, 219, 231–232, 235–
  236, 241–242, 251–252
Pericardial sinus
  oblique, 236–237, 266
  transverse, 232–233, 236, 241–242
Pericardium, 50–51, 220, 253–254, T4.1–T4.3
  bare area of, 217
  central tendon covered by, 215
  diaphragmatic part of, 236
Perichoroidal space, 116–117
Pericranium, 129
Pericyte, BP11
Periglomerular cell, 146
Perimysium, BP10
Perineal artery, 372, 380, 384, 405–407, BP93
  transverse, 407
Perineal body, 362, 368–369, 370, 378–379,
  383–385, 393, 397, 407, T6.1–T6.3, BP84
Perineal compartment, superficial, 398
Perineal fascia, 368, BP84
  deep, 371–372, 378–379, 382–384, 393, 397–
    398, BP86
    transverse, 385
  superficial, 371–372, 378–379, 380–384, 393,
    397–398, 406–407, 413, BP86
Perineal ligament, transverse, 359, 361, 366,
  368–369, 378, 384, 407
Perineal membrane, 274, 283, 343, 358, 362,
  364, 368–369, 371–372, 378–380, 382–385,
  393, 397–398, 406–407, 413, 415, 419,
  BP84, BP86, BP93
  anterior thickening of, 369, 384
Perineal muscles, 10
  deep, 393
    fascia of, 359
  transverse
    deep, 358, 364, 368, 379–380, 384, 385,
      398, 405–406, T6.4
    superficial, 358, 378, 379, 380, 382–384,
      393, 397–398, 406–407, T6.4
    superior, 407
Perineal nerves, 380, 394, 412–413, 415, 506,
  512
  deep, 413
    branches of, 380, 413, 415
  superficial, 413
    branches of, 380, 413, 415
Perineal pouch
  deep, with endopelvic fascia,
    BP84
  superficial, BP84
Perineal raphe, 377, 389
Perineal spaces, 398
  deep, 380
  male, 384
  perineal, 406
  superficial, 368, 372, 378–380, 382–383,
    406–407

Perineal subcutaneous tissue, membranous
  layer of, 368
Perineal vein, 407
Perineum, 2
  arteries of, 406
  deep, 379
  female, 364, 377, 379
    fasciae of, BP84
    lymph nodes of, 409
    lymph vessels of, 409
    nerves of, 415
    superficial dissection, 378
  interdigitating fibers of, 360
  male, 381–382
    arteries of, 407
    fasciae of, BP84
    nerves of, 413
    veins of, 407
  veins of, 406
Perineurium, BP4
Periodontium, 41
Periorbita, BP27
Periosteal vessels, BP9
Periosteum, BP9
Peripheral arteries, BP8
Peripheral nerve, features of, BP4
Peripheral nervous system, 4
Periportal arteriole, 304
Periportal bile ductule, 304
Periportal space, 304
Perirenal fat, 200
Perirenal fat capsule, 337
Perisinusoidal spaces, 304
Peritoneal cavity, 288–293
Peritoneal reflection, 393, 395
Peritoneum, 200, 364, 378, 301, 394, 402, 414,
  330, T6.1–T6.3, BP84, BP54
  inferior extent of, 417, 418
  paravesical pouch, 372
  parietal, 273–274, 363, 281, 368, 282, 369,
    371, 290, 292–293, 328, 343, 351, BP79,
    BP84
  pelvic part of, 368, 371
  visceral, 292
Perivascular fibrous (Glisson's) capsule, 303,
  304
Permanent teeth
  lower, 40, 41
  upper, 40, 41
Perpendicular plate, 25, 29
Pes anserinus, 498, 502–503,
  516–517
Pes hippocampi, 138–139
Petit, lumbar triangle of, 178, 195
  internal oblique muscle in, 195
Petropharyngeus muscle, 92
Petrosal artery, superficial, descending branch
  of, 123
Petrosal nerve
  deep, 64, 80, 150, 152, 158, 160
  greater, 33–34, 78, 80, 82, 113, 150, 152, 158,
    160, BP28
    in foramen lacerum, 33
    groove for, 32, 126
    hiatus for, 34
  lesser, 33–34, 77–78, 113, 122, 149–150, 152,
    161
    groove for, 32
    hiatus for, 34
Petrosal sinus
  inferior, 34, 130–131
    groove for, 29, 32
  superior, 126, 130–131
    groove for, 29, 32
Petrosal vein, 131, 171
Petrosquamous fissure, 35
Petrotympanic fissure, 31, 33

Peyer's patches (aggregate lymphoid
  nodules), 16, 297
Phalangeal bones, 468
  foot, 8, 535, 536
    distal, 535, 539
    middle, 535, 539
    proximal, 535, 539
  hand, 8
    distal, 466–468, 486
    middle, 466–468, 481, BP98
    proximal, 466–468, BP98, BP102
  middle, BP99
  proximal, BP99
  1st, BP101
Phallus, body of, 389
Pharyngeal aponeurosis, 94
Pharyngeal artery, 74
  ascending, 58, 74, 82, 91, 99, 123, 163, 165
    meningeal branch of, 34, 128, 163
    pharyngeal branch of, 90
    tonsillar branch of, 90
Pharyngeal constrictor muscle, 95, 108
  cricopharyngeal part of, 104
  inferior, 92, 96, 97, 104, 105, 108, 153, 256,
    T2.9–T2.14
    cricopharyngeal part of, 92, 255–256
    thyropharyngeal part of, 255
  middle, 54, 88, 90, 92, 94, 96–97, 104, 108,
    153, T2.9–T2.14
  superior, 73, 74, 85–86, 88, 90, 92, 94, 96–97,
    104, 153, T2.9–T2.14
    glossopharyngeal part of, 88, 96
  thyropharyngeal part of inferior, 94
Pharyngeal nerve, 149, 160
Pharyngeal nervous plexus, 91
Pharyngeal plexus, 98, 152, 153, 157–158
Pharyngeal raphe, 61, 92, 96, 104, 256
Pharyngeal recess, 61, 90, 93, BP22
Pharyngeal regions, nerves of, 98
Pharyngeal tonsil, 61, 63, 90, 92–93, 95,
  T2.1–T2.4, BP34-BP35
Pharyngeal tubercle, 31, 46, 63, 90, 92, 96
Pharyngobasilar fascia, 85, 88, 92, 96–97, 102
Pharyngoepiglottic fold, 92, 94
Pharyngoesophageal constriction, 255
Pharyngoesophageal junction, 94
Pharynx, 18, 93
  afferent innervation of, 84, BP26
  lymph nodes of, 102
  lymphatic drainage of, 102
  medial view of, 95
  muscles of, 92
    lateral view of, 97
    medial view of, 96
  posterior view of, 91, 104
  posterior wall of, 83
  zone of sparse muscle fibers, 104
Philtrum, 22, 83
Phrenic arteries, inferior, 216, 254, 258, 284,
  291, 308–310, 332, 342, 412, BP83
  left, 293
Phrenic ganglion, 216, 321
Phrenic nerve, 50–51, 55–57, 98–99, 103, 157,
  210–211, 214–216, 219–220, 231–232, 236,
  249, 251–254, 263–264, 276, 319, 341, 436,
  438–439, T4.1–T4.3, T2.9
  left, 253
  pericardial branch of, 214
  phrenicoabdominal branches of, 214
  relationship to pericardium and, 214
  right, 253
  sternal branch of, 216
Phrenic plexus, 319–320, 341
  left, 321
  right, 321
Phrenic veins, inferior, 285, 332
  left, 259

Phrenicocolic ligament, 291, 293
Phrenicoesophageal ligament, 257
Phrenicopleural fascia, 257
Pia mater, 127, 129, 189, 190
Pial plexus, 191
  arterial, 192
    peripheral branches from, 192
Pigment cells, 147
Pillar (rod) cells, 125
Pilosebaceous apparatus, BP1
Pineal body, 137
Pineal gland, 21, 133, 138, 140–141
Pineal recess, 133, 135
Piriform cortex, 146
Piriform fossa, 255
Piriform recess, 93, 107
Piriformis muscle, 283, 358–359, 498, 360–
    361, 404, 412–415, 504, 508, 511–513,
    T8.1–T8.2, T8.5–T8.8
  bursa of, 513
  left, 405
  nerve to, 506, 508, T8.3–T8.4
  tendon of, 513
Pisiform bone, 452, 456–458, 462–470, 472,
    475–476, BP98
Pisohamate ligament, 464, BP98
Pisometacarpal ligament, 464, BP98
Pituitary gland, 17, 21, 131, 133–134, 174–175
  anterior lobe (adenohypophysis) of, 167, 174
  cleft of, 174
  imaging of, 177
  posterior lobe (neurohypophysis) of, 167,
    174
Pituitary gonadotropins, BP95
Pituitary stalk, imaging of, 177
Pivot joint, 9
Plane joint, 9
Plantar aponeurosis, 540, 543–545, 549, 556,
    T8.1–T8.2, BP110
Plantar arch, 14, 522, 546–547, BP105, BP109
Plantar artery
  deep, 522, 532, 541–542, 547, BP105
  lateral, 533, 540, 545, 546–547, BP110
    cutaneous branches of, 543
    deep branch of, BP111
    superficial branch of, BP111
  medial, 533, 540, 545–546, BP110
    cutaneous branches of, 543
    deep branch of, 545–546, BP111
    superficial branch of, 543–546
Plantar calcaneocuboid ligament, 539
Plantar calcaneonavicular ligament, 539, 547
Plantar digital arteries, 543, 547
  common, 544–545, 547, BP109
  proper, 544
Plantar digital nerves
  common, 545, 553
  proper, 553
Plantar fascia
  lateral, 543, 544
  medial, 543
Plantar flexion, of lower limbs movements, 11
Plantar interosseous muscle, 546
Plantar ligaments, 539, 547
  of foot, 539
  long, 538, 547, 549
  short, 538
Plantar metatarsal arteries, 544, 546–547,
    BP109
Plantar metatarsal ligaments, 539
Plantar nerve
  lateral, 533, 540, 545–546, 549, 552–553,
    T8.3–T8.4, BP109-BP111
    cutaneous branches of, 543
    deep branch of, 545, 546
    plantar cutaneous branches of, 493
    plantar digital branches of, 544

Plantar nerve *(Continued)*
    proper plantar digital branches of, 545,
      546
    superficial branch of, 545, 546
  medial, 533, 540, 545, 546, 549, 552–553,
    T8.3–T8.4, BP110
    cutaneous branches of, 543
    deep branches of, 546, BP111
    plantar cutaneous branches of, 493
    plantar digital branches of, 544
    proper plantar digital branches of, 545–546
    superficial branches of, 543, 546
Plantar vein
  lateral, 523
  medial, 523
Plantaris muscle, 498, 501, 504, 511, 516, 520–
    521, 526, 528–529, 533, 552–553, T8.5–T8.8
Plantaris tendon, 504, 527–528, 533–534,
    BP106, BP110
Plasma, composition of, BP12
Plasma cell, BP11
Plasma proteins, BP12
Platelets, BP12
Platysma, 51, T2.9–T2.14
Platysma muscle, 48, 49, 50, 150, BP19, BP20
Pleura, 221, T4.1–T4.3
  costal part of, 17
  diaphragmatic part of, 17
  mediastinal part of, 17
  visceral, 227
Pleural cavity, 237, BP2
  costodiaphragmatic recess of, 215, 217–218,
    251–252
  costomediastinal recess of, 215
  left, 217
  right, 217
Pleural reflection, 217–219
Plica semilunaris, 110, 111
Polar body, BP87
Pons, 17, 133, 140–141, 161–162, 175, BP15,
    BP35
  imaging of, 177, BP32
Pontine arteries, 166–167
  basilar and, 168
Pontine vein
  lateral, 171
  transverse, 171
Pontomesencephalic vein, anterior, 171
Popliteal artery, 14, 504, 511, 515, 522, 527–
    529, 533, T8.1–T8.2, BP103, BP105
  pulse point, 14
Popliteal fossa, 491
Popliteal ligament
  arcuate, 518, 520–521
  oblique, 521
Popliteal nodes, 16
Popliteal vein, 494, 15, 511, 515, 523, 527–528,
    533
Popliteal vessels, 504
Popliteus muscle, 498, 504, 520–521, 526, 529,
    533–534, 553, T8.5–T8.8, BP106
Popliteus tendon, 517–519
Porta hepatis, 302
Portacaval anastomoses, 318
Portal arteriole, 304
Portal space, limiting plate of, 304
Portal tract, 303
Portal triad, 292, 296, 303, 306, 348
Postanal space
  deep, 398
  superficial, 398
Postcentral gyrus, 132
Postcentral sulcus, 132
Posterior chamber
  of eye, 110, 116, 118–119
  of eyeball, 117
Posterior cord, of brachial plexus, 433

Posterior cutaneous branches, 198
Posterior inferior cerebellar artery (PICA), 191
Posterior plane, 1
Posterior ramus, 190
  lateral branch of, 190
  medial branch of, 190
  meningeal branch of, 190
Posterior root, 160, 326
  rootlets of, 189
Posterior root ganglion, 419. *See also* Spinal
    sensory (posterior root) ganglion.
Posterior sacral foramen, BP81
Posterior segmental artery, 334
Postganglionic fibers, 6, 7
  of lower ureter, 419
  in reproductive organs, 417
    male, 418
  of urinary bladder, 419
Postnatal circulation, 247
Postsynaptic cell, BP3
Postsynaptic membrane, BP3
Poupart's ligament, 270–272, 280–283, 359.
    See also Inguinal ligament (Poupart's).
Preaortic nodes, 410, BP77
Prececal nodes, BP77
Precentral cerebellar vein, 171
Precentral gyrus, 132
Precentral sulcus, 132
Precuneal artery, 169
Precuneus, 133
Prefrontal artery, 166, 168
Preganglionic fibers, 6, 7
  of lower ureter, 419
  in reproductive organs, 417
    male, 418
  of urinary bladder, 419
Premolar teeth, 2nd, 41
Premolar tooth, 1st, 40, 40
Prenatal circulation, 247
Preoccipital notch, 132
Prepancreatic artery, 310, 312
Prepatellar bursa, subcutaneous, 521
Prepontine cistern, 136
Prepuce, 368
Preputial gland, 383
Presacral fascia, 366, 398
Presacral space, 366, 398
Presynaptic membrane, BP3
Pretracheal fascia, 51–52, 95, 225
Pretracheal (visceral) fascia, 50
Pretracheal lymph nodes, 103
Prevertebral anastomoses, 192
Prevertebral fascia, 86, 95
Prevertebral nodes, 260
Prevertebral soft tissue, BP34
Prevesical fascia, umbilical, 369
Prevesical plexus, 410
Prevesical space, 369. *See also* Retropubic
    (prevesical) space.
  fat in, 362
  of Retzius, 398
Primary oocyte, BP87
Primary spermatocytes, BP87
Princeps pollicis artery, 443, BP97
Procerus muscle, 48, 59–60, 150, T2.9–T2.14,
    BP19, BP20
Processus vaginalis, 387
Profunda brachii artery, 438, 441–442, 445
Profundus flexor tendons, 473
Progesterone, BP88
Promontorial nodes, 338, 408, 410
Promontory, 121, BP29
  sacral, 353, 183
Pronation, of upper limbs movements, 11
Pronator quadratus muscle, 450, 458–459, 461,
    473, 475, 486, T7.5–T7.8

Pronator teres muscle, 440, 442, 450, 455–456, 458–461, 483, 486, T7.5–T7.8
head of
deep, 428, 458
superficial, 428
Prostate gland, 6, 7, 20, 274, 368, 371, 384–386, 390–391, 393, 399, 405, 420, T6.1–T6.3, BP8, BP94-BP95, BP104
anterior commissure of, 385, BP92
apex of, 385
base of, 385
branch to, BP93
capsule of, 371, BP92
central zone of, 385
cross section of pelvis, BP92
lymphatic drainage from, 410
peripheral zone of, 385
primordium of, 390
transitional zone of, 385
Prostatic ducts, 385–386
Prostatic plexus, 412, 418–419, 339, T6.1–T6.3
Prostatic sinuses, 385–386, BP92
Prostatic urethra, 371, 386, 391, 420, BP85, BP92
Prostatic utricle, 385, 386, 390, BP92
Prostatic venous plexus, 368, 405
Proximal, as term of relationship, 1
Proximal convoluted tubule, BP71-BP72
Proximal interphalangeal (PIP) joint, 422, 467–468, BP98-BP99
Proximal palmar crease, 422
Psoas fascia, 337
Psoas muscle, 420, 505
major, 200, 216, 283–285, 287, 280, 296, 330, 331, 337, 351, 366–367, 502, 505, 507–508, 513–514, 550, T8.5–T8.8, BP38, BP94, BP80-BP81
nerves to, 506
tendon, BP94
minor, 351, 363, 367, 283, 505, 513, T8.5–T8.8
nerves to, 506
tendon, 283
muscular branches to, 507
Psoas tendon, 420
Pterion, 27, T2.1–T2.4
Pterygoid artery, 74
canal, 81
lateral, 74
medial, 74
Pterygoid canal, 81
artery of, 74, 82, 123, 164
nerve (vidian) of, 64, 66, 79, 80, 82, 123, 149–150, 152, 158, 160, 162
Pterygoid fossa, 38
Pterygoid fovea, 39
Pterygoid hamulus, 27, 29, 38, 73, 77, 85, 90, 96, 97, BP30
Pterygoid muscle
lateral, 72–74, 76–77, T2.9–T2.14, BP22, BP25
inferior head of, 52, 75, 78, BP25
nerve to, 98, 149
superior head of, 52, 75, 78, BP25
medial, 73–75, 77–78, 85–86, 92, 98, T2.9–T2.14, BP22
nerves to, 98, 149
window cut through right, 75
Pterygoid nerve, medial, 73
Pterygoid plate
lateral, 36, 38, 73, 97, BP30
medial, 38, 68, 73, 85, 90, 96, BP30
hamulus of, 36
right, 82
Pterygoid plexus, 78, 115
Pterygoid process, 31, 38
hamulus of, 27, 31, 38
lateral plate of, 27, 29, 31, 38
medial plate of, 29, 31, 38

Pterygoid process (Continued)
pterygoid fossa of, 31
scaphoid fossa of, 31
Pterygoid venous plexus, 65, T2.1–T2.4
Pterygomandibular raphe, 36, 73, 74, 85–86, 90, 97, T2.1–T2.4
Pterygomaxillary fissure, 27
Pterygopalatine fossa, 27, 38, 67
Pterygopalatine ganglion, 7, 64, 66, 79–82, 98, 148–150, 152, 158, 160, 162, BP7
branches to, 79
pharyngeal branch of, 64
schema of, 160
Puberty, BP95
Pubic arch, 353, 355
Pubic bone, 358, 359, 370, 384, 399, 421, 343
Pubic branch, 366
Pubic crest, 280
Pubic hair, BP95
Pubic ligament, inferior (arcuate), 353, 359–362, 366, 368–369, 378, 384
Pubic ramus
inferior, 353, 268, 495, 360, 364, 371
superior, 353, 268, 356, 495, 357, 496–497, 364, 505, 368, 508, 383, 555, T6.1–T6.3
radiograph of, 354
Pubic symphysis, 267, 280, 281, 283, 352–353, 268, 355–356, 359–360, 362, 364, 366, 368–369, 375, 378, 380, 381, 384, 404, 421, 513, T5.1–T5.3, T6.1–T6.3, BP94
radiograph of, 354
superior portion of, 420
Pubic tubercle, 267–268, 281, 283, 378, 383, 352–353, 356–357, 495, 271–272, 362, 502–503, 555, BP53
Pubic vein, 285
Pubis, 399
body of, 420, BP94
radiograph of, 354
superior ramus of, 361, 503
symphyseal surface of, 414
Puboanalis (puborectalis) muscle, 343, 360–362, 393, 397, BP94, T6.4
left, 358
Pubocervical ligament, 373
Pubococcygeus muscle, 358–360, 362, 372, 397, T6.4
Pubofemoral ligament, 496, BP104
Puboprostatic ligament, medial, 366, 369
Puborectalis muscle, 360–362, 393. See also Puboanalis (puborectalis) muscle.
left, 358
Pubovesical ligament, 373
lateral, 366
median, 369
Pudendal artery, 403
deep external, 276, 284
external
deep, 509, 522, BP103
superficial, 270, 272, 276, 284, 509, 522, BP103
internal, 284, 373, 380, 384, 400, 404–407, 420, 508, BP93, BP94
in pudendal canal (Alcock's), 406
Pudendal canal (Alcock's), 394, 400–401, 407, 413
pudendal nerve of, 415
internal, 407
pudendal vessels, internal, 407
Pudendal cleft, 377
Pudendal nerve, 186, 287, 380, 394, 412–420, 506, 508, 511–512, T6.1–T6.3, T8.3–T8.4, BP94
internal, 421
nerve to, 508
in pudendal canal, 415

Pudendal vein, 403
external, 277, 523
deep, 403
superficial, 492, 403
internal, 317, 401, 405, 420, BP94
Pudendal vessels
external, 285, 405
superficial, 270
internal, 285, 394, 421
Pudendum, 377
Pulmonary acinus, 226
Pulmonary arteries, 13, 227, T4.1–T4.3, BP45
left, 17, 220–221, 232–235, 242, 247, 252, 262, 265, BP50-BP51
right, 17, 220–221, 232, 234–235, 241–242, 247, 251, 262, 265, BP50-BP51
Pulmonary heart valve, 241, 246, 248
anterior semilunar cusp of, 241, 243
left semilunar cusp of, 241, 243
right semilunar cusp of, 241, 243
Pulmonary ligament, 221, 228, 251–252
Pulmonary (intrapulmonary) lymph nodes, 228
Pulmonary plexus, 153, 230, 261, BP6, BP7
anterior, 229
posterior, 229
Pulmonary trunk, 220, 232–234, 238, 241, 245–247, 248, 262, 265, BP51
Pulmonary veins, T4.1–T4.3, 13, 227, 234, BP45
inferior
left, 221, 235, 237, BP50
right, 221, 235, 237, 241, 266
left, 221, 235, 236, 242, 245, 247, 252
right, 221, 236, 242, 247–248, 251, 262
superior
left, 221, 232, 235, 242
right, 221, 232, 235, 241, 245
Pulse points, 14
Pulvinar, 134, 137–138, 140–141, 170
left, 171
right, 171
of thalamus, 166, 172
Pupil, 110
Purkinje fibers
of left bundle, 248
of right bundle, 248
Putamen, 135
Pyloric canal, 348
Pyloric glands, BP55
Pyloric nodes, BP75, BP78
Pyloric orifice, 297
Pyloric sphincter, 295
Pylorus, 294, 295, 296, 348, T5.1–T5.3
Pyramidal eminence, 122–123, BP28
Pyramidal process, 31, 35, 38
Pyramidal system, BP15
Pyramidalis muscle, 271, 281, T5.4
Pyramids, 140
decussation of, 140–141, BP15

Q

Quadrangular space, 436
Quadrate ligament, 448
Quadrate tubercle, 500
Quadratus femoris muscle, 415, 498, 503–504, 510–513, T8.5–T8.8
nerve to, 506, 508, 512
Quadratus lumborum fascia, 200
Quadratus lumborum muscle, 197, 216, 363, 505, 507, 283, 367, 284–285, 200, 287, 330–331, 337, 351, T5.4
Quadratus plantae muscle, 545–546, 553, T8.5–T8.8, BP110-BP111
Quadriceps femoris muscle, 498, 550
Quadriceps femoris tendon, 491, 502–503, 510, 516–517, 521, 530–532, BP108

Quadriceps muscle. *See* Rectus femoris muscle
Quadrigeminal cistern, 136
  imaging of, BP32

R

Radial artery, 14, 442–443, 455–458, 461, 464, 470, 472, 476–477, 479–480, T7.1–T7.2, BP96-BP97
  dorsal carpal branch of, 477
  palmar carpal branch of, 458, 475–476
  pulse point, 14
  superficial palmar branch of, 443, 457–458, 464, 470, 475–476, BP97
Radial collateral ligament, 448, BP98
Radial fossa, 427, 446
Radial groove, 427
Radial head, 483
Radial metaphyseal arcuate ligament, dorsal, 465
Radial nerve, 423, 433, 436, 438–439, 441, 445, 457–458, 482–486, 488–489, T7.1–T7.4
  communicating branches of, 478
  deep branch of, 457, 461, 483, 489
  dorsal digital branches of, 477, 482, 489
  dorsal digital nerves from, 423
  inferior lateral brachial cutaneous nerve from, 423
  posterior brachial cutaneous nerve from, 423
  superficial branch of, 423, 425, 454, 457, 461, 478–479, 482–484, 489
Radial recurrent artery, 442, 457, 458, BP97
Radial styloid process, 453, 462
Radial tubercle, dorsal, 462
Radial tuberosity, 440, 447
Radial veins, 15, 444
Radialis indicis artery, 443, BP97
Radiate ligament, of head of rib, 221
Radiate sternochondral ligaments, 221
Radiate sternocostal ligament, 429
Radicular artery
  anterior, 192
  great, 191
  posterior, 192
Radicular vein, 193
Radiocarpal joint, 463, 465
  articular disc of, 463, 465
Radiocarpal ligament
  dorsal, 465, BP98
  palmar, BP98
Radiolunate ligament
  long, 464
  palmar, 464
  short, 464
Radioscaphocapitate ligament, 464
Radioulnar joint, distal, 465
Radioulnar ligament
  dorsal, 465, BP98
  palmar, BP98
Radius, 8, 446–453, 455, 458–459, 460–465, 467, 475, 480, T7.1–T7.2, BP98
  anterior border of, 449
  anterior surface of, 449
  anular ligament of, 448
  articular surface of, 467
  body of, 447, 453
  head of, 440, 446, 447, 449
  interosseous border of, 449
  lateral surface of, 449
  neck of, 446, 447, 449
  posterior border of, 449
  posterior surface of, 449
  rotators of, 450
  styloid process of, 449, 467
  tuberosity of, 446

Rami communicantes, 416, 506
Ramus of mandible, 52
Receptors, 4
Recombinant chromatids, BP87
Rectal artery
  inferior, 400, 405, 406, 407, BP93
  middle, 366, 400, 402, 405, BP93
  superior, 366, 400, 402, 412
Rectal canal, 395
Rectal fascia, 366, 378, 385, 393–396, 398, BP84, BP90
Rectal nerves, inferior, 380, 412–413, 415–416, T6.1–T6.3
Rectal plexus, 323, 324, 412, 414, 339
  external, 401
  internal, 401
  left, 412
  perimuscular, 401
Rectal veins
  inferior, 401
  middle, 401
  superior, 401
Rectal venous plexus
  external, 317, 396
    in perianal space, 395
  internal, 395–396
  perimuscular, 401, 317
Rectalis muscle, 397
Rectococcygeus muscle, 283
Rectoperinealis muscle, 362
Rectoprostatic fascia, 368
Rectoprostatic (Denonvilliers') fascia, 362, 385, 393, 343
Rectosigmoid arteries, 400
Rectosigmoid junction, 301, 393, 395–396
Rectourethralis superior muscle, 362
Rectouterine fold, 363, 365, 335
Rectouterine pouch, of Douglas, 364, 365, 374, 375, 393, T6.1–T6.3, BP90
Rectovaginal spaces, 366
Rectovesical fascia, 385, 398
Rectovesical fold, 367, 293
Rectovesical pouch, 367, 368, 393, 343
Rectovesical space, 398
Rectum, 6–7, 283, 290, 293, 301, 330, 343, 358–360, 363–364, 366–369, 373, 375, 378, 385, 393, 399, 402, 414, 421, T6.1–T6.3, BP84, BP104
  ampulla of, 369
  arteries of, 400
  circular muscle of, 396
  longitudinal muscle of, 396
  muscularis mucosae of, 396
  radiology of, 182
  in situ
    female, 393
    male, 393
  transverse folds of, 395
  veins of, 401
Rectus abdominis muscle, 17, 202, 267, 271, 209, 272, 210, 273, 274, 212, 276, 278, 279–281, 363, 366, 368, 378, 399, 343, T5.1–T5.3, T5.4, BP43, BP104, BP94
Rectus anterior capitis muscle, 55, T2.9–T2.14, BP22
Rectus capitis muscle
  anterior, 102
  posterior
    major, 196–197, 199, T3.4
    minor, 196–197, 199, T3.4
Rectus femoris muscle, 491, 498, 501–502, 509–510, 515, 550, 420–421, T8.5–T8.8, BP104, BP94. *See also* Quadriceps femoris muscle.
  bursa of, 513
  origin of, 513
Rectus femoris tendon, 502, 515, 517, 531

Rectus lateralis capitis muscle, T2.9–T2.14, 55
  nerves to, 57
Rectus muscle
  inferior, 112, 114, 148, T2.9–T2.14, BP27
    branches to, 113
  lateral, 69, 112, 113, 114, BP27, T2.9–T2.14
    check ligament of, BP27
    tendon of, 116
  medial, 69, 112, 113, 114, 148, T2.9–T2.14, BP27
    branches to, 113
    check ligament of, BP27
    tendon of, 116
  superior, 112–114, BP27, T2.9–T2.14
    tendon of, 120
Rectus sheath, 270–271, 378, 351
  anterior layer of, 209–210, 271–273, 278, 280, 368, BP53
  cross section of, 273
  posterior layer of, 272–273, 276, 278, 280
Recurrent artery
  anterior, 443
  anterior ulnar, BP96
  of Heubner, 166–169
  posterior, 443
  posterior ulnar, BP96
  radial, 443, BP96
  tibial
    anterior, 522, 532, BP103, BP105
    posterior, 522, BP103, BP105
Recurrent laryngeal nerve, 51–52
Recurrent process, 146
Red blood cells, BP11-BP12, BP45
Red bone marrow, 16
Red nucleus, 134, 143–144
Referred pain, visceral, sites of, BP5
Reflexes, 4
Renal artery, 14, 284, 333–334, 336, 402–403, 419, T5.1–T5.3
  left, 311, 330, 332, 337, 342, 412
  pelvic branch of, 334
  right, 311, 325, 330, 332
  ureteric branch of, 332, 334
  variations in, BP70
Renal column (of Bertin), 333, BP73
Renal corpuscle, histology of, BP71
Renal cortex, 296, 349, BP72
Renal (Gerota's) fascia, 200, 337, 342
  anterior layer of, 337
  posterior layer of, 337
Renal ganglion, 418, 419, 339–341
Renal impression, 302, 307
Renal medulla, BP72
Renal papilla, 333
Renal pelvis, 333, T5.1–T5.3
Renal plexus, 419, 339–341
  left, 323, 412
  right, 319
Renal pyramid, 333
Renal segments, 334
Renal vein, 15, 333, 336, 343, 402–403
  left, 259, 285, 332, 337, 342
  right, 285, 259, 332
  variations in, BP70
Reposition, of upper limbs movements, 11
Reproduction, genetics of, BP87
Reproductive organs, innervation of
  female, 417
  male, 418
Reproductive system, 20, T4.1–T4.3
Residual body, 391
Respiration
  anatomy of, BP45
  muscles of, BP43
Respiratory bronchioles, 227, BP44
Respiratory system, 17, T4.1–T4.3
Rete testis, 391

Reticular fibers, BP11
  in arterial wall, BP13
Reticular formation, BP14
Reticular nuclei, 138
Retina
  ciliary part of, 116–117, 120
  iridial part of, 117
  optic part of, 116, 118, 120
  structure of, 147
Retinacula cutis. See Skin ligaments
Retinacular arteries, 514
  inferior, 514
  posterior, 514
  superior, 514
Retinal artery, 119
  central, 115, 116, 119, 120
Retinal vein, 119
  central, 116, 119, 120
Retrobulbar fat, 69, BP27
Retrocecal recess, 290, 298
Retrodiscal tissue (bilaminar zone), BP25
  inferior layer of, BP25
  upper layer of, BP25
Retromandibular vein, 24, 56, 65, 70, 86, 88,
      100, 115, BP22
  anterior branch, 56
  anterior division, 100
  posterior branch, 56
  posterior division of, 100
Retromolar fossa, 39, T2.1–T2.4
Retropharyngeal nodes, 102
Retropharyngeal space, 50–52, 95–96
Retroprostatic fascia, 398
Retropubic (prevesical) space, 343, 366, 369
  adipose tissue in, 272
  fat in, 362
  of Retzius, 398
Retropyloric node, BP74
Retrotonsillar vein, superior, 171
Retzius, retropubic space of, 272
Rhinal sulcus, 133–134
Rhomboid major muscle, 195, 198, 201, 212,
      263–265, 275, 428, 432, 488, T3.4
Rhomboid minor muscle, 195, 198, 428, 432,
      488, T3.4
Ribs, 8, 292, 233, 264–265, T4.1–T4.3
  1st, 17, 36, 45, 55, 103, 186, 203, 213, 217–
      218, 231, 251–252, 254, 429, 438–439,
      BP34, BP41
    lung groove for, 221
  2nd, 203, 205, 429, 52
  3rd, 203
  4th, 203
  5th, 203, 233
  6th, 203, 205
    posterior view of, 221
  7th, 203, 180, 266
  8th, 203, 237
  9th, 203, 307, BP36
  10th, 203
  11th, 203, 331
  12th, 186, 195, 203, 216, 268, 331, 337, 505
    radiology of, 182
  angle of, 203, 221, BP42
  associated joints and, 221
  body, 203
  costal groove of, 221
  false, 203
  floating, 203
  head of, 203, 221, BP42
    intraarticular ligament of, 221
    radiate ligament of, 221
  muscle attachment of, BP42
  neck of, 203, 221, BP42
  radiate ligament of, BP33
  superior articular facet on, BP33
  true, 203

Ribs (Continued)
  tubercle of, 203, 221, BP42
Right bundle
  Purkinje fibers of, 248
  subendocardial branches of, 248
Right colic plexus, 322–323
Right costal arch, 288
Right jugular trunk, BP76
Right lateral region, 269
Right lung, 337
Right lymphatic duct, BP76
Right subclavian trunk, BP76
Risorius muscle, 48, T2.9–T2.14, BP19-BP20
Rods, 147
Root canals, 41
Root ganglion, posterior, 155, 159
Root sheath
  external, 12, BP1
  internal, 12, BP1
Rotator cuff muscles, 434, T7.1–T7.2
Rotatores breves colli muscle, 197
Rotatores longi colli muscle, 197
Rotatores muscle, T3.4
Rotatores thoracis muscle
  brevis, 197
  longus, 197
Rotter's lymph nodes, 207–208. See also
  Interpectoral (Rotter's) lymph nodes.
Round ligament, 376
  of artery, 402
  of liver, 274, 288, 294, 302–303, 318, 335,
      351, BP83
  of uterine artery, 406
  of uterus, 364, 371–372, 375, 378, 402, BP86
Rubrospinal tract, BP40
Rugae, 295. See also Gastric folds (rugae).
Ruptured follicle, BP88

S

Saccule, 124–125, 151, T2.1–T2.4
Sacral arteries
  lateral, 191, 284, 404–405, 336, BP93
    spinal branches of, 191
  median, 365–366, 284, 314, 336, 400, 404–405
Sacral canal, 183, 361
Sacral crest
  intermediate, 183
  lateral, 183
  median, 183, 357–358
Sacral foramina, BP37
  anterior, 183, 356, 361
    radiograph of, 354
  posterior, 183, 356
Sacral ganglion, 1st, 6
Sacral hiatus, 183, T3.1–T3.3
Sacral horn, 183
Sacral kyphosis, 179
Sacral nerve, perineal branch of, 506, 508
Sacral nodes
  internal, 408
  lateral, 286, 408, 410
  median, 410
  middle, 286, 408, 410
Sacral plexus, 186, 287, 323, 339, 412, 417–419,
      506, 508
Sacral promontory, 268, 353, 355–357, 359,
      364–365, 367, 414, BP90, BP81
Sacral spinal cord, 324
Sacral spinal nerves
  S1, 186
    anterior ramus, 412, 416
    dermatome of, 188
    vertebral body of, 399
  S2, dermatome of, 188
  S3, dermatome of, 188
  S4, dermatome of, 188

Sacral spinal nerves (Continued)
  S5, 186
    dermatome of, 188
Sacral splanchnic nerves, 6, BP6
Sacral tuberosity, 183
Sacral veins
  lateral, 285
  median, 285, 317, 365, 401, 405
Sacral vessels, median, 402, BP93
Sacrococcygeal ligaments
  anterior, 356, 359, 361, 366
  lateral, 356–357, 184
  posterior, 356–357, 184
    deep, 356
    superficial, 356
Sacrogenital fold, 394
Sacroiliac joint, 355, 497, 361, T6.1–T6.3, BP81
  radiograph of, 354
  radiology of, 182
Sacroiliac ligaments
  anterior, 356
  posterior, 356–357, 184
Sacrospinous ligament, 184, 356–357, 358,
      360, 362, 412, 415, 504, 512
  left, 373
Sacrotuberous ligament, 184, 356–358, 360,
      362, 379, 397, 398, 404, 412–413, 415, 420,
      504, 511–512, BP104, BP94
Sacrum, 8, 178–179, 353, 268, 183–184, 186,
      360, 187, 361, 399
  ala of, radiograph of, 354
  apex of, 183
  arteries of, 191
  auricular surface for, 184
  base of, 183
  radiology of, 182
Saddle joint, 9
Sagittal plane, 1
Sagittal sinus
  inferior, 127, 130–131, 171–173
  superior, 127–133, 136, 171–172
    emissary vein to, 34
    groove for, 30, 32
    imaging of, BP31-BP32
Sagittal suture, 26, 30, 35
Salivary gland, 70, 86
  lingual minor, 83
  molar minor, 85
  sublingual, 86
  submandibular
    deep lobe, 86
    superficial lobe, 86
Salivary glands, 18
Salivatory nucleus
  inferior, 143–144, 152, 161
  superior, 80, 143–144, 150, 160
Salpingopharyngeal fold, 90, 93
Salpingopharyngeus muscle, 90, 92, 96, 153,
      T2.9–T2.14
Saphenous hiatus, 271, 277, BP53
  cribriform fascia and, 277, BP53
  falciform margin of, 282
Saphenous nerve, 492–493, 509–510, 515, 531–
      532, 534, 550, 553, T8.1–T8.2, T8.3–T8.4,
      BP106, BP109
  infrapatellar branch of, 492, 509–510, 531–
      532, 550
  medial crural cutaneous of, 550
Saphenous opening, 378, 492
  cribriform fascia and, 494
Saphenous veins
  accessory, 492–493, 523
  great, 15, 270–271, 277, 285, 381, 491–494,
      515, 523, 533–534, T8.1–T8.2, BP106,
      BP110-BP111, BP53
  greater, 421
  small, 491–493, 494, 511, 523, 527, 533, 534

Netter Atlas of Human Anatomy: Classic Regional Approach

Sarcolemma, BP10
Sarcomere, BP10
Sarcoplasm, BP10
Sartorius muscle, 420–421, 491, 498, 501–502, 504, 509–510, 515–516, 520, 527, 550, T8.5–T8.8, BP104, BP94
Sartorius tendon, 502–503, 516–517, 531–532, BP106
Satellite cells, BP3
Scala tympani, 124–125
Scala vestibuli, 125
Scalenus muscle, 49, 54, BP43
   anterior, 50–51, 54–57, 98, 99, 100, 103, 436–438, 219, 210–211, 214, 220, 231, 238, 251–254, T2.9–T2.14, BP41-BP43, BP50
   innervation of, 439
   medius, 51, 53, 55, 98–100, T2.9–T2.14
   middle, 50, 54, 56, 210, 253, BP41, BP42, BP43, BP50
   posterior, 50–51, 54–55, 210, 253–254, 436, T2.9–T2.14, BP42-BP43, BP50
   superior attachments of, 55
Scalp, 24, 48
   arteries of, T2.1–T2.4
   skin of, 48
   superficial arteries and veins of, 24
   superficial fascia of, 48
Scaphocapitate ligament, 464
Scaphoid bone, 462–463, 465–467, 477, T7.1–T7.2, BP98
   articulation with, 449
Scaphoid fossa, 122, BP30
Scapholunate ligament, 465
Scaphotrapeziotrapezoid ligament, 464
Scapula, 8, 201, 212, 263–265, 435–436
   acromion of, 203
   anastomoses around, 437
   coracoid process of, 203, 430
   glenoid cavity of, 431
   glenoid fossa of, 203, 427, 430
   inferior angle of, 178, 266, 427
   inferior border, 427
   infraspinous fossa of, 203
   lateral border of, 427, 430
   medial border of, 178, 427
   muscle attachment sites of, 428
   neck of, 203
   scapular notch of, 203
   spine of, 178, 195, 198, 203, 218, 432, 434–437, 490
   subscapular fossa of, 203
   superior border of, 434
   supraspinous fossa of, 203
Scapular artery
   circumflex, 436–437, 438, 440, 443, T7.1–T7.2, BP96, BP24
   dorsal, 56–57, 194, 437–438, T7.1–T7.2, BP24
Scapular ligament, transverse, 431
   inferior, BP24
   superior, 433–434, 436–437, BP24
Scapular nerve, 438
   dorsal, 439, 488, T7.3–T7.4
   lower, 438
Scapular notch, 427, 431, 436–437
   of scapula, 203
   superior, 436
Scarpa's fascia, 270, 381, 343, BP84
Scarpa's layer, 378
   of superficial fascia, 381–382
Schwalbe's line, 117
Sciatic bursa
   of gluteus maximus muscle, 513
   of obturator internus muscle, 513
Sciatic foramen
   greater, 184, 356–358, 552
   lesser, 184, 356–357

Sciatic nerve, 186, 504, 506, 508, 511, 512, 513, 515, 552, 415, 421, T8.3–T8.4, BP104, BP94
   articular branch of, 511
   left, 420
   muscular branches of, 511
   right, 420
Sciatic notch
   greater, 353, 268, 495
   lesser, 353, 268, 495
Sclera, 110, 116–120, BP27
   lamina cribrosa of, 116
Scleral sinus (canal of Schlemm), 116, 119–120
   veins draining, 120
Scleral spur, 116–117
Scleral venous sinus, 116–120
Scrotal arteries, posterior, 405, 407, BP93
Scrotal branches, posterior, 415
Scrotal ligament, 390
Scrotal nerves, posterior, 412–413, 512
Scrotal veins, anterior, 277, 405
Scrotum, 352, 267, 387–389
   Dartos fascia of, 270–272, 368, 381, 388, 392, 397, 407, BP84
   septum of, 368, 388, 392, 397, 407
   skin of, 388, 392
Sebaceous glands, 12, 110, BP1
   duct of, BP1
Secondary oocyte, BP87
Secondary spermatocytes, BP87
Segmental medullary arteries, 191, 192
Segmental medullary vein
   anterior, 193
   posterior, 193
Segmental motor nerve function, 11
Sella turcica, 28, 29, 32, 63, 175, BP34
   dorsum sellae of, 32
   hypophyseal fossa of, 32
   hypophysis in, 61
   posterior clinoid process of, 32
   tuberculum sellae of, 32
Semicircular canal
   anterior, 82, 124–126
      plane of, 126
   lateral, 124–126
      prominence of, 121, BP28
   posterior, 124, 125
      plane of, 126
Semicircular duct
   anterior, 124–126
   lateral, 124–126
      ampulla of, 151
   posterior, 124–126
      ampulla of, 151
   superior, ampulla of, 151
Semilunar folds, 90, 301
Semilunar hiatus, 67
Semimembranosus bursa, 516, 521
Semimembranosus muscle, 415, 498, 501, 504, 511, 515–516, 520, 526–527, 552, T8.1–T8.2, T8.5–T8.8
   bursa of, 520
Semimembranosus tendon, 516, 518, 520–521, 528–529
   groove for, 524
Seminal colliculus, 371, 385, BP92
Seminal gland, 274, 367, 368, 20, 385, 390, 391, BP95
Seminal vesicles, 385, 393, 405. See also Seminal gland.
Seminiferous epithelium, 391
Seminiferous tubules, convoluted, 391–392
Semispinalis capitis muscle, 195–197, 199
Semispinalis colli muscle, 197, 199
Semispinalis muscle, T3.4
Semispinalis thoracis muscle, 197

Semitendinosus muscle, T8.1–T8.2, T8.5–T8.8, 415, 491, 498, 504, 511–513, 515, 520, 527, 552
Semitendinosus tendon, 502–503, 517
Sensory cortex, 162
Sensory systems, 4
Septal papillary muscle, 241
Septate hymen, BP91
Septomarginal fasciculus, BP40
Septomarginal trabecula, 241, 245–246, 248
Septum, 392
   bulging, 61
   nasal, 26, 63
   of scrotum, 407
Septum pellucidum, 133, 135, 137–139, 168, 170, 172, 174
   posterior veins of, 173
   vein of, 172–173
Serosa (visceral peritoneum), 297, 307, BP57
Serous glands of von Ebner, 89
Serous pericardium, 236
   parietal layer of, 236
   visceral layer of, 236
Serratus anterior muscle, 428, 435
Serratus muscle
   anterior, 195, 201, 202, 205, 209–210, 212, 238, 263–267, 270, 271, 275, 278–279, 422, 432–433, 436, 438, T4.4, BP42
   posterior
      inferior, 275, 195–196, 200, 331, T3.4
      superior, 195–196, T3.4
Sesamoid bones, 8, 463, 466, 535, 536, 545–548
   foot, 539
Sherman's vein. See Perforating veins
Shoulder, 2–3, 10, BP15
   anteroposterior radiograph of, 430
   arthrogram of, 490
   computed tomography of, 490
   glenohumeral joint of, 431
   magnetic resonance imaging of, 490
   muscles of, 432
   nerves of, 488
   region of, 436
Sibson's fascia. See Suprapleural membrane (Sibson's fascia)
Sigmoid arteries, 284, 400, 314, 335
Sigmoid colon, 6–7, 288, 290, 301, 363, 365–367, 375, 393–396, 412, 414, BP60
   as site of referred visceral pain, BP5
Sigmoid mesocolon, 290, 301, 393, 314, 330, 335
   attachment of, 293
Sigmoid nodes, BP77
Sigmoid plexuses, 323
Sigmoid sinus, 82, 91, 126, 130–131, 176
   groove for, 29, 32
   imaging of, BP31
   in jugular foramen, 34
Sigmoid veins, 401, 317–318
Sinuatrial (SA) node, T4.1–T4.3, 239, 248
Sinus nerve, carotid, 157
Sinuses
   confluence of, 126, 130–131, 171–172, 176
   prostatic, BP92
Sinusoids, 303–304, BP78
Skeletal muscle, 4
Skeletal system, 8, T3.1–T3.3, T4.1–T4.3, T8.1–T8.2, T5.1–T5.3
Skene's ducts, BP86. See also Paraurethral (Skene's) glands.
Skin, 6, 50–51, 129, 273
   cross section of, 12
   papillary layer of, 12
   radiology of, 182
   reticular layer of, 12
   of scalp, 24, 48

Skin ligaments, 12
Skull
  anterior view of, 25
  associated bones and, 8
  lateral view of, 27
  midsagittal view of, 29
  of newborn, 35
  nuchal line of, superior, 196
  orientation of labyrinths in, 126
  radiographs of
    lateral view of, 28
    posterior view of, 26
    Waters' view of, 26
  reconstruction of, BP16
  superior nuchal line of, 197
Small arteries, 13
Small arterioles, 13
Small intestine, 18, 153, 288, 300, 343
  area for, 330
  autonomic innervation of, 322
  lymph vessels and nodes of, BP76
  mucosa and musculature of, 297
  as site of referred visceral pain, BP5
  veins of, 316
Smooth muscle, BP71
Smooth muscle cell, 4, BP13
Soft palate, 61, 63, 83, 90, 93, 95
  imaging of, 177
  muscles of, 96
Sole, muscle of
  first layer, 544
  second layer, 545
  third layer, 546
Soleus muscle, 504, 516, 526–534, 540, 552–
    553, T8.5–T8.8, BP106, BP110
  arch of, 504
Soleus tendon, BP110
Solitary lymphoid nodule, 297
Solitary tract, nuclei of, 143–144, 150, 152–153,
    162, 250, 340
Somatic fibers, BP8
Somatic nervous system, 4
Somatic sensory receptors, 4
Somatosensory system, trunk and limbs,
    BP14
Space of Poirier, BP98
Space of Retzius, 369
Special sense organs, 4
Spermatic cord, 274, 281, 382–383, 420,
    BP94
  arteries of, 276
  external spermatic fascia on, 270, 280, 407
  testicular vein in, 403
Spermatic fascia
  external, 271–272, 278, 280–281, 368, 381–
    383, 388, 392, BP54
    spermatic cord and, 407
    testis and, 407
  internal, 272, 281, 388, 392, BP54
    spermatic cord and, 282
Spermatids, 391, BP87
Spermatocytes
  primary, 391
  secondary, 391
Spermatogenesis, 391, BP87
Spermatogonium, 391, BP87
Spermatozoa, 391, BP87
Sphenoethmoidal recess, 61–62
Sphenoid bone, 25, 27, 29, 31–32, 36, 62–63,
    BP22
  anterior clinoid process of, 29, 32
  body of, 29, 32, 63, 68
    jugum of, 32
    prechiasmatic groove of, 32
    sella turcica of, 32
  carotid groove of, 32
  clivus of, 32

Sphenoid bone (Continued)
  crest of, 63
  foramen ovale of, 31
  foramen spinosum of, 31
  greater wing of, 25, 27–29, 31–32, 35, BP16
    groove for middle meningeal vessels, 32
  infratemporal crest of, 27
  lesser wing of, 25, 26, 29, 32
  of newborn, 35
  optic canal of, 29
  orbital surface of, 25
  pterygoid process of, 31, 38, 62
    hamulus of, 27, 29, 31, 38, 62
    lateral plate of, 27, 29, 31, 35, 38, 62–63
    medial plate of, 29, 31, 35, 38, 62–63
    pterygoid fossa of, 31
    scaphoid fossa of, 31
  sella turcica of, 29, 32
    dorsum sellae of, 32
    hypophyseal fossa of, 32
    posterior clinoid process of, 32
    tuberculum sellae of, 32
  sphenoidal sinus of, 29, 62–63
  spine of, 31, 38
Sphenoid sinus, BP34
Sphenoidal emissary foramen (of Vesalius),
    34
Sphenoidal fontanelle, of newborn, 35
Sphenoidal sinus, 17, 28–29, 61, 63, 69, 82, 90,
    95, 114, 131, 164, 175, BP23, BP27
  imaging of, 177
  opening of, 61, 62, 67
Sphenoidal vein (of Vesalius), 100
Sphenomandibular ligament, 42, 73–74, 76, 78
Sphenooccipital synchondrosis, 95
Sphenopalatine artery, 65, 74, 81, 99,
    T2.1–T2.4, BP21
  pharyngeal branch of, 81
  posterior lateral nasal branches of, 65
  posterior septal branches of, 65, 74
Sphenopalatine foramen, 27, 29, 38, 62, 66,
    74, BP21
Sphenopalatine vessels, in incisive fossa, 33
Sphenoparietal sinus, 130–131
Spherical recess, 124
Sphincter muscle
  external
    deep part of, 395
    subcutaneous part of, 395
    superficial part of, 395
  internal, 395
Sphincter pupillae muscle, 117, 148, 159,
    T2.1–T2.4, T2.9–T2.14
Sphincter urethrae, 364, 369, 378, 398, BP92
  muscle, 358, 362, 370–371, 379, BP86
Sphincter urethrovaginalis muscles, 369, 372,
    378–380, T6.4
Sphincters, female, 370
Spinal accessory nerve (CN XI), 57, 75, 98, 101,
    BP22, T2.1–T2.4
Spinal accessory nodes, 101
Spinal artery
  anterior, 191–192, 165–166, 168, 170
    anastomotic vessels to, 191
  posterior, 191–192, 166, 168, 170
    anastomotic vessels to, 191
    left, 192
    right, 192
    segmental medullary branches of, 165
Spinal branch, 192
Spinal canal, 345
Spinal cord, 4, 186, 250, 80, 263–264, 328–329,
    BP15, BP38
  arteries of, T3.1–T3.3
    schema of, 191
  central canal of, 136, 141, 144
  cervical part of, BP14

Spinal cord (Continued)
  cross sections, BP40
    fiber tracts, BP40
  descending tracts in, 230
  gray matter of, 189
  imaging of, 177
  lateral horn of, 189
  lumbar part of, 419, BP14
    upper, BP8
  principal fiber tracts of, BP40
  sacral part of, 419, BP8
  sections through, BP40
  thoracic part of, 159–161, 230, 326, BP8
  veins of, 193
  white matter of, 189
Spinal dura mater, 200
Spinal ganglion, 190, 419, 155, 212, T4.1–T4.3,
    BP14
  posterior root of, 189, 201
Spinal meninges, T3.1–T3.3
  nerve roots and, 189
Spinal nerve, 4, 51, 186, 189, 190, 201, 326,
    BP52
  anterior ramus of, 189
    C1, 98
    C2, 98
    C3, 214
    C4, 98, 214
    C5, 214
  anterior root of, 189, 279, 325
    C8, 186
    L4, 415
    L5, 415
    L5, posterior ramus of, 184
  lateral posterior branch of, 279
  medial posterior branch of, 279
  meningeal branch of, 279
  posterior ramus of, 189, 279
  posterior root of, 189, 279
  roots, 187
    S1, 415
    S2, 415
    S3, 415
    S4, 415
    T1, 186
Spinal nerve roots
  anterior root of, 189
  posterior root of, 189
    S1, 187
    S2, 187
    S3, 187
    S4, 187
    S5, 187
  vertebrae relation to, 187
Spinal nerve trunk, 279
Spinal nucleus, 149
Spinal sensory (posterior root) ganglion, 279,
    324–326, 328–329, 340
Spinal tract, 149, 153
Spinal vein
  anterior, 171, 193
  posterior, 171
Spinalis colli muscle, 196
Spinalis muscle, 196, T3.4
Spinalis thoracis muscle, 196
Spine, 427
Spinocerebellar tract
  anterior, BP40
  posterior, BP40
Spinocervical tract, BP14
Spinoglenoid notch, 427
Spinohypothalamic fibers, BP40
Spinomesencephalic fibers, BP40
Spinoolivary tract, BP40
Spinoreticular fibers, BP40
Spinothalamic fibers, BP40
Spinothalamic tract

Spinothalamic tract (Continued)
anterior, BP14
lateral, BP14
Spinous process, 180–181, 184–185, T3.1–T3.3, BP38, BP17
of C2, BP34
of C7, 196, 197, T3.1–T3.3, BP34, BP35
cervical, 43, 46
of L1, 182, 200
of L2, radiology of, 182
of L3, 181, BP37
of L4, BP37
of T, 12, 178, 195–196, 198
thoracic, 180
Spiral ganglion of Corti, 125
Spiral lamina
hamulus of, 124
osseous, 124–125
Spiral ligament, 125
Splanchnic nerves, 416
greater, 411, 416
thoracic, 6, 319–326, 328–329, 339, 341, BP6
least, 411, 416
lesser, 411, 416
thoracic, 6, 319–325, 339–341, BP6
lumbar, 6, 323–324, 411–412, 414, 416–417, 419, 339–341, T5.1–T5.3, BP6
fifth, 412
upper, 418
pelvic, 7, 323–324, 412, 414, 416–419, 339–340, 508
sacral, 6, 324, 339, 412, 414, 419, 508, BP6
thoracic, T5.1–T5.3, 412
greater, 417–418
least, 418
lesser, 417–418
Spleen, 16, 218, 291–292, 294, 234, 303, 306–307, 342, 345–346, 349, T5.1–T5.3, BP38, BP50-BP51, BP79, BP82
anterior extremity of, 307
area for, 330
inferior border of, 307
posterior extremity of, 307
as site of referred visceral pain, BP5
in situ, 307
superior borders of, 307
Splenic artery, 293, 296, 234, 306–313, 325, 328–329, 348, BP64, BP65, BP69, BP83
splenic branches of, 308
Splenic nodes, BP75
Splenic notch, 307
Splenic plexus, 319, 320, 321
Splenic red pulp, 307
Splenic trabeculae, 307
Splenic vein, 292, 293, 307, 315–318, 259, 342–343, 345, BP66
Splenic white pulp, 307
Splenius capitis muscle, T3.4, 5, 195, 196, 199
Splenius colli muscle, T3.4, 195–196, 199
Splenomesenteric trunk, BP67
Splenorenal ligament, 291, 293, 307, 330
Spongy bone, BP9
Spongy urethra, 391, BP85
bulbous portion of, 371, 386
pendulous portion of, 386
Squamous suture, 35, BP16
Stapedius muscle, 123, T2.1–T2.4, T2.9–T2.14
nerve to, 150
tendon of, 122, BP28
Stapes, 122, 123
footplate of, 122
limbs of, 121, BP28, BP29
Stellate ganglion, 229, 249–250, 261. See also Cervicothoracic (stellate) ganglion.
Stellate vein, 333
Sternal angle (of Louis), T4.1–T4.3

Sternal head, of sternocleidomastoid muscle, 22
Sternalis muscle, 209
Sternochondral (synovial) joint, 429, T4.1–T4.3
Sternoclavicular joint, 217, 263, 429
Sternoclavicular ligament, anterior, 429
Sternocleidomastoid muscle, 154, 156, 195, 219, 199, 49–52, 54, 56, 58, 70, 86, 202, 209, 432–433, 438, T2.1–T2.4, T2.9–T2.14, BP43
clavicular head of, 22, 202, 49, 54
medial margin of, 103
sternal head of, 22, 49, 54, 202
Sternocostal triangle, 211
Sternohyoid muscle, 49–54, 87, 155, 209, 211, 428, T2.9–T2.14
nerve to, 57
Sternothyroid muscle, 49–54, 108, 155–156, 209, 211, T2.9–T2.14
nerve to, 57
Sternum, 8, 17, 212, 215, 201, 234, 237, 255, 262, 432
angle, 203
body of, 202–203, 267–268, 209, 211, 213, 265, 266
jugular notch of, 203
manubrium of, 48–51, 54, 95, 211, 219, BP34, BP35
xiphoid process of, 202–203, 209
Stomach, 6–7, 13, 18, 217, 288, 291–292, 300, 238, 302–303, 305–307, 254, 258, 343, 345, T5.1–T5.3, BP50, BP79, BP82
air within, 234
area of, 330
arteries of, 308
autonomic innervation of, 320–321, 325
body of, 294–295
cardiac part of, 294, 255, 257
circular muscular layer of, 295
fundus of, 294–295, 255, 257, 347, BP51
greater curvature of, 294
longitudinal muscular layer of, 295
lymph vessels and nodes of, BP74
mucosa of, 295
oblique muscle fibers of, 257, 295
pyloric part of, 294
as site of referred visceral pain, BP5
in situ, 294
variations in position and contour of, BP55
veins of, 315
Straight arteries, 297, 313–314, 327
Straight conjugate, 355
Straight gyrus, 134
Straight sinus, 126, 130–131, 133, 136, 171–173, 176
imaging of, BP31
Straight veins, 316
Stratum basale, 12
Stratum corneum, 12
Stratum granulosum, 12
Stratum lucidum, 12
Stratum spinosum, 12
Stretch receptors (Hering-Breuer reflex), 230
Stria medullaris, 141
of thalamus, 133, 138
Stria terminalis, 133, 135, 138–139
Striate artery, long medial, 164, 166, 167–169
Striated muscle, BP8
Stroma, BP92
Styloglossus muscle, 74, 86, 88, 96–97, 155, T2.1–T2.4, T2.9–T2.14, BP22
Stylohyoid ligament, 36, 88, 90, 96–97
Stylohyoid muscle, 49, 53–54, 58, 70, 74, 77, 86–88, 92, 97, 150, T2.9–T2.14, BP22
Styloid process, 27, 31, 35, 36, 38, 42, 54, 58, 88, 97, 123, BP25
of temporal bone, 55

Stylomandibular ligament, 36, 42, 78
Stylomastoid artery, 91, 123
posterior tympanic branch of, 123
stapedial branch of, 123
Stylomastoid foramen, 31, 33, 150, 152, BP30
Stylopharyngeus muscle, 86, 88, 90–92, 94, 96–97, 152–153, T2.9–T2.14
Subacromial bursa, 431, 440
Subaponeurotic space, dorsal, 473
Subarachnoid space, 190, 116, 127, 129, 136
Subcallosal area, 133, 146
Subcallosal gyrus, 133
Subcapsular lymphatic plexus, 338
Subchondral bone tissue, 9, BP11
Subclavian artery, 14, 17, 55–57, 91, 98–100, 103, 105, 157, 163, 165, 191, 206, 210–211, 276, 436–438, 219–220, 231–232, 254, 258, 263, BP24
grooves for, BP42
left, 104, 194, 235, 252–253
lung groove for, 221
right, 104, 153, 194, 251, BP24
Subclavian lymphatic trunk, 228
Subclavian nerve, 439, T7.3–T7.4
Subclavian trunk, right, 286
Subclavian vein, 15, 17, 55–57, 100, 103, 210–211, 219–220, 228, 231–232, 254, 259, 277, 436, 438, T2.1–T2.4
grooves for, BP42
left, 104, 194, 252–253
right, 104, 194, 251
Subclavius muscle, 205, 209–210, 238, 251–252, 433, 435, 438, 429, T4.4, BP42
fascia investing, 435
groove for, 429
Subcostal artery, 284
Subcostal muscles, 279, T4.4
Subcostal nerve, 186, 253, 287, 330–331, 411, 416, 505–507, T5.1–T5.3
anterior branch of, 506
cutaneous branch of
anterior, 278
lateral, 275, 278, 287, 492
lateral branch of, 506
Subcostal plane, 269
Subcostal vein, 285
Subcutaneous artery, 12
Subcutaneous bursa, medial malleolus of, 540
Subcutaneous fat, radiology of, 182
Subcutaneous olecranon bursa, 448
Subcutaneous space, dorsal, 473
Subcutaneous tissue, 12, 51
deeper membranous layer of, 381
fatty layer of, 273
membranous layer of, 273
of penis, 368, 388
superficial fatty (Camper's) layer of, 381
Subcutaneous vein, 12
Subdeltoid bursa, 431
Subhiatal fat ring, 257
Subiculum, 139
Sublime tubercle, 446
Sublingual artery, 70, 87–88
Sublingual caruncles, 87
Sublingual fossa, 39
Sublingual gland, 6, 7, 69, 70, 77, 83, 87, 150, 160, BP7, BP23
Sublingual nerve, 77
Sublingual veins, 70, 88
Sublobular veins, 303, 304
Submandibular duct (of Wharton), 70, 83, 86, 88
Submandibular fossa, 39
Submandibular ganglion, 7, 70, 75, 77, 87, 88, 98, 149, 150, 158, 160, BP7
right, 75
schema of, 160

Submandibular glands, 6, 7, 22, 49, 52, 54, 70, 77, 87, 93, 99–100, 150, 160, BP7
Submandibular node, 101–102
Submental artery, 74, 99
Submental node, 101–102
Submental vein, 100
Submucosa, BP48
  esophageal, 257
  of ileum, 297
  of intestine, 327
  of jejunum, 297
Submucosal glands, 327
Submucosal (Meissner's) plexus, 327
Submucous space, 394–395, 398
Suboccipital nerve, 46, 199
Subpleural capillaries, 227
Subpleural lymphatic plexus, 228
Subpopliteal recess, 517–518
Subpubic angle, radiograph of, 354
Subpyloric nodes, BP74
Subscapular artery, 436–438, 443, BP24, BP96
Subscapular fossa, 427
  of scapula, 203
Subscapular nerve, T7.3–T7.4
  lower, 433, 436, 439, 488
  upper, 433, 438–439
Subscapularis muscle, 201, 212, 263–265, 428, 431, 433, 434–436, 440, 442, T7.5–T7.8, BP50
Subscapularis tendon, 431, 434, 436
Subserous fascia, 414
Substantia nigra, 134, 144
Subtendinous bursa, 521, 540
  of iliacus, 513
  of subscapularis muscle, 431
Sulcal artery
  central, 169
  precentral, 169
  prefrontal, 169
Sulcus
  calcarine, 132–134, 138, 147
  central
    of insula, 132
    of Rolando, 132–133
  cingulate, 133
  collateral, 133–134
  of corpus callosum, 133
  frontal
    inferior, 132
    superior, 132
  hippocampal, 139
  hypothalamic, 133, 141, 174
  lateral (of Sylvius), 132, 134
    anterior ramus of, 132
    ascending ramus of, 132
    posterior ramus of, 132
  lunate, 132
  marginal, 133
  median, 89, 141
  nasolabial, 22
  occipital, transverse, 132
  occipitotemporal, 133–134
  olfactory, 134
  orbital, 134
  paracentral, 133
  parietooccipital, 132–133
  postcentral, 132
  rhinal, 133–134
  temporal
    inferior, 132–134
    superior, 132
  terminal, 89
Sulcus limitans, 141
Sulcus terminalis, 235
Superciliary arch, 22
Superficial back, 10
Superficial capsular tissue, BP98

Superficial circumflex iliac vein, 267
Superficial dorsal vein, of penis, 267
Superficial epigastric veins, 267
Superficial fascia, 50, 378, BP84
  deep membranous layer of, 382
  fatty layer of, 343
  membranous layer of, 343
  penile, BP84
  of scalp, 24
  superficial fatty (Camper's) layer of, 378
Superficial fibular nerve, 534
Superficial glomerulus, BP73
Superficial investing cervical fascia, 51, 52
Superficial temporal artery, 60, 74
Superficial vein
  of forearm, 425
  lateral, of penis, 382
Superficialis flexor tendons, 473
Superior, as term of relationship, 1
Superior angle, 427
Superior anorectal plexus, 319
Superior anterior segmental artery, 334
Superior articular facet, of L5, BP36
Superior articular process, 181, 184
  facets of, 183
  of L1, BP37
  of L2, 182
  of L4, BP37
Superior border, 427
Superior bulb, in jugular fossa, 33
Superior cervical ganglion, 6, 230, BP6
Superior cervical sympathetic cardiac nerve, 158
Superior cluneal nerves, 275
Superior colliculus, 138, T2.1–T2.4
  brachium of, 138
Superior costal facet, 180
Superior costotransverse ligament, BP33
Superior fascia, of pelvic diaphragm, 369, 378
Superior gluteal cutaneous nerves, 275
Superior hypogastric plexus, 6, BP6
Superior labial artery, 60
Superior mediastinum, BP2
Superior mesenteric artery, 14, 284, 294
Superior mesenteric ganglion, 6, 7, BP6, BP8
Superior mesenteric plexus, BP6-BP7
Superior oblique muscle, 148
Superior orbital fissure, 37
Superior petrosal sinus, 82
Superior pharyngeal constrictor, 52, 99
Superior pubic ramus, 268
Superior rectus muscle, 148
Superior rectus tendon, 120
Superior segmental artery, 334
Superior suprarenal artery, 216
Superior vena cava, 13, 15
Superior vertebral notch, 180
  of L3 vertebra, BP37
Superolateral nodes, 409
Superomedial nodes, 409
Supination, of upper limbs movements, 11
Supinator muscle, 428, 450, 455, 457–461, 483, 489, T7.5–T7.8
Supraclavicular nerves, 23, 56–57, 156, 423, 424, T2.9
  intermediate, 156, 424
  lateral, 156, 424
  medial, 156, 424
Supraclavicular nodes, 101
Supracondylar ridge
  lateral, 427, 446
  medial, 427, 446
Supraduodenal artery, 308–310, 312–313
Supraglenoid tubercle, 427
Suprahyoid artery, 88
Suprahyoid muscles, 53, 87

Supralevator space, 394
Supramarginal gyrus, 132
Supramastoid crest, 27
Supraoptic recess, 133, 135
Supraopticohypophyseal tract, 174
Supraorbital artery, 24, 59, 60, 74, 99, 110, 115, 165
Supraorbital margin, 22, 26, BP16
Supraorbital nerve, 23, 59–60, 79, 110, 149
  lateral branch of, 113–114
  left, 114
  medial branch of, 113–114
Supraorbital notch, 25, 27
  of newborn, 35
Supraorbital vein, 24, 100, 115
Suprapatellar bursa, 517, 521
Suprapatellar synovial bursa, 518
Suprapineal recess, 133, 135
Suprapleural membrane (Sibson's fascia), 251–252
Suprapyloric nodes, BP74
Suprarenal androgens, BP95
Suprarenal arteries
  inferior, 284, 332, 334, 342
  middle, 284, 332, 342
  right, 332
  superior, 284, 332, 342
Suprarenal cortices, BP95
Suprarenal glands, 6, 7, 218, 19, 21, 296, 387, 302, 306, 307, 330, 332, 337, 341–342, T5.1–T5.3, BP8, BP38, BP50-BP51
  arteries of, 342
  autonomic nerves of, 341
  cross section, 342
  left, 291, 330, 332, 341, 349
  right, 293, 330, 332, 341, BP82
  veins of, 342
Suprarenal impression, 302
Suprarenal plexus, 319
  left, 323
Suprarenal vein, 342
  left, 259, 285, 332
  right inferior, 285
Suprascapular artery, 56–57, 103–105, 165, 436–438, T7.1–T7.2, BP24
  infraspinous branch of, 437
Suprascapular foramen, 437
Suprascapular nerve, 436, 438–439, 488, T7.3–T7.4
Supraspinatus muscle, 195, 198, 428, 431–437, 441, 488, 490, T7.5–T7.8, BP50
Supraspinatus tendon, 431, 434, 436, 441, 490, T7.1–T7.2
Supraspinous fossa, 427
  of scapula, 203
Supraspinous ligament, 46, 184–185, 200, 356–357
  radiology of, 182
Suprasternal space (of Burns), 49–52, 95
Supratonsillar fossa, 90
Supratrochlear artery, 24, 59–60, 74, 99, 110, 115, 165
Supratrochlear nerve, 23, 59–60, 79, 110, 113–114, 149
Supratrochlear nodes, 426
Supratrochlear vein, 24, 100
Supraventricular crest, 241, 245
Supravesical fossa, 274
Sural nerve, 493, 552–553, T8.1–T8.4, BP106, BP109, BP110
  calcaneal branches of, lateral, 493
Suspensory ligament, 387
  of axilla, 435
  of breast, 205
  of clitoris, 378, 380
  of ovary, 364, 374, 376, 402
  of penis, 271, 368

Tentorium cerebelli, 113, 130–131, 171–172
Teres major muscle, 178, 195, 198, 201, 264–
265, 428, 432–433, 435–438, 440–442, 445,
488
  lower margin of, BP96
Teres major tendon, 441
Teres minor muscle, 195, 198, 212, 263–264,
432–437, 441, 488, T7.5–T7.8
Teres minor tendon, 431, 434, 441
Terminal bronchiole, 227
Terminal ileum, 363, 367
Testes, 391, T6.1–T6.3, BP95
Testicle, 388
Testicular artery, 281–282, 284, 293, 330, 332,
351, 367, 388, 403, 405, 411, 418, BP104,
BP93
  covered by peritoneum, 281
Testicular plexus, 319, 411, 418
Testicular vein, 274, 280, 281–282, 293, 317,
332, 335, 367, 403
  covered by peritoneum, 281
  in spermatic cord, 403
Testicular vessels, 410
Testis, 20, 21, 387–388, 390, 392, 410, 418, 343
  arteries of, 403
  descent of, 387
  efferent ductules of, 391
  external spermatic fascia over, 407
  gubernaculum of, 387
  lobules of, 392
  mediastinum of, 388, 392
  rete, 392
  tunica albuginea of, 388
  tunica vaginalis, 388
    parietal layer of, 392
    visceral layer of, 392
  veins of, 403
Thalamic tubercle, anterior, 138
Thalamic veins, superior, 173
Thalamogeniculate arteries, 170
Thalamoperforating arteries, 167
Thalamostriate vein
  inferior, 171, 173
  superior, 127, 133, 135, 138, 172–173
Thalamotuberal artery, 167
Thalamus, 324, 137–141, 172, 174
  imaging of, 177
  left, 171
  pulvinar of, 166, 172
  stria medullaris of, 133, 138
  ventral posteromedial nucleus of, 162
Theca externa, BP90
Theca interna, BP90
Thenar eminence, 422
Thenar muscles, 467, 469–470, 486, BP100
Thenar space, 470–471, 473
Thigh, 2, 3
  anterior, 10
  anterior region of, 491
  arteries of, 509–511, BP105
  deep veins of, 523
  fascia lata of, 372, 378, 381–382
  medial, 10
  muscle of, 501–504
    bony attachments of, 498–499
    medial compartment of, T8.1–T8.2
  posterior, 10
  posterior region of, 491
  serial cross sections, 515
Thoracic aorta, 212, 215, 192, 237, 252, 254,
258, T4.1–T4.3
  descending, 262, 265, 266, BP51
  lung groove for, 221
Thoracic aortic plexus, 229
Thoracic aperture, superior, 213, T4.1–T4.3
Thoracic artery
  inferior, 258

Thoracic artery (Continued)
  esophageal branch of, 258
  internal, 104, 165, 206, 210–212, 214–215,
219–220, 231–232, 237, 254, 264–266,
276, 437, T4.1–T4.3, BP24
    anterior intercostal branches of, 210
    left, 252
    medial mammary branches of, 206
    perforating branches of, 206, 209, 211–
212, 436
    right, 251
  lateral, 206, 209, 210, 276, 436, 437, 438, 443,
BP96
    lateral mammary branches of, 206
  superior, 210, 437, 438, 443, BP96
Thoracic cardiac nerves, 229, 249, 250, 261,
BP52, T4.1–T4.3
Thoracic cavity, BP2
Thoracic duct, 16, 101, 210, 215, 219–220, 228–
229, 231, 237, 252, 254, 259–260, 263–266,
283, 286, 319, 344, 347, BP50, BP76, BP79,
T2.1–T2.4
  arch of, 220
Thoracic ganglion, 249–250, 261
  first, 6
  of sympathetic trunk, 229
Thoracic intervertebral discs, 215
Thoracic kyphosis, 179
Thoracic nerve, 436
  lateral, 438
  long, 206, 209–210, 278, 436, 439, T4.1–T4.3,
T7.1–T7.2, T7.3–T7.4
  posterior ramus of, 212
Thoracic skeleton, 8
Thoracic spinal cord, 230
Thoracic spinal nerves
  lateral cutaneous branches of anterior rami
of, 198
  T, 1, 186, 159–161
    dermatome of, 188
    vertebrae relation to, 187
  T2, 160–161
    dermatome of, 188
    vertebrae relation to, 187
  T3
    dermatome of, 188
    vertebrae relation to, 187
  T4
    dermatome of, 188
    vertebrae relation to, 187
  T5
    dermatome of, 188
    vertebrae relation to, 187
  T6
    dermatome of, 188
    vertebrae relation to, 187
  T7
    anterior ramus, 416
    dermatome of, 188
    lateral posterior cutaneous branch of, 275
    medial posterior cutaneous branch of, 275
    posterior rootlets of, 186
    vertebrae relation to, 187
  T8
    dermatome of, 188
    posterior rootlets of, 186
    vertebrae relation to, 187
  T9
    dermatome of, 188
    vertebrae relation to, 187
  T10
    anterior ramus, 412
    dermatome of, 188
    vertebrae relation to, 187
  T11
    anterior ramus, 416
    dermatome of, 188

Thoracic spinal nerves (Continued)
  T12, 186
    dermatome of, 188
    vertebrae relation to, 187
  typical, 201
Thoracic splanchnic nerve, BP52
  greater, 6, 279, 287, BP6
  lesser, 6, 279, 287, BP6
Thoracic sympathetic trunk ganglion, first,
159
Thoracic vein
  internal, 104, 210, 211, 213, 277, 215, 219,
232, 237, 264–266
    perforating tributaries to, 277
  lateral, 277
  thoracoepigastric vein, 277
Thoracic vertebrae, 179–180
  L1, body of, 184, 292
  T1, 36, 46
    anterior view of, 179
    inferior articular facet for, 36
    lateral view of, 179
    posterior view of, 179
    spinal nerve relation to, 187
    spinous process of, 218
  T2, spinal nerve relation to, 187
  T3, 263
  T3, spinal nerve relation to, 187
  T4, spinal nerve relation to, 187
  T5, spinal nerve relation to, 187
  T6
    lateral view of, 180
    spinal nerve relation to, 187
    superior view of, 180
  T7
    posterior view of, 180
    spinal nerve relation to, 187
  T7, body of, 266
  T8, 237
    posterior view of, 180
    spinal nerve relation to, 187
  T9, 182
    posterior view of, 180
    spinal nerve relation to, 187
  T10, spinal nerve relation to, 187
  T11, spinal nerve relation to, 187
  T12, 234
    anterior view of, 179
    lateral view of, 179, 180
    posterior view of, 179
    spinal nerve relation to, 187
    spinous process of, 432
Thoracic visceral nerves, BP6
Thoracic wall
  anterior, 209, 210
    internal view of, 211
  internal, veins of, 213
Thoracoacromial artery, 209, 210, 435, 437–
438, BP96
  acromial branch of, 437–438, 443, BP96
  clavicular branch of, 437–438, 443, BP96
  deltoid branch of, 432, 437, 438, 443, BP96
  pectoral branch of, 437, 438, 443, BP96
Thoracoacromial vein, acromial branch of, 424
Thoracodorsal artery, 436–438, 443, BP24
Thoracodorsal nerve, 433, 436, 438–439,
T7.3–T7.4
Thoracodorsal vein, 444
Thoracoepigastric vein, 270
Thoracolumbar fascia, 275, 195, 197–198, 200,
331, 351
  anterior layer, 200
  middle layer, 200
  posterior layer, 196–197, 200
Thoracolumbar spinal cord, 324
Thoracolumbar spine
  anteroposterior radiograph of, 182

Triangle
anal, 381
of auscultation, 178, 275, 432
cystohepatic (Calot's), 305
deltopectoral, 432
inguinal, 274, 280
transversalis fascia within, 280
lumbar (of Petit), 178, 275, 195
lumbocostal, 216
sternocostal, 211
urogenital, 381
Triangular aponeurosis, 474
Triangular fold, 90
Triangular fossa, 122
Triangular ligament
left, 293, 302
right, 293, 302
Triangular space, 436
Triceps brachii muscle, 178, 202, 422, 432–433,
437–438, 441–442, 445, 447, 460, 488,
T7.5–T7.8
lateral head of, 178, 264–265, 422, 432
long head of, 178, 422, 428, 432, 263–265
medial head of, 460
tendon of, 178, 422
Triceps brachii tendon, 445, 448, 454–455, 488
Tricuspid valve. See Atrioventricular valve; left
Trigeminal cave, imaging of, 177
Trigeminal ganglion, 77, 79, 84, 113–114, 131,
144, 159–162
Trigeminal impression, 32
Trigeminal nerve (CN V), 78–79, 82, 84, 126,
140, 144, 158, 160–162, 171, T2.1–T2.8
distribution of, 145
ganglion of, 149, 158
imaging of, 177
mandibular division of, 23
maxillary division of, 23
mesencephalic nucleus of, 143–144, 162
motor nucleus of, 143–144, 162
motor root of, 158
ophthalmic division of, 23
principal sensory nucleus of, 143–144, 149
schema of, 149
sensory root of, 158
spinal nucleus of, 143, 153
Trigeminal tubercle, 141
Trigonal ring, 370
Triquetrocapitate ligament, 464
Triquetrohamate ligament, 464–465
Triquetrum bone, 462–463, 465–467, BP98
Triticeal cartilage, 106
Trochanter of femur
greater, 178, 268, 353, 420–421, 496–497,
503–505, 511–512, 555, BP104
radiograph of, 354
lesser, 353, 268, 283, 496–497, 555
radiograph of, 354
Trochanteric bursae
of gluteus maximus muscle, 513
inferior extension of, 513
of gluteus medius muscle, 513
Trochanteric fossa, 500
Trochlea, 446, 112
Trochlear nerve (CN IV), 78, 82, 112–114,
131, 140–141, 143–144, 171, T2.1–T2.4,
T2.5–T2.8, BP27
distribution of, 145
schema of, 148
in superior orbital fissure, 34
Trochlear notch, 446–447, 449
Trochlear nucleus, 143–144, 148
True ribs, 203
Trunk, BP15
somatosensory system of,
BP14
Tubal artery, 123

Tubal branches
of ovarian vessels, 406
of uterine artery, 406
Tuber cinereum, 133–134, 140
median eminence of, 174
Tubercle, anterior, 138
Tuberohypophyseal tract, 174
Tuberosity, 537
tibial, 524
Tuberothalamic arteries, 167
Tufted cell, 146
Tunica adventitia, BP13
Tunica albuginea, 382, 386, 392
of ovary, BP90
of testis, 388
Tunica intima, BP13
Tunica media, BP13
Tunica vaginalis
cavity of, 387
testis, 388, 392, 343
parietal layer of, 388
visceral layer of, 388
Tympanic artery
anterior, 74–76, 78, 99, 123, 165
inferior, 82, 123
superior, 78
Tympanic canaliculus, inferior, 31, 33
Tympanic cavity, 80, 121–123, 125, 151–152,
BP28-BP29
labyrinthine wall of, BP28
lateral wall of, BP28
Tympanic cells, BP28
Tympanic membrane, T2.1–T2.4, 77, 121–123,
125, BP28-BP29
Tympanic nerve, 82, 122, 150, 152, 161
inferior, 123
Tympanic plexus, 123, 150, 152, 159, 161
tubal branch of, 152

U

Ulna, 8–9, 446–453, 455, 459–465, 467, 475,
480, T7.1–T7.2, BP98
anterior border of, 449
anterior surface of, 449
body of, 447
coronoid process of, 446
head of, 453
interosseous border of, 449
olecranon of, 422, 441, 446–447, 449, 451,
454, 455, 488
radial notch of, 446, 449
styloid process of, 449, 467
tuberosity of, 446
Ulnar artery, 14, 442–443, 456–458, 461, 464,
469–470, 472, 475–476, 483, T7.1–T7.2,
BP96-BP97
deep branch of, 464
deep palmar branch of, 457, 464, 469, 470,
476, 483, BP97
dorsal carpal branch of, 479
palmar branch of, 475
deep, 443
palmar carpal branch of, 458
pulse point, 14
Ulnar collateral artery
inferior, 442, 455, BP97
superior, 442, 445, 454–455, BP97
Ulnar collateral ligament, 448, BP98
Ulnar nerve, 423, 438–439, 441–442, 445,
454–458, 461, 464, 470, 472, 475–476, 479,
482–487, T7.1–T7.4
articular branch of, 487
common palmar digital branch of, 483
communicating branch of, 476, 478
deep branch of, 457–458, 469–470, 475–476,
483, 487

Ulnar nerve (Continued)
dorsal branch of, 423, 425, 454, 457, 458,
461, 478, 479, 483, 487
dorsal digital branch, 482
groove for, 427, 446
muscular branches of, 475
palmar branch of, 423, 425, 457, 469, 482,
487
common, 482
deep, 475
digital, 423, 482–483
superficial branch of, 457, 469–470, 476, 483,
487
Ulnar recurrent artery
anterior, 457–458, BP97
posterior, 454–455, 458, BP97
Ulnar styloid process, 462
Ulnar tuberosity, 440, 449
Ulnar vein, 15, 444
Ulnocapitate ligament, 464
Ulnocarpal ligament
dorsal, BP98
palmar, 464, BP98
Ulnolunate ligament, 464
Ulnotriquetral ligament, 464, 465
Umbilical artery, 247, 274, 400, 402, 405, 335–
336, 373, BP93
fibrous part of, 366
occluded part of, 404
patent part, 284
patent part of, 404
right, BP93
Umbilical fascia, 274, 281, 368
Umbilical fold
lateral, 367, 293
right, 274
medial, 273, 363, 375, 402
right, 274
median, 273, 293, 363
Umbilical ligament
medial, 247, 272–274, 363, 366, 280–282,
284, 367, 375–376, 402, 404, 335–336,
BP93
left, 274
median, 273–274, 363, 366–367, 281, 368,
282, 343, 369, 375, 378, 402, BP93
Umbilical prevesical fascia, 366
Umbilical region, 267, 269
Umbilical vein, 247
Umbilicus, 267, 274, 399, 318, BP83,
T5.1–T5.3
Umbo, 122, 125
Uncal vein, 172
Uncinate process, 61–62
cervical, 43
left, area for articulation of, 43
right, articular surface of, 43
Uncovertebral joints, 45, BP34
Uncus, 133–134, 146
Upper limb, 2–3, 422–490
arteries of, 443, 483
cutaneous innervation of, 423
dermatomes of, 5
electronic bonus plates, BP96-BP102
free part of, 2–3, 8
lymph vessels and nodes of, 426
muscles, T7.5–T7.8
nerves of, 483, 484
cutaneous, 424
superficial veins of, 424
surface anatomy of, 422
veins of, 444
**Upper limb movements, segmental
innervation of**, 11
Urachus, 378, 402
median, BP93

Netter Atlas of Human Anatomy: Classic Regional Approach

Ureter, 274, 282, 285, 363–369, 372- 376, 290, 293, 330, 333, 335, 385, 393, 394, 402–406, 412, 414, 419, T6.1–T6.3, BP80, BP90, BP93
in abdomen and pelvis, 335
arteries of, 336
autonomic nerves of, 339
left, 335
lower, innervation of, 419
in peritoneal fold, 367
right, 335
Ureteric orifice, 373, 375, 385, T6.1–T6.3, BP86
left, 371
right, 371
Ureteric plexus, 319, 412
Ureters, 19
as site of referred visceral pain, BP5
Urethra, 19, 283, 358–360, 362, 364, 369, 370–371, 373, 375, 379, 382, 385–386, 390, 421, BP8, BP85-BP86
beginning of, 420
cavernous venous plexus of, BP86
female, BP86
pelvis, 421
floor of, 386
hiatus of, 361
membranous, 384, 386, 391, BP85
musculofascial extensions to, 360
prostatic, 386, 420, BP85, BP92, BP94
roof of, 386
spongy, BP85
bulbous portion of, 371, 386
pendulous portion of, 386
Urethral artery, 384, 407
Urethral branches, BP93
Urethral crest, 385–386
Urethral fold
primary, 389
secondary, 389
Urethral glands, BP86
of Littré, 386
Urethral groove
primary, 389
secondary, 389
Urethral lacunae, of Morgagni, 386
Urethral orifice
external, 364, 368, 377, 380, 383, 386, 389, 397, BP86
internal, BP94
Urethral raphe, 389
Urethral sphincter muscle, 379
external, 274, 362, 368, 370, 378, 379, 384, 385, 386, 419, T6.4, BP84, BP93
female, 370
internal, 370–371, 386
loop of Heiss, 370
posterior loop, 370
trigonal ring, 370
Urethrovaginalis sphincter muscle, 358, 370, 372
Urinary bladder, 6, 7, 19, 182, 274, 281–282, 288, 293, 300, 323, 330, 335–336, 343, 363–367, 369, 375, 384–385, 387, 393, 399, 402, 405, 414, 419, T6.1–T6.3, BP8, BP84, BP85, BP90, BP94
apex of, 368–369
autonomic nerves of, 339
body of, 368–369
female, 371
fundus of, 368–369
innervation of, 419
interior of, 420
lymphatic drainage over, 410
male, 371
neck of, 368–369, BP86
orientation and supports of, 369
pulled up and back, 369
trigone of, 368–369, 385–386, BP86
ureteric orifice of, 369

Urinary system, 19, T5.1–T5.3
Urogenital diaphragm, 405
Urogenital hiatus, 370
Urogenital mesentery, 387, 390
Urogenital sinus, 390
Urogenital triangle, 352, 381
Uterine artery, 335–336, 375–376, 402, 404, T6.1–T6.3, BP90
branches of, 406
cardinal (Mackenrodt's) ligament with, 366
ovarian branch of, 376
vaginal branches of, 406
Uterine cavity, contrast medium in, 376
Uterine cycle, BP88
Uterine development, BP89
Uterine fascia, 366, 373
Uterine ostium, 374
Uterine tube, 20
Uterine (fallopian) tube, 363–365, 372–375, 390, 402, 414, 417, T6.1–T6.3
ampulla, 376
contrast within, 376
fimbriae of, 376
folds of, 374
infundibulum of, 376
Uterine veins, 285, BP90
superior, 401
Uterine vessels, 372, 406
Uterosacral fold, right, 375
Uterosacral ligament, 364, 366, 373–374, BP90
Uterovaginal fascia, 372, 378, BP84, BP90
Uterovaginal plexus, 414, 416
Uterovaginal venous plexus, 285
Uterus, 20, 372, 374–375, 378, 390, 393, 414, 417, T6.1–T6.3, BP84, BP95
arteries of, 406
body of, 364–365, 374
broad ligament of, 363, 365, 373, 390
cervix of, 364–366, 372–375, T6.1–T6.3, BP90
endometrium of, 399
external os of, 376
fascial ligaments, 373
fundus of, 363–365, 369, 374, 399
isthmus of, 374
ligaments of, round, 364, 371–372, 378, 402, BP86
myometrium of, 399
round ligament of, 285, 363, 373, 390
supporting structures, 372
veins of, 406
Utricle, 124–125, 151, T2.1–T2.4
Uvula
of bladder, 385–386
of palate, 83, 90
of vermis, 141

V

Vagal fibers, 153
Vagal trigone, 141
Vagal trunk
anterior, 153, 229, 254, 261, 283, 319–323, 325, 328–329, 339, 341, 412, BP7
gastric branches of, 153
hepatic branch of, 153
posterior, 261, 283, 319–323, 325, 328–329, 339, 341, 412, BP7
Vagina, 20, 358–360, 364, 369–375, 378–379, 390, 393, 397, 399, 402, 417, 421, T6.1–T6.3, BP86
musculofascial extensions to, 360
supporting structures, 372
vestibule of, 372, 377, 390
Vaginal arteries, 335–336, 372, 375–376, 402, 404, 406
inferior, 369
internal, 366

Vaginal epithelium, BP95
Vaginal fascia, 373, 393
Vaginal fornix, 364, 374
anterior, 364, 373–375
posterior, 375
Vaginal orifice, 364, 377, 380, 389, BP86
Vaginal vein, 401
Vaginal wall, 372
Vaginorectal fascial fibers, 366
Vagus nerve (CN X), 50–51, 56–58, 75, 82, 84, 91, 98–100, 103–105, 131, 140, 143–144, 152, 154, 157–158, 162, 214, 220, 229–232, 249–252, 254, 261, 263–264, 324, 326, 340, T2.5–T2.8, BP7, BP8, BP22, BP26
auricular branch of, 23, 33, 153
communication to, 152
bronchial branch of, 229
cervical cardiac branch of, 229, 261
inferior, 249–250, 153
superior, 249–250, 153, 157, 158
distribution of, 145
dorsal nucleus of, 143, 153
ganglion of
inferior, 82, 91, 153–155, 162
superior, 91, 153–154
inferior ganglion of, 230, 250, 261
in jugular foramen, 34
in jugular fossa, 33
laryngopharyngeal branch of, 157
left, 233, 104
meningeal branch of, 153
pharyngeal branch of, 91, 152–153, 157–158, 261, T2.5–T2.8
posterior nucleus of, 143–144, 250, 324, 340
right, 261, 104
schema of, 153
superior cervical cardiac branch of, 98–99, 157
superior ganglion of, 261
thoracic cardiac branches of, 153, 157, 214, 229, 249–250, 261
Vallate papilla, 89, 162
Vallecula, 89, 90
Valves of Houston, 395
Valves of Kerckring, 297, 305. See also Circular folds (valves of Kerckring).
Vas deferens, 371, 385, 393, 405, 411–412, 418
ampulla of, 385
arteries to, 400, 403, 405
Vasa recta spuria, BP73
Vasa vasis, BP13
Vastoadductor membrane, 502, 509–510
Vastus intermedius muscle, 153, 157, 498, 502–503, 515, 517, 550
Vastus intermedius tendon, 510
Vastus lateralis muscle, 421, 491, 498, 501–503, 509–511, 515–517, 531, 550, T8.5–T8.8
Vastus medialis muscle, 491, 498, 502–503, 509–510, 515–517, 531, 550, T8.5–T8.8
Veins, 12–13
of abdominal wall, anterior, 278
of anal canal, 401
of duodenum, 315
of esophagus, 259
of eyelids, 115
of forearm, 425
of iris, 120
of large intestine, 317
of pancreas, 315
of pelvis
female, 402
male, 405, BP93
of perineum, male, 407
of posterior cranial fossa, 171
of rectum, 401
of small intestine, 316
of spinal cord, 193

Veins *(Continued)*
  of spleen, 315
  of stomach, 315
  of testis, 403
  of thoracic wall, 213
  of uterus, 406
  of vertebral column, 193
Vena cava
  groove for, 302
  inferior, 13, 15, 200, 215, 220, 229, 234–236,
    238, 241–242, 251, 253, 254, 259, 262,
    285, 291–294, 296, 302, 306, 312, 315,
    330–332, 342, 345–351, BP38, BP79,
    BP82
    lung groove for, 221
    opening of, 244
    valve of, 241
  superior, 13, 103–104, 213–214, 220, 231–
    233, 241, 245–248, 251, 253, 259, 262,
    264–266, 344
    lung groove for, 221
Venae comitantes, 470
Venae rectae, 316
Venous arch
  dorsal, 492, 523
  plantar, 523
  posterior, superficial, 533
  superficial palmar, 470
Venous communications, body of C3, BP39
Venous palmar arch, superficial, 444
Venous plexus, 82, 369, 376, 388, BP93
  around vertebral artery, 194
  basilar, 130–131
  cavernous, of urethra, BP86
  external, 394, 401
  in foramen magnum, 33
  internal, 394
  pampiniform, 403
  pharyngeal, 91
  prostatic, 405, BP93
  rectal
    external, 395–396
    internal, 395–396
    perimuscular, 401
  vaginal, 421
  vertebral (of Batson), 130
    internal, 190
  vertebral artery with, BP39
  vesical (retropubic), 371, 405, BP93
Ventilation, anatomy of, BP45
Ventral posterolateral nucleus, of thalamus,
  BP14
Ventricles (cardiac), 245
  left, 232–239, 242, 245, 262, 266, BP51
    inferior vein of, 239
  right, 232–238, 241, 245–246, 262, 266, BP51
Ventricles (of brain), 141
  fourth, 133, 135, 141, 173
    choroid plexus of, 136, 141
    imaging of, 177
    lateral and median apertures of, 173
    lateral aperture of (foramen of Luschka),
      135
    lateral recess of, 135
    median aperture of (foramen of Magend-
      ie), 135
    outline of, 170
    rhomboid fossa of, 140
    taenia of, 141
    vein of lateral recess of, 171
  lateral, 133, 135, 137, 139, 166, 173
    body of, 135
    central part of, 133
    choroid plexus of, 133, 135, 136, 137, 138,
      166, 172
    frontal (anterior) horn of, 133, 135, 177
    imaging of, BP32

Ventricles (of brain) *(Continued)*
    lateral vein of, 173
    medial vein of, 173
    occipital (posterior) horn of, 133, 135, 137–
      139, 170, 173, 177
    right, 135
    temporal (inferior) horn of, 133, 135, 138–
      139, 173
  occipital horn of, 170
  third, 133, 135, 137–138, 141, 173
    choroid plexus of, 133, 135, 136
    imaging of, 177, BP32
    infundibular recess of, 69
    interthalamic adhesion, 135
    tela choroidea of, 135, 138, 172
Ventricular folds, 107
Ventricular veins
  hippocampal, 173
  inferior, 173
Venulae rectae, BP73
Venule, 13
Vermian vein
  inferior, 171
  superior, 171, 173
Vermiform appendix, 153, 290, 298–301,
  T5.1–T5.3, BP81
  orifice of, 299
Vermis, 69, 142
  imaging of, BP32
  inferior
    nodule of, 142
    pyramid of, 142
    tuber of, 142
    uvula of, 142
  superior
    central lobule of, 142
    culmen of, 142
    declive of, 142
    folium of, 142
Vertebrae, 8, 187
  L3, body of, 363, 367
  L4, spinous process of, radiograph of, 354
  L5, transverse process of, radiograph of,
    354
  lamina of, 262
  pedicle of, 262
  radiology of, 182
Vertebral artery, 34, 46, 91, 98, 104–105, 157,
  165–168, 170, 191, 194, 199, 249, 258, 437,
  BP18, BP24, BP50
  anterior and posterior meningeal branches
    of, 128
  cervical part of, 163
  in foramen magnum, 33
  imaging of, BP31–BP32
  left, 163
  meningeal branches of, 34
    anterior, 163, 170
    posterior, 163, 170
  right, 163, BP39
  venous plexus and, BP39
Vertebral body, 190, BP38
  articular facet for, 221
    inferior, 221
    superior, 221
  branch to, 192
  cervical, 43, 45, BP17
  of L2, 200
  of L3, BP37
  lumbar, 181
  posterior intercostal artery branch to, 192
  posterior surface of, BP33
  of S, 1, 182
  of T12, 182
  thoracic, 191
  fourth, 251
Vertebral canal, 180–181, BP2

Vertebral column, 8, 179
  veins of, 193–194
Vertebral foramen, T3.1–T3.3
  cervical, 42–43
  lumbar, 181
  thoracic, 180
Vertebral ganglion, 157, 249–250,
  261
Vertebral ligaments, BP39
  lumbar region, 185
  lumbosacral region, 184
Vertebral notch
  inferior, 180–181, BP36
  superior, 181
Vertebral plexus, 34, 157
Vertebral veins, 194, 259, BP39
  accessory, 194
  anterior, 194
Vertebral venous plexus, T3.1–T3.3
  anterior, BP39
  external, 193
    anterior, 194
  internal, 193
    anterior, 194
Verticalis linguae muscle, T2.9–T2.14
Vesical arteries
  inferior, 284, 335, 369, 400, 402–403,
    405
    posterior branches of, 405
  internal, 366
  superior, 274, 335–336, 366, 400, 402, 404–
    405, BP93
Vesical fascia, 366, 368, 371, 378, 393, 398,
  BP84, BP90
  urinary bladder in, 368
Vesical fold, transverse, 367, 375–376
Vesical nodes
  lateral, 338
  prevesical, 338
Vesical plexus, 323, 339, 412, 414, 416, 418,
  419
Vesical veins
  inferior, 401
    posterior branches of, 405
  superior, 285, 401
Vesical venous plexus, 285, 368, 405,
  BP93
Vesical vessels, inferior, 410
Vesicocervical fascial fibers, 366
Vesicocervical space, 366
Vesicouterine pouch, 364–365, 369, 375–376,
  393
Vesicovaginal space, 366
Vesicular appendix, 374, 390
Vessels
  blood, innervation of, BP52
  of neck, 56–57
Vestibular aqueduct
  internal opening of, 124
  opening of, 34, 126
Vestibular area, 141
Vestibular canaliculus, opening of, 29
Vestibular fold, 108
Vestibular fossa, 377
Vestibular ganglion (of Scarpa), 124, 151
Vestibular gland, greater, 379, 380, 406
  opening of, 377
Vestibular (Reissner's) membrane, 125
Vestibular nerve, 121, 124, 126, 151, T2.5–T2.8,
  BP29
  inferior part of, 124, 126, 151
  superior part of, 124, 126, 151
Vestibular nucleus, 143, 144
  inferior, 151
  lateral, 151
  medial, 151
  superior, 151